FOOD INDUSTRIES
MANUAL

FOOD INDUSTRIES MANUAL

Edited by

M.D. Ranken and R.C. Kill
Micron Laboratories, Luton, UK

and

C. Baker
ArvaTec (UK) Ltd, Wantage, UK
and
Chemical Engineering Dept, Kuwait University

BLACKIE ACADEMIC & PROFESSIONAL
An Imprint of Chapman & Hall
London · Weinheim · New York · Tokyo · Melbourne · Madras

**Published by Blackie Academic and Professional, an imprint of Chapman & Hall,
2–6 Boundary Row, London SE1 8HN, UK**

Chapman & Hall, 2–6 Boundary Row, London SE1 8HN, UK

Chapman & Hall GmbH, Pappelallee 3, 69469 Weinheim, Germany

Chapman & Hall USA, 115 Fifth Avenue, New York, NY 10003, USA

Chapman & Hall Japan, ITP-Japan, Kyowa Building, 3F, 2-2-1, Hirakawacho,
Chiyoda-ku, Tokyo 102, Japan

DA Book (Aust.) Pty Ltd, 648 Whitehorse Road, Mitcham 3132, Victoria,
Australia

Chapman & Hall India, R. Seshadri, 32 Second Main Road, CIT East, Madras
600 035, India

23rd edition 1993

This edition 1997

© 1997 Chapman & Hall

Typeset in 10/12pt Times by Acorn Bookwork, Salisbury, Wilts

Printed in Great Britain by T. J. International Ltd., Padstow, Cornwall

ISBN 0 7514 0404 7

A catalogue record for this book is available from the British Library

Library of Congress Catalog Card Number: 97-70332

∞

Printed on acid-free text paper, manufactured in accordance with ANSI/NISO
Z39.48-1992 (Permanence of Paper)

Contents

Contributors

Keith G. Anderson CBiol, FIBiol, MFC, FIFST, FRSH
Consultant
Ventress Technical Services Ltd
341 Reigate Road
Epsom Downs
Surrey KT17 3LT

David Arthey MSc, PhD, FIFST
Consultant
Spindlewood
Collin Close
Willersey
Broadway
Worcs WR12 7PP

Christopher G.J. Baker BSc(Eng), DIC, PhD, CEng, FIChemE
ArvaTec (UK) Ltd
Long Acre
East Hanney
Wantage
Oxfordshire OX12 0HP

R.G. Booth BSc, PhD, CChem, FRSC, FIFST, FRSMed
Consultant Food Scientist
19 Homewood Road
St Albans AL1 4BG

Roger Broomfield ANCFT, FIFST
Food Control Officer
Hereford & Worcester County Council
Trading Standards Service
28/30 Foregate Street
Worcester WR1 1DS

Alan J Campbell
Campden & Chorleywood Food RA
Chipping Campden
Gloucestershire GL55 6LD

M.N. Clifford PhD, MIFST
Reader in Food Science
University of Surrey
Food Safety Research Group
School of Biological Sciences
Guildford GU2 5XH

J.M. Conner BSc, PhD
Research Fellow
University of Strathclyde
Dept of Biosciences and Biotechnology
204 George Street
Glasgow G1 1XW

D.A. Cruickshank CBiol, MIBiol
Chocolate Development Manager
Cadbury Ltd
Bournville
Birmingham B30 2LU

John Gordon BSc, PhD, DIC
Consultant
Croft Cottage
71 Tilehouse Street
Hitchin
Herts SG5 2DY

S.D. Holdsworth BSc, MSc, CEng, FIChemE, CChem, FRSC, FIFST
Consultant
Withens
Stretton-on-Fosse
Moreton-in-Marsh
Gloucestershire GL56 9SG

Ron C. Kill BSc, PhD, FRIPHH, FIFST
Consultant Food Technologist
Micron Laboratories
Suite 3 Clarence House
30 Queen Street
Market Drayton
Shropshire TF9 1PS

Sri Parkash Kochhar MSc, PhD, FIFST
Consultant, Edible Fat and Oil Technology
SPK Consultancy Services
48 Chiltern Crescent
Earley
Reading RG6 1AN

E.R.E. Meersman
Laboratory Manager
Bavaria bv
P O Box 1
5737 Lieshout
Netherlands

G.G. Oliphant   MIFST, MInstM
Technologist
Micron Laboratories
10 Greenwood Court
Ramridge Road
Luton LU1 0TN

Fiona J. Palmer   BSc, PhD, MIFST
Analytical Laboratory Manager
Central Technical Services
Britvic Soft Drinks Ltd
Widford Industrial Estate
Westway
Chelmsford CM1 3LN

John R. Piggott   PhD, FIFST
Reader in Food Science
University of Strathclyde
Dept of Biosciences & Biotechnology
204 George Street
Glasgow G1 1XW

M.D. Ranken   BScTech, CChem, MChemA,
MFC, FRSC, FIFST, FInstM

(Retired, previously Consultant Food
Technologist
Micron Laboratories
10 Greenwood Court
Ramridge Road
Luton LU1 0TN)

29c Albert Road
Hythe
Kent CT21 6BT

D.A. Rosie   BSc, PhD, FIFST
Manager in Food Development Centre
Whitbread plc
Whitbread Technical Centre
Park Street
Luton LU1 3ET

S.P.E. Simon   MA(Cantab), MW
Director
Pat Simon Wines Ltd
15 Sutcliffe Close
London NW11 6NT

Derek Stansell   BSc, CChem, MRSC
Consultant
42 Maerdy Park
Pencoed
Bridgend
Mid Glamorgan CF35 5HX

Michael J. Urch   DipFoodTech
Editor
Seafood International
Meed House
21 John Street
London WC1N 2BP

Peter J. Wallin   BSc, MSc, PhD, MIFST
Technology Manager
Dalgety Food Technology Centre
Station Road
Cambridge CB1 2JN

David J. Wallington   BSc, MIFST
Chief Scientist
Weston Research Laboratories Ltd
Vanwall Road
Maidenhead SL6 4UF

W.E. Whitman   BSc, FIFST
Consultant
Food Solutions
19 Park Green
Great Bookham
Surrey KT23 3NL

Preface

It is a measure of the rapidity of the changes taking place in the food industry that yet another edition of the *Food Industries Manual* is required after a relatively short interval. As before, it is a pleasure to be involved in the work and we hope that the results will continue to be of value to readers wanting to know what, how and why the food industry does the things which it does.

For this edition we have made a major departure from the style of earlier editions by completely revising the layout of many of the chapters. Previously the chapters were arranged as a series of notes on specific topics, set out in alphabetical order in the manner of an encyclopaedia. This was useful to a reader seeking information on a specific topic but not very helpful for anyone looking for a general overview of even a moderately complex process. We have therefore chosen to present the same kinds of information but, wherever practicable, in an order which follows the logical sequence of the manufacturing process. To this end, as appropriate, each chapter begins with a flow diagram and the topics in the chapter are laid out accordingly. (Those who have any contact with the suddenly fashionable methods of HACCP (Hazard Analysis Critical Control Points) analysis will easily recognize the pattern.) We hope that any difficulties which may arise in searching for material no longer alphabetically arranged can be overcome by use of the comprehensive index.

The work has been revised and updated, and following the logic of the flow sheets there is some simplification and rearrangement among the chapters. Food Packaging now merits a separate chapter and some previous sections dealing mainly with storage have been expanded into a new chapter covering Food Factory Design and Operations.

There is one completely new chapter, entitled Alcoholic Beverages, divided into Wines, Beers and Spirits. There is a strain of thought which does not yet consider the production of those drinks to be a legitimate part of the food industry, but many of the processes and methods of manufacture and all those of production and quality control, are just the same in principle as anywhere else in the food industry and we believe that they have been excluded for too long.

It has been a pleasure to welcome a new member of the editorial team, Dr Christopher Baker. We wish also to record our deep thanks to the contributors for their hard work of revision and renewal, and to the Library and Technical Information staff at the Leatherhead Food Research Association for excellent help in finding and checking information.

M.D.R.
R.C.K.
C.G.J.B.

1 Meat and Meat Products

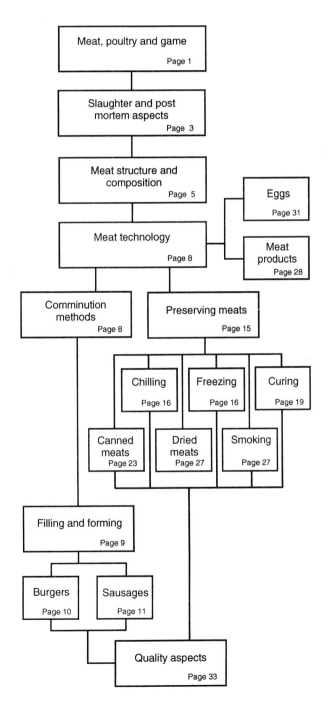

MEAT AND POULTRY

BEEF

Butchers' beef, for roasting or grilling, comes from male and female animals of about 15 to 24 months of age. Newer breeds from Europe such as the Charolais, Simmenthal and Belgian Blue have higher yields of saleable meat than traditional British breeds such as Aberdeen Angus and Hereford. As a prerequisite of annual milk production cows need to have one calf each year. Two-thirds of beef comes from dairy cross calves which are produced as a by-product of the dairy farmers' breeding programme for herd replacements. The remainder of the beef comes from suckled calves which stay with their mothers until six to eight months of age before being transferred to finishing units. Suckled calves are typically produced in the hills and uplands. Figure 1.1 shows typical names and cuts of beef.

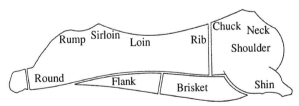

Figure 1.1 Composition of some typical cuts of beef.

Beef for manufacturing purposes consists partly of the forequarter meat from the beef animals noted above, but the majority comes from the carcasses of dairy and suckler cows at the end of their economic life which is typically 8 to 11 years of age. The meat is often known as 'cow beef'.

Veal calves are the same animals but taken at 3 to 4 months of age, 'bobby' veal at under 3 months.

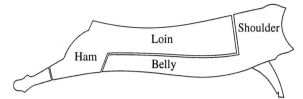

Figure 1.2 Main cuts of pork.

PORK

In the past pigs have been bred for their final carcass conformation, fat cover and weight. The requirements for pork and bacon were different in that the bacon manufacturer looked for a long back and a thin, even fat cover. These days both pork and bacon come from either the Large White/Landrace crossbreed or the Duroc. They are slaughtered from 140 to 200 days depending on the rearing regimes used. The final carcass weight is around 69 kg. The carcass typically has a fat cover of around 11 mm which suits both pork and bacon producers. Figure 1.2 shows typical cuts of pork.

LAMB AND MUTTON

Sheep (and goats, which are closely related) were probably the first animals domesticated by man. There is a large production in countries where wool production is important, elsewhere sheep tend to be kept mainly as a means of farming some marginal lands.

Mutton fat is less preferred than the other animal fats, mainly because of its high melting point which causes it to solidify unpleasantly in the mouth.

POULTRY

Broiler chickens are birds bred for high growth rate, good feed conversion of feed into flesh and high meat yield. In the USA, where the term means 'suitable for grilling', the birds are slaughtered at about 7 weeks of age, giving about 1.5 kg dressed weight. In Britain, birds up to 10 weeks old and correspondingly heavier are included.

The main source of chicken meat for manufacturing is 'spent' hens from egg-producing flocks. Since they were bred not for meat quality but for efficient egg production, they are small, of poor conformation and meat yield. They are usually about 18 months old, so the meat is also tough

unless slowly cooked, but they are cheap. A small proportion of these birds may be free range but the majority are grown intensively: flavour and texture differences due to growing conditions are negligible.

Turkeys for domestic sale may range down to 2 to 3 kg dressed weight. For manufacturing purposes males of 10 to 15 kg or heavier are used. The cost of the meat from these turkeys now compares favourably with meats such as pork.

The most efficient method of chilling poultry carcasses after slaughter and dressing is by immersion in ice-water, commonly done in rotating tanks called spin chillers. This water chilling can be hygienically operated, providing that water as close as possible to its freezing point is introduced into the exit to the tank and flows counter-current to the poultry, and that the water in the tank is agitated by having air blown through it. Some 2.5 l of water per bird of less than 2.5 kg is allowed and the water temperature is nowhere warmer than 16°C (60°F). Larger birds, such as turkeys, may require such a long residence time that after spin chilling they are kept overnight in tanks covered in ice. During the chilling process the carcasses absorb water. In the EU the average quantity absorbed is governed by regulation; the proportion absorbed by individual birds can vary greatly.

OSTRICH

All over the world the farming of ostrich for meat is becoming more widespread. In the UK ostrich farming is still in its infancy and the cost of meat is expensive. It can now be found in a wide range of supermarket outlets and as it becomes more accepted so the demand will increase. One of the limiting factors is the lack of specialist slaughter facilities (four in the UK). Farmers have the added cost and burden of moving animals over large distances for slaughter which in itself is less attractive to the would-be farmer. Farms are usually set up for breeding and rearing with two female and one male bird which will cost £1000 as an initial outlay. They also require specialist veterinary care. Birds are reared to a live weight of about 90 kg at the age of 8–14 months and will have an expected dressed weight of 50%.

The meat from ostrich has the appearance and texture of beef and a similar taste, although it is much lower in fat. As yet, due to scarcity and high costs, there is no manufacture of products from ostrich, products being solely devoted to premium cuts. It will be interesting to follow the growth of farming and consumption of this species in the future.

GAME MEAT

The EU defines farmed game to include wild land mammals and wild birds (except pheasants, quail and partridges which are poultry) which are bred, reared and slaughtered in captivity. The requirements for conditions in slaughterhouses and the conduct of slaughtering are similar in principle to those for conventional meat production.

Despite considerable difficulties in the management of such wild creatures in captivity, the farming of red deer in particular has increased in recent years. The product is mostly sold as venison for the table but there is a small production of venison sausages and pâtés.

SLAUGHTER AND POST MORTEM ASPECTS

SLAUGHTER AND DRESSING

There can be no doubt that the manner in which these operations are carried out has a profound effect on the quality of the meat obtained from the animals, and that the most humanitarian procedures are also those which give the best quality.

Animals should be well rested before slaughter and fed as recently as possible. Abattoir design and practice should be conducive to calm at the time of slaughter. It is required by law that the animals are anaesthetized in some way (with allowances made for ritual slaughter, see below). The captive bolt pistol is commonest for cattle and electric stunning for smaller animals and poultry. Carbon dioxide or electrical stunning is used for pigs.

The throat is 'stuck' with a knife and the unconscious animal is allowed to bleed to death, usually suspended from an overhead track to facilitate subsequent dressing operations. The knife may be a specially designed hollow one if it is intended to collect the blood for processing. After death there follows a sequence of operations, slightly different for each species, to remove the parts of the carcass not wanted for meat – hide, feathers, entrails – and in well ordered slaughterhouses there are appropriate subsidiary processes to deal hygienically with these offals. Some of the stages can be done fully or partly automatically, particularly in high speed poultry processing, but

there remains a large element of hand work. Both mechanical and hand operations have to be done hygienically but with good management this is quite possible.

Every carcass and its essential organs such as the heart and liver are inspected and approved by a veterinarian or qualified meat or poultry inspector.

Ritual slaughter

There are separate regulations in the UK governing religious slaughtering (Schedule 12, Additional Provisions for Slaughter by Religious Method). Under these requirements, only licensed practitioners are allowed, the animal must be upright in a restraining pen suitable to take the animal's weight. The knife used must be undamaged, sharp and of sufficient length.

The differences in the Kosher and Halal method are largely in the prayers given during the ritual. In Jewish ritual the rabbi will also inspect the lungs and diaphragm for adhesion or any lesions, the meat is 'porged' or stripped of its major blood vessels. In Halal the carcass is dressed in the same way as for the non-ritual animal.

POST MORTEM CHANGES

The chemical changes which take place on the death of an animal, changing its muscles into meat, are complicated but fairly well understood, and are fully described in standard textbooks, see Lawrie (1991).

When normal metabolism and the supply of oxygen to the bloodstream cease, any reserves of glycogen (the animal's energy supply) in the muscles are converted to lactic acid and the pH falls from its live value of 7.0–7.2 to a final 5.5–6.5.

The PSE (pale, soft, exudative) condition

If the pH falls very rapidly, with adequate glycogen present, the final low pH value is reached while the carcass is still warm. This leads to precipitation of sarcoplasmic protein, loss of water binding capacity and a paler colour due to light scattering by the precipitated material. The meat is soft and wet. The cause is to be found in the greater susceptibility of the animal to the stress of slaughter, and among pigs certain breeds including Danish Landrace and the Belgian Pietrain are notoriously badly affected.

The DFD (dark, firm, dry) condition

This condition is the reverse of the above, resulting from a failure of the pH to fall very far from its original value of about 7. The cause is in a deficiency of glycogen in the muscles at the time of slaughter, so that little lactic acid can be formed. That deficiency is usually the result of the exhaustion of the animals, for instance by prolonged fasting or excessive exercise before slaughter consuming the glycogen reserves. Associated with the high pH, the muscles retain water well and have a deep red appearance. Hence the descriptions 'dark cutting' in beef and 'fiery' or 'glazy' in bacon. The microbiological state may be adversely affected by the relatively low acidity.

The phenomenon of dark cutting beef has been known since the eighteenth century. Once caused, there is no known cure for the condition but it can be prevented by ensuring that animals are well fed, rested and gently treated at the time of slaughter.

Other changes

As the glycogen is exhausted rigor mortis sets in, to be resolved later on storage. These changes are described under AGEING. There are some advantages in processing the meat 'hot', before rigor mortis is complete. Hot deboning gives improved yield, more efficient cooling with savings in refrigeration costs and reduced drip and evaporation losses. If the meat can be immediately treated with salt and ground, as in large scale sausage or burger production, there are great improvements in water holding capacity compared with conventional cold meat processing.

AGEING

The object of ageing meat after slaughter is to make it more tender. When an animal dies, the adenosine triphosphate (ATP) in the muscle fibres, in the presence of magnesium, is decomposed by myosin ATP-ase. There is a large release of energy which is used up in contracting the muscle fibres: the actin filaments slide inwards between the myosin filaments, shortening the myofibrils. The heads of the myosin filaments then lock on to the actin, making the structure rigid. This is the well-known phenomenon of rigor mortis: opposing muscles contract and pull against each other and the whole carcass becomes stiff. If the meat is cooked when still in the rigor condition it is extremely tough and unacceptable.

When the meat is hung after slaughter the muscles gradually recover their extensibility and become considerably more tender. We say that rigor mortis has been resolved. The mechanism by which this occurs is not clear. Cracking of the muscle fibres occurs, probably a result of mechanical failure when the muscle was in the rigor state. Damage is also observed at the Z-lines of the muscle structure, where the individual fibrils are joined; this is probably caused by proteolytic enzymes including cathepsins and calpains. At ordinary ambient temperatures the approximate times for rigor mortis to commence and the times of hanging for adequate tenderization are:

	Time to onset of rigor	Time to resolution of rigor
Cattle/pigs	12–24 h	2–6 h[*]
Turkeys	1/2–2 h	6–24 h
Chickens	1/2–1 h	4–6 h

[*]Further slight increase in tenderness up to 14 days.

These differences in the hanging times necessary to achieve maximum tenderization are possibly due to different degrees of contraction of the myofilaments in bovine, porcine and avian muscles. Limited proteolytic changes have been observed in the sarcoplasm of the muscles but these do not appear to be the cause of the tenderization. Furthermore, very few micro-organisms are found deep within the intact meat after ageing, so neither the tenderizing nor the proteolysis is caused by bacterial action.

Accelerated ageing or conditioning

Ageing normally takes place after the meat is chilled but the process is greatly accelerated if chilling is delayed and the meat continues to be held at about 37°C. Undesirable bacterial and mould growth must be guarded against and there is an increased possibility of the production of pale exudative tissue. Beef carcasses held at 43°C become significantly tenderized in the 24 h after slaughter but the meat becomes pale and exudative. Times for achieving satisfactory tenderness, in typical experiments, are of the order of weeks at 0.5°C, 5 days at 13°C, 2 days at 18°C and a few hours at 29°C. After accelerated ageing at higher temperatures, the meat should be cooled and stored as necessary at 2°C.

Cold shortening or cold toughening

If beef (to some extent) or lamb (especially) is chilled rapidly after slaughter the muscles may undergo extreme contracture or 'cold shortening'. When cooked this meat is very tough. Under similar conditions pork is almost unaffected. The cause of the problem is that muscular contraction is triggered off by the cold conditions and it is mechanically possible because reserves of energy and of the energy-using ATP system still remain in the meat. Where the meat is cooled slowly these reserves become consumed and contraction is no longer possible when the meat is cold. The critical condition appears to be that if the temperature within the meat falls to 10°C (50°F) in less than 10 h then cold shortening is probable unless other precautions are taken – see below.

The tenderstretch process

If the contraction of the muscles on a carcass can be prevented then cold shortening is unable to occur even if the meat is cooled rapidly after slaughter. In the 'tenderstretch' process, also called 'hip-hanging', beef carcasses are suspended after slaughter not in the usual manner by the hind legs but by brackets which secure them by the aitch bone. By this means the muscles of the loin and back are kept under tension by the weight of the animal, contraction is inhibited and the meat can be rapidly chilled without undergoing cold shortening and consequent toughening. It is however necessary after evisceration to change the suspension of the carcasses by moving them from the special brackets which can interfere with the smooth flow of the slaughter line.

Because of their smaller size this process is less effective with lambs.

Electrical stimulation

When an electric current is passed through an animal carcass immediately after slaughter there is a considerable contraction of the muscles. The energy needed for this contraction consumes the remaining reserves of glycogen and ATP so that when the muscles relax on removing the current no further contraction can take place, and toughness from that cause, even if cooling quickly follows, can be eliminated. The process is effective both with cattle and lambs and is in widespread use wherever there are advantages in cooling rapidly after slaughter, for instance, with lambs to reduce the holding time necessary before freezing, or with beef to permit hot boning.

High voltage stimulation (700–800 V at peak) gives more rapid tenderizing than low voltage (80–100 V at peak) but the latter is safer to use in the abattoir.

There is some danger of reduced colour stability in the meat, possibly because of damage to the protein part of the myoglobin pigment by muscle enzymes released during the electrical contraction.

MEAT STRUCTURE AND COMPOSITION

MUSCLE STRUCTURE AND FUNCTION

Lean meat is the material which constituted the muscle of the live animal and was organized to do the mechanical work necessary for the motion of the animal when it was alive. Physiologically, each muscle is an organ capable of expansion and contraction, bounded by a muscle sheath and fastened at each end by means of tendons to those parts of the skeleton which it is responsible for moving.

The expansion–contraction mechanism is formed by a system of fibres running along the muscle: each fibre is encased in a sheath called the sarcolemma and consists itself of bundles of fibrils

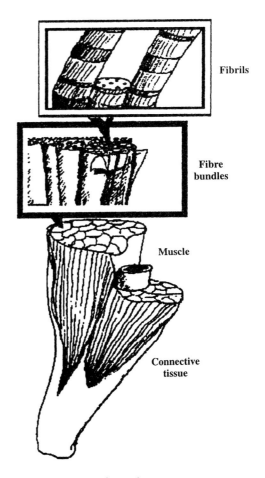

Figure 1.3 Structure of muscle.

with further subdividing sheaths called perimysium or endomysium (Figure 1.3). Finally, the fibrils are made of long, highly ordered chains of molecular protein, each chain running the length of the muscle fibre and surrounded by a complicated boundary structure called the sarcoplasmic reticulum. The proteins are mainly, though by no means exclusively, actin and myosin, arranged so that they can slide backwards and forwards against one another to provide the contraction and expansion of the muscle. At the same time, the regions in the fibril where these proteins overlap alternate with regions where they do not, making the familiar striated or banded appearance which is seen under the microscope or electron microscope. Surrounding and interspersing all these structures is a fluid called the sarcoplasm, the meat juice. Readers should refer to standard textbooks of meat science (e.g. Lawrie, 1991) for detailed descriptions of these structures.

Contraction of the live muscle comes about from biochemical changes in which the energy source, glycogen, is consumed producing energy which is transferred via the ATP–ADP system to cause myosin molecules to slide alongside the actin molecules so as to shorten the structure. The biochemistry and energetics of this process will also be found in appropriate texts.

MYOSIN

One of the major constituent proteins of the contractile mechanism of muscles, myosin is extracted from lean meat by salt solutions of moderate ionic strength, corresponding to 0.5 to 1.5 M NaCl or about 3 to 8% salt. It may be precipitated from solution by diluting the salt concentration, when it takes the form of fibrils spontaneously. The molecule consists of a long chain of 'light meromyosin' (LMM), molecular weight about 150 000, and a head of 'heavy meromyosin' (HMM), molecular weight about 400 000. The LMM tail provides the main structural element. The HMM carries an ATP-ase system which obtains the energy for contraction from muscle glycogen and also carries the mechanism which moves the whole molecule along neighbouring actin molecules and locks it in position when the movement is complete.

When lean meat is treated with salt and water some of the myosin may go into solution. If the solution is visible the technologist calls it 'exudate' or 'extract'. This myosin extract gels on heating, binding pieces of meat together and to other components in a meat product (see MASSAGING AND TUMBLING).

ACTIN

Actin is the major constituent of the thin myofilament, part of the contractile mechanism of muscles. The molecular weight is estimated to be about 43 000. It binds one molecule each of nucleotide (ATP or ADP) and divalent cation (calcium or magnesium). An unusual amino acid, N-methylhistidine, is found in actin as well as myosin, and may be used for the analytical determination of these muscle proteins.

COMPOSITION OF MEAT

A carcass consists of three kinds of tissue: muscular tissue (lean), fatty tissue and bone. The percentage of each varies with the overall fatness of the carcass. A lean beef carcass containing 20% of fatty tissue can be expected to contain 64% of muscular tissue and 15% of bone, whereas a fat one containing 40% fatty tissue would only contain 49% of muscular tissue and 10% of bone.

Muscular tissue

A muscle consists of

(i)   The main structural proteins, which also form the contractile mechanism of the living muscle, i.e. the system of actin, myosin, etc. (10.0%)
(ii) A light tubing or netting made of connective tissue, in which the structural proteins are encased (2.0%)
(iii) Muscle juice surrounding and permeating all the above, containing:
      Water (75.0%)
      Soluble protein (sarcoplasmic) (6.0%)
      Other solubles – myoglobin, salts, vitamins, etc. (3.5%)
(iv) About 3% of fat, finely dispersed in the above, with sinews, nerves, blood vessels, etc. (3.5%)

The moisture content of this material, from which all extraneous connective tissue and fat has been removed, is close to 77% on a fat-free basis and constant, as is the water : protein ratio of 4.8, allowing these values to be taken as constants for calculation of the muscle contents of mixtures when the protein contents are known. In real meat, of course, the situation is complicated by the presence of the variable amount of fat and connective tissue which accompany the muscles in different parts of the carcass or different cuts of meat. These are counted as meat under the definitions accepted in English law. In the UK, the lean meat content of a product such as a sausage is calculated from the protein content or the nitrogen content using the average factors for lean meat agreed by the Society for Analytical Chemistry, which take this variability into account. These factors include 3.50% nitrogen in lean pork and 3.65% in lean beef. To the lean meat content estimated in this way is added the fat content as determined analytically, to give the total meat content (see also under MEAT CONTENT, page 37).

Fatty tissue

Fatty tissue or adipose tissue, generally termed fat, is composed of round or polygonal cells of connective tissue in which the true fat (or lipid, consisting mainly of triglycerides) is stored. It has been shown that beef carcasses with about 20% of fatty tissue have about 70% fat in the fatty tissues while carcasses with 40% fatty tissue have about 86% fat in the fatty tissues. Some fat occurs in lean meat, not as gross fatty tissue nor as the minor amounts within the muscle structure (intramuscular fat) but as small deposits of intermuscular fat, also in cellular form. This is the fat known as 'marbling' which has long been considered a mark of high quality in meat for the table. That view was strongly contested by some scientific opinion a few years ago but recent work and current thinking appear to support the traditional view. There are similar quantitative relationships here: the more fatty the carcass the more marbling fat may be expected. A lean beef carcass with 20% fat may have 4.5% total intra- and intermuscular fat whereas a carcass with 40% fatty tissue may have 8.5%.

Connective tissue

Connective tissue contains the proteins collagen and elastin. When the collagen in the connective tissues of the meat of young animals such as veal or young chickens is hydrolysed during moist cookery it can be observed to form jelly in the gravies produced. In meat from older animals, however, the collagen is tough, relatively inelastic and not readily hydrolysed on moist cookery. Unhydrolysed collagen, with elastin, forms the well known gristle and gives rise to toughness in the meat. Different degrees of toughness in different cuts from the same animal are directly

related to differences in the amounts of collagen present. For instance, the psoas major (underfillet) of a beef animal contains about 0.27% collagen while the semitendinosus (eye of the round), a characteristically tough muscle, contains about 0.74% collagen. Muscles in the neck may contain over 1% of collagen. The decrease in solubility of the collagen with increasing chronological age of the animal is related to an increase in the degree of cross-linking among the collagen molecules which renders them both more stiff and more resistant to the action of a solvent such as hot water during cooking. Collagen contains about 12.5% hydroxyproline, an amino acid which is almost completely absent from muscle. Connective tissue can therefore be estimated analytically by measuring the hydroxyproline content. This is done colorimetrically after prolonged hydrolysis under standard conditions.

Proportions of lean and fat

The proportion of lean in a piece of meat can only be measured accurately by chemical analysis which may be slow and is usually impractical for large pieces. In commercial dealings recourse is most often made to an estimate of the 'visual lean' percentage by an experienced assessor. Such a subjective measurement is, of course, open to dispute in critical cases. The process can be somewhat refined in some cases, for instance, with boxed boneless frozen meat where the contents of several boxed may be sawn across, the cut surfaces examined under a plastic graticule and the lean and fat contents estimated by counting 'red' and 'white' squares.

Machines are now in use which make an essentially similar estimation on minced or diced meat passing on a conveyer belt below an optical scanner coupled with image analysis; the fat content is calculated from the ratios of the different colours detected. Another machine uses ultraviolet light and measures the fluorescence produced characteristically by the fat; it is said to be free from errors caused by smearing of the fat on cutting.

In the pork and bacon industry it is now commonplace to measure the thickness of the carcass back fat using optical or sonic probes with appropriate calculators built in. Back fat thickness is regarded as a good index of the fattiness of the whole carcass and the measure is commonly included in purchase specifications. Sonic probes can, of course, be used on the live pig; back fat thickness may therefore be included as a parameter in breed development programmes.

MEAT TECHNOLOGY

PRIMAL CUTS

Some years ago the fresh or frozen beef in the wholesale trade was almost all in the form of whole carcasses, sides or quarters, bone-in. It is now commonplace to bone out the meat soon after slaughter and to supply it in the form of 'primal cuts' which are intermediate in size between the quarter and the retail joint. This gives considerable savings in packaging and transport costs by elimination of the bones, and if the meat is deboned 'hot', of refrigeration costs also. The 'primals' may be vacuum packed and supplied unfrozen, usually for the retail trade, or frozen in polythene lined boxes for manufacturing. Other meat species are now supplied as primal cuts, but not yet to the same extent as beef.

'HOT' MEAT PROCESSING

There are advantages in processing carcass meat 'hot' after slaughter, before rigor mortis is complete. Hot boning gives improved yield and more efficient cooling of the meat with savings in refrigeration costs because the bones do not need to be cooled. Drip and evaporative losses are reduced. For the boners, warm meat is pleasanter to work with than cold meat. If meat for manufacturing can be immediately treated with salt and ground, as in large scale sausage or burger production, there are great improvements in water holding capacity compared with conventional cold meat processing.

COMMINUTION METHODS

As previously discussed, the high content of connective tissue in certain cuts of meat which makes them tough and less palatable has resulted in the manufacture of a wide range of comminuted meat products from the simplest, e.g. burgers, to the more complex emulsion-based sausages such as the cooked slicing varieties.

In order to produce these the meats need to be comminuted, or broken down, cutting the connective tissues into sizes more easily chewed. Some forms of comminution also result in the extraction of proteins and emulsification of fats.

Mincing

Mincers vary in size but not much in design. They consist of a worm feed screw that pushes pieces of

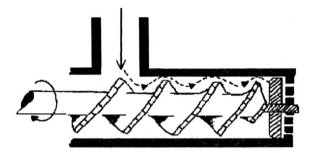

Figure 1.4 Cross-section of a mincer.

meat past flat blades attached to the end of the worm screw and press against a flat perforated plate such that as the meat passes across the knives and through the perforations of the plate it is cut by a scissor-like action (Figure 1.4).

The worm screw action and cutting method can result in significant tearing and cutting of fat cells and it is better to mince fresh rather than frozen meat and to have sharp plates and blades. In practice most factories will use tempered meats, especially where products such as free-flow mince are produced.

'Comitrol' shear cutting

This style of cutting is a much cleaner cutting method for lean meat and connective tissues. It consists of plates in the form of a ring made up from segmented knife-like blades (Figure 1.5, D), the segment gaps determining the size of the cut pieces. Rotating knife blades (Figure 1.5, B) similar to those of a mincer cause a scissor action against the ring plate (Figure 1.5, A). It operates at a much faster speed than a mincer but cuts more cleanly forming wedge-like slivers of meat. The fat cells suffer and are mostly all cut, especially when the plate segments are small (Figure 1.5, C). It works best with tempered meats.

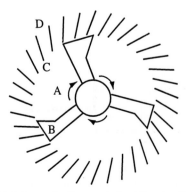

Figure 1.5 Schematic diagram of a 'Comitrol'.

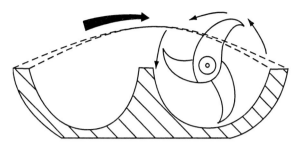

Figure 1.6 Diagram of a bowl cutter.

Bowl cutting

This method uses a bowl with a doughnut shaped profile into which a series of arc shaped blades are set mounted on a drive shaft such that they cut within the arc of the bowl against its edge. The speed of rotation of the bowl and knives can be varied to control the degree of comminution. Blades should be carefully set and sharp or excessive damage to fat cells will be caused and/or heat generated. There are models fitted with heating jackets and vacuum lids for producing pâtés etc. (Figure 1.6).

Milling

The meat mill works as a high speed cutter and pump producing very fine cutting from high speed rotating knives. The material that it cuts is usually premixed, minced and the milling process is designed to produce the final emulsion. This equipment is used to produce pâtés and emulsion-type sausages or 'brats' (Figure 1.7).

FILLING AND FORMING

Burger and shape formers

These can be anything in size from a simple hand operated press to large scale sophisticated machines. The principle is basically the same whatever the scale, the prepared meat mix is pressed to a desired shape. Pressing causes binding of the meat pieces. In large machines either gravity feed from a hopper or a worm screw drives the meat mix into a chamber where a pneumatic piston forces it into a die plate in the desired shape. The plate is then slid forward and the shaped meat knocked out.

Filling machines

Butchers and small manufacturers use piston-type fillers that extrude sausage mix though a nozzle into a casing (Figure 1.8). Larger manufacturers use vane pump machines which take sausage mix from a gravity fed hopper and by a series of vane plates set it on to an offset cam wheel which pumps it through the nozzle (Figure 1.9). Both types of machine can be fitted with automatic twist linking devices that can be indexed to a pulsed fill producing sausages of uniform weight.

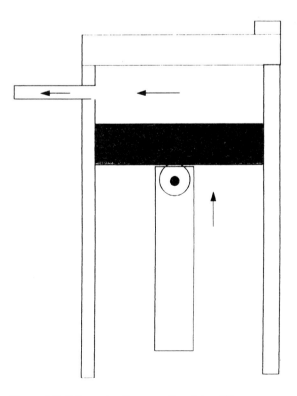

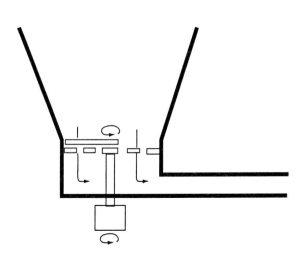

Figure 1.7 Cross-section of a meat mill.

Figure 1.8 Schematic diagram of a piston filler.

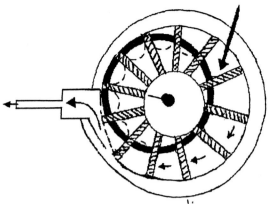

Figure 1.9 Schematic diagram of a vane pump filler.

BURGERS

In Britain this has become the generic name for raw meat patties with high content of lean meat. Originally the name derives from the German Hamburger sausage, made of beef and commonly cut into thick slices before consumption, but in English usage any other meat may be substituted for the 'Ham' part of the German city. So we may have lamburgers, porkburgers, baconburgers and so on.

In North America a hamburger is made with 100% beef, with about 20% fat content. In Britain this product is also common, especially in American-style fast food outlets. Other burgers are made containing 90% or 80% beef, the balance being made up with rusk and water.

The meat is prepared by mincing or flaking. A little salt may be added with the other ingredients if required, and the whole mixed under carefully controlled conditions: the degree of comminution and mixing largely controls the cohesive properties and the eating quality of the finished burger. The mixture is then fed to a patty former or burger press and either pressed or extruded into the required shape. In some of these machines with a sliding die plate there is a tendency to orient the fibres of the meat in one direction, which may lead to differential shrinkage on cooking and consequent distortion of the patties. This may be overcome, in the case of patties intended to be circular, by forming them into a slightly elliptical shape in the first place.

Maintenance of the desirable bright red colour of beef burgers sometimes causes problems, especially with frozen burgers. The cause of the trouble is frequently traceable either to defective microbiological quality of the meat or other ingredients or to oxidative rancidity in the fat which promotes browning of the meat pigments.

There is growing interest in the development of low fat burgers, particularly in the fast food area, for which burgers containing 10% fat or less have been developed. Carrageenan has been used along with hydrolysed vegetable protein to improve the palatability and flavour characteristics. It is likely that with the development of fat mimetics a wider range of low fat burgers and other meat products will become available in the future.

RE-FORMED MEAT

As generally understood, a re-formed meat has similar properties of colour, flavour and texture to those of good quality meat but is made from less expensive material. Usually this less expensive material consists of trimmings or other small pieces of otherwise satisfactory meat, re-formed or restructured into larger, more regular portions. The manufacturing processes include :

(i) Pieces of meat are bound into large units, for instance 'picnics' and related ham-like products, poultry rolls and some beef roasts. the technology used is based on massaging the meat followed by heat setting on cooking in appropriate moulds.

(ii) Meat is cut into regular flakes, for instance frozen meat tempered to −4 to −6°C may be flaked in a 'Comitrol' machine. This is an effective way of reducing connective tissues to a state in which they can be eaten, after cooking, without the sensations of gristliness associated with large pieces. Meat of moderate connective tissue content may thus be rendered more palatable. The flakes are allowed to stand for several hours with occasional stirring, then compacted under pressure into the desired shapes. Under good conditions a product made from forequarter beef will have a texture intermediate between a hamburger and a good steak.

(iii) Meat is intensely comminuted with salt, water and some phosphate to give a flowable paste which is extruded and heat set. It may be extruded in the form of fibres which can be formed into meat-like structures, perhaps with other fibrous materials to give interesting variations in texture. There are a number of patents in this area but it is not known how widely they may be applied.

In Britain, as a result of a legal test case, even if a re-formed steak were made entirely from best steak meat, it could not be called steak because it is no longer in the form that a consumer expects of a true steak. It should be called re-formed steak or something similar, or some neutral name such as just 'meat'.

MECHANICALLY RECOVERED MEAT

Mechanically recovered meat (MRM), also known as MDM (mechanically deboned meat) and in the USA as mechanically separated (species to be stated) meat (MSM), is the residual meat removed from bones which have already been more or less well trimmed by knife. Machines force the softer meat under pressure through perforated screens (e.g. Paoli, Beehive, Bibun machines) or through channels formed in other ways (as in the Protecon machine).

The material consists of the meat and fat which were on the bones, finely comminuted by passage through the machine. In some cases marrow from within the bones is included, in others the bones remain almost undamaged and little or no marrow is extracted. The connective tissue content is low. The content of bone fragments depends on the action of the machine and on the yield of MRM obtained, for instance later versions of 'bone press' machines like the Protecon produce few bone fragments. Obviously, greater pressure to extract meat increases the risks of making bone fragments. The EU limit for these is 0.05%, and is normally met without difficulty in commercial products.

Because of its finely divided condition and the nature of its origin, MRM is specially subject to microbial contamination and growth and extreme care is needed to keep it frozen until required and use it as rapidly as possible on thawing.

LOW FAT MEAT AND MEAT PRODUCTS

Breeding programmes in most of the developed countries since the early 1950s or so have been directed to the production of animals with lower proportions of fat than formerly, with significant effect in all the main edible species. This should be remembered when using tables of carcass composition, which if out of date, may be seriously misleading about modern meats.

The fat content of any meat product is a technical parameter which in practice can be varied over a wide range. It is not usually difficult to make acceptable products with different fat levels, informatively labelled, to suit the wishes of different groups of customers.

This subject is usefully discussed by Goutefongea and Dumont, from a French point of view,

in a chapter in the book *Reducing Fat in Meat Animals* (Wood and Fisher, 1990).

For consideration of fat substitutes see Chapter 8.

SAUSAGES

British-type fresh sausage

Fresh, uncooked, meat sausages are made in all countries. The British product is notable in that it is made with a meat content less than 100%, the balance being made up with a filler such as rusk and water. This gives the sausages a characteristic flavour and texture, preferred in this country to those of an all meat sausage. The sausages are made of various degrees of comminution or fineness, coarse cut sausages being commonly made by smaller producers and fine cut sausages by the larger ones. This may be partly at least because of difficulties in retaining the surface appearance of coarse cut sausages when these are filled using high speed equipment. Passage through such fillers tends to reduce the surface texture of the meat until it is like that of fine cut product.

In the UK pork sausages are required to contain not less than 65% meat, of which not less than half must be lean meat. Other sausages must contain not less than 50% meat, half of that being lean meat. 'Other' sausages include beef sausages and pork-and-beef sausages. These three varieties account for some 97% of current production in the UK. They are filled into various diameters of casings, between 20 and 30 mm. The majority are sold with casings but there is a significant manufacture of skinless sausages (see CASINGS).

The technology of comminution of these sausages is a matter of working the lean meat component with salt and added water so as to obtain the maximum meat binding and water holding properties, while treating the fat relatively briefly to avoid excessive damage to its cellular tissue structure. To the extent that fat can be kept separate from the lean and added to the mixture later, the comminution is fairly easy to manage, but with 'semi-lean' meat in which fat and lean are present together excessive damage to the fat cells will occur which may be difficult for the lean meat proteins to bind. Additional proteinaceous material such as soya isolate, caseinate or various milk derivatives or blood plasma may be added to assist the binding properties of the mix in such cases. Ice is used to keep temperature low, but protein extraction works more efficiently at warmer temperatures. For the British sausage this is less of a problem as it is comparatively open textured and chopped for a relatively short time.

Emulsion sausages

The range of such products is very wide. The common characteristic is that the meat mixture is finely chopped with added water (ice) and salt. Much of the fat therefore is liberated from the fatty tissue but remains bound or emulsified by the lean meat mixture. The result is a homogeneous paste which gels on heating to a firm sliceable mass which in a well made frankfurter will break with an audible 'snap'. The emulsion stability is important not only for the 'snap' of the frankfurter and to render the sliceability of other larger diameter varieties but to prevent cook out. Other materials may be added to the emulsion before the sausage is finished, for instance cubes of meat or fat, to give interesting appearances and textures. The meat may be cured by the addition of curing salts, or not, may be smoked or not, and the sausages may be filled into casings of a wide range of diameters. Permutations of all these variables lead to the wide range of sausage types to be found in countries such as Germany. Note how a basic emulsion component is common to them all. In manufacturing it is not unusual for a factory to make a single emulsion mixture (known in Germany as a brat) to go into a wide range of products. While this brat is mild in taste the main flavours of the various products made from it come from the other added meats which would have been preseasoned and cured with the seasoning typical of the product being made. There are as many different seasonings as there are product types with their own variations.

The durability of emulsion sausages depends upon what other treatments are given in the manufacture and varies widely. Some are fresh products with a shelf life of days or hours, some are cured and others which are canned have a long shelf life.

Dried fermented sausages

Salami and similar products are designed by a combination of factors to have a long shelf life at ambient temperatures. The main factors are:

(i) Nitrite is present from a very early stage in the manufacture, providing conservation and protection against pathogens. Traditionally, this nitrite is provided by a microbial fermentation of nitrate added to the sausage mix; a bacterial ferment containing micrococci and sometimes certain staphylococci is nowadays usually added to ensure that the fermentation takes place properly. The presence of nitrite gives the final product its typical deep red colour.

(ii) The presence of lactobacilli in the added bacterial culture ensures production of lactic acid, which may be further guaranteed by the inclusion of sugars in the sausage recipe. As acid is produced the pH of the sausage falls, reaching a value of around 4.5 in 1 to 2 weeks in a warm humid environment. In some varieties such as pepperoni the time is shorter and some manufacturers use accelerators such as glucona-d-lactone. These products are not required to have a long shelf life as they are used for pizzas and it is the final texture which is the desired characteristic, the flavours coming from the seasonings.

(iii) The above double fermentation is carried out for the first few days at a temperature of around 25°C, and later at a lower temperature of 15°C but with reduced humidity, so that the mixture loses moisture. The production process is usually considered to be finished when some 50% of the original weight has been lost. One corollary of this is that the conditions of chopping the original sausage mixture must be chosen to ensure that the material will lose water, not retain it, so chopping times are short, meats are tempered or frozen to ensure clean cut surfaces with no fat smearing which would inhibit the transfer of water out of the lean meat. Initial salt contents are low and phosphate is not needed. The percentage of salt increases proportionally as the sausage dries. The rate of drying must be carefully controlled to prevent the outside of the sausage drying out too rapidly creating a case-hardened surface which would inhibit the drying of the centre and hence cause spoilage. Humidity control is therefore important throughout the drying stages and is obviously more critical for the larger diameter sausages which have a much longer drying period.

Fillers for sausages

Rusk

The majority of manufacturers purchase proprietary brands of rusk. Rusk is made from wheat flour, chemically raised, baked and coarsely ground. The granules are supplied in specified ranges of particle size. The rusk absorbs about three to four times its weight of water when soaked. It may be added dry to a product during manufacture together with the requisite amount of water, or it may be presoaked before use. If this is done it should be noted that there is always a significant rise in temperature on hydration of the rusk and care must be taken not to hold quantities of wet rusk in bulk without arranging to cool them down.

Binders

Various other materials have good water holding or meat binding properties and have for a long time found application in sausage manufacture for these reasons. They are, however, more expensive than the cereal fillers mentioned above and should be used only when their functional properties are required. This is most likely to arise with cheap products of relatively low lean meat content. Since lean meat is the most important provider of water holding and meat binding qualities in a meat product, it follows that when the proportion of lean meat is low it may be advantageous to supplement it with some other food material capable of providing these properties. Such binders include dried or frozen whole egg or egg yolk, blood plasma, skimmed milk powder, caseinates and certain whey protein isolates, soya isolates, soya concentrates, soya flours and wheat gluten. Any of these may be used by direct addition to the product at the mixing stage.

CASINGS

Natural casings

Natural casings are processed from various parts of the alimentary tract of cattle, hogs or sheep. The types and amounts available are approximately:

	'Rounds' or 'runners' (small intestine)		'Middles' and 'bunds' (large intestine)	
	Length (mm)	Diameter (mm)	Length (mm)	Diameter (mm)
Cattle	36–40	36–46	9–12	45–60
Sheep	22–47	18–26	5–6	*
Pig	17–19	32–42	4–5	40–45

*Not usually used for casings, but sheep stomachs are used as casings for haggis.

They may be used as casings for sausages of various kinds, from chipolatas to large bologna types, according to diameter. About 1% of the sausage weight consists of casing.

Casings are normally packed in salt. Before use they should be thoroughly soaked and washed in lukewarm water at about 29 to 32°C. Too high a temperature is liable to ruin the texture of the casings which may burst on subsequent filling. After washing the casings should be kept wet at all times until they are filled.

The production and use of natural casings have diminished greatly in Western countries, as they have tended to be replaced with the more hygienic and more controllable artificial varieties. Some smaller manufacturers still prefer to use them.

Artificial casings

The advantages of artificial casings over natural casings lie in their uniformity of size, the absence of risk of contamination from improper preparation and the ability in most cases to be used without preliminary washing, soaking, etc. They are of several types.

Cellulose casings

These are made from cotton fibre or wood pulp, chemically dissolved, regenerated and extruded in the form of continuous tubing. The smaller diameters are widely used in the manufacture of skinless sausages. The raw sausage mix is filled into the cellulose casing in the usual way, then the surface of the sausage is cooked by scalding to a temperature of about 80°C (175°F), which produces a firm coagulated layer below the casing. The casing is cut lengthwise and peeled off mechanically, leaving the sausage skinless. For scalded sausages of which the frankfurter is typical, the scalding time is made long enough to cook the sausage right through, or the cooking is completed in a later stage, for instance if the product is canned.

Large diameter casings usually require soaking in water before use. They are used for a variety of bologna and large sausage types, which are sold with the casing intact although it is removed before consumption.

PVDC (Polyvinyl dichloride)

Casings made from PVDC provide an impermeable casing with good stretch for tight filling and reasonable appearance, although they wrinkle to some extent after cooking. They are available in a range of colours.

Fibrous cellulose casings

These are cellulose casings containing cellulose fibre for additional strength and are used for slicing sausage. They require soaking before use. They are porous and will allow the penetration of smoke. They attach well to meat and some are made with an easy peel coating on the inside. Their presentation is poor but they have the ability to shrink with the product during cooking.

PVDC/fibrous casings

These have the characteristics of both materials, good meat adhesion from the fibrous inner layer and enhanced presentation from the PVDC, however they are non-porous and do not take smoke. They are the most widely used casing for delicatessen presentation.

Collagen casings

These are made by a solution and chemical regeneration process similar in principle to the cellulose process but the starting material is animal collagen, usually in the form of cattle hides. They may be coloured for special purposes.

Reconstituted collagen casings have now largely replaced natural casings for most ordinary sausage manufacture in medium to large sized establishments because of their convenience in use and consistency of performance. Their behaviour in other respects is similar to that of the natural casing.

EMULSIFYING CAPACITY

The emulsifying capacity of a protein, a meat suspension or other emulsifying agent is the volume of oil which can be made into an emulsion by a solution or suspension of the protein, etc., under specified practical conditions. The best known experimental test is the Swift test. In the Swift apparatus, melted lard is added from a graduated separating funnel into a blender containing the meat suspension, protein solution or other emulsifier, with the blender running to give a high speed cutting/mixing action. An oil-in-water emulsion is formed which becomes increasingly more viscous with further addition of fat until the viscosity suddenly decreases when the emulsion 'breaks'. This is taken as the end point from which the emulsifying capacity can be calculated. Various modifica-

tions have been made to this basic method, notably in the application of electrical resistance methods to detect the end point and in the use of liquid vegetable oil instead of melted lard, but the principle remains unchanged. However, even with modifications and with great care in carrying out the experiments it is often difficult to obtain reproducible results.

Significant differences can be demonstrated in the emulsifying capacities of different manufacturing meats. However, attempts to relate these to differences in performance of the meats under manufacturing conditions have not met with conspicuous success. It is now clear that in many meat products the retention of fat is governed less by emulsification than by other factors related to the mode of breakdown of the cellular structure of the fatty tissue.

WATER HOLDING CAPACITY

The water holding capacity (WHC) of a piece of meat may be defined as its capacity to retain its own water content during such operations as cutting, pressing or heating; water binding capacity (WBC) refers to the same property in respect of added water; however the two terms are commonly used interchangeably.

It is well known that these properties vary among samples of meat and many aspects of this variation are still obscure. It has long been recognised that water is held in the muscle fibres mainly by chemical forces, especially those related to the charge state of the myofibrillar proteins and strongly influenced by the pH of the meat, but recently there has been a resurgence of the view that the forces are largely capillary in nature and related more to the spatial arrangement of the fibres than to their electrostatic charges directly. We still do not know.

WHC may be measured by a number of methods. The best known laboratory method is the Grau–Hamm press, in which a specimen of ground meat is pressed against a filter paper and the area of paper which is wetted is an inverse measure of WHC. Alternatively, some relevant technological property such as drip loss, cooking loss or weight of water absorbed by the meat may be measured directly under standardized conditions. However, although there are often fairly good statistical correlations among the results of these tests, they are almost never good enough to permit a single test to be used as a reliable predictor of the result of another test or important property.

POLYPHOSPHATES

All inorganic salts have some influence on the water binding and related properties of meat, which is similar in principle to the influence of sodium chloride. The size of the effect can be calculated in each case from the ionic strength of the salt in solution. However certain phosphates in the presence of common salt have considerably more effect than the calculations from ionic strength would predict. The phosphates with the greatest effects are the pyrophosphates and tripolyphosphates. The long chain polyphosphates such as 'Calgon' act rather slowly and orthophosphates are without any extra effect.

$(PO_4)^{3-}$	orthophosphate
$(P_2O_4)^{4-}$	pyrophosphate
$(P_3O_{10})^{5-}$	tripolyphosphate
$P_nO_{3n+1})^{(n+2)-}$	long chain polyphosphate

The polyphosphates act as catalysts to the 'salt' effect of sodium chloride, increasing its influence on water binding, cooking losses, meat binding, fat binding and texture. Strictly speaking, they only accelerate these effects but by permitting the results to be obtained in shorter times the practical outcome is as if the effects were greater. About 0.3% of effective phosphate is normally sufficient, though more is sometimes allowed to compensate for any unevenness of distribution in the meat. However at concentrations much above this a bitter taste becomes noticeable.

The explanation of the phosphate action is not yet clear. It is unlikely to be entirely due to the alkaline pH values of the phosphates in solution, for the pH of the meat is not greatly changed by added phosphate and similar effects on water binding properties are not observed among meats whose natural pH values differ in the same way. It has been suggested that the polyphosphates have an ATP-like effect on the actomyosin system, making it more readily soluble in salt, but this view is disputed.

PRESERVING MEATS

PREPACKING

Several distinct sets of conditions can be distinguished.

(i) *Fresh meat.* Butchers' meat is now commonly prepacked for retail display. The primary consideration is that the red colour and bloom of the meat should be retained. This requires generous access of oxygen to the meat surface (see COLOUR). The packaging film used must therefore have high permeability to oxygen as well as good appearance, sealability, and so on, and low permeability to water vapour to avoid weight loss. Coated cellulose films are made specially for this purpose and special grades of softer films such as PVC or polythene are also satisfactory.

(ii) *Vacuum packed fresh meat.* The removal of oxygen from a pack of fresh meat ensures a much longer preservation against microbial deterioration compared with packing with access to oxygen, but the colour of the meat becomes darker and purplish. On opening the pack, oxygen becomes available at the surface of the meat again and the colour reverts to the cherry red of oxymyoglobin. The purplish colour is not usually acceptable to retail customers, but the wholesale trade accepts it well and this form of packaging, in sealed heavy duty impermeable bags (e.g. Saran, Cryovac) is used for transport of boneless or bone-in primal cuts.

(iii) *Modified atmosphere packaging.* There are microbiological advantages if fresh meat is packed in sealed bags with an atmosphere initially rich in carbon dioxide. The carbon dioxide suppresses the growth of spoilage bacteria and thus extends the shelf life. However, the greater the carbon dioxide concentration the smaller is the concentration of oxygen and if the CO_2 level is too high then the colour will suffer. About 20% CO_2 mixed with air may be a useful compromise, particularly for products which also will be held at chill temperatures.

(iv) *Vacuum packed cured meats.* Here there is no problem with colour, in fact the absence of oxygen is beneficial to the preservation of the cured meat colour, cooked or uncooked. Retail and wholesale packs of bacon, ham, etc. are therefore prepared in impermeable transparent bags of various appropriate sizes, evacuated and sealed. The packs should be protected as much as possible from exposure to light, which is detrimental to the colour.

(v) *Frozen packs.* Frozen packs are widely supplied at both wholesale and retail, for all kinds of meat. With uncured meats shelf life is long provided correct temperatures are maintained and the packaging remains undamaged to avoid freezer burn. There are no special problems with colour. With cured meats, however, there may be more or less serious problems with rancidity. With frozen bacon, for instance, the product can become rancid in less than 6 months unless the vacuum packaging is absolutely perfect, with all oxygen excluded from all the packs. In such cases frozen storage can only be a short term expedient (see FREEZING AND FROZEN MEAT).

CHILLING AND COOLING

It is always important that warm food which is not intended for immediate consumption should be cooled down as rapidly as possible. When an animal carcass has been slaughtered and dressed its remaining body heat should next be removed in an efficient chiller. Some of the complications which may arise if this chilling is carried out excessively rapidly are discussed under AGEING, but too rapid chilling only occurs in exceptional circumstances – the more usual problem is that the chilling is not done soon enough, rapidly enough or thoroughly enough and that the microbiological condition of the meat suffers accordingly.

Meat products which have been cooked or otherwise processed in a warm state also need to be cooled as rapidly as possible. Sausages which may emerge warm from the chopping operation, and which tend to remain warm during filling and packing, may be passed thereafter through a continuous chilling unit to bring them to a centre temperature in the region 0–4°C (32–39°F). With freshly baked pies efficient chilling is even more important, partly because the temperature after baking is higher and it may therefore take longer for the available heat to be removed, partly because of the high probability of mould growth inside the pie or within the wrapping of the pie if conditions in those confined spaces remain warm and moist. Pies may be allowed to cool initially in the atmosphere, to bring their temperature down to near the ambient temperature, but the cooling should where possible be completed in a chiller using cold, filtered air. Alternatively, vacuum cooling units may be used. These are batch chambers into which racks of warm pies may be put, the door closed and a vacuum drawn. The vacuum causes the evaporation of some of the moisture in the pies and the loss of latent heat by evaporation reduces the temperature very rapidly. The process is efficient but the capital cost is high. The loss in weight resulting from evaporation can be allowed for in the original formulation of the pies.

It should be noted that the specifications for the performance and design of chilled stores and frozen stores usually make the assumption that the goods going into store have been properly chilled beforehand. If this is not done an extra burden will be placed on the refrigeration equipment for which it was not designed, and the overall performance of the store will be adversely affected.

FREEZING AND FROZEN MEAT

The subject of freezing is fully dealt with in Chapter 15; the purpose of this section is to draw attention to certain aspects which concern meat and meat products particularly.

Freezing processes

Any of the usual freezing processes may be used for meat or meat products. Blast freezing is the commonest for carcasses or cuts of meat, usually in bags or boxes lined with polythene, for boxes of various products or for more or less fluid material frozen in moulds and removed later as solid blocks (e.g. mechanically recovered meat). The air blast may be at -30 to $-40°C$ (-20 to $-40°F$) and the freezing time should be such as to freeze the material to a sufficient extent that the final equilibrated temperature will be close to $-20°C$ ($0°F$). Plate freezers are commonly used for rectangular blocks or boxes of products up to about 5 cm thick.

Liquid nitrogen vaporizes at $-196°C$ ($-320°F$) and therefore freezes the surface of a product very rapidly: with larger objects such as turkeys this crust may crack when the interior of the meat expands when it becomes frozen later. Liquid nitrogen is relatively expensive and nitrogen freezing is therefore normally used only for high value products.

Freezing the water in any sample of meat goes through three stages:

(i)   cooling the water to 0°C, from, say, body temperature, 37°C; this requires the removal of 155 kJ kg^{-1},
(ii) freezing the water without any change in temperature from 0°C; this requires the removal of 335 kJ kg^{-1},
(iii) cooling the ice to cold store temperature, say $-20°C$; this requires the removal of 42 kJ kg^{-1}.

Note that the true freezing stage, the removal of latent heat with no change in temperature, requires considerably more refrigeration than the other two stages which in this example brought the temperature down by 57°C. This point is of considerable significance for cold store operation, as is discussed below.

Frozen storage

To call a frozen store a 'freezer', as people often do, is a misnomer which may lead to difficulty because of the latent heat effect described above. A cold store is normally designed to keep the ambient air in the store at the required temperature, assuming that the contents are already at that temperature. It is not designed to freeze unfrozen material. If unfrozen material is put in the store, the refrigeration required to freeze it may well be beyond the capacity of the equipment of the store, which will be overloaded and unable to perform even its proper work of keeping the ambient cold.

Meat is usually held in frozen storage at −18 to −20°C (0°F). Some storage is commercially available at −30°C (−20°F) but the additional cost of this is not usually worthwhile. If properly packaged and handled, with consistent temperature control, the storage life of frozen unprocessed meat may be 1 to 2 years or more; for cured meats or comminuted meat products especially those with high fat content the storage life is shorter. Deterioration usually results from rancidity in the fat. Meats with more unsaturated fats, such as pork and poultry, are therefore the first to be affected.

Microbiological aspects

Microbiological activity effectively ceases at temperatures of −10°C (+14°F) and below, except that some moulds grow slowly at about −10°C. This is the cause of the condition known as black spot which was troublesome in meat cold stores up the mid-1960s when storage temperatures were not usually below −11 to −12°C (+10 to +12°F).

A small proportion of micro-organisms on the meat are killed by the low temperature of storage but the great majority survive and are capable of growth when the meat is eventually thawed out. The microbial quality of frozen stored meat is therefore not significantly better than that of the original meat at the moment when it was frozen.

Rancidity of fat

The oxidative chemical changes which lead to the rancid flavours of fats continue during frozen storage. This is apparently because although the rates of the chemical reactions are diminished at low temperature, those reactants which are dissolved in the water phase become highly concentrated as ice is formed and the proportion of liquid water becomes small. Also the increase in concentration has a greater accelerating effect than the slowing down caused by the lower temperature.

In fact, some liquid water is present in the tissues at all temperatures down to −30°C (−22°F). Rancidity due to this cause can be particularly troublesome in stored pork fats. In addition, the oxidative reactions are autocatalysed by oxidative changes in the haem pigments of the meat, so that rancidity in the fat and browning of the colour of the lean meat go together. This is sometimes observed in frozen stored burgers, where the fine comminution of the product ensures close contact between fat particles and the lean meat juices.

Freezer burn

In a frozen store with the doors closed the atmosphere becomes very dry because of condensation of the humidity in the air on to the coldest parts of the cooling system. Any exposed product in the store is therefore likely to suffer surface desiccation. Freezer burn is the name given to white patches on the surfaces of affected material. In the earlier stages the change may be more or less reversible when the surface becomes wetted again on thawing, but if the drying out continues the meat proteins become denatured, hard and leathery and quite unacceptable. The condition cannot then be cured. Its prevention requires the use of sound packaging impervious to water vapour for all but the shortest terms of storage. Polythene of a suitable gauge is usually satisfactory provided the pack is sealed, clipped or reasonably tightly closed against the passage of water vapour.

Of course, a further reason for close attention to the vapour-proof packaging of goods in store is that the loss of water can represent a significant loss of weight. Even from the cheapest of stored materials this potential loss of weight will usually justify the cost of simple polythene packaging.

Thawing

Thawing is more difficult to manage than freezing, for several reasons:

(i)   In thawing by heat transfer from a warmer medium such as water in ambient air, heat must be conducted inwards through an outer thawed layer whose thermal conductivity is lower than that of the ice within and which becomes thicker as thawing progresses; in freezing the conditions are the reverse, with heat conducted outwards through an outer layer of ice with higher thermal conductivity. Thawing is therefore bound to take longer than freezing even if the relevant temperature differences (between the centre of the meat and the ambient) could be the same.

(ii) Microbial growth is inhibited immediately the surface freezes, whereas in thawing at ambient temperatures, growth can occur at the surfaces throughout the whole of the thawing period. To minimize this, the surface temperature should be kept low, which is in conflict with the need for higher temperature differences to speed up the process (see (i) above).

(iii) Water is a better heat transfer medium than air, so thawing is faster in water. But air at the same temperature permits microbial growth less easily than water. Running water may be used to remove the microbial contamination but can be wastefully expensive.

Practical thawing conditions must usually be a compromise among these conflicting requirements. Prolonged thawing at a relatively low air temperature is the commonest solution, e.g. overnight at 5–10°C (40–48°F) for pieces up to 100–150 mm thick.

Thawing by microwave or dielectric heating avoids most of the problems. The processes work well on units of uniform size and shape such as blocks or boxes of frozen meat, but are not suited to irregular pieces such as whole or part carcasses.

Tempering

For many manufacturing processes it is necessary to temper deep frozen meat before use by bringing it to a uniform temperature suitable for cutting. Meat which is too warm is not firm enough to cut cleanly and straight; if it is too cold clean cutting may also be difficult and cutting tools become excessively blunted.

Bacon for slicing in high speed slicers is commonly tempered to a temperature close to −2°C (28°F); the temperature may be varied a little according to the softness (i.e. the melting point) of the bacon fat. Beef for hamburgers may be tempered in the range −3 to −8°C (27 to 18°F). The lower the temperature, generally speaking, the better the quality of the final product but the greater the wear on cutting equipment; variations of only one or two degrees may affect either of these considerably.

Because the final temperature to be achieved is so critical, tempering requires even more careful control of the operating conditions than thawing. The equipment and methods used are similar in principle to those for thawing; microwave tempering is favoured in appropriate cases because of its possibilities for fine control.

Detection of thawed frozen meat

When meat is frozen there is little disruption of the cellular structures, in contrast to, for instance, soft fruits, where the texture can be almost completely destroyed. However there is some disruption of the mitochondria, which are cells located in the muscle structure and rich in enzymes. The enzyme β-hydroxyl-CoA hydrogenase (NADH) is liberated on freezing and its presence free in the meat juices can be used to indicate whether a sample meat, presented ostensibly as chilled meat, has in fact previously been frozen. A small sample of juice is expressed and a colorimetric NADH test is performed; by use of a spectrophotometer the estimation may be made quantitative.

SALT

It can be argued that after the meat itself, salt is the most important single ingredient in meat product manufacture. It has a strong preservative action and a long history of use for this reason. In meat products a concentration of 4% salt in the water phase (a brine concentration of 4%) gives adequate conservation at refrigerator temperatures of products such as bacon; greater salt concentrations give longer storage or storage at higher temperatures. Perhaps equally important are the technological effects of salt in providing the water- and fat-binding properties, and the meat-to-meat binding of meat products. The greatest effects here are to be found at salt concentrations of 3–8% in the meat. However, the distinctive flavour of salt, though acceptable at low concentrations, is less so at values much over 4% in the water (say 3% in lean meat and 1.5 to 2% in a product containing much fat). The salt content of most products has for many years been tending to fall in response to this feature of consumer preference. This has meant an increased need for care and consistency in manufacturing practice as the safety margin between technologically effective concentrations and those acceptable for flavour has decreased.

The salt used in manufacturing is usually vacuum-dried crystalline salt of high chemical and microbiological purity. Rock salt was used in former days but the quality of this is poorer. Its content of calcium and other salts was the cause of 'white foots', a fine white precipitate in glass-packed tongues.

'Low salt' products can be made. Potassium chloride may be substituted for sodium chloride, within limits. First, the chemical and antimicrobial effects of given weight of the potassium salt are only 80% of those of common salt (the molecular weights are in the ratio 78.5/95.5); second the flavour of potassium chloride is much less agreeable.

CURING

The word curing means 'saving' or 'preserving' and food curing processes include sun drying, smoking and dry salting. All these methods have been used, or are still in use, to preserve meat. For meat products nowadays, however, the term 'cured' is usually taken to mean preserved using salt and nitrite.

When nitrites are present with meat there are changes in the meat colour due to the formation of nitrosyl pigments (see COLOUR) and the terms 'curing chemistry' or 'curing reactions' are often used to refer to the chemical changes occurring (see also NITRITES AND NITROSAMINES).

The use of salt for curing is ancient. Its main actions are:

(i) Preservative – at concentrations over about 4% in the water phase most of the ordinary meat spoilage micro-organisms are inhibited. This effect is obviously enhanced if the product is also dried, when reductions in moisture content lead to corresponding increases in salt concentration.

(ii) Flavour – in former times it was common to use high salt concentrations in cured meats and to remove the excess from bacon, etc. by prolonged soaking in water before cooking. In present day products the salt content is not usually as high as this, being limited by consumer preference for convenience as well as for milder flavours.

The main actions of nitrite are:

(i) Preservative – in addition to the preservative action of the salt, nitrite further restricts the range of organisms which are able to grow. Micrococci and lactobacilli are particularly selected: these will eventually spoil the cured meat but much more slowly than the normal fresh meat flora. In addition, when nitrite is heated in the presence of protein the so-called 'Perigo' effect occurs, causing an additional strong inhibition of the spores of *Clostridium botulinum*. This ensures the safety of canned cured meats such as ham even when these are given only relatively mild heat treatment during manufacture.

(ii) Colour formation – this is very characteristic of cured meats. For details see COLOUR.

(iii) Flavour – nitrite contributes to a minor extent to the characteristic flavour of cured meats.

Other factors contribute significantly to the shelf life of cured meats and may be critically important in products of marginal composition, where the salt content may have been restricted in response to consumer demand and the nitrite level may be kept low because of concern about nitrosamines. These factors include the hygiene of the manufacturing process, the maintenance of low temperatures during distribution and sale (by means of an effective cold chain) and vacuum packaging. Smoking the product, where this is appropriate, is also helpful. See also BACON, HAM.

Massaging and tumbling

This technology originated in Europe in the 1960s. The first production experiments were done using machines such as butter churns or cement mixers, but more suitable devices were soon introduced. A wide range of tumblers and massagers are now commercially available (Figures 1.10 and 1.11).

If pieces of meat containing injected salt brine,

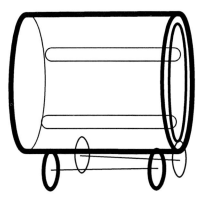

Figure 1.10 Meat tumbler.

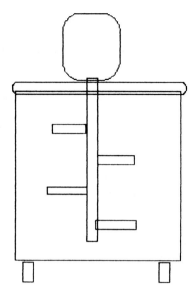

Figure 1.11 Meat massager.

or surrounded by added brine, are worked so that they rub against one another and against the surfaces of the containing vessel, penetration of salt into the meat tissues is accelerated. As the action continues there is some extraction of salt-soluble protein, mainly myosin, from the body of the meat to the surface and into any surrounding fluid. Two results follow:

(i)   The water holding capacity of the meat, which is improved by the interaction of salt with the structural proteins of the lean meat, is increased as a result of the improved penetration of salt.

(ii) The extracted salt-soluble protein forms a cement which not only renders the tumbled meat sticky, but also sets on cooking to bind the separate pieces of meat into an intact, sliceable whole. The presence of polyphosphates in the salt brine assists both of these effects.

The way is therefore open to make cooked cured meat products containing higher proportions of added curing brine than would be possible otherwise, and to make formed or re-formed products of standard shapes from irregular or small pieces of meat. Acceptable products are being made by these means, but they have also raised questions concerning the legal and commercial control of product composition.

The difference between tumbling and massaging is in the severity of the mechanical action. Generally, a long, gentle massaging action, for instance in machines with slow moving smooth paddles, is preferred to the vigorous action of the tumbler, where, for instance, pieces of meat are dropped through several metres inside a rotating drum. The effects are obtained more quickly in a tumbler but with much loss of texture in the meat due to structural damage.

Bacon

Bacon is cured pork. In the UK and in Europe generally it may be made from any part of the pig, but in North America the term usually refers specifically to cured pork bellies.

Wiltshire curing or tank curing

The original Wiltshire method of curing bacon has undergone many minor modifications but remains relatively unchanged in principle.

After slaughter and evisceration the pigs' heads are removed and the carcasses divided into halves. These are then chilled before curing. The curing process takes place in a curing cellar which is kept at a temperature $<5°C$ (42°F). First the sides are pumped with a brine or 'pickle', by injection under pressure through a hollow needle connected to a reservoir of pickle. The sites of injection are carefully controlled to ensure uniform distribution, as far as possible, throughout the meat. Where any bones have been removed in the preparation of the side, for example, the shoulder blade is usually removed, the pocket which remains in the meat and which is a probable source of infection, is stuffed with dry salt as an additional aid to the keeping quality of the final product. Next the sides are carefully stacked into large concrete or tiled tanks, immersed in curing brine and wedged with wooden beams to prevent floating. Extra salt may be sprinkled on the sides so as to keep the salt concentration in the brine high despite the diluting effect of meat juices seeping out from the pork. This immersion stage usually lasts 4 to 7 days. Finally the sides are removed from the tanks and stacked on the floor, skin side upwards, to drain, mature and, it is believed, to equilibrate further in composition. Traditionally, this stage might last a week but in recent times it has been shortened to only 2 or 3 days.

A typical Wiltshire brine as used nowadays in the curing tanks contains salt 24 to 25% by weight (a saturated solution contains, in theory, 26.3% salt at 8–20°C), sodium nitrate or saltpetre 0.5% and sodium nitrite 0.1%. There is also a high concentration of soluble proteinaceous material derived from pork previously cured in the same brine, so that a mature brine is a deep red colour. It has long been recognized that the proper management of this brine, to keep it in good condition, is essential for producing a satisfactory product. Before the chemical and microbiological principles were understood this could be achieved in practice only by strictly following the procedures laid down by generations of curers, but now it is possible to control the process more rationally. The essential features are those which control the microbial flora that convert nitrates into nitrites and that suppress the growth of other micro-organisms detrimental to the product. These factors are first, that the brine concentration must be maintained at or close to saturation. This enables a population of micrococci and lactobacilli to become established. These organisms reduce nitrate to nitrite in the course of their ordinary metabolism. The high salt concentration also assists in the suppression of other micro-organisms that could produce off-flavours or accelerate the later spoilage of the bacon. The maintenance of low temperatures is also important here. Second, it is essential that there should be a continuous supply of nitrate for reduction and as part of the food supply of the bacteria. This dependence of the whole process on the establishment and maintenance of the right microbial population throws light upon the old assertions that a curing brine should never be destroyed if it can possibly be salvaged, and that a new brine always performs much better if it is seeded by the addition of a portion of a good quality old brine.

With recognition of the nature of the microbiological activity of Wiltshire brines has come also appreciation of the need to monitor and control nitrite contents, so that the routine measurement of salt, nitrate and nitrite contents forms the basis of all bacon factory quality control testing. This in turn has led to new and simpler curing processes in which nitrite is added to the brine directly without any need to rely on microbiological processes to produce it from nitrate. These newer processes are discussed below (BLOCK CURING).

The third stage, the process of maturation, was previously considered to be the most important for the production of good bacon but was not well understood. During maturation the salt, nitrate, nitrite and any sugar and other ingredients present were believed to continue to diffuse through the meat, so that the composition became more uniform. It never becomes completely uniform however and all samples of bacon show differences in composition from point to point. During the maturation stage the characteristic bacon flavour does become more fully developed.

Dry curing

Instead of the tanking stage described above, the product may be made, after preliminary injection, by adding the requisite curing salts, not in the form of brine but in the dry state. Originally in this process the sides were sprinkled liberally with salt and saltpetre, built up in stacks of about eight sides, skin side downwards, and stored for about 10 days. If this operation was carried out in or near to a cellar in which 'live' brines from a traditional Wiltshire process were in use, similar microbial flora would become established in the dry cured bacon, with consequent production of nitrite in the bacon. Since no brine was added at this stage in the production, the moisture content of the product was lower and this was also helpful in ensuring good conservation. Furthermore, since the process was rather difficult to control, there was a tendency to use plenty of salt to ensure a good margin of safety, so that the product tended to be rather salty but for that reason it also had a good shelf life.

Machines are now available to provide a more controllable, more uniform distribution of the dry curing salts over the meat, but they are not in widespread use.

Block curing and other rapid curing methods

Two further developments have greatly speeded up the process of making bacon. The practice has grown of cutting the pork into blocks smaller than a whole side, for instance into backs or loins, bellies, fores or shoulders, etc. These cuts may also be more or less completely boned out before curing, in contrast to the Wiltshire method in which very few of the bones are removed until after the bacon is made. By these means the quantity of material to be handled is reduced by the elimination of bones and often by the diversion of some of the cuts of pork to products other than bacon. There are practical gains in convenience from working with smaller and more manageable pieces of meat.

The second development is the continuous automatic multi-needle injection machine. Applied to boneless pieces of meat this has not only greatly speeded up the preliminary injection stage of Wiltshire-type curing processes but because the injections can be made easily in many places, close together in the meat, it has become possible to achieve reasonable uniformity of composition in the product without need for the long tanking and maturing stages previously required. The brines used in these cures are almost all fresh brines made up as required, containing the necessary proportions of nitrite and not relying on microbial action to produce nitrite from nitrate. Indeed, nitrate is often omitted completely and there are good grounds to consider that unless curing is taking place in premises where live brines were previously in use and the appropriate nitrite-producing micro-organisms are present in the environment, it is better that nitrate should not be used.

Brines used in rapid curing processes typically contain 24 to 25% salt by weight, as in a Wiltshire brine. With, say, 10% weight uptake in the injection process this will give about 2.5% salt in the bacon. Similarly, 0.2% sodium nitrite in the brine will give bacon theoretically containing 200 mg kg^{-1}, in practice about 100 to 150 mg kg^{-1}.

The simplest version of these principles was probably 'slice curing', operated commercially for some years around 1970. This gave bacon of good uniform composition by dipping individual pork slices in a brine of appropriate composition. Satisfactory product was obtained in a few hours, less time than was required to distribute and sell it in pre-packed form. Unfortunately the handling of individual slices detracted from the neatly shingled appearance that customers expected in prepacked bacon and the process was later abandoned for that reason.

A method allowing curing to develop in the already prepacked product is now used with block cured bacon by a process known as 'cure-in-the-bag'. Injected blocks of meat are packed into strong plastic bags, sealed to prevent any loss or contamination. The curing reactions continue within the package, which again may be distributed to commercial outlets during the same time. Prepacked bacon blocks can be thus delivered to commercial outlets in the just-cured but fully cured condition, with significant savings in storage times and costs.

Sweet cures

Any of the cures described above may be modified by the addition of sugar, honey or other sweet substances. The sweetness modifies the sharp taste of the salt and is liked by consumers for that reason. If significant amounts of reducing sugars are present, as in honey or liquid glucose, brown colours and strong flavours will be produced by Maillard reactions during cooking and are favoured by some consumers. Reducing sugars also favour the formation of the red cured meat colours in the uncooked bacon and are of minor use for that reason. A content of 0.25% sugar in the final product is typical.

Ham

A 'ham' is the upper leg and buttock of a pig; ham is the cured meat product made from this. In the UK and North America it is commonly cooked but this is not essential and raw ham, shelf stable without needing to be cooked, is readily available in other countries. Ham may consist of pieces of cured ham meat formed into other shapes before cooking, but if the product is made from meat from other parts of the pig it may not be called ham.

A 'gammon' is exactly the same part of the pig, but cut from a cured side of bacon. It may then be cooked in the same way as ham, or sliced and treated as bacon.

Cured pork meat easily disintegrates when cooked and it is necessary to control the cooking of hams carefully to maintain the meat texture while giving sufficient heat to ensure shelf stability and safety from pathogenic spores. The presence of nitrite in the curing salts in the product permits the use of milder heat processes than are required for uncured meat products.

For detail of the manufacture and heat processing of canned hams, see CANNED MEATS.

The stability of uncooked hams referred to above depends on a relatively high salt-on-water content: such hams may be dry cured or allowed to dry out during the curing process. For cooked or pasteurized hams a high salt on water is not so necessary for stability or safety and products with higher moisture content can be made. There is evidence of consumer demand for milder flavoured, moister products, which may also presumably be cheaper. In most countries now there are regulations which either prescribe exactly which products may be made and sold, or lay down the names under which products of different moisture content must be described and other information which must be declared.

Picnic hams, shoulders, and so on

These are cooked cured pork products in which more of fewer small pieces of meat are formed into a solid, integral, sliceable whole by the use of massaging before cooking, then cooking after packing tightly into moulds or casings. The term 'ham' may only be used of products consisting entirely of pork leg meat.

CANNED MEATS

Meat and meat products have been canned since the beginning of that industry and nowadays are made in most countries where there are canneries. In recent years the largest quantities have been made in a few countries manufacturing especially for export – notably Denmark, Holland and Poland in Europe, Argentina and Brazil in South America.

Corned beef

Corned beef was originally a by-product of the process for making meat extract (q.v.). The boiled meat from which the extract is made is salvaged by mixing it with 'corns' or grains of saltpetre (sodium nitrate) and canning it. Sufficient nitrite is present in the saltpetre, either as impurity or as a result of adventitious microbial contamination, to produce a strong cured red colour in the product.

In modern processes, boned and trimmed meat is ground through 2–3 cm plates or cut into strips by rotary disc cutters, cooked for about 20 min in boiling water, then salt (about 1.5%) and sodium nitrite dissolved in a little water are added. The mixture is filled into cans in the usual way, closed under vacuum and processed for 2¼–2½ h at 115.5°C (240°F). Cooling with chlorinated water follows. The importance of scrupulous hygiene at this stage was demonstrated in the 1950s and 1960s when food poisoning outbreaks were traced to the use of untreated river water for can cooling in canneries in Argentina.

Despite the abstraction of much soluble nitrogenous material from the meat in the preliminary cooking process, the finished product has a high nitrogen content since water is also lost in cooking. If protein is calculated in the usual way from the nitrogen content, the apparent meat content is found to be 120% or more for normal corned beef of the traditional type. The fat content is fairly low, usually below 20%.

In recent years new products have been made which attempt to copy the earlier product without copying the process by which it was made. Such products are now made in various countries from unextracted beef, containing more or less fat and sometimes with more or less added water. Phosphates may be used together with fine chopping in order to retain the added water. Nitrite is added to obtain the cured colour as previously. Such corned beef products inevitably have apparent meat contents below 100% estimated from the nitrogen content, and their

texture is finer than that of the traditional South American version.

Corned meats other than beef are made by either of the above processes. Corned mutton is made in India and to some extent in Australia.

Ham

There are two distinct types of canned hams, a pasteurized product which must be kept under refrigeration and is frequently packed in large pear-shaped cans or long rectangular cans, and the commercially sterile product which is non-perishable at room temperature storage and is generally a piece of ham with a net weight of less than 908 g (2 lb). Because the latter shrinks considerably during the heat treatment necessary to produce a commercially sterile product, it generally has not as great a consumer acceptance as the larger pasteurized product.

Curing is carried out either by artery pumping or multiple injection (stitch pumping) by a machine with multiple needles; the latter method is becoming common practice for large scale production. After pumping, the hams are usually held in a cover pickle for 2 to 5 days at a temperature of around 3°C (38°F). The composition of the curing pickle for ham is similar to that described for tongue and it is usual to include polyphosphate materials to decrease shrinkage, the level of addition being probably about 5% in the pickle. When filling the cans, a little dry gelatine is usually added and, as with tongues, the larger cans are usually pressed in order to obtain a complete fill. They should than be vacuum closed. For the pasteurized product, e.g. as packed in long square-section 'Pullman' cans, it is necessary to heat process until an internal centre temperature of at least 66°C (150°F) is reached. Some packers find it preferable to use a two-stage process at temperatures of about 52°C (125°F) followed by 71°C (160°F). During retailing this product must be kept under refrigeration and should be labelled accordingly. The sterile product requires steam pressure processing in the normal manner and somewhere around 2–3 h at 110–115.5°C (230 to 240°C) will be required according to size. Rapid cooling of both types of product is absolutely essential and sometimes chilled cooling water is used. With the pasteurized product, cooling should be continued until the internal temperature is not greater than 21°C (70°F).

For small cans of product intended to be shelf stable at ambient temperature, the processing recommendations of Hauschild (see LUNCHEON MEAT below) should be followed:

where salt-on-water = 3.3% ensure F_0 = 0.3–0.5
$\qquad\qquad\qquad$ = 4.0% $\qquad\qquad$ = 0.1–0.2

For an explanation of F_0 values, see Chapter 15.

Luncheon meat

Canned luncheon meat consists of chopped cured meat with added seasonings and sometimes cereal. The most usual product is prepared from pork meat and marketed as pork luncheon meat or chopped cured pork, but it may be prepared from beef, veal or lamb meat. Various cuts of meat are used in preparing luncheon meat depending upon the type and quality of product. Butts, shanks and loins may be included and it is also usual to add a proportion of rinds, head meat and cured trimmings from the preparation of ham. The fat content may be up to 30%. A proportion of good quality frozen meat may be used. Individual formulae and methods of preparation will vary but the general basic procedure is as follows. The lean meat is chopped in a silent (bowl) cutter and seasoning, water and curing salts (also flour or cereal if included) are added, followed by the fat in strips. Cutting is continued until the required particle size is achieved. This procedure may be carried out in a vacuum chopper, but if it is not the next step is vacuum mixing at a high vacuum of some 26–28 inches of mercury (0.87–0.93 atm) for about 5 to 10 min. Efficient vacuum mixing is important to achieve the right texture and fill and to exclude air, but overlong mixing will cause the texture to become rubbery.

Curing salts consist of salt and sodium nitrite as the essential ingredients. Nitrate is now not recommended: the evidence is that not only is it unnecessary but it has a detrimental effect in subsequent can corrosion and product discoloration. In some circumstances it may also be microbiologically detrimental. Polyphosphates are almost always added for their beneficial effects in binding fat and water and minimizing separation, although they also accelerate corrosion (see CURING).

Small cans of luncheon meat, up to say 500 g, are shelf stable at ambient temperatures provided they are properly heat processed. The heat process is considerably less than the botulinum cook or F_0 = 3 which is essential for all other low acid canned foods, because in products of this type stability and safety from *Clostridium botulinum* are achieved by

(i) some destruction of spores by the mild heat process and

(ii) inhibition of the remaining spores, preventing them from germinating or forming toxin, by the salt, nitrite and Perigo-type factors (see NITRITES AND NITROSAMINES) in the product.

The minimum heat process required for any given product is therefore closely dependent on the concentrations of salt and the input of nitrite. As yet, no clear theoretical basis for calculation of the heat process has been worked out because of the complexity of the interactions among these and other factors, but there are guidelines based on good commercial practice and the proven good safety record of the products made thereby. The most recent guidelines, summarizing research and commercial practice in several important producing countries, are from Professor A.H.W. Hauschild of the Codex Alimentarius Commission (1994):

If input sodium nitrite is 150 mg kg^{-1}
and initial mesophilic spores are not more than 5×10^3 g^{-1},
then for 3.0–4.0% salt-on-water apply 1.0 to 1.5 F_0.
For 4.0–4.5% salt-on-water apply 1.0 F_0
and for 5.0–5.5% salt-on-water apply 0.5 F_0.

After vacuum mixing the next step is filling, which may be carried out semi-automatically using a sausage-type stuffer fitted with a horn shaped to the dimensions of the can, although fully automatic fillers are, of course, available. The cans should be closed under vacuum, heat processed as above and then cooled in water. Air pressure during cooling is advisable for rectangular cans, to assist them to return to the correct shape.

Rectangular cans weighing 7 and 12 oz (200 g and 340 g) used to be the most popular for retail packs, with 3 lb (1370 g) round cans common for catering purposes; because of the relatively high cost of manufacture of rectangular cans there is an increasing trend towards the use of round cans for retail packs also.

Minced beef in gravy

As well as the major constituent, minced beef, this product usually contains flour, tomato purée, salt, hydrolysed protein, caramel, monosodium glutamate and spices. The meat is boiled for a few minutes before adding salt, tomato purée, hydrolysed protein, caramel and monosodium glutamate, followed by the flour, previously mixed to a smooth paste with a little water. After further boiling, the spices are finally added just before the end of cooking and at this stage onions, if included, are also added. These may be reconstituted from the dehydrated form by soaking overnight. The whole is well mixed and usually filled hot without further exhaust, followed by processing for about 90 min at 115.5°C (240°F) and water cooling.

Poultry and game

Although there is no large scale production of this class of canned product, it is on the increase and some diversification of the types of pack has taken place. Also, some canners have a useful export business in high quality products. Game is usually packed whole, sometimes in a wine flavoured gravy. Part of chicken production is also in the whole form or as halves packed in oval cans. Other chicken products are boneless pieces in jelly or aspic (this is usually as breasts or fillets and is sometimes also packed in jars), boneless pieces in sauces, e.g. chicken supreme, curried chicken – sometimes with mushrooms, chicken fricassée, chicken à La King, and so on. Usually several of these products would be canned to use up the various portions of the bird and it is also likely that this type of production would be accompanied by preparation of chicken soup or spreads.

In the canning of whole poultry, after plucking and drawing and thoroughly cleaning, the bird should either be blanched for a few minutes in boiling water or, to obtain a brown roast appearance, be oven roasted for a short time or dipped into nut oil at a temperature of about 132°C (270°F) for about 2 min followed by draining and plunging into boiling water for 1 min. The birds are filled into cans by hand, topping up with a gelatin or agar brine and a little caramel may be added to give a slightly golden tint. After exhausting and closing, the cans are heat sterilized and, to avoid undue softening, the minimum safe processes are normally employed. Because of this it is necessary in order to avoid spoilage to ensure that the fill-in weight is carefully controlled and that the birds do not fit into the cans so tightly as to prevent circulation of the brine during processing. For the same reason, there must be passage through the body of the bird and it is important to ensure during filling that the body cavity does not become sealed against the can ends. Whole poultry cannot usually be stuffed before canning, except for small birds and with a carefully formulated process.

For the preparation of boneless chicken in jelly and other portion packs, precooking is necessary after the preliminary preparation. This may be carried out in boiling water or in steam-heated pressure vessels until the meat is sufficiently tender for it to be removed easily from the bone. The usual practice today, however, is to pack using preboned frozen chicken meat in blocks as the raw material. When packing in jelly this may be prepared from the filtered precooking water with the addition of salt, spices and gelatin or agar. The subsequent canning procedure will depend on the type of product but will entail the usual steps of filling, exhausting, closing, heat sterilization and cooling. Any gelatin used should be free from sulphur dioxide residues which would otherwise cause discoloration, as would contact with metals such as copper or iron during preparation.

Cans for poultry packs are usually fully lacquered but sometimes the product appearance in cans with plain bodies is preferred. For most products round cans are used but halves may be in specially designed flat, oval type cans and breasts or fillets are sometimes in small, oval cans.

Puddings and pies

Steak and kidney puddings are popular in 8 oz (227 g) and 1 lb (454 g) flat or basin-shaped cans, and pies in the 8 oz or 1 lb tapered pie cans. In each case the filling consists of appropriately chopped beef cubes and ox kidney pieces in a gravy of flour, salt, hydrolysed protein, monosodium glutamate, caramel and spices. The pastries used naturally vary considerably. Pastry for pies requires careful formulation according to the type of crust required. In filling it is usual to raise the pastry in the cans first, followed by filling the meat and gravy and sealing with the top section of pastry prior to exhausting and can closing. Some pies are formulated with top pastry only. Processes need careful evaluation since heat penetration through pastry is slow.

Stewed steak

This pack is often all meat, except for salt and possibly curing salts, or it may be canned as stewed steak in gravy. Beef sides are usually hung for about 24 h to give a firmer meat for boning out and trimming, with the separated materials being conveyed to different by-product departments for treatment. The boned-out meat is cut into appropriately sized pieces using disc cutters or a mincer, usually with 1–1¼ in (2–3 cm) plates. The neck meat, which is redder, is usually treated separately and divided amongst each batch. Each batch of meat is treated with salt

(0.75–1%), and the whole mixed in a paddle mixer for a few minutes. The meat is rough-filled into cans using vacuum fillers, which may be of simple construction although more sophisticated automatic fillers are available. The correct filling weight is normally made up and checked by hand and vacuum seaming is normal, although when the steak is in gravy this may be added hot, followed by steam flow closing. Heat processing for UT cans (73 mm diameter × 115 mm height) is 110 min at 115.5°C (240°F) or 80 min at 121°C (250°F) followed by water cooling.

Tongue

Cured beef tongues are canned as ox tongues and those of sheep and pigs as luncheon tongues. When received frozen they should be defrosted as slowly as is practicable to avoid excessive drip, and they must be thoroughly thawed prior to brining and curing which is usually carried out in two stages. First, the tongues receive a brine injection, in at least two places, followed by soaking in a brine of similar concentration for between 3 and 7 days. Composition of curing brines varies according to individual preference but the basic components are salt (around 20%) and sodium nitrite (see LUNCHEON MEAT, above). Other ingredients may be sugar, hydrolysed protein and monosodium glutamate, and some manufacturers continue to include potassium nitrate. After curing, the tongues are parboiled for 1 or 2 h which facilitates the skinning, rooting and trimming operations that follow. These are followed by filling into the cans whilst still hot, with the addition of a little gelatine, clinching, vacuum closing and processing. For large can sizes it may be necessary to compress the meat whilst in the can using a simple form of hand press. Care must be exercised in this case to avoid straining the can side seams. Heat processing times will depend on the can size and sometimes processing is carried out in two temperature stages when it is found that this gives a better product.

Internal can finishes for meat products

Because of the release of sulphur from protein during heat processing and because of the susceptibility to discoloration of certain products, a number of problems arise in specifying can finishes. For the straightforward non-cured or slightly cured semi-fluid packs, such as stewed steak, minced beef, Irish stew, etc., it is usually

sufficient to specify a can that is entirely coated with a lacquer that has reasonable resistance to sulphur staining. The same applies to products incorporating pastry. However, with products which retain their shape when removed from the cans, such as pork luncheon meat, corned beef, ham, etc., metal sulphide deposits may find their way onto the meat where this is adjacent to points of unavoidable metal exposure during can manufacture, particularly the body side seam. This type of corrosion is also aggravated by the inclusion of curing salts, particularly when nitrate is used in combination with polyphosphate additives. It is still traditional to pack corned beef into cans with plain bodies. Pork luncheon meat, hams, tongues, etc., are now packed in fully lacquered cans and because of side seam discoloration, the trend is to use cans with an internal side striped lacquer, applied after the side seam has been formed to cover this region. In the case of hams and tongues and other products where gelatine is added, it is important to ensure that this does not contain residual sulphur dioxide because this inevitably leads to black sulphide discoloration. These sulphide deposits are quite harmless but their appearance can be most unpleasant and is at all times undesirable. Another complicating factor is the difficulty in removing some solid meat products from the cans and lacquers are available incorporating an approved quick release or slip agent which facilitates product removal. It is important with all products, particularly solid types, to ensure that residual air in the cans at the time of closing is at a minimum by attention to filling, avoiding air pockets and exhausting, which is best achieved by vacuum closing.

INTERMEDIATE MOISTURE MEATS

Intermediate moisture foods have moisture contents in the approximate range 15 to 50%, have water activity (a_w) below 0.86, are shelf stable at ambient temperatures and can be eaten without need for rehydration. Some traditional foods, such as some dried sausages, fall within this definition but it is more usual to apply the term to newer formulations, especially those where the reduction in a_w is achieved by means other than drying off moisture.

Semi-moist pet foods have been made since the 1960s and are now widely available, and there has been some development for military purposes and for space travel, but there has not yet been a great deal of development for human food purposes.

The growth of pathogenic bacteria is effectively

stopped at a_w values below 0.86 and this is therefore taken as the upper limit of water activity for this class of foods. Some spoilage organisms, especially yeasts and moulds, grow at lower a_w's and may be controlled by other measures such as the use of fungistats, reduction in pH or heat pasteurization. Reduction in water activity is achieved by the addition to the food of humectants such as glycerol, sorbitol or other polyhydric alcohols, sodium or potassium chlorides or some organic acids (e.g. formic) or bases (e.g. urea). Even for animal foods, the number of suitable substances is not large and the fact that only a few of those are pleasantly edible by humans is a major impediment to large scale development of this type of food.

DRIED MEAT

In suitable climates strips of lean meat may be sun dried and are usually considered a delicacy. The product is known as Biltong in South Africa, Jerky or Charki in South America and Pemmican in Greenland.

For modern processes of dehydration, see Chapter 15, for the manufacture of dried sausages, see SAUSAGES.

COOKING

Meat and meat products are almost always cooked before consumption. When the temperature of any piece of meat reaches 65–72°C a number of changes take place.

(i) Microbial contamination is pasteurized, so that the meat is safer to eat. (Since 1991 the critical temperature has been considered to be 72°C, in order to control *Verotoxigenic E. coli* and *Listeria monocytogenes*. Of course, spores are not destroyed at these temperatures but they are not likely to be harmful if the food is eaten immediately after cooking.) Apparently greater exceptions are meats which are eaten after cooking to lower internal temperatures than this, such as underdone steak, where the microbiological condition at the centre may be presumed to be satisfactory even if not pasteurized, or a product such as the comminuted steak tartare which must be prepared rapidly and hygienically and consumed immediately to avoid the possibility of microbial growth, or products such as ham which are usually cooked at relatively low temperatures

but where the presence of nitrite in the curing salts ensures microbiological safety.

(ii) The colour changes from red to brown with uncured meats, from red to pink with cured meats, signalling that the meat is 'done'. The common kitchen test is to check the colour of the juice which runs from the centre when a skewer is inserted – it should not be red but colourless – but on this point, see the comment under COLOUR of cooked meat.

(iii) The 'binding' of sausages, rolls, picnic meats etc. is ensured, as the extracted protein on the meat surfaces gels.

(iv) Collagen begins to hydrolyse so that the meat is tenderized. For meat with little connective tissue, such as young poultry or good quality steak, brief cookery is sufficient but for the meat of older animals or the cuts with more connective tissue long cooking under moist conditions may be essential.

(v) The flavour of added herbs and spices is 'rounded' and blended into the mixture. Significant flavour changes occur at temperatures higher than 110°C (212°F). Of course, these cannot occur below the surface of the meat, where the temperature must remain at or below about 100°C for as long as any water is present.

(vi) Brown colours and strong pleasant flavours are developed at the surfaces of meat in contact with air at temperatures over about 150°C (300°F) – see COLOUR.

The widely different methods of cooking meat may be evaluated in terms of the extent to which they permit each of the above changes to take place.

SMOKING

Modern smoking processes are derived from the ancient practice of hanging meats in the chimney or fireplace to dry out. In addition to the preservative effects of the reduction in moisture content, the smoke has direct preservative and antioxidant actions and a strong flavour. The smoke of hardwoods is preferred. Present day applications include the use of special smoke generators burning sawdust or shavings and attached to cooking or combined cooking/drying cabinets. In this way the deposition of smoke can be controlled independently of cooking or drying processes and may be carried out hot or cold as required. Equipment also exists for purification of the smoke between generation and deposition, using water sprays or electrostatic precipitation to remove carcinogenic benzpyrenes and related tarry substances. Smoke concentrates and essences are also available, and in one modern application a concentrate in aerosol form is uniformly and efficiently deposited onto the product surface when the latter is electrostatically charged.

MEAT EXTRACT

This is the product obtained by extracting lean meat with boiling water and concentrating the liquid portion after removal of fat. Several portions of meat may be scalded in the same water until the soluble solids are brought up to a convenient concentration before commencing evaporation. Fifty or more parts of trimmed meat yield one part of extract containing about 18% moisture. Very little gelatin is extracted. Corned beef (q.v.) may be made from the residual cooked meat.

The first stage of concentration is in multiple effect evaporators, where the solids content of the extract is brought up to about 65% with little change in chemical composition. The final concentration is traditionally done in open steam-jacketed pans. Here the higher temperature and access of oxygen bring about important changes in composition. Insufficient heating gives a product with light colour, spongy texture and without a full meaty flavour. Excessive heat on the other hand leads to decomposition and unpleasant flavours. There is diminution of the creatine content and increase of creatinine, which may serve as an index of the heating undergone at this stage. On storage of the finished product further conversion of creatine to creatinine takes place slowly, about three years being necessary for alteration of about 75% of the creatine. In extracts with much residual moisture, clusters of creatinine crystals may be formed.

MEAT PRODUCTS

OFFALS

The quantities of different offals and animal by-products available are set out in Table 1.1.

In most countries the 'red' offals are fairly fully utilized for human food. The 'white' offals all find some application in human food, but in most cases the proportion actually used is small; the majority of the material is made into meat and bone meal. In the UK now the use of brains and

Table 1.1 Quantities of offals available

	Percentage of live weight		
	Cattle	Pigs	Sheep
Skin	6.9	–	17.0
Horns	0.09	–	1.3
Blood	2.2	3.0	4.1
Fats	5.8	1.4	5.3
Red offals			
Liver	1.3	2.9	1.0
Heart	0.41	0.3	0.5
Spleen	0.20	0.1	0.16
Kidneys	0.14	0.4	0.26
Brains	0.11	0.25	0.26
Thymus	0.05	0.16	0.19
Diaphragm	0.27	0.4	0.3
Tongue	0.65	0.4	0.3
Feet	2.0	2.1	–
Head	2.7	6.9	3.6
Tail	0.25	–	–
White offals			
Lungs	0.64	0.8	1.0
Oesophagus + trachea	0.27	0.35	0.58
Stomach(s)	2.34	0.5	0.5
Intestines	1.9	2.8	3.0
Udder	1.1	–	–
Sexual organs	0.06	0.32	0.26
Other glands	0.05	0.15	0.11
Stomach contents	17.0	13.92	9.05
TOTAL	46.43	34.23	50.4

(after S.P. Richards, Leatherhead Food RA Research Report No. 292, 1978)

spinal cord in human food is prohibited (see BSE in Chapter 19).

The large quantity of gut contents is notable. This is almost all thrown away and the handling of it poses severe hygienic problems. Similarly, the majority of the blood available is thrown away (see BLOOD below).

BLOOD

Blood sausages or black puddings are made in a number of countries. In most cases the quantities made are very small, only a tiny proportion of the available blood being used in this way. A certain amount of blood is carefully and hygienically collected at slaughterhouses for processing into blood plasma, to be used as a functional protein in meat products, particularly emulsion sausages (see REFORMED MEAT).

However, in at least all the countries of the EU the majority of the available blood is disposed of down the drain, with consequent loss of nutritious protein and increased burden on the waste disposal system.

PIES, PASTIES, SAUSAGE ROLLS, AND SO ON

Pastry and pie types

The usual kinds of pastry made with cold water (see Chapter 5) are used for products where strength in the pastry is not especially necessary. This includes pies to be eaten hot after cooking or reheating by the consumer, frequently packed in foil trays for strength and as a cooking aid, and also small goods such as sausage rolls and vol-au-vents. Puff pastry is used for all of these, short-crust pastry is often used for pie bottoms. Suet pastry is used for steak and kidney puddings and the like.

Where greater strength is required, the pastry may be produced by boiling the water in the recipe (hot paste) or the water and some or all of the fat (boiled paste) before addition to the flour. These pastes are used for cold eating pies such as the solid packed pork pie.

Pie making

The pastry is made in the usual ways (Chapter 5), with attention to the need for adequate relaxation after mechanical operations such as sheeting, cutting or forming pie bases, etc. There is usually a considerable production of scrap paste, for instance when circular tops are cut from a sheet of dough the unused dough round the edges is up to 25% of the total area of the sheet. The pastry recipes are normally robust enough to permit the incorporation of this worked material into subsequent batches without significant deterioration of final texture, but continual reintroduction of scrap paste (which itself contains scrap from earlier batches) causes a gradual reduction in microbiological quality. It is therefore necessary at intervals to dispose of the scrap in some other way so as to break the cycle of deterioration.

In the baking process there may be problems if the contents of the pie 'boil out' before the baking of the pastry is complete. Careful attention to the heating conditions, the geometry of the pie and the formulation of the filling is necessary.

After baking a jelly solution is added to solid pack pies (see GELATIN).

Baked pies must be cooled rapidly and efficiently to minimize microbial growth. Ambient air cooling may be used in the initial stages but a specially designed refrigerated air chiller is desirable. Vacuum cooling, where a rack of pies is taken into a chamber and a vacuum is drawn, cools the pies rapidly by the loss of latent heat as moisture is evaporated from them. Failure to cool properly at this stage leads to trouble later, especially in prepacked pies, when transfer of moisture within the pie or within the pack leads to condensation on later cooling, with fogging of packaging films and greatly increased dangers of mould growth.

Unbaked pies may be frozen without special difficulty, to be baked by the consumer. Baked pies may suffer in texture on freezing.

POULTRY PRODUCTS

Poultry products now available include a wide variety of chicken and turkey portions, sometimes precooked or partially prepared by coating with batter and breadcrumbs, and a range of products incorporating boneless poultry meat. These include pastes and spreads, canned products, sliceable chicken or turkey rolls and, in recent years, sausages and similar products. These last are usually made economically from turkey meat; since the turkey carcass carries relatively little fat, pork fat is used with it.

PÂTÉS, PASTES AND SPREADS

Spreadable meat products are made according to a wide range of recipes and presentations in almost every country. Their common features are:

(i) high meat content including relatively high fat, commonly 50% of the total meat; some pâtés have distinct layers of separated fat on top; meat pastes may contain non-meat ingredients particularly cereal but regulations in the UK, for instance, require a minimum 70% meat;
(ii) the meat is ground more or less finely depending on the texture required;
(iii) the meat is cooked before or during incorporation with the other ingredients; this ensures that no firm binding of meat to meat can take place (cf. picnic meats or most kinds of sausage, where binding is required); the paste is in contrast loosely bound by free semi-solid fat, does not flow but is easily displaced when a stress is applied on spreading.

These are precooked foods and unless they are fully heat processed in hermetically sealed containers (cans, jars or tubes) they must be stored and distributed under appropriate controlled chill conditions. Hygienic conditions during manufacture are essential. Contamination with *Listeria monocytogenes* is a real possibility which has been realized in commercial practice and was the reason for introducing the requirement in the UK (from 1995) for storage of pâtés (and soft cheeses) at temperatures not higher than 8°C.

GELATIN

Gelatin is formed by hydrolysis of collagen and is made from collagenous materials such as pigskin, cattle hides or bones. Meat from young animals, in which the collagen is not yet greatly cross-linked, yields gelatin easily under ordinary cooking conditions. Thus the cooked juices from veal, broiler chicken, etc., readily form jelly on cooling.

Dry gelatin may be added to canned hams and similar products to ensure that any liquid exuded from the meat on cooking sets to a firm sliceable jelly when the product is cold. The usual level of addition is 2%. Similarly, gelatin is usually added to cold eating pork pies to fill the space left when the meat filling shrinks on cooking. Here the gelatin is used as a 6% solution, injected hot into the pies immediately after baking. It is essential that care is always taken to keep the solution between 80 and 90°C (176 and 190°F): cooler than this and there is a risk of microbial growth, hotter than this the jelly strength is liable to deteriorate.

PET FOODS

The sale of prepared foods for domestic pets is a vast business. The UK production of canned cat and dog food is greater than any other canned product classification except for vegetables and about as great as the total canned vegetable production. It is mostly in the hands of a few large canners but there are a number of smaller ones with more local distribution. It is important to note that the law demands that premises for the preparation of animal food must be kept separate from those used for the preparation of food for human consumption.

Canned dog foods tend to be mixtures of meats or meat by-products with cereals, water and some material containing minerals and vitamins. Products prepared entirely from meat with nutritional additives, with meat visible in chunky form, are popular. Cat foods vary widely; some are very similar to dog foods and others contain principally fish and fish products with cereal, again usually incorporating vitamin carriers such as cod liver oil.

It is almost certainly impossible for any formulation to have universal appeal to either cats or dogs but it is essential to know that a prepared food in sufficient amounts will satisfy the growth requirements of the animals.

After mincing, where necessary, and mixing the various components, the product is heated and transferred by a screw conveyor to fillers which are usually of the rotary plunger type. Hot filling is normally practised and steam flow closure may be used as an additional exhaust when the product surface is sufficiently level for this to be beneficial. After closing, the cans are washed and processed as rapidly as possible to conditions which must be established by heat penetration experiments since a high sterilizing value is normally required for this type of product which may include material with relatively high bacterial counts. High temperature processing is common, around 127°C (260°F) for times which may vary up to 60 min or more according to can size and type of cooker. These products usually are excellent growth media for spoilage organisms. Rapid handling throughout the canning operation is essential and any 'gassy' meat or fish or contaminated cereals should be avoided. Line hygiene, especially on runways after processing, should be of a high standard.

Plain cans are usually used, with lacquered ends, although for certain products a fully lacquered container may be preferable. Two piece cans of the DWI (drawn wall ironed) type are always lacquered.

MEAT AND BONE MEAL

Meat and bone meal is the dried residue obtained from waste meat materials. Waste meat, bones, offals, etc. are usually collected from the meat factory or abattoir and transported to the rendering plant. In the older rendering process the material was ground, then treated under high temperature in a digester where the fat was melted and liberated from bones and fatty tissue. As much fat as possible was run off from the 'greaves' and the remainder was removed by centrifuging. This digester treatment has now been largely replaced by the more economical dry rendering process. Here the material is first completely dried in a closed retort. Most of the fat can then be run off, the remainder is centrifuged out and then the residue is ground.

Meat and bone meal is rich in calcium and phosphorus, with some residual fat and generally a high protein content. It is used in animal feeds, particularly for poultry. The nutritional value of the protein depends on the availability of lysine which can be reduced if the temperature conditions during rendering are too severe. Lower grades of meal may be used for fertilizer.

Prior to about 1980 in the UK, fat was removed from the greaves by solvent extraction, any residual solvent being removed with steam: the heat produced by that steam acted to give a final sterilization to the product. With the almost universal change to centrifuging, the sterilizing effect was lost and the subsequent production of meat and bone meal of lower sterility is now considered to have been a significant factor in the epidemic of bovine spongiform encephalopathy (BSE) in British cattle in the 1980s. Also highly significant was the then common practice of including meat-and-bone meal of bovine origin into cattle feeds, thus transmitting any surviving infection back to the host species: a ban on that practice has been imposed and appears to have been among the effective measures in controlling the disease (see BSE in Chapter 19).

EGGS

EGGS

The laying hen comes second only to the milch cow as a converter of feed protein into food protein. Despite the popular opinion that eggs from free range hens are better than those laid in batteries, practical tests have detected no differences in composition or functional properties.

Hens fed rations low in carotenoids may give pale yolked eggs, but this is easily remedied by feeding grass meal or synthetic carotenoids. The average overall percentage composition of the hen's egg is:

	Shell	White	Yolk	Whole egg
Proportion (%)	11	58	31	100
Water		88	48	74
Protein		11	17.5	13.5
Fat		0.2	32.5	11.5
Ash		0.8	2.0	1

The structure and main components of the egg are shown in Figure 1.12. The chalaza is a rope-like structure which permits the yolk to rotate and maintain a constant position within the egg.

The proteins in the egg white or albumen are:

Ovalbumin	56%
Conalbumin	14%
Ovomucin	12%
Globulins	18%, of which lysozyme 1%
	ovomucin 1.5%
	others 9.5%

The yolk proteins are mainly lecithin or ovolecithin, vitellin, phosvitin and livetin, present as phosphatides complexed with fat in lipoproteins. Cholesterol is associated with the lipoproteins, typically about 200 mg cholesterol per egg. The yellow pigments of the yolk are caroteboids, but have little vitamin A activity.

Candling of eggs

The interior of a shell egg may be examined when the egg is held against a bright light. The size of the air sac and the motility of the yolk can be

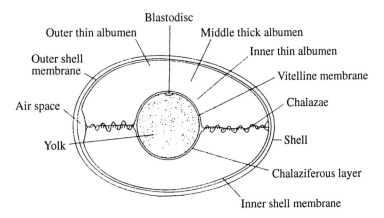

Figure 1.12 Structure of an egg.

seen, both of which can be correlated with the freshness of the egg. In commercial hatcheries it is usual to give the eggs a preliminary incubation for 7 days, then candle them to check their quality before transferring them to other incubators for the remainder of the incubation period. At this stage it is possible to identify infertile eggs by the absence of embryo development, and to reject them from the hatchery. The use of these 'incubator clears' for human food is now prohibited in the EU and many other countries.

Packing of eggs

Shell eggs in commercial practice are mechanically handled at all stages including candling (again) and the rejection of defects such as cracks, spots of blood or rot, size grading (by weight) and packing into plastic or paper cartons or moulded paper egg trays.

Storage of eggs

Eggs for storage should be clean, to ensure as low a microbiological load as possible, but should preferably not be washed. This is partly because of the difficulty of drying the surface thoroughly afterwards, partly because the pores in the shell absorb moisture and thus become permeable to bacterial infection from the environment. This means in practice that measures must be taken to ensure that eggs do not become dirty in the hen house, nesting boxes, batteries or elsewhere after laying.

Provided that the outside of the egg is kept dry, the contents remain stable for several weeks. The passage of micro-organisms through the dry pores in the shell is obstructed by the presence of inorganic material there. The optimum storage temperature is about 4–5°C; commercial packers usually aim to bring the temperature rapidly down to 10–12°C after packing at ambient temperature. EU regulations require that grade A eggs shall not be refrigerated at below 5°C. They may be transported to below 5°C provided it is for not more than 24 h. Lower temperatures than this may lead to condensation when the eggs are brought out of store, with possible mould growth. However under proper control, eggs may be deep chilled, e.g. in a blast chiller with air velocity of $1–2 \text{ m s}^{-1}$ and temperature of $-2°C$. Temperatures below $-3°C$ are undesirable as there is a possibility of freezing and shell shattering.

The proportion of thick white to thin white in the freshly laid egg is about 70–85%, varying among hens. Because of this high proportion, the egg 'stands high' when broken out. The proportion of thick white and the height of the broken egg diminish with time after laying: the height of the egg, measured with the Haugh instrument and expressed in Haugh units, gives a convenient estimate of the age of an egg.

Moisture is lost from the egg on ageing, with a corresponding increase in the size of the air cell. CO_2 is also lost, making the egg more alkaline: the pH rises typically from 7.6 in the fresh egg to 9.0–9.7 after a few days at room temperature. With the thinning of the white noted above, the yolk is no longer held central by the chalaza.

A 10% solution of water glass (sodium metasilicate) reduces moisture loss from the egg and a storage life of several months may be obtained; this process is now rarely used. The same effect may be obtained by oiling the shells; the process known as thermostabilizing consists of immersing the eggs in warm mineral oil for several minutes. Storage in closed containers, which reduces the loss of carbon dioxide, is also helpful.

Salmonella in eggs

Salmonellosis, infection with *S. pullorum*, is a well-known disease of poultry. For many years now it has been kept well under control in almost all countries and *S. pullorum* is not considered to be a public health problem. However, the sub-clinical infection of poultry with other salmonella species, and therefore the presence of symptomless carriers among poultry, especially chickens, is a major vector of human food poisoning worldwide.

The hen's egg is naturally constructed to provide a efficient barrier to the entry of anything harmful to the embryo, including microbes, and the inside of the egg as laid is therefore normally sterile.

Bacteria including Salmonellae are however transmitted down the poultry generations through the following sequence:

(i) salmonella is harmlessly present in the hen's intestines and caecum;
(ii) the outer egg shell surface is infected at the time of laying, surface infection is maintained through the time of incubation of the egg;
(iii) the chick hatches by pecking through its own shell, the digestive tract becomes infected with salmonella from the shell.

There are gradual changes in the serotypes of salmonella being carried by poultry flocks at any one time, as current serotypes die out slowly and new ones may be introduced from contaminated feed or other sources. These changes are commonly reflected in the serotypes identified from incidents of food poisoning.

Between 1987 and 1990 there was widespread infection with *Salmonella enteritidis* in the USA, Britain and elsewhere. The finding of an egg apparently infected on the inside with this organism led to considerable public alarm and concern for the safety of the UK egg supply. However, the finding does not appear to have been repeated, there was no epidemic of illness and the alarm has since diminished, though some level of public misgiving is likely to remain for a long period.

EGG PRODUCTS

Liquid egg products

Liquid whole eggs, and to a lesser extent separated egg yolks and whites, are important in the food manufacturing industries. Preparation consists of sorting, sometimes washing and disinfection, breaking and separation of the eggs either by hand or mechanically, individual inspection and rejection of any egg with a bad smell, screening to remove fragments of shell, followed by filtration, pasteurization and chilling or freezing.

Pasteurization is necessary to destroy pathogenic organisms, particularly Salmonella, and stipulated processes are laid down by government departments. For instance the USA requires pasteurization at 60–62°C for 3.5 min and the UK requires at 64°C for 2.5 min. With egg white there is a problem of coagulation and if the above temperatures are used the pH should be adjusted to 7.0 with lactic acid, with a protein stabilizer such as an aluminium salt plus triethyl citrate or triacetin. Alternatively it is possible to pasteurize at a lower temperature for a longer time. Pasteurized liquid whole egg has a shelf life of about 12 days at 2°C.

Pickled eggs

The eggs, in shell, are boiled for about 3 min until the yolks are just firm then cooled in running water as quickly as possible to minimize black discoloration (see below COLOUR). They may be shelled by hand whilst being cooled. The cooked shelled eggs are packed in jars with hot (over 70°C) spiced and salted vinegar, sealed and allowed to cool. Preservation is due to the low pH

and the direct effects of salt and acetic acid, as with vegetable pickles.

Frozen egg

Freezing has little adverse effect on egg white but the yolk proteins are coagulated, becoming thick, plastic and lumpy. Mixtures with sugar or salt (up to 10%) or liquid glucose may however be frozen satisfactorily.

The material may be blast frozen at −30 to −40°C in, for example, containers of 2 kg to give a frozen product with a shelf life of 2 years or more at −20°C.

Dried egg

It is essential for the liquid egg to be pasteurized before drying, as described above. The dried product is then tested for residual α-amylase content: this common enzyme has similar heat resistance to Salmonellae, so its absence indicates adequate pasteurization.

For details of the drying processes see Chapter 15.

QUALITY ASPECTS

COLOUR

Colour of fresh meat

Colour of lean meat

The colour of lean meat varies with the species, the age of the animal and with the cut of meat. These differences are related to differences in the concentration of the pigment myoglobin which is present in the muscles. Myoglobin is a large molecule consisting of a protein part, the globin, linked to a porphyrin ring structure containing an iron atom, which is the haem part. Its concentration is higher in beef than in pork, higher in old cow beef than in veal. There are also differences in the form in which the myoglobin is present. The bright red outer surface colour of fresh meat is due to oxymyoglobin, the oxygenated form which exists in the presence of plentiful oxygen, in this case in the air around the meat. Where oxygen is absent, as in the centre of the meat or in a vacuum pack, the pigment is in the reduced form, often called just 'myoglobin', which is a deep purple-red in colour. Between these two colours there is a brown layer, due to the oxidized form of the pigment, called met-myoglobin. This colour is more stable than the others and is formed where the oxygen concentration is positive but small. It therefore tends to accumulate as the oxygen concentration gradient in the meat alters due to consumption of oxygen by metabolic processes and by any microbial growth within the meat mass. So, as the meat gets older its colour becomes duller because of the increasing brown layer below the surface.

The colour is further influenced by the post mortem history of the meat. A high ultimate pH is associated with muscles with a closed structure, holding large quantities of water, appearing swollen and tightly packed. This muscle scatters relatively little of the incident light and appears dark red. At the other extreme is muscle with low pH, or ultimate pH reached quickly, where some of the protein is denatured, the structure is more open and the meat appears pale and wet.

It is not possible to do much about pale or dark colours resulting from the post mortem conditions (see POST MORTEM CHANGES), but the chemical changes in the pigments can be influenced to some extent. In prepackaging of fresh meat it is essential to select packaging materials with as high a permeability to oxygen as possible, to ensure maximum retention of the surface oxymyoglobin. On the other hand, reducing agents can slow down the rate of formation of the oxidized met-myoglobin and so delay the onset of the unacceptable brown colour. Ascorbic acid or ascorbates applied to the meat surface can give an extra day of apparent shelf life under normal storage conditions, but this is usually held to be to the detriment of the consumer since the microbiological deterioration of the meat continues during the period when the colour appears to be satisfactory. The addition of ascorbates to fresh meat is prohibited in most countries. Sulphur dioxide has a similar effect on the meat pigments but also has a preservative action against any microbial infection. It is not permitted in fresh meat and its use in meat products is restricted under the preservatives regulations in most countries.

The addition of oxidizing agents to the meat will promote the formation of met-myoglobin. One very common case is the accidental contamination of fresh meat or a fresh meat product with traces of sodium nitrite. This can turn sausages brown or grey almost instantaneously. Chlorine-based cleaning agents may have a similar effect.

Colour of fat

Beef, mutton and poultry fats may be of various shades from white to deep yellow. The yellow colour is due to carotenoids in the diet. In beef

and mutton, and in poultry in some countries, it is considered objectionable although it has no effect on palatability. Yellow fat in beef usually indicates either meat of old age or that from a dairy herd. When cows undergo a period of lower nutrition the fat deposits are depleted but the carotenoids remain behind. They consist of a mixture of the grass pigments carotene and xanthophylls, the carotene predominating. The majority of sheep have colourless fat but in some countries including Iceland and New Zealand a small number of sheep have yellow fat due to exclusive deposition of grass xanthophylls, predominantly lutein, lutein-5,6-epoxide and flavoxanthin. Chickens respond markedly to the carotene content of their diet and in countries such as the UK where white-skinned birds, with white body fat and leg scales, are preferred, care must be taken to control the use of grass meal or 'yellow' varieties of cereals in the feed.

Cured meat colour

The colour of cured meats is due to the pigment nitrosyl myoglobin or nitrosomyoglobin, formed from nitrite salts added or formed during the curing process. Nitric oxide is formed from the nitrite and combines with myoglobin in the meat. In the undenatured form, as in uncooked, unpasteurized bacon or raw dry sausage, the pigment is a deep red colour. On heating this is converted into a pink colour in which the protein is denatured. This is the colour of ham, canned luncheon meats or frankfurters. The red, undenatured, pigment is relatively unstable to air, becoming oxidized to brown met-myoglobin. It is not stabilized by the presence of ascorbate, on the contrary, at moderate ascorbate concentrations in the presence of air it may be completely destroyed. It is best preserved by vacuum packing or, in the case of unpacked bacon, by leaving the meat intact and unexposed to the air for as long as possible. The pink, denatured pigment on the other hand is less unstable to air but is very sensitive to light. It is, however, significantly stabilized by the presence of moderate concentrations of ascorbate. Ascorbate is commonly added to the curing mixtures for this class of product. Vacuum packaging is helpful for reasons other than colour stability, but care should be taken to minimize exposure of the packed product to light, especially to intense display illumination.

In all cured meats the stability of the cured colour is assisted by the presence of residual nitrite in the product. The movement in recent years to keep nitrite levels as low as possible, because of their implication in nitrosamine formation, has made it more difficult to manufacture cured meat products with reliable long term colour stability, but with careful control of both manufacturing and storage conditions it can be achieved satisfactorily.

Colour of cooked meat

The brown colour of cooked uncured meat is a heat denatured myoglobin or haemichrome, in which the links between the haem part of the molecule and the protein part become rearranged and even randomized during denaturation. The colour is relatively stable but fades on long exposure to air and light. The heat denaturation of the globin usually takes place at temperatures close to 68°C but the exact temperature may be found to vary somewhat. Until recently it was considered that the temperature of the colour change was also the temperature at which the meat becomes microbiologically safe – the scientific basis perhaps for the first elementary principle of meat cookery. Recent thinking in the Health Departments in the UK and the USA, influenced by questions about *Listeria monocytogenes*, has concluded that this relationship is not safe and that for full microbiological safety meat should be cooked to at least 72°C.

One question which is not yet clearly resolved is that in some specimens of meat a red colour may persist at the centre at temperatures much higher than this, or may be formed under anaerobic conditions at the centre after having apparently previously turned brown on heating. This red colour has a similar spectrum to oxymyolobin but its true nature and the exact conditions for its formation are not clear.

Deep brown colours are produced at the surface of roasted or grilled meat as a result of Maillard reactions. These occur mainly at temperatures in excess of 150°C (300°F) and therefore only at the surfaces from which all moisture has been driven off. They are due to reaction between the meat protein and unsaturated fat or reducing sugar. They can thus be enhanced by the addition of reducing sugars especially, as in products such as 'honey roast' ham or by the use of milk powders.

Added colours

Nicotinic acid

This is one of the B vitamins, also called niacin or PP (pellagra-preventive) factor. When used on meat it combines with the myoglobin colour of the meat to form a new, intense red colour. Such a treatment can deceive as to the true age and condition of the meat and is prohibited in most countries. The related substance nicotinamide is similarly prohibited.

Artificial colours

In most countries the addition of artificial colours to meat is not permitted. In the EU colour may be added to meat products but this appears to be confined to the use of certain colours in sausages and in canned luncheon meats. Black PN (E151) is also used for the identification of condemned carcasses.

Discoloration of cooked eggs

When whole hard boiled eggs are incorporated into products such as Scotch eggs or gala pork pies, they often suffer from a black discoloration of the yolk surface where it is in contact with the white. The chemical cause is hydrogen sulphide formed during the cooking of the egg white protein migrating as vapour through the white until it encounters iron present in the egg yolk, where black iron sulphide is formed. This is harmless but unsightly.

One method of prevention is to cool the eggs as rapidly as possible after cooking. This minimizes the movement of hydrogen sulphide and so reduces the formation of sulphide in the yolk. However, it is not always completely effective and a more certain treatment is to dip the cooked eggs immediately after cooking, cooling and shelling into a solution of sodium citrate (to chelate the iron and prevent it from forming sulphide) or hydrogen peroxide (to oxidize the hydrogen sulphide to sulphate whose iron salt is effectively colourless). Care is needed to keep the dipping solutions in good hygienic condition if they are to be used over long periods, and to maintain the concentration of the active agent.

FLAVOUR

This is perhaps the least understood of the characteristics of meat which influence quality. As animals grow older or when young cattle are put out to grass the flavour of the meat is improved. It is unfortunate that two of the most important characteristics of meat, tenderness and flavour, develop in opposite directions as chronological age increases. However, post mortem ageing improves flavour, partly because of the breakdown of nucleotides to ribose and hypoxanthine, as well as increasing tenderness. Some of the flavour of lean meat, and much of the characteristic flavour of each meat species, is due to the fat. Much of the improvement in flavour when animals are grass fed is due to fat-soluble constituents.

Many volatile compounds have been identified in cooked meat and broths by gas chromatography. These include carbonyl compounds, simple fatty acids, simple aliphatic alcohols, sulphur compounds and ammonia, together with many more complex substances. Short chain fatty acids contribute significantly to the flavour of fat and the role of some individual amino acids and small peptides appears to be important in lean meat flavour, but otherwise there do not appear to be any individual substances, or even small groups of substances, which are mainly responsible for meat flavours. It is likely that subtleties in the balance between many constituents are important, but the significant patterns have so far eluded discovery.

Added flavourings

In order to improve the flavours of various meat products, spices and aromatic herbs are employed, such as sage, thyme, coriander, pepper, mace, nutmeg, ginger, and so on, the combination being commonly referred to in the trade as 'seasoning'. The choice of a suitable combination is very important and many highly esteemed products owe their fame to success at this point.

Spices and herbs are frequently used in the form of fine or coarse powders; fine powders are usually preferable for producing more intense flavours but coarse ground spices or herbs sometimes contribute to the appeal of the product. Local tastes must be considered here.

Spice extracts are generally available as alternatives to crude spices, with a number of advantages. They consist of solvent extracts of the original herbs or spices, or of oleoresins, dispersed on a solid base of salt, rusk or sugar (dextrose). They are standardized during manufacture so that the variability in flavour which is one of the disadvantages of raw spices, is eliminated. Microbial and insect contamination associated with spices is likewise eliminated, the extracts are more convenient to handle and can indeed usually be made up in the particular mixtures required by the meat manufacturer, and in batch quantities. On the other hand, in some applications some of the subtleties

of flavour associated with the whole spice may be absent.

TEXTURE

Toughness or tenderness in meat arises from two causes, related to the amount and the hardness of connective tissues or related to the state of post mortem contraction of the muscle fibres. Toughness due to connective tissue may be ameliorated by tenderizing treatments. In the first place, meat with high connective tissue content, or from older animals with tougher connective tissue, is traditionally cooked by slow, moist methods (e.g. stews or casseroles) to permit maximum hydrolysis of collagen to gelatin. Second, such meat is made into products in which it is finely comminuted so as to break up the hard material into small pieces which are both less objectionable in themselves and also more readily softened on cooking. Third, the collagenous material may be softened by chemical degradation using enzymes. In cookery, if fresh pineapple is cooked with meat it serves as a source of bromelin for this purpose; in technology papain is used. In the patented 'Pro-Ten' process (widely used in the USA but now prohibited in the EU), papain is injected into animals just before slaughter when it is rapidly distributed throughout the carcass in the blood stream. Papain remains inactive in the muscles until it is activated by the heat of cooking, when hydrolysis and the tenderizing action take place.

Apart from the tenderizing that occurs during ageing (q.v.) whose nature is not yet understood, no methods are available to deal with toughness arising from muscular contraction. Proper management of slaughtering and post mortem cooling is therefore essential.

Various instruments may be used to measure meat texture, of which the Warner–Bratzler shear press, the Wolodkewitsch tenderometer and a number of applications of the Instron Tester are in common use. Sensory panels are widely used, scoring for tenderness or toughness or counting the number of chews required to disintegrate the sample. Good correlations can be obtained between panel scores and instrumental measurements.

In development work, texture profile methods can be used to reveal and quantify other features of the texture of a meat sample, for instance, its juiciness and elasticity. This method requires intense training of a panel and is not suitable for routine work.

FRESHNESS OF MEAT

The most useful test for the freshness of meat is to smell it. With a little practice it is not difficult to distinguish between:

(i)   putrid smells, characteristic of meat spoiling by mixed microbial action;
(ii) cheesy or acid smells, characteristic of lactic spoilage of cured meats;
(iii) rancid smells, characteristic of oxidative spoilage of fat.

This last is especially significant after frozen storage (see FREEZING AND FROZEN MEAT).

When microbial spoilage is well advanced the meat will often be slimy or sticky to the touch. An indication of the microbial numbers on meat at different stages of spoilage is given below under ANALYSIS OF MEAT PRODUCTS.

A detailed microbiological analysis may occupy several days; it will be useful for establishing the microbial quality of the meat before obvious spoilage takes place but will be of less value if spoilage has already occurred. Where a numerical index of the degree of spoilage is required, chemical tests may be more useful. Total volatile nitrogen (TVN) or extract release volume (ERV) may be measured, however interpretation of the results is difficult and there may be disagreement among experts about the degree of unsatisfactoriness represented by different values of TVN or ERV.

The principle of the TVN method has now been extended and its reliability greatly improved by Japanese workers with fish. The commercially accepted modern method for measuring the freshness of fish in Japan requires the rapid analysis of a number of nucleotides followed by calculation of an F-value. The method is said also to work satisfactorily with pork meat.

For evaluation of the quality of meat fat, the free fatty acid (FFA) and the peroxide or thiobarbituric acid value (PV or TBA) of the extracted fat should be determined.

ANALYSIS OF MEAT PRODUCTS

For most practical purposes in a meat factory quality control laboratory, the chemical analyses carried out on the products are for moisture, fat, protein, connective tissue and ash or salt. For cured meat products nitrite and nitrate determination will also be required. Rapid methods of carrying out these analyses are desirable.

Moisture may be determined rapidly by heating 5 g of the product for 30 min at 150°C (328°F).

Fat content may be determined by continuous extraction of the dried solids with mixed chloroform–petroleum ether (bp 40–60°C) in a Soxhlet extractor or in the Foss–Let or similar apparatus. In the latter method the sample is extracted with perchlorethylene which is then made up to a standard volume: the fat content is calculated from the refractive index of the solution. The proportions of saturated, unsaturated and polyunsaturated fatty acids may be determined by gas chromatography of the fat after hydrolysis and esterification with methanol.

Various methods of measuring protein content have been tested but modifications of the Kjeldahl process are still the most favoured. Semi-automatic systems are now in common use (e.g. Kjel–Foss, Kjel–Tec, Buchi) in which a first result may be available in 15–20 min from commencement and later ones at 3 min intervals.

Connective tissue may be determined by the method of Mohler and Antonacopoulos as follows. After hydrolysis of 4 g of the meat product, the hydroxyproline in the hydrolysate is determined by oxidation with hydrogen peroxide to give a compound which on acidification gives a red colour on heating with Erlich's reagent. Eight times the hydroxyproline content is reckoned as collagen or dry connective tissue, 37 times is wet connective tissue.

The estimation of nitrogenous substances other than meat protein may present severe difficulties (see MEAT CONTENT below).

Salt may be determined by titration with silver nitrate or mercuric nitrate. Multiple routine determinations may be done with a suitable calibrated salt probe or electrode. Salt in curing brines is commonly monitored by hydrometer.

Nitrite can be determined by the well-known diazo reaction followed by coupling with N-(1-naphthyl) ethylenediamine dihydrochloride (NED) to produce the water-soluble azo dye. Nitrates are usually determined after reduction to nitrite using metallic cadmium. The test solution is treated either by shaking directly with precipitated cadmium or by passing it through a column packed with the metal. The total nitrite content of the reduced solution is determined, the original nitrite deducted and the remainder calculated as nitrate. Nitrite and nitrate determinations should be carried out routinely on curing brines.

Meat content

Since the protein and moisture contents of pure muscle are relatively constant, this fact may be used in the analytical determination of the meat content of meat products. In real meat, of course, the situation is complicated by the presence of variable amounts of fat and connective tissue along with the muscles in different cuts of meat.

Meat content calculated from standard values for the nitrogen content of meat

In the UK the determination of the meat content of products such as sausages is done using average factors for the nitrogen content of lean meat, including the proportions of connective tissue and fat normally associated with the lean meat.

The following factors (percent nitrogen on the fat free basis) have been agreed between the Society for Analytical Chemistry, the Royal Society of Chemistry and others, and are used for control purposes in the UK:

Pork	3.50
Beef	3.65
Breast of chicken	3.9
Dark meat of chicken	3.6
Whole carcass of chicken	3.7
Ox liver	3.45
Pig liver	3.65
Liver of unknown origin	3.55
Tongue	3.0

The pork and beef factors are average values for all cuts of meat from the animal in question and may be incorrect for particular cuts whose composition (proportions of connective tissue and intermuscular fat) differs markedly from the average, as is the case with many of the cuts used for manufacturing.

The factors recommended by the Analytical Methods Committee of the Royal Society of Chemistry have been changed from time to time, the most recently recommended values for pork meat (with interstitial fat but without rind and subcutaneous fat) (1991) include: whole side 3.50, leg 3.49, neck or collar 3.38, hand 3.42, loin 3.66 and belly 3.50. Recommended values for beef (with interstitial fat) (1993) are: beef in general 3.65, clean beef 3.65, cull cow beef 3.70. The situation is complex and readers should consult the relevant publications for detailed information (Analytical Methods Committee, 1991 and 1993).

A further complication arises if added pork rind or gristle-containing material may have been used in the recipe of a product. The analyst may estimate the connective tissue content from a determination of hydroxyproline, deduct the connective tissue nitrogen content from the total nitrogen and calculate a connective tissue-free meat content. On the assumption that pork meat with the rind on, or beef flank meat, contains not more than 10% of connective tissue, a value for added connective tissue can be calculated.

Of course, meat products may contain nitrogenous substances other than meat protein and the detection and estimation of these may present difficulties. The Stubbs and More calculation applied to the analysis of British sausages assumes that the non-meat solids present consist of rusk with a nitrogen content of 2% and the appropriate deduction is made from the total nitrogen content before calculating an 'apparent' meat content. Soya, milk or other proteins may be estimated electrophoretically or by other means, provided the sample has not been strongly heated, and the appropriate corrections made. ELISA (enzyme-linked immunosorbent assay) methods can be used for cooked samples.

Attempts have been made to estimate meat content directly by measurement of the content of 3-methyhistidine, an amino acid which is characteristic of meat protein, but this process is not reliable unless the species of meat is known, the 3-methylhistidine content of muscle being rather variable.

Analytical control without calculation of meat content

In countries other than Britain it is common for control purposes to refer the composition of meat products directly to the nitrogen or protein content of the dry, fat free product, or to the water : protein or similar ratio. The analytical problems of determining the true meat nitrogen or protein content are of course the same.

From the figures considered above (see COMPOSITION OF MEAT), the water : protein ratio in muscle meat is close to 77% water : 23% protein = 3.35. This ratio is a pure number, independent of the fat content of the meat. The Feder number as used in Germany is closely similar, being defined as the ratio of water to organic non-fat in a sample, or:

$$\text{Feder no.} = \frac{\text{water \%}}{100 \times (\text{fat \%} + \text{water \%} + \text{ash \%})}$$

The organic non-fat consists of the protein of the meat plus other substances which are almost all nitrogenous; in practice this is close to the protein content as estimated from the nitrogen content.

In France the relationship HPD (humidité du produit degraissé or 'moisture of the defatted product') is used. For pure muscle, from our previous data, this value is 77%.

In USA regulations and in the FAO/Codex recommendation a related ratio is used, protein on fat free (PR_f) or:

$$PR_f = \frac{100 \times \text{protein } \%}{100 \times \text{fat } \%}$$

The limiting value of this expression is the protein content of fat free muscle, or 23%.

The expression ($100 \times \text{fat}\%$) in both HPD and PR_f is not the protein plus water content of the sample but protein plus water plus ash. It may therefore be affected by differences in ash content, as will be found for instance in cured meats.

Under systems such as these it is usual to provide a maximum tolerance for the presence of rind, gristle and connective tissues, calculated from the collagen content. French standards are expressed in terms of the ratio of collagen to total protein, the German ones as absolute or relative BEFFE values. The absolute BEFFE value is the percentage content of connective-tissue-protein-free meat protein (Bindegewebeseiweiss-frei Fleischeiweiss). The relative BEFFE is the percentage of BEFFE in the total protein. Of course, if non-meat proteins are present they will not be distinguished in these calculations and must be analysed for separately.

Meat species

The species of meat used in a given product, as long as the product has not been cooked, may be readily identified using immunological techniques. Simple test kits are commercially available.

With heated meats, immunological identification is very difficult but the species may still be determined by taking advantage of species differences in the composition of the muscle proteins, the myoglobin or certain enzymes including phosphoglucomutase. The individual components may be separated by gel electrophoresis with isoelectric focusing, then stained with Coomassie Blue. So long as mixed fats are not present, identification may also be possible from the fatty acid profile of the fat.

Rapid methods

To meet the increasing pressure from regulatory authorities for precise control of composition at the point of manufacture, and also in the manufacturer's own interest, a number of good and rapid analytical methods is now available. The initial cost of the equipment is high, but can be justified when large numbers of analyses lead to improved control of production. The systems presently available include those based on the following methods:

(i) *Infrared spectroscopy of a prepared sample surface (e.g. the Infralyser)*. This process was first developed for dry materials such as flour but has been adapted for meat. The instrument requires careful calibration for each type of sample analysed, so it is convenient only for long production runs on the same product. Another infrared method is the Foss Super-Scan, where light in the fundamental infrared region (2.5 to 10 μm) is transmitted through the sample and the energy attenuation is measured. The system consists of an electronic balance for accurate on-line transmission of sample weights, a reactor for sample preparation, a measuring unit where the fat, protein, carbohydrate and water infrared absorptions are measured and a desk top computer.

(ii) *X-ray method*. A large sample (e.g. 1–15 kg) may be required, but this may be an advantage as it makes it easier to ensure that the sample is representative; after analysis the sample may be returned to production without loss. The Anyl-Ray machine tests the sample by X-ray diffraction and gives a digital readout of the fat content.

(iii) *Microwave drying and solvent extraction*. Rapid destructive analysis by physical and chemical methods, as in the CEM Meat Analyser, can provide results in under 10 min. The CEM system consists of two units, an electronic balance fitted inside a microwave drying cabinet and a fat extraction unit with automatic solvent extraction and solvent recovery. This is an AOAC approved method.

(iv) *Video image and optical methods*. The Glafascan video image system is an on-line method for measuring the fat to lean ratio of fresh or frozen minced or diced meat. The system provides real time video measurement of the fat to lean ratio by continuously scanning the surface of the meat as it passes along a conveyer, calculating the fat content from the intensity and colour of the reflected light and displaying the accumulated results every 5 s. This information can be fed directly through suitable equipment to control subsequent manufacturing operations on the meat. An example of an optical device is the Lean Machine which consists of a hand-held optical device which is inserted into a meat sample; the device is linked to a microprocessor which gives a rapid calculation of the fat to lean ratio. Another optical method is the Trebor-99 Composition Analyser which is the laboratory-based instrument.

(v) *Automated heat and digestion systems*. A simple heating method called the Univex Hobart instrument consists of a heating element within which the sample, for instance minced meat, is placed. The heated fat melts and is collected in a graduated tube which is calibrated to give a direct readout of the fat content. The analysis takes 15 min. Although it appears rather crude, if the instrument is carefully calibrated it can provide accurate results.

(vi) *Chromatography*. The Dionex ion chromatography method is in principle capable of measuring the content of chloride, phosphate, nitrate and nitrite anions in a sample of meat in under 30 min. The system will also detect fluoride, benzoate, bromide, malate, sulphite and sulphate.

(vii) *Electrical methods*. The Dickie–John ground meat tester is a non-chemical method which relies on the generation of an electromotive force (emf) in a plug of meat; the emf generated is directly proportional to the lean meat content. A sample tube is loaded with around 750 g of meat and the instrument gives an immediate readout of the fat content. The accuracy is somewhat lower than that of the chemical methods but is generally acceptable in a factory environment.

(viii) *Density methods*. The specific gravity of fatty tissue is about 0.92 and that of lean tissue about 1.0. In theory therefore the fat content of a sample can be calculated if its density can be measured accurately. Machines have been designed to measure the density simply, for instance by compressing a known weight of minced meat and measuring the resulting volume, but they are not yet commercially successful.

The Foss–Let method uses automatic extraction of the fat from a sample of meat with perchlorethylene; this solvent has a specific gravity (SG) of 1.62, compared with about 0.92 for the extracted fat. Here the difference in SG is large and the SG of the solution of fat is easy to measure accurately; the Foss–Let instrument gives a reliable measure of fat content in meat. The results are available within 5 min and the method is AOAC approved.

FOOD POISONING BACTERIA

Food poisoning in general is dealt with elsewhere in this manual but it may be worthwhile to draw attention to the role of animals as an important reservoir of many of the food poisoning bacteria, and of meat as a vehicle of their transmission to humans.

Many of the bacterial species which cause human food poisoning are natural inhabitants of animal intestines. Salmonella, for instance, has been found in the intestines or the faeces of almost every warm blooded animal species which has been tested. In the great majority of instances the bacteria are harmless commensals with their hosts, causing no symptoms of illness whatever. Only when organisms move from one accustomed host to an unaccustomed one may illness sometimes result, usually in consequence of activation of the rejection mechanisms in the new host, causing vomiting and/or diarrhoea.

The organisms present in a healthy carrier host are commonly present only in low numbers, and usually only some of any group of animals will be carriers. Though bacterial numbers within an animal may be reduced considerably by a number of techniques (careful husbandry, appropriate medication), complete elimination has proved exceedingly difficult and only complete elimination can achieve what is required. However, recent developments in competitive exclusion show good promise of success. The newborn or newly hatched animal is fed as soon as possible with preparations containing high numbers of the types of organisms which are normal and harmless in the guts of its species, but containing none of the types to be eliminated; the normal organisms compete for space and nutrients with any harmful ones which may be present, outgrow and eliminate them. The technique, first developed with chickens, has found some commercial use.

Other food poisoning organisms are natural inhabitants of the soil or, like Listeria, are to be found almost everywhere. Animals for slaughter must therefore always carry potentially harmful bacteria on their feet and their skin, hair or feathers.

Even under laboratory conditions it is extremely difficult to kill and open an animal without contaminating any of the flesh from either the outer skin or the intestinal contents; in an abattoir it is practically impossible. Meat, from whatever source, is therefore an inevitable source of food poisoning bacteria. Meat technology therefore must attend to (a) minimizing initial contamination at the slaughter stage and cross-contamination later by good hygienic practice (but always expecting that some food poisoning bacteria will survive), and (b) minimizing the opportunities for bacterial numbers to increase, mainly by temperature control. By these means and under the best conditions, contamination levels can be kept very low, below the detection limits of most of the practical tests employed.

Bacteriological tests

There is much variation in the tests done in different laboratories and methods have not been standardized sufficiently for strict and enforceable bacteriological standards for meat and meat products to be laid down at international or, in most cases, national level. It is even considered by many that such standards are neither possible nor desirable. It is, however, common for large buyers to make their own microbiological specifications or guidelines and a supplier should, of course, follow these.

For raw meat, the following figures are a rough guide:

Microbial count, per g	
10^2	Excellent quality (laboratory conditions)
10^4	Good commercial quality
10^6	Rejection limit in many commercial contracts
10^8	Meat smells
10^9	Meat slimy

One recommended procedure for the examination of open pack meat products is as follows: make a plate count at 37°C, a presumptive coliform count at 37°C and follow if necessary with a faecal coliform count at 44°C. A staphylococcal count using the Baird Parker medium is also desirable. Salmonellae may be detected by enrichment in double-strength selenite F broth followed by plating onto to McConkey agar plate and a desoxycholate citrate agar plate. After incubation at 37°C, non-lactose fermenting colonies are inoculations are made into one tube each of Kohn's two-tube medium from the peptone water cultures. The Kohn's tubes may be read after 24 h at 37°C.

Clostridium perfringens screening tests may be carried out using litmus milk as in testing water; occasional false positives are obtained.

The 'agar' sausage may be used as a simple means of checking the level of microbiological contamination of equipment. The exposed cut end of the sausage is pressed on to the surface to be sampled, a thin slice is cut off and incubated.

Commonly used microbial media are now available in prepared forms such as ready poured plates or media impregnated pads, which greatly simplify routine laboratory procedure. Inoculation and counting can be simplified and speeded up with the Spiral Plate Maker. For intensive routine work, automatic systems such as the Bactometer or the Malthus Growth Analyser are becoming commonly used.

NUTRITION ASPECTS

Meat of all kinds and the products made from meat form excellent sources of protein and fat. Consumer pressure against the consumption of meat addresses either or both of these aspects but does not deny the truth of either.

The objections raised to animal protein may be summarized briefly here:

(i) animal protein is produced higher up the food chain than the protein of the plants upon which animals feed, therefore production of animal protein has only about 10% of the efficiency of plant protein (measured in land usage, human effort or other parameters);

(ii) there are moral objections to the killing of animals for human food;

(iii) the total consumption of fats in Western diets is considered too high by some; the ratio of saturated to unsaturated fats is high in meat relative to vegetable fats and this is considered to be related to certain disease conditions.

Meat is also, with less contention, a good source of the B vitamins thiamin (B1), riboflavin (B2) and niacin. Little destruction of riboflavin or niacin occurs on cooking but there are losses of thiamin, related to the length of the cooking time. Figures quoted for the losses of thiamin on cooking are: frying 10%, broiling 20%, roasting 30%, braising 35%, stewing 50%. There is also some loss of thiamin if the preservative sulphur dioxide is present.

The iron in meat, especially liver, is mainly present as haemoglobin and myoglobin, which are more readily absorbed than the iron from most other sources, hence the traditional value of liver in the treatment of anaemia.

MEAT SUBSTITUTES

In the 1940s and 1950s the reason for investigating substitute meats lay in a perceived world shortage of high class protein and a desire to upgrade the eating quality of vegetable proteins: that motivation has diminished with the understanding that the deficiencies are not specifically protein deficiencies but are part of general poverty-related shortages of food. In more recent years the motivation for meat substitutes has come rather from the 'green' and vegetarian markets in the wealthier countries.

Meat substitutes are almost entirely protein based. The starting materials are also very largely those of which there are large economic surpluses in the developed world, soya beans especially. Proteins from many other pulses have been investigated; few have come to commercial application. Proteins from wheat and milk have also been tried with somewhat more success. Very successful recently is Quorn, a patented product based on the mycelia of the mould *Fusarium graminearum* (hence mycoprotein).

Some applications are direct adaptations of traditional recipes, for instance iru made from locust beans (West Africa), tempeh made from soya or other beans (Indonesia), tofu made from soya (Japan).

In commercial applications, the proteins are of interest not only for their nutritional value but also for their effects on the texture of the meat substitute or the finished food containing it. The protein may be isolated from the parent food, commonly by alkaline extraction and reprecipitation with acid, or a simpler extract may be made, for example soya milk. In either case the protein may then be used as:

(i) Functional protein – The protein is mixed with the other ingredients of the product before cooking; cooking coagulates the protein which binds with the other constituents to confer a more or less cheese-like structure and texture to the final product;

(ii) Textured protein – For TVP (textured vegetable protein) and similar products the isolated protein is heated with or without other ingredients and made to coagulate under mechanical pressure, for instance in an extruder or spinneret, into pieces whose texture more or less resembles that of cooked meat.

For Quorn the mould mycelium is mixed with other ingredients including egg albumen which binds the mass together on cooking to produce a meat-like texture. The textured pieces may then be used as a meat substitute in manufactured or domestic recipe dishes.

The flavour of a substitute meat may approximate to that of meat if there is a high temperature cooking process to produce Maillard-type substances related to those of roasted meat. Otherwise the flavour may be enhanced by the addition of flavouring material, yeast extract or vegetable protein extract (from wheat gluten or from soya) are commonly used.

The use of blood and blood plasma as binding and texturizing agents may be noted here, though these may not be considered strictly to be meat substitutes (see BLOOD). Note also products such as surimi, based on minced fish, see Chapter 2.

NITRITES AND NITROSAMINES

Benefits of Nitrite

It has been known for almost a century that the essential agent in the production of cured meats is a nitrite salt, sodium or potassium nitrite. The nitrite has the following effects:

(i) Preservative. In uncooked cured meats preservative effect appears to persist for as long as residual nitrite remains in the product. In cooked cured meats there is an additional preservative effect due to an interaction between nitrite and protein on heating, by which some new preservative substance is formed. The nature of this is not exactly known; it is usually referred to as a 'Perigo-type' factor after J. Perigo who first discovered the effect in laboratory microbiological media. This factor has a strong action against spores, permitting products to be safe from formation of *Clostridium botulinum* toxin after only relatively mild heat processing.

(ii) Colour. Formation of the characteristic colour of cured meats (see COLOUR).

(iii) Flavour. Nitrite contributes to the characteristic flavour of cured meats.

Toxicity

(i) Nitrites. Nitrite in moderate doses is toxic. It reacts with the blood to form met-haemoglobin or nitrosyl haemoglobin and destroys the oxygen carrying capacity of the blood. The lethal dose for an adult is about 1 g. In all countries there are therefore regulations to control the maximum amount permitted in foods. UK regulations permit:

	mg kg^{-1} in the product, as NaNO$_2$	
	Nitrite + nitrate	Maximum nitrite
Cooked meat in sterile packs	150	50
Acidified or fermented cured meats	400	50
Bacon and ham, not sterile packed	500	200
Other cured meats	250	150

US legislation (9CFR) Part 172.175) permits sodium nitrite up to a maximum of 200 ppm and sodium nitrate up to 500 ppm in the finished product. In continental Europe nitrite is usually controlled by the use of curing salt containing 0.6% sodium nitrite, or in Germany since 1981, 0.4%.

(ii) Nitrosamines. Under mildly acid conditions and in the cold, nitrites react with primary and secondary amines to form nitrosamines which are now known to be strongly implicated in the formation of cancers. No link has yet been established between the consumption of cured meats and cancer, and it is likely that the quantities of nitrite or of preformed nitrosamines in cured meats are not very significant compared with the amounts found in humans from other sources. Nevertheless it is obviously prudent to restrict all uses of nitrites to the absolute minimum consistent with maintaining the safety of food from the well-known hazard of botulinum poisoning.

The quantities laid down in the Regulations quoted above were arrived at in consideration of all these factors. It does appear, to the best of present knowledge, that the concentrations of sodium nitrite commonly used in cooked and uncooked cured meat products in the UK can be considered satisfactory from these points of view. The concentration are, typically:
- 70 to 200 mg kg^{-1} added at the time of manufacture
- 20 to 150 mg kg^{-1} residual immediately after manufacture, depending on the product.

ASCORBIC ACID AND ASCORBATES

Ascorbic acid is the well-known vitamin C, occurring naturally in fruits and vegetables. Its optical isomer erythorbic acid (*D*-isoascorbic acid) has almost identical chemical properties but no vitamin activity, apparently because it is not phy-

siologically absorbed by living cells. Both acids are manufactured synthetically and are readily available. The use of erythorbic acid and its salts in food is permitted in the USA but not in the UK or most of the countries in the EU. There are no known ill effects from the consumption of moderate or even large amounts of ascorbic acid or its salts.

It is usually better in manufacturing practice to use sodium ascorbate than the free acid. If the acid is used in solutions such as curing brines, containing nitrites, free nitric oxide will be formed: on contact with the air this immediately forms brown fumes of nitrogen dioxide which are unpleasant to breathe and highly toxic.

In uncured meat the presence of ascorbate delays the oxidative processes which turn the red colour of the meat into brown. It therefore prolongs the apparent shelf life; an increase of about 1 day at ordinary temperature is typical. In the UK and certain other countries it is considered that to do this is to deceive as to the true age of the meat and the use of ascorbate in butchers' meat for this purpose is therefore prohibited (see also NICOTINIC ACID). It is however permitted in manufactured fresh meat products such as sausages and burgers.

In uncooked cured meat products such as dried sausage or unpasteurized bacon, the formation of the cured red colour is accelerated by the use of ascorbate but this is not usually of commercial significance because the time normally available for colour development is more than adequate. On the other hand if 200 mg kg^{-1} or more of ascorbate is used and the product is exposed to the air, hydrogen peroxide may be formed by reaction of oxygen with the ascorbate. This can form green choleglobin or further breakdown products which are colourless, thus destroying the red colour completely. The use of ascorbates is therefore not advisable in these products. The colour of cooked cured meats such as ham, pasteurized or hot-smoked bacon, luncheon meats and sausages of the frankfurter type is intensified and stabilized by ascorbates.

There appear to be at least three different effects:

(i) the yield of cured colour from the nitrite available is increased in the presence of ascorbate: this effect is significant where the concentration of available nitrite is low, so the uniformity of colour is improved in meats where the nitrite itself is not uniformly distributed;

(ii) the colour is formed more rapidly in the

uncooked meat so that more is available to be fixed by the cooking process;

(iii) the cooked colour is more stable to light, so long as some residual nitrite is also present.

In a period of intense activity in the early 1970s, American workers demonstrated the effectiveness of ascorbates in reducing the formation of nitrosamines (see NITRITES AND NITROSAMINES) in heated cured meats. The use of ascorbate was then required by the US authorities, at concentrations of 470 mg kg^{-1} (ascorbic or erythorbic acid) or 550 mg kg^{-1} (sodium ascorbate or sodium erythorbate).

Ascorbyl palmitate finds some use in proprietary mixtures as a fat antioxidant. It has the advantage of being both fat soluble and water dispersible. It can thus be added to a product by dispersion among the other water-soluble ingredients in order to perform its function in the fat phase in the product.

ANTIBIOTICS

Antibiotics are chemical substances, usually prepared from micro-organisms, which interfere with the metabolic processes of other organisms and retard their growth. The use of antibiotics as preservatives is prohibited in most countries, partly for the not very good reason that by checking spoilage they might reduce a major incentive towards good hygiene in processing factories. But public health considerations are more compelling.

In the late 1950s the authorities in a number of countries, including the Swann Committee (Joint Committee on the use of Antibiotics in Animal Husbandry and Veterinary Medicine) in the UK (1959), considered the dangers to human health posed by the carry-over of antibiotics from animals to meat, milk, etc. used as human food. Antibiotics may be administered to animals either for veterinary reasons (disease treatment) or as feed additives for growth promotion (achieved by suppression in the animal gut of bacteria which are harmful or which compete for the animal's food).

There is a double danger. The use of any antibiotic may be expected ultimately to lead to selection of resistant strains of the micro-organisms against which it will then be ineffective. If such micro-organisms from an animal were later to infect humans causing disease, the illness would not be treatable with the same antibiotic. Furthermore, this property of drug resistance is transferable between micro-organisms. Suppose that an animal is treated with antibiotic for a disease condition which is itself of no significance to human health, but that as a result of the treatment the bacteria responsible become resistant to the antibiotic. That resistance could then be transferred to other bacteria, for instance Salmonellae, and if the Salmonellae should cause human food poisoning then that illness would be resistant to treatment with the same antibiotic.

The Swann Committee therefore recommended that whilst the antibiotics used in human medicine might be used also in veterinary medicine with care and under proper supervision, antibiotics used as feed additives should be restricted to those without application in human medicine. This means in practice that antibiotic residues should not appear in meat. In milk, which might contain residues of penicillin following veterinary treatment of the cows, absence of antibiotic is interpreted in UK laboratory practice as meaning that the content of penicillin, or its equivalent of other antibiotic, should be below 0.05 i.u. ml^{-1}; a similar standard could be applied to meat.

PROBIOTICS

The term is used to cover two groups of agents or processes which produce modifications in the live animal body and influence the quality of the meat eventually produced.

(i) Competitive inhibition of harmful bacterial flora in the animal gut by cultures of other bacteria which are harmless to the animal or to later human consumers. Mixed cultures of lactobacilli and other organisms, derived from healthy poultry, can be introduced to day-old chicks via their drinking water and have been shown to outgrow and eliminate Salmonellae in the intestinal tract. Unfortunately they do not also eliminate Salmonellae in the caecum, so complete eradication from the poultry flocks is not yet possible, but the technique does show promise with other animals which do not have the avian digestive system.

(ii) β-Agonists or anticholinesterase drugs introduced in the animal feed as 'repartitioning agents', to increase the proportion of muscular tissue on the carcass. Clenbuterol (the so-called 'angel dust') and cimaterol are examples currently in use. These appear to work by stimulating the production of adrenaline, therefore increasing the metabolic production of energy which is consumed in a conversion of fatty tissue into protein, the chemistry of which is not yet explained. There is a tendency towards DFD meat.

HORMONES

Also referred to as anabolic agents, certain sex hormones are used in animal husbandry as growth promoters. Usually a mixture of androgen (male hormone) and oestrogen (female hormone) is administered by implantation or in the animal feed. Diethylstilboestrol, DES, is a synthetic compound with hormone-like properties which is not readily broken down in the animal or human body and its use is not now generally permitted. Others include zeranol and trenbolone. The natural hormones are broken down in the body and provided the quantity implanted is correct or there is a proper withdrawal period of a feed additive before slaughter, there are no residues in the meat.

Hormone treatment has been widely practised in the raising of steers (castrated males) for beef production, or for young animals for veal. Abuses have occurred, mainly involving excessive doses and inadequate withdrawal periods. Where bulls are the usual source of beef, as in France or Germany, the treatment of steers does not arise, but the practice is widespread in the USA, was so in Britain and was believed to be so in Italy. Since 1987, the use of anabolic steroids at any stage in the life of an animal has been prohibited in the EU.

REFERENCES

Analytical Methods Committee (1991) Nitrogen factors for pork: a reassessment, *Analyst*, **116**, 761.

Analytical Methods Committee (1993) Nitrogen factors for beef: a reassessment, *Analyst*, **118**, 1217.

Codex Alimentarius (1994) Volume 10, Annex D: *Preservation of Shelf-stable Cured Meat Products in Consumer Size Hermetically Sealed Containers*, Codex Alimentarius Commission, Rome.

Lawrie, R.A. (1991) *Meat Science*, 5th edn, Pergamon, Oxford.

Richards, S.P. (1978) *Research Report no. 292*, Leatherhead Food Research Association.

Wood, J.D. and Fisher, A.V. (eds) (1990) *Reducing Fat in Meat Animals*, Elsevier Applied Science, London.

FURTHER READING

Bailey, J.A. (ed.) (1984) *Recent Advances in the Chemistry of Meat*, Royal Society of Chemistry, London.

Bechtel, P.J. (1986) *Muscle as Food*, Academic Press, London.

Church, N. and Wood, J.W. (1992) *Manufacturing Meat Quality*, Elsevier Applied Science, London.

Cunningham, F.E. and Cox, N.A. (1987) *Microbiology of Poultry Meat Products*, Academic Press, London.

Flems, L.O., Cottyn, B.G. and Demeyer, D.I. (eds) (1991) *Animal Biotechnology and the Quality of Meat*, Elsevier, Amsterdam.

Institute of Refrigeration (ed.) (1987) *Meat Chilling: a Conference; Bristol September 1986*, AFRC Institute of Meat Research, Bristol.

Johnston, D.E., Knight, M.K. and Ledward, D.A. (1992) *Chemistry of Muscle-Based Foods*, Royal Society of Chemistry, London.

Kramlich, W.E., Pearson, A.M. and Tauber, F.W. (1973) *Processed Meats*, AVI, Westport, Connecticut.

Karmas, E. (1976) *Processed Meat Technology*, Noyes Data Corporation, Park Ridge, New Jersey.

Karmas, E. (1977) *Sausage Products Technology*, Noyes Data Corporation, Park Ridge, New Jersey.

Pearson, A.M. and Dutson, T.R. (1922) *Inedible Meat By-Products*, Elsevier Applied Science, London.

Price, J.F. and Schweigert, B.S. (1971) *The Science of Meat and Meat Products*, 2nd edn, Freeman, San Francisco.

Romita, A., Valin, C. and Taylor, A.A. (eds) (1987) *Accelerated Processing of Meat*, Elsevier Applied Science, London.

Solomon, S.E. (1990) *Egg and Eggshell Quality*, Wolfe Publications, London.

Wells, R.G. and Belyavin, C.G. (eds) (1987) *Egg Quality: Current Problems and Recent Advances*, Butterworth, London.

Wilson, N.E.P. (ed.) *Meat and Meat Products – Factors Affecting Quality Control*, Applied Science Publisher, London.

2   Fish and Fish Products

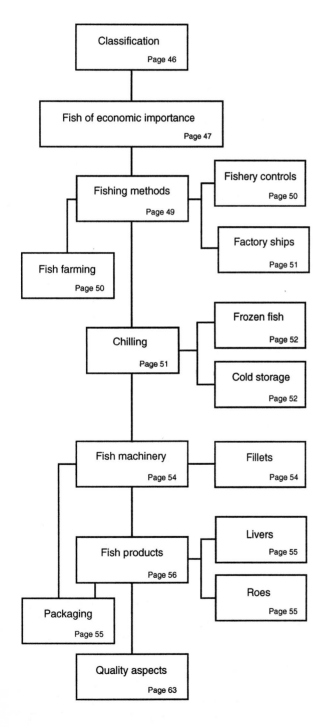

| Classification |
| Page 46 |

| Fish of economic importance |
| Page 47 |

Fishing methods	→	Fishery controls
Page 49		Page 50
		Factory ships
		Page 51

| Fish farming |
| Page 50 |

Chilling	→	Frozen fish
Page 51		Page 52
		Cold storage
		Page 52

| Fish machinery | → | Fillets |
| Page 54 | | Page 54 |

Fish products	→	Livers
Page 56		Page 55
		Roes
		Page 55

| Packaging |
| Page 55 |

| Quality aspects |
| Page 63 |

CLASSIFICATION

Seafood species come in a bewildering variety of shapes and forms which are the product of a long period of evolution. They can be grouped in various ways.

The first and most obvious division is between fish and shellfish, where the main biological difference is the position of the skeleton. True fish possess a recognized internal skeleton and swimming is the usual method of locomotion, whereas shellfish species are without a backbone and have a hard external shell or exoskeleton to compensate. Most shellfish, with the exception of the crustacea and cephalopods, remain stationary throughout their adult life.

FISH

Fish can be grouped into three main biological classes of which only two are of economic significance.

Cyclostomata

Apparently at one time there were many fish of this class of which only a few forms survive today. These are the hagfishes and lampreys which are eel-like in appearance. Most of them are parasites which live off other fish.

Selachii

Often referred to as the Elasmobranchii, this class comprises all the sharks, dogfish, skates and rays. These species are not bony in the accepted sense, having an internal skeleton composed of cartilage.

Pisces

By far the most important class commercially, comprising species such as cod, haddock, plaice and herring.

Fish can be differentiated according to whether they live in salt water (marine) or fresh water, whether they live on the bottom or near the surface of the sea (see below) and whether or not the flesh is fatty. They can also be divided according to shape :

(i)   tapering, or torpedo shaped, thicker at the head and slightly flattened out at the sides (e.g. cod)
(ii) arrow-shaped, with a fairly even cross-section and the fins set far back (e.g. pike)
(iii) flattened, either vertically (e.g. John Dory) or, more usually, horizontally (e.g. plaice)
(iv) serpentiform, long, round, slightly flattened at the sides (e.g. eel).

The shape of the fish is obviously important when it comes to the design of processing equipment, but so too is the biological state when it comes to yield – also the presence of roe can affect the economic value and eating quality of the species concerned. For catching purposes, fish are divided two broad groups, demersal and pelagic.

Demersal

This term means literally 'submerged' and is applied to all fish species living on or near the sea bed. Caught by trawls which are dragged along the bottom of the sea, or seine nets whereby the fish are encircled and trapped, demersal species are usually synonymous with white (non-oily) fish such as cod, haddock, whiting, plaice and sole.

Pelagic

Literally 'of or pertaining to the ocean' and applied to fish normally caught at or near the surface of the sea. Pelagic fish comprise, in the main, the migratory, shoaling, seasonal fish of families related to herring, mackerel and tunny.

Demersal species landed in western Europe tend to be much higher priced than pelagic fish. The average value of demersal species landed by UK vessels in 1994, for example, was £1067 per tonne (metric ton) compared with the average £128 per tonne for pelagic species.

SHELLFISH

Shellfish can be broadly divided into two groups, crustacea and molluscs.

Crustacea

This class comprises high value species like lobster, crab and shrimp or prawn. Their bodies are divided into segments with jointed appendages attached in pairs, while the exoskeleton or cuticle, composed mainly of chitin, is shed at regular intervals to allow for growth.

Molluscs

The soft fleshy body of these animals is a common feature. Species of commercial importance can be divided into three classes:

(i)   gastropods, which have a single piece external hard shell (e.g. whelk)
(ii) bivalves, which have a shell in two halves held together by an adductor muscle (e.g. mussel)
(iii) cephalopods, where the skeletal tissue has been reduced to an internal pen or cuttleshell (e.g. squid).

The flesh of shellfish, like that of fish, varies in composition and quality according to the time of year, but there is even greater seasonal variation which, in certain species, can dramatically affect their acceptability. For example, crab after moulting or shedding its shell in the summer can be rendered unusable as food. Shellfish also spoil more rapidly than fish.

Apart from shrimp and scampi (*Nephrops norvegicus*), the processing of shellfish is, by and large, unmechanized.

FISH OF ECONOMIC IMPORTANCE

Different species assume greater economic significance in some countries than in others, but in terms of fish (and shellfish) traded around the world, there are six main categories.

WHITE FISH

Known as groundfish in the USA, presumably because of their habitat (see previous section), species in this market sector are used in the manufacture of such popular products as fish fingers, coated fish portions, fish pies etc., which are sold around the world.

Cod is probably the single most important species in this sector in terms of value, but Alaska pollack is available in the greatest quantity. This latter species was used almost exclusively in the manufacture of surimi although more fish are now being sold as fillets or used for the kinds of products mentioned above.

Because of the growing scarcity of cod and other traditional white fish in some waters, formerly lesser known species such as hoki, caught in New Zealand waters, are being imported into Europe and the USA to make up the shortfall.

PELAGIC SPECIES

These oily fish are normally not eaten fresh in western Europe and the USA, but are sold preserved in some way such as smoked, cured, marinated or canned.

Markets often tend to be limited by country and seasonality in these areas – matjes (lightly cured herring) are eaten mostly in the Netherlands and Germany when the new season's herring are available in the early summer. There are bigger markets for some pelagic species, however, in Japan and Africa. The fact that pelagic species shoal and very large quantities can be caught in a short space of time, coupled with the lack of processing facilities or suitable markets, means that in some countries only relatively small quantities of pelagic fish are used for human consumption.

In Peru, among the top four of the largest fish catching nations in the world, around two-thirds of the catch may consist of anchovy. Most of this is used for the manufacture of fishmeal, with very little going into canning.

SALMON

The rapid growth of salmon farming since the mid-1970s means that salmon has moved from being a luxury food in some countries, into a worldwide traded commodity.

In 1994, 431 200 tonnes of farmed Atlantic and Pacific salmon were produced, according to the FAO information service GLOBEFISH. This compares with the 407 000 tonnes of wild Pacific salmon landed in the USA, the major catching country. Fresh farmed salmon can now be obtained, fresh, 52 weeks in the year, unlike the wild caught Atlantic or Pacific salmon which are only available as fresh fish on a seasonal basis.

SHRIMP

Shrimp, or prawns as they are called in the UK, have also become much more widely available due to the development of farming techniques in south east Asia, the Indian subcontinent and South America. The tropical, mostly Penaeid, species are now commonplace on the supermarket shelves of countries where the cold water *Pandalus borealis* was once the only species available.

Large shrimp are favoured for farming and three varieties account for most of the trade: the black tiger or giant tiger shrimp (*Penaeus monodon*), the Chinese white (*P. chinensis*) and the eastern Pacific white shrimp (*P. vannamei*).

There are about 50 species of shrimp which are commercially available and enter world trade channels. World catches (both cold and warm water varieties) have fluctuated between 2–3 million tonnes a year since the mid-1980s, while farmed shrimp have increased dramatically from 341 000 to 712 000 tonnes in the same period. The quantities fished appear to be stable but there has recently been a decline in farmed shrimp due to problems with disease.

TUNA

Tuna are pelagic species, but are treated as a separate product category because of their size – individual fish can weigh several hundred kilograms – and their importance to the industry.

There are five species, skipjack, yellowfin, albacore, bigeye and bluefin (both southern and northern). In most northern European countries and the USA tuna is rarely sold except in canned form. However this market is growing, and more tuna is starting to be sold in these areas as fresh or frozen fish.

By far and away the biggest market for tuna, is Japan, where the fish is regarded as a delicacy and, if the quality is right, is sold at very high prices to be eaten raw as sashimi.

It is interesting to note that the growing influence of the environmental lobby in the fishing industry has notably affected this market sector and most canned tuna labels carry a 'dolphin friendly' logo. (This is to confirm that the fishing method used to catch the tuna will not have caught any dolphins at the same time.)

CEPHALOPODS

The cephalopods, squid, cuttlefish and octopus, are actually molluscan shellfish where the 'shell' is an internal pen or cuttleshell, rather than an external means of protection.

Cephalopods are of more commercial significance in Japan and other countries in the Far East and in the Mediterranean, rather than in northern Europe and the USA, although squid is becoming more popular in these areas.

FISHING METHODS

It is not possible here to deal adequately with the different methods used to catch fish and shellfish around the world. This section will therefore be used to give a brief description of the main types of mechanized fishing employed.

Marine fin fish are mostly caught by nets, although lines with baited hooks are also used.

DEMERSAL TRAWL

The net, constructed of man-made fibres, tapers from wide shoulders down to the narrow tail or 'cod-end' where the fish are trapped. It is towed along the sea bed by two cables called warps which are normally 3–4 times longer than the depth of water being fished. A pair of otter boards or 'doors', one each side of the mouth of the net keep it open laterally, while floats on the headline and rollers (bobbins) on the footrope do the same in the vertical dimension.

Sometimes a bottom trawl will be towed by two vessels sailing on parallel courses abreast of each other, a system known as pair trawling, in which case otter boards will not be required to spread out the mouth of the net.

MID-WATER TRAWL

This method is most commonly used in the capture of pelagic species like mackerel, herring, sprats and pilchards, and can be operated by vessels working singly or in pairs. The main advantage in using a mid-water trawl is that the net can be fished at the depth distribution of the fish.

BEAM TRAWL

This is the oldest form of trawling in which the mouth of the net is kept open by a heavy, rigid beam which is towed in close contact with the sea bed. Beam trawls are used mainly for the capture of flat fish such as plaice and sole, but can also be used to catch shrimp.

SEINE NET

The shape of the seine net is similar to a trawl net, but it has much wider shoulders (wings). A rope warp is attached to one shoulder of the net and the other end of the warp to a floating buoy. The warp is paid out as the vessel circles away from the buoy until the net is cast into the sea and another warp is attached to the other wing of the net. (Depending on the depth to be fished and the nature of the sea bed, each warp may be 264–889 m in length.) While the second warp is being paid out, the vessel sails back to the buoy on a semi-circular course. The buoy is taken on board and both warps are attached to the winch which hauls in the gear as the vessel tows it. The movement of the warps drives fish into the path of the net. Alternatively, the vessel may be anchored after the encircling movement.

PURSE SEINE

This is probably the most efficient method of catching fish ever devised. It involves the complete encirclement of a shoal by a huge net in the region of 300 fathoms (550 m) long and 90 fathoms (165 m) deep. The net is cast into the shape of a lady's draw purse from which it derives its name, and the method of fishing is similar to that described above except, in this instance, one end of the net itself is attached to a buoy as the vessel goes round the shoal. When the vessel returns to the buoy, the net is closed at the foot and drawn in by means of a power block.

The concentrated catch can be scooped on board with the use of 'brailers', a smaller scoop-shaped net within the main net, or sucked on board through a fish pump.

OTHER NETS

Various stationary nets are used to trap fish which try to swim through them. Ground nets are anchored to the sea floor and rigged with floats to stand up from the bottom like a fence. A variation of the ground net is the trammel net usually consisting of three separate sheets of netting; the outer two have larger meshes than the inner net which 'catches' species like cod, haddock, ling, plaice, sole, skate, monkfish, etc.

Tangle nets are used, in some areas, to capture crawfish and other bottom-feeding species. Gill nets are used for many species and are sometimes set around wrecks where fish tend to gather.

POTS OR CREELS

These are traps to catch crustacean shellfish such as lobsters, crabs and langoustines or scampi. The traps vary in size and shape, and may be made of wood, metal or plastic with a netting cover. They are weighted so that they remain on the sea bed. Shellfish, attracted by the bait, enter through the 'eyes' and are unable to escape once inside.

In European waters pots may be fished singly or in a series of between 10 and 60 according to the capacity of the boat. Some vessels fish up to 700 pots.

FISH FARMING OR AQUACULTURE

The farming of fish in salt, brackish or fresh water is a growth industry in many parts of the world. Output has been predicted to reach 30 million tonnes by the end of the century, which is of great importance in the light of FAO predictions that world fish consumption may increase to 110 million tonnes in the same period, boosted mainly by population growth in the Third World.

Fish farming is not new. The Chinese have farmed carp for centuries; in Europe, in mediaeval times, the moats of castles were normally stocked with fish to provide food, particularly in time of war. It is only comparatively recently, however, that fish farming has become industrialized. In Scandinavia and other European countries Atlantic salmon and trout have been increasingly farmed since the turn of the century. Indeed, such is the success of the salmon farming industry in countries such as Norway, which is now producing more than 200 000 tonnes, Scotland and Ireland that the price of salmon may soon match that of cod on European markets.

In the USA catfish farming is of great importance in the southern states, and other fresh water species such as tilapia and striped bass are also reared under controlled conditions. Tilapias, which are omnivorous, are among the most widely cultured fish species in the world and are now grown in 75 countries. In Japan there is considerable cultivation of a number of fish and shellfish species, while the Chinese now intend to increase their total fishery production by nearly 50% (from 13 million tonnes to 18 million tonnes per annum) by the end of this century, with most of the increase to come from aquaculture. In Europe, including Scandinavia, special attention is being given to the farming of highly prized fish such as halibut, turbot, sea bass and even cod.

Molluscan shellfish such as mussels, Pacific oysters, scallops and abalone are also farmed in these and other countries with varying degrees of success. The most widely cultured shellfish, however, is the warm water shrimp.

Since the mid-1970s, the growth of shrimp farming has been extraordinary. The technology has been perfected for several species and production has been rising annually at a significant rate. Shrimp farmers contributed 712 000 tonnes to the world harvest in 1995. Production is concentrated in southern Asia and Latin America. There are many different systems of farming fish, the choice depending on biological, economic, technical and social factors. At one end of the scale, existing ponds or water courses are simply restocked with juvenile fish which grow by feeding on naturally available food. Such systems require little capital but their production per unit area of land or water is low. In Egypt, for example, some mullet ponds produce between 350 and 500 kg per hectare per year.

At the other end, modern hatchery and ongrowing systems are used with high stocking densities and specially formulated feed – Atlantic salmon reared in floating cages in western Europe or Scandinavia would be likely to have stocking densities ranging from 5 to 50 kg m^{-3}. In these systems natural food plays no part in the fattening of the stock; in both rearing and fattening there are straightforward analogies with animal husbandry practices.

In countries without a traditional aquaculture industry, farmed fish were at first regarded with suspicion by merchants and retailers. The advantages of constant supplies of fish of known size and quality, however, soon outweighed the possible disadvantages of the differences in colour and flesh texture, and irregularities in shape, which sometimes occur. Virtually all the trout now consumed in Europe is farmed, and sales of farmed salmon in most countries have long overtaken those of the wild-caught fish.

FISHERY CONTROLS

It became increasingly clear during the 1960s that the enormous growth taking place in fishery technology – purse seining, stern fishing, pelagic trawling, electronics, etc. – would make it necessary for fisheries to be regulated and managed if stocks were to be maintained at levels which were economically viable for fishermen to prosecute. This accelerated a worldwide movement, which had been in progress since the Santiago Declaration of 1952, away from the former doctrine of freedom of the seas and towards the concept of exclusive economic zones (EEZ). Most coastal states have now imposed an EEZ of 200 miles and have appropriated all the fisheries within that limit to manage and control as it sees fit. Depending on the species of fish involved, the state can either fish for them itself, or negotiate reciprocal rights with other countries in order to pursue its traditional fishing operations in foreign as well as domestic waters.

This policy has caused a number of political problems. Countries whose vessels have traditionally fished in waters within 200 miles of the coastline of other nations have found their vessels excluded, and members of the European Union (EU) have not ceased to argue about the theory of common access to European waters. To pave the way for eventual common access a Common Fisheries Policy (CFP) was established which effectively divided out the fish stocks among the various countries concerned. The CFP was drawn up in 1982 and a system of total allowable catches (TACs) and quotas was adopted in addition to various technical conservation measures such as the specification of mesh sizes in the nets used for catching the fish. TACs are determined annually by working groups of the International Council for the Exploration of the Sea (ICES) which collect the data from which they estimate the state of the fish stocks in the ICES sea areas. The Advisory Committee for Fisheries Management (ACFM) consolidates the findings of the working groups and tenders advice to the European Commission which, in turn, make proposals to the Council of Ministers which meets each December to agree the TACs for the following year. The agreed TACs are allocated to member states in the form of national quotas on a percentage basis. The CFP will last until 2002. What will happen then is uncertain for although it has been much criticised, both inside and outside the EU, no mutually acceptable alternative has yet been suggested.

Virtually all coastal states have instituted systems of fishing control based on catch limits calculated from scientific data although sometimes modified for political reasons. In most cases these systems have proved inadequate to prevent declining catches. It is only in New Zealand, where harvesting rights have been privatized through a transferable Quota Management System, that the ideal of sustainable fishing appears to have been achieved.

FACTORY SHIPS

Factory ships are vessels specially designed to carry out a modern mechanical fish processing operation, such as canning or freezing, entirely in the open sea. Some floating factories also make fishmeal in harbours or relatively sheltered waters. The concept is not new; Dutch 'busses' were used for salting herring in the sixteenth century.

The pioneer fish freezing venture on board ship was the attempt to freeze halibut at sea in 1928 to 1932 on a factory ship from Hull in the UK, supplied by small catchers. Soon after the 1939 to 1945 war, commercial experiments were started in the UK which led to a fleet of three Fairtry class freezing trawlers. Each of these returned with about 400 tons of frozen cod and haddock fillets, equivalent to about 1000 tons of whole fish every 70 to 80 days from fishing grounds such as the Grand Bank, off Newfoundland, which were otherwise normally inaccessible by vessels from British ports.

Subsequently, as a result of research and development work conducted by the UK government, partly financed by the fishing industry through the White Fish Authority, an impressive number of freezing trawlers were built for freezing whole white fish, chiefly cod and haddock in specially designed vertical plate freezers. The fish were stored at $-29°C$ ($-20°F$) on the vessel and thawed out at the Humber ports for subsequent filleting, etc. The next step, of course, was to carry out the filleting operation at sea. Several other countries have built freezer trawlers, particularly the USA and the former USSR.

Japan has a number of ocean going vessels for line catching and freezing tuna and this technique is also employed by countries such as New Zealand which supply the Japanese market. Japan also operates factory ships for on-board production of frozen surimi. These vessels were introduced in 1965 to overcome quality problems with land-based production that were attributed to the length of time between catching and processing. Production of surimi at sea expanded rapidly, surpassing land-based surimi in 1968 and accounting for 55% of the Japanese total in 1984.

CHILLING

One of the most perishable of all foods, fish begin to deteriorate immediately they are caught. The prime method of retarding this process is to reduce the temperature by chilling, or freezing, as soon as possible after capture. The simplest way is to surround the fish with ice, but mechanical refrigeration is now widely used and there is some usage of refrigerated sea water, particularly for pelagic species such as mackerel.

The bacteria responsible for fish spoilage are psychrophilic, i.e. cold loving, and even if fish are kept chilled at 0°C (32°F) under the best conditions of handling, the activities of these bacteria result in severe loss of palatability.

Species such as cod and haddock approach inedibility after about 14–16 days. At 5–6°C (42°F), a temperature at which many domestic refrigerators operate, spoilage will be about twice as fast, whereas at 11°C (52°F) it will be 4–5 times faster. Conversely, each successive degree of reduction in temperature towards the freezing point of the fish has a proportionately greater effect in reducing bacterial activity. This can be expressed quantitatively in terms of the temperature coefficient of growth, Q_{10}. Q_{10} values of marine bacteria of the types responsible for fish spoilage are shown in Table 2.1.

Table 2.1 Q_{10} values for marine bacteria

Bacterial species	Temperature range		
	°C	°F	Q_{10}
Pseudomonas fluorescens	20 to −5	68 to 41	3.7
	5 to 0	41 to 32	8.4
	0 to 3	32 to 27	9.3
Flavobacterium deciduosum	37 to 20	98 to 68	1.2
	20 to 5	68 to 41	1.2
	0 to −5	32 to 27	11.2
Achromobacter spp	25 to 7	77 to 45	2.3
	7 to −4	45 to 25	5.2

Chilled temperatures much below 0°C are to be avoided because of the detrimental effect of slow freezing on the texture of the flesh. The most satisfactory storage temperature is −2°C which results in an extension of the shelf life of white fish by several days. The success of this process, known as superchilling, requires accurate temperature control. Apart from the technical problems of maintaining this temperature exactly, a major disadvantage is that the material does become partially frozen and requires thawing out before processing can start. However, to avoid excessive wastage of ice, a temperature above the freezing point of water, but below 5°C, is probably best to aim for. It is important to stress that for efficient cooling the ice should surround the fish and not be used just as a cover.

Refrigerated chill rooms, cabinets and refrigera-tors are all designed to keep fish cool but, except for smoked fish and shellfish, it is good practice to use ice as well. The use of 'wet' ice may be unsuitable when fish is transported by air because the seepage of melt water into the aircraft must be avoided, but containers are now available to overcome this problem. Pending the acceptance of such containers by the major airlines the use of dry ice (solid carbon dioxide) or frozen gel packs, is common.

ICE

Various types of ice may be used, crushed, block, flake, or tube. The refrigeration effect of each type is more or less the same except when the ice is very slushy, when the effect is less. Flake ice, which is 'dry', light, has tiny pockets of air between the flakes and does not coagulate into lumps, is probably the best to use with fish, whether on board the catching vessel, in a processor's premises, or in a retail or catering establishment. Melting solid ice, in good contact with the surface of the fish because of the presence of liquid water, effects the cooling. In practice, cooling by ice is more effective the higher the temperature of the surroundings, because more melting takes place.

FROZEN FISH

Freezing as a general preservation process is dealt with elsewhere (Chapter 15), but in this section some of the features specially relevant to fish are noted.

Natural freezing of fish has long been known in cold climates such as Russia and Canada. The initial application of artificial refrigeration at the end of the nineteenth century gave poor results, largely due to the slow rates of freezing which were then the only ones possible. It is now well known that rapid freezing is essential for good quality. This is defined as taking no more than 2 h to pass from 0 to −5°C (32 to 23°F) in the thickest part of the fish, and freezing at this rate or faster is now normal good commercial practice.

Most fish begin to freeze at about −1°C (30°F), but multiplication of the putrefactive bacteria is only completely arrested at about −9°C (15°F). Although bacterial spoilage is then suspended and there is some reduction in bacterial numbers, not all the bacteria are destroyed by freezing. A proportion, depending on the time and temperature of storage and on other factors, remain dormant and survive. These resume growth and multiplication when the fish is thawed again.

The quality of the frozen product is closely related to the freshness of the raw fish: the fresher the fish before freezing the better it keeps in frozen storage. Whole fish of age equivalent to 3–4 days in ice will show some signs of deterioration on thawing and the staler the fish the more noticeable are the deteriorative changes that occur on storage.

Protection is required against evaporation in cold storage due to transfer of moisture from the air of the store to the refrigeration surfaces, resulting in unsightly and unpalatable freezer burn. Protection may be effected by moisture proof wrapping or packaging, or by glazing whereby water is sprayed on to the surface after freezing. The layer of frozen ice or glaze then serves as a protective coating. With small items, such as shrimp or scampi, the proportion of glaze may be as high as 50–60% of the finished product weight and obviously control is required to ensure that consumers are not defrauded. Commercial specifications and consumer labelling regulations commonly require the declaration of the quantity of glaze present, though this is not yet a legal requirement in all circumstances. The proportion of glaze is measured by weighing before and after thawing and allowing the liquor to drain away under standardized conditions.

Much fish is now frozen and cold stored pending further processing; this development has been accelerated by the development of freezing at sea on board the fishing vessel.

THAWING

Ideally frozen fish should be thawed immediately after removal from cold storage. The simplest methods are thawing in still air by simply leaving the frozen product out overnight at room temperature, or thawing whole fish in water. If fillets are thawed in water they may become waterlogged and lose much of their flavour. For air thawing the air temperature should not exceed 20°C to minimize bacterial growth on the surface, but very slow thawing must be avoided since bacterial growth may nonetheless cause the outer layers to spoil before the centre is fully thawed.

Thawing may be accelerated by mechanical means – air blast, heated water or vacuum methods – or electrically by dielectric resistance or microwave heating. The electrical methods are usually restricted to applications where the size and shape of the units to be thawed are constant, in other words to frozen fish blocks.

FISH BLOCKS

Fish is frozen in different forms, whole, headed and gutted (H&G) and as fillets, with or without skin. For the manufacture of many standard fish products, the fish block is a common starting point.

The international trade in frozen fish blocks has developed rapidly since the 1960s, and fish blocks are now the most important single source of raw material for large sectors of the fish products industry. Originally fish blocks were manufactured from layered blocks of skinless, boneless fillets and were termed laminated blocks. They were introduced into the UK in 1955 and by 1970, as the popularity of fish fingers and portions grew with the market for convenience foods, virtually all the cod frozen at sea and landed in Britain was made into blocks to satisfy demand at home and abroad, particularly for the Boston market in the USA.

In the 1970s, block production had become the largest single operation in frozen fish processing and the frozen block had become a commodity in international trade. A wide variety of sizes has been used; the nominal 7.4 kg block is now the most common.

As the industry expanded, mince, or fish flesh recovered by means of a bone separator, was added back into the block in the proportion naturally arising from the filleting operation (about 20%). Then blocks began to be made with higher proportions of mince, up to 100%, and it is now normal trade practice to control and specify the proportion of mince present in the fish block.

Polyphosphate solutions have been used to assist in producing blocks free from voids and to reduce drip losses on thawing or on subsequent cooking. This process can of course be abused to increase the content of added water in the block and it is now commonplace to specify also the polyphosphate and/or added water contents present.

COLD STORAGE

The normally recommended air temperature for the storage of frozen foods is −18 to −20°C. For frozen fish and a few other sensitive foods the temperature should be much lower and −30°C is recommended. Even at this temperature fish do not keep indefinitely. Microbial action ceases below about −10°C, but chemical reactions leading to irreversible changes in odour, flavour and appearance will continue slowly. In addition, unless the fish is properly protected against dehydration, physical changes will make it not only unattractive to look at – the condition known as 'freezer burn' – but also unpleasant to eat.

The most important factor limiting the storage life of frozen fish is the development of rancidity due to oxidation of the fat. This change is accelerated by the presence of small amounts of salt and for this reason fatty fish should not be brined before freezing. Good glazing excludes oxygen to some extent, and properly applied packaging films incorporating aluminium foil laminated with suitable plastics are effective. Smoked fish cannot be glazed, of course, but the storage life can be extended by packing them under vacuum before freezing. Table 2.2 shows the storage life of various kinds of fish when they are frozen and stored under good conditions; there will be variations according to fishing ground and season.

Table 2.2 Storage life of fish at $-30°C$

Commodity	Storage life (months)
White fish	
White fish, gutted	8
Smoked white fish	7
Fatty fish	
Herring, mackerel (gutted or ungutted)	6
Kippers	4.5
Shellfish	
Raw *Nethrops* (whole or shelled)	8
Cooked shucked mussels, raw shrimp,	
raw whole oysters, scallop meats	6
Cooked shrimp, whole crab and lobster	6
Extracted crab meat	4

(Source: Torry Research Station)

When fish are stored at temperatures above $-30°C$ deterioration is more rapid and storage life is reduced. Temperatures in distribution vehicles, secondary or distribution cold stores, frozen food rooms and display cabinets, are all likely to be above $-30°C$. In some retail display cabinets they may even be above the recommended level of $-18°C$.

It is important to remember that cold stores are designed to hold frozen goods at the specified temperatures, not to freeze the goods down to those temperatures.

FISH MACHINERY

Fish processing, in common with most manufacturing trades, is becoming increasingly mechanized, and equipment is now available for most basic operations such as gutting, heading (nobbing), filleting, splitting, skinning and bone separation (see below). Equipment is made in a number of countries including the UK, Germany, Scandinavia and North America. Computerization is playing an increasing role in fish processing as equipment becomes more sophisticated. For example, computers connected to electro-optical scanners are now used for the detection of bones and fins in fish fillets more reliably than by other methods.

Machinery also exists for many of the shellfish processing operations previously carried out by hand. Most of these relate to the peeling of shrimp and scampi, but some aspects of crab processing are also mechanized, particularly in North America.

BONE SEPARATORS

Various machines exist for recovering fish flesh from V-cuts (the anterior portion of a fillet containing the pinbones), from fish frames or even for separating flesh from the skin and bone in whole small fish. They operate mostly by forcing the softer flesh through a perforated cylindrical plate or screen, the harder bone and skin remaining behind. In some models the material to be separated is pressed to the outside of the cylinder by a moving belt, in others it is pressed from the inside by a rotating screw. In either case the separated flesh is produced in the form of a pasty or fibrous mass.

Care must be taken over the hygiene of the process and particularly to cool the product down immediately afterwards to minimize microbial growth since considerable heat is generated by the mechanical action.

FILLETS

Fillets are slices of fish muscle which have been removed from the carcass by cuts made parallel to the backbone, and from which all internal organs, head, fins, bones (other than pin bones) and substantially all discoloured flesh have been removed.

Block fillets, or cutlets, are paired fillets joined down the back from which all bones including the backbone (but excluding pin bones) have been removed; the fillets are defined as above.

The best quality fillets are prepared by hand; filleting machines are now in use which can give good results, though rarely as good as can be achieved by skilled hand operators. Fillets may be sold fresh or frozen; they may also be battered, breaded or otherwise treated as the major ingredient in further processed fish products.

ROES

Roes can be differentiated into the soft male and hard female roes which develop to edible proportions during the annual spawning cycle. The Japanese are probably the biggest consumers of fish roes (as distinct from caviar – see below) and export markets have been developed for the roes of several species. Cod roes are a delicacy when smoked, and so are salmon roes which can contribute as much as 35% of the total amount of the fish. There are existing or potential markets for the roes of other species.

CAVIAR

The name caviar should be strictly reserved for the roe of the various species of sturgeon. Pickled in salt, a controlled process of maturation encourages the development of specially characteristic flavours.

Of world production, 90% is shared between those countries which surround the Caspian Sea – Russia, Iran, Turkmenistan, Kazakhstan and Azerbaijan. Russia was, at one time, the largest producer of genuine caviar made from sturgeon caught in the Volga and other rivers, and processed on a large scale at Astrakhan on the Caspian Sea. Iran is the other major producer of caviar, sharing the resource of sturgeon from the Caspian Sea. China now exports keluga caviar from Manchuria.

Only sturgeon yields the top grade black caviar. A lower grade of pink 'caviar' made from salmon roe, called keta, has a milder flavour, and to people unaccustomed to caviar, perhaps a rather more pleasant flavour. It is very popular in Japan where most of the Canadian production is auctioned. A substitute caviar is manufactured in Scandinavia from the roe of the lumpsucker, or lumpfish (*Cyclopterus lumpus*), which has eggs comparable with, but larger than, those of the sturgeon. Scotland has also produced its own 'caviar' from trout. More recently a caviar substitute has been manufactured from seaweed in France.

LIVERS

Skate livers may reach 8–9% of the weight of the fish, cod livers 12–14% and shark 28–29%. In most other species they do not account for more than 3–4%, sometimes only 1–2%. Cod livers are served with the cooked flesh of cod in Norway, and some cod livers are canned, but fish livers are mainly used as sources of fish liver oils. This is particularly so where the content of the fat-soluble vitamins A and D is sufficiently high to justify the effort.

The livers are treated by steam injection, usually on board the trawler and the separated oil is decanted for subsequent purification, refining and destearinating (by cooling) at shore factories. The residual oil left in the 'foots' after steaming is also solvent extracted to furnish veterinary grades of oil.

COD LIVER OIL

Most of the vitamins A and D present in cod are stored in the liver, which is why cod liver was once regarded as a food supplement for children during and immediately after the Second World War. Recently, the product has made something of a comeback, although this time with adults concerned about their diets.

PACKAGING

In common with other foods, there was a swing to the prepackaging of fish for sale from self service counters in supermarkets in the USA, UK and elsewhere in the years around 1970. During this period, trials were carried out with chilled and frozen fish in packs consisting of expanded polystyrene base trays with simple printed or labelled film overwraps, or trays placed inside preformed pouches from which the air could be withdrawn if necessary. The trials were moderately successful in achieving consumer acceptance, but the potential wastage with chilled prepacks influenced fish processors and supermarket chains towards the frozen version. As a result of this work codes of practice were introduced for the prepacking of a number of wet fish and smoked fish lines and also for battered and breaded fish products.

Since that time the advent of the practice of gas flushing, or modified atmosphere packaging (MAP), has brought about a revival of the sale of chilled prepackaged fish by multiple retailers in the UK. There is an increasing use of MAP fish in Germany and other continental European countries. Vacuum skin packaging, which can be applied to both chilled and frozen fish, is another packaging system which attracted attention in the early 1980s in countries as far apart as France, Australia and New Zealand.

Both gas flushing and vacuum packaging offer advantages in shelf life extension, but both require strict control of temperature to ensure bacteriological safety. There are advantages in presentation as well as preservation. Gas flushed and vacuum skin packs are leakproof and odour free, and both provide excellent opportunities for merchandizing. One disadvantage is that of cost, particularly for the sophisticated equipment required and the packaging materials. However, the current consumer enthusiasm for fresh (i.e. non-frozen) fish and the fact that unusual and under-utilized fish and shellfish species can thus be presented in an acceptable form, have helped these and other forms of prepackaging become established.

FISH BOXES

The wooden barrels and boxes formerly used at fish auction markets have now largely been superseded by aluminium tubs, or kits and more recently by moulded plastic containers. These containers tend to be filled with different weights of fish at different ports.

For distribution from the port, the heavy, expensive, wooden box has been replaced by one made of newer and more hygienic materials such as fibreboard, polypropylene or polystyrene. Starting in Germany, there is now a definite move towards 'greener' fish boxes – returnable, recyclable, or even biodegradable.

FISH PRODUCTS

An extremely diverse range of products can be manufactured from fish and shellfish, depending on the species and the method of preservation of the raw material.

SALTING

Apart from the use of salt as a flavouring agent for smoked fish, its two principal uses as a preservative are brine pickle salting, chiefly of fatty fish such as herring, and dry salting, as applied to cod and similar species. Cod and other white fish can also be pickle salted, but fatty fish cannot be satisfactorily dry salted because the fat easily becomes rancid on exposure to air, whereas if immersed in a saturated brine pickle they are protected by a liquid barrier from the inroads of oxygen. In the case of products such as red herring, where fairly heavy salting is combined with drying in warm smoke, the antioxidants absorbed protect the fat.

It has been shown histologically that salt takes 24 h to penetrate into cod flesh to a depth of 17 to 18 mm. Penetration of salt through the skin and belly cavity is four times less than the rate of diffusion through a cut surface. Fat impedes diffusion, while freezing results in a 30% increase in the rate of penetration compared with unfrozen fish.

In brine pickling, the fish is first 'gripped', i.e. partly eviscerated in a special manner, the purpose of which is to remove the intestine but leave the pyloric caeca intact. These organs contain powerful enzyme systems which are responsible for the characteristic flavour of herring. The gipped fish are then packed in a wooden barrel, or tank, with solid salt which draws moisture out of the muscle and dissolves to form a saturated pickle which eventually covers the fish. The fish shrink and, after a week or two, the barrels have to be repacked and topped up before being filled 'bung-full' with saturated pickle. Once part of the staple diet in Europe, pickled herring have declined in popularity compared with the less highly flavoured products of the more expensive processes of canning and freezing.

Cod for salting (to produce 'klippfisk' as it is called in Norway) are usually headed, split and cleaned so as to leave a piece of the backbone at the tail end for strength. They are then packed in layers with solid salt to form stacks. The saturated brine formed is allowed to run away. The fish eventually become impregnated with a saturated salt solution (containing 26% salt) which amounts to a salt content of about 18% in the product. At a suitable stage in the curing process, the fish are removed from the stack, washed and allowed to drain, and are then hung up to dry for a few days at a time, with several intermediate periods of pressure in the stack to bring further moisture from the centre to the surface for the subsequent drying stage. Artificial dryers shorten this process and suitable plants are available (see DRYING). Various forms of microbial spoilage can affect salt fish, although the growth of most spoilage bacteria is stopped by sufficient salt.

CURING

A term widely used in the fish industry to denote the process of preservation by drying, salting, smoking, fermenting, acid curing or any combination of these.

MARINATING

The marinating of fish and shellfish involves its treatment with acid and salt. Not only does this process preserve the product by retarding the action of spoilage bacteria and enzymes (see AUTOLYSIS and BACTERIAL FLORA), it also gives it a characteristic texture and flavour.

The precise nature of the process is complex. The concentration of salt in a marinade will normally be lower than that required if salt alone were used as the preserving agent and the action of the acid, usually acetic or lactic, is essential in effecting preservation. At a pH of 4.5 or below, all food poisoning and most spoilage bacterial are prevented from growing and at such levels of acidity marinated products will keep for several months at 4°C. In practice, the shelf life will vary from a few weeks to several months depending on the nature of the product, the way in which it has been processed and packed and the storage temperature.

The main fish species associated with marinating is herring, although other species can be used including shellfish. Herring are usually 'cold marinated', but there are exceptions such as the German 'brathering' where the fish are fried as part of the process. Products are usually manufactured from boned or filleted fish, and a fat content of 5–15% is recommended. Too much fat is said to cause the fillets to gape and there are problems with oil separation, while too little can cause the product to be tough. Prolonged storage before processing can increase the incidence of blood discoloration of the flesh.

Methods of marinating vary, but the process usually consists of a two-stage process, the second of which may include the addition of sugar, spices and herbs depending on the final product. The process is more common in northern Europe and Scandinavia, but marinated products are becoming more popular in the UK and elsewhere.

Fish pickles

These include traditional craft products such as herring marinades, rollmops, pickled cockle and mussels, etc. In general, they are made by marinading in an acetified brine prior to packing in vinegar with appropriate spicing. Their microbial stability is related to the salt and acid contents (acetic with possibly some lactic acid); the general principles covering vegetable pickles apply. Acid hydrolysis of the protein during storage and the cooking effects of pasteurization may tend to shorten shelf life, but there is still a high demand for these products. In some cases salted herring are the raw material. They are partially desalted before rolling for final packing as rollmops.

Rollmops

Originating in Germany, rollmops consist of marinated herring fillets rolled up and held together with a thin sliver of wood with onions, gherkins and sometimes pieces of apple.

FERMENTING

Some European and Scandinavian fish products such as marinades, made from herring and other fatty fish, rely on enzymic action as part of the softening or maturation process. They must be recognizable as fish flesh at the end. In Asia, on the other hand, fish, small crustaceans and squid are deliberately liquefied to produce sauces or soft pastes. Products like nuoc mam, a fermented fish sauce used in Vietnam, Cambodia and neighbouring countries, are made from salted fish in which the action of the salt inhibits bacteria while allowing enzymic digestion to take place by autolysis. Some products consist of mixed fermentations of fish or shellfish with vegetables or cereals, in which the use of lactic acid bacteria or moulds may be involved. Yeasts, too, are used as fermenting agents in South East Asia.

SMOKING

The smoking of fish, as with meat, was originally a preservation process, almost certainly discovered by accident. Preservation results from a combination of partial drying caused by evaporation of moisture from the surface of the fish in the warm draught of air rising from a fire and the simultaneous deposition of aldehydes, phenols and other substances resulting from the partial distillation of the wood (cellulose yields mainly aliphatic compounds, lignin yields mainly aromatics). As a result, a smoked fish loses up to 20% by weight, but absorbs something like 10 mg of formaldehyde and other lower aldehydes, 100 mg of organic acids, 1 mg of ketones and 10 mg of mixed phenols per 100 g of product. In the process of the flesh becomes tougher, yellow or brown coloured and takes on a characteristic smell and flavour. Today, the reasons for smoke preservation are less important and the main reason for smoking is to produce a pleasant tasting alternative form of fish. The modern smoked products nevertheless do have somewhat longer shelf life than the unsmoked equivalents.

Most of the common white and oily fish can be smoked, including eels, and a number of shellfish such as mussels and oysters. Cold smoking, in which the smoke temperature does not exceed 30°C, gives a yellow colour to the fish but does not supply sufficient heat to cook the flesh. The best British example is the kippered herring, which requires cooking before consumption (see below).

In hot smoking the smoke temperature is 70°C or higher, thus the fish is cooked in the process. An example is hot smoked mackerel which is usually eaten cold; cold-smoked mackerel is available but is not nearly so popular. Fish are prepared for smoking by washing, usually removing the guts, beheading ('nobbing'), splitting, filleting and sometimes skinning, depending on the product. Salt is added by dry salting or immersing in a prepared brine. When the addition of colour used to be permitted, dye was included in the brine, but now this is rarely done. The fish may then be allowed to drain for up to several hours.

The traditional smoking kiln is simply a chimney. In modern mechanical kilns, which are usually computer controlled, smoke is generated outside the smoking chamber and moved over the fish by means of a fan. It is thus possible to exercise better control over the temperature of the smoke and the evenness of its distribution than in the simple chimney. Water sprays, or electrostatic precipitators, may be included in the smoke circuit to purify the smoke from benzpyrenes which are known to be carcinogenic.

Kipper

A kipper is a herring which has been split down the back to remove the guts and roes but leaving the backbone intact; an operation that is now almost always done by machine. After a light brining the herring is smoked.

Kipper fillets are herring similarly treated after preliminary filleting to remove the head and backbone. The term was originally applied to salmon which were caught after spawning and were usually smoked. Kippers in the UK, where they are mostly eaten, used to be coloured with the dye Brown FK, but this is now discontinued.

Bloater

A bloater is a whole, ungutted, slightly salted herring which is dried and then lightly smoked at a low temperature. The product therefore has a light smoke flavour and little alteration in colour.

DRYING

Natural drying in the sun and wind, and artificial drying over wood fires were the principal means of preserving fish in former times. Apart from the drying which occurs incidental to smoking, the chief processes in the North Atlantic area were the drying of stockfish and salt cod.

Stockfish (not to be confused with present day South African stockfish or stokvis, which is similar to hake) is still made in Iceland and Norway, and consists of plain dried, whole gutted cod or haddock. Large fish are usually split along the backbone into two portions, connected at the tail end. Drying takes several weeks, depending on the size of fish and the weather, and the final water content is about 15%. Above this water content moulds can grow and spoil the fish. Stockfish are mostly exported to hot countries such as Nigeria and Italy. There is also some limited demand in Scandinavia and the USA for the preparation of a dish called lutefisk, which requires prolonged soaking of the fish in water and alkali to soften it.

Salted cod is still dried to some extent in the open air in Norway, Iceland and Canada (Newfoundland). The water content is reduced from about 55% in the wet stacked state, to between 35 and 45% depending on the product and the market intended. In all the producing countries, mechanical dryers are now replacing natural drying for salt fish.

An appreciable proportion of the world catch of cod is still preserved as salt fish, with product mostly going to South America, the Caribbean area, Spain and Portugal, where bacalao is a popular product. (See SALTING).

CANNED FISH

More than a million tonnes of fish a year are canned throughout the world. Tuna and salmon head the list, but other pelagic species such as mackerel, herring, pilchard and sardine are also popular to some degree. Demersal white fish are not usually preserved by canning as the sterilization process usually produces a browning of the fish flesh. Shellfish species which may be canned include crab, shrimp, lobster, octopus, oyster and mussel meats.

Fish for canning are normally prepared by

nobbing (i.e. removing the heads), removing the tails, gutting, then either washing to remove loose scales or skinning completely. In some cases they are filleted. The prepared fish may be frozen at this stage and held in cold storage for completion of the canning process later, but this is a matter of local convenience and not essential. The fish may then be salted by soaking in 50–100% brine (13–26% salt by weight) and in some cases may be smoked. Brining has the effect of washing out the insides of the fish and getting rid of blood residues and most of any broken gut; it also reduces the moisture content of the flesh thus reducing the amount of cook out during sterilization and the dilution of any sauce.

The prepared fish may be packed into the cans by hand to ensure neat and efficient filling of the space with minimum damage to the fish or automatic filling machines may be used. The cans are topped up with brine, oil or prepared sauce, as required, then closed and sterilized in the usual way. It may be necessary to wash the seamed cans before sterilizing, using hot detergent followed by a water rinse.

Recent years have seen the emergence of the aluminium can which is now widely used for smaller sizes. The three-piece can is giving way to the safer two-piece can with one less seam to give trouble, and ring-pull ends are proving popular both for flat cans and for the smaller sizes of round can.

POTTED FISH

Potted fish use a recipe for short-time preservation of fish by cooking in vinegar, salt, spices, etc., often with a protective layer of butter. It is applied to herring, sardines, shrimp and salmon.

FISH PASTES

The usual starting material is cooked fish, normally canned. It is mixed with cereal (breadcrumbs), spices and usually with artificial colour, filled into jars or cans and sterilized by cooking in retorts under pressure. A 'botulinum cook' is necessary (see Chapter 15).

COATED PRODUCTS

There have been a number of interesting developments in coating technology since the first breaded products were introduced. Probably one of the best known coatings to come from Japan is the

tempura or puff-type speciality batter. These batters provide coatings of exceptionally high volume which are also light in texture, a far cry from the heaviness of some traditional batters.

Another innovative coating to originate in Japan, but which was subsequently developed in the USA, is the aptly named 'Japanese crumb'. The new coating was launched in the UK in 1980, and has since been followed by a variety of open textured, light eating crumbs. With long delicate rice-shaped particles giving a sponge-type texture described as a 'lighter bite', Japanese style breadcrumbs are manufactured in a variety of different flavours and colours.

Most coated products in the UK are now available with a three-way cook option and can be baked in a conventional oven, prepared under the grill or fried. Along with the removal of artificial colours from the coatings of frozen, and chilled, seafood products, other additives such as polyphosphate are being used less.

Fish fingers or fish sticks

The fish finger (fish stick in the USA) is the single product that has done most to popularize the eating of fish by children from pre-school to teenage. It is doubtful, however, if most children associate the product with fish that they see in whole or filleted form on a fishmonger's slab or supermarket fish counter. The UK is one of the leading markets for the development and consumption of coated fish products (fingers and portions).

Fish fingers, or sticks, and portions are regular sized pieces cut from rectangular frozen blocks of fish flesh, usually coated with batter, then crumbed before being flash fried and frozen. They may be packed in retail or catering size packs. Sometimes the fingers or portions are covered with batter alone, or portions may not be coated at all but be incorporated into products with sauces.

The typical British fish finger usually weighs about 1 oz (28 g) of which up to about 50% of the total weight may be batter and crumb. At the time of writing, there is no legal minimum fish content for the product, but the UK Government's Food Advisory Committee has recommended a minimum fish content of 55% for battered fingers and 60% for fingers coated with breadcrumbs.

Fish cakes

Fish cakes are made from a mixture of minced fish, potato and seasoning, coated with batter and breadcrumbs. White fish such as cod, haddock and saithe are normally used, but in recent years fish cakes have been manufactured from salmon (sometimes the offcuts from smoked salmon) and kippers. The composition of fish cakes in the UK is no longer controlled by legislation, however the compositional standard of not less than 35% by weight of the final product remains as an indication of a suitable standard.

Since fish cakes contain minced or flaked fish, trimmings from a processing line are commonly used, bone separators facilitate the preparation of bone free material, and potato powder has almost entirely replaced whole potato. Precooking is recommended when the history of the fish is uncertain and apart from fish cakes for local distribution, the product is mainly sold in frozen form.

FISH SAUSAGE

In Japan, fish hams and fish sausages are popular products. The sausage typically contains minced fish flesh with 10% pork fat, 10% starch and 2.5% salt, seasoning and preservatives. Chemical preservatives, such as furyl furamide and nitrofurazone, have been used to give a storage life at ambient temperature of a month or more.

The products are cooked and sliceable. Methods of making them were developed during the Second

World War when low cost sausage casings became available. The application of modern techniques led to a dramatic increase in sales of fish hams and fish sausages from 40 million lb (18 000 metric tonnes) in 1956 to a peak of 415 million lb (188 000 metric tonnes) in 1965.

Fish sausages were originally made from fish species such as tuna, but there was a switch to lower priced Alaska pollack surimi when tuna became expensive due to a combination of lower catches and an upgrading in quality which led to its use as sashimi. As well as the slicing sausage, the product could be canned after the fashion of frankfurters.

Several seafood processors are now selling fish sausages manufactured from surimi and the product is being manufactured in a variety of forms. Sales of fish hams and fish sausages have declined in recent years and were standing at 216 million lb (98 000 metric tonnes) in 1983.

SURIMI

Surimi is a washed, refined and stabilized fish mince which has been manufactured in Japan for about 900 years, first seasonally but later, after the advent of refrigeration, on a year-round basis. It is not consumed directly as such, but is used as the starting point for a whole host of seafood products, including kamaboko or kneaded fish pro-

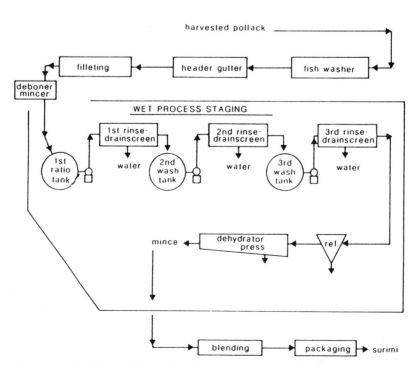

Figure 2.1 Diagram of a 'conventional' surimi manufacturing process (originally installed at the first land-based operation in Alaska, USA).

ducts which are available in a bewildering variety of shapes, colours and cooked forms. Kamaboko has a particularly elastic texture, known as 'ashi', which is much prized by Japanese consumers but strange to the Western palate. Therefore these products did not originally sell to any large extent outside Japan or the oriental communities in other countries.

In the 1970s, however, Japanese food technologists developed a method of manufacturing simulated crab leg meat from crab flavoured kamaboko, which came to be almost universally known as the 'crab stick'. This product became a spectacular success, not just in Japan, but in other countries, notably the USA, Canada, western Europe and Australia.

Surimi is mostly manufactured from Alaska pollack, (Figure 2.1) although other white fish species are now also used such as Pacific whiting, blue whiting and hoki. The Japanese originally fished Alaska pollack for its roe, which is considered a delicacy in Japan (see ROES). The flesh of the fish was also valued, but the quality on processing was poor due to denaturation of the protein after freezing. The use of cryoprotectants such as sugar, sorbitol and phosphates allowed the transition to a frozen product and subsequent industrialization of manufacture.

Land-based production of surimi began in 1960 and by 1964 there were 39 factories in Japan, all producing relatively low grade material. The poor quality was attributable to the length of time between catching and processing. To overcome this problem, factory vessels (see FACTORY SHIPS) were introduced in 1965 and sea-based production expanded rapidly, surpassing land-based production of frozen surimi in 1968 and accounting for 55% of the total in 1984. There was a five-fold increase in Japanese output of surimi in the 20 years between 1953 and 1973; in the peak year of 1973, surimi-based seafoods accounted for 37% of Japanese fish consumption on a live weight basis.

Small quantities of crab-flavoured kamaboko first appeared on the US market in 1978 as imitation crab legs, followed by lump and chunk forms of imitation crab meat and other shellfish analogues such as prawn/shrimp, scallop and lobster. In the late 1970s and early 1980s there was a great surge in popularity of the crab stick-type of product in Europe and Australia. Interest in some countries has since levelled off, but there now appears to be great potential for surimi-based seafood in France and southern Europe including Italy, Spain, Greece and Portugal.

Japan dominated the manufacture of surimi and surimi-based products in the early days, but many other countries such as South Korea have now started production. In 1986 it was estimated that nearly 20 countries were manufacturing surimi commercially or actively considering doing so.

There are other uses to which surimi can be put, both inside and outside the food industry. It is a versatile, functional protein material and development possibilities appear to be unlimited.

ANIMAL FEED

About a third of the world fish catch is converted each year to animal feed.

Fishmeal

Some species, mainly small bony and oily fish, such as sand eels and sprats, otherwise of little economic value, are fished specifically for conversion to fishmeal and oil. This operation is known as industrial fishing. The oil content of the species is important because fat is held at the expense of water, not protein, and therefore fatty fish give the same yield of fishmeal as lean fish and the valuable oil is an additional source of profit. Other sources of raw material for fishmeal are waste fish and trimmings from processing operations, condemned fish and fish withdrawn from sale for human consumption in times of glut.

Fishmeal from these sources is in demand for the manufacture of pet food. Concern is often expressed that much of the fish currently fed to animals could be upgraded and reclaimed for direct human consumption, particularly in Third World countries. Generally speaking, the major fishmeal producing countries in North and South America, Scandinavia and Africa, are those where pelagic species occur in such large quantities that there is no economically satisfactory means of processing them for human consumption. The fishmeal industries in these countries, however, have been subject to tremendous fluctuation. In Peru, for example, the presence of anchovies has risen and fallen dramatically due to over fishing on the one hand and natural biological phenomena on the other.

Controversy has risen over the catching of species suitable for human consumption by countries like Denmark with substantial agriculturally based economies, for conversion to feed for pigs and cattle. There is now EU legislation limiting the 'by-catch' when industrial fishing takes place.

Almost all fishmeal is manufactured by cooking, processing, drying and grinding the raw material in specially designed machinery. The drying stage, in particular, can be extremely smelly due to the substances emitted in the effluent from the dryers, and now what used to be tolerated as part of the odour of a fishing port has become a public concern. Fresh fish obviously causes less of a problem than stale material, so attention should be given to the thorough cleaning of equipment and unloading areas. Fishmeal made from fatty fish might contain 72% protein, 6% fat, 9% water and 13% minerals, whereas for meal made mainly from whole white fish and white fish offal, the analysis might read 66% protein, 4% fat, 9% water and 21% minerals. (The protein figure would be smaller if shellfish material had been included, or the edible flesh removed from the offal by a bone separator). The quality of the protein from fishmeal is high, as the constituent amino acids are present in a good balance for human and animal nutrition. Fishmeal is also a valuable source of the minerals calcium, phosphorus, magnesium and potassium, of four of the B complex vitamins and a number of trace elements.

Fishmeal should not contain food poisoning organisms which could infect the animals fed from it and thus find their way into human food. Salmonella is the most important organism in this respect (see BACTERIAL FLORA).

Prior to about 1910, the main use for fishmeal had been as fertilizer. Early in the twentieth century, the importance of fishmeal as an ingredient in animal feedstuffs was recognized and the pig and poultry industries are now large scale users. Up to 10% of the diets of pigs and poultry may now consist of fishmeal. At higher levels, with meal containing 10% fat, the fish oil may cause taint in the animal flesh. Fishmeal is also fed to mink, farmed fish, dogs, cats and cattle. It has been used in small amounts for prepared foods for humans.

Ensilage

In some countries, such as Denmark, fish or fish waste may be converted to liquefied feedstuffs for use in the pig, dairy and poultry industries. Liquefaction is achieved by adding acid to create the right environment for the bacteria naturally present in the viscera to work. The use of an organic acid such as formic, or a mineral acid such as sulphuric or hydrochloric, also halts the growth of spoilage bacteria and helps to break down bone. Although formic acid is more expensive than the mineral acids, preservation is achieved at a higher pH, about 4.5, and the silage does not need to be neutralized before adding it to the feed.

The composition of fish silage is similar to that of the material from which it is made and although there are some changes on storage, fish silage of the correct acidity appears to keep indefinitely at ambient temperatures. It is cheaper to produce than fishmeal and is without the problem of odour during manufacture. The disadvantages of fish silage compared to meal are, however, the higher cost of transport and the fact that it is still not yet an established commodity and therefore still requires marketing effort to encourage its acceptance. The most suitable outlet appears to be in pig farming, where it can be used in liquid feeding systems, but fish silage may also be blended with cereals to make semi-dry feeds. Silage from oily fish may have to be deoiled to ensure that the level of oil in the total diet does not exceed 1%, otherwise tainting of the animal or poultry meat may occur. Good results have been achieved by feeding fish silage to other animals apart from pigs; cow's milk and butter can be produced without taint and egg production from silage-fed hens is high.

The proteolytic action described above can be accelerated by the addition of plant enzymes, such as papain or bromelain, which are active at neutral pH and temperatures of 50–60°C. The liquid thus produced may be spray dried to prevent deterioration.

Fish hydrolysates made by these methods have been developed commercially for human use, principally in France. The high protein and low ash contents of fish hydrolysates meet the requirements for a casein substitute, but the high iron content of 40–50 mg kg^{-1} is a limiting factor for such a product.

Fish flour

Ethanol or isopropanol may be used to extract water, fat and some of the fishy flavoured substances from fishmeal, to produce fish flour with a protein content of 80–85%. The process is operated at ambient temperature, therefore there is minimum protein degradation and the product can be of suitable quality for human food use. It is more commonly used, however, for animal feed. For the manufacture to be economic it is essential to recover and re-use the solvent employed.

QUALITY ASPECTS

COMPOSITION AND ANALYSIS

Fish flesh contains chiefly water, proteins and fats, with traces of carbohydrates (several percent in certain shellfish), amino acids and other non-protein nitrogenous extractives, various minerals and vitamins (including the fat-soluble A and D). The elementary composition of fish is approximately as follows: O, 75%; H, 10%; C, 9.5%; N, 2.5–3%; Ca, 1.2–1.5%; P, 0.6–0.8%; S, 0.3%; together with traces of 60 other elements.

The enormous variation that can occur in the proximate composition of fish flesh from one species to another, and at different seasons, is indicated in the following table:

	Range (%)	Average (%)
Water	20–90	74.8
Protein	6–28	19
Fat	0.2–64	5
Ash	0.4–1.5	1.2

Edible portion

Fish carcasses vary considerably from one species to another in their proportion of edible flesh and, in addition, there is some seasonal fluctuation within species caused by variations due to feeding and breeding. Table 2.3 contains approximate average representative values for commercial species. Obviously shellfish have a low edible portion content, owing to the weight of shell. Whitebait, at the other extreme, are completely edible. Most white fish are landed in the gutted state, whereas pelagic fish are not usually gutted on board. The percentages given for edible portion are based on the landed weight.

Water content

Water content is generally between 80 and 85% in lean fish. For cod, it is usually about 81%, although in cases of starvation values as high as 89% have been observed. Even higher values have been recorded for diseased fish, e.g. 96% for a certain jellied condition of American plaice, the protein content being less than 3%.

In the case of fatty fish, water content is lower due to the higher fat content. At a rough estimate

Table 2.3 Approximate average yield of edible flesh and approximate composition of raw fish (UK species)

	Edible portion (%)	Composition per 100 g				
		Water	Protein	Fat	Carbo-hydrate	kcal
Demersal marine fish						
Gadoids						
Cod, haddock, saithe, whiting	45	81	16	0.5	0	99
Ling, tusk	50					
Pollack	60					
Hake	65	78	16	3	0	91
Flat fish						
Flounders	40					
Plaice	45	79	16	2	0	82
Lemon soles, megrims,* dabs soles, turbot, brill	50					
Witches						
Halibut	60	76	16	5	0	109
Elasmobranchs						
Skates and rays	45	80	16	1	0	73
Dogfish		78	13	5	0	61
Miscellaneous						
Monkfish	25	82	13	0.5	0	61
John Dory	40	80	16	1	0	73
Catfish Gurnards and latchets Bream	40	78	16	3	0	81
Redfish, red mullet Congers	50	76	16	5	0	109
Pelagic marine fish						
Herring, pilchards*†	55	66	16	15	0	199
Sprats		72	15	10	0	150
Grey mullet		78	16	3	0	91
Mackerel	50	73	16	8	0	136
Whitebait	100	78	16	3	0	91
Freshwater fish						
Salmon	65	67	17	14	0	194
Eels	50	69	13	16	0	196
Shellfish						
Crustaceans	40	73	20	4		116
Crabs		69	21	2	1	106
Shrimps, prawns* Lobsters, Norway lobsters Crawfish	45	73	20	4	2	124
Molluscs						
Oysters	10	78	10	2	6	82
Cockles*	20	60	11	0.5	6	73
Periwinkles		74	15	1	5	89
Scallops = clams		78	18	1	–	81
Mussels	30	82	12	2	2	74
Whelks	40	76	18	2	2	98

*Signifies no know edible portion data.
†Signifies no known analyses.

it can usually be taken as 80% minus fat content. Conversely, attempts have been made to predict fat contents of herring from a determination of water content.

Water, being the most abundant chemical component of fish, is perhaps the most frequently estimated constituent. Some commodities such as salt cod and fishmeal are often sold on a guaranteed maximum water content.

Methods of measuring water content in fish and fish products are various; the precision of the method selected depending on the accuracy required. For fishmeal, special quick methods have been devised depending on infrared heating of small samples directly on the balance pan. Electrical methods depending on conductance and capacitance have also been used. For general routine measurements, it is often accurate enough to use the Tate and Warren method of distillation with toluene and entrainent of the water driven off; a result being obtained in less than an hour. Where speed is no object, it is often good enough to dry a 50 g sample of minced fish in an air oven at 105°C (221°F) overnight. A precise method which has been used to show small seasonal differences in the water content of cod muscle, is to dissect out 8–9 myotomes (i.e. muscle flakes, numbers 8–16 counting from the anterior end of the fillet) and determine the loss on drying at 100°C for 7 days.

Oils and fats

The fat contents of fish are very variable, from below 1% in many white fish, while in pelagic species it can vary from less than 1% to over 30% (and certain US lake fish contain over 60% fat). The fat content of each species can fluctuate dramatically according to the season, although the difference is much more marked, and the upper limit much higher in pelagic species than with white fish. The table below shows the typical variation which can occur.

Species	Fat (%)
Cod	0.1–0.9
Hake	0.4–1.0
Halibut	0.5–9.6
Plaice	1.1–3.6
Redfish	3.2–8.1
Herring	0.4–30.0
Mackerel	1.0–35.0

Source: Torry Research Station

The fat of white fish is stored in the liver. In the case of cod livers, up to about 75% of the tissue may consist of fat in the form of oil, the liver amounting to 5% of the weight of the fish. Fat in oily species is distributed in globules in the flesh, most of which are concentrated in a layer just under the skin. All fish use fat to a certain extent as a source of energy, particularly when food is scarce during the winter months. Salmon use stored fat to supply the energy for swimming upstream and for jumping water falls.

The chemistry of fats is dealt with elsewhere in

this book, (see Chapter 8) but since there has been a great deal of publicity about the potential benefits to health from increasing the amount of fish oil in the diet, it is worth giving a brief overview of the situation.

The fats in fish are different from those of warm blooded animals in that they are usually liquid at room temperature. This is because they contain polyunsaturated fatty acids. Most vegetable oils also contain polyunsaturated fatty acids, and these acids have two or three double bonds in the hydrocarbon chain, while fish oils have a large proportion of fatty acids with five or six. The location of the double bonds is different in each case, giving rise to different names for the oils from the two sources. Fish oils primarily contain the ω-3 series of fatty acids which are also found in other forms of marine life. Two ω-3 fatty acids predominate in fish oil: eicosapentaenoic and docosahexaenoic, of which the former is the better known. The ω-3 fatty acids, and the ω-6 fatty acids in vegetable oils, cannot be synthesized by the body and must therefore be included in the diet. It now seems that ω-3 fatty acids have a role to play in the prevention of coronary heart disease, in fact medical evidence is accumulating for a curative as well as a preventive effect of fish oils. Sufferers from other apparently unrelated Western-style diseases such as rheumatoid arthritis, multiple sclerosis, asthma, diabetes and even breast cancer, may also benefit from diets with a higher seafood content than the average in most meat eating countries. The fact that fats in fish are unsaturated means, of course, that they react easily with oxygen from the air and go rancid. This will not cause problems with fresh fish stored for a few days in melting ice, but when frozen and kept in a cold store conditions are more conducive to oxidation. The higher the fat content, the greater the risk of rancidity and it is for this reason that fatty species have a shorter shelf life than those that are lean.

Phosphatides (phospholipids)

Fish flesh contains important phosphatides, including lecithin (phosphatidyl choline), cephalin (in which ethanolamine replaces choline), and sphingomyelin (in which sphingosine replaces glycerol). Some of the lecithin in fish tissues is free, but some is bound to protein in loose, easily dissociable complexes (lipoproteins). The total content of phosphatides in various species varies from 0.38–1.1%, including 0.21–0.65% of lecithin and

0.037–0.11% of sphingomyelin. The phosphorus content of lecithin is 5–10% of all the phosphorus in fish flesh.

Protein content

The protein content does not normally vary greatly and is usually around 16%, although tuna is an exception with over 20%. Actomyosin amounts to about 70% of the total protein content of fish muscle. Cod actin and cod myosin have been prepared separately, their electrophoretic and ultracentrifugal properties studied and amino acid analyses made. Thus G-actin from cod exists mainly in the form of a dimer of molecular weight of about 150 000, whilst cod myosin has a molecular weight of about 550 000.

Their respective isoelectric points are 4.75 and 5.3, which are practically the same as those for rabbit actin and myosin. In a number of respects, fish actin and myosin differ from those of rabbit. With regard to amino acid composition, however, cod myosin, cod actin and cod natural actomyosin closely resemble the corresponding rabbit proteins, and the ratio of myosin to actin in cod natural actomyosin lies between 2 and 4:1 in close agreement with the corresponding ratio in rabbit muscle.

When the various protein fractions of fish are examined electrophoretically, each species exhibits a characteristic pattern by which it can be identified. There is, however, no similarity between related species.

Fish muscle contains less connective tissue than, for example, lean beef, which probably accounts for its tenderness. In bony fish, the connective tissue nitrogen is usually 3–6% of the total protein nitrogen and in elasmobranchs 8–10%. About 80% of the connective tissue appears to be collagen and 20% elastin.

Amino acid analyses for food fish species are roughly similar. Protein fractions corresponding with tropomyosin and nucleotropomyosin have been isolated from fish muscle. So also has myoglobin. Myogens from fish exhibit different properties from rabbit myogen and seem even more heterogeneous. Cytochrome C is particularly active in the lateral brown band of fish; 0.1–2.0 mg of cytochrome C per 100 g has been reported in the white flesh definitely freed from dark muscle for a number of Japanese species, including mackerel and tunny; also 10–300 mg% of haemoglobin and myoglobin, including up to 100 mg% of myoglobin alone in the case of tunny. Over 90% of fresh fish protein is soluble in 0.5 M KCl, 5% NaCl, 7% LiCl, etc., and the reduction in salt solubility consequent upon processing by freezing or dehydration has often been used as a measure of denaturation.

Estimation of protein

For many industrial purposes an accurate enough figure for protein content is obtained by subjecting a sample of minced fillet to an ordinary Kjeldahl digestion and distillation and multiplying the resultant figure for nitrogen by 6.25. The 'crude protein' figure thus obtained, however, is too high because the total nitrogen includes anything from 10 to 40% of 'non-coagulable' nitrogenous constituents of a non-protein nature, quite apart from the accepted factor 6.25 which in any case is a little too high.

It is possible to coagulate proteins, peptones and proteoses by means of trichloracetic acid or by precipitation with cuprous oxide and so measure true or coagulated protein. For critical work, however, a more precise method is required than merely taking samples of bulked minced fillet.

For the greatest precision, a special part of the fish should always be selected for the sample, and the thirteenth and fourteenth myotomes, counting from the anterior end of the fillet, are regarded as the most consistent. Weighed samples are ground in 15% trichloracetic acid and centrifuged to remove non-protein nitrogen, which is then poured off. The precipitated proteins are then estimated by the micro-Kjeldahl procedure.

Nucleotides

Acid soluble nucleotides are found in cod muscle as follows:

	mg%
adenosine monophosphate (AMP)	2.4
adenosine diphosphate (ADP)	24.5
adenosine triphosphate (ATP)	272
inosine monophosphate (IMP)	44

Salt content

Although sea water contains about 2.5% salt, marine fish flesh contains only about 0.2%. Salt is introduced into fish in processing at two levels: either for flavouring at about 2%, as in kippers etc., or for long term preservation at about 18–20%, as in salt cod. Salt content is measured after extraction with water and titrating with standard silver nitrate solution, either using an indicator such as dichlorofluorescein, or potentiometrically using a platinum copper electrode system.

Minerals

Fish resembles meat in its content of useful minerals, and provides a well-balanced supply of those elements required in the diet. Sea fish, in common with other marine organisms, are an outstanding dietary source of iodine, a deficiency of which causes the disease goitre. The bones of fish afford a rich source of calcium and phosphorus in a favourable combination, and fish are the only animals of which some of the bones are likely to be eaten, for example the small bones in herring or the softened bones of canned fish. There is about 1–1.5% of mineral in fish muscle: figures above about 2% usually refer to samples containing skin or bone. Table 2.4 gives a rough indication of the mineral composition of fish muscle, all species being averaged together and no attempt

Table 2.4 Mineral composition of fish flesh (after Vinogradov, 1953) (mg per 100 g)

Element	Average	Minimum	Maximum	No. of readings
Na	72	30	134	48
K	278	19	502	50
Ca	79*	19	881*	51
Mg	38	4.5	452	51
P	190	68	550	49
S	191	130	257	26
Fe	1.55	1	5.6	48
Cl	197	3	761	13
Si	4	—	—	2
Mn	0.818†	0.0003	25.2	38
Zn	0.955	0.23	2.1	4
Cu	1.199	0.001	3.7‡	93
As	0.37	0.24	0.6	3
I	0.148	0.0001	2.73	201

*Most values lie between 20 and 40 mg%.
†Most values lie near 0.04%.
‡Some values recorded as 'trace'. (For further details see H.M. Stationery Office, 1959).

Table 2.5 Mineral elements in shellfish (mg per 100 g) (after Vinogradov, 1953)

Element	Suggested average	Minimum	Maximum	Suggested average	Minimum	Maximum
Ca	120	55	185	120	30	320
Mg	50	20	385	50	35	105
P	250	205	340	300	270	350
Fe	8.5	2.1	26.0	1.45	0.7	2.2
Cu	3.7	0.3	13.7	0.6	0.3	1.6
I	—	0.03	0.04	—	0.02	0.04

being made to distinguish reliable from unreliable determinations. All figures are mg% wet weight.

Shellfish, in general, contain rather more minerals than fish, as is shown by Table 2.5.

Vitamins

All the vitamins necessary for good health in animals, including man, are present to some extent in fish (Table 2.6), although amounts vary widely from species to species and throughout the year. The vitamin content of individual fish of the same species and even of different parts of the same fish can also vary considerably.

It is the parts of the fish not normally eaten,

Table 2.6 Vitamin contents of fish

Vitamin	Units	Range encountered among the main fish species
Fat soluble		
A	$IU\ g^{-1}$	0–1000 (in the flesh)
		200–800 000 (in the liver)
D	$IU\ g^{-1}$	0–1000 (in the flesh)
		5–30 000 (in the liver)
Water soluble		
B_1 (thiamine)	$\mu g\ g^{-1}$	0.2–2.4
B_2 (riboflavin)	$\mu g\ g^{-1}$	0.3–6.6
Niacin	$\mu g\ g^{-1}$	12–110
B_6	$\mu g\ g^{-1}$	0.3–9.8
B_{12}	$\mu g\ g^{-1}$	0.01–0.14
Panthothenic acid	$\mu g\ g^{-1}$	1.5–10.9
Biotin	$\mu g\ g^{-1}$	0.05–0.9
C (ascorbic acid)	$\mu g\ g^{-1}$	0.3
E		12

such as the liver and gut, which contain much greater quantities of the oil-soluble vitamins than the flesh. Almost all of the vitamins A and D in cod and halibut are located in the liver, hence the feeding of young children with cod-liver oil which was once commonplace. In contrast, in eels the same two vitamins are present mainly in the flesh.

Water-soluble vitamins such as B and C are more evenly distributed throughout the fish, and the flesh usually contains more than half the amount present. The roe is also a good source of these vitamins. In general, the vitamin content of white fish muscle is similar to that of lean meat, whereas herring is a good source of vitamins A and D, particularly the latter which is present in very few foods. The vitamin content of fish is not markedly affected by processing or preservation, provided storage is not prolonged.

pH

The pH of a live or newly killed fish is slightly alkaline (7.05–7.35). After death, there is a rapid fall as glycogen breaks down, with the formation of lactic acid. Fish flesh and extractives, particularly trimethylamine oxide, are fairly strong buffers and the pH of fresh fish is usually around 6.6 (compare 5.5 for meat). Only halibut seems normally to fall below 6.0.

As the fish begins to get stale, trimethylamine oxide is reduced by bacteria to trimethylamine which does not buffer in the physiological range so that the pH then falls slightly to 6.3. In the later stages, however, lactic acid disappears and more bases, including ammonia, are produced so that pH rises and can reach 8.0 or more in a really putrid fish. The natural variability of pH over this relatively narrow range renders it unreliable as an index of freshness or staleness.

MUSCLE STRUCTURE

It is obvious to the naked eye that fish do not possess the same kind of gross muscle structure that prevails in land animals, i.e. there are no discernible 'single' muscles sheated in connective tissue and attached by means of tendons to the bones. This is to be expected, as there are not the same kind of muscular activities to be performed. At the microscopic level, however, the basic structure within the muscle fibre is the same for both fish and red meat. (Meat muscle is tougher than fish due to the amount of connective tissue which is present, and also to the difference in pH which is lower in meat.)

There are two types of muscle tissue apparent in the body muscles of fish, which are differentiated by colour. In most species, the white tissue (in migratory species like herring, mackerel and salmon it is light brown or pink) makes up the bulk of the flesh. At the surface, and lying in a fairly broad band along the mid-line, is a strip of darker muscle varying in colour from red to brown.

The proportion of red muscle varies from species to species, seemingly related to the amount of activity. In some species it is not confined to a layer underneath the skin, but also penetrates inwards in a wedge towards the backbone. There are differences in the chemistry and physiology of the two types.

The musculature of fish includes three groups of lateral, striated muscles, head, trunk and fin. The greater part are trunk muscles or somatic muscles on either side of the vertebral column consisting of four longitudinal muscles, two dorsal muscles and two ventral muscles, separated from one another by strong connective tissue. Both the dorsal and ventral muscles are separated laterally by fine partitions of connective tissue, called myosepta, into a number of segments called myomers or myotomes (there are as many segments as there are vertebrae). Myotomes have the shape of hollow cones which fit one inside the other with their apices turned forward (towards the head of the fish). Myotomes consist of fasciae of longitudinal muscle fibres, covered by loose connective tissue (the endomysium). The ends of the muscle fibres are joined to the myosepta which in turn are connected, through intramuscular partitions and ligaments, to the skeleton. The primary muscle fibre, or cell, is the basic morphological and functional element of the trunk muscles. The surface of the muscle fibre is covered by a fine, elastic membrane (the sarcolemma) containing myofibrils which occupy most of the cell and sarcoplasm. Head and fin muscles consist of the same kinds of muscle fibres except that they are not separated into segments.

PHYSICAL PROPERTIES

It is often assumed in engineering calculations, as a satisfactory approximation, that fish behave in the same way as water with regard to latent heat, etc. For more discriminating requirements this is incorrect. Fish contain only about 80% water and less if they are fatty. Over the range 0–20°C (32–68°F) the mean specific heat of lean fish is about 0.80 ± 0.01 and of fatty fish about 0.7, depending on the relative fat and water contents. As the specific heat of fat can be taken to be 0.35, the specific heat of fish can be calculated approximately by the law of mixtures once its approximate analysis is known.

Obviously, the apparent specific heat of fish just below the freezing point will be much higher than that of ice (0.52 kcal kg^{-1}), depending on the temperature because of the latent heat involved in crystallizing ice from the freeze-concentrated tissue fluid. However, at $-30°C$ ($-22°F$) a value of 0.44 has been quoted.

Similarly, the thermal conductivity of fish is somewhat less than that of water (0.48 kcal m^{-1} h^{-1} °C^{-1}) as the conductivity of the dry matter of fish is only about 40% that for water. The thermal conductivity of ice (1.93 kcal m^{-1} h^{-1} °C^{-1}) is about four times that for water and for frozen fish is between 75 and 80% of the value for ice.

The thermal diffusivity (= thermal conductivity/specific heat × specific gravity, which is a measure of the rate of change of temperature of a body on being heated or cooled) of both fresh and frozen fish is about one-third of that for water $(0.00045\ m^2\ h^{-1})$ and ice $(0.00365\ m^2\ h^{-1})$, respectively.

The electrical properties of fish are of increasing importance with the development of dielectric and microwave heating, and electrical conduction thawing (see also FISH TESTER). The dielectric constant of wet fish is less than that of water (80), and that of frozen fish is slightly more than that of ice (2), owing to the residual liquid phase at ordinary cold storage temperatures. The electrical resistance decreases with increase in the frequency of the current transmitted and also with rise in temperature, and coagulation also leads to an increase. The following are recommended values for the principal physical properties of cod (from work at Torry Research Station):

	$-30°C$	$0°C$
Thermal conductivity (kcal $m^{-1}h^{-1}°C^{-1}$)	1.58	0.474[*]
Thermal diffusivity (m^2h^{-1})	10.4×10^4	1.5×10^4
Dielectric constant at 40 MHz	3	65
Specific resistivity at 50 Hz ($\omega \times cm$)	25×10^6	800

[*] Values of 0.39 and 0.40 have been quoted in the literature for other species.

FISH TESTER

The (now disestablished) Torry Research Station in the UK developed an instrument called the GR Torrymeter which measures certain dielectric properties of fish skin and muscle. These change in a systematic way after the death of the fish and the instrument thus gives an indirect measure of the freshness which it displays as a number. The latest commercial version of the instrument displays the average value from 16 readings to eliminate variation within a batch of fish and has the advantage of requiring little or no physical handling of the fish for the readings to be taken. The Torrymeter is more useful for some species than others, though it can be used for all species. It cannot be used for fish which have been frozen.

Whatever method is used for assessing the freshness of fish, the eyes, nose and palate of the fully trained assessor are probably the most reliable instruments overall.

FRESHNESS TESTS

Freshness is the most important single criterion of quality for most fish products and it is essential

for anyone concerned with the quality of fish products to be able to estimate freshness with an appropriate degree of accuracy. The circumstances under which freshness has to be estimated, however, will vary in the chain from the port to the retailer and different checks and tests will be applied accordingly. It must be borne in mind too that freshness is not an easy property to define or to measure. Loss of freshness followed by spoilage is a complex combination of microbiological, chemical and physical processes, and there is no single component to be measured which alone will give a freshness rating. What can be done is to measure some general feature of the spoilage process and use this an an index of the stage of freshness. Subjective sensory methods are still the most satisfactory (see QUALITY GRADING).

Of the numerous attempts to develop objective laboratory standards for fish, only two have stood the test of time as being reliable: determination of amines, particularly trimethylamine and determination of hypoxanthine. The former test relates to bacterial activity, the latter to enzymic. The two tests complement one another and have different ranges of acceptability and usefulness.

Trimethylamine and other volatile amines

Trimethylamine $(CH_3)_3N$ (TMA) is formed from marine fish as they become stale, as a result of bacterial reduction of trimethylamine oxide (TMO). Marine teleosts may contain 200–300 mg of TMO/100 g, depending on season, fishing ground, etc.; marine elasmobranchs may contain as much as 1.5%. TMO is found only in small quantities in freshwater fish.

TMA may be determined by the Conway microdiffusion technique, or by the picrate colorimetric procedure as in AOAC. Both methods require a high degree of laboratory skill and care. In species such as cod, the quantity of TMA is below 1 mg/100 g of wet muscle up to about 5 days in ice, increasing to about 5 mg/100 g after about 11 days in ice, then rises steeply to 10–20 mg/100 g after about 15 days in ice, by which time the fish is usually very unpalatable.

Although the estimation of TMA presents one of the most reliable objective tests for the freshness of fish, there are nevertheless a number of causes of variability and irregularity in the results, such as seasonal variations in TMO content and variable proportions of TMO-reducing organisms in the bacterial flora. Furthermore, TMO is water soluble and is liable to be leached out of the fish by melting ice. TMA itself is volatile, particularly at high pH. In spite of all this, a TMA value corresponds with the degree of freshness assessed by palatability in 70–80% of cases.

Other volatile amines and ammonia are formed during the deterioration of the fish, and the determination of total volatile bases (TVB) may also be used. Several analytical procedures are in use which differ mainly in the way the bases are released from the fish and distilled, and in comparing results it is essential to specify precisely the method used to obtain them. Changes in TVB content are similar to those of TMA, except that the initial value is high due to the presence of ammonia in very fresh fish. In non-elasmobranch species, the concentration of ammonia does not increase until the late stages of spoilage and virtually all the change in TBV is due to the TMA component; in many species there is also a small increase in dimethylamine, DMA.

In fresh fish, the TVB, expressed as nitrogen, should be below 20 mg N/100 g; in definitely spoiled fish the value may be greater than 30 mg N/100 g. However, because of variability and the fact that the starting values are high, TVB is not generally regarded as a sensitive index of freshness and the test is usually reserved for fish near the limit of acceptance.

Hypoxanthine

The measurement of hypoxanthine as an index of freshness is not as common as the measurment of TMA or TVB, but it is a better predictor over a wider range of qualities and is applicable to a wider range of species and products. Hypoxanthine is a normal component of fish flesh in only very low concentrations in the living animal. It results ultimately from the breakdown of adenosine triphosphate in normal glycolysis, but on death the balance of the enzymic reactions is disturbed and hypoxanthine, unlike TMA and TVB, increases soon afterwards. It can therefore be used to discriminate between batches of fresh fish.

Other tests

Other chemical tests that have been proposed include the estimation of tyrosine, H_2S, indole, volatile reducing substances and volatile fatty acids, with varying success. Physical tests which have been tried include the measurement of pH of the muscle or the skin, buffering capacity, electrical conductivity, refractive index of muscle juice or eye fluids and the optical density of the lens of the eye itself. Of them all, the most useful for some time was the Torry FISH TESTER (q.v.).

Finally, since bacterial activity is the prime cause of spoilage in unfrozen fish, various bacteriological tests have been investigated, but such tests are now basically used as checks on hygienic quality and are not well related to spoilage of fish. Dye reduction tests, like the methylene blue or resazurin tests used in the milk industry, have similarly failed to give reliable results.

AUTOLYSIS

Enzymes present in the flesh and various organs of a fish control a variety of processes after death which result in the breakdown of many of the tissue components. The process is called autolysis. The most obvious example is in ungutted fish when enzymes digest the gut walls and surrounding tissues, leading to burst bellies which are a noticeable feature of non-iced or poorly iced herring in the summer. Most fish apart from small pelagic species, and sometimes the larger herring and mackerel, are gutted soon after catching to avoid this problem. In some shellfish species such as shrimp and lobster, autolysis can occur extremely rapidly, the flesh being attacked by gut enzymes within a few hours after death, and therefore most shellfish must be processed as rapidly as possible.

The enzymatic changes lead to important changes in odour, flavour and texture, all of which can be reduced by lowering the temperature – hence the need to chill or freeze fish as soon as possible after capture. Of course, freezing also effectively stops bacterial growth, but unfortunately some autolytic processes such as the hydrolysis of fats can continue slowly in the frozen state. Mincing of the fish increases the rate of enzymatic reactions, both in the chilled and frozen states, making it imperative to handle minced fish with extra care.

RANCIDITY

Fish oils are relatively unsaturated and, as a result, readily react with atmospheric oxygen, forming a complicated mixture of oxidation and breakdown products which results in the development of an unpleasantly rancid flavour.

Pelagic fish, such as herring and mackerel which contain a high proportion of unsaturated fatty acids are subject to deterioration by the effect of atmospheric oxygen. The effects on odour and flavour are known as rancidity. Although unpleasant, rancidity does not markedly affect the nutritional value of the fish or cause it to become a hazard to health.

The rate of oxidation is affected by temperature, but the reaction can occur even in the frozen state. It is catalysed by the enzyme systems present in fish, particularly cytochrome C in the brown lateral band just underneath the skin. Traces of iron and copper ions also act as pro-oxidants. However, numerous attempts to apply antioxidants to fish have not met with much success. Frozen fish can be protected by keeping the storage temperature low and by restricting the access of oxygen by means of glazing, vacuum packaging or coating with batter, also by preventing dehydration since this increases exposure to oxygen by forming a porous layer on the surface of the fish. These methods, however, only retard the rate of oxidation and fatty fish always have a shorter life than lean or white fish.

In frozen herring, a peroxide value of more than about 5 in the oil corresponds with a degree of rancidity noticeable to the palate. (See also COLD STORAGE)

THIAMINASE

This enzyme is found in the flesh and viscera of aquatic animals including fresh water fish, herring and some invertebrates. It destroys thiamine by cleaving the pyrimidylmethylene group from the thiazole portion. It is of practical importance where raw fish, particularly herring, are fed directly to animals, as in the case of silver foxes and mink, where it may cause thiamine deficiency.

QUALITY GRADING

In spite of numerous attempts to develop objective laboratory standards for fish (see FRESHNESS TESTS, TRIMETHYLAMINE), the best method of assessing the degree of freshness is still subjective sensory (or 'organoleptic') examination. A marked disadvantage of such systems is that trained personnel are required to carry out the tests properly.

As fish spoils it goes through a sequence of changes which are readily detected by the human senses of sight, touch and smell. The spoilage patterns are constant for different species or closely related groups of species, but can vary to some extent between groups. The assessment of freshness can be simple or elaborate, depending upon the degree of accuracy required. A buyer in a

market, for example, will have different requirements to QC personnel in a fish processing factory.

Where there are facilities to do so, assessments of fish in its cooked state should also be made. Samples should be cooked in the simplest way so as not to impair flavour. Prepacked fish may be prepared by the recommended cooking method stated on the pack. Here again, odour and flavour follow definite patterns as fish spoil, but there are greater differences between species, at least in the initial stages. There is also a definite spoilage change of pattern during freezing and cold storage, particularly with regard to texture and taste.

Score sheets devised at the Torry Research Station, or variations derived from them, are in common use in the UK fish industry. A typical score sheet for round white fish such as cod, haddock, whiting and saithe is shown below.

Score sheet for white fish

Raw fish	*Score marks*
(1) General appearance (5 marks)	
Eyes perfectly fresh, convex black pupil, translucent cornea; bright red gills (colour depending on species); no bacterial slime, outer slime water-white or transparent; bright opalescent sheen, no bleaching	5
Eyes slightly sunken, grey pupil, slight opalescence of cornea; some discoloration of gills and some mucus; outer slime opaque and somewhat milky; loss of bright opalescence and some bleaching	3
Eyes sunken; milky-white pupil, opaque cornea; thick knotted outer slime with some bacterial discoloration	2
Eyes: completely sunken pupil; shrunken head covered with thick yellow bacterial slime; gills showing bleaching or dark brown discoloration and covered with thick bacterial mucus; outer slime thick yellow-brown; bloom completely gone; marked bleaching and shrinkage	0
(2) Flesh, including belly flaps (5 marks)	
Bluish translucent flesh, no reddening along the backbone and no discoloration of the belly flaps; kidney bright red	5
Waxy appearance, no reddening along backbone, loss in original brilliance of kidney blood, some discoloration of belly flaps	3
Some opacity, some reddening along backbone, brownish kidney blood and some discoloration of the flaps	2

Opaque flesh, marked red or brown discoloration along backbone, very brown to earthy-brown kidney blood, and marked discoloration of the flaps   0

(3) Odours (10 marks)*
Fresh 'seaweedy' odours   10
Loss of fresh 'seaweediness', shellfish odours   9
No odours, neutral odours   8
Slight musty, acetamide-like milky or caprylic acid-like odours   7
Bready, malty, yeasty, odours   6
Lactic acid, 'sour milk', or oily odours   5
Some lower fatty acid (e.g. acetic or butyric acids) or grassy, slightly sweet, fruity odours   4
Stale, sour, cabbage water, turnipy, or phosphene-like odours   3
Ammoniacal (trimethylamine and other lower amines) with strong o-toluidine-like odours   2
Hydrogen sulphide, other sulphide and strong ammoniacal odours   1
Nauseating, putrid, faecal odours; indole, ammonia, etc.   0

(4) Texture (5 marks)
Firm, elastic to the finger touch   5
Softening of the flesh, some grittiness on skin   3
Softer flesh, definite grittiness and scales easily rubbed off the skin   2
Very soft and flabby, retains the finger indentations, grittiness quite marked and flesh easily torn from the backbone   1

Cooked fish   *Score marks*

(About 6–8 oz (170–227 g) middle cut of fish steamed en casserole in resistant glass dishes (7 in (17.5 cm) diameter) over boiling water for 35 min)

(1) Odour (10 marks)*
Strong fresh seaweedy odours   10
Some loss of fresh 'seaweediness'   9
Lack of odour, or neutral odours   8
Slight strengthening of the odour, but no sour or stale odour; 'wood shavings', 'woodsap', vanillin or terpene-like odours; slight saltfish or cold storage odours   7
'Condensed milk', caramel or toffee-like odours   6
'Milk jug', 'boiled potato' or 'boiled clothes', or metallic odours   5
Lactic acid, 'sour milk' or toluidine-like odours   4
Some lower fatty acid (e.g. acetic or butyric acids), 'grassy', 'soapy', 'turnipy' or 'tallowy' odours   3
Ammoniacal (trimethylamine and lower amines) odours   2

Strong ammoniacal (trimethylamine, etc.) and some sulphide odours   1
Strong putrid and faecal odours (ammonia, indole, etc.)   0

(2) Texture (5 marks)
Firm thick white curd; bluish-white in appearance, no discoloration   5
Firm, but woolly; loss of bluish whiteness, some yellowing   3
Softer, cheesy; marked discoloration   2
Sloppy, soapy; very marked browning along the backbone   1

(3) Flavour (10 marks)*
Fresh, sweet flavours characteristic of the species   10
Some loss of sweetness   9
Slight sweetness and loss of the flavour characteristic of the species   8
Neutral flavour, definite loss of flavour but no off-flavours   7
Absolutely no flavour, as if chewing cotton wool   6
Trace of off-flavours, some sourness but not bitterness   5
Some off-flavours and some bitterness   4
Strong bitter flavours, some rubber-like and slight sulphide-like flavours   3
Strong bitter flavours, but not nauseating   1
Strong putrid flavours (e.g. sulphides) tasted with difficulty   0

*The descriptive terms, although capable of irrelevant associations, are those used spontaneously and agreed upon by the original panel at the Torry Research Station. Other panels may tend to use other, but probably similar, terms.

For grading purposes, freshness levels may be more or less elaborately defined depending upon the requirements of the purchaser or user of the fish. A good example of a simple system is that used within the EU, where batches of fish are graded at the quayside as E, A, B or unfit. Grade E is very fresh, Grade A is good quality fish that has lost its initial freshness but shows no sign of bacterial spoilage, Grade B is acceptable but not of good quality and covers the range from first signs of spoilage to condemnation. Features other than freshness which enter into quality grading include pieces of bones and membrane left by the filleters, blemishes such as blood marks and bruises and the presence of parasitic worms, particularly nematodes.

There are a whole range of specifications in use within the industry which cover the parameters required for grading within an organization or a company. They range from the *Codex Alimen-*

tarius standards issued by FAO, to in-house specifications used by processing or retail companies.

COLOURING MATERIALS

In most countries it is illegal to add colouring material to fresh fish offered for sale. Artificial colours were, at one time, quite widely used in the preparation of smoked fish, especially kippers and smoked white fish fillets. The colouring material was incorporated into the brine in which fish were soaked before smoking (see SMOKING). The colouring material originally used was annatto, then mixtures of various colours were adopted, principally amaranth, tartrazine and Brown FK (for kippers).

Since the mid-1980s when tartrazine and other azo dyes were withdrawn from use because of suggestions that they could be associated with hyperactivity in young children, the use of artificial colours has diminished considerably. Colouring materials are still used in the fish farming industry, where they can be included in the feed prior to harvesting in order to colour the flesh. Additives used in this way may remain undeclared, a point about which some consumer groups have protested. Natural or artificial colour added to any fish processed or prepared for retail sale is required to be declared on the label under the labelling laws in most countries.

BACTERIAL FLORA

Fish and shellfish carry bacteria on the skin, gills and in the intestines where in normal healthy fish they cause no harm. The flesh is generally sterile in a healthy fish. The surface bacteria are mainly of the Pseudomonas, Achromobacter and Flavobacterium groups, with some micrococci and, in the guts, some spore-bearing anaerobes. The total numbers of bacteria on the skin of freshly-caught fish have been recorded at $104-106 \, cm^{-2}$ and up to $106 \, g^{-1}$ in the gut. Fish landed at ports after 15–16 days in ice will have counts as high as $20 \times 10^6 \, cm^{-2}$, with Pseudomonas species preponderating, and it is this genus which in the main produces the typical odour and flavour changes associated with spoiled fish.

The broad pattern of changes in the bacterial flora developing on spoiling fish in ice is the same irrespective of species. Changes in the environment, however, will significantly affect the spoilage pattern by altering the growth rate and activities of the bacteria. The main environmental factors can be listed as temperature, pH, salt, water and toxic substances. By and large, chilled or frozen fresh fish are rarely incriminated in food poisoning outbreaks. Freezing significantly reduces the numbers of bacteria present and many of the surviving organisms die off or grow only slowly during cold storage. Other than freezing or sterilization the biggest factor in the prolongation of shelf life of fish is storage at ice temperature.

The changes brought about by the catching, handling and processing of fish will remove some bacteria, kill others and add others to the fish. Some of the bacteria added during processing operations may be of types which can cause food poisoning. These are generally not present in the fish before capture.

BOTULISM

Botulism is the disease, often fatal, which affects the central nervous system as a result of ingesting the toxin produced by the anaerobe *Clostridium botulinum*. There are seven types of *Clostridium botulinum*, labelled according to the first seven letters of the alphabet. Four types – A, B, E and F – consistently produce botulism in humans. Types B, E and F have been found in the sea and in freshly caught fish. Type E has mainly been responsible for outbreaks of the disease in places as far apart as Japan, Russia, Scandinavia and North America, usually resulting from the consumption of uncooked, partially cooked or fermented seafood products. Extensive research has shown since the mid-1980s that types B and F share the same properties as type E in that they have also been found in fish and sea mud and can grow and form toxin at temperatures as low as $3.3°C$. Surveys conducted as a result of a fatality caused by the consumption of smoked rainbow trout in Germany, have shown that types B, C, E and F are present in farmed trout and, furthermore, the incidence in farmed trout is higher than in marine fish, the organism being present in about 9% of farmed trout compared with less than 1% of marine fish.

Concern has been expressed about the safety of seafood packed in controlled or modified atmospheres (see PACKAGING) with regard to botulism, particularly those products which are eaten without further cooking, for example smoked salmon, hot smoked mackerel and trout, and cooked peeled prawns. The gas mixture recommended by the Sea Fish Industry Authority in the UK for use with fatty fish is one where oxygen is excluded. Even though some initital atmospheres for other species may contain oxygen, this gas will be particularly utilized by bacteria producing more anaerobic conditions within the enclosed pack.

Other types of packaging which cause anaerobic conditions, such as vacuum or vacuum skin pose potential problems when used for chilled seafood. Therefore, for all types of chilled prepacks, the fish or shellfish must be of high initial quality, hygienically handled and stored, and displayed at temperatures no higher than 3°C to ensure complete safety.

SCOMBROID POISONING

Scombroid food poisoning is usually associated with mackerel. The symptoms are those of an allergic response, including flushing of the face and neck, urticaria and dizziness. They do not usually persist longer than about 12 h.

This is an example of indirect bacterial action, the bacteria responsible do not themselves cause poisoning, nor do they produce toxin directly, but they do so by their action on normal constituents of the fish. Members of the Scombridae family of fish, to which mackerel belong, have high contents of the amino acid histidine. Some bacteria can convert this into histamine. It is not certain that histamine is the causative agent; it may be an indicator of the presence of other poisonous substances. The toxin is heat stable and not destroyed by normal cooking. Control is by proper attention to chilling the fish immediately after catching, and maintaining chilled or frozen conditions during storage.

PARASITES

Marine and freshwater fish are susceptible to a number of parasites, some of which can affect land animals. Epidemics of disease in marine fish are rarely seen because the vastness of the oceans prevents close contact and sick fish fall easy prey to predators. (Disease is of more consequence in freshwater fish, particularly farmed fish where, for economic reasons, high stocking densities occur.) Abnormal fish, however, are sometimes landed, and it is well known that fish may suffer from viral, bacterial, fungal, protozoan and tumorous diseases.

Round worms

Of the different types of parasite found in fish, round worms or nematodes are the most impor-

tant from a commercial and public health point of view. A number of fish species can be infected: cod, rockfish, pollack, Pacific halibut and sea bass among others. Round worms lodge themselves in the muscle and internal organs of the host fish where they can develop to a stage that can infect humans, sometimes causing distress. However, while the round worms can live in the bodies of fish, they cannot live long in the bodies of humans, making any distress temporary.

The two most common round worms are the 'cod worm' (*Phocanema decipiens*) and the 'herring worm' (*Anisakis simplex*). The former grows to about 4 cm long and varies in colour from creamy white to reddish brown. It is found in demersal inshore species, rather than pelagic offshore species, and can often be removed at the filleting stage – this is done by candling, i.e. examining fillets against a light. Fairly common in cod, as the name suggests, the incidence varies within wide limits depending on species, size, season and fishing ground. Cod from various distant fishing grounds may contain from 10 to 35% of fish infected with at least one nematode.

Anisakis simplex, the herring worm, is almost colourless and resembles a piece of coiled string about 2 cm long. It is found in hundreds of species of fish, about 80% of all Pacific salmon and rockfish, depending on the area, are estimated to contain them. The worms are almost always in the viscera of their hosts and generally only migrate into the flesh after the fish have been caught; therefore they are usually removed in the gutting process.

The life history of round worms is complex. It begins with the adult living in the stomach of a marine mammal; Phocanema in the grey seal and Anisakis mainly in dolphins, porpoises or whales. Excreted into the sea, the larvae are consumed by small crustacenas which, in turn, are consumed by fish. The larval worms are released into the stomach and bore into the guts or flesh of the host fish; the belly flaps can become infested. When the fish are eaten by a suitable host the life cycle is completed.

Cases of Anisakis infection in humans have been linked to the consumption of raw or lightly cured fish, and the 12 cases reported in the USA in 1985 may well be a result of the popularity there of sushi bars where uncooked fish is served. Round worms are readily killed by heating, the larvae are destroyed in one minute at a temperature of 60°C or more, by freezing and storage at −20°C for 60 h, or by salting and storing in 80° brine (21% salt) for 10 days.

Tapeworms

Of the tapeworms occurring in fish, *Grillotia erinaceus*, a common parasite of halibut and other species, is the most important from a commercial viewpoint. The encysted larvae are found in the belly cavity and flaps of the fish and infestation in halibut has been heavy in the past.

Protozoa

One of the protozoa, *Kudoa* (*Chloromyxum*) *thyrsites* is worth mentioning as it has in the past been troublesome in South Atlantic hake. The parasite can be present in the flesh in large numbers and after the death of the fish secretes an enzyme which softens and liquefies the flesh considerably duriing storage. This condition, known as 'milky hake', can only be controlled by rapid processing, freezing and cold storage.

MERCURY

It is now well known that aquatic life has great powers of concentrating certain substances. Contaminants present in the water, perhaps only in trace concentrations, are absorbed into fish or shellfish from the relatively huge quantities of water which pass through their bodies; they may not be excreted but accumulate to concentrations which may then be toxic to birds or people consuming the fish.

'Minamata disease' occurred in Japan in the early 1970s, among people eating fish from an estuary contaminated with mercury from the effluent of a factory upstream and resulted in 150 deaths. The cause was found to be methyl mercury compounds.

Following that tragedy, frequent analyses for heavy metals have become routine for fishery products worldwide. Studies published in Italy and the Faroe Islands in 1992 now indicate that mercury has become widely distributed among fish in many oceans and that traces of mercury may be found among fish eating people in many parts of the world. There is no evidence yet that the con-centrations found in humans approach toxic levels.

PHOSPHATE CRYSTALS

Crystalline disodium hydrogen phosphate dodecahydrate, or powdery dihydrate (containing 20% moisture), is sometimes found on smoked and dried fish products. This seems to originate from the fish muscle which normally contains about 0.2% total phosphorus, of which 0.16% is inorganic.

STRUVITE

Crystals of magnesium ammonium phosphate, glassy in appearance and texture, are sometimes found in canned fish and shellfish. They are formed when the necessary ions are present in sufficient concentration. If crystallization is slow, as when cans cool down slowly after retorting, the crystals may grow quite large and consumers may easily mistake them for small fragments of glass.

REFERENCES

Vinogradov, A.P. (1953) *The Elementary Composition of Marine Organisms*, Yale University Press, New Haven.
HMSO (1959) *The Chemical Composition of Fish Tissues*, H.M. Stationery Office.

FURTHER READING

Hall, G.M. (1992) (ed.) *Fish Processing Technology*, Blackie, Glasgow.
Hardy, R.W. (1992) Fish processing by-products and their reclamation, in Pearson, A.M. and Dutson, T.R. (eds) *Inedible Meat By-Products*, Elsevier Applied Science, Barking.
Lanier, T.C. and Lee, C.M. (1992) (eds) *Surimi Technology*, Marcel Dekker, New York.
Piggott, G.M. and Tucker, B.W. (1991) *Seafood: Effects of Technology on Nutrition*, Marcel Dekker, New York.
Regenstein, J.M. and Regenstein, C.E. (1991) *An Introduction to Fish Science and Technology*, Van Nostrand Reinhold, New York.
Sea Fish Authority and Torry Research Station (1984) *Specifications for the Purchase of Fish*, Sea Fish Authority, Hull.
Sikorski, Z.E. (1990) (ed.) *Seafood: Resources, Nutritional Composition and Preservation*, Wolfe Publishing, London.

3   Dairy Products

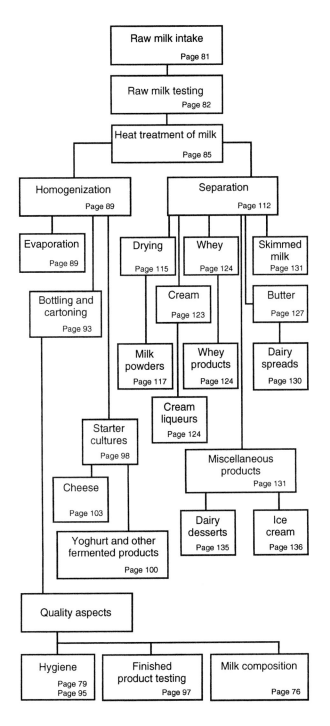

INTRODUCTION

Milk and dairy products have made a major contribution to the human diet in many different countries across the world since time immemorial. The dairy cow, the principal producer of milk, was domesticated over 6000 years ago and nowadays there are an estimated 280 million domesticated dairy cows in the world producing in excess of 400 million tonnes of milk annually.

Cows' milk represents around 90% of total world milk production but buffalo, sheep and goats produce 6%, 1.7% and 1.5% of total milk in volume terms and are also significant producers. Unless otherwise stated only cow's milk is considered in this chapter. Some 94% of the world's total milk supply is utilized as processed milk and dairy products. There are now over one hundred different varieties of domesticated cows with the most abundant type being Friesian–Holsteins.

Over the years the dairy cow has become a significant factor in world foodstuff production. In many areas of the world, particularly in Africa, the Indian subcontinent and in parts of central and South America the cow, or at most a few cows, is kept by single families to satisfy their daily needs for milk. Traditionally this was the pattern all over the world at one time. Over the past one hundred years however this pattern has changed dramatically as larger and larger farms have developed accompanied by larger herds. The chain of people involved in the milk industry in the most advanced and developed countries now extends from milk production on the farms, through transport to processing and on to retailers and consumers. The milk industry is now big business at all stages in the chain. It is on this sector that we will focus.

Wherever possible we have tried to follow a logical sequence, starting off with milk and ending up with products. Our aim is to provide easy to understand background knowledge and advice in as simple a way as possible. The chapter is not intended to be a comprehensive treatise since this would run into several volumes. For those wishing for further information a selection of key references is provided at the end of the chapter.

MILK COMPOSITION

The major constituent of milk is water (about 87.5%) with the remainder comprising mainly fat, protein and lactose. Smaller quantities of minerals are also present, together with non-protein nitrogenous components and vitamins. The average compositions of milk and related products are given in Table 3.1, but significant variations in milk composition occur as a result of differences in feeding regime, breed, stage of lactation and climate. The most marked change is with the flush of spring grass, when the solids-not-fat content rises by 0.2% or more and the fat content falls 0.1% or more as a result of the increased yield of milk.

MILK FAT

Approximately 99% of the fat in milk is in the form of triglycerides, which are esters comprising three fatty acid molecules with one molecule of glycerol. The distribution of the fatty acids in milk fat varies considerably with the seasons, mainly as a function of changes in diet. About 60–70% of the fatty acids are saturated (mostly palmitic, stearic and myristic acids), 25–30% are unsaturated (mostly oleic acid) and about 4% are polyunsaturated (linoleic and linolenic acids).

Milk fat, also known as butterfat, is characterized by the presence of a number of short chain fatty acids (4, 6, 8, 10 carbon atoms) which are present in only very small amounts in other fats. Butyric acid in particular accounts for about 4% of the fatty acids in milk fat but is not found in any other natural fat.

The presence of these short chain fatty acids is important in two respects. First, they have pronounced, characteristic odours which are important in determining the flavour and odour of certain dairy products, especially cheeses. Second, their relative abundance in milk fat also provides a means for indicating adulteration by foreign fats.

Physical form

Milk is in the form of a stable emulsion of fat globules dispersed throughout an aqueous phase containing the non-fat solids. Almost all of the fat is in the form of very small globules each surrounded by a fat-globule membrane comprising protein and phospholipid. The diameter of the fat globules varies between 0.1 and 20 μm, with an average of 3 to 4 μm. Under the influence of gravity, most of the fat globules are sufficiently large to rise on standing as a result of the difference in density between the milk fat and the aqueous phase, forming cream.

Hardness

The hardness of fat refers to the proportion of the fat which is solid at a particular temperature. It is the most important factor influencing the spreadability of butter and is determined by the fatty acid distribution in the fat and in particular by the relative amounts of the most abundant acids, i.e. palmitic, oleic, stearic and myristic acids. The hardness of milk fat decreases with an increase in the proportion of oleic acid: the iodine value (IV) gives an index of this. The iodine value of milk fat usually varies between 28 and 40, the lowest value being observed with hard fat in the winter and the highest value with soft fat in the summer.

Oxidation of fat

Oxidation of milk fat may occur at the double bonds in the unsaturated fatty acids, giving rise to undesirable flavours (see FLAVOURS). The addition of antioxidants to milk products is not generally permitted and it is therefore essential to minimize the oxygen content of manufactured milk fat products such as butter and anhydrous milk fat. The presence of bacteria in milk has an antioxidant effect, as does heating milk to temperatures above 80°C. The latter induces the formation of small quantities of sulphydryl compounds which act as reducing agents.

Table 3.1 Average composition (%) of milk and related products

Component	Whole milk	Skimmed milk	Butter milk	Cream (40% fat)
Fat	3.82	0.06	0.50	40.0
Protein (N × 6.38)	3.25	3.35	3.35	2.00
Casein	2.50	2.60	2.60	1.56
Whey protein	0.60	0.62	0.62	0.37
Lactose	4.80	4.95	4.95	2.90
Minerals	0.70	0.73	0.73	0.44
Water	87.4	91.0	90.5	54.6

SOLIDS-NOT-FAT

The solid constituents of milk, apart from the fat, are often measured together as the solids-not-fat content, i.e. the difference between the total solids content and the fat content. Solids-not-fat therefore includes the protein, lactose, minerals, vitamins and minor nitrogenous compounds in milk.

MILK PROTEIN

Milk proteins are classified into two major types, casein and whey protein, both of which are heterogeneous. Casein is usually defined as the protein which is precipitated from milk at pH 4.6, while whey protein is soluble under these conditions.

Milk contains about 3.3% total protein (total N × 6.38), of which about 2.5% is casein and 0.6% is whey protein. The remaining 0.2% comprises a number of nitrogenous compounds referred to as the non-protein nitrogen content.

Casein

Casein is a phosphoprotein and accounts for about 80% of the true protein in milk, where it is present in five major forms; α_s1, α_s2, β-, κ- and γ-caseins which on average account for 38, 10, 36, 13 and 3% of the total casein content, respectively. Most of the casein in milk is in the form of casein micelles, aggregates of several thousand casein molecules with a diameter of 10–300 nm. A number of minerals are bound within the casein micelle structure, the most important of which is calcium, without which the micelles dissociate. Calcium is present in casein micelles as bound calcium ions and as colloidal calcium phosphate (see SALTS); sodium, potassium and magnesium ions are also bound to protein in the micelle together with small amounts of citrate. The detailed structure of the casein micelle is not known with certainty but it is generally assumed that it comprises a roughly spherical aggregate of the α- and β-caseins surrounded by a coating of κ-casein.

The casein micelle is important industrially because it is destabilized, resulting in precipitation or coagulation, by acidification and rennetting (see RENNET).

Whey protein

The major whey proteins in milk are β-lactoglobulin, α-lactalbumin, blood serum albumin, immunoglobulins and proteose peptones, which account for about 50, 20, 6, 12 and 12% of the total whey protein respectively. The major whey proteins have a globular structure as a result of the fairly high number of disulphide groups in the whey protein molecules. Whey proteins are not precipitated by acidification or rennetting, but when milk is heated to about 65°C or above, they start to denature, and with the exception of the proteose peptones, become insoluble.

LACTOSE

Lactose, the milk sugar, is a disaccharide of glucose and galactose and is found only in the milk of mammals. The most important properties of lactose can be summarized as follows.

(i)   It is a reducing sugar and can therefore participate in the Maillard reaction when milk is heated.
(ii) It is considerably less sweet than sucrose.
(iii) It has only limited solubility (about 21 g per 100 g water) and crystallizes in concentrated milk products.
(iv) Lactose in solution is an equilibrium mixture of two forms, α-and β-lactose. At temperatures above 93.50°C, crystallization from a supersaturated solution yields the β-anhydride form while below this temperature, crystals of the α-hydrate are formed. The latter is present in dairy products in which the lactose is present as crystals, i.e. dry powders' and sweetened condensed milk.
(v)   It is fermented by lactic acid bacteria to produce lactic acid.
(vi) It is hydrolysed into its constituent monosaccharides, glucose and galactose, by the enzyme β-galactosidase (see LACTOSE HYDROLYSED MILK PRODUCTS).

SALTS

Milk contains about 0.8% minerals, mainly in the form of salts of Ca, Mg, K, Na, Cl, PO_4, bicarbonate and citrate. A number of other minerals are also present in small amounts, e.g. Cu, Zn, Fe. The concentrations of the individual minerals may vary significantly as a result of differences in breed, stage of lactation, season, feed, etc.

Apart from chloride which is present only in solution, the major minerals in milk are present in two forms: soluble and colloidal. Soluble potassium, sodium and chloride are present in ionic form. However, only about one-third of the soluble calcium and magnesium is ionized, the remainder is bound to citrate, phosphate and bicarbonate in a number of complex undissociated ions, e.g. Ca citrate$^-$, CaPO$_4$$^-$, CaHCO$_3$$^+$. Over 90% of the potassium and sodium in milk is soluble but only about 35% of the calcium and magnesium.

Colloidal potassium and sodium are bound to the negatively charged organic phosphate and carboxylic groups of casein. Approximately 30% of the colloidal calcium is also bound in this way but the remainder, about 50% of the total calcium in milk, is in the form of colloidal calcium phosphate, a complex compound also containing some citrate. The nature of the association between colloidal calcium phosphate and casein is not clear.

The minerals in milk are therefore in a complex equilibrium between a number of states. This equilibrium is commonly known as the salt balance. It is significantly altered by changes in pH, temperature and concentration. Increases in temperature and/or concentration result in the formation of colloidal calcium phosphate with a concomitant reduction in soluble calcium and a decrease in pH. Acidification on the other hand results in a progressive solubilization of colloidal calcium phosphate until at pH 4.9 all of the colloidal calcium phosphate is removed.

These changes in the salt balance, especially when they result from heat treatment and concentration, are important in milk processing and account for the deposition of calcium phosphate onto heating surfaces.

NUTRITIONAL VALUE OF MILK

Milk is a valuable element in human nutrition and supplies significant quantities of all five groups of nutrients; proteins, fats, carbohydrates, minerals and vitamins.

The average distribution of the nutrients in pasteurized milk is given in Table 3.2

The most valuable component of milk in terms of human nutrition is undoubtedly the protein. Milk protein provides an excellent balance of amino acids, and casein is actually used as the reference protein with which the nutritional quality of food proteins is generally compared. Milk and milk products provide about 20% of the total protein consumed across the world.

Milk and milk products are also the main dietary source of calcium and contribute significant amounts of vitamins A and D, and the B vitamins.

Table 3.2 Nutrients provided by pasteurized milk (average)

Nutrient	Quantity in milk	
	per pint	per litre
Protein	19.3 g	34.0 g
Fat	22.2 g	39.1 g
Carbohydrate	28.1 g	49.4 g
Calcium	702 mg	1.2 g
Iron	0.6 mg	1.1 mg
Vitamin A (retinol equivalents)	237 mg	417 mg
Vitamin D	0.17 mg	0.30 mg
Thiamin	0.23 mg	0.40 mg
Riboflavin	0.88 mg	1.6 mg
Niacin	5.3 mg	9.3 mg
Vitamin C	5.9 mg	10.4 mg

FLAVOUR

Flavour is one of most important attributes of milk and milk products, and correct handling of milk during collection and processing are important factors in maintaining the desirable flavour in milk and milk products. Fresh milk has an agreeable, slightly sweet flavour derived mainly from the soluble constituents of milk, i.e. lactose and minerals. However, milk fat is also an important contributor to flavour, together with small quantities of fatty acids and their condensation and oxidation products.

Off-flavours

Off-flavours may occur in milk for a variety of reasons.

(i) *Feed flavours*
 Feed taints may be detected in milk in winter and spring when grass is unavailable and alternative feeds are given, e.g. silage, turnip, beet. They are due to the adsorption of the volatile constituents of the feed via the bloodstream. Taints may also result when strongly flavoured weeds are present in the pasture. Feed flavours are not a serious problem in the UK but in some areas, such as the southeastern United States, the taints are so pronounced that the milk must be vacuum treated before consumption.

(ii) *Bacterial flavours*
 The production of lactic acid by lactic streptococci is responsible for the typical flavour of sour milk. A sour flavour is detectable when the total acidity reaches 0.2–0.3%.

However, with bulk collection and cold storage of milk, souring is uncommon except during hot periods in the summer. The major source of bacterial taints in milk nowadays is the group of psychrotrophic, mainly Gram negative, bacteria which are present in cold stored milk. The growth products of these organisms include a number of enzymes which may act on the protein and fat to produce a number of off-flavours variously described as unclean, bitter, etc.

(iii) *Enzyme-induced flavours*

In addition to the enzymes produced by psychrotrophic bacteria, milk also contains native lipases and proteases. These act on milk fat and protein, respectively and may result in the development of rancid flavours from the development of free fatty acids and bitter flavours through the development of bitter peptides. The latter, however, are very uncommon except in the case of UHT milk in which heat resistant proteases may give rise to bitter flavours developed during extended storage.

(iv) *Chemically induced flavours*

A burnt flavour may develop in milk in the presence of light and oxygen as a result of the oxidation of whey protein. This flavour is sometimes terms 'cabbagy' but is not generally considered to be a problem. The oxidation of milk fat is much more likely to give rise to off-flavour development. Two types of fat oxidation are recognized: (i) that induced by the presence of copper, and to a lesser extent iron, and (ii) that induced by the action of light. The first type is no longer a problem, presumably because copper surfaces are no longer present in milk-handling equipment. Flavour defects may also arise as a result of milk coming into contact with residual sanitizers in milking machines and bulk tanks.

(v) *Cooked flavours*

When milk is heated to approximately 80°C or higher, a cooked flavour develops as a result of the formation of sulphydryl compounds and other volatile compounds containing sulphur, such as hydrogen sulphide or mercaptans. These are produced from certain whey proteins. The Maillard reaction is also significant in the development of flavours in heated milk. This reaction also occurs to a limited extent in sterilized and UHT treated milks during extended storage.

HYGIENE

PREVENTION OF BACTERIAL CONTAMINATION

Milk drawn from healthy cows under hygienic milking conditions is relatively clean and free from bacteria. However, during milking the bacterial count in milk may increase substantially as a result of contamination by bacteria from the atmosphere (dust, etc.), dirty udders, unclean milking equipment and pipelines or if the cow is suffering from an infectious disease, e.g. mastitis. The major sources of contamination are the interior and exterior of the udder and the milking equipment. It is therefore important for farmers and their milking staff to be aware of sources of contamination and understand how they can be controlled.

Essentially in order to produce milk of a high hygienic standard the milker must prepare the cow thoroughly before milking, adopt a good milking technique and use a good post-milking routine. Adequate attention should be given to ensuring that the cows surroundings are clean, that their udders are thoroughly washed and that mastitic cows are milked separately and their milk discarded.

Hand milking used to be common on small farms and in fact it still is in many of the less developed countries in the world. Nowadays, however, it is much more the rule to see milking machines in use on farms in most countries. These machines suck out the milk from cow's udders under vacuum and are very efficient. If not properly cleaned, or if the teat clusters are allowed to drop on the floor or if operated under fluctuating vacuum conditions, they can be a significant source of bacterial contamination. Milk once collected is either poured into churns though a muslin filter or transferred by vacuum line to a bulk vat where it is cooled.

Having limited bacterial contamination of milk during milking, it is essential that contamination from equipment situated between the cow and the refrigerated storage tank is kept to a minimum. This means giving proper attention to the cleaning of the milking machine, milk pipelines, any interceptor vessels and the bulk storage tank. Cleaning regimes are therefore based on removing visible dirt, milk residues (fat, protein and milkstone) which harbour bacteria, then sterilization of the cleaned surfaces using heat or chemical sterilants such as sodium hypochlorite.

The multiplication of bacteria in milk is depen-

Table 3.3 Bacteria associated with milk and milk products

	Family	Genus
Gram-positive	Bacillaceae	*Bacillus*
		Clostridium
	'Coryneforms'	*Arthrobacter*
		Corynebacterium
	Lactobacillaceae	*Lactobacillus*
	Mycobacteriaceae	*Mycobacterium*
	Propionibacteriaceae	*Propionibacterium*
	Streptococcaceae	*Leuconostoc*
		Streptococcus
Gram-negative	Enterobacteriaceae	*Enterobacter*
		Escherichia
		Proteus
		Salmonella
		Serratia
		Shigella
	Micrococcaceae	*Micrococcus*
		Staphylococcus
	Neisserciaceae	*Acinetobacter*
		Moraxella
	Pseudomonadaceae	*Achromobacter*
		Alcaligenes
		Brucella
		Pseudomonas
	Vibrionaceae	*Aeromonas*
		Chromobacterium
		Flavobacterium

dent on both the temperature and time of storage. The storage temperature also influences the types of bacteria which grow and their spoilage characteristics. The growth rate of bacteria decreases at lower storage temperatures, and whilst bacterial numbers increase rapidly at 10°C and above, the growth rate of most bacteria is virtually zero at below 2°C.

Table 3.3 lists the most important families and genera of bacteria which have been isolated from milk and milk products. The control of bacterial levels in milk is important both in terms of destroying pathogenic bacteria and in the prevention of spoilage.

Pathogenic bacteria may derive from the milk of infected cows or from external contamination of the udder. The two most important diseases transmitted in milk are bovine tuberculosis (from *Mycobacterium bovis*) and brucellosis (*Brucella abortis*). Both diseases have now been virtually eradicated in most areas of the world.

Bacterial infection of the udder (mastitis) may also be a potential source of pathogenic bacteria (*Staphylococcus aureus*, *Streptococcus agalactiae*, *Escherichia coli*).

Pathogens which may arise from faecal contamination include Salmonella species, *Campylobacter jejuni* and *Listeria monocytogenes*. Yersinia usually derives from water or soil contamination. Listeria and Yersinia have come into prominence in recent years as more and more raw milk is refrigerated

on the farm. Both of these organisms are able to withstand and grow slowly at refrigeration temperatures but are destroyed at normal pasteurization temperatures, as are Salmonella and Campylobacter species. Spoilage of milk is practically always due to the growth of bacteria. The major groups of spoilage bacteria are as follows:

(i) *Lactic acid bacteria*
These are bacteria which cause souring by the fermentation of lactose to lactic acid. They include streptococci (e.g. *S. lactis*, *S. cremoris*) and lactobacilli (e.g. *L. bulgaricus*, *L. casei*) and form the basis for the starter cultures used in cheese and cultured products manufacture (see STARTER CULTURES).

(ii) *Coliforms*
These are facultatively anaerobic bacteria which ferment lactose to acid (mainly lactic) and gas (carbon dioxide and hydrogen) giving rise to an unclean flavour and smell. They can also cause 'blowing' in cheesemaking resulting in a large number of gas holes in the cheese and a bitter, unclean taste. Most of the coliforms in milk and milk products are of the genera Escherichia and Enterobacter. Since coliforms are killed by pasteurization, their presence in pasteurized products is used routinely as an indication of insufficient pasteurization or of post-pasteurization contamination.

(iii) *Butyric acid bacteria*
These are anaerobic, sporeforming bacteria of the genus Clostridium (e.g. *C. butyricum*, *C. tyrobutyricum*) which ferment lactose and/or lactate to butyric acid. Carbon dioxide and hydrogen are usually produced as well as acid. Because of their anaerobic nature, butyric acid bacteria are not a problem in milk, but in some cheeses, especially less acid varieties, 'blowing' occurs as a result of the butyric acid fermentation (see CHEESE).

(iv) *Enzyme producing bacteria*
These bacteria produce extracellular enzymes which hydrolyse protein and fat giving rise to bitter and rancid flavours, respectively. Spores of bacteria of the genus Bacillus (e.g. *B. cereus*, *B. subtilis*) survive pasteurization and can germinate on storage. These bacteria produce extracellular proteases and lecithinases which give rise to sweet curdling and bitty cream in pasteurized milk and cream (see BITTY CREAM). Another important group of bacteria which produce extracellular enzymes are the psychrotrophic bacteria which are capable of growth at 70°C or

below. Raw milk must be stored at 40°C or below in order to control the growth of these organisms.

The importance of psychrotrophic bacteria has grown substantially since most raw milk is now stored on the farm in refrigerated bulk tanks. Most psychrotrophic bacteria are Gram-negative but some Gram-positive Bacillus species are also reported to be psychrotrophic. The most important psychrotrophs are of the genus Pseudomonas, and organisms of the genera Flavobacterium, Acinetobacter, Alcaligenes and Achromobacter. Whereas psychrotrophic bacteria are destroyed by pasteurization, the enzymes produced by these organisms are only partially deactivated. Residual proteases and lipases act on the protein and fat in milk giving rise to bitter and rancid flavours, respectively.

COOLING AND STORAGE OF MILK AT THE FARM

Milk leaving the cow emerges at around 37°C, an ideal temperature for the growth of many bacteria. Since bacteria multiply themselves by binary fission, often at 20 min intervals at their optimum growth temperature, it is possible for one single bacterial cell to produce over 10 million similar bacterial cells in 12 h. It is therefore important that milk be cooled as soon as possible after milking to a temperature below 4°C in order to suppress bacterial growth.

Cold water, chilled water or preferably refrigerated units may be used for cooling milk. Immersion coolers with chilled water circulation are often used for churn milk. Where milking machines and refrigerated bulk farm tanks are used, an ice bank in the bulk vat rapidly cools milk to below 4°C.

RAW MILK INTAKE

The frequency at which milk is collected from the farm depends on various factors, for example the refrigeration temperatures employed (if refrigeration is available) and the storage capacity at the farm. Nowadays, with fuel costs being so high in many countries, there are significant transport cost savings to be made by less frequent collection of milk. In practice, milk produced under clean conditions is stored for collection every 2 days with the more costly alternative of every day collection becoming less common.

Cooled milk in churns or cans is usually taken to a convenient collection point, where they are protected from direct sunlight. The collection vehicle takes the milk either to a collection centre or to the nearest dairy where each individual churn or can is inspected and providing the milk is acceptable the volume or weight is measured and recorded, samples taken and the churns are emptied into a bulk vessel for cooling and storage or for immediate heat treatment.

Milk stored in refrigerated bulk tanks is collected by being pumped into an insulated transport tanker, after the driver has recorded the milk's temperature and assured himself by sight and smell that the milk is satisfactory. Milk volumes are measured by means of a calibrated dipstick or, more often nowadays, by use of an automatic flow meter on the collection vehicle. Since the individual farm supply is blended in the tanker with supplies from other farms, it is normal practice to take samples of each consignment at the point of collection to enable faults to be traced back to source. The collection tanker will then deliver the milk to the dairy where the contents are sampled on receipt, inspected by smell and taste and tested for bacteriological and compositional quality. Once the tanker load of milk is accepted it is pumped into a silo to be used within a few hours or to be stored for later use.

STORAGE OF MILK AT THE DAIRY

When milk is stored before pasteurization or further processing bacteriological spoilage can occur. Many countries now lay down strict standards either in legislation or codes of practice for the maximum desirable temperature for acceptance of milk at the dairy. Despite this, milk spoilage will occur if the temperature becomes too high. Various methods are employed by dairies to prevent raw milk spoilage in silos. These include deep cooling to 2°C, the addition of carbon dioxide (subsequently removed under vacuum prior to processing) and the addition of starter cultures to milk destined for cheesemaking. In the latter case the slow growth of the starter cultures inhibits the growth of psychrotrophic bacteria, thereby extending the shelf life of the stored milk until cheesemaking can begin.

In some countries thermization is used to extend preprocessing storage life. Thermization is a subpasteurization heat treatment which may be used to treat raw milk on reception at a dairy or creamery. Thermization involves heating the milk at a temperature of 60 to 66°C for 5 to 20 s. This destroys most of the psychrotrophic bacteria in milk. The production of extracellular enzymes during cold storage is thus prevented and thermized milk can be held for up to 4 days at 40°C without significant increase in the total bacterial count.

This heat treatment is not sufficient to eliminate phosphatase activity (see PASTEURIZATION) and has no significant effect on the suitability of the milk for further processing. Thermization is therefore a useful means of extending the storage life of raw milk without restricting the possibilities of its final use. It may be used for milk prior to cheese manufacture. In some countries thermization is used to treat milk for weekend storage so that a 5 day week can be operated.

RAW MILK TESTING

The range of tests applied to milk on reception at the dairy depend on a number of factors. Essentially milk is tested to satisfy the dairy that the milk which they are purchasing from the farm is of a satisfactory quality and suitable for the end use to which it is to be put. There are also minimum legal standards laid down in all countries which must also be met. Without exception tests will be conducted to analyse for the hygiene and compositional quality and normally to detect any unlawful additions or adulterants in the milk, for example, added water, antibiotics, etc.

The type and complexity of the test methods employed will depend to a large extent on the source and condition of the milk, especially where hygienic quality testing is concerned. Once the dairy staff have satisfied themselves that the temperature, general appearance, cleanliness, colour and smell of incoming milk is satisfactory they will proceed to more detailed testing.

HYGIENE AND BACTERIOLOGICAL TESTS

In many countries milk is still collected in churns, is often uncooled, and can arrive at the dairy at temperatures around 20°C or higher. In the majority of countries, however, milk is stored in refrigerated bulk vats on the farms and transported to the dairies in insulated tankers. Uncooled and cooled milks often have a different bacteriological profile and thus need to be subjected to different tests.

For milk collected in churns as unrefrigerated milk bacterial spoilage can readily occur, especially if the milk is kept at higher ambient temperatures

for any length of time. Spoilage is due to lactic acid production often associated with high levels of bacteria in the milk. Hence the quality control tests are focused on measuring this spoilage factor.

Normal milk is slightly acid, with a pH between 6.6 and 6.8, and exhibits a significant buffering capacity due to the presence of salts and protein. The developed acidity (lactic acid) is negligible in fresh milk and rises as bacterial action attacks lactose. The bacterial group responsible for spoilage belong to a group called 'mesophiles', namely those which grow best at temperatures between 10 and 37°C. Any of three simple tests can be used to check for developed acidity, the 'clot on boiling' test, the alcohol stability test or the titratable acidity test.

The 'clot on boiling' test involves heating a small volume of milk to boiling point in a test tube. If the milk has developed more than 0.1% lactic acid it will not clot and should be rejected. This test is cheap, rapid and easy to perform but suffers from being relatively insensitive since any milk failing the test will already be too sour for processing.

The alcohol stability test is slightly more sensitive but still not very reliable or quantitative. Essentially a very small volume of milk is shaken in a test tube with an equal volume of alcohol. If no flocculation occurs a further equal volume of alcohol is added. If no flocculation occurs the milk is adjudged to have an acceptable acidity.

The acidity of milk can also be measured by titration with N/9 sodium hydroxide using a phenolphthalein indicator. The titratable acidity of fresh milk is typically 0.14 to 0.16%, expressed as lactic acid. The actual value is determined by the buffering capacity of the milk and not by the lactic acid content only. Values higher than 0.16% are an indication of the development of lactic acid by bacteria, i.e. that the milk has soured to some extent. However, the degree of souring can only be accurately assessed by subtracting the initial acidity from the measured acidity, and the former can vary substantially with season, feed, breed, etc. For this reason, titratable acidity can only be used as an indication of milk quality, not as an absolute method.

All of the above methods give a measure of developed acidity. Since mesophilic bacteria also change the redox potential of milk they will reduce dyes such as methylene blue and resazurin which change colour and hence these can also be used as an indicator of bacterial activity. These tests are easy to carry out, use the simplest of equipment but suffer from being imprecise, since some bacteria have a high rate of reduction and others low. Despite this both tests are still in much use since they offer a useful general grading method for uncooled milk.

For milk collected from bulk farm vats as refrigerated milk 'psychrophilic' bacteria (i.e. those which grow best below 10°C) tend to grow and thus a different set of test methods are required. The most commonly used method nowadays is the total bacteria count (TBC) although direct microscopic counting can be employed but is not very accurate.

The TBC method is an empirical method of judging milk quality which entails incubating predetermined dilutions of the raw milk in nutrient agats for 3 days at 30°C. At the end of incubation the number of bacterial colonies formed (derived from individual live bacteria) is counted and, after taking account of the levels of original dilution, the total number of bacteria calculated. There are many different variations of the TBC method in use, the most common one involves incubation of the inoculated agars on Petri dishes although one can also use the Astell roll tube method or a more recent development, the Spiral Plate Count. An American corporation, the 3M Company, has also introduced its own variant of the TBC test which is called 3M Petrifilm and this is finding increasing application in many laboratories.

Like all analytical methods the TBC methods have advantages and disadvantages. They are simple and in skilled hands can offer a reasonably reliable indication of the bacterial content of milk. On the other hand they are labour intensive, lack real precision and there is sometimes poor reproducibility between different laboratories.

Since about 1980 there has been a significant drift away from techniques which involve growing and counting bacterial colonies towards techniques involving physical concentration of bacteria, staining and then counting the stained cells. The first such development occurred in the UK with the introduction of the direct epifluorescence filter technique (DEFT). In DEFT a small aliquot of the milk is treated with the enzyme trypsin and then a surfactant to disperse the fat and cellular debris. The resulting mixture is passed through a very fine pore filter to retain the bacteria which are then stained with Acridine Orange and counted using microscopic techniques. The number of live bacteria per millilitre can then be calculated. The DEFT method is quick (20 min) and sensitive to 1000–10 000 bacteria/ml. It is however expensive to buy and operate, is labour intensive and needs skilled operators to achieve consistent results.

The DEFT was the first of what are now known as direct counting methods. Other automated tests which have followed on from this test include the Biofoss, Bactoscan and Autotrak which, in principle, are very similar to DEFT. They offer high levels of accuracy, are rapid but are expensive. There are however many such machines now in use, particularly in laboratories in the more advanced dairying nations.

A further significant development in rapid bacterial analysis has taken place in recent years with the introduction of test methods based on the detection of the metabolic products of bacteria, now commonly known as indirect methods of analysis. Included amongst such methods is a method based on bioluminescence. All living cells contain adenosine triphosphate (ATP) which acts as a substrate in the biolumeniscent firefly enzyme system luciferin – luciferase, and results in the emission of light. This very sensitive reaction can be used as a measure of low levels of bacteria via their ATP content.

$$ATP + luciferin + oxygen \xrightarrow{luciferase} reaction\ products + light$$

There are some problems with the test (e.g. somatic cells which also contain ATP have to be removed before bacterial ATP can be measured) but nonetheless the test is rapid and sesitive down to 10 000 bacteria/ml. Competition in the field of bioluminescence testing is intense and some manufacturers are claiming that their instruments are sensitive down to 100–1000 bacteria/ml with a result being available in 2 min. Other applications in addition to raw milk testing are also possible, including the use of these instruments for the rapid testing of UHT milk and for assessing process plant cleanliness after swabbing. Examples of equipment now on the market and in use include Biotrace (Brigend, UK) Bactofoss (Foss Electric, Denmark), Bio-Orbit (Finland) and Hy-Lite (Amersham International, UK). Further information on these instruments and of their capabilities can be obtained from the manufacturers. Other indirect methods for the rapid assessment of bacteria in milk include Electrical Impedance methods (Celsius UK) and those based on Flow Cytometry (Chemunex, Paris).

COMPOSITIONAL TESTS

The total solids (TS) of milk are made up of butterfat (usually around 3.9%) and solids-not-fat (SNF, normally around 8.6%). The principle components of the SNF of milk are proteins (caseins and whey proteins, immunoglobulins, etc.) and lactose. Additionally the fat phase will contain the fat-soluble vitamins, whilst the aqueous phase has water-soluble vitamins and salts. Occasionally the aqueous phase will also contain antibiotic residues which have entered the milk accidentally after treatment of the cow for mastitis or it can contain additional water, normally as a result of failing to drain milking pipelines or the bulk farm vat of water after cleaning operations.

The range of compositional tests applied to incoming milk can be quite wide, although most dairies nowadays are selective both in the type of component to be analysed and in the frequency of testing. It is also important to remember why dairies carry out compositional testing. Their first priority from a commercial standpoint is to ensure that they are getting value for money since the compositional quality of milk has a direct bearing on the yield of products manufactured. Another important reason, especially for dairies processing milk for direct supply to their customers as drinking milks, is that the final product must meet strict minimum legal compositional standards.

Smaller dairies and many of those in less well-developed countries are often only equipped to carry out the simplest of compositional testing. Sometimes only an acidity test (by titration, as previously described) and a test for BUTTERFAT using the Gerber or Babcock methods will be carried out. This particularly applied in the less well-developed countries. Normally, in addition, a simple test for SNF by the use of a hydrometer or lactometer would be carried out. Detailed descriptions of the protocols for carrying out these tests and also the standard oven-drying reference method for assessment of TS in milk can be found in any classical dairy analytical textbook. Likewise, the internationally recognized reference methods for protein (Kjeldahl) and lactose (by polarimetry) are well documented in the same textbooks.

Reference testing of milk for compositional quality is lengthy, labour intensive and expensive, making it unsuitable for routine use in quality control or for quality payment purposes. The need for lower cost, quicker, high throughput tests led to the development of instrumental methods for compositional analysis. Since the 1960s there have been many new developments in this field, ranging from the introduction of both manual and automatic machines using turbidimetric principles for the quantitative measurement of butterfat through to the emergence nowadays of instruments that will rapidly measure the fat, protein and lactose content of incoming milk and finished products. All these methods are based on infrared absorp-

tion and operate on the principle that these constituents, because of their chemical bonding structure, absorb infrared light at distinctly different wavelengths.

In order to be certain that only light of the desired wavelength is detected, optional interference filters which allow only the deisred wavelength to pass are used. The filters are arranged in a filter wheel which rotates to bring each of the filters in turn, two per component, into position in the path of the infrared beam. Two different wavelengths, 5.7 µm, can be used to determine the fat content of milk, whilst proteins are measured at 6.5 µm and lactose at 9.5 µm.

The major manufacturer of these infrared absorption instruments is Foss Electric of Denmark but there are others, such that now there is a wide range of instruments available of varying costs and versatility capable of meeting the industry's specific needs, from manual sample presentation for small laboratories through to fully automated instruments with automatic computer data capture.

None of these instrumental methods are independent measuring techniques and all have to be calibrated against reference methods on a regular basis. Calibration is achieved by making sure that the instrumental methods give the same results as the reference methods, that is, Gerber for fat, Kjeldahl for protein and polarimetry for lactose. Further details on the method of use of infrared absorption instruments and of their capabilities can be found in the manufacturers' operational manuals. Many diaries, especially those manufacturing fermented products such as yoghurts and cheeses also carry out periodic checks on milk for the presence of antibiotics.

Antibiotics

Antibiotics are used in the treatment of cows suffering from diseases caused by bacteria, principally for udder infections. Residual antibiotic is present in the milk for several days after treatment and such milk is usually discarded but the exact time required to guarantee milk free from antibiotic (3–4 days) varies with different antibiotic preparations and milking conditions.

The presence of antibiotic residues in milk is undesirable for two important reasons:

(i)   A small proportion of the population is sensitive to antibiotics, particularly penicillin.
(ii) The manufacture of cultured products, e.g. cheese and yoghurt, relies on the growth of

lactic acid bacteria which are severely inhibited by antibiotics.

The presence of antibiotics is routinely determined using a variety of proprietary test kits which detect any inhibition to the growth of selected strains of bacteria. Supplies of milk from farms are regularly tested in this way and the positive identification of antibiotics results in a substantial decrease in the price the farmer receives for his milk.

A conflict of standards can sometimes exist in that the upper limit set by national Regulations is that set for toxicological reasons. However, lower concentrations may inhibit yoghurt or cheese starter cultures and for this reason many of the more recently developed assay techniques are capable of detection at levels well below the legal maximum.

The presence of antibiotics is routinely determined using tests which in principle work by detecting inhibition of growth of selected strains of bacteria when the milk sample contains antibiotics. The original and much used method until 20 years ago was the International Dairy Federation Disc Assay method in which zones of inhibition of growth were observed on a Petri dish seeded with *Bacillus stearothermophilus* if the milk sample placed in wells on the plate contained antibiotics. Since then antibiotics testing has advanced considerably both in terms of the length of time taken to obtain a result and of the sensitivity of the test.

In the 1970s refinements on the earlier disc assays took place enabling test results to be obtained in 2½ to 3 h. One such method, still used today, is the Delvotest, developed by Gist Brocades in Holland. There are other competing tests, but essentially they are all based on microbial inhibition. The length of time taken to complete these tests is not a disadvantage if speed is not of the essence. However, in the 1980s growing interest emerged in more rapid testing to meet the needs of a platform acceptance/rejection test at the dairies and to conform to new European Union Regulations. In this way the dairies can minimize the risk of contaminated milk and dairy products reaching the consumer.

These rapid tests are termed competitive immunoassays and employ highly specific antibodies to detect antibiotic residues in milk. Very low levels, down to parts per billion, of antibiotics in milk can be detected often in as little time as 10 min. There is now a wide choice of test kits available to dairies from different manufacturers, although none are cheap either to buy or to operate. Examples of detection systems available include LacTek (produced by the American company Idetek), Delvo-X-Press (Gist Brocades, Holland) and AIM-96 (Charm, USA); the latter takes longer to complete but has the advantage that it can detect a much wider range of antibiotics than those based only on derivatives of β-lactams (Penicillins are the most frequently used β-lactams). Further details of operating instructions can be obtained from the manufacturers.

HEAT TREATMENT OF MILK

In dairies milk undergoes specific forms of heat treatment depending on the use for which it is intended. The most common forms of treatment of milk destined ultimately for the liquid milk consumer market are pasteurization, sterilization or ultra heat treatment (UHT).

PASTEURIZATION

Milk and milk products are pasteurized to destroy pathogenic organisms, particularly the tubercle bacillus, and to reduce the number of non-pathogenic organisms which might adversely affect product quality. In most countries there are Regulations specifying minimum time/temperature combinations which much be used for pasteurization of milk:

(i) *Batch pasteurization*: this is normally referred to as the holder process in which milk must be held for at least 30 min at a temperature not less than 62.8°C and not more than 65.6°C.

(ii) *Continuous pasteurization*: this is normally referred to as the high temperature short time process (HTST) in which milk must be held for at least 15 s at a temperature of at least 71.7°C.

Batch or holder pasteurization

The holder process is now installed only in very small dairies processing less than about 2500 l of milk per day. The heating, holding and cooling operations are carried out in a jacketed vat fitted with an agitator.

HTST pasteurization

This method of pasteurization is now used almost universally, even in relatively small dairies. The heating and cooling operations are carried out in a plate heat exchanger which gives very good economy of energy by using hot milk as the medium for heating incoming cold milk, and vice versa (regeneration). The HTST process comprises five stages.

(i) *Regenerative heating*
     Cold milk from a storage tank is pumped via a balance tank to the regeneration section of the heat exchanger where it is warmed to 60–65°C by heat transfer from the freshly pasteurized milk. In modern plants, up to 92% of the heat required to warm the milk to pasteurization temperature can be recovered from the pasteurized product.

(ii) *Heating*
     In this stage, the milk is further heated to the required pasteurization temperature (e.g. 72°C). The heating medium is hot water.

(iii) *Holding*
     The holding section can either be a chamber in the heat exchanger itself or an externally mounted holding pipe. The latter is usually preferred and the pipe is suitably dimensioned to give the required residence time at the specified flow rate. Close control of product flow rate is necessary to achieve a constant holding time. The milk temperature at the end of the holding pipe is continuously monitored to ensure that the milk has been held at the required temperature. Should the temperature fall below a set temperature, say 71.7°C, a flow diversion valve closes and the milk is diverted back to the balance tank. A record of the pasteurization temperature is kept on a chart recorder.

(iv) *Regenerative cooling*
     The hot milk from the holding section passes to the regeneration section where it is cooled to 10–15°C by heat transfer to the incoming cold milk (see (i) above).

(v) *Cooling*
     The regeneratively cooled milk is then further cooled to 4–5°C, first with cold water and finally with chilled water or refrigerant. The temperature of the cold product is also recorded.

The operation and cleaning of the pasteurization plant are described in depth in the Society of Dairy Technology (1983).

The phosphatase test

The enzyme phosphatase, which is present in raw milk, is inactivated under the conditions prescribed for HTST pasteurization and its detection is therefore used in routine quality control of pasteurized milk. The phosphatase test, which indicates the presence of residual phosphatase activity, should be carried out on the day of production since the enzyme may become reactivated during storage. Reactivation is more common with cream products.

STERILIZATION

Sterilized milk still accounts for over 5% of the liquid milk consumed in many countries and its cooked flavour and rich texture are particularly popular with its consumers. The most important feature of the manufacture of sterilized milk is the fact that the milk is heat treated in the bottle (glass or heavy duty plastic) and recontamination is therefore prevented. The description 'sterilized' is in fact somewhat misleading since the heat treatment involved is insufficient to destroy all heat resistant spores. However, in-bottle sterilized milk has a shelf life of at least seven days and should remain fit for consumption for several months.

The milk Regulations of most countries specify that sterilized milk shall be filtered or clarified, homogenized and heated to at least 100°C for a sufficient length of time to comply with a turbidity test. This test gives an index of the degree of whey protein denaturation during sterilization and is based on the fact that the major whey proteins (β-lactoglobulin, α-lactalbumin and blood serum albumin) are completely denatured when heated to 100°C for a few minutes. The test involves two stages. Casein and denatured protein are first precipitated by the addition of ammonium sulphate. The filtrate is then boiled for 5 min to precipitate any residual undenatured whey protein. The latter should be absent in sterilized milk and the boiled filtrate should therefore remain clear if the milk has been adequately sterilized. The turbidity test only indicates that the milk has received the minimum required heat treatment; it cannot be taken as a measure of sterility or shelf life.

In the usual manufacturing process, clarified milk is heated, homogenized at 2500 psi (170 bar) (to retard fat separation during storage) and further heated to 80°C before filling into bottles. It is necessary to leave a head space when filling in order to allow for expansion of the milk during heat treatment. Traditionally, glass bottles have been used, hermetically sealed with crown corks. However, in the past few years, plastic bottles have been introduced; these are usually blow moulded on site from high density polyethylene and closed with a plastic foil laminate which is heat sealed to the top of the bottle.

The batch sterilization process, by which bottles are filled into crates and heated in a batch autoclave at 105–115°C for 20–30 min, has now been largely superseded by continuous sterilization techniques, i.e. the rotary valve sealed sterilizer and the hydrostatic sterilizer. The main feature of these continuous sterilizers is the provision of a pressure lock system which allows the bottles to pass continuously into a sterilizing section operating at a relatively high pressure.

The sterilization section of the rotary sterilizer is in the form of a horizontal chamber maintained at the required sterilization temperature, typically 125°C (110–140°C). The bottles are loaded onto a chain conveyor and pass into the chamber via a mechanical rotor. The residence time in the chamber is typically 10 min and during heat treatment the bottles are continuously rotated to improve heat transfer. After sterilization, the bottles are precooled under pressure and finally returned to atmospheric pressure via the rotary valve.

The rotary sterilizer is used mainly for the sterilization of plastic bottles and with these it is important to provide an overpressure in the sterilization chamber in order to avoid bursting or distorting the plastic. This is achieved by using a mixture of steam and compressed air as the heating medium.

Sterilization of glass bottles is usually carried out in a hydrostatic sterilizer, in which the pressure in a central sterilization zone is maintained by a column of water at the inlet and outlet. The head of water necessary to maintain the required sterilization conditions is 4–6 m high giving the hydrostatic sterilizer the appearance of a tower. The bottles are loaded onto a chain conveyor and fed downwards through the inlet column of water, the temperature of which is increased as the bottles progress through it. At the bottom of the inlet column, the direction of the bottles is reversed and they pass upwards into the sterilization section (108–112°C). At the top of this section, the direction of the bottles is again reversed and they pass downwards to the bottom of the sterilization section before passage through the outlet column of water in which they are progressively cooled. The residence time in the sterilization section is between 20 and 30 min.

Two-stage sterilization

In the conventional manufacture of sterilized milk, raw milk is simply heated to 80°C before filling into bottles and sterilizing at, for example, 110°C for 30 min. On some occasions this regime of heat treatment has been insufficient to destroy all spores and most manufacturers have adopted the practice of giving the milk a UHT treatment, e.g. 140°C for 2 s (see UHT), before filling into bottles and sterilizing conventionally. In this case, the in-bottle sterilization treatment may be less intensive since it is only required to destroy those organisms on the internal surface of the bottle together with those which entered the milk during the filling stage.

ULTRA HEAT TREATMENT (UHT)

Ultra heat treatment of milk was developed to provide a sterile milk with a flavour similar to that of pasteurized milk, rather than the caramelized flavour of in-bottle sterilized milk. The UHT technique is based on the fact that at temperatures greater than 130°C, the heat treatment required for the complete sterilization of milk gives rise to much reduced colour formation and flavour deterioration than conventional sterilization treatment, e.g. 110°C for 30 min.

Most Regulations require that UHT milk should have received a heat treatment at least 132°C for at least one second. In fact, UHT treatments usually fall within the range 135–150°C with holding times of 2–5 s; these temperatures are achieved either by indirect heating, with pressurized hot water or steam, or by direct heating, whereby the product is mixed with high pressure steam. Alternative methods of heating are also available, e.g. electric heating and friction, but these are not yet commercially significant.

After UHT treatment, the sterile product must be protected against recontamination and UHT products are therefore filled into containers under aseptic conditions.

Indirect heating

The main feature of the indirect heating method is that there is no contact between the product and the heating medium. Heat exchangers for indirect UHT treatment may be of the plate or tubular type but economic considerations favour the former for the production of UHT milk. The heating medium is usually pressurized hot water but steam may also be used.

Milk is pumped from a balanced tank to a regeneration section of the heat exchanger where it is heated to about 85°C. The milk is then homogenized and returned to the heat exchanger where it is heated, first by regeneration and finally by pressurized hot water, to the required sterilization temperature (e.g. 138°C). The hot milk is held for the required length of time (e.g. 4 s) in a suitably dimensioned holding tube and cooled in the regeneration section before aseptic filling.

Automatic safeguards are incorporated into the plant to ensure that if the product sterilization temperature is not reached, the flow is diverted and underheated product is returned to the balance tank. Prior to UHT processing, it is of course necessary to sterilize the plant itself. This is achieved by recirculation of hot water at the sterilization temperature for at least 30 min. After presterilization, water is run through the plant to establish stable operating conditions before switching to product.

The high temperatures involved in UHT treatment give rise to deposition of protein in the heating, holding and cooking sections. These deposits increase the pressure drop and reduce the rate of heat transfer across the plates and must therefore be minimized, particularly in the heating section, if long run times are to be achieved. Some reduction in deposit formation can be brought about by holding partially heated milk at an intermediate temperature (e.g. 80 to 85°C) for up to a few minutes before heating up to the sterilization temperature. Recently, however, equipment manufacturers have developed modified heating and regeneration systems to minimize the temperature difference between the product and the heating or cooling medium, for example by using a closed water circuit as the regeneration medium rather than milk. This not only reduces fouling but also reduces the level of cooked flavour in the product.

Direct heating

Heating of milk by direct mixing with steam requires the latter to be of the highest quality. A direct heating system should therefore be equipped with its own steam generation plant which must be carefully controlled to eliminate contaminants.

Direct heating can be achieved either by injecting steam into milk or by spraying milk into steam in an infuser. There appears to be little difference between these two methods and the overall process is essentially identical for each.

Milk is first preheated to 80–85°C either by regeneration or with hot condensate or with low

pressure steam, before heating to sterilizing temperature (e.g. 140–145°C) by direct contact with steam. The latent heat of the steam required to heat the milk to sterilization temperature results in the condensation of a significant amount of steam, diluting the product by about 10%. This added water is subsequently removed by evaporation in a vacuum chamber, the conditions in which are controlled so that the quantity of water evaporated is exactly equal to the quantity of steam condensed during sterilization. The evaporation in the vacuum chamber rapidly cools the milk which is then asepticallly homogenized before further cooling and aseptic filling.

With the direct heating method, it is usually necessary to homogenize the milk after sterilization because direct heat treatment causes reagglomeration of fat globules. The homogenizer must therefore be capable of aseptic operation and sterilization.

Presterilization of the plant is by means of high pressure steam. The plant is heated at the sterilization temperature for 30 min before operation on water and finally product.

Comparison of heating systems

The indirect heating system is generally the least expensive in terms of capital and running costs but is more prone to fouling by deposit formation on the heating surfaces. In the past, it has been generally accepted that milk processed by direct heating has a better flavour than indirectly heated milk but recent developments in the design of indirect heating plant have overcome this.

Aseptic balance tank

In many instances, it is possible to match the throughput of the UHT plant with that of the aseptic filling machines. An aseptic balance tank may then be used for short term storage of sterile product prior to filling. The tank is sterilized with high pressure steam before use and the product valves are surrounded by steam barriers to maintain sterility. Air entering the tank is sterilized by filtration and a slight overpressure is usuallly maintained to avoid the entry of contaminating bacteria.

UHT treatment of milk-related products

Cream containing up to 40% fat can be processed satisfactorily by UHT treatment but homogenization and heat treatment must be carefully controlled in order to prevent fat and serum separation during extended storage. Viscous products, e.g. custards, desserts, can also be UHT processed, but highly viscous products, and products containing particulates, require the use of a scraped surface heat exchanger.

Age gelation in UHT milk and creams

In some instances, UHT milk or cream may coagulate on prolonged storage to form weak, custard-like gels. This is usually the result of chemical and physical changes in the milk protein. UHT treatment is usually sufficient to destroy most of the enzymes in milk but some psychrotrophic bacteria produce proteases which survive UHT temperatures. Such proteases act in a similar way to rennet and result in gelation. Gelation may also be induced through interactions between protein and calcium ions and the Maillard (browning) reaction has also been implicated.

Instances of age gelation are rare, especially when the raw milk is of good quality. Age gelation in UHT cream can be controlled by the use of permitted stabilizing salts (phosphates, citrates, carbonates).

HOMOGENIZATION

The fat in untreated milk floats freely and rises to the surface to form a creamline on the top. Centrifugal separation of the milk is simply an accelerated form of the same phenomenon. A creamline on top of market milk when packed in bottles used to be considered attractive in most countries of the world and indeed it still is where bottles are used.

However, an increasing volume of market whole milk is now packed in cartons where the consumer cannot see the creamline before opening the carton. In such milks and in long life UHT milks and also in modern dairy products like chocolate milks and coffee milks a creamline is undesirable. Such milks are therefore homogenized.

Homogenization eliminates the creamline problem. During homogenization the fat globules are subjected to mechanical treatment which breaks them down into smaller globules that are then uniformly dispersed in the milk. The rising velocity of these small fat particles is extremely low and thus homogenized products are very stable.

The main advantages of homogenizing market milk and cream are:

(i) Uniform distribution of the milk fat globule sizes
(ii) No creamline
(iii) Whiter, cleaner colour
(iv) A more even and full bodied flavour.

The efficiency of homogenization can be increased by two-stage treatment. In this case two homogenizing heads are installed in series in the same machine. The homogenization head comprises a high pressure positive pump (usually a piston pump) which forces the milk through a narrow gap in a specially designed valve. The high pressure on the inlet side of the valve (usually 2000–3000 psi (136–204 bar)) causes the fat globule in the milk to break up under the influence of mechanical shear and cavitation forces. The degree of globule breakdown increases with the pressure drop across the valve and this is controlled by adjusting the gap width in the valve. Thus homogenization takes place principally at the first homogenizing head. Those homogenizers with a second head (the majority) operate this head at a lower pressure (about 500 psi (34 bar)) which serves to break up any clusters of fat globules which may have formed after the first homogenizing stage. The two-stage process is common in UHT processing.

Market milk is the most commonly homogenized dairy product, now that most milk is sold in cartons or polybottles. The milk is either totally or partially homogenized. In total homogenization the full flow of milk is passed through the homogenizer, while in partial homogenization only the fat and a part of the skim milk are treated. UHT processes always include homogenization which counteracts some adverse effects of heat treatment.

Pretreatment of milk for yoghurt and fermented milk processes always includes an homogenization stage to serve to give the product a much better body as well as a smoother texture. Homogenization is also included in some cheese production lines, especially for rennet cheeses to speed up coagulation.

EVAPORATION

Evaporation is a unit operation widely used in the dairy industry both in the production of concentrates, e.g. evaporated and sweetened condensed milks, and in the preconcentration of milk prior to drying. The cost of water removal using modern evaporation techniques is less than 10% of the cost of water removal by spray drying.

The degree of concentration of milk and milk products by evaporation is, however, limited by the viscosity of the concentrate and the tendency of concentrated milk to gel at high solids concentrations. In the case of skimmed milk, the maximum solids concentration achievable by evaporation is about 55%.

For some years, the dairy industry has almost exclusively used highly automated, continuous evaporators. A detailed assessment of the different evaporator types, i.e. falling film, climbing film or plate is given by Kessler (1981). However, the falling film evaporator is by far the most widely used with dairy products because the high rate of heat transfer allows low temperature operation with a relatively short residence time.

FALLING FILM EVAPORATORS

The essential feature of the falling film evaporator is the calandria, which comprises a vertical bundle of tubes (diameter 25–80 mm, length 4–10 m) surrounded by a steam jacket. The milk feed is evenly distributed over the tubes and flows down the inside of the tubes as a thin film. The water is rapidly evaporated by the heat transferred from the steam jacket through the walls of the tubes. The mixture of concentrate and vapour is discharged from the base of the calandria to a vapour separator where it is separated into concentrate and vapour. The final vapour from the evaporator is then condensed using either an indirectly cooled surface condenser or a directly cooled spray condenser.

Evaporators are operated under reduced pressure in order to allow low temperature operation (70–80°C) which minimizes the heat treatment given to the milk and avoids the formation of deposits on the surface of the evaporator tubes.

MULTIPLE EFFECT OPERATION

Commercial evaporator installations comprise a series of calandria, or effects, whereby the vapour from the first effect is used as the heating medium for the second effect and so on. This mode of operation is termed multiple-effect evaporation and industrial units may comprise between three and seven effects. In order to maintain a temperature difference between the heating vapour and the milk in each effect, it is necessary to operate the effects at progressively lower temperatures (down to 40°C) and hence lower pressures. The production of viscous concentrates with high solids con-

tents necessitates the inclusion of a specially designed final effect, called a finisher, which must be heated using raw steam.

A six effect evaporator operates with a steam economy (kg steam required to evaporate 1 kg water) of approximately 0.25. Further economies are gained by compressing the vapour for re-use (see below).

VAPOUR RECOMPRESSION

Thermal vapour recompression involves the compression of vapour from the second or subsequent effects in a thermocompressor. In the thermocompressor, vapour from one of the effects is drawn into a suction chamber and mixed with a jet of live steam. The mixture then passes through an orifice which causes it to be compressed and therefore to be suitable for use as the heating medium in a previous effect. Thermal vapour recompression may be carried out over a number of effects and can reduce the steam economy to as little as 0.12 with a six-effect evaporator. Vapour may also be recompressed mechanically using electrically driven compressors. The recompressed vapour is then used to provide a large proportion of the steam required to heat the first effect.

PREHEATING

Preheating of the raw milk product is necessary both to pasteurize the feed and to impart specific characteristics to the final product, e.g. viscosity, resistance to oxidation, heat stability. Initial preheating is usually carried out in spiral heat exchangers housed in the calandria. Further heating is then carried out either directly by steam injection or indirectly in tubular heat exchangers, which are usually duplexed to allow heater changeover should one heater become fouled during continuous operation.

Milk is usually preheated to 80 to 100°C but other temperatures may be used in the manufacture of specialized products (see SKIMMED MILK POWDER; WHOLE MILK POWDER). A number of commercially important products are produced by evaporation including evaporated milk and sweetened condensed milks. It is also an essential prerequisite in the production of milk powders.

EVAPORATED MILK

Evaporated milk is a sterilized milk product containing a minimum of 9% fat and 22% solids-not-fat. The manufacturing process comprises the following operations.

Standardization

Clarified or filtered milk is first standardized to make the fat:solids-not-fat (SNF) ratio required in the final product, e.g. 9:22. This usually involves reducing the fat:SNF ratio by the addition of a calculated quantity of skimmed milk to the raw milk.

Heat treatment

The standardized milk is then given a heat treatment to stabilize the milk protein against coagulation when the evaporated product is subsequently autoclaved. Traditionally, milk was heated to about 95°C and held in a hot well for 10 min. Modern plants utilize higher temperature treatments, e.g. 120°C, with shorter holding times, e.g. 2 min. This type of treatment is not only more effective in stabilizing the protein but also allows continuous operation, since the holding period can be carried out in a suitably dimensioned tube.

In spite of the stabilizing effect of the heat treatment, milk produced at certain times in the year, e.g. late spring, is inherently unstable because of abnormalities in the salt balance (see SALTS). Some manufacturers therefore screen raw milk by the ALCOHOL STABILITY test. Milk for the manufacture of evaporated milk should have an alcohol stability of at least 70%.

Evaporation

The milk is evaporated at low temperature (below 65°C) to a solids level slightly in excess of that required in the final product. Falling film evaporators are used (see EVAPORATION, main section above), usually with three or four effects.

Homogenization

After evaporation the milk is homogenized to prevent fat separation during the storage life of the product. Homogenization must be sufficient to disperse the fat effectively, but excessive homogenization has a destabilizing effect on the protein. Two-stage homogenization is preferred, with pressures of about 2500 and 500 psi (170 and 34 bar) on the first and second stages, respectively.

Stabilization and standardization

The evaporated milk is cooled to 5–8°C and held at this temperature while the composition is checked (and adjusted if necessary) and the optimum level of stabilizer addition established. Stabilizing salts are used further to protect the milk protein against coagulation during the subsequent sterilization process. Phosphates, citrate and bicarbonate are permitted stabilizers and may be used at concentrations up to 0.2% in total. The optimum level of stabilizer addition is established by sterilizing a number of sample cans of the evaporated milk containing different quantities of stabilizer. After sterilization, the can contents are examined and the level of stabilizer addition determined as that giving a non-coagulated product with the desired consistency. The required amount of stabilizer is then added to the bulk evaporated milk. Vitamins are also added at this stage.

Filling and sterilization

Filling should be carried out at a low temperature, to avoid excess foaming at the filler, and around 4°C is usual. Filling is done automatically followed by immediate closing of the cans. Heat sterilization of evaporated milk is always carried out in rotating sterilizers. An end-over-end rotating batch cooker may be used but the larger canners use continuous sterilizers which impart rotation and these may be of the hydrostatic type (see STERILIZATION, main section above). The normal heat processing for evaporated milk is around 118°C with 15–20 min holding time after the come-up period. This is followed by rapid water cooling.

Two types of can are used for evaporated milk. The conventional food can with double seamed ends has now almost entirely replaced the older traditional type known as the vent hole. The internal finish of cans for evaporated milk was traditionally plain but there is a trend toward overall lacquering, mainly for the sake of appearance.

SWEETENED CONDENSED MILK

Sweetened condensed milk is an evaporated milk product containing added sugar. This product is not sterilized but is preserved by the high concentration of sugar which raises the osmotic pressure of the product to a level where it is microbiologically stable.

Most condensed milk and dried milk Regulations specify three types of sweetened condensed milk. For example in the UK:

(i) Sweetened condensed skimmed milk: this must contain a minimum of 24% milk solids and not more than 1% fat.

(ii) Sweetened condensed partly skimmed milk: compositional requirements for this product are different for retail and industrial sales. The retail product must contain a minimum of 28% milk solids and between 4.0 and 4.5% fat and the non-retail product must contain a minimum of 24% milk solids and between 1 and 8% fat.

(iii) Sweetened condensed milk: this product also has different specifications for retail and industrial sales. The retail product must contain a minimum of 31% milk solids and 9% fat and the non-retail product a minimum of 28% milk solids and 8% fat.

The manufacturing process comprises the following operations.

Standardization

Clarified or filtered milk is first standardized to have the fat:solids-not-fat ratio required in the final product. This step is not necessary in the manufacture of sweetened condensed skimmed milk.

Sugar addition

Sugar is either added to the milk before evaporation or is introduced into the evaporator in the form of a sugar syrup. The quantity of sugar added is fixed within narrow limits by the need to achieve a sufficiently high osmotic pressure without the risk of sucrose crystallization in the final product. The parameter used to define the sucrose content in the final product is the sugar ratio, which is the weight of sucrose divided by the combined weights of sucrose and water. The sugar ratio must be at least 62.5% to prevent bacterial spoilage and less than 64.5% to avoid sucrose crystallization; a sugar ratio of 63.5% is usually chosen. The amount of added sugar is calculated on the basis of this ratio, the quantity and composition of the milk and the required total solids content in the final product.

Heat treatment

After sugar addition, the milk is subjected to a heat treatment which not only acts as a pasteur-

izing step but also inactivates most of the enzymes which might give rise to spoilage. The heat treatment also serves to modify the milk protein and control the viscosity of the final product (20–50 P). Traditionally, the heat treatment consisted of holding at 80 to 95°C for 15–30 min but modern plants utilize higher temperatures (110–115°C) with shorter holding times (1–2 min).

Evaporation

The sweetened milk is evaporated at low temperature (below 60°C) to the solids level required in the final product. Falling film or plate evaporators are generally used (three to four effects) and because of the high viscosity of the product, a finisher is necessary as the final effect (see EVAPORATION, main section above).

The total solids content of the product is controlled indirectly by measurement of the specific gravity or refractive index of the concentrate.

Cooling and seeding

After evaporation, the sweetened condensed milk must be cooled and seeded to promote the crystallization of a large number of very small lactose crystals. At ambient temperature, sweetened condensed milk is supersaturated with respect to lactose and if crystallization is allowed to proceed slowly, large lactose crystals will form and the product will be gritty.

Cooling can be carried out in vacuum coolers, tubular coolers, or more rarely in plate coolers. After rapid cooling to about 30°C, seed crystals of lactose (350 mesh) are added to promote rapid crystallization. Crystallization may be allowed to proceed for up to one hour before final cooling to ambient temperature, e.g. 15°C. The product is then held in storage tanks for at least one day until crystallization is complete. Seeding is not necessary with tubular coolers since the product is continually seeded by the film of product on the surface of the cooling tubes.

Packaging

Cans of condensed milk receive no post-filling heat sterilization process and the product is preserved by its total solids content. Therefore, because the empty cans may contain mould spores they should be presterilized. This may be carried out by inverting the cans over a series of superheated steam jets followed by hot air jets to dry them or by passing them over gas jets. Top ends should be similarly treated. It is, of course, essential that the cans be completely dry before filling, which should be done in a sterile area. Filling is carried out to leave an absolute minimum of head space and specially contoured can ends are used for the same purpose. The presence of air in the head space might allow mould or yeast growth, oxidation of the product or rusting.

Conventional food cans with double seamed ends are used for this product, with the exception that the ends are specially contoured, and they are usually plain. Sweetened condensed milk for industrial use, such as confectionary manufacture, is either filled into drums (250 kg) or is delivered in road tankers.

BOTTLING AND CARTONING OF MILK

In this section the information presented relates principally to the ways in which heat treated milks are packaged for sale to the ultimate consumer. The chain of events within the dairy following the various heating processes (pasteurization, UHT, sterilization) are critical and great care has to be taken to ensure that the customer receives a finished product which not only satisfies the minimum compositional and hygiene standards laid down in legislation, but also is of the desired quality.

The choice of package is an important consideration both for the dairies and the consumer. Close attention must also be paid to the hygiene and cleanliness of the processing and filling plant and to the environment under which the heat treated milk is packaged. Reference is made to these aspects later in this section. Essentially milk for sale to consumers is packaged in either of two types of containers, bottles or cartons.

BOTTLING

Glass bottles were the preferred container for pasteurized milk for well over half a century in most countries. In spite of the problems associated with the returning and washing of glass bottles, and the development of plastic and paper cartons, bottles are still widely used for pasteurized milk in the UK because of the continuing popularity of the doorstep delivery system. However, the latter has declined sharply in recent years, with sales of plastic bottles and cartons especially from supermarkets taking a larger share.

Filling pasteurized milk into glass bottles comprises two main steps, i.e. washing and sanitizing the dirty bottles, followed by hygienically filling the bottles with the correct quantity of milk.

Bottle washing

The average life of a milk bottle in the UK is approximately 20 return trips, so a key feature of the hygiene of bottled milk is the washing of the bottles prior to use.

Bottle washing machines are of either the sprayer type (jetting or hydro) or the soaker–sprayer type. The former system relies totally on internally spraying the bottles for cleaning and rinsing while the soaker–sprayer type incorporates soaking tanks which wet and soften the residual soil in the bottles and thus make it easier to remove in the subsequent spraying stages. Whatever the system, the bottles are first prerinsed, treated with caustic detergent at up to 70°C, attempered and given a final rinse with good quality water.

The detergent is based on caustic soda and would normally be a proprietary blend with other chemicals which act as sequestrants (to avoid scale), wetting agents and others which aid in holding particles in suspension. Apart from the degree of hardness of the water being used, the choice of detergent obviously depends on the type of bottle washing machines, especially the choice of wetting agent which must be low foaming in the sprayer type. The detergent alkalinity used is also somewhat lower (0.5–1%) in the sprayer type than in the soaker–sprayer type (1.0–1.5%). The effectiveness of cleaning obviously depends on good control of detergent concentration and temperature, but if this is achieved the bottles will be almost sterile when they leave the detergent section.

The final rinse is therefore crucially important with regard to the hygiene of the bottle: soft water should be used to prevent scaling up and the water should be chlorinated to a low level (1–2 ppm) to ensure that the bottles are not recontaminated.

Satisfactorily washed bottles should contain less than 200 colony forming units (cfu), and it is possible with care to achieve below 50, but it has to be said that counts of several hundred are not very rare.

Bottle filling

Most filling machines are of the rotary type, whereby milk held in a large bowl is gravity filled into the bottle. The filling bowl is under a slight vacuum and the level of milk within it controlled by a float. Bottles are conveyed from the washer to individual pedestals on the filling machine. As the bottle travels around, a cam lifts the pedestal until the top of the bottle presses against the filler valve rubber (which is spring loaded). As the bottle is lifted further it opens the valve and milk flows in.

After filling, the bottle is lowered to close the valve and excess milk is removed through a central suction (the filler bowl operates under a slight vacuum). The bottle is then lowered and transferred to a capping machine where caps are applied and crimped on.

Modern filling machines are designed for CIP (cleaning-in-place, see CLEANING AND SANITIZING) but even with hygienic cleaning there is ample opportunity for product contamination via the air. This is obviated in the conventional bottle filler, since air from the environment is actually drawn into the filler as a result of the vacuum suction devices. Also, the aluminium cap provides nothing like a hermetic seal. So, packaging of milk into glass bottles exhibits several areas where post-pasteurization contamination needs careful control.

CARTONING

Cartons have grown in popularity in recent years as the sales from retail outlets have increased. Their filling is relatively straightforward and typically the cartons are supplied in the form of blanks which are held in a magazine at the end of the cartoning machine. The blanks are opened and fed on to a mandrel which moves to a heating position, here the bottom flaps are heated to soften the polyethylene coating on the sealing area. At the next mandrel position, the bottom flaps are tucked and closed, the sealing being achieved through the rehardening of the plastic coating. The carton is then dropped on to a conveyor which takes it to a filling head. After filling, the top flaps are heated and finally sealed.

In recent development, plastic lidded cartons with a ring-pull for opening have been introduced. An example of this is the 'Tetratop' system where the lids are formed from reels of material within the unit which is forming and filling the cartons.

Usually, cartons have a lower degree of contamination than bottles and since air is not actually being drawn into the filler, the opportunities for post-pasteurization contamination are less than in bottle filling (see BOTTLING). However, air contamination does occur, and with the piston-type

filling mechanisms used, contamination can also occur as a result of the difficulty of effectively cleaning the sliding areas of the piston.

In 1983, one carton filler manufacturer (Liquid Pak) introduced some interesting developments in machine design to enable more hygienic operation, resulting in the following equipment modifications:

(i) The filler bowl is totally enclosed to avoid product contact with air. Changes in liquid content in the bowl are accommodated by a flexible steel diaphragm.
(ii) A plastic bellows is used in a positive displacement mode for filling. This avoids the cleaning difficulty previously encountered with a piston filler.
(iii) Air contamination in the filling zone is avoided by the provision of a flow of sterile air blown against the direction of carton travel, inside a stainless steel chamber.
(iv) Finally an inexpensive and safe way of pre-sterilizing the cartons was found in the use of low concentrations of H_2O_2 in conjunction with UV radiation which have a synergistic sporicidal effect; originally discovered by the Food Research Institute, Norwich.

The major manufacturers of cartons, the treated cardboard rolls from which cartons are often produced, and filling machinery (both aseptic and non-aseptic) on which cartons are filled are the Tetra-Pak Division of Tetra-Laval, Elopak, Combipak and Purepack. Comprehensive descriptions of the filling machinery models available and their hygienic methods of operation, along with preventive maintenance instructions are provided by all of these manufacturers and reference should be made to these for further details.

There are advantages and disadvantages to both types of container, bottles and cartons, but in recent years it is the carton that has taken by far the greatest market share worldwide. As such, compared with returnable bottles, the unit and processing costs of cartons are higher than those of bottles, but cartons offer many advantages, for example they are lighter, take up less space in the storage room, in transport and on the retailers' shelves. They also offer the opportunity for better labelling.

Plastic bottles

Just since about 1985 a further competitor to bottles and cartons has emerged in Europe, namely the blow-moulded plastic bottle. These containers have the advantage of price over glass bottles, they are non-returnable and are supplied in a whole range of volumes to suit the needs of shops and consumers. Most of these are now resealable, unlike bottles and thus have another advantage over glass bottles. These plastic bottles are now taking an increasing market share mainly at the expense of cartons.

HYGIENE ASPECTS OF MILK PACKAGING

The primary requirement for hygiene in the packaging of milk arises from the microflora of the milk itself. However, there are different requirements for packaging hygiene depending on the type of milk (e.g. fresh, UHT, etc.) and the shelf life required from it. These requirements can be broken down into three groups:

(i) Packages for doorstep delivery, i.e. bottles and cartons, which will be delivered and consumed within days of manufacture.
(ii) Packages for the retail trade, i.e. cartons and plastic bottles, where a shelf life of up to 2–3 weeks may be required (under chilled distribution).
(iii) Packages for ambient storage of UHT milk and milk products which must be aseptic if the required shelf life of several months is to be achieved. By aseptic, we mean less than one defective container in 10 000 packages.

The ability to achieve these shelf lives is dependent not only on the number of bacteria in the heat-treated milk, but also on the degree of recontamination of this milk during its packaging. The shelf life of pasteurized milk under commercial conditions (e.g. a few days) is normally restricted by post-pasteurization contamination. Several studies in recent years have shown that the aseptic filling of freshly pasteurized milk into sterile containers can extend the shelf life of the pasteurized milk to over 3 weeks at 4°C.

FACTORS AFFECTING POST HEAT TREATMENT CONTAMINATION

The following factors may be sources of post-heat treatment contamination.

Heat exchangers

Product direct from heat exchangers has been known to be contaminated with untreated milk where pinholes have appeared in the heat exchanger plates in the regeneration or cooling sections. In the case of UHT milk, where it is necessary for the product to be sterile, even the minutest pinhole can give rise to unacceptable levels of contamination. Regular examination of the cooling sections of the heat treatment plant should therefore be carried out, and in some pasteurization systems a special pressure booster pump is installed in the product line to ensure a positive pressure differential on the pasteurized product side in the regeneration and cooling sections.

Pipework, valves, and so on

The most obvious source of contamination from pipework, valves, etc., arises from inefficient cleaning and sanitation. Cleaning is obviously facilitated by good plant design, e.g. by ensuring that fittings are crevice free, that pipelines have the correct fall for drainage, that dead ends or pockets are eliminated and correct flow rates for cleaning fluids are maintained.

However, aseptic processing lines are always presterilized by heat (either steam or pressurized hot water) to ensure sterility even if small milk deposits are not removed during cleaning.

For the hygienic processing of pasteurized products, cleaning of process pipework is equally important. After sanitizing, it is important that the final rinse water contains a few ppm of available Cl_2. A common source of contamination is the use of untreated water for rinsing downstream of the pasteurizing plant.

In aseptic processing, the need to avoid contamination totally means that special care must be taken where there is the possibility of contact between the sterile product in the pipelines and the surrounding air, e.g. valves. This is overcome by the use of what are known as sterile barriers whereby a small pipeline between two valves connects product flow between them when required, but when they are disconnected, low pressure steam is introduced to create a barrier against contamination.

Filling equipment

The filling equipment itself, and particularly the filling and sealing areas, is obviously a major source of contamination and there is a distinction between the equipment surfaces and the surrounding air. As with pipework, the most important factor in avoiding contamination in equipment is good design which, together with efficient cleaning regimes must remove a major source of problems – residual soil. Some examples of these design principles are:

(i) Product contact surfaces to be smooth, impervious, corrosion resistant and non-toxic.
(ii) Internal angles should have minimum radii.
(iii) Product contact surfaces self draining.
(iv) Bottoms of filler bowls and hoppers require a slope towards the outlet.
(v) All mechanical cleaning systems for bowls or hoppers should be designed so that solution is applied to all product contact surfaces.
(vi) Design of valves is another important feature; wide radii must be incorporated throughout the dead-end pockets eliminated to prevent product accumulation.
(vii) Drip deflectors should be installed at each filling valve to prevent condensation and drippage from falling into the container.

Airborne contamination

Air in dairies is a major source of product recontamination – one report quotes a figure of 3.2 cfu l^{-1} in a milk-processing area. In some factories in the USA and Germany, equipment has been installed to supply sterile filtered air around filling machines, and as a result an increase in shelf life of pasteurized milk to 10 days has been observed.

The provision of a positive pressure of sterile air in the filling area of the packing machine was a part of the development of fully aseptic packing machines in the 1960s and 1970s, the small positive pressure ensuring that non-sterile air from the surrounding atmosphere is not drawn into the filling area. In aseptic processing, air is sterilized by either filtration, heat treatments (300°C) or both. Depth filters are preferred, based on compressed mats of glass fibre, metal or ceramic. They can be steam sterilized *in situ*, but care must be taken to ensure air flow is maintained within a recommended range to ensure their effectiveness.

For the less stringent requirement of hygienic (not aseptic) processing, high efficiency particulate air (HEPA) filters are adequate (giving a retention of 99.99% of particles larger than 0.3 mm), and some of the newer carton filters use these.

Packaging materials

The original microbial count of carton packing materials has been estimated at 0.5 organisms

cm^{-2}. This is equivalent to about 25 organisms in a half-litre carton. In fact this is quite a low level and this level of hygiene is sufficient for many market requirements. However, aseptic packaging requires the packaging material to be sterilized.

CLEANING AND SANITIZING

Milk processing inevitably results in the formation of deposits on the surface of dairy equipment and effective cleaning and sanitizing are therefore essential in maintaining high standards of hygiene. A standard procedure for cleaning and sanitizing dairy equipment would comprise the following steps:

(i) Recovery of product residues by drainage, flushing with water or expulsion with compressed air.
(ii) Rinsing with warm water to remove soluble and loose deposits.
(iii) Cleaning with an alkaline detergent solution.
(iv) Rinsing with warm water to remove loosened deposits and to flush out the detergent solution.
(v) Sanitizing with a chemical solution, e.g. hypochlorite, iodophor.
(vi) Final rinsing with potable water.

While the above procedure is satisfactory for general purposes, the cleaning of heat exchange surfaces requires more vigorous treatment to remove denatured protein and 'milkstone', i.e. deposits of calcium phosphate. This is carried out by including an acid (usually nitric) cleaning stage after the second rinse.

The delicate membranes used in the ultrafiltration of milk and milk by-products need special low heat treatments to remove any fat and protein deposits blocking the molecular size pores. Successful enzyme preparations have now been developed which ensure minimum structural damage to the membranes.

CLEANING SYSTEMS

Manual cleaning has now been largely superseded by cleaning-in-place (CIP) in which the cleaning fluids are held in tanks and recirculated around the plant at high velocity. This procedure is ideal for the cleaning of pipes, heat exchangers, separators, homogenizers, etc. In the case of vats and tanks, cleaning solutions are directed against internal surfaces by spray nozzles, the design of which can be modified to give direct impingement of cleaning liquid onto areas requiring particular attention, e.g. agitators, manways, etc. CIP systems are available with large variation in the degree of automation, depending on size of plant. Larger systems also incorporate a facility for recovering detergents for re-use.

DETERGENTS

The most commonly used detergents are alkaline based and purchased as proprietary blends containing alkali, surfactant, sequestrants and phosphates. The optimum level of each of these depends on the cleaning duty required and the hardness of the water. The functions of the detergent components are as follows:

(i) The alkali causes protein deposits to swell and dissolve. At high temperatures, the alkali also saponifies fat, coverting it to water-soluble salts.

(ii) Surfactants reduce surface tension and have a dispersing effect on deposits.

(iii) Sequestrants prevent scale formation in hard water and help disperse calcium phosphate deposits.

(iv) Phosphates are effective dispersing and emulsifying agents and also have a sequestering effect.

Acid detergents are used less frequently when scale is a problem. Proprietary acid detergents usually contain a surfactant only in addition to nitric acid or sometimes phosphoric acid.

FINISHED PRODUCT TESTING

Great emphasis is nowadays placed in all the advanced dairying countries on quality. Above a description was given of the range of compositional and hygiene tests applied by dairies to incoming raw milk supplies. Exactly the same level of attention is required with regard to their analytical testing of finished milk products, namely pasteurized, UHT and sterilized milks.

PASTEURIZED MILK

Since pasteurized milk has a relatively short shelf life because it is not a sterile product there is a need for routine quality assurance of the plant. Thermographs monitoring and recording pasteurization times and temperatures are inspected and kept on file for verification for at least 6 months in case a public health problem subsequently arises.

Samples are taken from the filler approximately every half-hour to check that butterfat has not separated out, extraneous water is absent (by the freezing point test) and that the milk has been correctly pasteurized (by the phosphatase test). Close control is also kept on the temperature of milk held in the cold store prior to despatch.

The keeping quality of correctly pasteurized milk may be markedly shortened by post-pasteurization contamination which can occur anywhere downstream of the pasteurizing plant, normallly at the filling stage in the case of glass bottles. Contamination can occur from the surrounding atmosphere or from the plant. Even very low levels of post-pasteurization contamination, especially with Gram-negative micro-organisms, can quickly cause spoilage if the temperature of the finished product is allowed to rise during storage and distribution. Likewise, thermoduric bacteria, which survive pasteurization treatment may also cause spoilage problems especially during the warmer summer months.

The most commonly used test to assess pasteurization efficiency is the phosphatase test. Alkaline phosphatase occurs naturally in raw milk but is destroyed by normal pasteurization processes (71.5°C for 15 s). Measurement of residual phosphatase after pasteurization therefore offers a means of checking that correct pasteurization has occurred and that raw milk has not been reintroduced accidentally downstream of the heating cycle in the pasteurizer.

Phosphatase tests are undertaken on every batch of pasteurized milk to conform with legislation. The basis of the reaction to measure residual phosphatase involves splitting a phosphate radical from a phosphatase ester by means of the enzyme phosphatase which is naturally present in milk. Since the first use in the early 1930s the test has undergone several modifications mainly to improve the speed or sensitivity by changing the substrate and hence the reagent released by the action of the enzyme phosphatase.

Nowadays the most commonly used test for phosphatase is the Aschaffenburg and Mullen method. In this method *para*-nitrophenol is liberated and measured colorimetrically using a photometer. Recently a new technique for measuring residual phosphatase has been developed and introduced by Advanced Instruments in the USA. It is called 'Fluorophos' and involves the use of a proprietary substrate as the phosphate monoester. This proprietary substrate is not of itself fluorescent, however, the liberated phenol derivative, the concentration of which is directly related to the amount of the alkaline phosphatase present in milk, is highly fluorescent and can be measured at very low levels.

'Fluorophos' has many advantages over all other methods, principally in terms of speed and sensitivity, and it is rapidly gaining favour in the industry as the preferred test for alkaline phosphatase.

In addition to the phosphatase test most dairies carry out a limited range of bacteriological tests on representative samples of packaged pasteurized milks. These could include either carrying out simple Petri-dish plate counts for the total bacterial count, or for Coliform bacteria or for Pseudomonads. The latter are psychrophilic bacteria which grow well at the refrigeration temperatures used to distribute and retail milk and thus this check offers arguably a more representative assay of milk quality than simple total bacteria count.

All of these bacteriological methods take two or three days to complete and thus, at best, can only offer a retrospective assessment of pasteurized milk quality.

UHT MILK

The only significant quality problems which can occur with UHT milk are the development of off-flavours and gelation in storage due to the survival of heat stable protease enzymes (derived originally from psychrotrophic bacteria present in the original raw milk) and the risk that bacteria from the environment may enter the finished package if the latter has not been properly sealed.

A rapid and sensitive method has been developed for heat stable protease assay using a modification of the bioluminescence technique. The principle of the test is that luciferase enzyme catalyses the reaction of ATP and luciferin to produce light. Luciferin originates from fireflies. ATP is found in all living cells and supplies the energy to power biological reactions. Psychrotrophic protease is incubated in the presence of luciferase which degrades and causes a light output when luciferin and ATP are added. This rapid (5 min) method is now available in kit form and some dairies have now started to use this technique.

Aseptic filling is an essential part of the UHT process and to ensure that sterility has been achieved a representative sample of containers is taken and stored for 15 days at 30°C followed by a standard bacteriological plate count. The aseptic failure rate should be lower than 1 in 100 000 and certainly not higher than one in 10 000. This is the

most commonly used test method by dairies on UHT milk.

STERILIZED MILK

Sterilized milks are normally checked using the Aschaffenburg turbidity test which is a measure that milk has received the correct time–temperature heat treatment. This test involves precipitation of casein and heat denatured whey protein from milk. In the case of sterilized milk a test for residual non-heat denatured whey protein would prove negative whereas UHT milk would give a positive test for whey protein since it would not have been totally denatured by the less severe UHT heat treatment.

LIPOLYSIS

Lipolysis of milk fat by native and bacterial lipases gives rise to rancidity as short chain fatty acids (butyric, caproic) are released. Lipolysis is not often noticed in liquid products but may cause defects in some butter products. Two major factors control the extent of lipolysis in milk.

(i) *Heat treatment.* Native milk lipase is inactivated by pasteurization but certain bacterial lipases are very heat resistant and may even survive UHT treatment. However, a large proportion of the lipase activity in milk is destroyed by high temperature pasteurization and residual lipase activity does not often present a problem.

(ii) *Fat globule membranes.* Most of the fat in milk is protected from lipolytic attack by the membrane surrounding the fat globules (see FAT). However, agitation of the milk, especially in the presence of air, may lead to disruption of the fat globule membrane and thus render the fat susceptible to attack. Great care is therefore taken in the handling of unpasteurized milk to avoid the risk of development of lipolytic rancidity.

Lipolysis is also a problem sometimes in whipping creams since the free fatty acids released by lipolysis can often suppress the whipping powers of the product.

STARTER CULTURES

Acidity and flavour development in cultured milk products such as cheese, lactic butter and yoghurt, is achieved by inoculating milk, cream or skimmed milk with selected bacteria in the form of a starter culture, or bulk starter.

The most important functions of a starter are the controlled development of acidity and flavour characteristics. Starters also serve to suppress the growth of undesirable bacteria and in some cases are considered to impart therapeutic properties to the final product. In cheesemaking, the acidity developed by the starter is a necessary aid to coagulum formation and subsequent expulsion of whey from the curd (syneresis). The starter bacteria also govern the development of the cheese flavour and texture during maturation.

STARTER ORGANISMS

The most important bacteria in starter cultures are the lactic streptococci, particularly *S. lactis* and *S. cremoris*, which ferment lactose to lactic acid. These organisms are homofermentative and ferment lactose only. *S. thermophilus* which has a higher optimum growth temperature is used in the manufacture of high scald cheeses, e.g. Emmenthal, and yoghurt.

S. diacetylactis belongs to a group of bacteria which produce aroma compounds as well as lactic acid. *Lactobacillus lactis*, *L. helveticus* and *L. bulgaricus* may also be included in this group. These organisms are heterofermentative and are capable of fermenting the citric acid in milk, as well as lactose, to produce volatile aroma compounds, e.g. diacetyl, acetaldehyde and acitic acid. Leuconostoc strains, e.g. *L. dextranicum* and *L. citrovorum*, are also used for flavour development in the manufacture of a number of cultured milk products.

Recently much interest has been shown in starter cultures exhibiting 'health giving' properties. Examples of starter cultures in this probiotic category include *L. acidophilus* and species of Bifidobacterium all of which are natural inhabitants of the healthy human intestine.

Requirements of a starter culture

The requirements of a starter culture are summarized as follows:

(i) It should contain bacteria of the required type only.
(ii) It must produce acid rapidly and at a consistent rate.
(iii) It must produce a clean acid flavour and,

where appropriate, aroma. It must not produce off-flavours, taints or gas, unless the latter is required in an open-textured cheese, e.g. Swiss.
(iv) It must be capable of growing under the conditions of manufacture of the particular cultured product.
(v) It should show high resistance to bacteriophage attack.

Types of starter culture

The starter bacteria may be used in one of three ways:

(i) The single-strain starter, which is a pure culture of only one strain of bacteria, e.g. *S. cremoris*, *S. lactis*.
(ii) The defined multiple-strain starter, which is a defined mixture of two or more strains of lactic acid bacteria and flavour producing bacteria.
(iii) The mixed-strain starter, which is a mixture of unknown proportions of two or more strains of different species.

The susceptibility of starter bacteria to bacteriophage attack, especially in cheesemaking, has led to the practice of using starter cultures which include at least two strains of bacteria, i.e. types (ii) and (iii) above. Thus if bacteriophage attack, then the others can continue to produce acid. In cheesemaking, bacteriophage problems may also be reduced by rotating the starters so that consecutive vats do not contain the same strains of bacteria.

Yoghurts traditionally contain a mixed strain starter culture comprising *L. bulgaricus* and *S. thermophilus* but nowadays, in addition, 'bio' yoghurts contain species of Bifidobacterium and sometimes *L. acidophilus*.

Bulk starter preparation

The milk used for the propagation of a bulk starter culture must provide a suitable medium for growth and steps must therefore be taken to ensure that inhibitory factors are absent. For this reason, most manufacturers now use skimmed milk powder which has been specially screened to ensure that it is substantially free of antibiotics. The powder is reconstituted to 12% solids and heated at 85–95°C for 30–60 min to destroy bacteriophage and vegetative bacteria.

Alternative media are also available for starter

culture propagation. These include bacteriophage inhibitory media, which give protection against bacteriophage by chelating the calcium necessary for bacteriophage propagation, and buffered media which reduce the rate of fall of pH with lactic acid production and hence allow the development of greater bacterial numbers.

The traditional method of bulk starter production is carried out in two or more stages. The starter culture, which is purchased either frozen or freeze dried, is first used to prepare a mother culture. This is grown in a glass bottle (about 100 ml) which is first sterilized in an autoclave. The bottle is then three-quarters filled with milk or reconstituted milk and heated to about 90°C for 1 h before cooling to the incubation temperature (e.g. 21°C for cheese, 43°C for yoghurt). The starter is then inoculated into the milk and the bottle incubated until the correct acidity has been reached, usually 0.7–0.9%. It is then cooled to 10–15°C before use in order to minimize any loss in activity. A portion of this mother culture can be used to inoculate another bottle to produce an active mother culture for the next day.

The mother culture is then used to inoculate the bulk starter, although an intermediate culture may be prepared in larger factories. The bulk starter is grown either in churns or in specially designed starter vessels which incorporate facilities for heating and cooling the milk, mixing, temperature control and aseptic operation. It is very important that the transfer of cultures is carried out aseptically in order that contamination by bacteriophage and undesirable bacteria is avoided.

This method of starter propagation is very time consuming, and developments in starter production in the 1960s have resulted in the now widespread use of concentrated starters which can be used to inoculate the bulk starter directly. These concentrated starters are either supplied freeze dried or liquid-nitrogen-frozen in ring-pull cans and a single container is sufficient to inoculate up to 1000 l of bulk starter.

More recently, highly concentrated cultures have become available for direct inoculation of milk, thus avoiding the need for bulk starter preparation. These are significantly more expensive than the cultures used for direct bulk starter inoculation, but because they offer a choice of pure, defined bacterial strains and cut out the need for the user to produce his own starters at the bulk vat stage, they now find wide usage in large dairy plants.

There has been an interesting development in France recently where some cheesemakers complained that there is a possibility that the new direct-to-vat concentrated cheese starters (and possibly the milk and processing plant) were 'too clean' to allow the old-fashioned cheese flavours to develop. Development of non-lactic ripening cultures (*L. casei, L. helveticus* and micrococci) which enhance flavour development for addition with convenient starters has virtually opened up a new market sector. These bacteria have served to reintroduce the small ancillary fermentations in the ripening cheese and have thus improved the taste and flavour of the matured cheese in a beneficial way.

Testing of starter cultures

Routine monitoring of starters is particularly important in cheesemaking where bacteriophage attack is more common than with other cultured products. The two most important tests establish the ability of the starter to produce acid and detect the presence of bacteriophage.

Bacteriophage

Bacteriophage, or phage, are viruses which may attack bacteria used in the manufacture of cheese and other cultured dairy products. The point of entry of bacteriophage is not known with certainty but it is most likely that it is present in raw milk. A pronounced reduction in the rate of acid development by starter bacteria is almost always traced back to the presence of bacteriophage, and the selection and use of starter cultures are carefully planned in order to minimize this problem.

YOGHURT AND OTHER FERMENTED PRODUCTS

Yoghurt is a cultured milk product which owes its characteristic flavour and texture to the growth of a mixed culture of *Streptococcus thermophilus* and *Lactobacillus bulgaricus* in specially prepared milk. The fermentation process was originally used as a means of extending the shelf life of milk, since the low pH of yoghurt inhibits the growth of undesirable bacteria. However, yoghurt is nowadays consumed for its pleasant organoleptic properties. Also, in recent years, the reputed therapeutic properties of yoghurt have been established through scientific research.

The manufacture of yoghurt is covered by differing legislation in different countries. Often the main stipulation in legislation relates to composition (percentage of butterfat and milk solids-not-fat) and to labelling of the packaged product.

Most of yoghurt produced in the UK and in America is of the stirred type which is manufactured in bulk before mixing with fruit and packaging. Set yoghurt, which is packed into retail containers before incubation, is manufactured extensively on the Continent. In recent years yoghurt has also become available in the form of drinking yoghurt, long life yoghurt and strained yoghurt.

MANUFACTURE OF STIRRED YOGHURT

The manufacture of stirred yoghurt comprises six major steps:

(i) Preparation of the milk
(ii) Homogenization
(iii) Heat treatment
(iv) Fermentation
(v) Fruit addition
(vi) Packaging

Milk preparation involves the standardization of the fat and solids-not-fat content of the milk and the addition of stabilizers, where used. The fat content of milk for yoghurt manufacture is usually between 0.5 and 1.5%, although some very low fat products (approx. 0.1% fat) are manufactured and at the other extreme, the 'thick and creamy' products contain between 3 and 10% fat.

The solids-not-fat content of the milk is usually increased by between 1.5 and 3% (either by evaporation or skim milk powder addition) to increase the viscosity and stability of the yoghurt curd. Stabilizers may also be added at this stage (e.g. gelatin, pectin, modified starches) to aid in water binding and thus increase viscosity and reduce the tendency for syneresis (whey separation).

The homogenization step is carried out at 55°C and at a pressure of around 3000 psi (204 bar). Homogenization is not only important to reduce the fat globule size, and thus retard creaming, but also in causing changes in the structure of the casein micelles (see MILK PROTEIN) which improves their water binding properties.

Probably the most important operation of the prefermentation steps is the heat treatment, which not only pasteurizes the milk but also induces whey protein denaturation and the formation of complexes between β-lactoglobulin and κ-casein (see MILK PROTEIN). These changes again significantly increase the water binding properties of the milk proteins, thus increasing the viscosity of the

final product and reducing the risk of syneresis. The heat treatment involves holding the milk at 90–95°C for about 5 min in continuous flow processes, or at 85°C for 30 min in batch processes.

After the high temperature treatment, the milk is cooled to 43–45°C and pumped to fermentation tanks. Here the milk is inoculated with a mixed culture of *Lactobacillus bulgaricus* and *Streptococcus thermophilus*. The inoculation rate is about 2% and the starter culture is prepared in the same way as cheese starter cultures. The ratio of *L. bulgaricus* to *S. thermophilus* is between 1 : 2 and 1 : 1.

The milk is incubated for 3–4 h in the final stage of the fermentation (e.g. when the pH has fallen to 4.5) the yoghurt is cooled to around 15°C to retard any further acid development. Some yoghurt tanks have the facility for integral cooling, but usually the yoghurt is cooled in tubular or plate heat exchangers. The structure of the yoghurt at this stage is extremely sensitive to shear, and plant is therefore designed to ensure gentle stirring in the fermentation tank itself and minimal shear in downstream processing. The latter is achieved by the use of positive displacement pumps and by using short lengths of wide diameter pipeline.

After cooling, the addition of fruit preparation where used is via in-line mixers. Again, these are designed to give good mixing with minimum shear. Finally, the yoghurt is filled into plastic pots (125 or 150 g), closed with heat sealed foil and transferred to cold stores, where the yoghurt is held prior to despatch and where the temperature falls to 4–5°C.

Stirred yoghurt should have a thick, smooth consistency and be sufficiently viscous to hold the fruit in suspension during the shelf life of the product. Actual shelf life depends very much on the type of packaging material used and the general hygiene of the ingredients and the manufacturing process itself. Deterioration can occur through excessive acid production, yeast and mould contamination, or syneresis, and an expected life of 14 days under chilled storage is normal.

MANUFACTURE OF SET YOGHURT

The selection, standardization and heat treatment of the milk are the same as for stirred yoghurt manufacture. Bulk starter is then metered into the milk and, where appropriate, colour and flavour are added.

The inoculated milk is first filled into the retail

containers which are then packed into trays and incubated in a temperature controlled room for the required period, e.g. at 42°C for 4 h. The yoghurt is then cooled to 15–20°C in another temperature controlled room, then stored at 4–5°C.

Set yoghurt has a firm, gel-like consistency which leaves a clean surface when cut.

STRAINED YOGHURT

Strained 'Greek-style' yoghurt is manufactured by removing some of the whey from yoghurt, either by straining in bags or by using a centrifugal separator. Typically, strained yoghurt contains about 25% total solids and 10% fat and is particularly popular in Eastern European and Middle Eastern countries.

DRINKING YOGHURT

Drinking yoghurt is a low viscosity drink made by blending yoghurt with fruit juice and sugar. It has a sharp, refreshing quality and usually contains very little fat.

Yoghurt is first manufactured as described under MANUFACTURE OF STIRRED YOGHURT above. When the fermentation is complete, the coagulum is broken by stirring, and fruit juice, sugar and a stabilizer (e.g. pectin) are added. The stabilizer is important in preventing protein aggregation during the subsequent heat treatment which may otherwise cause a chalky mouthfeel and protein precipitation on storage. Good dispersion of the stabilizer is ensured by homogenization. The drinking yoghurt is then heat treated and packed. It can either be pasteurized and filled into cartons to give a fresh product with a shelf life of a few days, or pasteurized and aseptically filled to give a long life product with a shelf life of several months.

LONG LIFE YOGHURT

Fresh yoghurt has a shelf life of 2–3 weeks at 4°C, but this can be extended to a few months if the yoghurt is pasteurized. This can be achieved in one of three ways:

(i) Pasteurization of yoghurt in a scraped surface heat exchanger at, for example, 75°C for 15 s, followed by hot filling.
(ii) Pasteurization as in (i) followed by aseptic cooling and filling.

(iii) In-pot pasteurization where the yoghurt is first filled into pots which are then held in pasteurizing rooms at 75–80°C for 10–20 min.

All of these types of heat treatment and filling techniques are aimed at eliminating the presence of yeasts and moulds in the final product. These organisms are capable of growth at the acidic pH of yoghurt and can lead to off-flavour development, gas production and syneresis (whey separation).

YOGHURT PRODUCT DEVELOPMENT

Extensive segmentation of the yoghurt market has taken place in western Europe and the USA over the past few years. Conventional yoghurts have been targeted at specific market sectors, e.g. breakfast yoghurts (containing muesli, oats, etc.), thick and creamy yoghurts (relatively high fat, with smooth creamy texture) and savoury yoghurts (aimed particularly at people from the Indian subcontinent).

Frozen yoghurts made their appearance in the American and European markets in the late 1970s and are now popular products in many countries, largely as a result of marketing efforts by the ice-cream manufacturers.

The introduction of newer types of starter culture such as *Lactobacillus bifidus* has been exploited in health conscious areas, particularly Germany and Scandinavia. There are now many different variants of yoghurt available in which manufacturers have used a whole spectrum of micro-organisms reputed to have health giving and special nutritional and protective properties. As a group such yoghurts are known as 'Bio yoghurts' and the 'Probiotic' starters used in their manufacture were originally isolated and purified from the intestinal flora of healthy human beings. Types of probiotic bacteria include *L. acidophilus* and species of Bifidobacterium as well as *Enterococcus faecium*, the latter being one of the starters employed in a successful health yoghurt 'Gaio' which originated in the Ukraine and which is now manufactured under licence by a Danish company.

ACIDOPHILUS MILK

Acidophilus milk is the name given to milk cultured with *Lactobacillus acidophilus*. It is claimed to have therapeutic properties in the treatment of digestive disorders because the bacteria can flourish in the intestine and produce acid which inhibits the growth of putrefactive bacteria.

Lactobacillus acidophilus grows only slowly in milk and it is important to eliminate competing organisms. Milk, either whole or skimmed, is homogenized and sterilized before cooling to 37°C. A bulk starter of *Lactobacillus acidophilus* is then added at a level of 2 to 5% and the milk incubated until an acidity of 0.7% is reached. The product is bottled and held between 5 and 10°C during storage. Plain (natural) and fruit flavoured varieties may be made.

KEFIR

Kefir is a cultured milk product which originated in the Balkans. In addition to lactic acid, kefir also contains alcohol and carbon dioxide which contribute to its flavour.

The starter culture for kefir manufacture is in the form of coarse particles known as kefir grains. These contain coagulated casein together with lactose fermenting yeasts (*Torula kefir*, *Saccharomyces kefir*, lactobacilli and *Streptococcus lactis*). The yeasts account for up to 10% of the microorganisms.

The starter culture is prepared in two stages: the grains are first inoculated into milk and incubated at 20–25°C for 20 h. The grains are then separated, inoculated into another batch of milk and incubated under the same conditions to produce a bulk starter.

Milk for kefir manufacture may be partly skimmed or whole milk and this is homogenized at high pressure before heat treatment at 95°C. Both of these operations are important in developing the desired consistency in the final product. The milk is then cooled to 23°C and inoculated with about 3% of the bulk starter. After 20 h, the pH has fallen to 4.5–4.6 and the kefir is cooled to 40°C and held before packaging.

Kefir is a smooth viscous fluid with a fresh acidic taste. It usually contains about 1% alcohol as a result of the yeast fermentation.

KOUMISS

Traditional Koumiss originated with the Tartars and was made from mare's milk. Its present day equivalent is normally made by fermenting skimmed cow's milk with *Lactobacilllus acidophilus*.

Traditional Koumiss has a uniform consistency with no tendency to whey-off. The protein of mare's milk is somewhat different from milk of other species because when fully renneted, it forms no visible curd. Rennet is not used to make Koumiss, but acid produced during fermentation results in the milk protein forming a fine precipitate which remains in suspension.

The starter culture used in Koumiss preparation in the home by Kazakhs Tatars is thought to contain a Candida species of yeast and a species of *L. delbreukii*.

CULTURED BUTTERMILK

Cultured buttermilk was originally manufactured from fresh buttermilk, but difficulties in maintaining a consistent supply of high quality buttermilk have led to skimmed milk being used as the raw material.

Skimmed milk with a fat content of about 0.5% is first pasteurized and inoculated with a starter culture (1%) containing acid producers, e.g. *Streptococcus lactis*, *S. cremoris* and flavour producers, e.g. *S. diacetylactis*. The milk is incubated at 22°C until an acidity of 0.7–0.8% is achieved and the product then cooled to 4–5°C and filled into plastic cups or cartons.

Cultured buttermilk has a mild lactic flavour and is considerably more viscous than fresh milk. It is usually consumed as a refreshing drink.

CHEESE

The Food and Agriculture Organization has given the following definition of cheese: 'Cheese is the fresh or matured product obtained by the drainage of liquid after the coagulation of milk, cream, skimmed or partly skimmed milk, buttermilk or a combination thereof'.

Cheesemaking provides a very elegant solution to the problem of preserving the protein and fat in milk. Cheese may, therefore, be thought of as a product in which the most important constituents of milk are converted into a smaller bulk which can be stored for considerable lengths of time. This is undoubtedly the main reason for the historical development of cheese as one of the major dairy products (the other being butter). In addition, cheese can be produced with a variety of textures, flavours, etc. which appeal to a wide range of palates. It can either be eaten in its own right or used as an ingredient in domestic cuisine or commercial food processing.

Cheese is also a very important source of nutrition, accounting for about 5% of the protein con-

Table 3.4 Important nutrients in cheese (per 100 g)

Cheese	Energy (kcal)	Protein (g)	Fat (g)	Calcium (mg)	Vitamin A (mg)	Riboflavin (mg)
Cheddar	410	25	33	800	310	0.5
Camembert	300	20	25	150	150	0.4
Cottage	95	14	4	60	32	0.3

sumed in the diet. It is a valuable source of energy and contains a number of the minerals and vitamins important in the human diet (see Table 3.4).

Cheese represents a concentrated food. It is normally made from cow's milk, but can also be made from the milk of goats, sheep or buffalo, and consists of casein, variable amount of the fat, mineral salts, and part of the milk serum. In the small amount of serum present are dissolved lactose, whey protein and vitamins. The cheese curd may be formed by addition of rennin or by acidification of the milk. The cheese may be unripened, or ripened by bacteria or by moulds.

Although more than 400 varieties of cheese have been described, many of them are similar and there are only about 10 distinct types. Cheese may be classified according to texture, moisture, ripening agent and method of manufacture. Examples of several different cheese categories and types are shown below:

(i)   Low fat soft unripened, e.g. cottage cheese.
(ii)   High fat soft unripened, e.g. cream cheese.
(iii) Firm unripened, e.g. Mozzarella.
(iv)   Soft, ripened by surface bacteria, e.g. Limburger.
(v)   Soft, ripened by surface moulds, e.g. Camembert.
(vi)   Semi-soft, ripened by internal bacteria, e.g. Port du Salut.
(vii) Semi-soft, ripened by internal moulds, e.g. Stilton, Roquefort.
(viii) Firm, ripened, e.g. Cheddar, Provolone.
(ix)   Hard, ripened by eye-form bacteria, e.g. Emmentaler.
(x)   Very hard, ripened by bacteria, e.g. Parmesan, Romano.

The moisture content in cheeses can vary between 79% maximum in cottage cheese down to 30% in cheeses such as Parmesan. Butterfat contents range from a low of 0.4% in cottage cheese up to 37% in high fat cream cheese. Protein contents also vary, ranging from a low of 8.8% in high fat cream cheeses up to 36% in Parmesan. The remaining components comprise mainly minerals, salt, calcium and lactose.

CHEESEMAKING

The cheesemaking process varies greatly between the many types of cheese produced. However, in general terms, cheesemaking comprises three essential operations:

(i)   Production of a milk coagulum, or curd, through the action of rennet and/or lactic acid on the milk protein casein (see CASEIN).
(ii)   Separation of the curd from the whey.
(iii) Manipulation of the curd to produce the desired characteristics of the final cheese.

The important unit operations which make up the cheesemaking process are summarized in Table 3.5 (see also CHEDDAR CHEESE). For any particular cheese variety, the manufacturing process may include only some of the operations listed and, in some cases, these may be carried out in a different sequence from that indicated in Table 3.5. Many of the operations were traditionally performed manually but in most cases, especially in developed dairying countries, are now highly mechanised and, wherever possible, automated.

Continuous coagulation

In spite of major developments in the mechanization of cheesemaking, it is only in the last 20 years that industrial equipment has become available for continuous coagulation. Earlier methods relied on the fact that rennetted milk does not coagulate in the cold. Most of these systems therefore provided a means of continuously heating cold rennetted milk, and in one case concentrated cold rennetted milk (Stenne–Hutin process), to induce coagulation. However, the curd produced by these systems possesses different characteristics from conventional curd and for this reason it has not gained acceptance.

A relatively new development has been the Alpma Coagulator which is now installed in several factories in Europe. This coagulator comprises a slow moving flexible belt of semicircular cross-section, in which the various stages of cheesemaking are carried out in different sections

Table 3.5 Unit operations in cheesemaking

Unit operation	Purpose
Pretreatment of milk:	
(i) standardization of composition	Control of fat and protein content of final cheese.
(ii) pasteurization	Destruction of pathogenic bacteria.
(iii) additions	Various: e.g., calcium chloride to assist curd formation, sodium nitrate to inhibit butyric acid bacteria.
Starter culture addition	Production of lactic acid by bacterial fermentation of lactose. Development of flavour and aroma compounds (mould spores may be included to develop mould growth in mould ripened cheese).
Rennet addition	Coagulation of milk protein to produce a coagulum or curd, entrapping milk fat.
Cutting	Cutting of curd into small cubes (5–30 mm) to facilitate release of whey (syneresis).
Cooking	Raising temperature accelerates syneresis to reduce moisture content of curd further.
Whey drainage	Drainage releases bulk of whey from curd.
Curd texturization	Manipulation of curd, e.g. Cheddaring, develops required texture in cheese.
Milling	Milling curd into small pieces to facilitate distribution of salt and convert curd into a suitable form for pressing.
Salting (dry or brine)	Salt addition retards acid production and helps curd fusion through solubilization of protein. Salt also enhances flavour.
Moulding	Filling curd particles into moulds which decide final shape of cheese.
Pressing	Removal of residual whey and control of texture of final cheese.
Packing	Suitable wrapping applied to cheese to provide protection during storage.
Ripening	Storage at defined temperature and humidity, to allow development of texture and flavour through enzymic breakdown of protein and fat.

along the length of the belt. The coagulation section is divided into compartments by spacer plates in order to minimize unwanted movement of the milk during coagulation. Starter culture and rennet are metered into the milk feed which coagulates in the segmented compartments. The coagulum passes through a cutting section and the curds and whey are then conveyed to a drainage section before filling into moulds. This coagulator does not incorporate the facility for scalding and

is therefore restricted to the manufacture of soft cheeses.

A continuous tubular coagulator (Alcurd) has also been developed for the coagulation of milk concentrated by ultrafiltration.

Ultrafiltration

The use of ultrafiltration in cheesemaking was developed in the 1970s and many countries now use this technique in the manufacture of soft cheeses.

Ultrafiltration of the cheesemilk removes the necessary amounts of lactose, salts and water before rennetting. The protein and fat in the milk are concentrated by ultrafiltration to produce a 'pre-cheese' which has the same composition (fat, protein, moisture) as a drained cheese. The pre-cheese is then incubated with a starter culture and after a suitable period of acid development, rennet is added. The concentrate is filled into moulds where coagulation takes place without loss of whey. Brining and further processing then proceed conventionally.

The major advantage of ultrafiltration in cheesemaking is that the whey proteins are retained in the cheese and the cheese yield is therefore increased by up to 20%. Starter and rennet requirements are also reduced and the elimination of vats gives savings in equipment costs and floor space. However, the texture and flavour of ultrafiltered cheeses tend to be slightly different from those of conventional cheeses and the commercial application of ultrafiltration has therefore been slower than was originally expected.

The application of ultrafiltration in cheesemaking is presently limited to the production of cheeses with a moisture content of over 50%, e.g. Camembert, Feta, because of the inability of existing ultrafilteration systems to concentrate milk to a solid content in excess of about 50%. However, systems have recently been developed whereby a 'cheese base' can be produced with the composition of Cheddar and other hard cheeses.

In these systems milk is first concentrated to a solids content of 40% by ultrafiltration. In the latter stage of concentration, water is added (diafiltration) to reduce the lactose content and thus control the final pH of the cheese. The concentrate is then cooled and a starter culture added. After a short ripening period, salt is added and the concentrate evaporated at low temperature to a solids content of 60 to 65%. The cheese base is packed into plastic drums and after a final ripening period the finished product is transferred to a cold room for storage. As an alternative to this procedure, the ripening can be completed prior to evaporation and the concentrate pasteurized to give an essentially sterile product.

The yield of cheese base is up to 20% greater than that of conventional cheese because of the incorporation of whey proteins. However, the product is very bland and is not comparable with a fully ripened hard cheese. The main application for cheese base is in the manufacture of processed cheese and cheese spreads where it is an economic alternative to unripened conventional cheese.

To illustrate the process of cheesemaking a few examples of cheese manufacture can be considered. Detailed descriptions of the manufacture of most cheeses can be found in any classical textbook on the subject and thus for the purposes of this chapter only the manufacture of Cheddar, probably one of the best known of all cheeses, and of 'Quarg' and 'Fromage Frais', two soft cheeses which have grown rapidly in popularity in recent years will be outlined.

CHEDDAR CHEESE

Traditional manufacture of Cheddar cheese was a batch process in which nearly all the cheesemaking operations were carried out in the same cheese vat. However, this has largely been superseded through the mechanisation of many of the stages in cheesemaking so that the cheese vat is used only for the production of curd. Texturization and subsequent operations are then carried out separately in highly mechanised equipment.

Pretreatment of milk

Modern cheesemaking practice aims at producing cheese of a consistently good quality. In order to achieve this, it is a prerequisite that the cheesemilk be of a consistent quality and composition, and for this reason milk is usually pretreated before cheesemaking.

Cheesemilk is usually pasteurized to destroy pathogens and other bacteria that may adversely affect cheese production. Standard HTST conditions are used, i.e. 72°C for 15 s, since higher temperatures cause whey protein denaturation and precipitation of calcium phosphate, both of which have a deleterious effect on curd formation and syneresis.

The ratio of casein:fat in the cheesemilk may also be standardized to what is considered an optimum value (usually 0.7 : 1) in order to overcome seasonal variations in milk composition. The purpose of standardization is twofold; first to produce a cheese of consistently good texture and body, and second to maximize the yield of cheese from a given quality of milk.

Standardization usually involves an increase in the natural casein:fat ratio, either by the removal of fat (in the form of cream) or by the addition of skim milk solids (in the form of skim milk or skim milk powder). When milk is standardized, it is important that the standardized milk produces a cheese conforming to the legal standards regarding composition.

Pretreatment may also involve the addition of calcium salts (chloride, phosphate) at levels up to 0.02%, in order to improve the coagulation characteristics of the milk. The natural dye, ANNATTO, may also be added in the manufacture of coloured cheese.

Starter culture addition

After pasteurization, the cheesemilk is cooled to 30°C and delivered to the cheese vats. Traditionally, open rectangular vats with swinging stirrers were used. However, these have now been largely replaced by round-ended vats with rotating stirrers and more recently by enclosed cylindrical vats (up to 25 000 l capacity), with rotating stirrers which, when reversed, act as knives in the cutting process.

The starter bacteria (see STARTER CULTURES) are added as an active culture of lactic acid and streptococci, usually a mixture of *Streptococcus lactis* and *S. cremoris*. The level of addition is usually 1–2% of the quantity of cheesemilk.

The amount of lactic acid developed by the starter bacteria is simply measured by titration and the titratable acidity is widely used to fix the different stages of the cheesemaking process. The rate of acid development in the milk, and later in the curd, determines the moisture, texture and flavour of the finished cheese, so control of acid development is one of the most important aspects of good cheesemaking practice.

Rennetting

When the acidity of the milk has risen to 0.20 to 0.22%, usually after 30 to 45 min, the milk is ready for rennetting in order to produce a coagulum (see CHEESEMAKING INGREDIENTS – RENNET). The exact time of rennet addition may be simply defined by the acidity of the milk, but in some cases as empirical test may be used to assess the suitability of the milk for rennetting, e.g. the test cup. Rennet is added at a level of 22 to 25 ml per 100 l of milk, usually in a diluted form (five times). The milk is stirred for a few minutes after rennetting in order to prevent the fat rising as a cream layer. However, it is important that the coagulum remains undisturbed once the milk has started to set in order that the clot forms as a gel, entrapping the fat, rather than as a granular precipitate from which the fat is lost.

Cutting

When the coagulum is sufficiently formed, usually after 30 to 45 min, it is cut into small cubes using a set of vertical and horizontal knives. The coagulum is usually considered ready for cutting when it breaks cleanly over an inserted finger or rod. However, a number of instruments are now available whereby the firmness of the coagulum can be measured in the vat. In the future, these may be used to give an objective measure of the suitability of a coagulum for cutting, but at present they are not in widespread use. After cutting, whey is released from the curd particles (syneresis) and the curds and whey are gently stirred.

Scalding (cooking)

The expulsion of whey from the curd particles is accelerated by heating the curds and whey in a process known as scalding. Steam or hot water is applied to the heating jacket of the vat so that the temperature is raised to 39–40°C over about 40 min. The rate of heating is initially slow (1°C per 5 min up to 35°C) to avoid case hardening of the surface of the curd particles; if the surface of the curd hardens too quickly, it reduces the rate of syneresis and the curd particles retain too much whey.

The maximum scald temperature is determined by the thermal sensitivity of the starter bacteria; it is important that the starter bacteria are not substantially inactivated in order that acid development in the curd is not retarded. When the maximum scald temperature has been reached, the stirring is continued until the curd reaches a condition suitable for the texturization or cheddaring process (acidity 0.20%). In traditional manufacture, however, curd was first 'pitched' or allowed to settle, while the titratable acidity developed from 0.17% up to 0.20%.

Whey drainage

Traditionally, the curd was pushed to one end of the vat and the whey drained through a tap. In modern systems, the curds and whey are pumped or fed by gravity to a screen which in some cases forms part of the texturizing equipment.

Texturization (cheddaring)

Differences in manufacturing technique between traditional and mechanized processes are greatest in the texturization stage of Cheddar manufacture. The object of this process is to allow the curd to consolidate or fuse and, by subjecting the curd to a stretching action, to convert the curd from a rubbery mass to a structure with a tearability similar to that of chicken breast.

In the traditional system, the curd mass was cut into blocks and piled up two or three high. The blocks were turned every 15 min and this squashing of the curd, together with acid induced fusion of the curd particles, gave the required texture in the cheese when the acidity reached 0.65 to 0.75% (pH 5.2–5.3).

A number of mechanized plants are now used to carry out the cheddaring process; the two most important types are those in which cheddaring takes place in a tower or on a series of flexible conveyor belts.

In one tower system, the separated curd is first delivered to a perforated conveyor belt on which the curd is stirred by rotating pegs. This dry stirring promotes further dewheying of the curd. The curd is then pneumatically conveyed to the top of a cheddaring tower (1.5–2 m diameter) where the curd fuses under the influence of its own weight and the developed acidity. Stretching of the curd is brought about by a change in cross-section, from round to square, in the lower part of the tower. The curd is discharged from the base of the tower and cut into blocks.

The other major type of mechanized cheddaring system comprises two or more conveyor belts arranged above one another. After whey drainage and dry stirring, the curd reaches the end of the top conveyor and is automatically turned upside down onto the conveyor belt below. The curd mass is stretched by running the two conveyor belts at different speeds.

Milling and salting

The blocks of curd are broken into small pieces using a mill, Salt is added to the milled curd at a level of about 2% and the curd is then stirred to ensure even distribution of the salt. A short time is allowed for the salt to dissolve before the curd is pressed (mellowing). In mechanized systems, the milling and salting stages are usually incorporated within the cheddaring equipment and the curd mellowed on a conveyor belt.

Pressing

Salted curd is usually filled into moulds or hoops for pressing. Traditionally, the cheese was pressed in hoops under increasing pressure for 2–3 days. However, Cheddar is now usually pressed in rectangular moulds for 24–36 h to give an 18 kg block. The pressing time is further reduced in some systems by the application of a vacuum before pressing. The need for individual moulds has been obviated by the 1 tonne presses which have been installed in some factories or by continuous block-forming systems in which the curd is pressed under its own weight under vacuum.

Wrapping (packaging)

After pressing, the cheese must be treated to give protection against moisture loss and mould growth during storage. Traditional round cheeses were wrapped in a bandage which was either pasted or waxed to prevent mould growth. The paste was a flour and water mixture containing a preservative, e.g. sorbate or pimaricin, while wax was applied by dipping the bandaged cheese into a bath of melted wax.

However, the traditional round cheeses have now been largely replaced by rectangular 18 kg blocks. These are a convenient shape for cutting into consumer portions and wrapping in plastic film which gives more effective protection against mould growth and mite damage. Wrapping in a plastic film also prevents rind formation and therefore reduces wastage. In addition to the normal requirements for food packaging materials, the plastic film wrapping must be relatively impermeable to water vapour (to prevent moisture loss) and to oxygen (to prevent mould growth). Two major types of plastic wrapping materials have become popular: waxed cellulose laminates, and films based on polyvinylidene chloride copolymer (PVdC). Alternative plastic materials are also in use, as single layer films or as laminates.

Waxed cellulose films, e.g. Pukkafulm, Parakote, are sealed around the cheese by the application of heat and pressure in a suitable press, e.g. the Flower's press. In this operation, the air between the cheese and the film is squeezed out and the fat in the cheese melts and becomes attached to the wrap. This type of film can also be used in a vacuum wrapping operation where the seal is produced by application of pressure and heat to the wrapper fold overlap. In this case no heat is applied to the surface of the cheese. PVDC materials, e.g. Saran, can be used as a film wrap, which is then protected by a paper overwrap, or as a vacuum pack.

The tendency nowadays is towards vacuum packing in PVdC barrier bags, e.g. Cryovac. The cheese is filled into the bags, a vacuum applied and the bag sealed with a metal clip. The sealed bag is then heat shrunk onto the cheese. After wrapping, the cheese blocks are stored in rigid containers, e.g. plastic or wooden boxes, in order to prevent deformation on storage.

Ripening

Cheddar cheese is matured in ripening rooms maintained at a temperature of 4–8°C. The ripening or maturing stage permits the controlled breakdown of the protein and fat in the cheese to produce the desired flavour and texture characteristics in the final product.

During the early stages of ripening the starter bacteria (lactic streptococci) are inhibited while lactobacilli develop and bring about changes in the flavour and aroma of the cheese. Residual lactose in the cheese is quickly fermented to lactic acid and other organic acids. The major change in the texture of the cheese, from a rubbery to a soft, creamy consistency, is brought about by the breakdown of the protein to peptides and amino acids by the rennet and bacterial proteases. Fat is also broken down to glycerol and fatty acids by lipolytic enzymes. Amino acids and fatty acids are then broken down to a number of lower molecular weight compounds through the action of a number of bacterial enzyme systems. These breakdown products account for the characteristic flavour and aroma of Cheddar cheese and include amino acids, amines, fatty acids, ketones, diacetyl, hydrogen sulphide, methanethiol, aldehydes and lactones. Mild cheddar is generally ripened for 3–6 months; mature cheddar requires 6–12 months.

QUARG

Quarg is a soft unripened cheese with a refreshing lactic flavour. It originated in Germany but is becoming increasingly popular in the rest of Europe. In France it is called fromage frais (q.v.).

Quarg is manufactured from pasteurized skimmed milk inoculated with a bulk starter of *Streptococcus lactis* and *S. cremoris*. A small amount of rennet is also added (0.5–1 ml/100 l milk) to improve coagulation. Acid development (to pH 4.5–4.6) is carried out in an insulated tank and takes 5–6 h 32°C (short method) or 16–20 h at 22°C (long method).

Traditionally, the coagulum was then cut and the curd drained in bags, although any medium-to-large plant would now incorporate a centrifugal separator to concentrate the curd. Quarg leaving a separator typically contains between 17 and 20% solids and is cooled before storage in a silo. Quarg produced by this process utilizes only the casein fraction of the skim milk. Two other methods of quarg manufacture are also in use in Europe whereby some or all of the whey proteins are recovered in the quarg, thus increasing yield.

The most widely used process for quarg manufacture in Europe is the thermoquarg process. Here, the skimmed milk is first heated to about 90°C and help at this temperature for 5 min to denature the whey proteins (see MILK PROTEIN). The milk is then cooled to the fermentation temperature and the acidification process proceeds as described above. Following fermentation, the coagulated skimmed milk is given a second heat treatment, this time at 60–65°C for 3–5 min. The quarg is then separated from the whey in a separator, cooled and held in a silo prior to use.

The other process which incorporates whey proteins into the quarg involves the use of ultrafiltration (see above) and there are two variants of this. First, the ultrafiltration process can be used instead of a separator to separate the fermented skimmed milk into quarg and whey. In this case, the skimmed milk would be fermented to pH 4.6 and then ultrafiltered: the concentrate from the ultrafiltration plant would be the quarg (incorporating all the whey proteins) and the permeate would be the aqueous phase of the milk containing the lactose and the minerals. Alternatively, the quarg is separated conventionally and the ultrafiltration process is used to concentrate the protein in the whey to the same protein level as that in the quarg. The concentrated whey stream is then heat treated at 90°C for 2–5 min to denature the proteins, homogenized, cooled and then blended with the main quarg stream.

The use of ultrafiltration in quarg manufacture enables the achievement of higher yields than those obtained with the thermoquarg process. However, the product resulting from the latter process is much closer in organoleptic terms to conventional quarg and for this reason the thermoquarg process is currently preferred over ultrafiltration processes.

The fat content of quarg can be increased by blending with cream in specially designed contin-uous mixers. The incorporation of fruit and sugar is also becoming popular as a means of extending the use of quarg as a dessert.

FROMAGE FRAIS

Fromage frais is a type of soft cheese which has become very popular recently and which is produced with a number of different fat contents (0–8%) with and without added fruits.

Essentially, fromage frais is the French equivalent of quarg, i.e. a fresh, acid ripened soft cheese (see QUARG, above). In France, it is sold in containers of up to 2 kg as well as in individual portions, and is used as a general cooking ingredient and as a base for other ingredients, such as fruits, nuts, sugars, etc.

Fromage frais can be manufactured in one or two ways:

(i) By blending quarg and cream: here it is very important to control the composition and pH of the quarg in order to control the texture of the final product and avoid graininess.
(ii) By ultrafiltration: here milk is first fermented by a starter, as in the manufacture of quarg, and then ultrafiltered to the required composition, e.g. 20% solids, 4% fat.

Fromage frais produced by these two processes has different organoleptic properties, the ultrafiltration product generally being considered less desirable. However, because the ultrafiltration process recoveres the whey proteins, the yield is increased by some 15–20% over the traditional method.

In most countries, fromage frais is usually sold in individual portion pots and most of the products are flavoured (10–20% fruit, 5–10% sugar). Recent refinements to the process of manufacturing fromage frais by ultrafiltration have led to a much more acceptable soft cheese being produced and the final enormous sales; for example, in tonnage terms the market for this product in the UK alone now tops 50 000 tonnes per year and grew by 22% in 1994 alone.

CHEESE INGREDIENTS

Rennet

Rennet is the enzyme preparation isolated from the fourth stomach (abomasum) of the calf. Commercial rennet preparations contain varying proportions of the proteases chymosin and pepsin; chymosin accounts for about 90% of the proteolytic activity in rennet extracted from young calves but the proportion of chymosin decreases with increasing age of the animal to only 10%. Commercial rennets usually contain 70 to 80% chymosin.

Rennet is usually supplied as a liquid containing salt, propylene glycol and sodium benzoate in addition to the enzymes. Stored at 4°C, liquid rennet loses 1–1.5% of activity per month. Powdered and paste rennets may be used in hot countries where liquid rennet will not keep.

Rennet is normally diluted with cold tap water before use; water prepared for sanitizing and containing hypochlorite should not be used since this has an adverse effect on rennet activity.

Rennet action

The coagulating effect of rennet is the result of a highly specific proteolytic reaction which renders the casein micelles (see MILK PROTEIN) susceptible to aggregation in the presence of calcium ions. Rennet cleaves the phenylalanine–methionine bond (105-106) in γ-casein releasing a soluble glycomacropeptide. The remainder of the κ-casein molecule, termed *para-κ-casein*, is no longer able to stabilize the casein micelles and these become linked via calcium bridges to form a coagulum.

The first stage of rennet coagulation, the proteolysis of κ-casein, can take place at low temperatures but the second stage, calcium-induced aggregation, is not significant at temperatures below 20°C. In cheesemaking, rennet action takes place at about 30°C in order to achieve the desired rate of coagulation.

Overheating milk during pasteurization has a deterimental effect on rennet coagulation due to a reduction in the soluble calcium content in the milk and to whey protein denaturation. The former can be overcome to some extent by the addition of a small quantity of calcium chloride (up to 0.02%) to the cheesemilk in order to restore the calcium balance. The pH of the milk is also important and a firm clot is only achieved at pH 6.6 or below; acidity development by starter bacteria is therefore an important factor in the coagulation of milk by rennet.

Approximately 6% of the rennet added to milk is retained in the cheese after the separation of whey and its proteolytic activity is in part responsible for the development of the flavour and texture of the cheese during maturation.

Standardization of rennet

The coagulating power of rennet is usually determined by measuring the length of time required to coagulate a sample of milk under defined conditions of temperature (e.g. 30°C) and pH. This is known as the rennet coagulation time (RCT). The substrate is usually low heat skimmed milk powder reconstituted in 0.01 M calcium chloride, adjusted to pH 6.5. The RCT is inversely proportional to the rennet concentration over a fairly wide range of concentration and this relationship can be used to calculate the strength in rennet units (RU), i.e. the number of millilitres of milk which can be clotted in 40 min by 1 ml of rennet. Rennet in most countries is usually standardized to a strength of 15 000 rennet units per ml.

Alternative coagulants

The decreasing supply of calf rennet has led to the development and use of a number of alternative coagulants. Mixtures of bovine and porcine pepsine with calf rennet (50 : 50) have proved satisfactory in use, as have some fungal enzymes (e.g. from *Mucor miehei*). Enzymes derived from plants and bacteria have good coagulating strength but exhibit excessive proteolytic activity during maturation and give rise to flavour and texture defects.

The microbial coagulants derived from Mucor spp. exhibit similar proteolytic specificity to calf rennet and are used commercially in a number of countries around the world. More than one-third of the cheese made worldwide utilizes microbial coagulants. Microbial coagulats are standardized in a similar manner to rennet and are usually offered as single strength (15 000 RU ml^{-1}) or triple strength (50 000 RU ml^{-1}) products. A number of thermolabile products are also available which are completely inactivated in whey by pasteurization.

Annatto

Annatto is the natural colouring agent used in the manufacture of some types of cheese and butter. It is extracted from the fruit of the annatto tree, *Bixa orellana*, and is used in the form of a paste in butter manufacture and in the form of a dilute alkaline solution as an additive to cheesemilk. The colour of annatto is dependent on pH; between pH 6 and 7 (i.e. that of sweet cream butter) it is predominantly yellow, while at pH below 5.5 (i.e. most cheeses) it is predominantly red.

In the UK, annatto is used in the manufacture of Leicester, Double Gloucester and coloured Cheshire cheeses. Coloured Cheddar is not usually manufactured in England but is particularly popular in some parts of Scotland.

CHEESE PRODUCTS

Cheese powder

Cheese powder is a convenient form of cheese solids for use as an ingredient in food manufacture. It is mainly used in the flavouring of snacks, where a powder is ideal for coating and in the manufacture of dips, sauces, fillings, etc. where it can simply be mixed with the other ingredients.

The most popular form of cheese powder is based on Cheddar cheese, but smaller quantities of semi-hard cheese may be incorporated to reduce cost. Blue cheese and quarg may also be used to manufacture cheese powders with specific flavour characteristics.

The raw material for cheese powder manufacture is usually a carefully selected blend of cheeses of different ages. The cheese is comminuted and mixed with water to produce a slurry containing 35 to 45% solids. Stabilizing salts (phosphates, citrates) are added in small amounts (about 2%) to modify the physical characteristics of the milk protein so that the fat is effectively emulsified and the slurry melts evenly on heating to give a smooth body in which grains of unmelted cheese are absent. The slurry is pasteurized and homogenized (2000 psi (136 bar) first stage; 500 psi (34 bar) second stage) before spray drying to produce a powder containing 3% moisture. There is usually a significant loss of cheese flavour during spray drying and cheese powders generally have a much milder flavour than the original cheese. This can be overcome to some extent by correct choice of raw materials and the flavour can also be improved by the inforporation of enzyme modified cheese or cheese flavour enhancers into the slurry.

Cheese analogues

Cheese analogues are cheese-like products manufactured from milk protein (casein) and vegetable fats. These products are considerably less expensive at present than natural cheese, not only because of the use of vegetable fat in place of milk fat, but because the pricing system for casein within the EU is such that casein in the form of acid or rennet casein or caseinate is less expensive than the equivalent amount of casein in raw milk. The milk protein source for manufacture is either a blend of sodium and calcium caseinates, or rennet casein. When the latter is used, it must first be solubilized by the addition of emulsifying salts.

The manufacturing procedure for cheese analogues is very similar to that used for processed cheese, using a steam heated kettle. The casein or caseinate is added first with a small quantity of water to produce a slurry. The vegetable fat is then added together with small amounts of salt, lactic acid and emulsifying salts. The ingredients are then heated and mixed to ensure effective distribution and emulsification of the fat. The heat treatment must be sufficient to pasteurize the mix, e.g. 70–85°C for 4–15 min, which is then filled into moulds or formed into slices. The flavour of cheese analogues may be improved through the incorporation of a proportion of natural cheese or enzyme-modified cheese.

Cheese analogues usually possess a fairly mild flavour and a close texture similar to that of processed cheese. The textural characteristics can be altered by varying the type and quantity of casein and fat used and by modification of the pH. In this way, it is possible to produce cheese analogues with similar texture and melting characteristics to a number of popular cheeses, particularly Cheddar and Mozzarella.

Cheese analogues find greatest application in the catering sector, especially for use as pizza toppings and in the form of slices for use in hamburgers. The flavour of cheese analogues is not yet sufficiently well developed to compete with natural cheeses for retail sale.

PROCESSED CHEESE

Processed cheese is manufactured by pasteurizing a blend of hard cheeses with added emulsifying salts to give a prolonged shelf life. The emulsifying salts serve to control the pH of the processed cheese and to bind calcium so that the protein can effectively stabilize the fat.

There are two major types of processed cheeses:

(i)   Block cheese; this is a firm, slicing cheese with a relatively low moisture (40%) and a high pH (5.7–6.3). This type of processed cheese can also be produced in the form of slices.
(ii) Cheese spread; this is a soft cheese with a higher moisture content (50%) and a lower pH (5.4–5.8).

Manufacture

The raw material for processed cheese manufacture is usually a blend of hard pressed cheeses of varying age (a few days up to over one year old). The cheese is scraped, washed, cut into blocks and ground between rollers to a fine consistency. Emulsifying salts (citrates, phosphates, polyphosphates or a mixture of these) are then added together with common salt, skimmed milk or whey powder and water. Colour and flavour may be added together with a preservative. After thoroughly mixing the ingredients, the mixture is transferred to the cooker.

Cooking can be carried out batchwise, in agitated kettles or continuously in scraped surface heat exchangers. Batch cooking is achieved by direct steam injection and may take place under vacuum. The mixture is held at 70–85°C for up to 15 min and then discharged to the filling machine. Continuous cooking is carried out at higher temperatures, e.g. 115°C, with a shorter residence time. Hard processed cheese is either filled into blocks and wrapped in plastic film or aluminium foil, or formed into slices. Processed cheese spread can be portion packed in aluminium foil or filled into glass jars.

The popularity of processed cheese derives from its good keeping quality, even without refrigeration. It is also a good base for incorporation of flavours and various additives, e.g. pieces of fruit, vegetables, meat, etc.

SEPARATION

In the dairy industry, separation refers to the recovery of cream from milk by means of a centrifuge or separator. The latter was invented in 1878 by Gustaf de Laval and has been much modified from those early days of small, farm-based machines with capacities of a few litres per hour to the modern high speed, high efficiency, self-cleaning machines used today processing up to $55\,000\,l\,h^{-1}$. The principle remains exactly the same however.

The separator comprises a stack of conical discs, with a spacing of 0.5 mm between each, inside a bowl. Milk is fed to the base of the disc stack and distributed via holes near the centre of the discs. Under the action of centrifugal force (about 4000 g), the fat globules separate towards the centre of the centrifuge where they are removed as cream, while the skimmed milk is forced outwards and is removed from the top of the separator via the gap between the disc stack and the inside of the bowl. Any insoluble material in the milk is also thrown outwards and accumulates as sludge in a space between the disc stack and the bowl. Modern separators are self-deslud-

ging; at predetermined intervals, the lower part of the bowl is allowed to drop so that the accumulated sludge is thrown from the bowl with a small amount of skimmed milk.

The milk feed to a separator is usually at 50–60°C to maximize the difference in density between the fat globules and the skimmed milk and thus maximize the rate of separation. The performance of a separator is assessed as the skimming efficiency, that is, the fat content of the skimmed milk. This is usually about 0.05% which means that 98.5–99% of the fat in the milk has been recovered in the cream. The efficiency of separation is determined to a large extent by the size of the fat globules in the milk; fat globules with a diameter less than 0.6–0.7 mm are not separated and remain in the skimmed milk. Milk must therefore be handled gently prior to separation in order to minimize fat globule disruption which would impair separation efficiency. It is also important to minimize the enterainment of air in the milk since this enhances the susceptibility of fat globules to disruption under shear.

Conventional separators are fed with milk at atmospheric pressure and a certain degree of air entrainment is therefore inevitable. However, in the past few years, hermetic separators have become available in which the milk is fed under pressure to an air tight bowl. The reduction in air incorporation gives an improvement in skimming efficiency of some 0.01–0.02%.

Hermetic separators are intrinsically more efficient than conventional separators principally because of their design and operational features leading to less fat globule membrane breakdown during operation. This in turn leads to less free fat being produced which can spoil the appearance of liquid milk in bottles or can result in a higher level of unremovable free fat in the skimmed milk phase. It is possible to separate up to around 70% fat cream using an hermetic machine.

Today, separators are found at the centre of every liquid milk dairy and are expected to produce the entire range of milks for packaging in glass, plastic and cartons. The use of these machines for the production of whole and standardized 3.5% milk for bottling without homogenization is entirely possible.

In the dairy industry worldwide there are several other forms of separation process used to separate out different components of milk or milk by-products. Some of these use centrifugal separation principles like the conventional separator (e.g. clarification, standardization), whilst others (e.g. ultrafiltration, reverse osmosis) make use of physical membrane separation processes. Each is now considered in turn.

CLARIFICATION

Clarification is often used as an alternative to filtration in order to remove dirt, cells, etc. from raw milk. The clarifier is a continuous disc centrifuge similar in design to the separators used to separate cream from milk, with the exception that the clarifier does not separate the cream. Clarifiers are usually self-desludging, whereby the sludge in the centrifuge bowl is ejected at preset intervals in order to permit fully continuous operation.

STANDARDIZATION

The requirements of today's production environment are such that many product changes are required in order to satisfy demand. This means that to be competitive and to offer a full range of products of various fat levels it is necessary to use some form of standardization where one may change products quickly, often at the touch of a button.

Blending systems, either in-line or in batching tanks, require the preparation of skim milk prior to mixing with whole milk and pasteurizing. Thus they are inherently less flexible and also tie up vessels and piping which would otherwise be used for raw milk. To overcome this deficiency a number of manufacturers have developed direct on-line systems. One such company, the Alpha Laval company amongst others, has developed what it calls the 'Direct On Line' system which offers many advantages over the in-tank blending systems. The 'Direct On Line' system can yield the highest level of accuracy in milk and cream fat contents thus allowing the processor to maximize revenue from the production of surplus cream.

It is possible to produce semi-skim products with a minimum fat content of 1.5% with the standardizing system set at 1.51% without any risk of producing product below the legal limit of fat in semi-skim in many countries. This is a major achievement since the market for semi-skimmed milks in recent years has witnessed significant growth largely as a result of consumer nutritional pressures.

Hermetic separators are better suited to changes in operating conditions than paring disc separators and respond more accurately to changes initiated by the standardizing unit. Such changes can include product flow rate changes, separator dis-

charge, finished milk tank changes, raw milk silo changes and cream tank changes. The accuracy of the standardization equipment is a function of the ability of the equipment to cope with all such changes but nowadays most in-line systems are greatly assisted in their accuracy by the use of specific computer software for each individual change in operating conditions.

Standardization systems are available to standardize the fat content of the cream, the fat content of the milk and cream, the ratio of fat to SNF and for the enhancement of the milk stream by the addition of cream and additives or removal of skim milk.

Protein standardization of milk is not yet officially recognized for liquid milk in the UK but it is commonly practised in many parts of the world. In central Europe and in Australia it is regularly used in evaporated milk and powder production and for standardizing cheesemilk. More recently, protein-standardized liquid milk is being used for UHT production. Particularly in southern Europe, all the permeate produced as a by-product of ultrafiltration plants (see ULTRAFILTRATION below) is utilized to standardize the raw material.

In plants manufacturing products that require preconcentration such as fromage frais, ultrafiltration is an integral part of the process. The permeate discharged is low in protein but approximately the same in other components so when added to bulk raw milk only the overall protein is adjusted downward. Most of the milk protein is retained in the retentate so incorporation into the milk supply raises the protein level. Standardization is achieved by controlled in-feed directly from the ultrafiltration plant or from holding tanks and the amount to be added is calculated by a simple equation.

MEMBRANE SEPARATION

As the name implies membrane separation processes differ radically from other separation processes in that membranes are used to filter out particular components of milk or whey rather than centrifugal force as in a separator or clarifier. Membrane filtration was originally developed for desalination of salt or brackish water some years ago.

The original membranes were made of cellulose acetate. In the intervening years membranes have become tougher and more adapted to the more rigorous conditions experienced in industrial processing as the types of material which can be used have increased and the range of applications has expanded.

Cleaning techniques also have been improved. Many membranes now are able to withstand harsher cleaning chemicals (even chlorine-based disinfectants in some cases) and higher temperatures (up to 150°C) and thus steam sterilization becomes feasible.

The advantages conferred by the use of membrane separation processes depend on particular applications and situations. In general, certain processes can be performed that were not previously possible. Bacteria, yeasts and moulds and even viral particles can be removed without product damage. Yields can be increased either by reducing losses or by upgrading low quality materials. Membranes can be used to concentrate, clarify or separate at low temperatures so that there is little or no change in phase and materials are protected from damage or change in characteristics.

Membrane systems in use nowadays are based on four main types; reverse osmosis, nanofiltration, ultrafiltration and microfiltration.

Reverse osmosis

Reverse osmosis (RO) uses the tightest membrane and smallest pore size. Reverse osmosis membranes are semi-permeable, that is they only allow the passage of water while retaining the solid constituents of milk, i.e. fat, protein, lactose and salts. When a pressure greater than the osmotic pressure of milk is applied, water molecules diffuse through the membrane, thus increasing the solids concentration of the milk above it. The reverse osmosis processes in operation today operate typically at pressures from 15 to 60 bar.

The design and operation of reverse osmosis membranes and equipment are similar to those used in ultrafiltration (see ULTRAFILTRATION). The single pass mode of operation used in earlier plants has now been superseded by multistage recirculation systems which allow the achievement of higher levels of concentration.

The degree of concentration by reverse osmosis is limited since the rate of water removal decreases with increasing solids content. In practice, modern plants are used to concentrate skimmed milk up to 18% total solids and whey up to 30% total solids. In future one can envisage the reverse osmosis process being used for polishing condensate and also RO permeate.

The capital and running costs of RO are significantly lower than the equipment costs of heat eva-

poration and quite a lot of companies in the western world now use RO to preconcentrate skimmed milk and whey prior to final evaporation and drying.

Nanofiltration

Nanofiltration (NF) or 'loose RO' is a more recent advance of the technique of reverse osmosis brough about by the development of special types of membranes. This technique retains most chemicals but allows through a proportion of the low molecular weight materials and water. Operating pressures range from 10 to 30 bar.

The NF technique has been brought about only relatively recently by the development of special types of membranes by various manufacturers. As yet there are no known major installations of NF in the dairy industry although it is known that a number of membrane plant manufacturers are developing systems which in the near future will be capable of desalting whey to produce specialized whey protein concentrate products; in the longer term manufacturers are also in the process of developing systems which could be used for the concentration and desalting of waste waters.

Ultrafiltration

This is another type of membrane separation process which can be applied to milk, skimmed milk and whey. It is probably the most widely used of all the membrane separation techniques in the industry. Ultrafiltration (UF) retains suspended particles, bacteria and the larger molecular weight materials (e.g. whey protein) but allows through salts, lactose, other sugars and materials with a molecular weight less than the pore size. UF operates within the range of 2 to 10 bar.

Ultrafiltration membranes are manufactured from synthetic polymers with a defined pore size such that when pressure is applied to milk or whey above the membrane, compounds with molecular weights in excess of 10 000 Da (proteins, fats) are retained (the retentate) while lower molecular weight compounds (lactose, salts, water) are able to pass through it as the permeate.

In practice, ultrafiltration membranes are used in a number of configurations e.g. flat sheets, spiral-wound sheets, tubes, fibres, etc. Ultrafiltration equipment usually provides a porous support for the membranes in order to protect them against physical damage. The rate of ultrafiltration increases with temperature and pressure but decreases with increasing concentration and with the presence of undissolved solids, e.g. fines in whey, which cause the membranes to become fouled. The feed to an ultrafiltration plant must therefore be effectively clarified in order to main high rates of filtration. Maximum rates (flux rates) are achieved in practice by operating sets of membranes in two or five stages in series, each stage operating at a progressively higher concentration.

Within each stage, the product is recirculated by means of a pump in order to prevent the accumulation at the membrane surface of a protein layer (membrane polarization) which would otherwise reduce the flux rate. Each stage operates at a constant concentration and product is continuously bled off as the feed for the next stage.

In this way skimmed milk and whey can be concentrated to 60–65% protein in dry matter. Higher protein concentrations can only be achieved by diafiltration, whereby water is injected into the final stage to reduce the concentration of lactose. Diafiltration permits the production of concentrates with up to 80% protein in dry matter.

Ultrafiltration plants are operated at 50–55°C in order to minimize microbial growth while ensuring that the protein remains undenatured. The concentrated product should either be used immediately or cooled to 4°C for storage.

Commercial application for UF have in the past been mainly restricted to their use in the production of whey protein concentrates. Nowadays however processing plants are available for the protein standardization of milk, for the production of milk protein concentrate, for cheese bases and for soft cheese production (concentration of milk prior to production of fromage frais). Although not yet legally permitted in any country, protein standardization of drinking milk and preserved milk products could well become reality in future years and in this area UF processing will have a major role to play.

Microfiltration

Microfiltration (MF) is the latest addition to the membrane technology field. It uses a more open membrane and retains all particles in the range of 0.05 to 8 μm; hence membranes with pore size up to 1.5 μm offer the possibility of removing bacteria and bacterial spores from milk and the process is now being applied in a number of markets. Operating pressues are typically 0.3 to 5 bar.

MF is used mainly for cheesemilk to avoid the necessity for the addition of nitrates and other preservatives used to prevent defects caused by bacterial spores, to remove lactobacilli and other bacteria causing problems in Mozzarella and to obtain controlled bacteriological quality in the milk used for raw milk cheese production.

It is known that a number of major plant manufacturers have further MF applications under development including the use of MF for fat reduction in whey so as to obtain very high protein (up to 90% w/w) whey protein concentrate products.

DRYING

The drying of milk products and skimmed milk in particular has become increasingly significant as the level of milk production within the world has increased. Milk and milk products may be dried using either spray or roller drying techniques, although spray drying is now used almost exclusively because of the superior quality of the spray dried product in terms of flavour, colour and solubility (see also MILK POWDERS).

ROLLER DRYING

Roller dryers comprise two counter rotating rollers with a clearance of 0.1 to 0.5 mm. The rollers are heated internally by steam at 140–150°C and rotate at 5–30 rpm.

Preconcentrated milk is either fed to the trough formed between the rollers or sprayed on to the rollers by nozzles. The latter method allows a greater utilization of the drying surface. A film of milk is formed on the surface of the rollers and the water is quickly evaporated. The rate of drying is controlled by the thickness of the film, the temperature of the rollers and the speed of rotation of the rollers. The dried film is continuously removed with a doctor knife and the flakes of dried product fall to a grinder. After grinding to a power, the product is sieved to remove burned particles.

A variant of roller drying is the vacuum roller drying process whereby the rollers are enclosed in a chamber which can be operated under vacuum. This enables a lower temperature (e.g. 60°C) to be used for drying, giving a product with a lower degree of heat treatment.

SPRAY DRYING

Almost all dried milk products are now produced by the spray drying process. A spray drier comprises a chamber in which preconcentrated milk is atomized to produce a fine spray which is then dried by contact with a steam of hot air. The essential elements of a typical spray drying process are as follows:

Hot air supply

Filtered, heated air is drawn into the drying chamber by fans and distributed by a series of vanes around the top of the chamber to provide a downward flow of air. The pattern of air flow within the drier is designed to avoid the deposition of powder particles on the roof and walls of the drying chamber. Fresh air is heated to a temperature of 180–220°C before entry into the drier. The heating system is indirect using oil or gas fired heaters. The outlet temperature of the air is usually in the range 80–100°C.

Atomization of the feed

The atomization systems used to produce finely dispersed droplet sprays may be classified into two types: rotary atomizers and pressure nozzles. The rotary atomizer comprises a disc or bowl rotating at high speed. The liquid feed is distributed to a series of jets, orifices or slots at the periphery of the atomizers and the peripheral speed is sufficient to break up the liquid into droplets which are thrown out as a spray.

Pressure nozzles can themselves be classified into two distinct types: high pressure nozzles and two-fluid nozzles. The first of these involves pumping the liquid feed under high pressure through an orifice. The entry to the orifice (swirl insert) is designed to impart a rotary motion to the liquid which leaves the orifice as a conical sheet. The flow of hot air in the drying chamber is then sufficient to break up the liquid into fine droplets.

The two-fluid nozzle also involves the flow of liquid through an orifice but in this case the nozzle comprises two concentric tubes. The liquid feed enters the central tube while compressed air is fed to the outer tube. The compressed air is directed at an angle across the central orifice and breaks up the liquid into a conical spray.

Most of the spray driers used in the dairy industry utilize the rotary atomization technique but newer plants generally have the facility for interchanging rotary and nozzle atomizers in order to achieve maximum flexibility. The nozzle atomization system has advantages in the production of compounded products and high density powders while the rotary atomization system permits a greater throughput.

The drying chamber

The shape of the drying chamber depends on the type of atomization system used. Rotary atomization requires the drying chamber to be of cylindrical design, either with a conical or flat bottom. The conical-based chamber is designed to allow powder removal by gravity but it is usual to incorporate externally mounted hammers to aid powder flow. Flat bottomed chambers (e.g. Scott drier) require a means for powder removal and this can be pneumatic or mechanical, e.g. rotating scrapers or brushes.

Chamber design for nozzle atomization can either be of the cylindrical types mentioned above or a rectangular box (e.g. Rogers drier).

The most common chamber design in industry is the conical-based cylinder (Niro, Stork, Anhydro). Where nozzle atomization is used alone, the aspect ratio of the chamber is higher (tall-form drier) because of the lower cone angle of the spray produced by a nozzle atomizer. Where interchangeable atomizers are used, the chamber must be designed for rotary atomization. The nozzle atomizer then comprises a cluster of nozzles designed to produce approximately the same spray dimensions as the rotary atomizer.

Powder recovery

The main bulk of the powder is removed from the base of the chamber and either pneumatically conveyed to a bulk storage silo or fed to a secondary drier (see below TWO-STAGE DRYING). However, a significant quantity of powder is entrained in the exhaust air and this is recovered using a series of cyclones. Powder losses in the air from the cyclones should be less than 0.5% of the total production.

Further powder recovery from the exhaust air may be justifiable on economic or environmental grounds. In such cases, exhaust air from the cyclones is passed to a bag filter or wet scrubber, but powder recovered in this way does not usually conform to the specifications of first grade powder and may have to be sold for animal feed. Scrubbers are also available in which raw milk is used as the recovery medium. In this case the recovered powder can be converted to first grade product.

The exhaust air from a secondary powder recovery system, e.g. bag filter or scrubber, contains a significant amount of heat, and some heat recovery systems are now in use whereby the exhaust air is used to preheat the cold air feed to the drier.

Drying temperatures

The thermal efficiency of the spray drying process is estimated as follows:

$$\text{Thermal efficiency} = \frac{(T_i - T_o)}{(T_i - T_a)}$$

where T_i, T_o are the inlet and outlet air temperatures, respectively and T_a is the ambient temperature.

The inlet temperature should therefore be as high as possible but there is a practical limitation in that the temperature of the powder particles must not exceed the point where protein denaturation, and hence powder insolubility, becomes significant. The outlet temperature is determined by a number of factors, the most important of which are the required moisture content of the powder, the temperature and humidity of the drying air, the total solids content of the feed, and the design of the drier and its components.

The following equation predicts the change in ΔT_o (in °C), required to compensate for changes in ΔT_i (in °C) and feed solids content, ΔTS (in %):

$$T_o = \frac{\Delta T_i}{10} + \Delta TS - KM$$

where K is a constant (about 5 in the case of skimmed milk) and M is the moisture content of the powder (in %).

This equation implies that any change in the moisture content of the powder brought about by a change in for example the solids content of the feed, can be compensated by an appropriate change in the outlet temperature. The latter is used as the parameter by which the final moisture content is controlled.

Two-stage drying

An important feature of the spray drying process is the decrease in drying efficiency as the powder particles become progressively drier. This has led to the introduction of two-stage drying systems in which the final drying stage is carried out using a fluid bed.

The feed is first spray dried to a moisture content of 5–8%, compared with 3.5–4% in the conventional system, using a lower outlet temperature (e.g. 85°C, cf. 95°C) and discharged to a vibrating fluidized bed drier located near the base of the spray drier, where the moisture content is further reduced to 3.5–4%. The use of a lower outlet temperature in the two-stage drying system gives an energy saving of some 15 to 20% compared with the simple spray drying process and powder quality is also generally improved because of the lower temperatures used in the final stages of drying. Most modern spray drying plants now utilize the two-stage drying system.

New developments in spray drying

Further energy economy and gentler drying can be achieved if the powder is spray dried to a moisture content of 10–15% followed by fluid bed drying. However, powder at this moisture content is too sticky to transport to the second stage of a conventional two-stage system so modified plants have developed to overcome this.

The Filtermat (Damrow) drier incorporates a slow moving conveyor belt at the base of the spray chamber. The powder builds up as a cake on the belt and secondary drying takes place on the belt before discharge.

The multistage drier (Niro) is a similar concept in which the powder falls on to a stationary fluid bed drier at the base of the spray drying chamber. The powder is further dried on this fluid bed before discharge to a vibrofluidized bed drier for final drying.

MILK POWDERS

The most important milk powders are skimmed milk powder, whole milk powder, fat-filled powders and whey powder (see individual sections). Smaller amounts of more specialized powders are also produced, namely buttermilk, caseinate, cheese, coffee whitener, infant formula, whey protein and yoghurt powders. Several aspects of milk powder quality are routinely measured to determine the suitability of the powder for its particular end use and to establish a basis for agreed standards of quality which are understood both by the manufacturer and the customer. Quality aspects can be classified into four groups: organoleptic, compositional, physical and microbiological quality. Milk powders are usually packed in 25 kg bags with polyethylene liners which may be tied manually or heat sealed.

ORGANOLEPTIC QUALITY

The taste and odour of a powder should be consistent with its description, e.g. milk, cheese, yoghurt, etc. Undesirable flavours, such as those resulting from overheating, fat oxidation or bacterial growth, should be absent, although in lower grade powders they may be acceptable to a slight degree.

COMPOSITIONAL QUALITY

The three most important aspects of compositional quality are the titratable acidity, moisture and fat content. A minimum protein content may also be specified and in powders prone to oxidation during storage, e.g. whole milk powder, maximum levels of copper and iron may be prescribed.

The titratable acidity is measured as an index of the microbiological quality of the milk product before it was dried while the moisture content is important in determining the storage life of a powder. Fat content is often used as a means of differentiating and defining different types of milk powder, e.g. from skimmed or whole milk, and is therefore an important aspect of grading. Free fat, i.e. that which can be extracted from a powder by washing with a non-polar solvent, affects the reconstitution properties of the powder and a maximum level may be specified for certain products.

PHYSICAL QUALITY

The physical quality of a powder relates to its suitability for reconstruction or incorporation into food products. In the first place, the powder should be free from lumps, except those that break up readily under slight pressure, and exhibit satisfactory flow characteristics. Other frequently used indices of physical quality are solubility and the extent of contamination with burned particles. The bulk density of the powder may also be important where a minimum weight of powder is required to be filled into a container of given volume. Instant powders are specifically manufactured to give ease of reconstitution and a number of empirical tests have been devised to assess this property, e.g. dispersibility, wettability, sinkability. Ease of reconstitution is particularly important in powders manufactured for retail sale and in calf milk replacers.

Certain physical characteristics are also sought in reconstituted powders which are to be further processed, e.g. viscosity and heat stability (see RECOMBINATION).

MICROBIOLOGICAL QUALITY

The microbiological quality of a milk powder is a function of the quality of the raw material and the level of hygiene during manufacture. The most important index of microbiological quality is the total viable plate count, although specifications usually include maximum levels of coliforms, yeasts and moulds. There has been a tendency in the past few years for microbiological specifications to become more exacting, and the absence of specific groups of bacteria may be demanded in some cases, e.g. Salmonellae, Staphylococci, thermodurics.

SKIMMED MILK POWDER

Skimmed milk (SMP) powder is by far the most important of all the milk powders produced and, apart from a small quantity which is roller dried for use in chocolate manufacture, is almost exclusively manufactured by the spray drying process.

Skimmed milk can be dried in any of the various types of spray drier available, but for reasons of economy and quality, it is usually dried by rotary atomization in a cylindrical chamber with a conical base. Before drying, the skimmed milk is first evaporated to a concentration of about 50% total solids. Ordinary, or regular, skimmed milk powder is manufactured using either a spray drier with a simple pneumatic conveying system (single-stage drier), or a two-stage drier with secondary drying in a vibrofluidized bed (see DRYING).

Use of SMP

Skimmed milk powder is a valuable source of protein (35%) and carbohydrate (lactose, 50%) and is used both for reconstitution and as a food ingredient. There are three major outlets for skimmed milk powder:

(i) *Food ingredient*
Skimmed milk powder performs three major functions as a food ingredient: it imparts a desirable dairy flavour, it contributes to food texture and enhances the development of desirable colour and flavour compounds. Skimmed milk powder is the most widely used form of milk protein in the food industry. It functions effectively in terms of water binding, fat emulsification and structure formation. This accounts for the well established uses of skimmed milk powder in the manufacture of baked goods, confectionery, ice-cream and other dairy desserts, processed meats, dry mixes, sauces etc.

(ii) *Reconstitution and recombination*

A substantial quantity of skimmed milk powder is sold for reconstitution in institutions (hospitals, schools etc.) and catering establishments as well as for domestic use. Increasing quantities of skimmed milk powder are also being traded around the world for use in the manufacture of recombined dairy products.

(iii) *Animal feed*

A large proportion of the skimmed milk powder produced in the EU, America and Australasia is used in the manufacture of calf milk replacers (see also SPRAY MIXING under FILLED MILK POWDERS). Similar products are manufactured for feeding lambs, pigs, etc.

Specifications for SMP

Grading requirements vary slightly with different end users but the basis for a specification for skimmed milk powder is usually taken as the 'Extra Grade' requirements defined by the American Dried Milk Institute (ADMI). Powder for sale into EU Intervention must comply with specifications laid down in EU Regulations. In addition to powder quality, specifications for skimmed milk powder also include packaging requirements.

The ADMI Extra Grade specification is listed in Table 3.6. The Intervention Board specification undergoes fairly frequent revisions but follows a similar form to the ADMI specification, albeit more stringently. Skimmed milk powder for use in the manufacture of cheese and yoghurt must, of course, be essentially free of antibiotic in order to avoid inhibition of the starter bacteria.

Table 3.6 ADMI specification for 'Extra Grade' spray skimmed milk powder

	Not greater than
Milkfat	1.25%
Moisture	4.00%
Titratable acidity	0.15%
Solubility index	1.25 ml
Bacterial estimate	50 000 per g
Scorched particles	Disc B

Heat classification of SMP

The heat treatment given to skimmed milk before evaporation and drying has a significant effect on some of the characteristics of the resulting powder. For this reason, skimmed milk powder is usually classified according to the degree of heat

treatment received. A number of criteria may be used for this purpose, the most widely accepted being the whey protein nitrogen index (WPNI) used by the ADMI; the amount of undenatured whey protein in the powder is inversely related to the degree of heat treatment. According to this system, low and high heat powders are defined as containing more than 6 mg and less than 1.5 mg of undenatured whey protein nitrogen per gram of powder, respectively. Powder which falls between these classifications is termed medium heat.

The heat treatment used in the manufacture of low heat powder must be at least equivalent to HTST pasteurization, e.g. 72°C for 15 s or flash heating to 90°C. However, some manufacturers may use higher temperatures in order to ensure microbiological quality but great care is necessary to ensure that the whey protein nitrogen index does not fall below six. This is particularly important since the original level of undenatured whey protein in skimmed milk varies significantly through the year. Low heat powders are mainly used in the manufacture of reconstituted or recombined milk products such as milk, cheese and yoghurt.

The heat treatment used in the manufacture of high heat powder must be sufficient to denature a large proportion of the whey protein. Traditionally, skimmed milk was heated to about 95°C and held for about 10 min in a hot well. Modern plants, however, utilize higher temperatures (e.g. 120°C) with shorter holding times (e.g. 2 min). High heat treatment improves the heat stability of skimmed milk (see EVAPORATED MILK) and high heat skimmed milk powder is therefore suitable for use in the manufacture of recombined evaporated milk (see RECOMBINATION). High heat powder is also used in breadmaking where the use of low or medium heat powder depresses the loaf volume.

The medium heat classification covers a wide range of heat treatments and medium heat powders may, therefore, exhibit substantial variations in certain characteristics, e.g. flavour, viscosity. If the latter is an important factor, it may be necessary to specify a particular range of WPNI or a viscosity test to ensure consistency in powder supply. Such characteristics are controlled during manufacture by manipulation of the temperature and holding times used in the heat treatment.

The effect of heat treatment on the WPNI and other powder characteristics varies significantly throughout the year as a result of changes in milk composition. The production of powders with specific properties therefore requires close attention to the heat treatment required at different times of the year.

Instant skimmed milk powders

Ordinary skimmed milk powder is easily reconstituted by mixing in warm water with efficient agitation. However, in cold water, skimmed milk powder is difficult to reconstitute because the powder particles become wetted very quickly and form into lumps surrounded by a gelatinous layer of concentrated skimmed milk. This layer hinders the wetting of the powder inside the lumps, and because air is also entrapped the lumps are not sufficiently dense to sink down in the water and are therefore unable to dissolve.

The need for skimmed milk powders with good reconstitution properties in cold water, e.g. for domestic use, led to the development of instant powders, where the powder particles are produced as porous agglomerates. When these are reconstituted, the water flows into the pores and wets the powder evenly; when the water has displaced the air the agglomerates become sufficiently dense to sink down into the water where they disperse and finally dissolve.

The instant properties of a powder are usually assessed by measuring a property termed dispersibility. Several tests are used but each involves reconstituting the powder under defined conditions of temperature, concentration and agitation. After a specified time, e.g. 15 s, the reconstituted powder is sieved and the dispersibility determined as the proportion of powder passing through the specified sieves. Instant powder should have a dispersibility of at least 85%.

Skimmed milk powders with improved instant properties are manufactured in two ways:

(i) *Straight-through instantizing*

The skimmed milk is dried using the two-stage process and the small powder particles (fines) entrained in the main drying air and the fluid bed drying air are separated in cyclones (see DRYING). The fines are then conveyed back to the drying chamber and introduced near the atomization zone where they adhere to the moist, newly formed droplets. In this way agglomerates are formed and the gentle secondary drying in the vibro-fluidized bed ensures that most of them maintain their integrity.

(ii) *Re-wet instantizing*

Superior instant properties are achieved by re-wetting previously dried powder in a process known as surface agglomeration. This is carried out in a chamber of similar design to a spray drier, and in fact some spray drying chambers can be used for this purpose.

The powder is fed through a tube at the top of the chamber where hot, moist air is also introduced. The moisture condenses onto the powder particles, causing them to become sticky and form agglomerates which are then dried by contact with hot air. The powder agglomerates are cooled in a vibrofluidized bed and sieved to separate fines which are then recycled to the agglomeration tube, together with fines separated from the fluid bed cooling air.

Re-wet instantized powders have a significantly lower bulk density than straight through instantized powders, reflecting a greater degree of agglomeration and hence improved instant properties.

WHOLE MILK POWDER

Whole milk powder is available in both spray dried and roller dried form, although the latter is produced only in very small quantities.

Spray dried whole milk powder can be manufactured using a single stage drier with pneumatic powder conveying (see DRYING) but it is important that the powder is cooled rapidly after drying in order to minimize the free fat content. For this reason, whole milk powder is best manufactured using a drier equipped with a fluid bed cooler.

Milk is first standardized to give a fat content of 26.5% (minimum 26.0%) in the final powder. The milk is then pasteurized, evaporated to a total solids content of 40 to 50% and homogenized in two stages (3000 psi (204 bar) first stage; 500 psi (34 bar) second stage).

The drier may be equipped with either rotary or nozzle atomization and operates with an inlet air temperature of 180 to 200°C. The powder leaves the chamber at its final moisture content (2.5%) and is discharged to a fluid bed cooler before packaging.

Instant whole milk powder

Whole milk powder usually contains between 0.5 and 3% free fat and this is sufficient to cover the powder particles and make them difficult to wet, except in warm water at 40 to 50°C. Whole milk powder can be instantized, i.e. made cold water soluble, by coating the particles with a surface active agent such as lecithin.

The process first involves the production of an agglomerated powder in a similar manner to the straight through process for instantizing skimmed milk powder (see SKIMMED MILK POWDER). The agglomerated powder is then conveyed to a vibrofluidized bed, where it is sprayed with a solution of lecithin in butter oil before conditioning in a second fluid bed to disperse the lecithin evenly throughout the powder. The lecithin content of the powder is about 0.2%.

Instant whole milk powder is packed in nitrogen flushed cans immediately after production in order to prevent fat oxidation during storage.

Fat oxidation

The shelf life of whole milk powder is limited by off-flavour development as a result of fat oxidation during storage. Oxidation can be retarded by storage under nitrogen but this is only applied to retail packs. The shelf life of bulk whole milk powder, packed in 25 kg bags, can be extended by preheating the milk prior to evaporating. This results in the production of sulphydryl compounds which act as antioxidants. The heat treatment may comprise holding at 90°C for about 10 to 15 min, but better results are achieved by heating at 110°C for up to 2 min.

FILLED MILK POWDERS

Filled milk products contain vegetable fat, in place of milk fat, together with non-fat milk solids. Two major types of fat-filled milk powder are manufactured in practice.

(i) *For human consumption*
    These powders contain 26 to 28% fat and are used for reconstitution, either domestically or by caterers. In addition to skimmed milk solids and vegetable fat, fat-filled powders for human consumption may incorporate emulsifying salts and small amounts of added carbohydrate, e.g. lactose, glucose solids.
(ii) *For animal feeding*
    These powders contain 30 to 50% fat and up to 2% of an emulsifying agent e.g. lecithin, monoglycerides. They are blended with a number of other ingredients (whey powder, starch, minerals, vitamins, antibiotics) to produce milk replacers, containing 15 to 28% fat, suitable for the feeding of calves, pigs, lambs, etc.

The choice of fat for these products is important in imparting a desirable mouthfeel. The fat should have a melting point slightly below body temperature and for this reason the lauric fats, i.e. coconut oil and palm kernel oil, are preferred.

The manufacture of both types of fat-filled milk powder is essentially the same and involves the homogenization of a mixture of fat and skimmed milk concentrate prior to spray drying. Skimmed milk is evaporated to a solids content of about 45%, and in the case of animal feed production, the conditions of preheating and evaporation are carefully controlled to minimize whey protein denaturation which can cause scouring in young animals. The skimmed milk concentrate is then mixed with the appropriate amount of melted vegetable fat and other ingredients and homogenized at high pressure to produce a stable emulsion in which most of the fat globules are less than 1 mm in diameter. The mixing process may be carried out batchwise or continuously. In the latter case, a piston pump is used to deliver defined proportions of skimmed milk, fat and emulsifier to the homogenizer via a mixing device. Homogenization conditions are carefully controlled to ensure effective dispersion of the fat and thus minimize the amount of free fat in the final powder.

A maximum free fat content is included in most specifications for fat-filled powders and drying conditions are therefore chosen carefully in order to minimize free fat levels. Drying temperatures are somewhat lower than in the case of skimmed milk drying (inlet temperature 160 to 190°C compared with 180 to 230°C for skimmed milk) and the powder is cooled as rapidly as possible in a fluidized bed on exit from the base of the drying chamber.

Spray mixing

With the exception of whey powders (see below) a type of process called spray mixing is often used for various types of powder production operations. Spray mixing is a process whereby a powder is brought into contact with a liquid spray for the purpose of mixing, agglomeration or carrying out a chemical reaction. Spray mixing is used extensively on the European continent for the manufacture of fat-filled powders for animal feed.

The mixing operation is carried out by conveying powder through a mixing zone where a series of nozzle atomizers produce a fine liquid spray. The powder particles are deposited on to the liquid droplets forming agglomerates which are then dried and/or cooled in a fluid bed.

This technique can be used in the production of:

(i) Fat-filled powders, by spray mixing skimmed milk powder and vegetable fat, followed by cooling

(ii) Instantized powders, by spray mixing powder and water, followed by fluid bed drying and cooling

(iii) Sodium caseinate, by spray mixing casein and sodium hydroxide solution, followed by fluid bed drying and cooling.

WHEY POWDER

Whey can be evaporated and spray dried in a similar manner to skimmed milk (see SKIMMED MILK POWDER) but the resulting product is very hygroscopic because of the high concentration of lactose (70 to 75%). Whey powder in this form is not suitable for use as a food ingredient because it is very sticky and absorbs moisture during storage to form hard lumps. Non-hygroscopic (or non-caking) whey powder is produced by precrystallizing the lactose before drying so that most of the lactose is present in the α-crystalline form (see LACTOSE) which is non-hygroscopic.

Whey is first evaporated in a falling-film evaporator to a solids concentration of about 50%. The concentrate is cooled in a plate heat exchanger to 30°C and transferred to a jacketed, stirred tank where a small amount of whey powder is added to promote crystallization. The temperature is gradually reduced to 15–20°C over a period of about 6 h during which a large number of very small lactose crystals are formed. The precrystallized concentrate is then spray dried to a moisture content of 4% to give a free flowing powder.

Higher quality whey powder is produced by incorporating a secondary crystallization step after spray drying. Powder is removed from the drying chamber at 8–14% moisture and conveyed to a moving belt or a fluidized bed. The moisture remaining in the powder permits almost complete crystallization of the lactose and the residual moisture can then be removed in a secondary drying system, e.g. fluid bed drier, before cooling the powder and packaging.

Whey powder is finding increasing use as an economic source of milk solids in food manufacture. The main applications are in bakery products, ice-cream and dry mixes with smaller amounts used in confectionery and margarine. Whey powder may also be mixed with other dairy ingredients, e.g. caseinates, skimmed milk powder, to produce specialized blends for use as ingredients in various sectors of the food industry. The other important use for whey powder is in the formulation of animal feeds, e.g. calf milk replacers.

RECOMBINATION

Recombination refers to the manufacture of dairy products from dry ingredients, such as anhydrous milk fat and skimmed milk powder. The manufacture of recombined milk products is most significant in areas of the world without an indigenous dairy industry, e.g. Africa, South America and the Middle and Far East, and the supply of the raw materials is important to the European Dairy industry. Large quantities of dry dairy ingredients are used. The most important recombined products are sweetened condensed milk and evaporated milk, which have good storage stability. However, recombined UHT products are growing in importance and small quantities of butter, cream, cheese and yoghurt are also made by recombination.

Raw materials for recombination are usually skimmed milk powder and anhydrous milk fat. Whole milk powder is sometimes used, although there are problems with its storage, and buttermilk powder incorporation (up to 10%) is reported to improve flavour and emulsion stability. Indigenous vegetable oils are sometimes used wholly or partly to replace the milk fat.

Standards for the anhydrous milk fat for recombination are similar for all the recombined products, maximum levels being specified for moisture, free fatty acid (FFA), peroxide value, copper and iron (see ANHYDROUS MILK FAT). Standards for skimmed milk powder relate to the end product use but the composition (fat, moisture, acidity) and quality (solubility, scorched particles, bacteriology) specifications do not vary greatly. The major differences among the skimmed milk powders used for recombination lie in their physical properties which are related to their processing characteristics and the properties required in the particular recombined product. In the manufacture of recombined evaporated milk, for example, it is important that the recombined product does not coagulate during the sterilization process (see EVAPORATED MILK). The skimmed milk powder used in its manufacture must therefore be heat stable and this is achieved by giving the skimmed milk a high heat treatment before evaporation and drying. In the case of recombined sweetened condensed milk, however, the viscosity of the recombined product is more important (see SWEETENED CONDENSED MILK) and a medium heat skimmed milk powder is more appropriate.

CREAM

Creams are manufactured with a wide range of fat content, shelf life and physical characteristics. Additives are not usually permitted, with the following exceptions in the UK and most other countries:

(i) Whipping cream and whipped cream may contain alginates, carboxymethylcellulose (CMC, E466), carrageenan or gelatin as stabilizers (up to 0.3% of any or all of these). Sugar may also be included (up to 13%).

(ii) Sterilized cream, including UHT cream, may contain up to 0.2% stabilizing salts including calcium chloride and the sodium or potassium salts of carbonic, citric or orthophosphoric acids.

MANUFACTURE

The manufacture of all types of cream involves the following steps:

(i) Separation of fresh milk (see SEPARATION) to produce a cream containing a slightly higher fat content than necessary.

(ii) Standardization of the fat content of the cream to the desired final concentration by the addition of skimmed milk.

(iii) Homogenization of the cream to retard fat separation during storage and to control the product viscosity.

(iv) Heat treatment of the cream, by pasteurization or sterilization, to provide adequate shelf life. Holder pasteurization (63–66°C for 30 min) is suitable for small scale production but most manufacturers practise HTST pasteurization (74°C for 15 s or 80°C without holding).

(v) Packaging into suitable containers. Bulk supplies to the food industry are delivered in road tankers, while smaller bakers and caterers may receive cream in churns (10–50 litre) or bag-in-box containers. Cream for retail sale is usually packed into plastic pots or cartons while some sterilized cream is sold in glass bottles and cans.

Particular details of the manufacture of some of the different types of cream are given below.

Single and half cream

In the UK, single and half cream contain a minimum of 18 and 12% fat, respectively. In the USA the corresponding product is called 'light cream', with 18–30% fat content. Single cream is homogenized at pressures between 2000 and 3000 psi (136 and 170 bar) depending on its use. In most cases it is desirable for single cream to have as high a viscosity as possible and this is achieved by the use of higher homogenization pressures. Cream used in coffee, however, sometimes gives a feathering effect because the acidity and temperature of the coffee destabilize the protein in the cream. Coffee cream therefore requires homogenization at lower pressures in order to minimize the destabilizing effect of homogenization on the milk protein.

Double cream

Double cream contains a minimum of 48% fat and is only homogenized at very low pressures (500 psi (34 bar)) if at all. In the USA the corresponding product is called 'heavy cream' (minimum 36% fat content) or 'dry cream' (40–75% fat content).

Whipping cream

Whipping cream contains a minimum of 35% fat, but in order to ensure consistent whipping properties the fat content is usually closer to 40%. Whipping cream is not homogenized as this impairs whipping properties. After standardization, the cream is pasteurized and cooled to 4°C. The cream must then be held at this temperature, or below 5°C, for at least 24 h in order to allow the fat to crystallize. After packaging, the cream should be held at 5°C or below prior to use. Cream should be whipped at 8°C or below.

Sterilized creams

Two types of sterilized cream are manufactured; aseptically packaged UHT cream (see UHT) and in-container sterilized cream. Sterilized cream must contain a minimum of 23% fat and sterilized half cream a minimum of 12% fat.

In-container sterilized cream is packed in either glass bottles or cans. Canned sterilized cream usually contains 23 to 24% fat while bottled sterilized cream usually contains 35 to 48% fat. Some manufacturers sterilize cream by UHT processing and then use the in-container sterilization step to destroy bacterial contamination from the filling operation.

Before filling the cream is homogenized; a pressure of 2500 psi (170 bar) is suitable for 23% fat cream for sterilization in cans while a lower pressure (300 psi (2 bar)) is used for double cream for sterilization in bottles. In the case of canned cream it may also be necessary to add stabilizers (bicarbonate, citrate) to control age thickening.

Either type of container may be sterilized in static retorts (e.g. for 45 min at 105–110°C), but to increase heat penetration and to prevent caramelization, agitating retorts are preferred. A speed of rotation of 6 to 8 rpm is satisfactory for batch rotary retorts (e.g. for 15–30 min at 116–121°C) but continuous cookers are preferred, especially for high speed production. The fat content of the cream should not be so great that it solidifies during the sterilization process and about 25% butterfat is suitable.

Clotted cream

Clotted cream was originally prepared from milk but it is now commercial practice to use mechanically separated cream (55% fat). The cream is first filled into shallow aluminium or stainless steel trays to a height of 1 to 2 cm. The trays are then placed on a hot water bath (90 to 100°C) and left for up to an hour by which time the cream should have attained a temperature of about 80°C. The heating not only pasteurizes the cream but causes it to thicken giving the typical consistency of clotted cream and forms a crust on the surface. A certain amount of evaporation also occurs raising the fat content by 2–4%. The cream is then cooled to 7°C or below and filled into retail containers, usually by hand.

Aerosol cream

Aerosol cream usually contains sugar, a stabilizer, usually carrageenan, and a propellant, usually nitrous oxide. The aerosol is designed so that the cream is dispersed in aerated form as a convenient alternative to whipped cream. The volume of cream dispensed is high, up to ten times the original volume of cream, but the texture is quite fluffy compared with conventionally whipped cream.

Recombined cream

According to the prevailing prices for milk, cream, etc. there may be a significant economic advantage in recombining cream from water, skimmed milk solids and unsalted butter. Machines are now available for the small scale manufacture of recombined creams in which the constituents are reconstituted, homogenized, pasteurized and cooled. Good recombined creams are difficult to distinguish from fresh product.

Filled cream

In recent years a number of cream-type products containing vegetable fat have been developed and marketed in the western world. Some butterfat may be present, or the product may consist wholly of a vegetable oil mixture combined with emulsifiers.

CREAM LIQUEURS

The market in cream liqueurs has grown dramatically since their introduction in the mid-1970s and is now worth some £600 million sterling worldwide, with over 100 commercially available varieties. These consist of mixtures of cream, sugar and alcohol. The alcohol can be in the form of neutral spirit, in which case the product would incorporate a flavouring system, or a spirit, e.g. whisky or brandy, which would also give the desired flavour.

In addition to the three major ingredients, sodium caseinate (3%) is used to aid the formation of a very fine fat emulsion. The caseinate is first dissolved in hot water and the required amount of sugar added. This mix is then pasteurized and mixed with the required quantity of pasteurized cream (40–50% fat content). At this stage, it is necessary to add an ingredient which will stabilize the milk protein against precipitation by the alcohol; this can either be sodium citrate or a proprietary emulsifier of the glyceryl monostearate (GMS, E471) type. The mix is then pumped to a homogenizer where the alcohol is added in-line upstream. The homogenizer is operated under high pressure (4000–5000 psi (272–340 bar)) and the cream liqueur is then cooled and filled into glass bottles.

As an alternative to the above method, the cream/sugar/caseinate mix can be first homogenized and subsequently mixed with the alcohol. Cream liqueurs typically contain 15% fat, 20% sugar and 17% alcohol.

WHEY

Most whey is derived from cheesemaking and as such contains small quantities of curd particles (fines) and fat which must be recovered for reasons of economy. Fines are usually recovered using vibrating screens or rotary filters and may either be returned to the cheesemaking process or treated separately to produce a lower quality cheese. Fat separation is carried out in centrifugal separators similar to those used to separate cream from milk (see SEPARATION). The whey cream may either be sold as such or churned to produce whey butter.

Classification of whey types

Whey is produced during the manufacture of different types of cheese and casein and four major types of whey may therefore be identified:

(i)   Sweet cheese whey from the manufacture of cheeses coagulated by rennet at pH 6.0 to 6.4, e.g. Cheddar.
(ii) Acid cheese whey from the manufacture of cheeses coagulated at pH 4.5 as a result of lactic acid production by starter bacteria, e.g. cottage.
(iii) Acid casein whey from the manufacture of acid casein (pH 4.5).
(iv) Rennet casein whey from the manufacture of rennet casein (pH 6.6).

In view of the similarities between different types of whey, a simpler classification differentiates acid wheys (acid cheese whey or acid casein whey) from sweet wheys (sweet cheese whey or rennet casein whey).

Composition of whey

Whey from different sources shows some variation in composition but the most important type can be taken to contain about 6.5% solids of which about 4.8% is lactose, 0.6% protein, 0.6% minerals, 0.15% lactic acid, 0.25% non-proteinaceous nitrogen compounds and 0.1% fat (after separation). Whey therefore contains about 50% of the total solids in whole milk.

WHEY PRODUCTS

Whey has always posed a considerable disposal problem to cheese factories and large quantities of whey are still either sold for stockfeeding, discharged to effluent plants or used as a fertilizer. However, the past 15 years have seen a change in the philosophy of whey utilization with the recog-

nition that whey is a potentially valuable source of protein and lactose which may be used as ingredients in food manufacture. The conversion of whey to lactose and whey products has increased substantially and a number of new technologies have been applied commercially to convert whey into saleable products. The most important of these are described below.

WHEY BEVERAGES

A fermented whey beverage called 'Rivella' is produced and sold in Switzerland but its success there does not appear to be repeated elsewhere. A number of whey 'champagnes' are produced in Eastern Europe as by-products of cheese manufacture.

Alcohol fermentation

The production of alcohol from whey is a well-established fermentation process but it is only in the last 20 years that whey alcohol production has become commercially viable.

As a carbohydrate source for alcohol production, whey has two major disadvantages: lactose is fermented only by a limited number of yeasts and the low concentration of lactose in whey gives rise to low ethanol concentrations and high distillation costs. However, the identification of efficient lactose fermenting yeasts (strains of *Kluyveromyces fragilis*) and the development of less energy intensive distillation and rectification equipment have enabled the commercial production of whey alcohol. Whey alcohol is of very good quality and can either be used in the manufacture of high grade potable spirits, e.g. gin, vodka or as an industrial product. It is produced from whey which has first been deproteinized, either by ultrafiltration or by heat precipitation.

DEMINERALIZATION

Reduction of the mineral content of whey increases the range of opportunities for its use as a food ingredient. Demineralization is either carried out by ion exchange treatment or by electrodialysis and the demineralized whey is spray dried in the same way as whey powder.

The main use of demineralized whey powder (90% demineralization) is in the manufacture of infant foods, where it is used in conjunction with skimmed milk powder to give the same ratios of casein : whey protein : lactose : minerals as is found in human milk. Another important application is in chocolate manufacture, especially chocolate coatings.

LACTOSE HYDROLYSIS

(See LACTOSE HYDROLYSED MILK PRODUCTS)

Hydrolysed whey syrups are manufactured by hydrolysing the lactose in whey or deproteinized whey, followed by evaporation to a solids concentration of 60–75%. The whey may be partly demineralized (80–90%) before evaporation, depending on the intended application. Approximately 80% of the lactose is hydrolysed in order to avoid lactose crystallization in the resulting syrup.

Lactose can be hydrolysed by hot acid or by enzymes (β-galactosidase), but the latter is almost always preferred and may be used in a batch process or in immobilized form. The most successful immobilized system to date comprises a fungal β-galactosidase attached to porous glass beads. The hydrolysis process is a continuous one with whey or deproteinized whey being pumped through a column packed with the glass beads. Hydrolysed whey products may be made either from whole whey or from deproteinized whey, e.g. the permeate from ultrafiltration. The most important product, lactose hydrolysed whey, is usually evaporated to 70–75% solids and contains about 10% protein. It is used as an economic form of milk protein and sweet carbohydrate in the manufacture of products such as toffees, ice-cream, mousse and bakery products. Lactose hydrolysed, deproteinized whey is evaporated to only 60–65% solids and may be used as an alternative to glucose syrup.

LACTOSE MANUFACTURE

One of the oldest methods of whey utilization is the manufacture of lactose. Whey is first evaporated to a solids concentration of 60 to 65% and then transferred to crystallization tanks where lactose seed crystals are added to promote crystallization. The lactose crystals are recovered by centrifugation, washed to remove residual protein and salts, and dried. The liquid by-product remaining after centrifugaion is known as mother liquor and contains about 20% protein in dry matter. It is sold either for animal feed or as an ingredient in pet foods, but its high mineral content precludes use in human foods. However, the mother liquor may be demineralized and dried to give a product similar in composition to skimmed milk powder.

Lactose manufactured as described above is suitable for food use (edible grade) but for pharmaceutical use (e.g. in tabletting) it is necessary to purify the washed crystals further. To this end the crystals are redissolved in hot water and boiled with calcium phosphate and charcoal; this causes the protein to precipitate while coloured compounds are adsorbed by the charcoal. The solution is then filtered hot and the lactose crystallized and recovered as described above.

METHANE PRODUCTION

Whey, or deproteinized whey, can be fermented anaerobically to convert the organic matter to methane. The methane produced is equivalent in energy terms to about 0.5 kg fuel oil for each kg of whey solids converted and the gas can be used to fuel the factory boilers.

WHEY PROTEIN PRODUCTION

Whey proteins are superior to most other food protein in terms of nutritional value and possess a number of physical properties which make them very useful as food ingredients.

Commercial whey proteins are usually manufactured by ULTRAFILTRATION and are available as spray dried powders containing between 25 and 75% protein. In food products, whey proteins perform in a similar manner to egg proteins; they are highly soluble, even at acid pH and they foam and form gels on heating. Whey protein powders are therefore used in the manufacture of bakery and meat products, where their gelation properties are particularly useful.

Whey proteins are also extracted from whey using ion exchange chromatography techniques. These products may contain up to 90% protein and are mainly used in nutritional and dietetic products. However, production of this type of protein has been limited to date.

See also MILK PROTEIN.

MILK FAT PRODUCTS

In this section a number of milk fat containing products, including butter and dairy spreads, will be considered.

Milk fat combines a natural quality image with a highly desirable flavour, an attractive colour due to the carotenoid (β-carotene) content, desirable nutritional value (vitamins A and E, high monounsaturated fatty acid content, etc.) and a range of functional characteristics which make it suitable for many food applications, Notwithstanding this, butter consumption fell significantly in most of the major butter consuming countries during the 1970s and 1980s due to a combination of health scares and high prices. During this period, after first virtually ignoring the decline in butter consumption largely in the hope that it would reverse itself, the once highly conservative dairy industry has reacted vigorously and produced some significant developments in the market place particularly in relation to new technology and the generation of new products in respect to market demand.

BUTTER

Buttermaking is synonymous with churning of cream in batch or in continuous buttermakers, the latter having been established since the 1930s. While both processes have been optimized over the years, batch churns are mainly used in smaller dairies and cheese factories (where the cream recovered from whey is churned into whey butter) and the continuous plants are used in large scale operations productions producing 1–4 tonne h^{-1} of butter.

The high value and high levels of production of butter have led to the development of sophisticated and highly controlled equipment for its manufacture. The butter Regulations of most countries require that butter must contain not less than 80% fat, not more than 2% solids-not-fat and not more than 16% water.

The major proportion of the butter consumed in many countries is sweet cream butter, that is, butter manufactured from fresh cream. Lactic butter, which is more popular in some other countries, is manufactured from cream which is first cultured to develop aromatic flavour compounds, e.g. diacetyl. Both types of butter may be salted (up to 2% salt) depending on market requirements.

FRESH OR SWEET CREAM BUTTER

Cream was traditionally churned in batch churns but these have now been almost entirely superseded by continuous buttermakers based on the Fritz process which combines the operations of churning, drainage, salt addition and working in one machine.

Manufacture

Cream is first separated (see SEPARATION) to a fat content of 38–42% and pasteurized at 85–95°C. A vacuum deodorization treatment may also be given where the cream is cooled and held at 3–5°C in insulated tanks for a minimum of 4 h (usually overnight) to allow the fat to crystallize; the development of an extensive network of fat crystals is crucial for efficient churning. After this ageing period, the cream is heated to churning temperature (7–13°C) and pumped to the buttermaker.

Continuous buttermakers (Contimab, Silkeborg, Westphalia) vary slightly in design but each incorporates three major sections:

(i) A churning section comprising a horizontal cylinder in which high speed rotating beaters churn cream into butter grains
(ii) A drainage section where the buttermilk release during churning is separated from the butter grains
(iii) A working section in which augers force the butter grains through a series of perforated plates to ensure efficient moisture distribution and develop the required texture.

In the churning section, the fat globule membrane is disrupted allowing the fat to aggregate into grains. With efficient churning, over 99% of the fat in the cream is converted into butter and the aqueous portion of the churned cream, the buttermilk, should contain less than 0.5% fat.

Churning temperature is chosen to minimize fat losses in buttermilk and is determined by the fat content of the cream and the hardness of the fat. Approximate formulae are available for the calculation of churning temperature from the cream fat content, but the exact temperature is determined by practical experience.

After churning, butter grains and buttermilk fall by gravity into the working sections which comprise a cylinder inclined upwards at approximately 20° to the horizontal. Two counter rotating augers carry the grains upwards through the various stages of the working section. The lower part of the working cylinder is perforated to allow buttermilk drainage (where a secondary churn is used, this incorporates the buttermilk drainage screen). The level of buttermilk is controlled by a variable-height siphon through which the buttermilk is drained; the level of the buttermilk together with the auger speed are the two parameters most often used to control the final moisture content of the butter.

The working section comprises a number of stages in which the following operations are performed:

(i) *Washing/cooling*

The grains are cooled by spraying with chilled water or buttermilk. Washing is nowadays considered unnecessary (and in fact a potential source of contamination) and it is general practice to use chilled buttermilk to cool the grains.

(ii) *Working*

Working is achieved by forcing the grains through a series of perforated plates between which are mixing vanes. The degree of working is controlled by a throttle plate which can be adjusted to vary the back pressure on the butter.

(iii) *Salting*

Salt is injected into the working section as a 50% slurry in water. Microfine salt is used to ensure even distribution and solution.

(iv) *Water injection*

In the case of unsalted butter in particular, it is common practice initially to produce a low moisture butter and control the final moisture content by injection of water.

(v) *Air removal*

Some buttermakers incorporate a vacuum working section to reduce the air content and produce a denser butter.

(vi) *Final working*

A final working section gives the butter the required texture before it is extruded as a ribbon and conveyed to the packaging unit. The moisture should be very well dispersed after this stage, with most of the water droplets less than 5 mm in diameter.

The most important aspect of process control in continuous buttermaking is the final moisture content of the butter. The moisture content must not exceed the legal maximum of 16% but must be as close as possible to this in order to maximize yield. Moisture control can either be achieved by manual adjustment of process conditions (e.g. buttermilk level, auger speed) following rapid measurement of the final moisture content, or by the use of automatic control systems. These are based on the continuous monitoring of the final moisture content of the butter using dielectric moisture meters. The output from the meter regulates the injection of water into the working section of the buttermaker (see (iv) above) and is reported to control the final moisture content of the butter to an accuracy of 0.1%.

Packaging

Butter is either packed into retail containers or into bulk packs for storage or export. Bulk packaging is necessary where butter is to be stored for long periods but a significant quantity of butter is also bulk packed and stored prior to repackaging for retail sale.

Butter from the buttermaker can be pumped directly to the packaging unit but it is usual to provide intermediate storage as a buffer against differences in the rate of butter production and the rate of packing. Trolleys equipped with augers are used to transfer butter from the machine to the packing unit and are available with capacities of 1–10 tonnes. Larger installations may use butter silos (10–30 tonnes capacity) for intermediate holding. These are totally enclosed tanks which accept butter directly from the buttermaker and are equipped with one or more butter pumps capable of transporting butter through a pipeline system of the packer. Packaging machines can either be fed directly from a butter trolley or silo or can have their own feeding units comprising a hopper and a filling auger. Bulk butter is packed into 25 kg lined cardboard boxes. The lining material used to be parchment but this has now been replaced by low density polyethylene.

The degree of automation of the packing operation is very much a function of throughput. In smaller units, box forming and lining are carried out manually. The packing machine pumps butter into the box and when the box is filled, an automatic cutter give a smooth finish to the surface of the butter.

The rest of the packing operation, i.e. weighing, weight correction and closing is then carried out manually. Larger units may be completely automated and include automatic box forming and lining. After filling, the box contains less than the final required weight of butter and an automated weighing device determines the difference between actual and desired final weight. The electronic output from this device is then used to control a weight correction unit which adds the necessary amount of butter to obtain the exact final weight. Box closing is also automated; the liner is folded over, the surface of the butter flattened to give a good finish, and the box closed and taped.

Butter for retail sale is usually in the form of 250 g or 500 g packets but a number of machines also incorporate the facility to pack in smaller sizes, e.g. 100 g. Another range of packaging equipment produces individual catering portions of various sizes between 6 and 20 g.

Wrapping materials for retail packs should

provide a barrier to light to prevent oxidative deterioration, as well as being impermeable to water in order to prevent surface drying and discoloration. Parchment fulfils the latter criteria but aluminium foil and other laminates have now gained wide acceptance because of their light barrier properties. After moulding and wrapping, retail packets are packed in cardboard boxes and palletized prior to storage.

Freshly packaged butter is still quite soft since a large proportion of the fat is in the liquid state. The typical consistency of butter is only reached after cold storage.

Storage

Butter is always stored at a low temperature in order to ensure a satisfactory shelf life. For short periods, storage at 0 to 4°C is adequate but generally butter is stored at −10°C or below. For long term storage (up to 12 months), butter should be held at −25°C. During the early stages of storage the fat crystallizes and the structure 'sets' giving the solid consistency of finished butter.

Grading

Butter grading is not compulsory but many manufacturers use a grading system. Butter is examined at 12–13°C and points are awarded for flavour, body and texture, colour and appearance and for absence of free moisture. The most important property of butter is its flavour and this is continuously monitored. Bacteriological specifications (e.g. total bacterial count, coliforms) are also set by most customers.

Blending and repackaging

Butter blending provides a facility for combining the various attributes of a number of batches of bulk-stored butter. For example, it is advantageous to blend butter with extremes of colour, hardness, moisture content, etc. to produce a homogeneous butter which meets market requirements.

Repackaging of bulk packaged butter is relevant not only to blended butter but also to butter manufacture during periods where production exceeds demand for retail butter, e.g. during peak milk production in the spring. The first stage in butter blending is the preparation of the bulk butter in a suitable form for blending. This is carried out in a shiver in which a rotary cutting drum cuts blocks of cold butter into thin slices. In the blender itself, the butter is worked vigorously to break down the 'set' structure of stored butter and improve its plasticity so that it can be handled by the packaging equipment. During blending, salt may be added and the moisture adjusted as required. The blended butter is then discharged into a trolley, or the feeder, of a packaging unit, prior to moulding and wrapping in retail packets.

LACTIC ACID BUTTER (RIPENED CREAM BUTTER)

While most of the butter consumed in the UK is of the sweet cream variety, a number of countries in Europe, especially France and Denmark, show a preference for lactic butter. Small amounts of lactic butter are now manufactured in England.

Lactic butter is manufactured from cream which has been ripened by lactic acid bacteria to produce lactic acid and aromatic compounds, principally diacetyl, which give lactic butter its characteristic strong flavour. Lactic butter is normally unsalted but may be salted at levels up to 0.5%. After pasteurization, the cream is cooled and inoculated with a starter culture (see STARTER CULTURES) which incorporates acid producers (e.g. *Streptococcus lactis*, *S. cremoris*) and flavour producers (e.g. *S. lactis* subsp. *diacetylactis*, *Leuconostoc cremoris*). The incubation temperature depends on whether a thermal treatment is being applied to the cream to modify the hardness of the final butter (see BUTTER HARDNESS, below). In the UK, this is not the case and a single incubation period of 2 to 4 h at 16 to 21°C would be sufficient for the ripening process. After ripening, the cream is cooled to 7°C and held for several hours before churning.

The acidity of ripened cream gives rise to several differences between the manufacture of ripened and sweet cream butter. The low pH of ripened cream weakens the fat globule membrane, giving faster churning and reduced fat losses in the buttermilk, i.e. increased yield. However, the acidity also causes migration of copper to the fat in ripened cream making lactic butter more prone to oxidative rancidity. Lactic butter also gives rise to an acidic buttermilk which is difficult to utilize.

These disadvantages of traditional lactic butter production have led to new developments in the manufacturing process. The most popular of these was developed in NIZO in Holland and is in operation in many factories in Europe. The raw material is sweet cream, and after churning and buttermilk drainage a culture concentrate (containing 11–12% lactic acid) and an aromatic starter culture are injected into the buttermaker in the first working section. This gives the butter the required acidity and aroma but the butter has a lower copper content and therefore greater oxidative stability and the buttermilk is not acidic. The product is filled into retail packs which are then packed in cardboard boxes and transferred to a cold room for storage.

BUTTER HARDNESS

Seasonal variations in the melting point of milk fat give rise to significant variations in the hardness, or spreadability, of butter made at different times of the year. However, the hardness of butter can be reduced to a certain extent by careful control of fat crystallization while the cream is being cooled prior to churning. This involves subjecting the cream to a series of tempering heat treatments designed to produce relatively large crystals of fat in a continuous phase of liquid fat (Alnarp process). Such treatment is generally considered uneconomic in the manufacture of sweet cream butter but is more compatible with the manufacture of lactic butter where the tempering programme can be carried out while the cream is being ripened.

DAIRY SPREADS

Some 25 years ago only two kinds of yellow fat spreads were available, butter and margarine. During the late 1970s the fuller potential of buttermaking technology was realised in two important ways. First, vegetable fats were added to cream to form an artificial cream and then the mix was churned successfully using standard buttermaking equipment to produce a spreadable butter with vegetable fat. Second, the moisture content of these products was increased from the normal 16% maximum up to 25% the apparent upper limit of churning technology. These developments were stimulated by both convenience and health demands from the marketplace.

Butter fat – vegetable fat spreads

The demand for a butter product which spreads at refrigerator temperature has led to the development of spreads based on blends of butter with lower melting point vegetable oils, usually soya. The first of these products, Bregot, was successfully marketed in Sweden and equivalent products are now manufactured in the rest of Europe. The composition of these products is similar to that of butter with the vegetable oil comprising 15% or more of the total fat content.

Manufacture is similar to that of butter itself using continuous buttermaking techniques. The vegetable oil is mixed with cream in the appropriate ratio to give a final fat content of about 38%. The cream mix is then aged and churned in a continuous buttermaker using a slightly lower churning temperature than in conventional buttermaking.

Low fat spreads

Driven by consumer demand for increased convenience and the high price of butter, but particularly by increasing awareness of the link between health and diet, the first low fat spread was developed and introduced in Britain and western Europe in the late 1960s. It was produced by Unilever and was called 'Outline'. This product is still available today but has now been joined by over 30 varieties of other low fat spreads either produced with a butterfat component only or, more usually, with mixtures of butterfat and vegetable oils. Fat levels in different types of spreads range from 80% down to 5%. These health sector spreads derive their consumer appeal from reduced fat levels, enhanced polyunsaturated but reduced saturated fats and restricted cholesterol and salt contents.

Spreads are emulsions of water-in-oil above about 15% fat and normally emulsions of oil-in-water at lower levels. Compositionally the fat can be derived from milk fat, non-milk fat or blends of the two.

In water-in-oil spreads the main function of fat is to provide structure by an interlocking network of small, micron-size crystals. A secondary but essential function is to stabilize the water droplets by forming solid thin coats of fat around the droplets. The amount and the size and shape of fat crystals are related to the level of solid phase in the fat blend, the physical structure of the solid phase and the degree of chilling and working imposed during the manufacturing process. Fats also provide taste and desirable flavour notes and lubricate foods during the chewing process. Individual fats can be modified by a variety of processes which alter the solids/temperature profile and crystal properties. These include blending, hydrogenation, fractionation and chemical or enzymatic interesterification.

As the fat level in spreads is reduced so the technical significance of the aqueous phase increases. In water-in-oil products the aqueous phase is designed to inhibit water droplet coalescence during processing and on spreading. Various methods are used to achieve this most of which involve the use of edible emulsifiers and stabilizers such as milk proteins or plant-derived polysaccharides.

The manufacture of low fat spreads is not possible using cream churning techniques and the fat/water mix is emulsified in a continuous working unit normally, similar to that used in the manufacture of margarine. Essentially there are at least two ways of processing low fat spreads and these can either involve:

(i) standard, margarine-type procedures involving crystallization and plasticizing water-in-oil emulsions by use of scraped or swept-wall heat exchanger units;

(ii) mechanical inversion of part-crystallized oil-in-water emulsions to water-in-oil spreads using high speed tubular pin mixers.

Nowadays there are a number of other butter-derived products and some of these are included in the section on MISCELLANEOUS PRODUCTS.

SKIMMED MILK

(See also MILK POWDERS)

The separation of cream from whole results in the production of a quantity of skimmed milk equivalent to about 90% of the milk processed. The pattern of skimmed milk production has changed dramatically since the mid-1960s. Traditionally the major outlet for skimmed milk was for stock-feeding of animals on farms and in some of the less developed countries of the world this is still the case. In most countries however this practice has long been superseded by either the conversion of liquid skim into skimmed milk powder or by direct sale to the public as a retail product.

SKIMMED MILK POWDER finds many industrial and domestic uses and is a major item in international trade. Most skimmed milk powders are produced in the European Union, Australasia, Eastern Europe and the United States.

As public opinion has turned against consumption of diets with too high a saturated fat content, consumers have sought lower fat products for nutritional and health reasons. Despite the fact that the saturated fat theories are still disputed by the medical profession, consumers have erred on the side of safety and reduced fat milks are now much in demand. There has been a large increase in the domestic consumption of pasteurized milk (with less than 0.1% fat) and especially of semi-skimmed milk (with 1.5–1.8% fat), both of these being in replacement of the full cream milk previously consumed. In the UK alone sales of semi-skimmed milks now represent more than half of the total domestic sales of pasteurized milks. Skimmed milk is also used extensively in the manufacture of cottage cheese, yoghurt, casein, etc.

Skimmed milks are produced to strict legal standards and can either be produced by batch mixing methods (by predetermined volumes of whole milk and skimmed milk), by in-line blending systems or by direct in-line standardization.

MISCELLANEOUS PRODUCTS

In this section a number of products are described which the dairy industries around the world have developed over the years for various reasons. Products were produced either in response to a market demand (e.g. CHOLESTEROL-REDUCED PRODUCTS), to meet a particular functional need (e.g. MILK PROTEINATE), to meet a particular set of climatic conditions in certain parts of the world (e.g. GHEE) or simply to find a commercial application for a by-product which hitherto had little realisable value (e.g. LACTOSE HYDROLYSED MILK PRODUCTS).

Each of these products add value to the original milk and serve to demonstrate just how entrepreneurial the industry has become. Examples of some such products follow.

LACTOSE HYDROLYSED MILK PRODUCTS

Lactose hydrolysis has been of interest to the dairy industry for a number of decades but it is only in the past 10 years that commercial preparations of β-galactosidase have become available in a technically suitable, economic form. There are two major areas of application for the hydrolysis of lactose.

(i) *Avoidance of lactose intolerance.* A significant proportion of the world population is lactose intolerant, i.e. unable to digest lactose. This arises from a deficiency of β-galactosidase in the individual and the ingestion of lactose by such an individual can result in diarrhoea, stomach cramps, etc. Prior hydrolysis of the lactose in milk therefore provides a means of supplying milk to a sector of the population who are presently unable to consume it.

(ii) *Production of sweet whey products.* The glucose and galactose produced on hydrolysis of lactose are significantly sweeter and more soluble than lactose itself. This enables the production of sweet whey syrups which can be used as ingredients in food manufacture (see WHEY PRODUCTS).

Lactose hydrolysis

A number of commercial β-galactosidase preparations are available in a range of purities; technical grades are suitable for lactose hydrolysis in whey products but purer preparations must be used in milk where coagulation could result from the presence of contaminating proteases.

Commercially available β-galactosidases are usually derived from yeasts (*Kluyveromyces fragilis, K. lactis*) and fungi (*Aspergillus niger, A. oryzae*). The former exhibit maximum activity in the range pH 6–7 and are therefore suitable for use in milk and whey, while the latter exhibit maximum activity in the range pH 4–5, and are therefore suitable for use in whey only. Both types of enzyme have been immobilized onto a number of supports, although these are mainly used for the hydrolysis of lactose in whey (see WHEY PRODUCTS).

Lactose hydrolysis is usually carried out at the optimum temperature of the enzyme, 30–40°C, or in the cold (5–10°C). Actual conditions are usually specified by the enzyme supplier.

BUTTERMILK

Buttermilk is the liquid released when cream is churned into butter. Its composition is similar to that of skimmed milk (see MILK COMPOSITION) apart from a slightly higher fat content (0.5%). Buttermilk also contains a large proportion of the fat globule membrane material which is rich in phospholipid.

Buttermilk from the buttermaker may be separated to recover part of the fat and either sold for stock feed or spray dried to produce buttermilk powder. The phospholipid content of buttermilk has valuable emulsifying properties and this has led to the increased use of buttermilk powder in bakery products, ice-cream and recombined dairy products. See also CULTURED BUTTERMILK.

CASEIN

Casein is the major protein in milk, accounting for 80% of the true protein content (see MILK COMPOSITION). Recovery of casein from skimmed milk is carried out commercially by precipitation using either acids or rennet, giving acid casein and rennet casein, respectively.

Casein manufacture has become a significant feature of the European and American dairy industries and manufacturing processes are now very efficient (up to 99% recovery) and conform to the highest standards of hygiene. The latter has been achieved by ensuring that the entire manufacturing process is totally enclosed.

Acid casein

Several acids may be used in the manufacture of acid casein, e.g. hydrochloric, sulphuric, lactic, but hydrochloric acid is used almost exclusively. Acid precipitation can also be carried out by using starter bacteria to develop lactic acid. Casein made in this way is called lactein casein.

The precipitation process is a continuous one in which skimmed milk is first heated to about 30°C and acidified to pH 4.5 by in-line injection of the appropriate amount of hydrochloric acid (1.5 M). The temperature of the acidified milk is then raised to 45°C by direct steam injection and at this temperature the casein precipitates as aggregates (2–10 cm) suspended in the whey. The curd and whey stream then passes through a syneresis tube designed to give a residence time of up to 1 min to allow the curd to consolidate and expel more whey.

Separation of the casein curd from the whey can be carried out in one of two ways; traditionally, inclined screens have been used, but horizontal decanter centrifuges give a much more efficient separation and reduce the extent of washing required downstream.

Washing is necessary to remove residual whey solids from the curd; for most applications, casein must contain less than 0.1% lactose. Washing takes place in two to four stages in a continuous operation with countercurrent flow of wash water. The wash water requirement is high (up to 40% of the volume of skim milk treated) and contributes significantly to the effluent load. Washing stages have traditionally comprised stirred vats but some factories now use specially designed washing towers. One of the stages must be maintained at a temperature of 70–75°C in order to pasteurize the curd. The washed curd is then pressed in order to remove as much water as possible before drying. Drying is usually carried out in multideck vibro-fluidized beds to a moisture content of less than 12%. More recently, attrition driers have been introduced whereby the friction generated by a rotating peg mill is used to dry the curd. Dry casein is then ground and sieved before packaging in 25 kg bags.

Rennet casein

The manufacture of rennet casein was traditionally a batch process but this has now been superseded by continuous systems. The process involves the preparation of a rennet coagulum by the action of rennet on skim milk. The curd is then broken by agitation, heated to 55–65°C by steam injection and held for 1–2 min. Separation of the curd from the whey, washing and drying are then carried out in a similar way to acid casein.

Caseinate

Caseinate is the soluble form of casein prepared by reacting acid casein curd with an appropriate alkali, e.g. NaOH, Ca(OH)$_2$. The caseinate dispersion containing 20–25% solids is then spray dried or roller dried to a moisture content of about 5%. Sodium caseinate is the most widely used caseinate and a number of grades with different viscosities can be manufactured by manipulation of manufacturing parameters.

Uses of casein and caseinates

Acid casein is by its nature insoluble and is therefore not often used in food products in this form. However, an increasing number of food processors now prefer to buy acid casein and convert it to the sodium form themselves; this not only gives economic advantages but provides a more effective means of preparing a caseinate solution rather than trying to dissolve the very fine, poorly dispersible, sodium caseinate powder.

Acid casein also finds some use in the non-food industries and the two major applications are in paper coatings and adhesives.

Rennet casein has not generally been associated with food use because of its insolubility, even at natural pH. However, rennet casein can be rendered soluble using polyphosphates and in this form rennet casein is finding use in the manufacture of cheese analogues. Another use for rennet

casein is the manufacture of plastics, especially blanks for button manufacture and the manufacture of artificial horn and ivory.

Caseinates are used almost exclusively in the food industry. Sodium caseinate is superior to most food proteins in its emulsifying and water binding ability and is used for this purpose as an ingredient in the manufacture of meat products, coffee whiteners and whipped toppings.

CO-PRECIPITATES

Co-precipitates of casein and whey proteins may be manufactured in a similar way to CASEIN by adjusting the precipitation conditions so that the whey proteins are denatured and recovered with the casein. The additional recovery of whey protein gives an increase in yield of about 20% over that obtained for casein alone from the same milk.

The manufacture of co-precipitates involves first heating skimmed milk to 9°C in order to render the whey proteins susceptible to precipitation. The protein is then precipitated by a combination of pH adjustment and calcium chloride addition. Curd recovery and washing procedure are then similar to those used in the manufacture of acid casein. The extent of calcium chloride addition and pH adjustment can be used to modify the properties of co-precipitates substantially and Table 3.7 shows the conditions used in the manufacture of the three most important types. Co-precipitates can be dried as such, using a conventional casein drier, or converted to dispersible or soluble forms by the addition of sodium hydroxide or tripolyphosphate and spray dried. The properties of co-precipitates do not offer significant benefits over those of the various caseinates available. In spite of the increased yield of protein obtained in the manufacture of co-precipitates, pricing systems within the EC have not favoured their production over casein and only small quantities are manufactured.

Table 3.7 Precipitation conditions used in the manufacture of co-precipitates

Co-precipitate type	CaCl$_2$ addition (% of skim milk)	Precipitation pH	Calcium content of co-precipitate (%)
High-calcium	0.2	Natural	3.0
Medium calcium	0.06	5.3	2.0
Low calcium	0.03	4.6	0.5

ANHYDROUS MILK FAT

Anhydrous milk fat, as its name suggests, is a very pure form of milk fat containing less than 0.1% moisture, allowing it to have a long shelf life and when suitably packaged it can be stored for several months at ambient temperature. This gives anhydrous milk fat a significant advantage over butter which must be stored at −10°C or below.

The International Dairy Federation has defined three types of purified milk fat as follows:

(i) *Anhydrous milk fat* is the product obtained from fresh raw materials (milk, cream or butter) to which no neutralizing substances have been added. Anhydrous milk fat must contain a minimum of 99.8% milk fat and no more than 0.1% moisture.

(ii) *Anhydrous butter oil* is the product obtained from butter or cream, which may be of different ages. The compositional specifications for anhydrous butter oil are the same as those for anhydrous milk fat.

(iii) *Butter oil* is the product obtained from butter or cream, which may be of different ages. Butter oil must contain a minimum of 99.3% milk fat and no more than 0.5% moisture.

For each of these three products, specifications are also set for free fatty acids, peroxide value, copper and iron concentrations, to minimize the extent of lipolytic and oxidative rancidity.

Butter oil, anhydrous milk fat and butter oil (including GHEE or clarified butter, which is a product in this class) have superior keeping properties to the butter from which they are made. In the absence of water the hydrolysis of triglyceride fats to fatty acids does not take place. (When hydrolysis does occur in ordinary butters on storage at ambient or elevated temperatures, short chain fatty acids such as butyric and caproic acids are liberated and it is these which are largely responsible for the smell and flavour associated with rancid butter.)

The major use for anhydrous milk fat is in the manufacture of recombined dairy products (see RECOMBINATION). Butter oil is used in the UK and Europe as a food ingredient, particularly in the manufacture of bakery products, ice-cream and chocolate.

Anhydrous milk fat and butter oil may be manufactured from either cream or butter. It is generally more economic to use fresh cream as the raw material since the cost of churning into butter is eliminated. However, manufacture from butter does give added flexibility in terms of converting excess butter into a form which is less expensive

to store and allows the use of butter which may be out of specification for other markets, e.g. with poor body, high moisture.

Manufacture from cream

Cream (about 40% fat) is first pasteurized, cooled to 55–60°C, and concentrated to a fat content of 75–80% in a centrifugal separator. The concentrated cream is then pumped to a specialized separator equipped with a serrated disc which disrupts the fat globules, thus liberating free fat.

The liberated fat is fed to a further separator which removes residual fat globules and fat globule membrane material. The fat phase at this stage contains between 99 and 99.5% fat. The purified fat is then heated to 90 to 95°C and dehydrated to a fat content of 99.8 to 99.9% in a vacuum drier. The product is cooled to about 40°C prior to packaging.

When the original cream is of good quality, this method of milk fat recovery gives a high quality product which conforms to the standards for anhydrous milk fat.

Manufacture of anhydrous milk fat and butter oil from butter

Blocks of butter are first melted in a steam-heated tank and the melted butter heated and pumped to a holding tank where the protein is allowed to agglomerate. The fat is then separated from the aqueous phase in a centrifugal separator; the fat phase at this stage contains 99% fat.

This degree of purity is sufficient for the manufacture of butter oil, but if anhydrous milk fat is required, the fat must be washed (by the addition of water) and then reseparated. It may also be necessary to neutralize the fat if the original butter contained a high level of free fatty acids; treatment with warm alkali converts the free fatty acid to soluble soaps which can then be removed by washing. The purified fat is then dehydrated in a vacuum drier and cooled prior to packaging.

GHEE

Ghee is a type of clarified butter originally prepared in India and popular in Asian communities where it is primarily used in cooking. Ghee may be manufactured in a number of ways and different ghees may have completely different flavour and textural characteristics. Manufacture usually involves heating the butter, during which characteristic flavours are produced as a result of Maillard-type reactions involving the solids-not-fat components. Moisture is removed either by evaporation or by decanting the fat layer after allowing the aqueous and fat phases to separate. The clear oil is then filled into cans and allowed to cool to a solid, grainy mass.

Packaging

Anhydrous milk fat is usually required to have a long shelf life and is therefore packed into small cans or lacquered mild steel drums under an atmosphere of nitrogen. The latter is necessary to prevent fat oxidation during storage. Butter oil, on the other hand, is usually packed into plastic lined cardboard boxes (25 kg) and held chilled prior to use.

MILK PROTEINATES

Milk proteinates are milk protein preparations in which the casein and whey proteins are recovered from milk in their naturally occurring ratio. Unlike CO-PRECIPITATES, milk proteinates are completely soluble at neutral pH.

Milk proteinates are manufactured in a similar way to CASEIN, only the skimmed milk is first heated under alkaline conditions (e.g. 55°C at pH 10, or 90°C at pH 7.5) to induce an interaction between the casein and the whey proteins. The proteinate is then recovered by isoelectric precipitation as in the manufacture of casein. Milk proteinates are reported to have particularly good functional properties, but their use seems to have been restricted to the USA to date.

CHOLESTEROL REDUCED PRODUCTS

Cholesterol consumption in industrialized countries is almost twice that recommended in the WHO Report (1990) *Diet, Nutrition and the Prevention of Coronary Diseases*. Interest in the reduction of cholesterol content of foods has increased significantly in the last few years and various methods for its destruction or extraction are being developed.

Destructive or degradative methods tend to be based on the use of micro-organisms (e.g. *Arthrobacter, Bacillus, Nocardia* or *Streptomyces* species) or enzymes such as cholesterol reductase found in the leaves of plants such as cucumber or lucerne.

However none of these methods has yet been operated on a commercial scale although Japanese workers are thought to be well advanced with a microbial conversion process.

Most researchers have focused their efforts on extraction processes, seeking to remove the cholesterol from milk without otherwise modifying it. Cholesterol can be extracted with non-polar organic solvents or alcohols but this is necessarily limited to analytical techniques. Techniques using supercritical carbon dioxide, fatty acid polyglycerol esters or molecular distillation are all now in the development stage.

One such process is now operating on a commercial scale. This is the Asterol process, developed by a joint industrial partnership between Sanofi and Entremont. The method consists essentially of bringing the food product into contact with β-cyclodextrin then separating the cholesterol–cyclodextrin complex which is formed. The equipment is simple (mixing vats and centrifuges); a factory has been built in Quimper, Brittany.

By mixing cholesterol-extracted butterfat with skimmed milk, a wide range of diary products (e.g. butter, light butter, fermented milks, soft cheeses) with reduced cholesterol content can be manufactured.

DAIRY DESSERTS

The term 'dairy desserts' refers to a range of sweet products, based on milk and/or cream, which are thickened or set by the incorporation of a suitable gelling agent or a stabilizer. They are almost always sold in pots in individual portions. Dairy desserts may be either short shelf life products, with a chilled shelf life of up to two weeks, or UHT products with a shelf life of several months at ambient temperature.

Examples of dairy desserts include mousses, puddings, trifles, syllabub, cheesecake, etc., and these are often multilayer products in which up to three different components are filled sequentially into the pot (e.g. fruit, pudding, cream) to give a layered effect.

In spite of their diversity, the ingredients in dairy desserts are common to most products and, of course, the major ingredient is milk (with added cream and/or skimmed milk powder). Sucrose or glucose solids are used to sweeten the products and the texture is achieved through the use of thickeners and stabilizers. Modified starch is commonly used to thicken and provide a pudding-like texture while gelling agents (gelatin, pectin) are used to control consistency and provide firmness.

Stabilizers are also used to prevent milk protein precipitation and syneresis.

Emulsifiers, usually in the form of mono- and diglycerides of fatty acids, generally improve eating quality through providing a smoother consistency and aiding in the even distribution of fat. Finally, desserts are usually characterized by their flavour, and fruits, chocolate, cocoa powder, caramel, etc., may be used on their own or in combination with food flavourings and/or colours. Manufacture of dairy desserts is a relatively straightforward process in which the milk components and dry ingredients are first dissolved at 60–70°C. Special care is usually necessary to ensure complete dispersion of the stabilizer and a high shear mixer is usually used for this purpose. The product is then homogenized, for example, at 1000 psi (68 bar), and either pasteurized or UHT processed before filling into plastic pots. The shelf life of fresh desserts is dependent on the degree of post-pasteurization contamination that occurs, and good hygiene and cleaning is therefore very important.

Some desserts may also be aerated or whipped. Continuous equipment (e.g. Mondomix) is available for this purpose and the aeration is usually carried out with cold product. The whipping of UHT desserts requires the use of an aseptic whipping device downstream of the UHT plant.

Dairy desserts represent an enormous 'added value' to milk for the dairy industries worldwide. From their original humble beginnings, normally in the form of custards or trifles made traditionally at home in the kitchen, they now represent a significant percentage of the dairy products sales of most retailers.

ICE-CREAM

Traditionally, ice-cream was manufactured from milk or cream, sugar and eggs. Ice-cream labelled as 'dairy ice-cream' must contain milk fat as the sole source of the fat but much ice-cream now contains vegetable fat in place of milk fat.

COMPOSITION

The composition of ice-cream may vary considerably between different manufacturers and different countries. The major constituents are as follows.

(i) *Fat*
Ice-cream usually contains between 9 and 12% fat, and in dairy ice-cream it may be in the form of milk, cream, butter or butter oil. However, many ice-creams now contain vegetable fat, usually hardened palm kernel oil.

(ii) *Milk solids-not-fat*
Ice-cream in the UK must contain a minimum of 7.5% milk solids-not-fat (MSNF) and although the latter term is not defined in the ice-cream Regulations it is usually taken to mean skimmed milk solids. However, ice-cream usually contains between 9 and 12% MSNF (see FORMULATION, below) and while some manufacturers may supply all of this in the form of skimmed milk solids, it is usual to use the latter only to satisfy the legal minimum requirements (i.e. minimum 7.5%) and to make up the desired SNF by the incorporation of less expensive whey materials such as whey powder, demineralized whey powder or delactosed demineralized whey powder (see WHEY PRODUCTS).

In the United States the minimum MSNF is 20%, whilst the minimum butterfat in plain ice cream is 10%. In the USA a product made with vegetable fat cannot be called ice-cream. It is called mellorine by law.

Skimmed milk solids are used either in the form of concentrated skimmed milk (30 to 40% total solids), skimmed milk fortified with skimmed milk powder or skimmed milk powder alone.

(iii) *Sugar*
Ice-cream contains between 14 to 16% sugar, usually in the form of sucrose. However, fructose, glucose or corn syrups may be used in place of some of the sucrose in order to reduce the sweetness and to reduce the freezing point of the ice-cream.

(iv) *Emulsifier*
An emulsifier is incorporated to facilitate the formation of a stable fat emulsion and to aid the dispersion of air when the ice-cream mix is frozen. The emulsifier is usually a mono- or diglyceride and is incorporated at a level of 0.3–0.5%.

(v) *Stabilizer*
The stabilizer functions to bind water and to promote the formation of small ice crystals. Various stabilizers are used, including carboxymethyl cellulose, alginates, gelatin, pectin and vegetable gums. The level of incorporation is usually 0.2–0.4%. A number of ingredient suppliers offer proprietary blends of emulsifiers and stabilizers which give consistent performance together with convenience of use.

(vi) *Colour and flavour*

These ingredients are only added in very small amounts but their selection and level of incorporation are very important in providing an attractive appearance and taste.

FORMULATION

The quantities of fat and sugar required in a recipe are chosen on the basis of the desired texture and flavour of the product. Legislation might demand a minimum of 5% fat in ice-cream in some countries but actual values are higher than this (9–12%) for quality reasons.

The content of solids-not-fat, however, must be carefully calculated to avoid crystallization defects in the ice-cream. A low content of milk solids-not-fat is likely to result in the formation of large ice crystals while a high content of milk solids-not-fat is likely to result in the formation of lactose crystals. It has been established by experience that a ratio of one part MSNF to 7 parts of water gives an ice-cream free of crystallization defects.

MANUFACTURE

The ingredients are preweighed and blended in a mixing tank heated to 50–60°C. Liquid ingredients may be metered directly via a positive pump or weighed into the tank using a load cell. The mix is then heated to 70–75°C and homogenized using pressures of 2500 and 500 psi (170 and 34 bar) on the first and second stages, respectively. After homogenization, the mix must be pasteurized and because the sugar gives some protection to the bacteria in the mix, it is necessary to use a higher temperature than would be the case with milk. HTST pasteurization is generally used and the mix is held at 80–85°C for 2–15 s. It is then cooled to between 2–5°C and held in an insulated tank for up to 24 h (minimum 4 h). This ageing period is necessary to allow the fat to crystallize and the milk protein and stabilizer to hydrate fully. Colour and flavour are also added at this stage.

The method of freezing the ice-cream mix depends on the scale of operation. Smaller manufacturers may use a batch freezer in which the mix is frozen in a refrigerated cylinder fitted with scraper blades. The mix is first cooled to −6°C in about 10 min and a revolving beater then incorporates air into the mix to achieve the desired light texture and increase in volume. The increase in volume is described as the 'over-run'. An over-run of 80–100% may readily be achieved in a batch freezer.

Most ice-cream manufacturers use continuous freezers which inject compressed air into the mix as it is continuously frozen to about −6°C in a refrigerated cylinder fitted with scraper blades. The residence time in a continuous freezer is of the order of only 30 s and the rapid rate of freezing produces a large number of very small ice crystals. The quality of the ice-cream depends upon the smoothness recognized when eating the product. In general, the smaller the ice crystals and the more uniformly distributed, the smoother the texture and the better the mouthfeel.

The over-run can be controlled more closely in a continuous freezer by adjustment of the pressure of the compressed air feed. An over-run of 100% is normal but values up to 180% may be achieved. The frozen mix from the freezer is filled into cups, cartons or plastic containers and frozen further by holding in the deep freeze.

Future prospects for ice-cream sales around the world look good. Ice-cream manufacturers are real innovators and have taken their products a long way since the early days when ice-creams were perceived mainly as children's products.

The manufacturers have segmented the market very well and ice-creams are available to suit all tastes at all times. In America alone the per head annual consumption is a massive 22 l. Even cholesterol-reduced ice-creams are now available.

ICE-CREAM POWDER

Ice-cream powder is used to make reconstituted ice-cream mixes for use in the preparation of soft serve ice-cream for dispensing in cones. Ice-cream powder is manufactured from concentrated skimmed milk to which fat, emulsifier and stabilizer are added. The concentrate is then homogenized, pasteurized and spray-dried before dry blending with ground sugar to produce a powder containing about 40% sugar, 25% milk solids-not-fat, 30% fat, 2% emulsifier and stabilizer and 2% moisture. The powder is reconstituted in water in a ratio of 1:2.

REFERENCES

Society of Dairy Technology (1983) *Pasteurizing Plant Manual*, Huntingdon.
Kessler, H.G. (1981) *Food Engineering and Dairy Technology*, Verlag A. Kessler, Freising.
WHO (1990) *Diet, Nutrition and the Prevention of Coronary Diseases*, WHO Report, Geneva.

FURTHER READING

Early, R. (ed.) (1991) *The Technology of Dairy Products*, Blackie, Glasgow.

Harding, F. (ed.) (1995) *Milk Quality*, Blackie, Glasgow.

Kosikowski, F. (1977) *Cheese and Fermented Milk Products*, 2nd edn, Edwards Books, Ann Arbor.

Marshall, Valerie M. (1993) Starter cultures of milk fermentation and their characteristics *J. Soc Dairy Technol*, **46**(2).

Milk Marketing Board (1992) *Code of Practice for the Hygienic Manufacture of Ice Cream*, Thames Ditton, Surrey, UK.

Moran, D.P.J. (1991) Spreads and other products from fats, in

Rajah, K.K. and Burges, K.J. (eds) *Milk Fat: Production, Technology and Utilization*, Society of Dairy Technology, Huntingdon.

Moran, D.P.J. (1993) Yellow fat spreads. *J. Soc Dairy Technol*, **46**(1), 2–4.

Rajah, K.K. (1994) Milk fat developments. *J. Soc Dairy Technol*, **47**(3), 81–3.

Robinson, R.K. (ed.) (1990) *Dairy Microbiology*, 2 vols, 2nd edn, Elsevier Applied Science, London.

Robinson, R.K. (ed.) (1986) *Modern Dairy Technology*, 2 vols, 2nd edn, Elsevier Applied Science, London.

Robinson, R.K. (1991) *The Therapeutic Properties of Fermented Milks*, Elsevier Applied Science, London.

4 Fruit and Vegetable Products

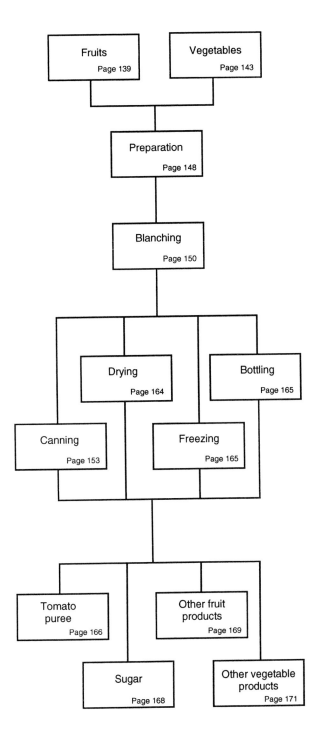

INTRODUCTION

This chapter begins with a description of the principal varieties of fruit and vegetables, classified by their end use rather than according to strict biological definitions (i.e. fruits contain seeds, vegetables do not). This is followed by descriptions of the industrial processes used in the canning, drying and freezing of fruit and vegetable products, together with the specific processes used for products such as tomato paste and sugar. Then a number of particular products are described in some detail.

RAW MATERIALS

FRUITS

Apples

All eating and cooking apples, which number over 1000 cultivars, belong to the species *Malus pumila*. They are all derived from crab apples and are probably complex hybrids of several species of this wild apple. The varieties commonly available as dessert apples include Cox's Orange Pippin, Golden Delicious, Granny Smith, Starking, Red Delicious, McIntosh, Spartan, Laxton Fortune, James Grieve and Egremont Russet. The dominant variety for cooking is Bramleys Seedling because of its sharper flavour.

Apricots

This pitted fruit, *Prunus armeniaca*, originated in China and is now grown extensively. The sweet yellow flesh makes it both a good dessert fruit and a good fruit for canning. The almond-like kernels may also be used as an almond substitute in certain products.

Avocados

The avocado, *Persea americana*, is an evergreen tree native to central America. The edible berry is pear shaped and has thick yellowish flesh and a large stone. The cultivated varieties have a smaller fruit than the wild tree and they are grown widely in warm regions. Now very popular as a fresh vegetable, avocados have some industrial use in dips and sauces.

Bananas

The banana, *Musa sapientua*, is the fruit of a giant herb which superficially resembles a tree. It is native to southeast Asia and cultivated bananas are grown throughout the tropics. Bananas are used in canning, mainly as a constituent of fruit salads.

Bilberries (blueberries or blaeberries)

The bilberry (*Vaccinium myrtillus*) is a native European shrub. It has a bluish fruit and is also known as huckleberry.

Blackberries

The blackberry (*Rubus ulmifolius*) is a shrub, also known as bramble, and is native to Europe.

Blackcurrants

The blackcurrant (*Ribes nigrum*) is a native European and Asian plant and is widely cultivated. Its name derives from its superficial resemblance to dried currants.

Cherries

There are several species and varieties of cherry, all members of the genus *Prunus*. They are deciduous trees mostly found in northern temperate climes.

Cranberries

This is an evergreen shrub (*Vaccinium oxycoccus*) which grows in boggy soils. Indeed the berries are sometimes harvested by floating in water. Cranberries are mainly used for making cranberry sauce.

Dates

The fruit of a palm, dates (*Phoenix dactylifera*) have been cultivated since 6000 BC. Each tree is very productive and can produce up to 250 kg fruit per year for up to 100 years. Dates are usually available in dried form.

Exotic fruits

Exotic fruits and vegetables generally include those from tropical climes which have become more available in northern Europe as fresh items in recent years and more in demand as tastes have widened. However, the term exotic is now used very much as a marketing ploy and is no longer applied, for example, to pineapples or bananas which are not now considered exotic because of their familiarity. Some examples are given here.

Fig

Figs (*Ficus carica*) are perhaps more familiar than most 'exotics' in fresh, canned and dried forms. There are many species of fig tree (about 800) some deciduous, some evergreen.

Guava

Guavas (*Psidium guajava*) are sold as a fresh item but are also commonly available as a canned fruit. They are sharp tasting, with either pink, yellow or white flesh.

Kiwi fruit

Kiwi fruit (*Actinidia sinensis*) is a native, not of New Zealand where it is widely cultivated, but of China and once called the Chinese Gooseberry. As well as a fresh item it is used in canning and as an ingredient.

Kumquat

Kumquats (*Fortunella margarita*) are small oranges, usually eaten whole, hence they have a bitter flavour. They are not usually used industrially.

Lychee

Lychees (*Litchi chinensis*), which may also be spelt litchi, are the fruit of an evergreen tree native to China. Available fresh, they are also commonly canned.

Mango

Mangoes (*Mangifera indica*) are the fruits of an evergreen tree native to southeast Asia. The fruit is large and fleshy and is available fresh or used in canned fruit salads.

Okra

Okra (*Hibiscus esculentus*) is also known as 'ladies fingers' or 'gumbo'. The fruit is a pod which, when cooked, becomes glutinous in texture. It is eaten whole and commonly used in Indian cookery. It is not usually used in industry.

Passion fruit

The passion fruit (*Passiflora edulis*) comes from a large genus of plants mainly originating in America. The fruit have a leathery skin with a soft edible pulp. The flavour is very characteristic and while the fruit is available fresh, the industrial use is often as a flavour, e.g. in soft drinks (see also Chapter 6 – FRUIT JUICES AND SOFT DRINKS).

Pawpaw

Pawpaw (*Carica papaya*) is also known as 'papaya' or 'papaw'. It is a large fruit with pink flesh and black seeds and, while its origin is uncertain, it is cultivated throughout the tropics. In addition to its food properties, it is also a source of papain, an enzyme used in the meat and leather industries.

Plantain

Plantain (*Musa paradisiaca*) is a relative of the banana and its green banana-like fruit is used in Caribbean cookery.

Pomegranate

Pomegranate (*Punica granatum*) is a large berry from a deciduous shrub native to southwest Asia. The fruits are eaten fresh or used in Persian cookery but have little or no industrial use.

Star fruit

Star fruit (*Averrhoa caramboli*) is also known as 'carambola' and is available fresh with limited industrial use, usually for its decorative appeal.

Gooseberries

The gooseberry (*Ribes grossularia*) is a deciduous shrub native to Europe; varieties produce green, yellow or reddish fruit. The berries are often harvested in an immature condition, especially for canning.

Grapefruit

Grapefruit (*Citrus paradisi*) is a citrus fruit which continues to be popular in fresh and canned forms. There have been an increasing number of varieties in recent years which offer new colours of flesh (pink) and sweeter flavours. New hybrids also continue to appear, e.g. ugli, a tangerine–grapefruit hybrid and jaffa sweetie, a lime–grapefruit hybrid. Grapefruits are also important for juice production and in soft drinks.

Grapes

Grapes (*Vitis vinifera*) are divided into those for the table, those for winemaking, and those for drying. Most European vines are derived from North American species because of their resistance to *Phylloxera*, an aphid insect pest which devastated northern European vines in the 1860s and 1870s. Grapes vary in colour, size and degree of sweetness. The presence of tannin in the peel and stalks is important in wine production, especially red wine. Grapeseeds are also used to produce oil and breakfast cereal.

Lemons

This citrus fruit (*Citrus limon*) is widely grown in warmer temperate regions and widely used in industry as a flavour in cookery, as a juice in soft drinks or as a juice to be sold both sweetened and unsweetened.

Limes

Limes (*Citrus aurantifolia*) are smaller, rounder and greener than lemons. Their uses are similar to this fruit but not so extensive.

Mangoes

See EXOTIC FRUITS

Melons

These are the fruits of trailing or climbing vines (*Cucumis melo*). There are many distinctive varieties including honeydew, watermelon, cantaloupe and gallia. Some are used industrially in, for example, canned fruit salads.

Nectarines

See PEACHES

Olives

Olives are commercially available in two types, black and green. This colour difference arises from picking times – green olives are picked at an early ripe stage when the fruits are pale green or straw coloured, but black olives are picked when the fruit is fully mature and purple or black in colour. Olives destined for oil production are always picked when black.

Freshly harvested olives are unacceptably bitter in taste and the bitterness is normally removed by soaking green olives in 1–2% lye solution at ambient temperature until the lye has penetrated to within a short distance of the pit or stone, leaving a small and acceptable amount of bitterness in the fruit. The lye is washed out of the olives by a succession of soakings in water and then the olives are brined. Initially, a dilute salt solution is employed but this is increased in stages to 7.5–9% salt. In warm weather, it may be necessary to increase the final salt concentration. A small amount of lactic fermentation takes place which lowers the pH and assists in preserving the olives; this is frequently encouraged by the addition of glucose or other sugar.

A newer method for storage of lye-treated olives is by means of a solution containing 0.67% lactic acid, 1% acetic acid, 0.3% sodium benzoate and 0.3% potassium sorbate. In many cases, this produces a better final product and avoids most of the problems attached to the disposal of strong brines. Black olives are frequently preserved without initial lye treatment, the brine soaking being relied upon to remove most of the bitterness.

Olives stored in barrels or in vats in favourable and well-controlled conditions may be good for up to one year. If, however, olives are canned or processed in glass, a much longer shelf life can be expected.

For canning, olives are frequently pitted (or stoned) and in some cases may be stuffed with pimiento or other material of contrasting colour, texture and flavour.

See also OLIVE OIL in Chapter 8.

Nuts

See SNACK FOODS in Chapter 12.

Oranges

Oranges are widely grown in all warmer temperate regions and are thus an important crop. Sweet dessert oranges (*Citrus sinensis*) are generally eaten as a table fruit, used in juices and soft drinks, and canned as segments. Other species include *Citrus aurantium*, the seville orange, important in marmalade manufacture, and *Citrus bergamia*, the bergamot orange which is grown as a source of bergamot oil obtained from the rind and Neroli oil from the flowers. Bergamot oil is used as a flavouring, e.g. in speciality tea.

Peaches

Peaches (*Prunus persica*) are the fruits of a deciduous tree possibly originating in China. They are used widely as a table fruit and in canning and to a lesser extent in drying. Canned peaches are produced in syrups and juices in whole, halved and sliced form. They are usually an excellent fruit for canning although size variation caused by growing conditions can lead to problems. Peaches are also an ingredient in fruit salads and in 'Two fruits' (peaches and pears).

Nectarines (*Prunus persica* var *nectarina*) are a smooth-skinned variety which are not commonly canned.

Pears

Pears (*Pyrus communis*) are native to Europe and Asia. There are numerous varieties of which the most popular dessert ones are Doyenne du Comice, Conference and Williams bon Chretien (Bartlett).

Pineapples

The pineapple (*Ananas comosus*) is a native of South America now also cultivated in other areas, e.g. parts of Africa. It is a multiple fruit with a superficial cone-like appearance, hence the name. It is available fresh, canned and dried and is also used to produce juice. It may also be an ingredient of fruit salads in cans.

Plums, damsons and prunes

There are numerous varieties of plum (*Prunus domestica*) which vary greatly in size and colour. Important table varieties include Victoria, Golden Cape, Red Ace and Greengage. The Prune plum may be eaten raw but, of course, is best known as a dried plum, which may then be canned. Another species is the damson (*Prunus institia*), which may also be canned.

Raspberries and loganberries

Raspberries (*Rubus idaeus*) are the fruit of a deciduous shrub native to Europe and Asia and are included amongst the cane fruits. The larger, dull-red fruits of the loganberry are very acidic in flavour and have occasionally been canned in syrup.

Rhubarb

Rhubarb (*Rheum rhaponticum*) is a long lived perennial plant with edible stalks, originally native to Siberia. The leaves have a high oxalic acid content and, because of their toxicity, should not be consumed.

Strawberries

The strawberry (*Fragaria ananassa*) is a perennial plant which propagates itself by producing runners. It is widely grown and produces a false fruit of a swollen receptacle.

VEGETABLES

Artichokes

The globe artichoke (*Cynara scolymus*) is a thistle-like plant with edible flower heads. Artichoke hearts are available in cans. The Jerusalem artichoke (*Helianthus tuberosus*) is completely unrelated to the globe artichoke. The similarity in the flavour of its edible tuber to the globe artichoke is the reason for the name.

Asparagus

Asparagus officinalis is native to Europe and Asia and it is the young shoots which are the edible part. There are varieties in green, and purple. White spears are produced by 'forcing' the plants with the exclusion of light; these are popular in mainland Europe. The green variety is the one commonly canned and is most popular in the UK.

Aubergines

The aubergine (*Solanus melongena*) is a bushy perennial native to New World tropics. The fruit is variable in shape and colour but commonly purple when ripe. It is also known as the eggplant because originally the shape of the fruit was more egg-like. It is used in casserole-type dishes and its industrial use is mainly limited to complete meals on trays or in canned vegetable mixes. It is also known as brinjal.

Bamboo shoots

Bamboo shoots (*Bambusa vulgaris*) are more usually available in cans from China.

Beans

Beans are pulses belonging to the family Leguminosae. A wide range is available in dried form. Alternatively green beans are grown for their edible, fleshy pods and are used in fresh, canned or frozen forms. Some of the dried beans available are as follows.

Adzuki bean

Phaseolus angularis is a small bean from China and used in Chinese and Japanese cookery.

Black-eyed bean

Phaseolus vulgaris is a characteristic bean from Africa, now grown in North and South America with over 50 varieties.

Broad bean

Vicia faba, also available fresh in the pod, is considered to be the first bean used by man and is universally popular.

Butter or lima bean

Phaseolus lunatus is known as the butter bean when dried or canned and as the lima bean when fresh.

Flageolet bean

Phaseolus vulgaris is a small green bean, an immature haricot, popular in French cookery and sometimes canned.

Haricot bean

Phaseolus vulgaris is a popular bean used domestically and in canning. It originates in South America. It is also widely canned.

Mung bean

Phaseolus aureus, widely cultivated in China and India, this small bean is used domestically and is of importance as beansprouts (see BEANSPROUTS).

Red Kidney bean

Phaseolus vulgaris, a very popular bean, especially used in Mexican cookery. Also commonly canned.

Soya bean

Glycine max also known as soy or soja bean. The beans are rich in protein and have several uses including the production of soya flour, soya oil, meat substitutes and soy sauce.

Among the green beans, the commonly-used species are the following.

Broad bean (see above)

French bean

Phaseolus vulgaris is probably from South America; a variety, Bobby Bean, is grown in France.

Runner bean

Phaseolus coccineus is the most popular green bean in the UK, originating from South America and also called the stick bean.

Beansprouts

Beansprouts are germinated mung beans, eaten in the early stages of germination when only the first shoot is present.

Beet and beetroot

There are several important varieties of beet, all derived from the wild *Beta vulgaris*. They can be divided into root crops and leaf crops.

Among the important root crops are beetroot, mangel-wurzel and sugar beet. The latter contains up to 20% sugar and has largely replaced sugar cane as a source of sugar in western Europe.

The leaf beets include spinach and chard.

Brassicas

Brassicas are a large genus of important food plants which include cabbages and similar plants but also turnip and mustard. The types discussed below are almost all derived from the wild cabbage *Brassica oleracea*. They are important sources of vitamins A, B and C and provide significant amounts of iron and calcium.

Sprouting broccoli and calabrese

These varieties of *Brassica oleracea* are cultivated for their edible immature flowering heads, which may be white, green or purple. They are rarely used in canning but calabrese (broccoli spears) are commonly seen as a frozen vegetable.

Brussels sprouts

This is a variety of *Brassica oleracea* which produces tight round shoots or sprouts in the axils of the leaves on the main stem. They can be harvested over a long period, characteristically in winter. They are said to have originated near Brussels in the thirteenth century. They are used in both the canning and frozen foods industries.

Cabbage

There are numerous cultivars of cabbage (*Brassica oleracea*), which have a range of features. One or another cultivar can be harvested all year round. Examples include spring greens, January King, Savoy, Roundhead, white cabbage and red cabbage. They are used in the frozen foods industry and in prepared salads, e.g. coleslaw.

Cauliflower

Like broccoli this type of *Brassica oleracea* is grown for its edible immature inflorescence. The plant probably originated in the Orient. Most common is the white headed variety, but there are also green headed cauliflowers. Cauliflower is used in the frozen foods industries and as an ingredient, e.g. in pickles.

Chinese leaves

This cabbage (*Brassica pekinensis*) is usually eaten raw as a salad vegetable. It has a subtle flavour and characteristic shape. Not usually used in industry.

Kale

Kale is a non-hearting type of cabbage (*Brassica oleracea*). It is a hardy plant and grown both as a food crop and a fodder crop. Not usually used in industry.

Kohlrabi

This variety of *Brassica oleracea* is a turnip-like plant with a swollen, edible stem, although the leaves can also be eaten. Not usually used in industry.

Capers

Capers are the flower buds of a nasturtium-like plant, *Capparis spinosa*, mainly obtained in brine from France or Algeria. Somewhat delicate, they are carefully washed and drained prior to packing into glass jars and covering with liquor, usually distilled malt vinegar or wine vinegar. They may or may not be pasteurized, depending on final acidity.

Capsicums

The species *Capsicum annuum* includes sweet peppers and chillis, other species include paprika and cayenne. The hot flavour notable in chillis is due to the chemical capsicin, contained in the placenta. The plants are native to the tropics and are related to the tomato and the potato. There are many varieties of chilli pepper which range in colour from green through yellow to red and in degree of hotness. Sweet peppers also range in colour but partly this is due to the degree of maturity. They have a wide range of industrial uses as general ingredients.

Carrots

The carrot (*Daucus carota*) is a herb with an edible orange tap root. It is native to Europe and

temperate climes and extensively cultivated. Widely used in canning and in salads.

Cauliflower

see BRASSICAS

Celery

Celery (*Apium graveolens*) is a herb, widely cultivated for its edible swollen leaf stalks. It is available in green and white varieties.

Celeriac is a variety of celery (*A. graveolens* var *rapaceum*) with an edible tuberous base to the stems.

Chickpeas

Chickpeas (*Cicer arietinum*) have been cultivated since ancient times, although mainly in the past as fodder. They are available in pale coloured or sometimes black forms, are used in canning and may also be milled into flour.

Chillies

see CAPSICUMS

Chinese leaves

see BRASSICAS

Courgettes

see SQUASHES

Cucumbers

see SQUASHES

Garlic

Garlic is a perennial bulb, native to Asia but now widely cultivated. Its use lies solely in its strong flavour; as a food it is available in fresh, paste, dried flake, dried powder and garlic salt forms. In recent times there have been claims that it has an antiseptic effect and that it has health giving properties. A market for garlic in tablet form therefore also exists.

Gherkins

see SQUASHES

Herbs

see Chapter 14

Horseradish

Horseradish (*Amoracia rusticana*) is mainly used as a base for horseradish sauce.

Kale

see BRASSICAS

Leeks

see ONIONS

Lentils

This is a leguminous plant (*Lens culinaris*) which produces pods containing 1–2 of the disc-shaped seeds. There are several varieties of differing colour. They have a high protein content (about 24% raw) and are very popular especially in vegetarian cookery.

Lettuce

Lettuce (*Lactuca sativa*) is widely cultivated in several varieties which differ in shape and texture, e.g. Butterhead, Iceberg, Cos. This popular vegetable has a short shelf life and does not lend itself to processing. Thus industrial use is limited to chilled salad mixes.

Marrows

see SQUASHES

Mushrooms

There are a large number of edible fungi that are esteemed for their own flavour and for improving or influencing the flavour of other foods. There are also poisonous varieties of fungi that are not always dissimilar from those that are edible and naturally it is imperative to be able to distinguish between them. Commercially this problem does not normally arise since cultivation is carried out under carefully controlled conditions and the only important fungus is that known as mushroom, or champignon, which is botanically named *Agaricus campestris*.

Onions

There are several edible species of the genus *Allium*. The common onion (*Allium cepa*) includes spring onions and the Spanish onion as varieties; other well-known species are leeks (*Allium ampeloprasum* var *porrum*) and shallots (*Allium ascalonicum*). All are used both domestically and industrially.

Parsnips

Parsnip (*Pastinaca sativa*) is a root vegetable native to Europe but introduced all over the world. The tap root is also used as fodder.

Peas

Peas (*Pisum sativum*) are members of the Leguminosae family, native to Europe and Asia. They have been widely cultivated since prehistoric times and now have many cultivars. The pods contain up to 10 seeds and these are available fresh, dried, canned or frozen. Mangetout (*Pisum saccaratum*) is a variety in which the pod contains small immature peas and is eaten whole. Petit pois are types of *Pisum sativum* which naturally produce small peas and are now also available in canned or frozen form. It should be noted, however, that some peas canned as petit pois are no more than small peas graded out from large varieties and may not have the same tender texture.

Peppers

see CAPSICUMS

Potatoes

Solanum tuberosum is the well-known tuber crop which is grown throughout temperate regions and is an immensely important food plant both historically and economically. The tubers are an important source of vitamin C and starch and the thousands of varieties, some of which are listed below, vary in physical appearance and eating quality. 'New' potatoes are those tubers harvested in an immature condition; 'old' potatoes are those which have been lifted when mature and have developed a second skin layer. Old potatoes may be stored until the first lifting of tubers in the following season.

Historically, diseases of potatoes were of great importance as they were so damaging. Modern varieties are more disease resistant.

In industry, potatoes are important in canning, freezing and in snack production.

Some important varieties currently used are: Maris Piper, Maris Peer, Wilja, Record, Golden Wonder, Estima, Pentland Squire, Desiree. There are many others.

The sweet potato (*Ipomoea batatas*) is not related to the potato nor the yam with which it is sometimes confused. It is available fresh but has no industrial use.

Pulses

A general term meaning seeds of leguminous plants, i.e. peas, beans, lentils and so on.

Radish

Raphanus sativus, a tuberous root vegetable used since ancient times. It has little industrial use other than as a salad ingredient.

Seaweeds

Seaweeds are eaten in various parts of the world and are important notably in Japanese cookery. Some seaweeds are available in canned form. Important ones are nori, wakame and hiziki. Nori is made into sheets in Japan and dried and used in sushi dishes. Other well-known seaweeds are laver, used in Welsh cookery, and carrageen which is also a source of carrageenan gum.

Spices

see Chapter 14

Spinach

Spinach (*Spinacea oleracea*) is a leaf vegetable available in fresh, frozen and canned form.

Squashes

Squashes are members of the family Cucurbitaceae, native to America; indeed the term 'squash' derives from a native American word. Apart from cucumbers, all the following are varieties of the species *Cucurbita pepo*.

Courgettes or zucchini are immature marrows. Like marrows and pumpkins, they are mainly available as a fresh item with limited industrial use.

Cucumbers (*Cucumis sativus*) have been cultivated since early times. Baby cucumbers are used as gherkins in pickling and are a variety, not an immature form.

Swedes

see TURNIPS

Sweetcorn

Sweetcorn (*Zea mays*) is the only cereal native to the New World. It is an important crop and available fresh as corn cobs, and is frozen and canned both as cobs and as kernels. Canned sweetcorn is often packed with little added water or brine and consequently the cans usually have a higher vacuum relative to other canned vegetables. Immature cobs are also available fresh, frozen and canned. Sweetcorn is also important as the principal ingredient in cornflake breakfast cereals.

Tomatoes

Lycopersicon esculentum, the tomato, is native to South America and widely cultivated. There are many varieties which range in size and shape from the small cherry tomato to the large beefsteak tomato. Plum tomatoes are important and grown in large quantities for canning. The final pH on canning is sometimes critical as canned tomatoes fall on the borderline between low and high acid foods. To achieve the correct balance, citric acid is sometimes added during the canning process. Tomato forms the base of many types of dishes and food products and industrially, tomato puree is often the most useful form.

Turnips

The turnip (*Brassica rapa*) is widely grown as a vegetable and as fodder. It has whitish skin and flesh. The swede (*Brassica napus*) or Swedish turnip or rutabaga is similar with reddish skin and yellowish flesh. Both are canned and frozen, usually diced and as part of vegetable mixes.

Zucchini

see SQUASHES

MANUFACTURING PROCESSES

PREPARATION OF FRUIT AND VEGETABLES

Because the aim of all preservation processes is to offer a convenient product with no extraneous matter or waste, coupled to economic processing at optimum yield and minimum deterioration from fresh, preparation is a quality control procedure. It follows a common pattern, varying only slightly according to the product being processed. Bruising and stoppages must be avoided. Root vegetables require the most attention.

Prewashing

The evidence of the field not eliminated at the purchasing stage has to be washed away first, usually with process water still sufficiently clean for the purpose. Items such as celery and leeks require special attention.

Most root-type vegetables can be dealt with whole in a horizontal drum or rod washer consisting of slats carried on galvanized rings. A noncorroding scroll inside the slats gives rise to a propellant action which conveys the vegetables along, and a perforated pipe supplies jets of wash water. The combined washing and tumbling action removes most of the dirt in a slurry, which leaves through the slats to a bunker and drain; a drum rotational speed of 8–10 rpm is typical.

Stone traps, based on buoyancy difference, are regularly incorporated in the line, usually with some form of elevator from the sump, containing water to which salt may have been added; preferably, another screw conveyor discharges the stones continuously. Cinders and small stones can prove problematical, so this stage is followed by a rough-sorting inspection to safeguard the down-

stream size cutters. Riffles are used to remove stones from peas and beans but they are not completely effective.

Peeling

There are several systems, each with a particular merit.

Abrasive peeling

In this method the vegetables are rotated against rough surfaces such as carborundum, which rubs off the skin. Machines cannot adapt to the irregular contour of the vegetable, so the method is savage, although otherwise simple and cheap. Losses may amount to as much as 30% but, by careful control, this may be kept down to around 12%; losses tend to be higher when the roots are mature.

A batch-type abrasive peeler consists of a drum coated internally with carborundum, with a rotating undulating base plate, also carborundum covered. A measured amount of vegetables is fed into the drum and thrown by the spin of the base plate against the rough inner surface, the loosened peel being washed away by a powerful water spray. Batch peelers are usually mounted in tandem and operated sequentially by means of a timer. A check has to be kept on wear.

Onions can be dealt with by a not dissimilar style of machine, having knives instead of abrasive, or on specialized machines which can also top and tail.

Continuous abrasive peelers contain rows of spindles with abrasive materials fused to their surface. These rotate at high speed and are arranged in sets to give a forward and sideways movement to the root vegetables being peeled. The vegetable are fed in at one end of the peeler, travelling in zigzag fashion through a series of compartments before being discharged at the opposite end. Water sprays in each compartment wash away the loosened peel. The degree of peeling is regulated by adjusting the openings of gates between the compartments. Such machines can handle potatoes, carrots, beetroot and can even snib gooseberries.

Lye peeling

Lye peeling has a number of drawbacks, not least being effluent disposal problems. It introduces certain hazards, particularly the danger of handling hot caustic solutions, but it has definite merit as a method of peeling root vegetables. The best techniques will completely lye-treat the eyes as well as the skin of potatoes and a properly functioning washer will remove practically all but the deep-seated eyes at the ends of the potatoes. Shallow-eyed varieties of good quality are lye peeled and washed so completely that only stem ends and occasional eyes on the ends require hand removal.

Lye-bath peeling requires caustic soda and a generous supply of water and steam. The steps involved are:

(i)   A preliminary water wash in standard equipment
(ii) A short immersion in a hot caustic soda solution in the lye peeler itself
(iii) A vigorous water rinse and tumbling in a washer to remove all skins and chemicals
(iv) Neutralization in an acid bath (used occasionally)
(v)   Hand trimming and cutting to give perfect finish.

The skins of new potatoes treat easily at low caustic concentrations and a temperature between 82°C and the boiling point. The temperature and time of immersion (2–4 min) vary with the type and condition. Stored potatoes require a higher concentration of lye or need a higher temperature, thus shortening what would otherwise be a long immersion time.

Fluctuations in rate of feed cause variations in conditions and treatment. If the temperature falls, the production and quality of the peeling drop. The boiling point of a contaminated solution of caustic soda may not remain the same as the original, especially at higher concentrations. A 5% solution will boil at about 102°C, but a 50% solution boils at about 143°C.

When high concentrations are used, at high temperature and short immersion time, there will be less cooking of the vegetable flesh. Apples, apricots, pears, beet, carrot, celery, cucumber, garlic, gherkin, horseradish, paprika, potato, rhubarb, tomato and turnip have all been peeled in this way. The operator adjusts the strength of the solution, the time of immersion and the temperature accordingly. Chemical tests of the solution are invaluable as a control procedure.

Potatoes should emerge from the lye bath with the skin still adhering, but in a soft somewhat gelatinous condition loose enough to be easily washed away. Removal of parts of the skin in the lye indicates too long exposure or too much mechanical action. Overexposure is corrected by speeding up the passage through the lye peeler to give a shorter immersion time. If the potatoes emerge with too little softening of the skins these will not be in a condition to be properly removed in the washer. It is important not to increase the time of immersion to the point of removing a greater depth of flesh than is desirable for a proper balance between weight loss and trimming labour. Machines to be used for more than one type of crop need greater flexibility in terms of speed than is required for potatoes alone, particularly if the range includes fresh as well as stored produce.

Freshly dug carrots are lye peeled with short exposures of 1½ min upwards in very weak (3% and upwards) caustic soda solutions, with the temperature maintained at 80°C up to boiling point. Weight loss is only a fraction of that encountered in abrasive peeling and the finish of the carrot is better. Turnip requires a lye concentration of 4 to 10%, with the temperature close to boiling point. Beetroot is more difficult; discoloration and bleeding necessitate a pronounced scalding as part of the peeling operation. Lye concentrations for beets are around 7% and the temperature is best at or near boiling point. The time of immersion depends on the condition of the beet skin, but tends to be longer than for other root crops. Freshly dug beets can be lye treated in 2–3 min.

Skin removal following the lye bath can be accomplished with various tumbling combinations plus water jets, all of a fairly gentle nature.

Lye bath peeling has tended to give way to the dry-lye process, in which lye application is kept to a minimum, heat is supplied by infrared lamps and a buffing action removes the dry peel.

Steam peeling

Subjecting root vegetables briefly to high pressure steam is now a major method of peeling. The principle involves vapour formation just under the skin to loosen it, and the timing has to be such that overcooking is avoided. The steam vessel may be of the batch type, mounted to rotate on trunnions. The steam expels the air just as the quick-acting lid is closing and the contents are tumbled for a short time. When the steam is automatically shut off, cold water is injected for rapid cooling. The water is then shut off and the exhaust valve opened. The batch is subsequently discharged and the vessel uprighted to receive the next load. Alternatively, the system may be continuous in the form of a cylinder containing a helical conveyor fed and discharged by rotary valves at each end. Exposure to steam at 620 kPa for 25 s is sufficient for potato. A steam:product ratio of 1:7 by weight is typical. The loosened peel is washed off afterwards under high pressure water jets in a rotary slatted drum.

Flame and radiant heat peeling

In this method, vegetables are brought either into direct contact with a live flame or are subject to high levels of radiant heat from the refractory lining of rotating furnaces through which the vegetables pass. Temperatures as high as 1000°C can burn out the eyes of potatoes as well as charring away the skins which are subsequently removed by high pressure water jets. Hand trimming is largely eliminated and wastage is remarkably low. Energy consumption during warm up, loss during cool-down and intermittent loading costs detract from the advantages.

The root vegetable is now in a surface-clean condition, ready for general inspection and the trimming out of blemishes prior to mechanical subdivision into the regular shaped pieces in which form it is to be processed. The preparation of leafy vegetables begins at such a hand trimming stage, as of course does that of soft fruits and carcass meat. A deft hand, a critical eye and trained judgment play an important part here in maintaining high standards of quality assurance and yield.

Cutting to regular size

This is probably the most telling part of the preparation for drying. Pieces having different paths for moisture removal will not dry uniformly together. Anything other than clean cuts across the cells will fail to produce a satisfactory final appearance. The machines must be protected from damage to their knives, which operate in gangs to produce true strips, dice or shreds. Strips 5 mm × 8 mm or 3 mm × 8 mm cross-section are common; dice or slices have similar limits, although 12 mm sections are possible. The desired rate of reconstitution will also dictate the size. Sliver removing machines are used when necessary.

Leaf vegetables are washed after cutting. Root vegetables containing starch also need a surface wash, but most other products are not washed after cutting.

Strip washing

Strip washing machines remove surface starch after certain root vegetables (e.g. potato) have been stripped or sliced. They take the form of a perforated stainless steel cylinder, the lower half of which dips into a trough where water may be maintained at any desired level.

The strips are fed down a chute at the end of the machine and are then propelled through the water by a stainless steel scroll. Discharge is effected by an arrangement of lifter plates, perforated to allow the water to run back, while picking up the strips and discharging them through a residual water-draining outlet chute. The drive gives low rpm and is variable.

BLANCHING OF FRUIT AND VEGETABLES

The term 'blanching' originally denoted a culinary process, defined by the famous Mrs Beeton as 'placing in cold water and bringing to the boil'. In that terminology the word also bears the sense of 'to whiten' (as with almonds and celery) but one of the purposes in modern practice is just the opposite – to retain colour. For this reason the word 'scalding' is considered proper in the dehydration industry; however, since the term 'blanching' persists both words are used interchangeably here.

Vegetables are now normally blanched by immersion in hot or boiling water or in steam, as soon as possible after preparation and immediately before processing. This has a fivefold purpose:

(i)   to cleanse the vegetables;
(ii) to inactivate enzymes which might otherwise produce undesirable flavours or discoloration, or cause loss of vitamin C;
(iii) to bring about shrinkage of the vegetables without which it would be difficult to pack sufficiently high filled weights in cans;
(iv) to expel gases from the intercellular spaces of the vegetables and thus prevent an avoidable increase in the pressure developed in the cans during processing;
(v)   to improve the colour and flavour.

An additional advantage when canning is that if the product is filled immediately, the closing temperature of the canned product will be higher. However, if immediate processing cannot be ensured, water cooling should follow the hot blanch to reduce heat degradation and especially to prevent the bacterial multiplication which can take place rapidly in a warm product.

Good canning practice demands a high level of hygiene in blanching equipment and in storage hoppers and fillers handling vegetables hot from blanching. This is essential in order to prevent the undesirable build up of thermophilic bacteria.

Some fruits such as apples and rhubarb are blanched before canning, particularly if they are to be solid packed, because the softening and shrinkage permit better filling into the cans. Hot filling also reduces the processing time in such packs where heat penetration can otherwise be slow.

The advantages of blanching are partly offset by the high loss of nutritive value which occurs. Blanching times should therefore be kept as short as possible. Although blanching in steam entails more practical difficulties than blanching in water, it does give lower losses of nutrients.

The identities of the enzymes which need to be inactivated have not been completely established. In the case of peas for freezing, for example, it is necessary to inactivate catalase or peroxidase in order to obtain good quality and prolonged storage. However, for other frozen vegetables, catalase or lipoxygenase must be inactivated. The use of peroxidase as an indicator has achieved popularity because of the availability of the simple hydrogen peroxide guaiacol test, and also because it is the most heat resistant enzyme commonly encountered in vegetable systems.

Water blanching

Continuous water scalding may be employed, the vegetable being passed through a machine akin to a rotary pea blancher, in which the temperature of the (minimal) scald liquor is held at about 95°C. The residence time is usually about 2–3 min, the exact time varying with the vegetable and the variety.

If the vegetable is to be dehydrated, care must be taken to avoid overscalding, which leads to mushiness and difficulty in subsequent drying. For dehydration of certain vegetables, it is usual to add sodium sulphite or metabisulphite during scalding in order to improve colour, storage life and vitamin C retention.

In the course of scalding, leaching of solutes from the cells of the vegetable tissue occurs. Within certain limits, the concentration of solutes in the scald liquor should be allowed to build up, as this will minimize leaching losses. The liquor is serially replenished. Recommended solute concentrations are 3% for carrot, 1% for cabbage and 1% for potato. At higher soluble-solids contents, the quality of the product will be adversely affected and the storage life reduced, due to the accumulation of the products of non-enzymatic browning reactions in the scald liquor. This is par-

ticularly pronounced if the liquor, with its content of dissolved solids, remains at high temperature for some hours without replacement.

Since the cell contents of plant tissues are acidic, the pH of the scald liquor falls during blanching. When green vegetables, e.g. cabbage, are being scalded, it is necessary to add sodium carbonate solution to maintain the pH of the liquor at 7.3–7.8, in order to avoid the degradation of chlorophyll that takes place when green tissues are heated in the presence of acid. Phosphate may be added in certain cases where discoloration by iron–tannin reactions appears, e.g. in certain stocks of carrot.

Steam blanching

Blanching in steam has certain advantages and is often preferred to water scalding. The main benefit is the smaller leaching loss, but for some vegetables the product gains an enhanced storage life and there is no need for handling between blanching and the subsequent processes. For delicate products such as cauliflower florets, steam blanching using a belt and tunnel system is preferred, since the product can be handled more gently.

Microwave blanching

This process uses heat generated within the food by subjecting it to high frequency radiation. The advantage claimed is that there is no leaching of solutes. Hence the losses of flavour and nutritive value are kept to a minimum. Against this has to be set the high initial cost of the equipment.

A variety of blanching equipment is available, as described below.

Rotary water blancher

This device incorporates a perforated rotating cylinder with an internal helical scroll. It is enclosed within a cylindrical shell, the lower half of which contains steam coils to heat water to a thermostatically controlled temperature. Alternatively, steam injection can be employed. For dehydration, the shell is continuously dosed with sulphite, carbonate or phosphate. These solutions are fed at controlled rates from tanks by means of variable delivery pumps or other devices. All contact surfaces should be of stainless steel. The rotating cylinder should preferably be supported on external bearings to avoid introducing metallic contamination into the scald liquor.

Steam blancher

Most designs are based on a conveyor taking the vegetable, spread in a thin layer, through a tunnel at atmospheric pressure into which steam is injected (Figure 4.1). The heating zone may be followed by a deep-layer holding zone and a cooling zone. If the material is subsequently to be dried on trays, there are advantages in employing a system in which the conveyor takes the trays upon which the fresh produce is already spread. Unscalded vegetables are cool and crisp, and much more easily spread into even layers on the tray than are hot, limp, scalded vegetable pieces. Furthermore, spreading before scalding avoids handling the material after the partial sterilization effected by the blanching process. Despite any possible leaching, weight is sometimes even gained.

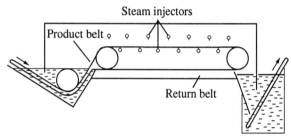

Figure 4.1 Water-sealed steam blancher. Reproduced from Arthey and Dennis (1991).

Another form of steam blancher consists of a helical screw conveying heaped vegetable pieces, with some friction and tumbling, along a covered trough into which steam is injected through perforations in the walls and the hollow screw shaft. A modified version of this allows scalding in water. Here, the relatively small volume of scald liquor permits it to be replaced more frequently than in the earlier rotary water blancher described above. This again leads to an improvement in the storage life of the product.

Sulphite or other solutions which may be required for dehydration can be applied by sprays to the material emerging from the tunnel.

Water spray blancher

In this device, the hottest water is sprayed over the product in the centre of the blancher, collected and resprayed in sequential steps towards both ends of the blancher. Thus the material is warmed up towards the centre and then progressively cooled.

Tubular blancher

Particulate material is conveyed through a long tube for the specified time by pumped water, which is heated by steam injection.

After scalding or blanching some cooling may occur naturally by evaporation, but a separate product cooling system is almost always necessary. It is also very important that the product is thoroughly washed in potable water to reduce microbial contamination and to remove concentrated blancher liquids. Direct water spray reels or countercurrent coolers are in general use.

Modern blanchers incorporate water cooling as an integral part of the system, so that heat can be recovered from the precooling sections and indirectly used to heat the blancher water (Figure 4.2). Advantage is also taken of the heat-hold process in which the product is heated and then adiabatically held at a particular temperature prior to cooling. The general principle is referred to as individual quick blanch (IQB) (Figure 4.3) and involves short exposure of a thin layer of product and then equilibration in deep bed conditions. Evaporative cooling has also been incorporated in more recent designs.

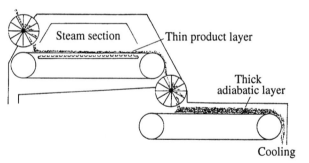

Figure 4.2 Layout of blancher using heat regeneration. Reproduced from Arthey and Dennis (1991).

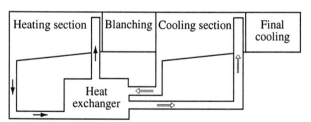

Figure 4.3 IQB blancher. Reproduced from Arthey and Dennis (1991).

Some typical blanching times (minutes) in water maintained at 95°C are: artichokes, 5–9 depending on size; asparagus, 2–5; green beans, 2–3; broccoli, 2–3; Brussels sprouts, 4–5; carrots, 2–5, depending on style; cauliflower florets, 3–4; celery, 2; corn-on-the-cob, 6–11; peas, 1–2; spinach, 2.

PRESERVATION BY CANNING

Fruit

Apples

The best English apple for canning is the most extensively grown cooking variety, the Bramleys Seedling; another variety that has been used is Newton Wonder. Golden Delicious will be used more in the future, chemically adjusted to resemble the Bramleys Seedling. Cox's Orange Pippin has also been used for sliced or diced apple in syrup.

The apple used must be of good size, shape and quality and should be graded so that after peeling and slicing the segments are approximately the same size. Windfalls or other damaged fruit should not be accepted, as the removal of the bruised portions will result in badly shaped sections in the finished pack. The quality on arrival at the factory should be such that washing and sorting are unnecessary, and the fruit is conveyed straight to peeling machines which, in one operation, remove the peel and cores.

The peeled apples should drop straight into a tank containing dilute brine (about 1%), as it is essential that the apples are not exposed to the atmosphere after peeling. This applies throughout the whole of the canning operation to prevent brown discoloration by oxidation or enzyme activity. The peeled apples are then passed through a machine which removes the seed cells. After this, some hand trimming will be necessary to remove any blemishes and the apples are cut into sections, the number depending on the size of the apples; sixes and eights are commonly employed. This work can be carried out automatically or by hand. It is desirable that the sectioned apples should be held in dilute brine for several hours – overnight is common. Brining improves the texture, colour and keeping properties of the canned apples and, by removing air occluded in the fruit tissue, renders them less liable to cause internal rusting or other corrosion of the cans.

Prior to use, the segments are turned out onto a conveyor belt where they are well sprayed with fresh water and may be given a final inspection to remove any remaining discoloured patches. The next operation is blanching, either by passing the sectioned apples through a steam tunnel or by immersion in hot water for a sufficient length of time to render the segments pliable. This will usually mean anything from 3–10 min at temperatures of 77–93°C (170–200°F).

The majority of apples are canned in solid pack form, frequently in the A10 size for the catering trade. Filling should be carried out immediately after blanching since it is essential that the product is not allowed to cool appreciably. The filling tables are usually constructed so that the empty cans are supported under a surface in which holes with diameter slightly less than the cans have been cut. The apple sections are scooped into the cans and pressed in firmly by hand operated plungers. Not less than the minimum weight per can laid down by the UK Code of Practice for Canned Fruit and Vegetables (1986) should be filled. The cans are topped up with boiling water and, because the pack is susceptible to internal rusting of the headspace, this should be kept to a minimum – 8 to 10 mm is recommended and 13 mm should never be exceeded. The filled cans should then be clinched, since during exhausting the apples tend to rise above the top of the can. Provided that the cans are filled hot, an exhaust of 8–10 min at 88°C (190°F) should be sufficient to ensure that the centre temperature reaches not less than 60°C (140°F). The cans are then seamed. This is followed by processing as quickly as possible for a time sufficient to raise the temperature of the centre of the contents to 68–71°C (155–160°F). This will usually mean a period in boiling water or steam of 16–30 min, according to can size. The cans should be well cooled in cold water.

The following points should be noted.

(i) Although it is essential to store apple segments under brine during handling, an alternative method of air removal can be employed in which a high vacuum is applied to the product within an appropriate vessel, for example a cannery retort. The vacuum is subsequently relieved with either steam or water and this is followed by a normal blanching operation. In commercial practice, the cut apples may be filled into shallow trays, which are run on a rail into a large box-like vacuumizing and steaming chest in which the whole operation can be carried out and at the end of which the hot apple segments may be filled straight into the cans. The advantages of this method claimed for solid pack apples are a better quality product and in particular a higher drained weight.

(ii) A method of holding prepared apples or apple segments for some months at refrigerated storage temperature prior to canning involves a short dip in 0.25% sulphurous acid followed by soaking in 0.20% potassium phosphate. This raises the pH to around 8, at which level it is reported that browning does not take place. To avoid subsequent microbiological spoilage, the storage temperature must not exceed 1°C (34°F) and reacidification is necessary immediately prior to canning. Alternatively the apples may be treated with malic acid, adjusting the pH to 3, when prolonged storage under refrigerated conditions is achieved with the addition of sorbates.

(iii) The can specification for all apple products is a plain body with lacquered ends, except for small cans, e.g. 50 mm diameter cans where a fully lacquered can is used.

(iv) It is very important that apples do not come into contact with iron or copper at any time because discoloration will result. All preparation and filling machinery, knives, etc. should be of stainless steel construction or alternatively for some of the equipment, monel metal or aluminium.

Bilberries

These are not canned to any great extent except in certain districts, where they are in demand by confectioners and are then usually packed in water in A10 cans. The fruit grows wild on moorlands and is very tedious to pick; a 'bilberry rake' which is a scoop shaped instrument provided with tines or forks set closely together may be used.

The canned pack is very corrosive and excellent canning practice is essential to obtain reasonable shelf life. To preserve the colour of the product, packing in fully lacquered cans is recommended.

Blackberries

Blackberries are cultivated for canning; suitable varieties are Bedford Giant and Himalayan Giant. They should be canned when fully ripe as the seeds are less noticeable at this stage and the flavour is at its best. The fruit should be handled in shallow baskets or trays to prevent crushing and should be canned as quickly as possible after picking, using standard fruit practice and lacquered fruit cans.

Blackcurrants

Blackcurrants have a very high vitamin C content and the canned product generally contains about 100 mg of vitamin C per 100 g of canned material. Thus, in theory, an A2 can would provide the full daily requirement for an average family. The best variety of blackcurrants for canning is Baldwin. The ripe fruit should first be prepared by

removing the stalks, an operation known as 'strigging', which on a commercial scale is carried out by a machine consisting of an inclined reciprocating screen through the interstices of which mechanical fingers rotate at high speed. The currants are fed on to the higher end of the screen and are freed from the stalks by the gentle trapping of the rotary fingers. The stalks are then drawn through the screen by the rotating fingers into a receptacle below, whilst the fruit rolls down the inclined screen into another vessel at the bottom of the machine. This operation may be carried out more easily if the fruit is frozen. Before being filled into cans, the fruit should be carefully picked over to remove any small strigs not detached by the machine. The fruit fill should be according to the UK Code of Practice (BFVCA, 1986) and completed with sugar syrup. Can closing is followed by processing at 100°C (212°F) for 9–30 min according to can size and whether a stationery cooker or a rotary cooker is employed. Fruit lacquered cans must be used.

Blackcurrant pulp intended for storage for subsequent juice manufacture is prepared differently. The whole fruit including stalks is reduced to pulp by heating with a minimum of water and the pulp is filled as hot as possible into A10 cans, which are rapidly sealed and processed at 100°C (212°F) for about 15 min and quickly cooled. Every precaution must be taken to prevent oxidation of the vitamin C and no metal equipment other than stainless steel should be used.

Cherries

There are three types of preserved cherries. The first two are the sweet fruit which is usually the white type and the red sour fruit that is most often pitted. Both of these may be canned. The third is the characteristically flavoured Maraschino which derives its name from a Dalmatian sweet-sour liqueur originally prepared from Amarasco bitter cherries. This pack is usually bottled with sulphur dioxide preservative being added to extend the shelf life after opening, since these cherries are often used for cocktail drinks.

The sweet cherry varieties most suitable for canning are Napoleon Bigarreau, Kentish Bigarreau and Elton Bigarreau, all of these being 'white' varieties. None of the 'black'-coloured sweet cherries has been found to can entirely satisfactorily although Hertfordshire Black (or Caroon) is sometimes used. The large Morello is the most suitable variety of sour cherry, other fairly good ones being May Duke, Kentish Red and Flemish Red.

The sweet 'white' cherries usually require added dye to colour them red. Since they are packed in cans with internally plain bodies, the only satisfactory colouring material is erythrosine, which is relatively stable to dissolved tin. The dye may be applied as a simple addition to the syrup but, if the cherries are to be subsequently used for addition to fruit salad packs, it is necessary to control the amount of dye added very carefully to prevent subsequent 'bleeding' of colour which would affect the other constituents of the fruit salad. Alternatively, the cherries may be blanched in dye solution to which citric acid is added halfway through in order to 'fix' the dyestuff in the cherries, since it precipitates in acid solution. In other respects, standard fruit canning practice is applied with fill and syrup strengths according to the grade required and in compliance with the UK Code of Practice (BFVCA, 1986).

All the 'black' cherries, that is those with dark red or purple-coloured flesh, are packed in lacquered cans and when they are canned whole, that is non-pitted, it is essential to ensure that enzymes in the stones are inactivated, otherwise excessive corrosion with rapid formation of hydrogen swells may result. A thorough exhaust is recommended, preferably coupled with steam flow closing, and an adequate heat process, which should be carefully established. Again because of their corrosive effect, it is essential that canned cherries are stored under cool conditions. If Maraschino cherries are to be canned, it is essential that any sulphur dioxide employed as preservative in the storage containers should first be leached out in order to prevent subsequent severe corrosion.

Gooseberries

The most important green variety for canning is Careless. Berries should be picked when they have reached full size but before they have ripened sufficiently to become soft or change in colour. The stalks and blossom ends are removed in a carborundum peeler or snibber in which a small jet of water is emitted from the side. This prevents deep cutting and also washes the fruit. The revolving disc brings the berries in contact with the side walls, and the stems and tails are thus rubbed off. The outer skin of the berries is also slightly punctured during the operation which is an advantage since it allows the sugar syrup subsequently to penetrate the fruit and prevent shrinkage or wrinkling of the berries.

Fill of syrup and berries should be according to the UK Code of Practice (BFVCA, 1986), and standard fruit canning procedure is used. Processing will be from 5–30 min according to can size, type of exhaust and whether in stationary or automatic agitating cookers. Cans with plain bodies are recommended. Occasionally canned gooseberries are artificially coloured, in which case a can lacquered throughout is required.

Pears

The Bartlett (Williams bon Chretien) variety of pears is the most popular and best suited for canning. It is of uniform size and shape with a fine texture, bright colour and good flavour. The Conference pear, sometimes imported, is also canned and it has been found that this variety may be sub-acid so that, if canned untreated, the final pH may be above 4.5, up to 4.9. Some acidification is therefore required and if packing of the Conference variety is contemplated, it is suggested that the Campden & Chorleywood Food Research Association (at Chipping Campden, GL55 6LD, UK) first be consulted.

Pears for canning should be allowed to grow to full size on the trees but should be picked while still green and hard. They should then be allowed to ripen in lug boxes or baskets in a well ventilated store with daily inspection to determine suitability for canning. Sometimes storage is at cool temperatures in order to slow the ripening process and thereby increase the length of the pear canning season. A special pressure tester may be used to decide when the fruit is ready for canning.

An approximate mechanical grading for size precedes peeling, coring and halving which is usually done by hand using guarded curved knives. After trimming, the halves must be immediately immersed in weak (1%) brine to prevent browning up to the time when they are finally hand graded and filled into cans with fill weights and syrup strengths according to the appropriate UK Code of Practice. Processing will vary between 12 and 35 min at 100°C (212°F) according to can size and whether a stationary or agitating cook is used. Sometimes pears need longer processes to soften than to sterilize them. Cooling should be thorough and complete otherwise the edges of the pieces may 'fray' and stacking away hot will cause some pears to turn a pink colour which, as tin dissolves during storage, may even go to purple. Plain cans with lacquered ends are normally used for this product.

Plums

The principal canning varieties are 'Yellow Egg' Pershore, Victoria and Damsons (Shropshire

Prune is the only good variety, Common Damson is moderate). High grade fruit properly prepared and processed gives a canned product of excellent qualities which is quite popular. The raw material must be good sound fruit of optimum maturity. On receipt at the cannery, the plums are washed and the stems removed, normally by machine. This is followed by size grading, inspection and then filling directly into cans, according to the UK Code of Practice. Syrup is added. Exhausting should be thorough because all plums have a considerable corrosive effect and heat processing should be sufficient to inactivate enzymes in the stones which will otherwise accelerate corrosion. Cans with plain bodies are used for the yellow or green varieties, but fully fruit lacquered cans are necessary for the red or purple types. Heat processes will vary from 10–30 min at 100°C (212°F) according to method of exhausting, can sizes and type of cooker.

Some varieties of plums, and in particular Victoria, show a tendency to form pockets of gum close to the stone, the origins of which may be physiological changes during growth and enzyme activity associated with the stone. There may also be surface spots which do not extend into the flesh but are merely protective seals produced by the plums to cover insect punctures on the skin. Practically the only thing known with certainty about gum formation in plums is that its extent increases with the amount of rainfall during the period during which the plums are maturing on the trees. Extensive research into the prime cause of the defect have not so far shown any practical means of eliminating it. The spots of gum swell during canning so as to form relatively large lumps of jelly which, in extreme cases, may cause the plums to burst and spoil the appearance of the canned product. The only way to avoid this is in selection of raw material, and trial cannings may be advisable. Problems should be referred to the Campden & Chorleywood Food Research Association.

Prunes

These makes an excellent canned product and are second only to fruit salad as an out-of-season canned fruit, being prepared from the dried prune imported usually from the USA. Prunes selected for canning are usually 80 to 90 grade (representing the number of fruit per pound or per 454 g). They are picked over for broken or defective fruit, then washed and blanched. Alternatively, they may be soaked overnight in cold water, but much of the natural quality may be lost

in the water that is discarded. The prunes are filled, with allowance being made for the syrup that will subsequently be taken up during processing and storage. The UT (16 oz/454 ml) can will need about 20 prunes according to the actual grade. The fill is made up with a relatively low Brix syrup and a thorough exhaust is necessary to minimize subsequent corrosion. The heat process will vary according to the previous treatment and quality of the prunes. The texture of the product will invariably require a process longer than that needed to prevent fermentation. The process should, however, be as slow as possible, again to reduce corrosion and pressure processing is frequently employed.

The choice of container specification will depend upon the flavour preferred, the processing temperature and the shelf life required, and the container supplier should be consulted. This pack is extremely corrosive and at various times problems have arisen both with excessive tin pick-up and hydrogen swells. Close attention should be paid to canning practice and carrying large stocks of packed cans should be avoided. The current trend is to use fully lacquered containers, usually with a white acrylic enamel.

Raspberries

The raspberry is one of the finest flavoured English fruits and quite popular as a canned pack. The berries should be packed when they are ripe but firm and should never be used for canning purposes when they have become soft. Rather, such fruit should be used for jam making or the production of fruit syrup. Malling Jewel and Glen Clova are in wide use. Loganberries, which originated as a cross between the wild blackberry and the raspberry, are also moderately successful and are handled in a similar manner.

The best method of conveying the fruit to the factory is in the shallow punnets or chips used for marketing the berries, contained in larger crates or boxes. Other types of shallow container may be used, but the former method has been found very satisfactory in preventing bruising or crushing and in subsequent ease of handling.The fruit should be canned promptly on arrival because it tends to mould quickly. When the berries arrive in punnets, a worker can take a punnet in one hand, shake a few berries into the other hand, pick out the defectives and roll the good berries direct from hand into the can. By this method, the good berries are never picked up between the thumb and finger, which tends to break the seeds apart so that they subsequently separate during processing and spoil the appearance. Some canners prefer to wash the berries, but where they are grown in a clean area and on canes tall enough to prevent sand splash, this should be unnecessary.

Standard fruit canning practice is employed. A good vacuum is necessary because this product is corrosive. The heat process will vary between 5 and 12 min at 100°C (212°F) according to can size and type of cooker used. Fully fruit lacquered cans are specified.

Rhubarb

This leaf stalk is classified as a fruit for food purposes because it is canned in syrup and, as such, it used to be one of the largest UK canned fruit packs and the first of the season. One of the best varieties for canning is Champagne, but many others are suitable, e.g. Sutton, Timperley, Victoria. In pulling rhubarb, the leaves are stripped but the bottom end is not cut, but broken from the plant to avoid drying out and splitting of the stalk. The first operation on receipt at the cannery is to cut off the top and base; this is followed by thorough washing. The fruit is then inspected and trimmed and cut into lengths of about 1 inch, which may be done by machine. Soaking in water after cutting may reduce the corrosivity of the product but overlong soaking causes curling and splitting. After filling a thorough exhaust is essential to minimize subsequent corrosion. Rhubarb is sometimes canned as a solid pack for catering purposes.

Because of the unusual combination of acids, including abnormal amounts of oxalic acid that rhubarb contains, it exerts an extraordinary detinning action. Fully lacquered cans are essential, but, even so under-film detinning can cause lacquer stripping with a consequent unsightly appearance on storage. Spray lacquered cans have been shown to give some improvement but were rejected by the industry on the grounds that they made the operation uneconomical. Thus the shelf life of rhubarb, calculated as the time until the cans appear unsightly, may not be much greater than 6 months, although some packs have survived for more than 18 months. Tin-free steel ends for cans used for packing rhubarb give superior performance over tinplate ends with respect to corrosion resistance. Tin-free steel can bodies are likely to replace tinplate ones with the advent of welding. It is hoped that superior corrosion resistance will then give an increased shelf life.

Strawberries

The delicate colour and flavour of most strawberries deteriorate appreciably as the result of heat processing in canning and the research associations are continually experimenting to find more suitable varieties.

The strawberry requires careful and rapid handling to produce good results and it is desirable that the fruit be delivered in shallow chips or punnets. It may be possible to avoid washing the fruit if it arrives in first class condition but washing may be essential to remove adherent soil and sand, particularly after rainy weather and if there is any possibility of agricultural residues which could affect the flavour. Any residual calyces should be removed and the fruit graded for size, after which standard fruit canning procedure is used. Processing will be 6–10 min at 100°C (212°F) according to can size and type of cooker. Rapid processing in rotary cookers, followed by quick and thorough cooling, is an advantage. Fully fruit lacquered cans should be used.

Vegetables

Beans

Broad beans. The common, purple-flowered varieties are not suitable for canning because the heat process turns them brown and the surrounding liquid rapidly takes on a muddy appearance. Some white-seeded varieties have been specially developed for canning and the most widely grown that gives a good result is Medes.

On a small scale, podding is done by hand, but pea-viners may be modified to deal with large quantities of broad beans in the field. They are washed and screened to remove debris, then blanched in hot water for about 3–4 min, followed by further washing and inspection to remove diseased or blemished beans or skins. They are filled into lacquered cans, covered with a simple 2% brine, steam-flow closed, processed at 115°C (240°F) for 25–65 min according to can size and rapidly cooled.

Butter beans. This is a fairly popular out-of-season pack using dried beans, at one time Madagascan but now American 'white' lima beans. Supplies should be carefully selected, the beans being not more than a year old, for old beans do not readily absorb the covering liquor and consequently tend to give a hard pack. During long storage, beans may also develop defects which become evident on canning, such as a tendency to shed their skins and a defect known as 'pink eye' which refers to a pink spot that develops after blanching. A trial canning is recommended.

The beans should be of a reasonably uniform size and a pleasant light colour. They are sorted to remove those with visual defects and then soaked overnight for not less than 16 h, although if the temperature of the water is too low a longer soak may be necessary. The beans should approximately double in weight after soaking. They are blanched for 5–15 min at 88–100°C (190–212°F), the exact time varying with the nature of the bean, washed and again inspected for visual defects including splitting during blanching. After filling to the prescribed weight (UK Code of Practice; BFVCA, 1986), they are covered with a brine containing about 2% salt and 2–3% sugar. The process depends to some extent on the quality of the beans but will be around 40–60 min at 115.5°C (240°F) according to can size. Cans with plain bodies are sometimes used, although this pack should normally be regarded as having a limited shelf life due to product colour deterioration (greying of beans and liquor). The use of cans with a sulphur-absorbent lacquer throughout gives the best result and is essential where artificial colour is added, although in any event the colour will probably start to deteriorate after about 6 months. It is particularly important with this product that neither the beans nor the brine should be allowed to come into contact with iron or especially copper during preparation, since this will accelerate colour deterioration.

The UK Code of Practice (BFVCA, 1986), which states the minimum dry solids content of canned butter beans and how it shall be determined, should be consulted before canning this product.

Green beans. The dwarf and runner beans commonly grown for the fresh market are not satisfactory for canning owing to the tough nature of the fibres which join the two halves of the pod together. Therefore, varieties have been specially developed for canning. Information on these varieties is available from the Campden & Chorleywood Food Research Association.

All green beans are mechanically harvested and should be canned as soon as possible. If they are dry, they may be held in the unsnipped condition in a cool place for about 2 days. They may be canned whole, sliced or cut according to size and requirements. After the preliminary washing, they are snipped, graded if necessary and then sliced as required. Automatic machinery is available for these operations. They are then blanched for 2–3 min at 82–88°C (180–190°F) and apart from the

usual advantages, this imparts sufficient pliability to the beans to obtain the correct fill weight. The beans are then washed in cold water and thoroughly drained. This may be accomplished by using a reel-type washer immediately following the blancher. They are covered with hot brine usually containing 2% salt. The cans are steam-flow closed and processed for 17–45 min at 115.5°C (240°F), or equivalent, according to can size. The cans should preferably be coated throughout with sulphur-absorbent lacquer in order to preserve the product colour for as long as possible.

Beans in tomato sauce. This is an immensely popular product in Britain. It can be canned all the year round avoiding the times for seasonal fruits and vegetables. Minor variations are beans with pork, beans with bacon, and sausage and beans, all in tomato sauce. The primary raw materials are dried beans, known as pea or small white navy beans, imported mainly from the North American continent (e.g. Michigan and Ontario) and from Chile and canned using tomato purée. The UK Code of Practice No. 5 (BFVCA, 1986) should be referred to before canning beans in tomato sauce.

The beans must first be sorted for size and to remove those that are discoloured. In modern high-speed canneries the discoloured are now removed by machinery using electronic eyes. Further sorting may be carried out by fluming the beans over a series of waterfalls or 'riffles' so that any remaining foreign bodies sink and are drawn off.

The next stage in conventional practice is to soak the beans in cold water overnight for 12–18 h, according to the nature of the beans and hardness of the water, followed by blanching for 10–20 min at 88–90°C (190–210°F). This results in the beans approximately doubling their original dry weight. Substitution of a shorter soak at higher water temperatures leads to microbiological problems and should be avoided, but newer methods combine soaking and blanching in continuous procedures, one of which consists of passing the washed and sorted beans through pipes with hot water into which steam is continuously injected. This results in far less moisture pick-up and the subsequent fill-in weight of beans must be lowered with an increased amount of diluted sauce, but the beans swell further during heat processing of the cans and sauce penetration is assisted.

After blanching the beans are again inspected thoroughly and this may be followed by a short oven bake. They are filled as hot as possible into the cans and the sauce is added immediately afterwards. If beans with pork or bacon are being canned, a small piece of pickled pork fat or smoked bacon is added to each can. Approximately equal amounts of beans and sauce will be filled, the exact amount to be determined by experiment. Beans and sauce are added as hot as possible to give a temperature around 77°C (170°F). This is necessary to obtain a reasonably high final vacuum in the can since this product is corrosive. In order to obtain a good quality product, the tomato purée used should be carefully selected to have a good bright colour and be free from any harshness or bitterness of flavour. Apart from tomato purée, the sauce will probably contain sugar, salt, flavour and spices.

Processing may be carried out at 115.5°C (240°F) to 127°C (260°F) according to can size and type of cooker used. The lower temperature is essential for static processing of large can sizes. Agitation improves the quality of the product as it increases the rate of heat penetration. Hydrostatic cookers are commonly used for UT cans, rotary cookers for A6 and A10 cans.

Normally cans with plain bodies and lacquered ends are used, some plain plate being required to confer sufficient shelf life. However, it should be noted that a product with a high cured bacon content may cause excessive detinning of plain tinplate due to the curing salt content of the bacon and a suitable can recommendation should be sought.

Beetroot

Beets for canning must be of a uniform deep red colour and small ones of about 1–1¼ inch (2.5–3 cm) diameter are preferred although larger ones may be used for a sliced or diced pack. An American variety, Detroit Dark Red, cans successfully. Internal black spotting in beetroot is due to boron deficiency in the soil. The seeds should be sown late in the season to avoid the plant running to seed and so that the crop will be ready when the rush of the summer canning season is over.

After lifting, the beet tops are screwed off by hand or by mechanical topper and the roots transported to the cannery in sacks or boxes. Clamped beets may be used later in the season but the final product is not of such high quality. On arrival at the cannery, the first step is a thorough wash to remove adhering soil. It may be necessary, for example with beets grown in clay soil, to soak in tanks prior to spray washing. Care should be taken that the skin of the beetroot is not broken or damaged during this operation. The beets should be graded into sizes before blanching,

which consists of steaming in a retort at about 104°C (220°F) for 10–20 min according to size, followed by peeling by hand or mechanical abrasion peeler. They should be canned as quickly as possible after blanching in order to avoid discoloration, particularly if the product is sliced. After filling the beets, boiling 2% brine which may contain a little sugar is added to complete the fill; it is important that the vegetable is entirely covered. If the brine is filled at a sufficiently high temperature, exhausting may be unnecessary. The cans are closed and processed for 30–45 min at 115.5°C (240°F), according to can size and rapidly cooled.

Because of their proneness to discoloration it is important that beets should avoid contact with iron during preparation: stainless steel equipment is recommended.

Sliced beetroot in vinegar is a popular product, which is similarly prepared but with a covering liquor of vinegar of about 4% acetic acid, spiced or not, according to requirements. Because the beet is not fermented before packing and contains a fairly high proportion of soluble sugars, the finished pack is susceptible to spoilage by yeasts and moulds and a pasteurizing process must be given. This usually consists of raising the temperature to 82°C (180°F), maintaining this for 20 min and then rapidly cooling. The final acetic acid content should be 1.8–2%, and the pH must not be greater than 4. For canning, 2% acidity must not be exceeded because of the corrosive effect on the container. This product is often bottled in glass jars with hermetic metal closures.

Cans for beetroot are specially lacquered and therefore the product should be clearly stated when ordering: this is particularly important for beetroot in vinegar.

Carrots

Carrots may be packed whole, sliced or diced. Red-cored varieties of the Chantenay type, which have conical roots, are used almost exclusively for the whole pack in the UK. Amsterdam types are used elsewhere. The roots are lifted when they have a maximum diameter of about 2.5 cm. Berlicum types are used for slicing and dicing.

If carrots are not topped mechanically at the canning factory, they are topped by hand at special on-farm locations before delivery. At the factory an efficient spray wash is usually sufficient to clean the roots, but if much dirt adheres they should be soaked beforehand. Various methods are available for peeling, the standard carborundum vegetable peeler being quite successful. Lye peeling is also used, but the modern tendency is to use steam pressure peelers operating at around 80 psi (5.5 bar). This is followed by mechanical dicing or slicing for these types of pack. Blanching is unnecessary with steam peeling but otherwise it should be carried out either before peeling, before dicing or slicing or immediately prior to filling, according to the condition of the raw material. The small whole pack is normally tumble filled; sliced or diced carrots are also tumble filled. Hot brine is then added, followed by steaming and processing for 15–60 min at 115.5–127°C (240–260°F) according to can size and type of product. The larger can sizes will require pressure cooling. Cans with plain bodies and a lacquered end are recommended.

Celery

Celery must be selected as firm and free from disease. 'Rust' marks and any form of internal breakdown of the hearts should be avoided. Imperfect bleaching, that is the presence of green patches on the outer stalks near the base, will give rise to an unsatisfactory colour in the canned product. However, green celery is sometimes canned because of its superior flavour. It is packed in the form of trimmed hearts cut to fit the length of can used, the outer stalks being canned separately, sometimes in the form of chopped pieces for use in soups. The bunches are broken apart or cut across just below the crown and then thoroughly washed under high pressure water sprays to remove sand and soil. They are then cut to within 0.4 inch (1 cm) of the height of the can. This is followed by blanching in hot water containing 0.15% citric acid for 3 or 4 min. The acid blanch assists in maintaining the crispness and improves the whiteness of the pack. The hearts are packed carefully into cans with plain bodies and lacquered ends and covered with boiling 2% brine for processing at about 115.5°C (240°F) for a time which is primarily dependent on the diameter of the heart. Process recommendations are made on this basis, with times being between 25 and 46 min at 115.5°C (240°F), followed by rapid cooling.

Clean waste may be cooked in a steam kettle for a short period, then run through a pulper and canned in A10 cans to be used later in the preparation of soup.

Mushrooms

Mushrooms are canned in brine, in butter or cream sauces, as constituents of a great variety

of products such as savoury risottos, curried meats, meat pies and puddings, and, of course, in mushroom soup. The cultivated mushrooms are gathered daily and for canning on their own, usually as button mushrooms, they are picked just before the veil breaks at the stem. They are conveyed to the cannery in punnets or chip baskets. They may be graded by hand as they are picked. Alternatively, this operation may be carried out mechanically after trimming the stems. Next they are thoroughly spray washed. This is followed by a short soaking in running water to remove any embedded contamination, and then blanching in boiling water for up to 3 min. The blanched mushrooms are subjected to inspection prior to filling, topping up with brine and steam flow closing and processing for about 30–45 min at 115.5°C (240°F), according to can size. Cans with plain bodies and lacquered ends are used for mushrooms but other finishes may be necessary where mushrooms are being canned as only one component of a product.

Damaged and oversized mushrooms and those which are too open for canning are removed during inspection operations and may be used for other purposes such as inclusion in combination savoury products or to prepare canned or dehydrated soup mixes.

Parsnips

This is not a popular product, which is surprising because it makes an excellent canned vegetable. Parsnips for canning should be pulled before they are fully mature to avoid being tough and stringy. The root is seldom an even shape and usually looks best if it is sliced before canning. Most of the common market garden types are suitable for canning. The roots should be thoroughly washed and a short blanch in boiling water containing about 1% of citric acid improves the brightness and colour, but further washing is required after this treatment. Peeling is carried out by scraping or by using a steam pressure peeler. This is followed by slicing lengthwise, usually into four quarters, and preferably cut to the length of the can. The pieces are filled vertically to weights according to the UK Code of Practice (BFVCA, 1986) and made up with 2% brine. The cans are then steam-flow closed and heat processed for about 30 min at 115.5°C (240°F), the time varying with can size and finally water cooled. Plain cans with lacquered ends are recommended.

Peas

Fresh or garden peas. The processing of fresh peas is a highly specialized industry, probably involving more capital outlay than any other canned or frozen vegetable pack for large scale production. Crops are grown specially. Peas mature very rapidly as they reach the optimum condition and sophisticated machinery is necessary to obtain high production rates during the very short season of only a few weeks in the height of summer. Some of this machinery of necessity remains idle for the rest of the year. Because the recommended varieties are constantly changing it would be pointless to name any here and the Campden & Chorleywood Food Research Association should be consulted for such information.

The stage of maturity of the peas is determined frequently during their ripening by examining sample lots on a machine called a 'Tenderometer', which registers the force necessary to shear a sample of the peas. When a crop reaches the Tenderometer grading that the canner or freezer has calculated as optimum, the crop is harvested with a pea harvester which first combs the pods from the plants then shells the pods to discharge the peas. The complete vining operation is carried out in the field using mobile machines. The peas are chilled and transported to the processing plant where they should arrive within only a few hours of being mown.

On arrival at the plant, the peas are passed through a winnower, which removes light debris by air blasting, then washed in a continuous flotation washer and riffle. This is followed by grading, blanching, and a second wash. They then proceed to a filler, which automatically measures the correct quantity of peas and volume of covering liquor. All of these operations are continuous and automatic. The brine contains salt and sugar in roughly equal proportions and artificial green colouring matter to prevent the peas turning yellow after canning. Although an increasing quantity of canned peas without added colour are now being marketed, such products are not normally acceptable. Because of the speed of the whole operation and the consequent high filling temperature, exhausting is usually by steam flow closure, after which cans are heat processed at 115.5°C (240°F) or above for times which will vary not only according to can size but also the type of cooker employed. Although some stationary processing is still carried out, many canners now use continuous cookers, often of the hydrostatic type. Cans should be cooled efficiently and dried as quickly as possible since this product is prone to leaker spoilage if wet cans are allowed to cause a build-up of infection on runways. Fully lacquered vegetable cans should be used.

The UK Code of Practice states that no size description is necessary for ungraded peas, although they may be described as 'as from the pod'. Where the words 'small' (or petit pois), 'medium' or 'large' (or standard) are used on the label for canned fresh garden peas, the sizes of the peas are defined and the UK Code of Practice (BFVCA, 1986) should be conformed with.

Processed peas. This is the name given to canned reconstituted dried peas. Popular varieties are the Lincolnshire or Dutch Blue, the Marrowfat and the Alaska. Peas intended for canning should be free from any which are dark in colour or turn black during processing. The best method of determining the suitability for canning of a parcel of peas is by a trial canning. A mixture of peas, particularly if old and new season peas are included, will lead to variation of texture within the pack. The trial canning will also indicate water uptake during soaking, which is vital to the economics of the pack and give a lead to the process requirement for optimum final texture.

The first operation is sorting of the dried peas which in large scale operations may be carried out using machines with electronic eyes to separate defective and discoloured peas. They are then soaked for a period of 12–16 h, usually overnight, in cold water to which sodium bicarbonate may be added if the peas are unduly resistant to softening or the degree of hardness of the water is high. Alternatively, if oversoftening takes place, the addition of calcium chloride may be beneficial, although this is apt to impart a bitter flavour. Blanching is carried out in a standard rotary pea blancher, with times and temperature that must be determined by experiment; these will be about 5 min at 99°C (210°F). This is followed by washing in warm water and immediate mechanical filling of peas and brine, which will contain salt, sugar, colouring material and sometimes mint flavour. The cans, which should be vegetable-lacquered throughout, are steam flow closed and heat processed at 115.5°C (240°F) or above for a time that is dependent upon can size, type of cooker used and sometimes upon the final product texture.

The UK Code of Practice (BFVCA, 1986) stipulates the minimum dry solids content of canned processed peas and how it shall be determined.

Potatoes

Variety. Varieties that have been canned successfully include Royal Kidney, King Edward, Home

Guard and Duke of York. Maris Peer is the most important variety and Carlingford has possible potential for the future. Different varieties are usually used for crisps and chips; only one or two varieties (such as Maris Piper) are sometimes used for both. The 'Pentland' series are frequently used for chips and Record is a principal maincrop used for crisps, but preferences vary and change with time.

However, many factors other than variety affect performance. Extraneous factors such as the location of the factory and the type of soil nearby may well influence or even dictate varietal choice. Preliminary canning trials should therefore always be carried out.

Early planting. To ensure early maturity, early planting is necessary as late harvesting brings a chain of problems in its wake, such as easier bruising and damage, high sugar content, etc.

Size. Grading is necessary as only medium to large ware is suitable – over 45 mm for chips and over 40 mm for crisps.

Shape. A regular shape with shallow eyes is necessary for both products; in addition crisps need very clean skinned potatoes.

Flesh. For chips, cream or white flesh with no blemishes is necessary. For crisps, flesh colour is less important and a higher level of blemishes is permissible though undesirable.

Disease and greening. Freedom from rot, cracking, splitting and disease is generally demanded, together with only a minimum of greening, if any.

Solids content. For chips the specific gravity (SG) should be over 1.080 (= minimum 20% solids contents) whereas for crisps a slightly higher SG of 1.085 (= minimum 21% solids content) is required. For canning a specific gravity of 1.08 or less is preferred.

Reducing sugar content. This must be low for frozen or dehydrated products because, if it is high, it leads to excessive colour in the finished product (caramel production) and also to higher production losses by leaching, etc. For chips a maximum of 0.25% is generally accepted and quoted. For crisps, whilst the same maximum level may be acceptable, a figure of below 0.1% is much preferred. These levels have to be maintained during storage.

Storage. Controlled storage of potatoes after harvest is critical if they are to be suitable for crisp or chip manufacture for up to seven months or so. To avoid an increase in sugar content, potatoes for manufacture of chips (French Fries) are usually stored at around 8°C, whereas those for crisps are held at around 10°C. However, tempera-

ture and conditions of storage may frequently need to be modified, e.g. to 'burn off' sugar, and controlled environment storage is necessary to minimize losses and to maximize the product quality. Sprout inhibition (chemical spray or fog usually) is necessary in order to avoid major loss late in the storage period. The maintenance of a high relative humidity, which will at the same time minimize shrinkage yet be inadequate to encourage mould growth, is also desirable. There must never be any risk of freezing. This would immediately damage the crop, which is composed of cells which as long as they are alive have appropriate defence mechanisms against many forms of microbiological attack. Freezing disrupts the living cells and kills them; this is followed by severe losses.

Constant in-store vigilance and skilled control is required to run a potato chip or crisp factory on main crop, home grown stock from late September until April/May, whilst maintaining final product quality. This, however, minimizes the time when the factory has to be run on imported potatoes (if at all) and then briefly on earlies.

Potatoes to be used for dehydration are much less constrained in their quality characteristics in that size, shape, blemishes and flesh colour are not serious problems. This is just as well because the yield of dry potato from fresh is only about 1 : 7 and therefore potato cost is most important (the yield of crisps from potatoes is about 1 : 4, but crisps contain about 35–40% of fat and this accounts for the different ratio.)

Some potato used in snack foods may be in the form of dehydrated mashed potato powder (q.v.). The potato required for this is very different and its raw material requirements much more exacting, particularly in respect to the intrinsic quality of the mashed potato it yields on boiling and the storage conditions necessary for it to retain this quality. These approximate closely to those employed for crisp and chip potato storage.

Canning operations. The operations in canning are size grading, which is important for obtaining a good out-turn appearance, cleaning, peeling, inspection and trimming, filling, closing, processing and coding. Both abrasion and lye peeling have been used but the preferred method is to use a high pressure steam peeling machine operating at about 550 kPa (80 psi), followed by rod washing. The brine contains about 2% salt and may sometimes include a little mint flavour. Steam-flow closing is the usual method of exhaust and processing will be around 40 min at 115.5°C (240°F) according to can size. Plain cans with lacquered ends may be used although sometimes, because of a localized black discoloration effect that potatoes cause when in contact with plain tin-plate, vegetable cans lacquered throughout are preferred.

Because potatoes darken readily through enzyme action or exposure to air, any which must be held after peeling should be immersed under dilute brine solution. Some potatoes disintegrate on processing and attempts to overcome this have been made by using calcium chloride, either in the brine or as a soaking or blanching liquid, but this has been found to impart a bitter flavour. Calcium lactate may be better although more expensive, but it is preferable to select the right quality of raw material.

Spinach

Spinach should be harvested before the seed stalks are well developed, whilst the leaves are young and tender. It requires careful sorting before canning and all roots and discoloured leaves should be removed. Thorough washing is essential to remove adhering soil, grit and insects. Blanching is carried out for some 2 min in hot water until the spinach is tender but not mushy or disintegrated and, at this stage, it should have a bright green colour. After further inspection, it is filled, topped up with brine, closed and processed. Heat penetration through leaf spinach is slow, since there is practically no convection, and processing times are longer than for most other vegetables. Adequate sterilization depends on efficient blanching, correct filling weights of vegetable and brine and a high initial temperature. Since blanched spinach tends to become stratified horizontally in large cans, heat penetration into these may be more rapid when they are processed on their sides.

Spinach purée or pulp may be prepared by passing the blanched spinach through a pulper using a screen of $\frac{1}{8}$ or $\frac{3}{16}$ inch mesh. Processing times for purée will be even longer than for whole leaf.

Fully lacquered cans should be used but it is essential that sulphur-absorbent lacquers are not included since spinach may discolour them.

Factors affecting quality of canned products

Discoloration of fruit

If the raw material is unsuitable for canning because it is under-ripe or of an unsuitable variety, the canned product may well also be of poor colour. Green Victoria plums, for example, will not give a satisfactory product however much artificial colouring matter is added because the natural green substance turns brown when the plums are heated. Similarly, most varieties of pear grown in the UK turn red when heated.

Some fruits and vegetables also contain natural substances which undergo chemical change when heated and the result may well be an objectionable colour. Modern methods of selection of varieties are making increasing use of the identification of substances known to undergo such changes. Apples and pears tend to turn brown when peeled, and these fruits may be of poor colour before they are packed into the cans; this browning is generally prevented by placing the fruit in a dilute brine until ready for canning. Plums and greengages occasionally turn slightly brown on the top of the cans during exhausting, but correct regulation of the exhaust process should prevent this. Certain fruits, such as strawberries and some varieties of raspberries and blackberries, lose their colour badly in the cans during processing; in the case of strawberries artificial colour is added, but several good varieties of raspberries and blackberries do not require added colour.

Another discoloration problem in canned fruits is the bluish purple tint found in raspberries, loganberries, blackcurrants, damsons and other red and purple fruits. This is due to the action of the fruit acids on the tinplate through scratches in the lacquer. Modern fruit cans have two coats of lacquer and a stripe of lacquer over the side seam to minimize this effect.

Enzyme-related changes

Many changes which occur in fruits and vegetables after they have been gathered result from enzyme activity and these often lead to loss of quality as a raw material for preserving. For example, the browning of the cut surfaces of apples, pears and potatoes is a result of enzymic changes, as is also the flavour deterioration of raw podded peas. Rapid loss of vitamin C may occur through enzymic activity in fruit and vegetables which have suffered tissue damage during preparation. The rate of the chemical changes is dependent upon the storage temperature, being much slower in frozen food, although still proceeding at a perceptible rate. Enzymes may be inactivated or destroyed by exposure to heat which constitutes one of the important advantages of blanching. For frozen vegetables in particular, adequate blanching is essential for a good quality final product, and checks for enzyme survival should be included as a control.

Survival or revival of inactivated enzymes may occur in canned fruits or vegetables if 'short-high' processes are used which, while producing microbiological stability, may be inadequate to destroy all the enzymes deep in the tissue. Such survival may lead to undesirable flavour changes on storage and must, therefore, be guarded against. Similarly, fruits containing stones (e.g. cherries and plums) require sufficient heat treatment to inactivate enzymes in the kernel of the stone, since the rate of can corrosion may be affected by the products of changes induced by these enzymes.

PRESERVATION BY DRYING

A very wide range of fruits is dried, either naturally in the sun or in dehydration chambers. Sulphur dioxide is often added, especially in vine fruit, to prevent the fruit becoming too dark on drying. Table 4.1 summarizes the principal varieties of fruit that are dried on an industrial scale.

Table 4.1 Dried fruit

Dried fruit	Drying method	Principal countries of origin
Sultanas	Solar	Greece, Turkey, Australia
Currants	Solar	Greece, USA, Turkey
Raisins	Solar	USA, South Africa, Afghanistan
Figs	Solar	Turkey, Greece, USA
Prunes	Tunnel dryer	USA (California), Chile, Argentina, Romania, former Yugoslav Republics
Apples	Band dryer + SO_2	USA, Italy, Spain, Chile
Peaches/ Pears	Solar	USA, Australia, South Africa
Apricots	Solar + SO_2	Turkey, USA, South Africa

Imports of dried fruits tend to be in compressed bulk packs, which make quality inspection difficult except in the parent country. The subsequent processing centres on separating the consignments into individual fruit pieces for scrutiny, followed by thorough washing to remove extraneous material such as twig fragments. The moisture content may be adjusted in vacuum steam chests to give a soft, plump, texture, followed by drying off to around 17–20% moisture content to obtain free flow.

Berries

Single-sac berries such as cranberries are rarely dried with a view to reconstituting them fully. Rather, they are often perforated for faster drying or precooked in a steam-jacketed screw conveyor to soften the skin. A variety of dryers have been used to bring the moisture content down from 88% (wet basis) to 5% (dry basis), when the product may be screened as flake or ground to powder.

Multi-sac berries, such as strawberries (halved with stob removed) and raspberries (plug and calyx removed) can be dried with good colour and flavour retention, but texture is a problem: best results are by freeze drying and such products have been used successfully not only in breakfast products, but also in yoghurts.

Peas

Like other pulses, peas lend themselves to drying as a way of preservation. Often these are later canned. Dehydrated immature peas are processed by a unique method which involves piercing each testa.

Raisins

The drying of ripe grapes is much facilitated by alkali dipping of the fruit, usually using a hot lye bath. The dip may be 0.2–0.3% NaOH at 94°C (200°F) for a few seconds, followed by a rinse, but longer dips at lower temperature and with higher caustic content may be used. The dip has a dual purpose, namely to remove the slight waxy coating found on fresh grape skins and to convert the skin from a semi-permeable to a permeable membrane. Both these processes enable the moisture in the grape to evaporate more quickly and readily.

After lye treatment, the grapes are usually 'sulphured' for about 4 h. They are spread out on trays in a space which is poorly ventilated and in which 1.8–2.3 kg of sulphur per tonne of fruit is burnt. Alternatively, the fruit is exposed to an equivalent amount of liquid SO_2. The grapes may then be sun dried or tunnel dried and should finally contain 1500–2000 ppm of SO_2 and 12–13% moisture. To aid in handling and packing, the dried fruit is frequently provided with a very thin coating of light edible oil.

The principal variety of grape used for making dessert raisins is the Thompson Seedless, though the European Muscat (with seeds or 'stones') is regarded by many as being superior in flavour.

PRESERVATION BY FREEZING

Fruit

Freezing has largely supplanted sulphiting as a means of preserving berries intended for further processing into jams, conserves, pie fillings, etc. Many fruits suffer substantial damage on freezing, the changes occurring as a result of ice formation destroying the semi-permeable membranes without which the turgor which is responsible for the characteristic texture of fruit, cannot be maintained. Fruits in general do not require blanching before freezing and can be individually quick frozen (IQF), packed in sugar or syrup or puréed before freezing.

Frozen fruit juices or concentrates are also important commodities in international trade. The consumption of frozen orange juice concentrate is very high in the United States. Brazil is now the largest exporter of frozen orange juice concentrate, which is gaining popularity in countries where citrus cannot be grown.

The large scale production of frozen citrus concentrates demands mechanical picking of the ripe fruit which is then washed, sorted for size and has the juice extracted mechanically before it is filtered, pasteurized and vacuum concentrated. Some 10% of raw cut-back juice and a quantity of pulp is then added to replace the volatiles lost during concentration. The juice can be blended (early season juice is high in acid, low in sugars; late season juice is low in acid, high in sugars) before being packed for distribution, generally as a 64 concentrate, which is a plastic solid even at $-18°C$ (0°F).

Vegetables

Most vegetables preserve their intrinsic characteristics well during freezing, storage and thawing. Frozen vegetables are generally regarded as preferable in quality to their canned counterparts and occupy an important place in the frozen food industry.

Almost all vegetables need to be blanched by steam or hot water before freezing. Failure to inactivate enzymes, especially in peas, green beans, sweet corn etc., results in objectionable flavours developing during cold strorage as well as a loss of colour.

Mechanical harvesting, particularly of peas, broad beans and green beans, results in bruising or damage followed by flavour loss, development of off-flavours and discoloration. To minimize these undesirable changes, the delay between harvesting and blanching must be carefully controlled; a delay of only some 3–4 h in peas can result in detectable off-flavours.

Most vegetables benefit from quick freezing which causes less cellular damage and results in a crisper texture in the finished product. The nutritional value of vegetables is well preserved by freezing; the only nutrients which suffer important losses are water-soluble, comparatively unstable vitamins such as vitamin C. Providing storage temperatures are maintained at $-18°C$ (0°F) or colder, some 80% of this vitamin will be present after a year in storage.

Before being blanched, cooled and frozen, vegetables are washed, sorted, graded, trimmed, peeled or otherwise prepared so that the frozen product is edible in its entirety.

The most popular frozen vegetable is potato chips or French fries. The potatoes are lifted in the autumn when the skins have set and stored in warehouses. In cold weather, a conditioning period to reduce the content of reducing sugars, which give rise to dark-coloured chips, is required before processing. Colour control of the finished product is achieved not only by controlling the level of reducing sugars by conditioning but also by washing surface sugars out of the raw chips in a warm water blanch and then deliberately adding back a low level of reducing sugar (dextrose) in order to achieve the desired light golden yellow colour after frying. Chelating agents, such as sodium acid pyrophosphate (SAPP), are used in a preblanch to prevent the formation of grey compounds between the chlorogenic acid in the potato and any iron compounds present.

A high solids content (typically 18–24%) in the raw material is important. This will produce an opaque, firm and even floury texture while a low solids (12–16%) potato results in a soggy, translucent chip which is far less acceptable. Some chips are partially air dried before freezing which imparts a crisper texture to the outer layer.

PRESERVATION BY BOTTLING

Bottling of fruit is not a large industry. Vapour vacuum sealing and compound-lined caps are used. Choice and preparation of fruit follow similar lines to that for fruit canning and have been described under the individual fruits. Since the package is transparent, uniform size grading and freedom from blemishes are obviously very important. Because the fruit shrinks on cooking, it should be packed as tightly as possible and layers are frequently pressed in to minimize free space. Filling is assisted if the surface of the bottle is wet so that the fruit slips easily; consequently bottles are usually used straight from the washing machine. Except for the hot-filled packs closed with steam injection, where a controlled headspace is essential, the jars are usually filled completely with water or syrup.

Maraschino, otherwise known as glacé or cocktail cherries, are best prepared from the Royal Anne variety, the fruit normally being received in barrels preserved with sulphur dioxide. This product is most frequently bottled, with up to 100 ppm sulphur dioxide being added as a preservative to extend the shelf life of the opened bottles, where the product is likely to be used over a period of time. The fruit is pitted and impregnated with sugar syrup to give a total solids content of around 75% and coloured either red with erythrosine, or green with a mixture of permitted yellow and blue dyes.

Table 4.2 lists typical processing times and temperatures for bottled fruit. It may be found necessary to vary these conditions slightly according to the condition of the fruit and the details of manufacture, but technical advice should be obtained before any other procedure is adopted.

The heat treatment may be carried out either in water in an open tank heated by steam coils or in a specially designed steam heater. Either way, the temperature may be brought up slowly from cold to the processing temperature (see Table 4.2).

Table 4.2 Typical processing conditions for bottled fruit

Fruit	Processing conditions	
Apples	Stage 1:	Heat up to 60°C (140°F) in 1 h.
	Stage 2:	Heat up to 79°C (175°F) in next ½ h; hold for 10 min.
Blackberries	As Apples	
Cherries	Stage 1:	Heat up to 66°C (150°F) in 1 h.
	Stage 2:	Heat up to 85°C (185°F) in next ½ h; hold for 15 min.
Currants	Stage 1:	Heat up to 66°C (150°F) in 1 h.
	Stage 2:	Heat up to 85°C (185°F) in next ½ h; hold for 10 min.
Damsons	Stage 1:	Heat up to 66°C (150°F) in 1 h.
	Stage 2:	Heat up to 82°C (180°F) in next ½ h; hold for 10 min.
Gooseberries	As Damsons	
Loganberries	Stage 1:	Heat up to 66°C (150°F) in 1 h.
	Stage 2:	Heat up to 79°C (175°F) in next ½ h; hold for 10 min.
Plums	As Damsons	
Raspberries	As Loganberries	
Rhubarb	As Loganberries	

Water cooling is not usually employed, although it may be carried out by replacing the hot water slowly with cold. If water cooling is used, the water should not be allowed to cover the caps in case any is drawn into the bottle by the vacuum produced.

The filled bottles should not be handled more than is absolutely necessary until at least 24 h after heating. During this period the fruit tends to firm up and, if it is moved too soon, the appearance of the pack may be spoilt by broken fruit. The store room should always be protected from bright light, because the colour of the red fruits in particular tends to be spoilt by exposure to sunlight.

MANUFACTURE OF TOMATO PURÉE

Tomato purée is an important ingredient used in the manufacture of ketchup, relishes, pickles, sauces and many other products.

Considerable quantities of tomato purée are manufactured in the USA and the countries of southern Europe, the eastern Mediterranean and North Africa. In the UK the Italian product was once considered the best, but other growers have succeeded in breeding into their tomatoes the desirable 'Italian' characteristic, and there are now well-equipped factories producing high quality product in many other countries.

Canned tomato purée may be in the form of 'single concentrate' (18–20% total solids), 'double concentrate' (28–30% total solids) and 'triple concentrate' (38–40% total solids). These traditional descriptive terms are misleading and represent approximately fourfold, sixfold and eightfold concentration of the original pulp. At one time most of the tomato purée produced was double concentrate but now triple concentrate is more widely used. Although triple concentrate is the more economical way of obtaining a given amount of tomato solids, it may tend to be of inferior colour and flavour. However, triple concentrate tomato purées are available with satisfactory colour and flavour characteristics.

The standard pack for industrial tomato purée was the so-called '5-kilo' can, usually internally unlacquered. The '5-kilo' designation is another traditional but misleading one, as the actual net weight can be significantly less than 5 kg. Aseptic packaging, in large drums or bag-in-box, is widely used. For direct retail sale, smaller cans and flexible tubes are used.

Manufacture

Different manufacturers have their own ideas on techniques, procedure, equipment, layout, etc. at each stage of the manufacturing process and many variants exist. The following is a general description of the method employed.

Mouldy and defective tomatoes are first removed. This is essential if a low Howard Mould Count (see below) is to be achieved. After pulping and seed removal, carefully controlled heating is aimed at inactivation of pectolytic enzymes without adversely affecting colour and flavour. Screening removes skin, hard and under-ripe pieces and any residual seeds, and finally reduces the particle size of the pulp.

Vacuum boiling may be carried out batchwise or as a continuous process. There is now widespread use of multiple-effect evaporators; these result in lower concentration temperatures, particularly in the later stages when the purée becomes more concentrated and thickens, and consequent improvements in colour and flavour. Copper contamination used to be a serious problem, largely arising from the use of copper and brass equipment, but nowadays, with the almost universal use of stainless steel equipment, it is rare to find a tomato purée from a reputable source containing even as much as 5 ppm copper.

Quality

The assessment of quality and suitability for use falls under six main headings.

Colour and appearance

Appearance may provide clues in relation to texture and consistency, especially how 'smooth' the purée appears to be and to what extent separation of liquid has occurred. The purée should be of a good, bright red colour and free from black specks, insects or other foreign matter. Different approaches to colour assessment have been practised, including the Munsell spinning disc method, the Lovibond Tintometer, Hunter, reflectance spectrophotometry and absorption spectrophotometric characterization of petroleum ether and aqueous extracts of the tomato purée. Perhaps the most useful method is reflectance spectrophotometry, measuring the X, Y and Z tristimulus values and thence the x and y coefficients on the CIE Chromaticity Chart. The Lovibond Tintometer serves adequately for routine purposes, despite its subjective weakness. A good tomato purée colour would comprise 20 or more Lovibond red units, seven or less yellow units and two or less blue units.

Flavour and odour

Foreign and abnormal flavours and odours should be absent. Bitter, caramelized, 'cardboardy' and 'metallic' off-flavours are occasionally encountered, though less frequently than in the past.

Assessment of flavour suitability will depend on the nature of the product in which the tomato purée is incorporated, the extent of the incorporation, whether it is a mild or highly flavoured product, the nature of the processing involved and the manufacturer's assessment of the customer acceptance of the resulting finished product. To illustrate with opposite extreme cases, the flavour of the tomato purée used has a far greater bearing on the flavour of the resulting product when it is incorporated at a rate of 40–50% in a mild flavoured, vacuum boiled tomato ketchup, than when it is incorporated at a rate of 5–10% in a 'hot', highly spiced sauce containing other highly flavoured ingredients and over 3% acetic acid, prepared by prolonged boiling at atmospheric pressure.

Composition

The compositional characteristics should be determined on the well-mixed contents of each can to be examined, as soon as possible after opening. Determinations include refractometric solids (and optionally total solids by a vacuum oven method), insoluble solids, pH, acidity as citric acid, salt content, total reducing sugar content, copper content and other trace metals (tin, lead, arsenic, iron).

The limiting values shown in Table 4.3 have been included in trade specifications for tomato purée, paste and concentrate (Codex Standard 57-1981, Vol. 5A, 1994).

Microbiological quality

Unless there has been underprocessing, defective seaming, post-process seam leakage or can damage in transit, canned tomato purée will not contain non-sporing bacteria or yeasts. The pH of the tomato purée is normally low enough to prevent any problem with pathogenic organisms. There has been some tendency towards higher pH values in Italian tomatoes in recent years, which could cause difficulties in this regard and it is always prudent to check the pH value.

Table 4.3 Trade specifications for tomato paste products

Specification	Limiting value
Minimum content of total reducing sugars (as invert sugar), expressed as a percentage of the total tomato solids	45%
Maximum titratable acidity (as hydrated citric acid), expressed as a percentage of the tomato solids	10%
Maximum water-insoluble inorganic impurities, expressed as a percentage of the tomato solids	0.1%
Maximum lead content of purées of 15–20% solids	3 ppm
Maximum lead content of purées of 25% solids or more	5 ppm
Maximum arsenic content	1 ppm

Acidity – Not referred to in the trade specification, but the pH of good quality tomato purées normally lies within the range 4.0–4.4.

Mould contamination is another source of concern for the microbiological condition of tomato purée and other tomato products. The Howard Mould Count is obtained by a standardized microscopical technique. Using a special graticule, a large number of separate fields is viewed through the microscope; those which are seen to contain a specified amount of mould hyphae are counted positive, the others negative and the Mould Count is returned as the percentage of positive fields. This is taken as an index of mould contamination, or lack of it, in the tomatoes from which the product was prepared.

Many have questioned the validity of the Howard Mould Count, but it has been the basis of many official and commercial standards since its widespread introduction in the 1950s. The current standard in the UK, under an informal agreement between the authorities and the trade (1995, LACOTS Circular CO 9 949), is 50% maximum positive fields. The USDA standard is 40% maximum.

Texture and consistency

The main factors involved in texture and consistency are the pectinous nature of the purée (partly dependent on effective control of degradation by pectic enzymes in the early stages of manufacture of the purée), the insoluble solids content and the particle size of the insoluble material.

Various laboratory tests may be carried out to assess texture and consistency. These include pectin determinations (of doubtful value), insoluble solids content, the 'blotting paper' test (which measures seepage of fluid from the purée),

apparent viscosity determination by flow measurements or torsion viscometer and sedimentation tests on a dilute suspension. Results, however, should be interpreted with caution, and always in the light of experience in correlating past results with product performance, particularly where any of the critical factors may be subject to modification during processing. For example, sample A, on examination, might appear preferable in consistency to sample B, due, in fact, to the smaller particle size of the insoluble matter in A; but when incorporated in a tomato ketchup subjected to a high pressure homogenization during manufacture, the consequent reduction in particle size of the tomato solids in B led to B performing better in the product than A, from a consistency point of view. Similarly, differences due to pectin may appear significant on examination of samples of tomato purée, but may well be insignificant when the purées are incorporated in products containing substantial thickening or stabilizing ingredients (e.g. brown fruit sauce).

Performance in the product

The final critical test should always be performance in the product. This need not necessarily mean on full production scale, which is sometimes impossible, as when assessing a relatively small quantity of a buying sample. A laboratory scale or pilot scale trial procedure should, however, be followed, simulating as closely as possible the formulation and processing conditions of production. Complete simulation is unlikely to be achieved, if only due to 'scale' effects, but this is less important so long as there is saftisfactory correlation between performance in the simulated trial and performance in actual production.

Specifications

Under the FAO/WHO Codex Alimentarius Commission, the Committee on Processed Fruit and Vegetables has produced a Codex Standard (1981) for Processed Tomato Concentrates and this should be referred to for specification details.

SUGAR PRODUCTION

Sucrose (sugar) is produced from two principal sources – sugar beet, which is grown principally in Europe and sugar cane which emanates from hotter climates such as the West Indies, the southern parts of the United States, Hawaii, Jamaica, Cuba, Brazil and Queensland in Australia.

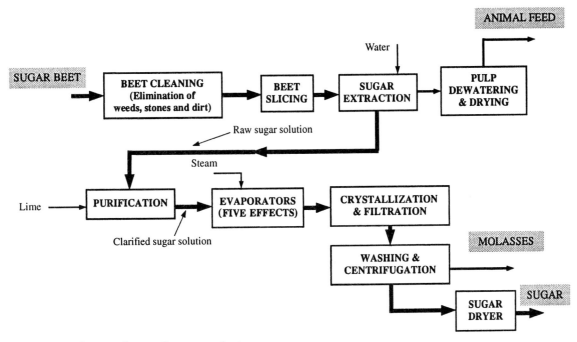

Figure 4.4 Manufacture of sugar from sugar beet.

In the United Kingdom, sugar production from beet accounts for over half the country's requirements; the remainder is met mostly by cane sugar imports. During each annual 'campaign', which lasts from September to January, around 1.14 million tonnes of sugar is extracted from 8.5 million tonnes of beet, which is grown on 170 000 ha in East Anglia and the West Midlands.

In a typical manufacturing process in which sugar is extracted from beet (Figure 4.4), the crop is delivered to the factory in lorry loads according to a timetable that must be strictly adhered to. Every load is sampled for sugar content, dirt and remaining top leaves. The beets are then washed and separated from dirt, stones and other contaminants. They then enter the factory where they are sliced into thin strips known as cossettes. Sugar is extracted from the cossettes by water at around 70°C in countercurrent diffusers. The resulting sugar solution is purified by adding milk of lime, which causes the impurities to settle out. The clarified liquid is subsequently concentrated in a multieffect evaporator. This syrup is then 'seeded' in vacuum pans with very finely ground sugar crystals, thereby causing the sugar to crystallize out. The white sugar is separated from the syrup in a centrifuge, washed, dried in a rotary dryer and transported to storage silos. The residual syrup is returned to the process where it undergoes two further crystallization operations. The final syrup is termed molasses.

The residual pulp leaving the diffusers is pressed to remove as much as possible of the residual liquid. It is then dried further in rotary dryers to yield animal feed.

The cane sugar process is as follows. After the cane is cut, the stalks are trimmed, washed and cut into short lengths or shredded. The cane is then crushed between heavy rollers while, at the same time, being sprayed with water. The resultant juice is dark grey or greenish in colour and is acid to the taste; milk of lime is therefore added to clarify it. This raw juice is heated immediately and run into tanks, where the impurities settle out. The clear juice is subsequently evaporated until it forms a mixture of sugar crystals and molasses. The remainder of the process broadly follows that of the manufacture of sugar from beet.

OTHER FRUIT PRODUCTS

Apples in syrup

Blanching is carried out as for solid pack apples followed by filling to the weights recommended in the UK Code of Practice (BFVCA, 1986). Syrup should then be added at 82°C (180°F), varying from 15–50° Brix according to the UK Code of Practice for the appropriate label designation. Cans are steam-flow closed and processed at 100°C (212°F) for between 10 and 25 min, according to can size and water cooled.

Apple sauce or purée

It is not possible to prepare a good quality sauce or purée from the waste produced during the canning of apple segments. It is, however, a useful outlet for fruit of a size inappropriate for the other products but which is otherwise in good condition. After peeling, the fruit should be sliced or chopped into pieces about 1 cm thick which are then cooked until soft in steam or water. Continuous belt steamers or closed vessels may be used, and after cooking the apples are passed through a pulper with screens having perforations of 3 mm or smaller according to the texture required. Dry sugar is added to the cooked, finished pulp, followed by mixing and either reheating in steam-jacketed pans or a continuous screw conveyor, or flash pasteurizing, which will give a better colour. If sugar or syrup is added before cooking, the mixing and reheating may be omitted, giving a better colour, as cans may be filled straight from the pulper. However, there is some loss of sugar using this method. If the apple sauce is heated in a continuous cooker and filled into the cans at 88°C (190°F) or higher, no processing is required after closing, it being sufficient to invert the cans and hold them for 3 min before water cooling as quickly as possible. If filled into cans at a lower temperature, further processing will be required in steam or boiling water. Continuous-agitating cooling equipment permits more rapid cooling and better colour retention of this product. The UK Code of Practice (BFVCA, 1986) states that apple purée (sweetened) shall contain not less than 8% of added sugar.

Apple pie

This is a product that is finding favour in the 603 diameter tapered pie can. Since it is a product that requires pressure processing, there are twin problems: retaining a good quality of filling and preventing moisture absorption and subsequent sogginess of the pastry. A suitable filling comprises apple segments, rusk, cornflour and salt-based clove spice. The pastry also requires careful formulation and it is advisable to carry out product development and storage tests before attempting to market this product. It is essential to maintain pH control of the filling, which should not be greater than 4, and special types of margarine and rusk are required. Processes for the 1 lb (454 g) can of 40 min at 115.5°C (240°F), 35 min at 121°C (250°F) or 30 min at 127°C (260°F) have been used, with the higher temperature being favoured.

Apple pudding

This product is easier to prepare than apple pie and is usually marketed in 8 or 16 oz (227 or 454 g) cans. A slab of pastry is raised in the can by a dough-forming machine, and after adding a fill of sliced apples with a little syrup and one or two cloves, a pastry lid is applied. The cans are then clinched, exhausted and processed for about 70 min at 110°C (230°F). Careful selection of apples is required in order to avoid pink discoloration due to the high temperature process; Bramleys are usually most suitable.

Fruit leathers

Fruit leathers are extruded mixtures of dried fruit purées and other ingredients (sugar, starch glucose, acid and pectin), which are intended as a snack food. These are not yet common in the UK but are available in the USA and there appears to be no limit to the type of fruit which can be used to produce them. They are available in apple, strawberry, apricot, banana, cherry, blackcurrant, grape, pineapple, mango, kiwi and others.

Fruit salad and fruit cocktail

These are popular packs which because not all of the fruits employed are in season together, usually consist partly or entirely of recanned fruits. Special packs are prepared in A10 or larger cans for use as salad stocks and the fruit is carefully selected to withstand the double processing it is to receive. This applies particularly to the pears and apricots which must be properly ripened but not to the extent that would cause them to break down during the reprocess. In the UK, the Code of Practice (BFVCA, 1986) states that canned fruit salad should contain fruits in the following proportion of the total filled weight of fruit: peaches 23–46%, apricots 15–30%, pears 19–38%, pineapples 8–16%, cherries or grapes 5–15%. The syrup drained from the canned fruits may be used in the preparation of the syrup to be used in the salad after filtering and adjusting to secure the necessary degree of Brix to conform with the UK Code of Practice (BFVCA, 1986) for the label designation used.

The fruits are usually prepared to give the same count of each in the can and a common fill for the A2½ size of can is six units each of peaches (quarters), pears (quarters), apricots (halves), pineapples (pieces) and cherries (whole). The cherries used for this pack (sometimes Maraschino type) must have been carefully dyed with erythrosine to prevent 'bleeding', that is transference of colour to other fruits in the salad. Pears are particularly prone to this and, for this reason, it is advisable to fill cherries and pears at opposite ends of the can. In order to avoid overcooking, exhausting, if by heat, and processing should be kept to the minimum necessary to ensure freedom from spoilage. This should not be difficult with entirely recanned fruit provided it is not left open to infection for long periods.

Cans with plain bodies and lacquered ends are recommended. As the product is entirely recanned, the headspace gas balance will favour rusting, even with lacquered ends. Consequently, even with good canning practice, it may sometimes be necessary to invert the cans not less than 4 days after packing. It is possible to use cans that are fruit lacquered throughout, with the only difference being a slightly darker colour pack and possible minor differences in flavour over that obtained from a pack in plain cans. 'Fruit cocktail' is the name given to the similar pack with diced fruit. The UK the Code of Practice states that fruit cocktail should consist of diced peaches 30–50%, diced pears 25–45%, pieces of or diced pineapple 6–25%, cherries 2–15% and (optional) seedless grapes 6–20%.

OTHER VEGETABLE PRODUCTS

Mixed vegetables or macedoine

Macedoine consists of a mixture of vegetables: peas, carrots, turnips, potatoes, stringless beans, celery and sometimes dried, resoaked white beans. The relative proportions may vary and it is not necessary or even usual to include all of these ingredients. The UK Code of Practice (BFVCA, 1986) states that canned macedoine or mixed vegetables shall consist of at least four vegetables of which no single vegetable should exceed 40% of the total filled weight of vegetables. In the case of green beans and peas, it is usual to include previously canned products since only in this way can the vegetables be artificially coloured with green dye. The others are prepared separately using the methods that are described for them individually elsewhere. The large vegetables are diced into about 3/8 inch cubes and celery and stringless beans are cut into small pieces. They are either filled separately into cans in layers or mixed thoroughly in a drum-type mixer before machine filling. The fill is made up with hot 2% brine followed by steam-flow closing and heat processing for between 30 and 45 min at 115.5°C (240°F) according to can size.

Cans with plain bodies and lacquered ends are used, but even with lacquered ends the inclusion of recanned vegetables may cause a headspace rusting problem by altering the normal headspace gas composition. To avoid this, headspace should be minimized and a good final can vacuum achieved.

REFERENCES

Arthey, D. and Dennis, C. (eds) (1991) *Vegetable Processing*, Blackie, Glasgow.

BFVCA (1986) *Code of Practice on Canned Fruit and Vegetables*, British Fruit and Vegetable Canners Association, London.

FURTHER READING

Baker, C.G.J. (ed.) (1997) *Industrial Drying of Foods*, Blackie, Glasgow and London.

Eskin, N.A.M. (ed.) (1989) *Quality and Preservation of Vegetables*, CRC Press, Inc., Florida.

Mallett, C. (ed.) (1992) *Frozen Food Technology*, Blackie Academic and Professional, Glasgow.

Masefield, G.B., Wallis, M., and Harrison, S.G. (1969) *The Oxford Book of Food Plants*, Oxford University Press, London.

Pattee, H.E. (ed.) (1985) *Evaluation of Quality of Fruits and Vegetables*, AVI, Westport, Connecticut.

Selman, J.D. (1987) 'The blanching process', in Thorne, S. (ed.) *Developments in Food Preservation*, 4, 205–49, Elsevier Applied Science, London.

Weichmann, J. (ed.) (1987) *Postharvest Physiology of Vegetables*, Marcel Dekker, Inc., New York.

5 Cereals and Cereal Products

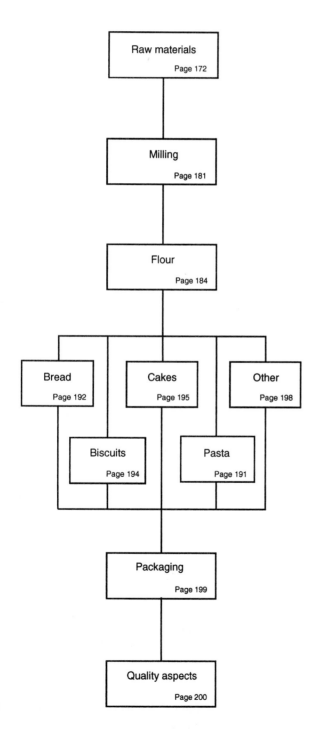

Raw materials
Page 172

Milling
Page 181

Flour
Page 184

Bread	Cakes	Other
Page 192	Page 195	Page 198

Biscuits	Pasta
Page 194	Page 191

Packaging
Page 199

Quality aspects
Page 200

INTRODUCTION

In this chapter, the progress of the raw materials, cereals, is followed through the various processing steps, such as milling and baking, into finished products. General quality factors are grouped at the end of the chapter.

Since the last edition, the Bread and Flour Regulations 1995 have removed the old definition for brown flour, based on crude fibre content, but have included hemicellulases as ingredients permitted in flour and bread. It is expected that all enzymes will be permitted as ingredients by the end of 1996 and that the bleaching agents benzoyl peroxide and chlorine dioxide will cease to be permitted treatment agents by July 1997.

RAW MATERIALS

WHEAT

Characteristics

Wheat grains are the fruit of the wheat plant, which is able to grow in most kinds of soil and under widely differing climatic conditions. The principal wheats of commerce belong to the botanical species *Triticum aestivum* and *Triticum durum*; *Triticum compactum* or Club Wheat is not widely grown now, but in the USA it is milled to produce a soft flour for the manufacture of confectionery and biscuits.

A grain of wheat is ovoid in shape and bears at one end a number of short fine hairs, the beard. Down one side of the grain runs a deep longitudinal crease. The grain consists of three main parts, the enveloping skins (bran), the embryo (germ) and the endosperm. The relative proportions of these parts vary with the plumpness of the grain, but the average composition of wheat is 83% endosperm, 2.5% embryo, 14.5% enveloping skins.

The enveloping skins are:

(i)   The pericarp, which consists of the epidermis, the epicarp and the endocarp. The pericarp represents about 6% of the grain.
(ii) The seed coat consisting of the testa and the hyaline layer, which comprises about 2% of the grain.
(iii) The aleurone layer representing about 6.5% of the grain.

The composition of wheat varies quite widely, which is not unexpected in view of the many varieties that are grown and the very different conditions under which they are cultivated. The ranges of data normally encountered for commercially traded wheats are as follows:

Moisture	10.0–16.0%
Protein	8.5–15.0%
Fat	2.0–2.5%
Fibre	2.0–2.5%
Mineral matter	1.5–2.0%
Sugars	2.0–3.0%
Starch	63.0–71.0%

Wheats are categorized as 'spring' or 'winter' according to sowing and growing conditions. Winter wheat is sown in the autumn and because of its longer growing period, will generally out-yield spring-sown varieties. However, spring wheats need to have been 'cold' before they will germinate, but they are generally recognized as the best breadmaking wheats, despite their short growing cycle.

Wheats can be classified as either red or white according to the colour of the bran. White wheats are susceptible to premature sprouting, with consequent high amylase activity. They are not, therefore, used widely in the northern hemisphere in breakmaking grists.

The protein content and protein quality, and hence the breadmaking quality, of different wheats vary widely. Some wheats, such as Canadian spring wheats, contain a high quantity of strong proteins. Others, such as English wheat, contain a low quantity of weak protein. The British miller blends together strong and weak wheats so as to obtain a mixture suitable for milling into bread flour.

Aleurone layer

The aleurone cells of the wheat grain lie between the endosperm, that is the starchy parenchyma, and one of the surrounding wheat skins known as the nucellar layer. The aleurone layer in the wheat grain consists of a simple layer of cubical cells. They are heavily walled and contain protein, but not protein which gives rise to gluten when admixed with water. The outer bran layers of the wheat grain may have detrimental effects in bread-making, but there is evidence that the aleurone layer can be included with endosperm in the milling of breadmaking flours.

Embryo of wheat

The embryo is that portion of the wheat grain from which the new plant originates. It lies within the branny skins of the grain and is separated from the endosperm by a membrane known as the scutellum. It is customary to separate the embryo during the milling process and the product thus obtained, which consists of embryo plus branny skins, is known as germ.

Endosperm

The endosperm constitutes the bulk of the interior of the wheat grain. It serves as the food supply for the new plant until the rootlets and leaf shoots have developed sufficiently to be able to obtain the required nutrients from the soil and by photosynthesis in the leaf. About two-thirds of the endosperm is starch. The object of the milling process is to remove the endosperm from the wheat grain with the minimum contamination of bran powder and germ.

OTHER CEREALS

Barley

The cultivation of barley (*Hordeum* spp.) reaches far back into human history. Barley was used as a bread grain by the ancient Greeks and Romans. Roman gladiators were known as the hordearii because their general food was barley. Bread made from barley and rye flour was the staple diet of the poor in England in the fifteenth century, while nobles ate wheaten bread.

Barley has a husk which is difficult to remove and is largely indigestible. The husk and bran are usually removed by abrasion to produce pearled barley. Barley flakes made from pearled barley can be used as a flavouring ingredient in speciality breads.

Barley is widely used for malting, brewing into beer and for distilling in whisky manufacture. The controlled germination of barley produces malt for brewing but the malted barley can be dried and milled into a malt flour for use in the production of malt bread and other baked goods.

Maize

Maize (*Zea mays*) is known as 'corn' in America. It originated in America, where it is still widely grown and is now cultivated as a crop in Africa, India, Australia and the warmer parts of Europe. The grain of maize is much larger than those of other cereals and it is used industrially as a good source of starch, which is extracted by a wet milling process. The principal types of maize are: dent (*Z. mays indentata*) and flint (*Z. mays indurata*), pod (*Z. mays tunicata*), pop (*Z. mays everta*), soft or flour (*Z. mays amylacea*), sweet (*Z. mays saccharata*) and waxy (*Z. mays ceratina*).

Dry milling of maize aims to obtain the maximum yield of grits (endosperm chunks), containing the least possible contamination with fat and black specks from the tip cap. Dry milling should also recover the maximum amount of germ in the form of large particles with the maximum oil content (maize is steamed and then degermed in a pin mill); it should also produce the minimum quantity of fine maize flour. This maize flour should not be confused with 'corn flour', the term used in the UK for maize starch obtained by the wet milling process.

Maize grits from dry milling are used in the production of breakfast cereal and in extrusion cooking in the snack foods industry. In Italy maize porridge, made from fine grits or coarse meal and flavoured with cheese, is called 'polenta'. Fine meal or maize cones are used for dusting bakery products like muffins, in bakery mixes or in infant foods.

Maize has a characteristic flavour which is popular in Mexico and the southern states of the USA but is only just being introduced to the British public in Mexican restaurants specialising in tortillas and tacos.

Millet

Millet is a name given to numerous small seeded grasses which are of Asian or African origin. In 1981, the world crop was estimated at 29 million tonnes of which 40% was Pearl millet (*Pennisetum americanum*) and 24% was Foxtail millet (*Setaria italica*). Another 25% of the world crop is made up of two further types of millet, Proso (*Panicum miliaceum*) and Finger (*Eleusine coracana*).

The millets have been used as human food for thousands of years in Asia, Africa and Europe. Their cultivation can be traced back into prehistoric times, and in parts of India and Africa, they are still a staple food. The traditional method for the preparation of millet for domestic consumption is simply pounding in a mortar to loosen the husk and to reduce the grain to a wholemeal or semolina, followed by winnowing. Preparation of large quantities of meal is not practised because in hot climates the undegermed meal becomes rancid very quickly.

The flour or millet meal can be used to make unleavened bread, or the ground product may be used to make a beverage. Millet may be consumed in the form of porridge, or it may be cooked with sugar, peanuts or other foods to make desserts. The protein contents of pearl, foxtail and proso millet are comparable with that of wheat and barley but finger millet has a somewhat lower protein content, similar to the level in rice. Although the protein quality of millet is good and has a higher lysine content than wheat, it does not form gluten. The lack of gluten makes it difficult to make unleavened pancakes (roti, chapatti or tortilla) from millet meal and cold water. The use of boiling water to make the dough results in potential gelatinization of the starch and allows the dough to be rolled out into thin pancakes without breaking or tearing.

Oats

Oats (*Avena sativa*) are a crop of the temperate regions, but are not suitable for particularly dry or prolonged sub-zero Celsius conditions. The kernel, which is called a groat, retains its husk on harvesting like rice and barley and has the characteristic cereal structure of endosperm, bran and germ. Compared with other cereals, oats are characterized by a high protein and fat content and a low carbohydrate content. The high fat content and naturally occurring fat-degrading enzymes provide potential for rancidity in milled oat products. However before oats are used for human consumption, the tightly adhering husk has to be removed. This is normally carried out by an abrasion process after the oats have been heat treated (kilned). This causes the husk to become friable and also inactivates the fat-degrading enzymes, thereby eliminating most of the potential for enzymatic rancidity. After removal of the husks, the oats may be flaked, milled into flour or cut into smaller pieces.

In the United Kingdom, oats have traditionally been used for porridge, oat cakes and biscuits.

Since the mid-1980s, the market for oat flakes has increased with the increasing consumption of mueslis. In more recent years, oats and oat bran have been used in an increasing number of products. One of the reasons for this has been a claimed link between the intake of soluble fibre, from oats and particularly oat bran, and a reduction in blood cholesterol level. This, in turn, reduces the chances of developing cholesterol-related heart diseases. There is, however, still much debate about the validity of the cholesterol-reducing claims for oats.

Naked oats are recently developed varieties in which there is no closely adherent husk on the kernel. Whilst various beneficial claims have been made, heat treatment will still be required to avoid rancidity. Wild oats are a different species to *Avena sativa* and can be a serious weed in crops of cultivated oats and other cereals.

Rice

Rice (*Oryza sativa*) is generally considered to be a semi-aquatic annual grass, typically associated with paddy fields and tropical countries. However, there are varieties that have adapted to a wide range of environmental conditions from dry hillsides to the foothills of the Alps and Andes, although rice is not grown in the extreme latitudes where wheat and barley are cultivated. Together with wheat and maize, rice is one of the three major cereals of the world and is the basic food for over half the world's population. In some countries, it provides 80% of the daily food intake.

After harvesting, rice kernels usually retain their husk and are known as paddy or rough rice. When the husk is removed it is called cargo or brown rice. The kernel is of the typical cereal form, having endosperm, bran and germ, but is different from wheat, in not having a longitudinal crease. When rice is milled it is not ground up to produce a flour, but goes through various degrees of abrasion to remove the bran and germ to give white rice. The white rice may then be polished to give it a brighter and smoother appearance and to enhance its cooking properties. As with other cereals, the bran and germ contain a significant proportion of vitamins and minerals which are lost on milling.

Rice is varied in its properties and hence has a wide range of uses and different quality preferences throughout the world. In the UK, rice is principally used as a vegetable, as a dessert and for breakfast cereals. There are three major types

of rice, long, medium and short grained, which are classified by the grain shape and tend to have different properties and hence different uses. In the UK, long grain rice in white or brown form, usually from the USA, is used as a vegetable and short grained rice, usually from Italy, is used for rice puddings. Other categories of rice are:

(i) *Basmati rice.* An aromatic long grained type, which only comes from the Himalayan foothills in India and Pakistan.

(ii) *Parboiled rice.* This is rice that has been steeped, steamed and dried whilst still in its husk. This process carries soluble vitamins and minerals from the husk into the kernel, which following dehusking is more nutritious than usual. The process also decreases the likelihood of the kernels being sticky when cooked, but they will tend to be darker and require a longer cooking time than usual.

(iii) *Easy/quick cook rice.* This has been processed to increase the rate of penetration of water into the kernel, hence decreasing the cooking time. The extreme of this is instant rice which only requires the addition of boiling water.

(iv) *Wild rice.* This is not a true rice, but an aquatic grass with long slender black-to-brown grains; it is used to add to long grain rice for its visual effect.

Rye

Rye (*Secale cereale*) tends to be grown on land just outside the belt which gives the most satisfactory return for wheat, such as areas of northern and eastern Europe that have a temperate climate. As a bread grain, rye is second only to wheat in importance and is the main bread grain of Scandinavian and eastern European countries. Although nutritious and palatable to some people, rye bread is not comparable with wheaten bread as regards crumb quality, large loaf volume or taste. Cultivated rye was derived in about 400 BC from the rye grass that occurred as a weed in wheat and barley crops.

Rye is particularly susceptible to infection by the fungus *Claviceps purpurea*, which can also attack barley, oats and wheat. The sclerotia of this fungus are known as ergot. Ergot is a toxic contaminant of cereals, which has a maximum limit of 0.05% in wheat and rye traded within the European Union.

The grains of rye are longer and thinner than those of wheat and have a greenish-grey colour. Rye milling differs from wheat milling and usually involves the use of eight of more break rolls and comparatively few reduction rolls. Rye meal and flour have a characteristically high amylase activity and its protein produces little gluten which limits its use in bread production. The high pentosan content of rye allows batters to be whipped from water and rye meal, which form a stable foam. These stable foams can be baked at high temperature into crispbread.

In order to avoid excessive amyloytic breakdown of the starch in breadmaking, the pH of the rye dough is lowered by acid modification in a 'sour dough' process. A starter 'sour dough' is prepared by allowing a rye dough to stand at 25°C for several hours to induce a natural lactic acid fermentation caused by species of *Lactobacillus*. Although rye bread is still not popular in the United Kingdom, the successful new range of 'soft grain' breads, such as Mighty White, contain presoaked kibbled wheat and rye grains.

Sorghum

Sorghum probably originated in Africa. Egyptian frescoes suggest that it was in cultivation in 2200 BC. All the cultivated sorghums are known as *Sorghum valgase* or *Sorghum bicolour*. Grain sorghum is a coarse grass which bears loose panicles containing up to 2000 seeds per panicle. It is an important crop and the chief food grain in parts of Africa, China and the Indian subcontinent, where it forms a large part of the human diet.

In the so-called 'waxy' type of sorghum, the starch is composed almost entirely of amylopectin. These varieties have starch with physical properties similar to those of cassava, from which tapioca is prepared.

The whole sorghum grain may be cooked like rice or fried. Acceptable bread products, such as muffins or griddle cakes, can be produced from a blend of strong wheat flour and sorghum flour. Rollermills are not appropriate for processing the round sorghum grain. The bran and germ of sorghum pulverize easily to a fine powder which is difficult to separate from the fine flour. Dry milling is accomplished in two stages: an abrasive pearling to remove bran is followed by pulverizing the pearled grains. Wet milling of sorghum to produce starch uses methods similar to those employed for maize but the process is more difficult. Varieties with dark coloured outer layers are not satisfactory for wet milling because some of the colour leaches out and stains the starch.

Triticale

Triticale is a polyploid hybrid cereal derived from a cross between wheat (*Triticum*) and rye (*Secale*). The first wheat–rye hybrid was described in 1876 but it was nearly ninety years before the first hexaploid triticale varieties were released in Canada, Spain and Hungary.

The objectives in making the cross were to combine the grain quality, productivity and disease resistance of wheat with the vigour and hardiness of rye. Varieties of triticale in commercial production prior to 1990 did not offer much promise for milling and baking; they resembled their rye parent more than their wheat parent, particularly in a tendency towards undesirably high α-amylase activity and the inability to form an elastic gluten protein. Triticale can be milled on rollermills used for wheat milling, and baked products made from triticale flour–wheat flour mixtures can be purchased in Europe and North America.

ADDITIVES

Ascorbic acid

Ascorbic acid (vitamin C, E300) acts as a flour improver, and if used in the proportion of 20 to 200 ppm of flour, it improves the baking quality materially. It differs from other flour improvers in that it is a reducing agent; however it is readily oxidized by air to dehydroascorbic acid and it is this compound which exerts the improving action. It is the only oxidative improver for use in the manufacture of bread, allowed by the UK Bread and Flour Regulations 1995. It finds its chief use in the shorter breadmaking processes, such as the Chorleywood process, and is added by the baker in a proprietary bread improver. In the UK it may also added to the flour by the miller.

Benzoyl peroxide

The organic peroxide $(C_6H_5CO)_2O_2$ has been widely used as a flour bleacher. It gives an effective bleach when added in the proportion of 1 part to 30 000 parts of flour, but the reaction is slow and needs about three days to reach completion. Benzoyl peroxide has been used as a proprietary mixture in which it is diluted with about six times its weight of inorganic fillers.

Until July 1997, all bread and flour other than wholemeal may contain benzoyl peroxide in any proportion not exceeding 50 ppm. However, from 1996 onwards, most white flour milled in the UK will be unbleached.

Bleaching agents

At one time flour was always 'aged' for several weeks before it was despatched to a baker and during this period it improved significantly in baking quality and became bleached. For many years both the improving and bleaching effects have been attained rapidly by adding small amounts of oxidizing agents to flour. The bleaching agents used had no action on the branny particles and consequently did not make a dull or inferior flour seem to be of better grade, they only removed the yellowness, which is present in even the highest grade of flour, and so yielded the white bread which was thought to be demanded by the public. In 1996, the substances which are permitted in the UK as flour bleachers are chlorine dioxide, benzoyl peroxide and chlorine, although the latter chemical is legally restricted to use in cake flours. Both benzoyl peroxide and chlorine dioxide will be banned from July 1997.

Chalk

When the rate of extraction of flour was raised to 85% during World War II, millers were instructed to add prepared chalk of British Pharmacopoeia (BP) quality to all flour in order to compensate for the calcium-immobilizing effect of the phytic acid present in the longer extraction flour. The BP name for such material was 'creta praeparata' and some millers still refer to 'creta'. The UK Bread and Flour Regulations 1995 require the addition of prepared chalk of BP quality and of a stated degree of fineness to all flours other than self-raising flour with a calcium content of not less than 0.2%, wholemeal and wheat malt flour. The flour should contain not less than 235 mg and not more than 390 mg of chalk per 100 g of flour. The addition is now no longer made to counteract the effect of phytic acid, which is very small in white flour, but as a means of supplementing calcium in the national diet.

Chlorine dioxide

Chlorine dioxide is a gaseous flour improver and bleach that superseded Agene (nitrogen trichloride) in Britain and in the USA. Plants were

available that enabled it, greatly diluted with air, to be added continuously and automatically to flour in a precisely controlled proportion. A normal rate of treatment in Britain was about 15 ppm. However the major production streams from a flour mill were treated with different rates of chlorine dioxide. The 'patent stream' required least treatment and the 'low grade' stream required most bleaching and improving.

The reaction against the use of additives described under BENZOYL PEROXIDE also applies to chlorine dioxide and by July 1997, at the latest, chlorine dioxide (Dyox) will not be used by British millers.

Chlorine

Chlorine is used in the production of speciality cake flours. These are finely dressed flours milled from a weak wheat of low protein content, which have been treated with chlorine at the rate of 1500 ppm.

Fungal amylases

For many years it was customary for the miller to correct diastatic insufficiency in a bread flour by the addition of malt flour as a source of α-amylase. During the last few years, however, there has been a growing tendency for malt flour to be superseded by fungal α-amylases for this purpose.

Fungal α-amylases are obtained by growing *Aspergillus oryzae* in deep culture and extracting and purifying the α-amylase that accumulates in the culture medium. The purified enzyme is suitably diluted with a carrier so that the necessary diastatic correction can be attained by supplementing a flour with about 100 ppm of the mixture.

Fungal α-amylase preparations have several advantages over malt flour as diastatic correctives. These include:

(i) Freedom from other and possibly undesirable enzymes, e.g. bacterial amylase, which can survive the baking process.
(ii) Standardized potency.
(iii) Low temperature of inhibition, which avoids stickiness in the crumb in the event of overdosage.

Iron

The UK Bread and Flour Regulations 1995 require that in the UK white flour shall contain not less than 1.65 mg of iron per 100 g. At present iron may be added as ferric ammonium citrate of BP or BP Codex quality or as finely divided iron. Some forms of ferric ammonium citrate cause the development of rancidity in the master mix on storage and for this and other reasons the iron is almost invariably added as finely divided iron. However, evidence is accumulating that the assimilation of finely divided iron is far from complete and in future salts of iron may be the preferred means of enrichment.

RAISING AGENTS

Acid calcium phosphate

Acid calcium phosphate (ACP, $Ca(H_2PO_4)_2$) has been used effectively as a preventive of the bacteriological spoilage of bread known as 'rope' (see ROPE). This compound has an inhibiting action on amylases and has been recommended as an ameliorative measure in combating sticky bread crumb and problems encountered when slicing bread in plant bakery. Acid calcium phosphate is widely used in the milling industry as the acid ingredient in self-raising flour. Its neutralizing value, i.e. the parts of sodium bicarbonate that would be neutralized by 100 parts of the phosphate, varies with the sample, the usual value being about 80.

Acid calcium phosphate coated with sodium phosphate or other phosphates is widely used in the USA. It acts more slowly on sodium bicarbonate in the cold than does ordinary acid calcium phosphate and hence gives rise to less loss of gas from self-raising doughs during the prebaking period. In Northern Ireland, acid calcium phosphate is added to 'plain' flour to cater for local taste and particularly for the production of hot plate goods such as farls.

Acid sodium pyrophosphate

Acid sodium pyrophosphate (ASP, $Na_2H_2P_2O_7$), which is used as an acid ingredient in baking powders and also in self-raising flours, is available undiluted. It is also marketed diluted with a filler (usually dried starch) to a concentration of 64%, at which level 2 parts of the mixture will neutralize 1 part of sodium bicarbonate. Acid sodium pyrophosphate reacts more slowly with sodium bicarbonate than does acid calcium phosphate and hence leads to less loss of gas from self-raising doughs during the prebaking period. Baked goods produced from self-raising doughs containing acid sodium pyrophosphate may, however, possess a

tang or 'bite'. There is, therefore, much to be said for the use of a mixture of acid sodium pyrophosphate and acid calcium phosphate and such a mixture is available on the market.

Baking powders

Baking powders are mixtures containing chemicals used for the aeration of various types of bakery products. The active ingredients are sodium bicarbonate and an acid substance, which in the presence of water, and with or without heat, will react to produce carbon dioxide. The relative portions of bicarbonate and acid are such that they almost exactly neutralize each other but an adjustment may be made so that the resulting product is intentionally slightly acid or alkaline. The two chemicals are dispersed in an inert filler such as starch.

The acid components mainly used in baking powders are acid sodium pyrophosphate, acid calcium phosphate, cream of tartar, glucono-δ-lactone and sodium aluminium phosphate. Each of these has its own characteristic effect, e.g. on flavour or speed or slowness in working off.

Cream powder

A cream powder is a preparation containing an acid phosphate to replace cream of tartar in baking powders. Its neutralizing value is adjusted by dilution with flour to be equivalent to that of pure cream of tartar, which is much more expensive. The common phosphates used are acid calcium phosphate and sodium pyrophosphate. They have the advantage that they do not act on the bicarbonate in the cold. It is claimed that sodium aluminium phosphate has an even more delayed action than the other two phosphates mentioned. The disadvantage of the acid calcium phosphate is that when it is present in excess of the sodium bicarbonate it produces an acid bite, and unless they are very finely ground, the particles will tend to produce brown spots on the surface of sugar-containing products.

Sodium pyrophosphate is said to improve gluten by making it more extensible.

Cream of tartar

Cream of tartar, potassium hydrogen tartrate, is used as an acid ingredient in some baking powders. It has the disadvantage of working off quickly and has been replaced by a range of cheaper acid phosphates which, however, do not produce such a satisfactory flavour in the finished product.

One hundred parts of cream of tartar will neutralize 45 parts of sodium bicarbonate so that the usual 2:1 mix for baking powder will give a product slightly on the alkaline side.

Glucono-δ-lactone

Glucono-δ-lactone is the lactone of gluconic acid. It is used as an acid in raising agents for two reasons, first, it does not have the distinctive bite of sodium pyrophosphate and second it is only converted slowly to the active state when it comes into contact with moisture and thus delays the working off of mixed batters, for example. It is, however, more expensive than the more commonly used acidulents.

Sodium bicarbonate

Sodium bicarbonate is the gas producer in baking powders and in self-raising flours. Carbon dioxide is released when it reacts with an acid and the balance of bicarbonate to acid needs to be adjusted so that normally the final product finishes up just on the acid side of neutral. Sodium bicarbonate will release its carbon dioxide simply by the action of heat, but in the absence of acid, the residue is very alkaline. This not only produces an unpleasant flavour but causes the crumb of products to turn yellow.

PRESERVATIVES

In bread, the following substances may be added as preservatives: acetic acid, monocalcium phosphate (ACP), sodium diacetate and propionic acid and certain of its salts. All these help to prevent the onset of 'rope' but the propionates also have an antimould effect.

Acetic acid

A 12% aqueous solution of acetic acid, at the rate of 0.5 l per 100 kg of flour, may be used as a preventive against the development of 'rope' in bread. The purpose of adding the acid is to reduce the pH of the crumb to below 5.4.

Propionic acid and propionates

In the EU, propionates are allowed in bread (but not in biscuits) up to 3000 ppm of the flour weight and they are used at this, and sometimes higher levels, in other countries also. There is a flavour breakthrough at this level and in the UK it is more usual in practice to use propionates at about 1000 ppm. Under normal conditions, this will give an extra 24 h shelf life to wrapped bread before it goes mouldy. Propionates have a delaying effect on yeast and where they are used, this must be allowed for by adding extra yeast.

During 1986, the pressure for 'healthier' foods led to bakers removing propionate from bread so that they could add the claim 'free from preservatives' to their wrappers. However, the shortened mould-free shelf life has not pleased all consumers.

Sorbic acid and sorbates

Sorbates are more effective mould inhibitors than propionates. They also have a delaying effect on yeast which must be allowed for in the manufacturing process.

PROCESS INGREDIENTS

Ammonium carbonate

Commercial ammonium carbonate, which is a mixture of ammonium bicarbonate and ammonium carbamate, is known as 'Vol' in the baking trade. Under the action of heat it decomposes to carbon dioxide, ammonia and steam and thus acts as an aerating agent. It is used mainly in biscuit manufacture. It has the advantage of leaving no residue in the finished product but the disadvantage that freshly baked goods made with it have a smell of ammonia although this quickly disappears.

Ammonium chloride and sulphate

Ammonium chloride and ammonium sulphate have been used as yeast foods, being readily available sources of nitrogen for the organism. One or the other was used as an ingredient in most commercial bread improvers, but yeast foods are not necessary in a modern dough conditioner which contains mixed enzyme systems and emulsifiers.

Improvers or dough conditioners

The breadmaking quality of freshly milled white flour improves with age due to the action of atmospheric oxygen; the gluten is rendered more stable and stronger and also more elastic. At one time millers never despatched flour to a baker until it had been stored for several weeks and thus had undergone this natural 'ageing'. Later, it was discovered that this delay could be avoided and the effects of natural ageing attained promptly by adding a very small amount of ammonium persulphate to the flour. Subsequently it was found that a number of other oxidizing substances were able to age flour artificially. It became common practice in the milling industry to treat all breadmaking flour with improvers, as they are called, in order to obtain the improvement in breadmaking quality which previously used to be obtained by ageing freshly milled white flour for several weeks. The improvement consists of rendering the gluten more stable and stronger and also more elastic. Some improvers bleached the yellow pigments of flour as well as improving the baking quality, but others had no such bleaching effect. When the latter were used, the miller added a bleaching agent to the flour, in order to reproduce the change from creaminess to whiteness that accompanies natural ageing. Since 1989, when potassium bromate was banned, the only oxidative improvers that are used in Britain are ascorbic acid and chlorine dioxide. Chlorine dioxide also has a bleaching effect, but ascorbic acid acts only as an improver. From July 1997, chlorine dioxide will not be allowed in British mills.

The mode of action of improvers has been the subject of numerous investigations and several theories have been advanced to explain their beneficial effect. The weight of the evidence is in favour of an increase in the cross-linkages between neighbouring protein molecules by oxidation of contiguous sulphydryl groups to form new disulphide bonds.

Commercially nowadays the term 'improver' is applied to an all-in mix containing all or some of the following: flour improvers, yeast foods, fats, emulsifiers, enzyme preparations, soya flour, a flour filler and possibly calcium propionate. The mix of ingredients will depend on the intended use. It may also contain calcium sulphate and ammonium sulphate as yeast food. The hard fat essential for the process is suitably dispersed, possibly on an enzyme preparation such as malt flour or, more commonly, fungal amylase, hemicellulase and some inert filler. Bread improvers are usually designed to be used at 1% or 2% of the flour weight.

Glyceryl monostearate

Glycerol (glycerine) is a trihydric alcohol and when each molecule of glycerol is combined with three molecules of fatty acid, either the same or different acids, the result is a fat. Glyceryl monostearate (GMS) is obtained when glycerol is combined with only one molecule of fatty acid, stearic acid. This produces a molecule with glycerol, which is water soluble (hydrophilic) at one end, and a fat, which is oil soluble (lyophilic), at the other. This configuration confers emulsifying properties on GMS. To be effective in emulsification and for crumb softening, the GMS should be presented in the correct crystalline form and as a hydrate. Commercially available solid GMS is generally a mixture of GMS (33%), glyceryl distearate and stearin, but distilled GMS (90%) is available, and also prepared stabilized hydrate.

Lecithins

Lecithins are emulsifiers occurring naturally, for example, in egg yolk and the soya bean. Commercial lecithin is extracted from soya bean as a very sticky yellow paste which is difficult to handle and it is now more usual to buy it dispersed on a suitable support. When lecithin is used at 0.5% of flour weight in bread, a silkier softer crumb is claimed. It is used more on the continent than in the UK where it appears to have lost ground to the newer emulsifiers and crumb softeners.

Shortenings

When vegetable or animal oils are hydrogenated, their melting points are raised, producing fats or shortenings. The degree of hydrogenation and therefore the melting points of the product may be varied. Like oils, shortenings are a mixture of many triglycerides of different melting points.

Ideally the proportion of fat with a melting point above that of body temperature needs to be controlled, otherwise, on eating, the presence of the high melting point fraction becomes apparent in the mouth as a 'palate cling'. This is not obvious when a product is eaten hot. Some tailored shortenings for puff pastry have greater proportions of these higher melting point fats so that the layers of fat are not dispersed in preparing the pastry. Shortenings used in recipes containing higher weights of sugar and liquor than flour, so called high ratio shortenings, have emulsifying agents such as glyceryl monostearate incorporated in them. Because of the hydrogenation, shortenings are more stable and resistant to air oxidation than their parent oils.

Soya flour

Soya flour is milled from the soya bean, a legume widely grown around the world. The UK is too far north for existing varieties of soya bean to crop successfully and beans milled in the UK usually come from North America. Commercial soya flour exists in two forms, raw enzyme-active flour and an enzyme-inactive flour milled from heat treated beans.

Raw soya was once added (up to about 2%) to most plant made bread, but now it is normally only added as an ingredient of commercial bread improvers. Various advantages for its addition to bread have been claimed but possibly the only important one is its whitening effect on the crumb due to the bleaching effect of the lipoxidase present; because of this and the pressure to remove chemical bleaching agents, e.g. benzoyl peroxide, an increase in its use for this purpose may be anticipated. The enzyme-inactive flour is used in cake pastry and biscuit formulations.

Soya flour is rich in protein (about 40%) and oil (about 23%) and the defatted flour is used as feedstock in the preparation of texturized vegetable protein.

Salt

Salt is an important ingredient of bread, without which the product appears to have little flavour. However, the 1991 COMA (Committee on Medical Aspects of Food Policy) report has recommended that the salt intake of the UK population should be reduced. Experiments showed that salt levels in bread could be reduced by up to 12.5% without it being detectable by taste and most bakers have taken this action. This does not mean that all bread has the same salt content and regional differences still exist, e.g. southern England 1.1% and Scotland 1.2%.

Yeast

There are several hundred species of yeast. Baker's yeast is the species *Saccharomyces cerevisiae*. It is a single cell organism roughly oval in shape and measuring about 4 mm in length.

Commercial yeast is grown on a sterilized molasses feed stock, first anaerobically and finally aerobically with air being blown through the fermenters. It is normally sold in a compressed block form containing about 30% solids. Granular yeast, which contains rather less moisture, is now not so popular in the UK although it is still used in some European countries. Cream yeast, a liquid containing about 20% yeast, has the advantage that it can be pumped and metered in continuous processes. Dried yeast might be considered a useful standby in case supplies of fresh yeast should be interrupted but considerably more yeast has to be used to achieve the same gassing activity. Commercial yeast is the result of much development work to obtain an all-purpose product and that in the UK is generally faster acting than those used elsewhere.

Yeast works by converting the fermentable carbohydrates in the dough into alcohol and carbon dioxide, the latter producing the required aeration in the production of bread. The contribution that yeast makes to the flavour of bread is important. The activity of yeast increases with temperature to an optimum at about 40°C and above this temperature the activity quickly drops off until 50°C when all the yeast cells are killed. Yeast is a live product and must be treated as a perishable commodity – ideally it should be stored between 1°C and 4°C. Yeast stored in a warm bakery will quickly lose its acitivity. More detail is given by Brown (1982).

Baker's yeast contains a very complex enzyme system. The enzymes of importance to the baker are invertase, converting sucrose to fructose and glucose; maltase, converting maltose to glucose; and zymase, converting glucose and fructose into ethyl alcohol and carbon dioxide. Recently an unwanted enzyme appeared in some commercial strains which reduced the flavouring principle of cinnamon, cinnamic aldehyde, to styrene, the monomer of polystyrene, thus producing a strong plastic flavour in spiced buns.

PRIMARY PROCESSES

MILLING

The wheat grain consists of three parts, the enveloping skins, the embryo or germ and the endosperm. The object of the milling processs is to separate most of the endosperm in a manner that leads to minimum contamination with powdered

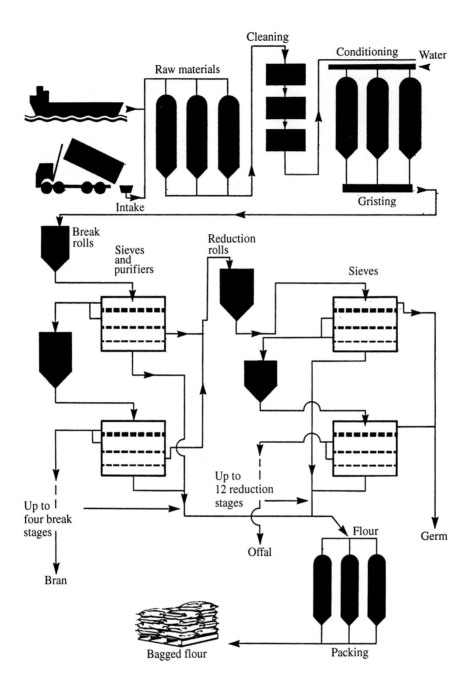

Figure 5.1 Flour milling process.

skins and germs. The operations included in the conversion of wheat into white flour are cleaning, conditioning, breaking, scalping, purification and reduction. A schematic diagram of a typical milling process is shown in Figure 5.1.

Cleaning

The wheat is subjected to a number of sieving operations to remove impurities larger or smaller than the grain. It is passed through disc separa-

tors to remove other cereals such as barley and oats and other impurities which are similar to wheat in diameter but different in shape. The wheat is then scoured and brushed in order to remove dirt and the thin papery pericarp which is called 'beeswing'. During these operations the grain is subjected to aspiration to remove light grains and seeds, 'beeswing' and dust. Heavy material such as stones which are of similar size to wheat grains are removed on a gravity table and metal detectors or magnets remove iron and swarf.

Conditioning

The objects of the conditioning process are to toughen the bran, to enable the bran and endosperm to be separated easily, to facilitate the pulverization of the separated endosperm, to enable the products of the flour dressers to be sieved easily and accurately and to arrive at a correct moisture content in the finished products. The moisture content of dry wheats has to be increased and this is accomplished by preliminary damping and subsequent conditioning. Proper distribution of the added moisture is achieved by allowing the moist wheat to lie in a bin for 4–36 h, but more commonly nowadays water is added in a machine which tumbles the grain through high pressure water sprays and this efficient wetting process enables the conditioning time to be reduced. British wheat is often too moist for optimum milling and hence requires drying before it can be accepted into the mill.

Breaking

The breaking stage of the milling process is performed by a series of pairs of corrugated rolls, known as break rolls, which revolve in opposite directions at a differential speed of 2.5 to 1. It is common practice to have four pairs of break rolls, each pair having finer corrugations and a smaller clearance than the preceding pair.

The corrugations and the differential speed give the break rolls a shearing action and their purpose is to split open the grains and scrape the endosperm from the branny skins. After passing through the first break rolls, the material is sieved and the coarsest fraction, consisting of split-open grains from which a little endosperm has been removed, goes to the second break rolls. Here the scraping away of endosperm is carried further; the mixture leaving the rolls is sieved and the coarsest material sent to the third break. More endosperm is scraped away and removed by sieving and the residue then goes to the fourth break rolls for another, and often the final, scraping. The material that remains when the last scraping has been performed consists of flakes of the outer skins of the wheat bearing a small proportion of endosperm. This product is bran, which can be further cleaned in a drum sieve or bran finisher.

Scalping and grading

Scalping is the name given to the sieving operation that is performed upon the stock or chop, as it is

often called, that leaves a pair of break rolls. The object of this sieving is to separate the stock into three main products: (i) coarse particles, which are the remains of the grains and from which endosperm can still be obtained, (ii) flour and (iii) particles of intermediate granularity, which are nodules of endosperm and are known as semolina or middlings, according to their particle size and purity. Scalping is usually performed by plansifters which consist of a number of superimposed enclosed horizontal sieving surfaces to which is imparted a rotary motion. The nests of sieves are hung on sets of flexible canes and they are given their rotary motion by a vertical shaft attached to the middle of the machine and supported by an overhead bearing. Most mills also employ plansifters as flour dressers at the end of the main flour collecting conveyor.

Purification

The semolina and middlings that have been removed from the various break stocks by scalping are treated on machines known as purifiers. These consist of enclosed reciprocating sieves through which a current of air is blown. The combined effect of the aspiration and the movement on the sieves is to remove much of the loose branny material with which the stock may be contaminated and to grade the stock on the basis of particle size and, to some extent, of purity.

Reduction

The purified semolina and middlings are sent to smooth rolls, known as reduction rolls which revolve at a differential speed of 1.5 to 1. The clearance of these rolls is adjusted to suit the granularity of the stock fed to them. There will be from 10 to 15 sets of reduction rolls in a mill. They are designated by consecutive letters of the alphabet (excluding I), and the later the code letter of a roll occurs in the alphabet the lower the grade of the stock with which the roll deals.

The crushing action of the reduction rolls reduces the size of the nodules of endosperm in the stock, reducing some of them to the fineness of flour, and at the same time tends to flatten out branny and germy particles. The stock leaving a pair of reduction rolls is sieved, or dressed as it is called, on plansifters, so as to remove the flour that has been produced by the rolls. The residue is separated into two fractions; the finer and purer of these goes to one of the succeeding reduction rolls, while the coarser and more branny material goes still further down the reduction system. The rolling action of the smooth, head reduction rolls, whilst flattening the bran particles in the desired manner, does inevitably produce some endosperm flakes. In modern mills flake disrupters (high speed pin mills) are installed after the rolls and before the sifters to increase the flour release.

By the time the last reduction roll of the series has done its work, most of the endosperm has been converted to flour and most of the coarser branny fragments have been removed. The residue is an intimate mixture of endosperm and equally fine branny matter from which it is no longer practicable to separate good flour.

Divides

The dressing machines in the mill furnish streams of flour which differ from one another in colour and baking quality. All these streams may be mixed together to constitute straight-run flour, or they may be segregated into several groups or 'divides' which differ in grade and can be sold by the miller as flours for different end products. For example, a brightly coloured high grade 'Patent' flour for the production of high quality bread and rolls could be milled on a 40% divide from a breadmaking grist of 12% protein content, with the remaining high protein flour being used to produce malt bread.

Entoletion

The Entoleter is designed to destroy all forms of insect life in flour and other cereal products. Internally it consists of two horizontal steel discs, each of which carries at right angles to its surface a number of cylindrical steel rods. One of these discs remains stationary, while the other revolves at 2000–3500 rpm. The stock to be treated is fed into the middle of the upper and moving plate and thus thrown by centrifugal action toward the periphery. On its passage to the periphery the flour is violently impacted against the cylindrical rods whereby all forms of insect life are destroyed.

Flaking

Cereals and malted grains can be milled into flakes by crushing the moist grain through heated metal rolls. Flaked maize can be used as animal feed. Flaked cereals are used extensively in the production of breakfast cereals.

Roasting

Dry drum roasting of cereals or malted grains not only dries the product but also imparts a characteristic taste and flavour. Roasting cereals under infrared heaters is known as 'micronization' of grain and results in expanded products which are easier for animals to digest and which can be used in human foods such as muesli.

Wet milling

Maize is dry milled to produce grist for brewing or for the production of breakfast cereals, but a wet milling operation is used to extract starch (cornflour) from maize.

Pearling

Grains with a tightly adhering husk, such as rice, oats and barley, normally have the husk removed by an abrasion process known as pearling.

PRIMARY PRODUCTS

FLOUR AND RELATED PRODUCTS

Wholemeal

Wholemeal must contain the whole of the product derived from the milling of cleaned wheat.

Wheat flour, white flour

Flour is the primary product obtained from the milling of wheat by the gradual reduction system. The object of this milling process is to separate, as cleanly as possible, the endosperm of the grain from the enveloping skins and embryo. How effectively this separation can be performed in a given mill is a measure of its efficiency.

The term flour extraction rate is used to denote the proportion of flour obtained from wheat during the milling process. The numerical magnitude of the degree of extraction will depend on the basis upon which it is calculated. The extraction can be based upon (i) dirty wheat to silos, (ii) cleaned wheat to first break, (iii) total end products including screenings, and (iv) end products less screenings. Method (iv) is the most usually employed in Britain.

The wheat grain contains about 82% of white

starchy endosperm and so theoretically it is possible to produce white flour of 82% extraction. In practice 80% is about the limit of white flour extraction and most commercial white flours are around 77% extraction.

Flour is produced from wheat in stages and when all the separated fractions are mixed together the resulting mixture is known as 'straight run flour'. A divide consisting of a few of the flour fractions separated early in the milling system, thus only slightly contaminated with bran powder, is known as a 'patent flour' and sells at a higher price than the 'straight run flour'. Bread-making flour in the UK is milled from selected home grown wheats alone or blends containing a proportion of strong wheat from North America, and has a protein content of 10.5–12.5%. Its diastatic activity should be sufficient to provide ample sugar for the yeast throughout the fermentation of the dough. It has been customary to add an oxidative improver to bread flour to enhance its bread-making quality and to bleach it to meet the public demand for white bread.

Flour sold for domestic use may be less strong and less proteinous and consequently may be produced from blends of European or other weak wheats. A protein content of 9.5–10.5% would be suitable. Such flour is not always bleached and often contains no oxidative improver.

Biscuit flour

A biscuit flour should be weak and have a low protein content, 8.5% or less. The protein should have good extensibility. For some formulations in which the biscuit dough needs to be very extensible, flour is treated at the mill with sulphur dioxide. It can best be made from a weak British wheat.

Cake flour

Cake flours used for recipes with high sugar and high liquor contents are designed to contain the majority of particles in the 15–55 μm range, thus maximizing the amount of large undamaged starch granules. To achieve this narrow size range, flour particles are reduced in size by milling in a pin or impact mill which minimizes the amount of starch damage. A modern cake flour plant would normally consist of a system of pin mills with air classification stages following each milling process. Speciality cake flours may also be treated with chlorine to aid cake stability.

Brown flour

Brown flour differs from white flour in that it contains a relatively high proportion of bran, but it does not represent the whole of the grain from which it was produced. If all the endosperm, germ and all the skins of the grain were present, the product would be a wholemeal. Brown flours differ not only in bran content but also in granularity; in some brown flours, the bran is finely ground and in others it is present as relatively large fragments.

Most meals are produced in the mill by running together in suitable proportions stocks that in admixture give a product with the desired characteristics. Whatever the method of manufacture, the wheat blend should be a strong one if the meal is to be used for the production of brown breads. Meals milled from home grown wheats are used for the production of 'wheatmeal' biscuits.

The UK Bread and Flour Regulations 1984 required that brown flours should contain not less than 0.6% of crude fibre on dry basis. In the 1995 version of the Regulations, brown breads and flour have been removed as defined products. However, custom and practise would refer back to the 1984 definition.

Patent flour

Flours milled from the head of the reduction system, which have been processed from the cleanest semolina, representing the centre of the wheat endosperm, are known as 'patent flours'. These patent flours are very bright in colour and are used in the production of high class confectionery and morning goods (rolls, etc.) where brightness of crumb is desired.

Air classified flour

Wheat flour may be further processed into speciality flours of defined particle size and protein content by air classification. Conventional sieving cannot be used to separate particles below 80 μm in size but fine particles can be classified by a process in which the effect of centrifugal force is opposed by the effect of air drag on individual particles. The cut size at which the separation is made can be adjusted, for example, by altering the volume of air or its direction of travel.

Free wedge protein, in which starch granules are embedded in the wheat endosperm, is found in flour as approximatley triangular pieces which are normally smaller than 17 μm in size. This interstitial protein can be removed from flour by air classification, together with small B starch granules and fragments of damaged starch granules. Cake flours, particularly those used for the manufacture of sponges, need to be low in protein content and low in starch damage. Therefore large and fine particles may be removed from a flour by air classification to improve its sponge-cake making properties.

Semolina

Semolina is nodules of wheat endosperm, some of which carry particles of adhering bran. This mill stock is produced on the break rolls and after purification and grading is ground into flour. Purified semolina is sold as a commodity or for processing into pasta. The best quality pasta is made from semolina milled from durum wheat.

Self-raising flour

Self-raising flour is flour into which sodium bicarbonate has been incorporated and a water-soluble acid substance in such proportions that if they are permitted to react will yield sufficient carbon dioxide for aeration purposes without leaving a marked excess of either ingredient. Most self-raising flours contain 1.16% sodium bicarbonate and about 1.52% of 80% acid calcium phosphate. In some instances a mixture of acid calcium phosphate and acid sodium pyrophosphate is used as the acid ingredient. The presence in the self-raising flour of 1.16% sodium bicarbonate and sufficient acid body to neutralize this gives the flour an available carbon dioxide content of about 0.59%, which is well above the statutory minimum requirement of 0.40% and provides a good margin for loss during storage.

Since chemically aerated goods are baked shortly after the dough has been made, there is no opportunity for the gluten to 'ripen' hence self-raising flours need to be weaker and to contain less proteinous material than bread flours. A protein content of about 10% is sufficient. The flour should contain not more than 14.5% moisture and preferably only 13.5%, as there is then little fear of undue deterioration during storage. The use of sprouted native wheat in the blend is to be avoided, otherwise the flour will have a high α-amylase acitivity. This may not prove detrimental when baked goods are made, but may well cause trouble with steamed or boiled goods. The flour should be bright (colour grade

3.0 or less) and consequently in the past it used to be bleached.

Heat treated flour

The physical properties of wheat protein can be altered by the application of heat, the effect of the treatment lying in the direction of increased strength and stability and diminished extensibility. Properly applied heat, therefore, can lead to an improvement in the baking quality of flour milled from the wheat. The total effect of heat treatment is dependent upon three factors, the temperature, the duration of the heating and the moisture content of the flour being heated. The higher the moisture content, the greater the effect of a given degree of heating. A time and temperature of heating which cause a beneficial change in the protein quality of a flour containing 14% moisture may completely ruin the baking quality of a flour containing 16% moisture.

The heat treatment of flour can be used totally or partially to eliminate its enzyme activity and to denature the gluten proteins. Such inactivated flours are used for gravy thickening or soup manufacture, or as the basis of commercial batter mixes.

Malt flour

Malt flour is produced by milling barley or wheat that has been germinated and then kiln dried. α-Amylase is produced during the germination but the final diastatic activity of the malt flour is influenced by the temperature to which the grain is exposed during kilning. Malt flour was used for many years as a corrective for low diastatic activity in bread flour, a common rate of addition being 800 ppm. There is now a tendency for standardized fungal α-amylase preparations to be used in place of malt flour for this purpose, since fungal amylase has a lower thermal death point and will not continue to work in the baking oven, as was the case with malt flour. Malt flours of medium-to-high diastatic activity can be used in the production of sticky malt breads of characteristic flavour. Heavy kilned or roasted malt flours can have a bitter taste and are used mainly to colour bread.

Bran

The object of the modern flour milling process is to split open the wheat grains and to scrape out the endosperm. A perfect separation is not possible and some of the skins become powdered up in the flour, while some endosperm remains attached to the residual skins. The co-product of the flour milling operation, which consists of separated skins plus adhering endosperm, is known as bran. Its dietary fibre content is in the region of 40% (10% crude fibre). Bran is much used for stockfeeding but wheat bran for human consumption is used in brown breads and breakfast cereals, giving the consumer the benefits of increased fibre content in their diet.

Germ

The commercially separated embryo of the wheat grain is known as germ. Commercial germ is relatively high in protein content, fat and enzyme activity. In Britain the proprietary germ breads are made with commercial germ which has been steam cooked.

Wheatfeed and screenings

Wheatfeed is the name used for the by-products of the milling process other than germ. They comprise bran, the coarse residue from the break grinds and fine wheatfeed, the accumulated residues from the purifiers and the reduction grinding.

The term 'screenings' is used to describe the usable impurities that are removed from wheat during the cleaning processes and hence excludes such contaminants as stones, string and pieces of metal. Screenings consist of cereals other than wheat and foreign seeds and are ground and mixed into wheat feed. In Britain the screenings removed during cleaning of the wheat are generally ground, and unless contaminated with ergot, added to the fine wheatfeed.

Polentas

Maize polentas are coarse ground particles of maize corresponding to wheat semolina. The term 'polenta' is also given to a porridge made from them.

Rice cones

These are granular rice particles with sizes resembling that of sand. They correspond to wheat semolina.

Breakfast cereals

(See Chapter 12)

Snack products

(See Chapter 12)

SECONDARY PROCESSES

BREADMAKING

Bread is produced by making a dough from wheat flour and aerating this with carbon dioxide produced by yeast fermentation. The proportion of water in the dough mixture varies with the type of equipment used but is normally within the range 55–65% of the flour weight.

About 2% SALT is added to give flavour to the bread. It has a depressant effect on the yeast and the addition of salt to the dough is often delayed to the 'knock-back' before dividing and moulding in bulk fermented bread doughs. This is the so-called delayed salt method.

The proportion of yeast required varies according to its activity and the proposed fermentation time. Various factors have been suggested for calculating the correct amount. Commonly it varies between about 1% for bulk fermented doughs to 2% for short time doughs.

Developments in the technology of breadmaking have been concerned with speed and the elimination of bulk fermentation and its concomitant weight loss due to flour being converted to carbon dioxide by the yeast. About 80% of the bread eaten in the United Kingdom is made on plant bakeries.

One of the most important developments in breadmaking technology has been the Chorleywood Bread Process (CBP), illustrated in Figure 5.2, in which the dough is put into the ripe condition, previously only obtained by long bulk fermentation, by the input of a large amount of mechanical work within a short time. The method has been so successful that the bulk of the bread manufactured in the UK is now made by this method.

The characteristics of the CBP method are as follows.

(i) More yeast is used than in the bulk fermentation method because of the much shorter total fermentation time of about 1 h. Special

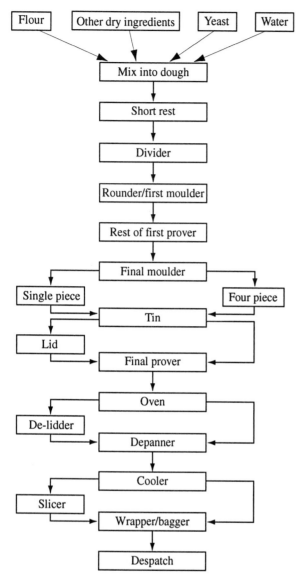

Figure 5.2 Chorleywood bread process (reproduced from Gedye *et al.*, 1981, with permission of the Royal Agricultural Society of England).

yeast strains have been developed for this purpose which are more active than the normal continental yeasts.

(ii) A very small percentage of fat is required (about 0.1%) with a melting point above the final dough temperature.

(iii) Ascorbic acid is used as an oxidizing improver.

(iv) An extra 3–4% of water can be added to the dough.

(v) The mechanical work input is 11 W h kg^{-1} of dough during the 5 min mixing period.

The method was originally taken up by plant bakers but it is now used in one variation or another by many small bakers. In the larger bread plants, doughs of 160–240 kg are mixed within 3

min. This requires very powerful mixers. In warm weather, because of the heat generated during the mixing, the dough water is chilled before addition. The dough is divided straight from mixing, given a 5–10 min rest or intermediate proof, moulded, tinned, proved for typically 45 min and then baked.

Activated dough development (ADD) is a quick alternative method of producing fermented products, e.g. bread and rolls, without using bulk fermentation or mechanical dough development. The essential feature of the method is the use of L-cysteine hydrochloride (at about 35 ppm of flour weight) as a reducing agent and ascorbic acid as an oxidizing improver. The addition of at least 1% fat is regarded as essential to improve the gas retention properties of the dough. As with the mechanical development of dough, extra yeast must be added to obtain the same degree of final proof as obtained in the longer fermentation methods. The addition of L-cysteine hydrochloride up to 75 ppm of flour is allowed in the UK.

Flour handling

The flour is delivered in bulk to bakeries in batches of up to 25 tonnes. It is then transferred pneumatically from the tanker to the flour silos, which commonly hold 50 tonnes each.

Dough mixing

The flour is drawn pneumatically from the silo to the mixing station, where the amount required may be called for by an automatic weigher, and the correct amount of water at the correct temperature metered into the mixer. The other ingredients may be metered in or, more usually, added by hand. A small proportion of doughs are still mixed in the traditional steel open-bowl mixer with one metal arm performing a kneading action as the bowl rotates. The bulk of doughs, however, are mixed on high speed mixers requiring a source of chilled water to counteract the increase in dough temperature caused by the rapid input of energy.

It has long been known that the improving action of ascorbic acid requires its oxidation by atmospheric oxygen to dehydroascorbic acid before it becomes effective. It is also known that during mixing oxygen is rapidly removed from the dough by the yeast. Thus, it is perhaps not surprising that the addition of oxygen as the sole oxidant to the mixer during mechanical mixing with ascorbic acid has a remarkable improving effect on the bread produced, comparable to the use of ascorbic acid with potassium bromate. This effect had been known for some time, but it was only in 1986 that the pressure to remove potassium bromate as an ingredient in bread resulted in a successful attempt to employ oxygen commercially.

The previous fears of an explosion resulting from passing a gas containing 60% oxygen by volume through the mixer headspace have now been dispelled. In the presence of raw soya flour, which contains a lipoxidase, the method has an added bleaching effect on the xanthins in the unbleached flour, producing a very bright white crumb.

Dough dividing

Dough dividers produce dough pieces of equal weight by a volumetric method. Ideally this should be done by weighing, but this is only practical for the small baker. Volumetric machines furnish dough pieces of constant weight only as long as the density of the dough remains unchanged. In practice this is very difficult to achieve. It is therefore essential that the weights of these dough pieces are checked frequently. Because of the difficulties associated with bread production and the requirements of the Weights and Measures Act, it is general practice to record the time and weight of each piece and to retain the records.

Handing-up (umbrella moulder)

This is a machine for moulding dough pieces into round shapes as they leave the divider. The best known type consists of a hemisperical shaped tunnel that spirals around a finely corrugated cone. This cone revolves and in so doing carries the dough pieces through the tunnel and discharges them at the top in a nicely rounded condition. A stream of air blown over the cone helps to prevent losses of pieces of dough due to sticking.

First proof

In order to relax the dough after dividing and handing-up, it is customary to allow the dough pieces to rest for a short period, usually 4–8 min.

Final moulder

The most common type of final moulder consists of a vertical train of sets of rollers through which the round piece of dough drops, exiting onto a moving belt in the form of a pancake. The leading edge of the pancake is flipped up and back by passing it under a dragging chain curtain. The resulting roll of dough passes under a pressure board designed to push any entrapped air out of the ends of the roll. This roll is deposited into the greased bread tin, either as a whole piece to give single piece bread, or cut into four pieces for the production of four piece bread.

There has always been a desire to produce bread with a bright white crumb and in large part, this is achieved by removing bran from the flour and bleaching the natural pigment remaining in the endosperm. However, there is a residual darkening effect produced by the natural gas holes in the slice, which is similar to that observed on looking into an unlighted cave. In single piece bread, this effect is accentuated because the standard method of moulding produces gas cells along the length of the loaf and these are cut across when the bread is sliced. Thus in cut bread one looks down deep holes into relative darkness.

In four pieced bread, the moulded doughpiece is cut into four and the pieces laid in the tin side by side. In this way the gas holes are oriented across the loaf, so that, when the loaf is cut, the cave-like holes are cut along their length and their maximum depth is their width and not their length. They now act, not as dark holes, but as concave reflecting surfaces with a consequential apparent increase in the brightness of the crumb face.

Final proving

The tinned dough pieces pass into a prover, where the tins are loaded onto travelling shelves so that, as they leave, a push bar can discharge a whole shelf full of tins across the width of the travelling oven. The temperature and humidity of the prover atmosphere are controlled, 42°C, 70–75% relative humidity (RH) being typical. The internal temperature of the dough piece on leaving the prover after 45–50 min is about 35°C.

Baking

The two main types of oven are the forced convection ovens in which hot flue gases are circulated around the tins and the radiant ovens where hot flue gases pass through tubes which radiate heat to the tins. This latter kind of oven also has some forced turbulence to help distribute the heat evenly. Typical baking times vary between 23 and 30 min. The bread leaving the oven is dislodged from the tins by an air blast and sucked out and deposited on the feed to the cooler.

Cooling

The cooler is basically a tunnel through which the bread is passed against a countercurrent stream of air of an adequate humidity to prevent too high a weight loss due to evaporation. Cooling is completed in 2–3 h.

Cooling of baked goods between baking and packaging requires an intermediate storage space and is a stage in processing during which there is a chance that the product will become contaminated with mould spores. It is also a time during which the collapse of some products occurs and in which the warm product is more susceptible to damage.

Vacuum cooling offers the possibility of avoiding these disadvantages. The hot baked product enters a sealed chamber and is subjected to a vacuum. The latent heat of evaporation, lost as some of the hot moisture within the product quickly evaporates, cools the product accordingly. It may then be rapidly moved onto the packaging line. One disadvantage of this method of cooling is the extra loss in weight incurred as a result of the evaporation of moisture and some allowance has to be made for this in the formulation of the product.

Slicing

The cooled bread should have an internal temperature of 27°C in order to avoid tearing during slicing. Although band slicers do exist, in which the knife consists of a continuous band, the most common kind of slicer used in the UK consists of a reciprocating frame of scalloped-edged blades through which the loaf is passed. The bread is packed either in waxed paper or heat sealable film or, more commonly, placed in plastic bags with a neck-tie.

BISCUIT MAKING

The major ingredients of a biscuit dough are flour, sugar and fat. Compared with a bread dough, very little water is added and the doughs are very firm. In mixing, the wheat gluten is not fully developed and an extensible gluten is required for processing either in a rotary moulder or in sheeting and cutting the biscuit shapes. Sodium metabisulphite (up to 200 ppm of sulphur dioxide on flour weight) may be added to biscuit doughs to increase their extensibility. The cut biscuit shapes are baked rapidly in a hot oven to give a dry product.

Continuous dough mixing

During the 1960s a considerable amount of work was carried out on developing a continuous mixer for bread dough. Essentially, a continuous mixer is barrel shaped and consists of a primary mixing chamber, where the flour and solutions or aqueous dispersions of the other ingredients are mixed, followed by a 'work' chamber in which the dough is subjected to sufficient mechanical work to ripen it. The dough exiting from the mixer as a continuous ribbon could then either be fed into conventional dough dividers and through the usual bread plant or it could be divided at the mixer head and dropped straight into the baking tin. This method gives a remarkable consistency of dough piece weight, something which is difficult to attain otherwise.

Continuous biscuit dough mixing is used very successfully by the UK biscuit industry but continuous dough mixers have almost entirely disappeared from the UK bread industry.

CAKE MAKING

The basic ingredients of cakes are flour, shortening, eggs, sugar and milk. However, the variety of cakes that may be produced by varying the proportions of these ingredients and the types of flour and shortening, plus the addition of other ingredients, is such that no attempt can be made to list them here.

A typical recipe is as follows:

Flour	1000 g
Butter	590 g
Shortening	90 g
Eggs	900 g
Sugar	700 g
Baking powder	15 g

Cake batters are usually prepared by either the sugar batter or the flour batter method. the former procedure, which is also known as the creaming method, is preferable when high class cakes are being made, but the flour batter method

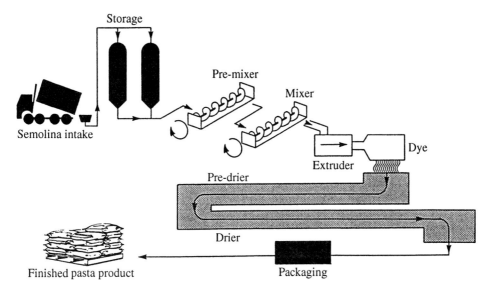

Figure 5.3 Pasta manufacture (reproduced from Gedye *et al.*, 1981, with permission of the Royal Agricultural Society of England).

is more suitable for the production of cheaper cakes with a low egg content.

The sugar batter method is performed by creaming together the fat and the sugar at medium speed for about 10 min, adding the egg in stages while mixing is continued, and finally making several alternate additions of flour and milk. The egg should be brought to a temperature of about 21°C before being added to the mix or the batter may curdle.

In the flour batter method, the shortening and the flour are creamed together until a fluffy mass is obtained. The eggs and sugar are whipped together and the resulting foam is them carefully blended into the creamed fat and flour. Milk is then added in small portions.

Cake batters are baked at temperatures between 149 and 208°C depending upon their richness, their weight and their moisture content. Richer batters of high sugar content require less heat than those made on leaner formulae.

RUSK AND CRUMB MANUFACTURE

Sausage rusk and crumb is baked from a tight chemically raised dough to give an open textured biscuit or bread which is ground into a coarse pale crumb, which has the ability to absorb large amounts of water without becoming sticky.

PASTA MAKING

Durum semolina is mixed into a firm dough with a small amount of water, and sometimes whole egg if yellow pasta is desired. The dough is either sheeted or extruded into a range of strips, strands or complicated shapes all known by Italian names such as spaghetti, lasagne, vermicelli etc. The cut or extruded pasta shapes are dried slowly in air to prevent the pasta from cracking. Figure 5.3 illustrates a typical pasta manufacturing process.

GLUTEN AND STARCH SEPARATION

Coarsely ground flour is milled from wheat in a simple roller milling process. The flour is either mixed into a dough or batter from which starch and fibre are washed to leave strands of gluten protein. The gluten is agglomerated into lumps, dewatered and dried very carefully in a ring dryer to minimize any heat denaturation of the vital wheat gluten. The starch is washed in hydrocyclones or decanters before being sold either in slurry form or as a dried product. The process is illustrated in Figure 5.4.

SECONDARY PRODUCTS

BAKED PRODUCTS

Bread

Bread is a baked aerated dough, the primary ingredients of which are flour, salt and water. The aeration is normally obtained by fermentation with yeast but may be obtained by other means.

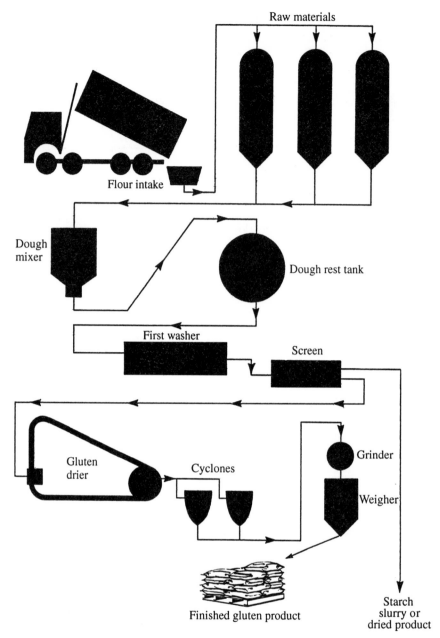

Figure 5.4 Starch–gluten separation (reproduced from Gedye *et al.*, 1981, with permission of the Royal Agricultural Society of England).

Fat is sometimes included in the dough mix. The yeast ferments the sugars natural to the flour and those produced by diastatic activity and thereby evolves carbon dioxide, which aerates the dough.

The majority of bread produced in the UK is made from white flour, but the proportion of bread made from wholemeal, brown and other non-white flours has increased rapidly over recent years. Speciality breads, such as malt, germ and protein-enriched breads are also available, usually under proprietary names. The composition of breads of all types is controlled by the UK Bread and Flour Regulations 1995.

Batch bread

This traditional type of bread is produced mainly in Ireland and Scotland. The loaf is of similar weight (800 g) to normal bread baked in a tin, but does not have any side crusts, having been produced in a fashion that resembles a giant bun round.

Although traditionally made by the bulk fermentation process, a greater proportion is now being made by mechanical development methods. This tends to give a less dense, silkier bread crumb which in spite of these uncharacteristic qualities appears to be acceptable, possibly more so to the younger generation. In both processes the dough is divided after mixing and moulded. The well-oiled dough pieces are packed together on the sole of a travelling oven, the first part of which acts as a final prover. The side dough pieces are traditionally retained by wooden setter boards. The touching dough pieces have no way to develop but upwards, producing a loaf of similar size to that constrained by a tin. However, as the dough pieces are touching throughout the bake, when they separate on leaving the oven, only the top and bottom of the bread have crusts. The bake required, about 1½ h, is considerably longer than for normal plant bread.

Part-baked loaves

The production of fresh bread for sale first thing in the day traditionally required bakers starting work in the early hours of the morning. These times are now regarded as unsocial and great efforts are made to avoid them. This may be accomplished by the use of part-baked bread. This is bread that has been made in the conventional manner up to the point just before the crust starts to caramelize, but after oven spring has occurred and the crumb has set. Oven spring is the sudden increase in loaf volume due to the warmth of the oven expanding the gas in the dough. This type of loaf, which has the same shelf life as ordinary bread, had a brief spell of popularity some years ago in the retail trade. It enabled freshly baked loaves to be baked in the home for immediate consumption.

This method is now being used on a large scale, particularly by plant bakers, for the production of speciality breads, e.g. baguettes, not easily produced on existing plant and by hot bread shops who wish to have hot crusty bread for sale within 15 min of the baker arriving to a preheated oven. For these commercial uses, the part-baked loaves are deep frozen immediately after baking and distributed deep frozen. Any staling that occurs is reversed during the subsequent bake-off.

Baguettes

Made in France traditionally, a baguette is a long (about 0.3 m) thin, round loaf with four to six diagonal cuts across the surface. The crust is very crisp and the cuts have a typical ragged edge. The crumb is open and non-uniform. Such bread, despite its comparatively short shelf life, has become increasingly popular in the UK.

Many early attempts to make such bread in the UK were unsuccessful for a variety of reasons, not the least being the use of too strong a flour. The result was often a loaf only resembling the traditional bread in shape and with an uncharacteristic 'blind' cut on the crust, the edge of which was almost continuous with the surface of the loaf. In French baguettes, the edges of the cut are irregular and sharp. Ideally such bread should be made with a flour having a low starch damage and a protein content between 9.5 and 10.5%. The dough should be relatively cool and, after moulding, the dough piece should be given a long cool proof (2–3 h) before being lightly cut.

In France, great efforts are being made to preserve the handmade baguette, where each dough piece is proved in an individual cloth lined basket before being hand tipped onto a flour sprinkled 'peel', a 2 m long handled wooden board, to be placed on the oven bottom. However, plants now exist that can produce baguettes indistinguishable from the handmade products.

Croissants

The croissant is now a major breakfast food item in countries far beyond its native France. It is manufactured using techniques similar to Danish pastry and puff pastry. Products termed 'croissant', however, vary hugely in texture and overall quality.

Croissants always contain flour, sugar and a fat (frequently butter), and may also contain egg and milk. Most are made from a yeasted dough. However, some are not and may rely on dough lamination with water-containing fat for their expansion. Altogether, the croissant is a very voluminous, lightweight expanded product. A typical recipe for croissants is as follows:

Flour	1000 g
Butter or margarine	600 g
Sugar	50 g
Yeast	50 g
Water or milk (warm)	0.5–1 l
Liquid egg	for glazing

The yeast should be blended with a little water and mixed with the flour and sugar. More water (or milk) should then be added to achieve a soft elastic dough. This should then be allowed to rise in a warm place for about 1 h to about double its original volume. The dough should then be placed on a floured board and lightly rolled. The softened butter or margarine should then be placed on it, and enclosed within the small piece of thick dough, which is then rolled and folded twice before being left to rest in a warm place. After three more rolling and folding operations, the dough should then be cut, first into wide strips and then into large triangles, rolled into crescent shapes, and placed on a baking tray to prove for about 20 min in a warm place. After brushing over with egg, the croissants should be baked for 10 min in a very hot oven.

Malt bread

There are many types of malt bread, ranging from the lightly malted to the heavily malted loaf with a dark sticky crumb. It is this latter type that would appear to be most popular and it is normally sold containing fruit. The stickiness of the crumb is obtained by the use of diastatically active malt flour or malt extract.

Protein-enriched breads

Bread containing not less than 22% protein (dry basis), produced by adding more gluten or other protein, may be called high protein or protein-enriched bread. Such bread normally has a high specific volume and although of the same calorific value weight-for-weight as ordinary bread, has a lower calorific value volume-for-volume. It therefore has some appeal in slimming diets as bread is normally eaten by volume, i.e. by the slice. The actual claim that the bread contains extra protein would appear to have little impact in the UK market.

Vienna bread and rolls

The characteristic features of Vienna bread and rolls are an open crumb and a thin, crisp glazed golden-brown crust. Traditionally this is achieved by a long cool fermentation with either a straight dough or a sponge and dough process using a stronger flour or, alternatively, by the addition of dried gluten to bread flour. Fat, malt flour, milk powder and sugar are common ingredients producing bread with a good flavour. Baking was traditionally done on the sole of a sloping oven designed to trap the steam within it. Steam is essential to obtain the characteristic appearance of Vienna goods. However, in modern electric ovens, the controlled provision of steam from a suitable boiler achieves the desired conditions much more easily.

Wheatgerm bread

In the milling of white flour the wheat germ, about 2% of the grain weight, is removed and should an excessive amount remain, bread from the flour will be of poor volume. Wheatgerm bread is prepared from a meal made by taking the separated wheatgerm and inactivating the enzymes present by moist heat treatment, then adding it back to white flour at not less than 10% by weight.

Wholemeal bread

The UK Bread and Flour Regulations 1995 state that wholemeal bread must be made from 'flour' milled from the whole wheat grain.

Biscuits

The range of biscuits made is clearly too large to enumerate here. Their individual characters are determined by the type of flour used, the proportion of sugar and fat, the condition of those ingredients when added to the mix (e.g. crystal size), the method of mixing (e.g. batch, continuous, creaming, all-in), the treatment of the dough and the method of baking. In this manner, biscuits varying from water biscuits to butter shortcake, and marie to puff biscuits, may be obtained.

Most types of sweet biscuit call for a low protein flour with an extensible gluten. In the UK, these flours are obtained almost entirely from British wheats. If the gluten in the biscuit dough retains too much of its elasticity, the cut biscuit shape will contract in the direction of the oven band producing, for example, an oval biscuit from a circular cut piece of dough. This may be overcome by treating the flour with sulphur dioxide or by the addition of sodium metabisulphite to the dough. An alternative method involves the use of proteolytic enzymes. Exceptions to the above include the cream cracker and puff biscuits, which sometimes require stronger flours.

Shortbread biscuits

A recipe for a good quality shortbread is:

Soft flour	1000 g
Butter	500 g
Caster sugar	250 g
Salt	20 g
Egg	100 g

The egg should be mixed with the sugar, then the butter rubbed in and the flour added by degrees. The baking temperature depends upon the size of the pieces which are being baked and normally ranges from 195–225°C.

Butter is the essential flavouring ingredient of shortbread. The flour used should be low in protein and of a granular nature.

Buns

A breadmaking flour is required for making fermented buns, basic recipes for which are as follows:

	Bulk fermentation (g)	Chorleywood bread process (g)
Flour	1000	1000
Yeast	70	85
Skimmed milk powder	30	30
Sugar	125	125
Fat	125	125
Egg	75	75
Salt	10	10
Colour (as required)		
Water	500	510
Commercial bread improver	–	10

The dough should be fermented for 1–1.5 h at 24–28°C, with a knock-back about three-quarters through the fermentation. After moulding, the buns should be proved to about double their size in a cool proof with enough steam to prevent skinning. Baking is at 210–220°C.

Cakes

Flour, sugar and egg are the essential ingredients for a cakemaking batter. Air is whisked into the egg to form a stiff batter which is poured gently into a mould or frame. The batter is baked to gelatinize the starch in the flour and to set the egg.

Sponge cakes

Traditionally the recipe for a good quality sponge was referred to as pint, pound, pound, i.e. a pint of egg, a pound of caster sugar and a pound of soft flour. Taking care to use grease-free equipment, the warmed sugar (32°C) is gently whisked into the egg to form a stiff batter and the flour is then blended in very lightly. The resulting batter is piped into tins and baked at 204°C. A small proportion of glycerine (5% of flour weight) may be incorporated into the recipe to extend the shelf life of the sponge. Such sponges rely on their egg content for their aeration.

Cost saving may be achieved by cutting the amount of egg used and using baking powder. A typical recipe is as follows:

Soft flour	1000 g
Caster sugar	1000 g
Whole egg	400 g
Milk	700 g
Baking powder	50 g
Glycerine	85 g

In this case, one batter is made with the egg and an equal weight of sugar and another with the milk beaten with an equal quantity of flour and the rest of the sugar and the glycerine. The two batters are blended together and the remaining flour, in which the baking powder has been sieved, is folded in by hand. The resultant batter is piped and baked as before. Commercially an 'all-in' method is used and all the ingredients are beaten together in a pressurized whisk for a much shorter period to produce the batter.

Crispbread

Although regarded as bread in continental European countries, rye crispbread is classified as a biscuit in the UK. It could very well stand in a class of its own.

Its manufacture relies on the high pentosan content of the wholemeal rye flour used, which allows a batter made from the flour and water to be whipped up to a stable foam. This is spread out evenly over a wire oven band, docked and baked off at a high temperature. Rye crispbread contains a high proportion of dietary fibre, 11.8% on a dry matter basis. Crispbreads may be made from wheat flour, sometimes with the addition of gluten, to produce the required structure more easily obtainable with rye flour.

Crumpets

Most crumpets are now made on a continuous plant. A typical recipe is as follows:

Flour (medium strength)	1000 g
Salt	30 g
Sodium bicarbonate	7 g
Cream powder	14 g
Yeast	11 g
Water	1200 g
Preservative	as required

In plant manufacture, water at about 38°C is used and the batter is allowed to ferment for 30 min. It is then deposited from a holding tank into hoops on polished hot plates. The traditional turnover to lightly bake the tops of the crumpets when they have set may be replaced by a hot plate lowering down onto their surface.

Pikelets are made from a similar batter but cooked on a hot plate without hoops. When the surface of the pikelet loses its wetness, it is turned on the hot plate to brown the top.

Crumpets have tended to be a seasonal (winter) product by tradition and, having a high water content, they are subject to mould growth. Efforts are being made to extend the shelf life.

Doughnuts

A doughnut is a fermented bun cooked in fat rather than being baked in an oven. The fried doughnut is drained and rolled in caster sugar, flavoured or unflavoured. Traditionally doughnuts contain jam which was folded into the unproved dough. The more modern practice is to inject the jam into the cooked product.

Muffins

Although baked like crumpets on a hot plate, muffins must not be confused with them. Muffins are thick well-aerated dough cakes with a soft moist crumb and are designed to be split in half for toasting before eating. Crumpets, by comparison, are made from a batter and are much tougher. A typical recipe is:

Strong flour	1000 g
Water	780 g
Yeast	25 g
Salt	15 g
Sugar	6 g
Skimmed milk powder	40 g
Shortening	10 g

The water level may need to be adjusted, but the dough should be soft. The dough is well mixed and should be moderately warm, e.g. water at 38°C should be used. After an hour the dough is knocked back and allowed to ferment for a further 30 min. The dough is scaled, moulded and set on a board dusted with rice cones or maize polentas and allowed to prove for about another 30 min before baking on a hot plate at a moderate temperature. When baked sufficiently on one side the goods are turned, but care is needed to ensure that the outside of the muffin is not burnt before the centre is baked sufficiently.

Puff pastry

Whilst much puff pastry is still made by hand in the traditional manner, considerably more is now being machine-made and distributed deep frozen.

In this form, it is much more economical in use for caterers and in-store bakeries. Puff pastry consists of alternate layers of dough and butter or pastry margarine. The pastry margarines are especially tailored for the purpose, having a sufficiently high melting point for working without oiling, but having a plasticity comparable to that of the dough so that they will both roll out evenly. When baked in a hot oven (232°C), the steam generated between the layers of butter or margarine 'puffs' the layers of dough apart.

Essentially, a recipe should contain margarine, about 75% of the flour weight, although fat contents range between 50% and 100%. The flour should be of medium strength. A cool dough is made using about 10–15% of the margarine. The dough is rolled out to a rectangle about 5 mm thick and the remaining plasticized fat is spread over two-thirds of the surface of the rectangle. The remaining area of dough is folded over the fat and then folded over again to give two layers of fat between three layers of dough. It is rolled again to form a rectangle of the previous size and folded in three once more. The paste is allowed to rest for a few minutes and the process repeated another 5 times with rests between folds or 'turns'. When the pastry is cut it should still be possible to distinguish the numerous layers of fat and dough with the aid of a magnifying glass. Much puff pastry is now made mechanically (see DANISH PASTRY).

Short pastry

Although the name 'short paste' is used for both unsweetened and sweetened pastes, it traditionally described the unsweetened paste used to make savoury products. In making a short paste, it is important not to overmix, and so develop the gluten, thereby producing a tough product liable to shrink on baking. Thus in the lean recipes where water provides the liquid, the flour and fat should be rubbed or mixed together first, before the other ingredients dissolved in the water are slowly mixed in. A typical recipe is:

Soft flour	1000 g
Baking powder	10 g
Shortening	250 g
Margarine	250 g
Fine caster sugar	125 g
Water	125 g

Danish pastry

Danish pastry is made by layering fat between sheets of lightly yeasted dough. The flour used is normally a high protein breadmaking flour. The fat may be butter or margarine. A typical recipe is:

Flour	1000 g
Yeast	70 g
Butter or margarine	600 g
Milk	50 g
Flour	150 g
Sugar	115 g
Egg	170 g

Danish pastry is rolled and folded in the same way as puff pastry. The bulk of such pastry is now made mechanically and deep frozen in appropriately sized squares. In this manner, even the small baker may benefit by taking out the deep-frozen squares and allowing them to prove before baking off, almost to demand, without having to go through the rather long procedure of making the paste.

Much work has gone into the development of machines to carry out this layering of fat and dough, the main problem being in rolling out the dough in a gentle enough way to avoid the dough layers splitting, as they will do if the reduction of dough layer thickness is effected too quickly or severely. These developments also apply to PUFF PASTRY.

Pittas

Pittas are traditional Greek flat breads which are gaining popularity in the UK. Of an elliptical shape, the two surfaces may be opened to form a pouch into which a filling may be put. Thus pittas may be described as an early convenience food.

They are now mass produced mechanically. The doughs used to make pittas contain a low level of yeast and can be mixed on a high speed mixer. Divided dough pieces are moulded into round balls, relaxed for about 15 min and then pinned first in one direction and then at right angles to give elongated flat dough pieces. After a short rest, these are baked for a few minutes on hot plates in a dome-roofed travelling oven. The dough pieces blow up to a rugby-ball shape before collapsing on cooling, thus giving the product its characteristic structure.

Scones

The aeration of scones is achieved through the use of baking powder, the choice of which is important from the point of view of taste in the final product. Cream of tartar is recognized as superior

in this respect but has the disadvantage of working off quickly in the cold. The use of acid phosphates gets over this problem, but often leaves the scones with a phosphate 'bite'. The use of glucono-δ-lactone as the acidifying agent helps to produce a sweet scone. Overworking of the scone dough, particularly in the commercial production of tea scones cut from a sheeted dough, produces a tough product.

Plain scones may be prepared from variations of the recipe given below. The fat is creamed with the sugar and the egg worked in. The remaining ingredients are added and the whole mixed to a dough of medium consistency. The dough is scaled, the pieces moulded round and flattened to an inch thick. Each round is half-cut into four pieces, given an egg wash and baked at 230°C.

	Rich recipe	Lean recipe
Flour	1000 g	1000 g
Baking powder	50 g	65 g
Fat	250 g	125 g
Salt	1 g	1 g
Sugar	220 g	125 g
Egg	150 g	nil
Milk	approx. 480 g	approx. 625 g

Soda bread

Soda bread is a chemically aerated bread, characteristic of Eire. The leavening is obtained by carbon dioxide liberated from sodium bicarbonate, traditionally by the action of the lactic acid in buttermilk. However, cream of tartar is commonly used as an acidifying agent for the bicarbonate, ensuring that the resultant bread is not too alkaline. Soda bread requires a weaker flour than white bread and a lower baking temperature is recommended. The flour used should be low in α-amylase acitivity.

Pasta

Products such as spaghetti, macaroni and noodles are extruded from very firm doughs made with semolina. The semolina may have been milled from a selected strong common wheat (*Triticum aestivum*) but the best quality pasta is made with semolina milled from the hard flinty durum wheats (*Triticum durum*). Pasta paste is extruded at high pressure without expansion into sheets or shapes, depending on the type of pasta and then carefully air dried. In the dried state, pasta goods such as spaghetti have a shelf life of many months. Chilled pasta products are becoming increasingly popular.

PRODUCTS FOR USE AS FOOD INGREDIENTS

Breadcrumb

Breadcrumbs, used for coating a variety of food products, have been manufactured for over 60 years. Whilst the component materials that go into breadcrumbs may be similar, the end results can be very different, according to application and end use requirements.

Whilst breadcrumbs were used for stuffings, scotch eggs and croquette-type products, the first real growth explosion in their use occurred with the advent of fish fingers and similar products. This type of crumb, still widely used today, is often referred to as conventional crumb. Its function is to add weight and to provide a total coating, which protects the substrate and plays a significant role in improving cooking and eating qualities.

In the late 1970s, new types of crumb were introduced which differed from conventional crumb. The new products were larger in particle size, irregular in appearance and much less dense in structure. These materials gave products which looked more 'homemade', had lighter eating textures and created products which were more versatile in terms of application and cooking, i.e. products that could be oven baked, grilled or fried. This second generation of breadcrumb was, and still is, referred to as 'Japanese style', 'J' type, or 'oriental'. Throughout the 1980s, variants on these products were introduced to include flavours, colours, herbs and spices.

The growth of fast food in the 1980s and 1990s has resulted in the development of newer breadcrumbs and 'breaded' products, which are used in coating systems to deliver particular texture and flavour characteristics, and to withstand the rigours of fast food and food service demands. Whilst most breadcrumbs are based on wheatflour, alternatives are available such as wholemeal, maize, potato and rice crumbs.

Breadcrumb is made from wheatflour, salt and yeast. The process is similar to the breadmaking process, except that the baked bread is dried to a lower end moisture content. The particle size ranges of the breadcrumbs depend on the end product application.

Breadcrumb of any type is only part of a food coating system. The design of a food coating will need to take into account the substrate, the method of processing, the storage cycle, the method of cooking, etc. This can be simple or complex, from a simple batter/crumb coating to a multilayered system, which may include predusts, more than one batter layer and even more than one type of crumb. For this reason, a food coating should be treated as a system, with each element contributing to the final result.

Rusk

Chemically raised biscuit rusk or bread dough is baked to form a hard dry product, which is ground into an open textured, pale coloured material used as a binder in the manufacture of sausages. Rusk has been manufactured on a commercial scale for about 100 years. Its use as a food ingredient has applications in many food products. The product most people readily associate rusk with is the British sausage, which it characterizes in terms of texture and cooking performance. The use of rusk is largely confined to markets in the UK, although there are exceptions.

Rusk is a biscuit-type material, usually made from wheat flour, salt and a chemical raising agent. The wheat flour needs to be of a consistent type as it is the major constituent. Therefore protein quality is as important as quantity and equally the starch content must also be consistent. Salt is included for flavour and preservative action. Salt can also affect colour and texture.

Fast acting chemical raising agents can be used and contribute to product structure and porosity. They leave little or no residuals in the end product. Yeast is not used as it produces a different texture and flavour and it can potentially give rise to a product with a higher residual yeast count. This would be a distinct disadvantage in the meat products in which rusk is primarily used.

The general process stages are: make a dough, form a sheet, bake evenly, dry the baked product to a consistent final moisture content and finally grind the material. The end product application will determine the particular particle size range required.

The role of rusk in a product such as sausages is to soak up and hold free water during the manufacturing stage and then to firm up the sausage in its casing. Its role during the cooking process is to characterize the product in terms of shape retention and texture and to retain meat juices and fat, thus giving a sausage which does not shrink or shrivel in the pan/grill and which has a succulent mouthfeel.

Rusk has a wider application than sausages. These can include pies, pastries, sausage rolls, small goods, vegetarian foods, stuffings and retorted products. It can also be used in puddings and some non-savoury products. The value of rusk as a material is that it has a bland colour and flavour, and delivers bulking and texturizing properties, thus providing food manufacturers with a versatile food ingredient.

Starches

Starches for food and industrial use are washed from cereal doughs or batters on an industrial scale in starch factories. These starches can be dried and/or chemically modified to give a range of materials for thickening and texture modification of foods.

Cornflour is not a flour but separated maize starch. Custard powders are coloured and flavoured cornflours. Cornflour is also used as a minor ingredient in some proprietary sponge mixes and cake flours.

Gluten

Gluten is formed when the endosperm wheat proteins (gliadins and glutenins) are kneaded in the presence of water. This is normally undertaken by mechanically manipulating a ball of dough under running water until the starch has been washed away. Gluten washed out from flour in this way contains about 70% water. When wheat flour is mixed with about half its weight of water, the proteins present absorb the water and unite to form an interwoven network of gluten strands. This network serves as the skeletal structure of the dough and confers upon it those elastic properties characteristic of a wheat dough. No other cereal, except rye to a very limited extent, behaves in this way.

The separation of wheat gluten is carried out on a large scale, the basis of the process being similar to that above. The wet gluten obtained is dried and ground. Drying must be carried out carefully and at a relatively low temperature otherwise the dry gluten will be devitalized. In other words, it will not reconstitute with water to form a coherent elastic and extensible mass analogous to fresh wet gluten.

Separated gluten, mostly in the dry form, is used principally in the production of high fibre and high protein breads. The latter are no longer popular and dried vital gluten is employed more frequently to supplement the low protein contents of the European and English wheats used almost exclusively in the grists of flour for ordinary white bread.

PACKAGING

PRIMARY PRODUCTS

Flour

Most flour milled in the UK is delivered in bulk road tankers but 25% of the production is packed into paper sacks in quantities ranging from 50 kg to 500 g.

Whether any material tends to dry out or to absorb moisture depends upon the relative humidity of the atmosphere to which it is exposed, and for each substance there is a relative humidity with which it is in equilibrium. Packaged flour will tend to lose moisture and hence weight when stored in very dry air, but will pick up moisture, and therefore gain weight, if exposed to air of high relative humidity.

Milling co-products

Bran, germ, semolina and wheatfeed may be delivered in bulk road tankers, 1 tonne bulk bags or a variety of different sizes of paper sack. Wheatfeed is usually pelleted before being delivered in bulk.

SECONDARY PRODUCTS

Breads

Bread is either heat sealed in wax wrapping paper or placed in a plastic bag sealed with a sticky tag. Crusty breads are best packed in perforated plastic bags to retain a dry crust.

Biscuits and crispbreads

Very dry products like biscuits and crispbread are sealed in moisture proof wraps and may also be boxed.

Metal detection

The chance inclusion of small pieces of metal in baked goods is not only dangerous for the consumer but can carry severe penalties for the manufacturer by way of fines, court appearances and adverse publicity. Such pieces of metal can arise, for example, from broken sieve wires, wear of wire bands, nuts and bolts falling from machinery, etc. Magnets will remove most of the ferrous contamination but not stainless steel – an argument against its use in sieves.

Non-ferrous metals are much more of a problem and their presence may be detected by their effect on the electric field produced by the metal detecting device through which the contaminated product passes. Once detected, the product may be diverted from the production or packing line or be blown off the belt into an appropriate receiver. All product thus rejected should be examined to find the source of the contamination, which should be corrected.

QUALITY ASPECTS

WHEAT

Intake tests

Every lorry load of home grown wheat is tested at the mill intake against stringent quality standards. The wheat is usually contracted as a known variety and typical tests are as follows:

(i)   Visual appearance (varietal purity, disease, infestation, etc.)
(ii) Screening/admixture
(iii) Specific weight
(iv) Hardness
(v)   Moisture content
(vi) Protein content
(vii) Hagberg falling number

Insect pests associated with wheat

The common field insect pests which may occasionally arrive at the flour mill with the wheat are:

(i)   Thrips (*Thysanoptera*). These are small insects, 1–1.5 mm in length, which are commonly trapped between the chaff and the grain or within the grain itself.
(ii) Wheat bugs (*Hemiptera*). The wheat is attacked shortly before maturity and the insects enter the grain while it is still soft. The proteolytic activity of the saliva of the insect can cause bread doughs to be very runny and the bread to collapse in the oven.
(iii) Wireworms (larvae of click beetles)

(iv) Leatherjackets (larvae of crane flies)
(v)   Wheat bulb flies (*Delia coarctata*)
(vi) Frit flies (*Oscinella frit*)
(vii) Clover weevils (*Sitona* species) which are active at harvest time and spread to the cereal crop from adjacent grassland.

Wheat stored in silos or passing through the mill screenroom is also subject to attack by a small range of active insects. The common wheat-infesting insects are the saw-toothed grain beetle (*Oryzaephilus surinamensis*), the red-rust grain beetle (*Cryptolestes ferrugineus*), the grain weevil (*Sitophilus granarius*), the rust red flour beetle (*Tribolium castaneum*), the white shouldered house moth (*Endrosis sarcitrella*) and occasionally the warehouse moth (*Ephestia elutella*). Where cleaning has been long neglected, the scavenging insects, e.g. the yellow mealworm beetle (*Tenebrio molitor*), the brown house moth (*Hofmannophila pseudospretella*), the Australian spider beetle (*Ptinus tectus*) and the globular spider beetle (*Trigonogenius globulus*) are also often present.

Hagberg falling number

The falling number method is applicable to both wheat and flour and is a measure of the α-amylase activity. An aqueous suspension of ground wheat or flour is stirred in a precision-made test tube with a stirrer of standard design. During this time the test tube is immersed in a boiling-water bath. At the end of exactly 60 s, the stirrer is lifted to its extreme vertical position. The time taken for the stirrer–viscometer to fall a standard distance is recorded. Hagberg quotes the falling number as the time in seconds for the viscometer to fall plus 60 s. The higher the falling number, the lower the α-amylase activity. Provided that all conditions are standardized, the test is particularly useful in that it gives a rapid indication of which wheats or flours are liable to yield a sticky crumb when used for breadmaking.

Moisture content

A knowledge of the natural moisture content of the wheat is necessary to enable the miller to decide on the amount of water he will need to add in the conditioning process. The moisture content of the final blend determines both the efficiency with which the wheat can be milled and the moisture content of the milled product.

Whether the moisture content of flour remains stationary or changes during storage depends upon the relative humidity of the atmosphere to which it is exposed. A moisture content of 13.5–14.5% would be in equilibrium with a relative humidity of about 65%, an average figure for the climatic conditions prevailing in Britain.

The figure obtained in a determination of the moisture content of wheat or flour varies with the method employed. A British Standard method exists which the milling and baking industries normally use as a reference method. This is an air oven method in which 5 g of white flour are heated for 1.5 h at 130°C. The moisture content of grain and flour can be determined more rapidly but usually with less accuracy by means of electric moisture meters, which depend upon the change in conductivity or dielectric constant with moisture content. Recently near infrared reflectance (NIRR) instruments have been introduced into cereal laboratories, and this technique can give a good correlation with moisture content determined by oven drying. NIRR methods must be calibrated against an oven method.

The moisture content of bread for analytical reference purposes is normally quoted at 40%. This figure applies more to tinned bread than oven bottom bread. Because of its greater exposed area during baking, oven bottom bread may have a moisture content as low as 36%.

Near infrared spectroscopy

Prior to 1965, Carl Norris working in the USA on the development of a grain moisture meter had investigated the use of near infrared reflectance (NIRR) spectroscopy. Difficulties were encountered in the moisture measurements from interference as a result of absorption due to other constituents such as protein, starch and oil. This problem was solved by the use of a computer correlation transform technique, which not only enabled the interferences from protein, starch and oil to be eliminated but also permitted the determination of these constituents. The first commercially produced NIRR instruments, which were launched in the USA in 1971 and 1973, were crude to operate and needed repeated calibration against known standard samples. However, in the next five years, developments in optics, detector systems and computing power resulted in the production of a range of very successful NIRR spectrometers that were simple to use and had stable calibrations. Near infrared reflectance spectroscopy is now well established in the food industry as a rapid means of quality control on raw materials and products. The milling industry in the UK

has used NIRR measurements of moisture and protein in wheat and flours since the mid-1980s. Analysis of moisture, protein and fat in bread, biscuit and cake has been investigated by many workers using NIRR techniques. Methods and equipment based on NIRR spectroscopy for on-line control of moisture and protein have been developed for use in the flour milling industry.

Robust near infrared reflectance calibrations for moisture and protein in wheat and flour rely on the selection of the most appropriate wavelengths in the spectra. Most commercial NIRR instruments have six filters, with wavelengths chosen on other grounds, but the most reliable calibrations are those based on specific wavelengths known to correlate with prime vibrations, or overtones, of chemical bondings in the constituent being monitored. The NIRR technique is affected by the particle size of the sample being scanned, so wheat samples must be ground to a fine powder and all powder samples presented to the optics of the machine must be compressed in a standard manner to avoid errors due to difference in packing density. Since about 1986 it has become possible to use infrared transmission spectroscopy to assess moisture and protein in whole cereal grains. However, the falling number test requires ground grain for testing, so millers have not adopted whole grain techniques as readily as have cereal merchants and shippers.

Protein content

The determination of protein content is an important routine test of the cereal laboratory. Nitrogen content is determined by the Kjeldahl method and for wheat this figure is multiplied by 5.7 to give the protein. (It should be noted that the factor 6.25 is used for wheat products that are intended for animal feedstuffs and that many European countries use the factor 6.25 for all wheaten products).

Wheat bran and germ contain high levels of protein but little of the gluten-forming proteins which give a bread dough its elasticity and ability to hold the gases produced in fermentation.

Specific weight

The specific weight of wheat is determined in an instrument known as a chondrometer, which consists of a metal cylindrical receptacle attached either to a spring balance or to a pivoted beam which carries a sliding weight. The dial attached to the spring balance, or the beam on which the weight slides, is calibrated in kilograms per hectolitre. This measure replaced the Imperial Bushel weight in 1973. In North America, the Bushel weight is still used for cereals.

The specific weight will be influenced by the presence of foreign material in the wheat, by the plumpness of the grain, by the proportion of damaged kernels and by the moisture content of the wheat. A wet English wheat may weigh only 68 kg to the hectolitre, whereas a dry North American wheat may weigh as much as 84 kg to the hectolitre.

FLOUR

Quality tests

Alveograph

This is a dough-testing instrument developed by Chopin from his earlier extensometer. An unyeasted dough prepared under standardized conditions is rolled into a wide ribbon from which four circular discs are cut. These are allowed to rest for 20 min in a constant temperature chamber and then, in turn, are clamped across an orifice through which air is passed so as to blow the disc into a bubble. The inflation is continued until the bubble bursts. Throughout the inflation process, the pressure within the bubble is automatically recorded. From the mean of the four curves thus produced, certain measurements are obtained: stability (maximum curve height), area under the curve, length of the curve. These figures will depend on the quantity of water added. The International Association for Cereal Chemistry (ICC) has suggested 50% for laboratory milled flour.

For export flour an extra figure, the W value, is often required by former French colonies still operating older French standards. This is derived from the area under the extensometer curve multiplied by 6.54×10 ergs. In practice only the factor 6.54 is used, giving a range of W from 50 to 250. For breadmaking flours destined for export, a requirement exists that the W value should not be less than 130.

Amylograph

The Amylograph, which is in effect a temperature-controlled rotating-bowl viscometer, was designed to assess the α-amylase activity of flours. It provides a continuous record in gra-

phical form of the changes in viscosity which occur as the temperature of an aqueous suspension of the flour under test is raised at a uniform rate. The viscosity will increase as the starch gelatinizes but the magnitude of the increase will be related to the α-amylase activity of the flour. The higher the α-amylase activity the smaller the rise in viscosity.

The amylograph will only measure the effects of natural cereal α-amylase (from the wheat grist or any supplement with malt flour). The presence of fungal amylase supplements is not measured by either the Amylograph or the HAGBERG FALLING NUMBER method.

Ash

The endosperm of the wheat grain has an ash content of 0.3–0.4%, whereas the skins which envelop the endosperm have an ash content in the region of 9%. Hence, the natural ash content of a flour is an index of the extent to which it is contaminated with powdered skins, i.e. bran powder, and therefore a measure of the efficiency with which the milling process has been performed.

The brightness of a flour, its 'colour grade', is related to the proportion of bran powder present and hence to the natural ash content, and for many years the milling industry accepted ash content as a measure of flour colour. The ash content lost its significance in Britain when it became obligatory for millers to add chalk to flour, since it no longer represented the ash natural to the flour. It is for the same reason that the ash content of self-raising flour, i.e. flour to which aerating chemicals have been added, is no index of the colour of the flour.

The ash content of flour can be determined by igniting a weighed quantity in a silica or platinum dish at 550–900°C until the loss in weight does not exceed 1 mg h^{-1}. The ash of flour is mainly potassium phosphate. The following figures are typical of the results of an analysis of flour ash:

Potassium (as K_2O)	37.04%
Magnesium (as MgO)	6.12%
Calcium (as CaO)	5.53%
Iron and aluminium (as Fe_2O_3 and Al_2O_3)	0.36%
Phosphorus (as P_2O_5)	49.11%
Sulphur (as SO_2)	0.40%
Chlorine	trace

Damaged starch

During the milling of wheat, some of the starch granules in the endosperm become damaged. If a flour is wetted with a dilute solution of Congo Red, the damaged starch granules will take up the stain and can be identified under the microscope. Hard wheats suffer more starch damage than soft wheats during milling and the starch of any one wheat will undergo a degree of damage that is related to the condition and mode of operation of the milling plant.

The breadmaking properties of a flour are influenced in two ways by the proportion of damaged starch that it contains. A flour that contains a small proportion of damaged starch may fail to produce sufficient sugar for the yeast despite an adequate α-amylase acitivity because α-amylase cannot act on undamaged starch. The proportion of damaged starch also affects the water absorption of a flour; the higher the content of damaged starch, the higher the water absorption. By increasing the pressure on the reduction rolls which operate on very pure stock, it is possible to increase the amount of starch damage and hence the water absorption, without detriment to colour grade. Too high a degree of starch damage results in loss of loaf volume. The damaged starch content of a flour can be assessed by microscopic examination or by enzymic digestion methods. In the United Kingdom, the Farrand method is widely used. This method assesses the amount of reducing sugars produced after a suspension of flour has reacted with excess α-amylase under specified conditions.

In contrast to breadmaking, the manufacture of biscuits, cakes and pastry requires a flour with a low level of starch damage. Biscuits and cakes are relatively low moisture products which are made from doughs or batters containing little water.

Extensograph

In the Brabender extensograph, a dough-testing instrument, a cylinder of dough is supported in a horizontal position at its two ends and is then deformed by a moving arm which pulls the middle of the dough piece downwards. The stress imposed upon the dough and the degree of extension are recorded continuously throughout the test upon a sheet of paper affixed to a rotating drum. The curve affords numerical measures of the resistances and extensibility of the dough under test. The area under the curve gives an indication of the strength of the flour and of its breadmaking potential. Doughs for this test are preferably prepared by the FARINOGRAPH, which determines the water addition to be used. Oxidative improvers such as chlorine dioxide and ascorbic acid affect the shape

of the curve and their presence may be inferred from the results.

Farinograph

The Brabender farinograph is a dough-testing instrument. The dough under test is mixed in a twin-bladed water-jacketed mixer by means of a freely suspended electric motor to the housing of which is attached a pen, which records its movement via lever systems on a sheet of moving paper. The torque exerted on the motor as a result of the resistance offered by the dough to the movement of the mixer blades causes the whole motor and its housing to rotate through a small arc and this is reflected in a movement of the pen. The consistency of the doughs is adjusted, by varying the water added, to bring the pen always to the same chosen value.

The farinograph is a useful instrument in a mill or bakery laboratory since it can be used to assess the water absorption of a flour and to prepare doughs for the extensograph test.

Fermentograph

The fermentograph is designed to measure the gas producing power of a dough. A weighed amount of a yeasted dough is placed in a rubber balloon, which is then suspended from a pivoted counterpoised arm in a tank of water, the temperature of which is thermostatically controlled. As gas is produced in the dough, the balloon becomes more buoyant, and as it gradually rises in the water the counterpoised supporting arm moves. In so doing, it causes a pen to trace a line on a sheet of moving paper. At the end of each hour, the balloon is removed from the water and thoroughly pummelled, so as to knead the dough within it. The gas expelled from the dough by this treatment is allowed to escape via a tap and the balloon is then resuspended in the tank of water for the next hour of fermentation. In this way a series of curves is obtained, which portrays the amount of gas produced each hour and the rate at which it is produced. This equipment is widely used for testing the gassing power of commercially produced yeasts.

Filth test

The filth test is the name given to a technique which originated in the USA for the determination of the extent to which a foodstuff is contaminated with rodent hairs and insect fragments. It is a measure of the hygiene attendant upon the storage of the raw materials and the finished product and upon the manufacturing operations. The rodent hairs are not harmful in themselves, but their presence is an index of contamination with rodent excreta, since rodent faecal pellets are highly charged with hair.

Wheat may become contaminated with rodent excreta on the farm, and some of the faecal pellets may escape the cleaning processes to which grain is subjected at the mill. In this event, the flour will contain some fragment of rodent hair. Although no effort should be spared to attain the highest possible standard of purity in a foodstuff, it may also be borne in mind that flour is rarely consumed in the uncooked state.

The filth test is performed by digesting a foodstuff so as to free the hair and insect fragments and then floating these contaminants off in light mineral oil. The oil is filtered and the collected debris examined microscopically.

Grade colour

Grade colour of flour is determined by:

(i)   the proportion of branny matter present,
(ii) the proportion of unoxidized yellow pigments present,
(iii) the presence of smut or dirt,
(iv) granularity.

That aspect of flour colour which is determined by the bran powder content is the brightness factor, known as the 'colour grade'. The smaller the proportion of branny matter in the flour, the brighter is its appearance. The colour grade is independent of the degree of creaminess, which is governed by the proportion of unoxidized yellow pigments present. In order, therefore, to express the colour of a normal sound flour of average granularity, it is necessary to be able to measure separately the two independent factors of brightness and creaminess.

The brightness, or the colour grade, of a flour is evaluated instrumentally on a Kent-Jones and Martin colour grader. The reflecting power of a suspension of the flour in water is measured photoelectrically with a standard surface as reference. Attempts to measure the colour grade in the same way but continuously on the flour stream have not been successful. However, automated image analysis systems which detect the presence of bran particles in white flour are being developed.

Granularity

The granularity or particle size of a flour is determined by the severity of the grinding and the mesh sizes of the sieves through which it is sifted during manufacture. Variations in granularity encountered in normal commercial bread flours have little effect upon baking behaviour. Excessive overgrinding, however, will have a detrimental effect since high levels of starch damage reduce the tolerance of the dough to changes in fermentation conditions. The finer a flour, the whiter its appearance, but the effects of granularity disappear when the flour is made into bread.

Some guide to its granularity can be obtained by submitting a flour to a standardized and controlled sieving test employing a range of flour sieves. More detailed information and a better picture of the distribution of particle size are provided, however, by sedimentation or more sophisticated methods such as laser scanning or electrolyte displacement (Coulter counter).

Maltose figure

The maltose figure has been used as a measure of the ability of a flour to produce maltose from its starch by diastatic action. It is now very rarely used, having been superseded by direct measurement of α-amylase and starch damage.

Mixograph

This American dough-testing instrument, designed by Professor Swanson in 1933, is not as widely used as the farinograph or extensograph. Doughs are mixed in a high speed pin mixer and a mixing curve is drawn whose shape can be related to flour or wheat strength. The mixograph can differentiate between very tough stable doughs in less than half the time needed for a farinograph test. However, this type of tough dough is unlikely to be encountered with British milled flours.

Amylases

The amylases are enzymes that attack damaged starch and thereby produce reducing sugars. Two are present in wheat. α-Amylase attacks 1:4 α-glucosidic linkages in both amylose and amylopectin at random points within the starch molecule. The main product of the reaction under normal conditions is dextrins of low molecular weight. β-Amylase also attacks 1:4 α-glucosidic linkages, but only from the non-reducing end of the starch molecule. Maltose units are split off until a branch point (e.g. a 1:6 linkage) is approached, when action stops. Thus, β-Amylase can degrade amylose completely to maltose, but its action on amylopectin ultimately yields maltose and a high molecular weight dextrin, the so called β-limit dextrin. A combination of the two enzymes accelerates and extends the action of β-amylase because the α-amylase exposes additional straight chains on which the β-amylase can react. Wheat flour contains high levels of β-amylase, but its total diastatic activity is controlled by the α-amylase activity.

An excess of cereal amylase will produce bread with a 'sticky' crumb, which has a tendency to be torn by the high speed slicers in modern plant bakeries.

Three sources of amylase have been used in the manufacture of bread: fungal (see FUNGAL AMYLASES), cereal and bacterial. They are differentiated by their thermal death points, about 73°C, 78°C and 100°C, respectively. Thus for white bread, fungal amylase is the safest to use being inactivated before excessive degradation of starch has occurred. Bacterial amylase, although considered very powerful, is not suitable for most bread because it can survive the baking process. It is however useful in the manufacture of malt bread where a sweet sticky crumb is being sought.

Ageing of flour

White flour stored under normal atmospheric conditions changes with time in both baking quality and in colour. The colour gradually whitens as a result of oxidation of the yellow pigments present. The baking quality is improved, the protein characteristics being altered in such a way that the flour gives a stronger, more stable and more resilient dough. If flour is stored for a protracted period, the initial improvement in baking quality is followed by deterioration, the flour becoming rancid. Although white flour may improve in breadmaking quality over a period of a year, wholemeal flour, which contains the wheat germ, will go rancid more quickly and baking properties can deteriorate after two months' storage at ambient temperature.

Because of the beneficial changes that occur during storage, it was at one time the practice of millers never to send flour to bakers until it had been aged for several weeks. Later it was found that the desirable improvement in baking quality

could be achieved promptly by the addition to the flour of a small amount of ammonium persulphate. Subsequently it was found that other oxidizing agents could exert a similar effect and these became known as 'improvers'. They were adopted by the milling industry for the artificial and rapid ageing of flour. It was also discovered that bleaching, which was a feature of natural ageing, could be accomplished rapidly by certain oxidants and these too were adopted by the industry. Full artificial ageing, that is, the accomplishment of both bleaching and improvement of baking quality by the addition of chemicals, has been common practice in British flour mills for many years. Benzoyl peroxide and chlorine dioxide have been used as bleaching agents. The latter chemical also had an improving action and could assist the oxidizing improvement of ascorbic acid in a long fermentation breadmaking system. By July 1997, at the latest, British millers will not use either benzoyl peroxide or chlorine dioxide and all flour will be unbleached.

Elasticity of doughs

A dough made from wheat flour has elastic properties, whereas the doughs furnished by other cereal flours are plastic. It is because of the elasticity of a wheat flour dough that it is possible to inflate the dough with gas and retain the gas in bubbles which give bread crumb its characteristic texture.

Phytase

This is an enzyme present in flour which hydrolyses phytic acid into inositol. Phytase is located in the bran and germ of wheat; relatively little occurs in the endosperm. Phytase is active during the fermentation and in the early stages of baking and may hydrolyse 80% of the phytic acid.

Strength

For a long time, an accepted definition of the strength of a flour was its ability to produce large, shapely loaves. This, however, is an unsatisfactory definition, because the nature of the bread produced from a flour is not necessarily a reflection of the strength of that flour; bread quality can be markedly influenced by the diastatic activity of the flour, the oxidative treatment of the flour, the manipulative skill of the baker and the bakery technique. It is only when the diastatic activity of the flour is adequate and the dough has been correctly fermented, manipulated and baked that loaf volume and shapeliness are reliable indices of strength. A more satisfactory method of assessing flour strength is by appraisal of the physical properties of the dough during fermentation, e.g. its stability, resilience, etc. This method excludes the gas production factor and enables a flour to be classed as strong, although it could yield a loaf of small volume owing to insufficient diastatic activity.

Water absorption of doughs

The amount of water in any particular system will depend on the type of end product being made. Biscuit doughs are relatively stiff compared with bread doughs and the consistency of pastry doughs is generally between these two extremes. A bread flour will absorb 65% water but a biscuit flour will absorb only 50% water.

By mixing a fixed quantity of flour with water in a dough mixer to a set dough consistency, the amount of water absorbed by the flour can be assessed. This can be measured accurately in the laboratory by the farinograph. In modern plant bakeries, the dough mixers work in a similar way with weighed additions of water and flour being monitored against dough consistency during the mixing cycle.

Enrichment

The UK Bread and Flour Regulations 1995 require that all flour shall contain not less than prescribed proportions of vitamin B, nicotinic acid and iron, these proportions being:

Vitamin B	0.24 mg/100 g
Nicotinic acid	1.60 mg/100 g
Iron	1.65 mg/100 g

A flour of 80% extraction could meet this specification but flours of lower extraction must be appropriately enriched with a mixture of these nutrients. The composition of the master mix employed is such that compliance with the statutory requirements is achieved by adding it to 75% extraction flour at the rate of 22 ppm. These Regulations also require that all flour except wholemeal, self-raising flour with a calcium content of not less than 0.2% and wheat malt flour shall contain between 235 and 390 mg of chalk per 100 g.

Hygiene

Insect pests in flour and cereal products

Flour and cereal products also attract and support the rapid development of a number of prolific insect pests. Failure to clean flour silos, storage bins and flour handling equipment on a regular basis and to apply, where necessary, approved insecticides will rapidly result in infestations.

The pests most commonly encountered are the Mediterranean flour moth (*Ephestia kuhniella*), the confused flour beetle (*Tribolium confusum*), the broad-horned flour beetle (*Gnatocerus cornutus*), the flat grain beetle (*Cryptolestes turcicus*), the Drugstore or bread beetle (*Stegobium paniceum*), the Cadelle (*Tenebroides mauritanicus*), mites and in recent years, the booklice (*Psocids*).

The Mediterranean flour moth is the most common and most troublesome pest in the flour mill and bakery. The larva produces a mass of silken threads and excreta (frass) which often collects flour dust and forms a matted web. This may cause obstructions to the flow of stocks or contaminate products. The adult moth is grey in colour with a wing span of about 2.5 cm. The beetle pests (*Coleoptera*) are found in flour and cereal products in both their larval and adult forms.

Mites infest most types of stored products, the species usually found in flour and wheat products being *Acarus siro*. The adult mite is about 0.5–1 mm in length. The eggs are only about 0.1 mm in length and hence will pass through a No. 14 flour sieve. The usual climatic conditions in the United Kingdom are not sufficiently favourable for the rapid development of mite eggs which may be present in stored flour. If, however, the storage is unduly prolonged and the humidity of the surrounding atmosphere is high, then infestation may occur. A relative humidity below 60% however, does not permit the eggs to develop even if the temperature is favourable.

The presence of living mites in flour can be detected by smoothing the surface of the sample and examining it after a few minutes. If living mites are present the surface will be roughened by their emergence. The presence of dead mites in flour can be detected by microscopical examination, but only readily by carrying out an extraneous matter (filth) test involving acid hydrolysis of the flour.

The psocids found associated with stored flour and cereal products usually comprise the two species *Liposcelis bostrychophilus* and, less commonly, *Lepinotus patruelis*. Both are tropical insects which have established themselves in Europe since the early 1940s. They seek locations having a temperature between 25°C and 30°C with a relative humidity of 75%. Because of their small size (smaller than a pin head), their pale brown colour and lack of mobility, they are very difficult to detect. *Lepinotus* can tolerate cooler conditions and is common in grain silos, where it can be found on dusty walls and in warehouses, where it occurs on wooden pallets. *Liposcelis* is parthenogenic; the eggs are not fertilized and all adults are female. Under favourable conditions, a single female will lay on average 110 eggs during her 6 months life and frequently 75% of the juveniles will survive to become egg-laying adults. Although they present little hazard to health, they are annoying psychological pests, particularly when present in large numbers. Contamination of foods frequently takes place in the home, the insects commonly being present in warm humid kitchens. It has not yet been established whether or not they are semi-permanent occupants of households or are introduced with foodstuffs, possibly in small numbers which increase rapidly under the favourable environmental conditions. Recent evidence suggests that *Liposcelis bostrychophilus* is not easily killed with safe domestic insecticides and control techniques may well have to rely upon environmental conditions, i.e. the cold, dry storage of susceptible foodstuffs.

Fumigation

Flour mills, on occasion, may be fumigated in order to keep in check the common mill pests, particularly the Mediterranean Flour Moth, which produces a sticky webbing during the larval stage and which may cause blockage of spouts and elevators.

The most commonly used fumigant is methyl bromide. The mill premises are carefully sealed, strips of adhesive tape being stuck over all cracks and crevices and methyl bromide is then released into the building. The gas is fed from cylinders outside the building into pipe lines bearing outlet nozzles which have been installed throughout the premises. The building is kept under gas for 24 h and is then opened up and thoroughly aired by natural or forced ventilation for a further 24 h. This method is now very little used in the UK, where more emphasis is being placed upon regular detailed inspection of the mill situation in which infestatin could develop. The Montreal Protocol advocates a worldwide ban on the use of methyl bromide early in the next century, but as yet no safe or effective alternative exists for the flour

miller to use as a fumigant. Where necessary, localized application of insecticide is used to combat insect infestations.

Rodents and rodent control

Contamination of wheat and flour by rats and mice will clearly affect the hygiene quality of the final product. The miller can only ensure that the wheat being purchased has been harvested and stored satisfactorily by buying from reliable sources which maintain adequate control measures, by laying down relevant specifications and by examining the wheat on delivery. The miller also needs to ensure that the mill perimeter is clean and rodent free, that the premises are adequately rodent proofed and free from rodent harbourages, that storage conditions and housekeeping are satisfactory and that efficient perimeter and internal rodent-baiting techniques are employed.

Similar precautions are also essential in bakeries where adequate inspection of incoming ingredients, packaging materials and equipment, screening of raw materials before use, good storage and housekeeping and efficient rodent proofing and rodent baiting, including the possibility of employing a specialist contractor, should form part of a continuous rodent control programme.

BAKED GOODS

Crust colour

The colour of the crust of bread is due in part to the reaction between reducing sugars and proteins. Any factor which affects the amount of sugar remaining in a dough when the dough goes to the oven will affect the crust colour of the bread. Dextrose, milk or whey powder is often added as an ingredient to intensify crust colour, the latter two ingredients giving a more 'foxy' hue.

Crumb softeners

It is said that water is the best bread improver and certainly any method of incorporating more water into the final crumb will improve its softness. Water absorption of the flour can be increased by increasing the level of damaged starch but if this is taken to excess and without a parallel increase in protein content, poor quality bread will result.

A decrease in loaf density, i.e. a greater volume,

will give a softer crumb; this may be achieved by the addition of emulsifiers, with or without fat. The addition of fat alone will soften the crumb and 'bank holiday bread', which has to last over the long weekend, often has extra fat added for this purpose.

Of the emulsifiers permitted in bread, the mono-glycerides are thought to produce crumb softening by complexing the starch and delaying retrogradation, the diacetyl-tartaric esters of monoglycerides do the same by complexing with the protein and giving bread of higher specific volume and the stearoyl-2-lactylates achieve crumb softening by complexing both starch and protein but not to the same degree as either of the former emulsifiers. The stearoyl-2-lactylates are now more widely used in the United Kingdom than previously. Stearyl tartrate, a permitted emulsifier, is rarely if ever used.

Flavour

Bread flavour arises from two main sources in addition to the obvious flavour ingredients, such as salt and sugar. These are the yeast fermentation and the caramelization reactions in the crust. The former source is rather more subtle and although an expert may be able to distinguish bread produced by bulk fermentation from that produced from mechanically developed dough, the differences are not great.

The flavour developed during fermentation is mainly the result of organic acids and alcohols and although losses occur during baking, some reactions between these compounds take place at this time. Maillard reactions, in which the reducing sugars present in the dough react with free amino acids produced by the yeast, also occur during baking and result in the brown coloration. In this reaction, highly flavoured carbonyl compounds are produced giving bread its characteristic flavour. The trend to produce underbaked loaves to enable them to pass the so-called 'squeeze' test consequently tends to produce insipid bread.

The addition of flavour to bread is not prohibited by the UK Bread and Flour Regulations 1995. The main problem is that of acceptable labelling.

Staling

Bread staling is commonly thought by the consumer to be the result of drying out but it has long been known that even when not allowed to lose moisture, bread will develop the characteristics that the consumer associates with staleness. These are mainly a dryness and a firmness. Scientific measurements of crumb firmness correlate closely with subjective estimates of staleness. X-ray diffraction analysis shows unbaked starch to have a crystalline structure which is lost when it is baked in a loaf and becomes gelatinized. However, as an increase in staleness is subjectively perceived, the crystallinity of the starch can also be seen to increase. It is this change which is generally accepted as being the major contributor towards bread staleness. Support for this theory is given by differential scanning calorimetry (DSC), which shows up a difference in the heat absorption of fresh (non-crystalline) and stale (crystalline) bread when heated in the range 50–70°C.

The rate of staling of bread depends on its storage temperature. Above 55°C the starch will not become crystalline, but it is impractical to store bread above this temperature because of bacteriological problems, ROPE (see below) in particular, and a gradual discoloration of the crumb. As the storage temperature falls, the rate of staling also increases to a maximum at about 0°C, accounting for the increase in complaints of stale bread in cold weather. Below −10°C, the rate of staling is zero and bread may be kept fresh for long periods at these temperatures. Delaying of perceived staleness may be achieved by baking lower density bread. Certain additives also delay staling, e.g. the permitted emulsifiers and in particular glyceryl monostearate, which is thought to complex with the gelatinized starch and delay crystallization. The gain in shelf life from such additions will rarely exceed two days.

Rope

Rope is a bacterial infection of bread and fermented products caused by bacteria of the *Bacillus subtilis* group. Infections in bread are characterized by a distinctive fruity odour reminiscent of pineapple. As the infection develops, the colour of the crumb becomes yellow and then brown and its develops a stickiness so that when touched, strands or 'ropes' may be drawn away from it.

Bacillus subtilis is a common organism occurring in large numbers in soil, and hence it easily finds its way onto the grain and into the flour. Similarly it may be introduced into the dough by soya flour or, more rarely, via the yeast. The bacteria produce heat-resistant spores, which can survive baking, the number of spores surviving depending on the degree of the original infection. Bread is always likely to contain these spores and thus precautions should be taken to prevent rapid growth of the bacteria. These precautions include thorough baking, rapid cooling of the loaf after baking and avoiding storing the loaf in warm, humid conditions. Rope is effectively controlled through the use of calcium or sodium propionate at 0.15–0.2% of the flour weight or propionic acid (12.5% solution) at 0.8% of flour weight. Now that propionates are being used less frequently, bakers are falling back onto an older but less effective remedy, the use of vinegar or a 12% solution of acetic acid at 0.5 l per 100 kg of flour. The rope bacteria cannot thrive at a pH below 5.

Resistant starch

When starch is heated with water, a proportion of it becomes resistant to break down by the digestive enzymes of the human gut. The reason for this change has not yet been determined.

If this so called resistant starch were to pass through the gut unchanged, there would be no doubt that it should be included in the total dietary fibre figure. However, it is broken down in the lower gut by the bacteria there and an opinion has been expressed that it should not be included in the unassimilable carbohydrate or dietary fibre figure. Nevertheless, it is argued that because it has been shown to have the same physiological effect as dietary fibre, it should be included in this figure. This disagreement has not been resolved.

Yield of bread

The yield of bread obtained from a flour is determined by the amount of water used in the dough and the losses which occur during fermentation, baking and cooling. The amount of water that may be used in the dough is determined by the water absorption of the flour, which in turn depends on its moisture content, the chemical treatment received by the flour, the degree of starch damage and the amount and quality of the protein present. Attempts to increase yield by simply increasing starch damage will result in poor quality bread. Careful control of cooling conditions, particularly in respect of humidity, will prevent unnecessary loss of weight due to evaporation.

Enzymes

Wheat flour contains many enzymes, the most important of which are α-amylase and β-amylase, which have the power of splitting off maltose units from damaged starch thus providing nutrient for the yeast. Flour that is made from sprouted wheat contains, in addition to β-amylase, another diastatic enzyme, α-amylase, only present to a small extent in sound wheat. α-Amylase can attack across the branching point of an amylopectin chain and thus open up more sites for attack by both amylases. Thus, the level of α-amylase activity controls the extent of starch breakdown, even though all flours are high in β-amylase activity.

Proteolytic enzymes which can break down protein are occasionally used in the biscuit-making industry to relax doughs and to make them more extensible during the sheeting and cutting operations.

Lipolytic enzymes, i.e. enzymes that can break down fat, sometimes occur naturally, producing rancid flavours in for example coconut products.

Dietary fibre and crude fibre

It is now appreciated that the non-assimilable, mainly carbohydrate, fraction of food, far from being useless has a valuable part to play in human physiology. This fraction, generally termed 'dietary fibre', consists of the pentosans, hemicelluloses, celluloses and lignin present in the food.

The determination of dietary fibre is complex and the result obtained depends on the method used. Much collaborative work is still in hand to develop a rapid reliable method. Currently the most popular method is the AOAC technique. A definition of dietary fibre has still to be agreed as there are differing schools of thought as to whether resistant starch should be included in the dietary fibre content or not.

The simple term 'fibre' may be taken to indicate 'crude fibre', as determined by the method of the UK Feeding Stuffs (Sampling & Analysis) Regulations (S.I. 1982:1144). This is an empirical method which measures the residue left after boiling a food or feeding stuff with acid and then alkali. The residue consists mainly of cellulose and lignin but does not equate to the total cellulose and lignin present in the sample.

It is interesting to note that white bread commonly regarded as containing very little or no fibre, because its crude fibre content is only 0.2%, in fact contains about 1.6% dietary fibre.

	Crude fibre (%) (Fertilizer & Feeding Stuffs Regs)	Dietary fibre (%) Englyst method
White bread	0.2	1.6
Brown bread	0.7	5.1
Wholemeal bread	1.5	7.2

STARCH

Cereals contain 60–70% starch. The shape and size of the starch granules are characteristic of the cereal from which they are derived. The granules of wheat, barley and rye starches are round and those of maize, oats and rice starch are polygonal. Wheat starch granules fall into size groups; these are termed A-starch, the larger granules (25–35 µm diameter) and B-starch, the smaller granules (2–8 µm diameter). Some of the starch granules of a cereal undergo damage during the milling process and these damaged granules are susceptible to attack by amylase, whereas the undamaged granules are resistant to this enzyme.

Starch is insoluble in cold water, but if a starch suspension is heated the granules swell and eventually burst. The released inner portion of the granules is soluble in hot water and when the solution is cooled it sets to a stiff paste. This effect of heat upon starch in the presence of an excess of water is termed gelatinization.

Starch consists of two fractions, amylose and amylopectin, each of which is a polysaccharide composed of glucose units. In amylose, the glucose units are linked together in straight chains but amylopectin contains branched chains.

VITAMINS

Wheat contains a number of vitamins but they are concentrated in the fractions of the grain that are removed during the milling of white flour. Vitamin B is concentrated in the scutellum, riboflavin in the germ and aleurone layer, nicotinic acid in the skins and the adjacent layer of endosperm and vitamin E in the germ. Mean figures for the amount of these vitamins occurring naturally in flour of 72% extraction are:

Vitamin B	0.11 mg/100 g
Riboflavin	0.035 mg/100 g
Nicotinic acid	0.72 mg/100 g
Vitamin E	1.5 mg/100 g

The UK Bread and Flour Regulations 1995 prescribe minimum limits for the amount of vitamin B and nicotinic acid in flour and as these correspond to the amount occurring naturally in flour of about 80% extraction, it is necessary to add these two vitamins to flours of lower extraction.

The vitamin content of bread is less than that of the flour from which it was made; about 20% of the vitamin B and about 45% of the vitamin E are destroyed during baking.

REFERENCES

Brown, J. (1982) *The Master Baker's Book of Breadmaking*, Turret Press, London.
COMA (1991) *Committee on Medical Aspects of Food Policy no. 7*, HMSO, London.
Gedye, D.J., Doling, D.A. and Kingswood, K.W. (1981) *A Farmer's Guide to Wheat Quality*, NAC Cereal Unit, Royal Agricultural Society of England.
UK Bread and Flour Regulations S.I. (1995) *Regulations Governing Composition, Labelling and Descriptions of Bread and Flour*, HMSO, London.

FURTHER READING

Cotton, R.T. (1963) *Insect Pests of Stored Grain and Grain Products*, Burgess Publishing Company, Minneapolis.
Kent, N.L. and Evers, A.D. (1994) *Technology of Cereals. An Introduction for Students of Food Science and Agriculture*, 4th edn, Pergamon/Elsevier Science, Oxford.
Kent-Jones, D.W. and Amos, A.J. (1967) *Modern Cereal Chemistry*, 6th edn, Food Trade Press, Orpington.
Lauer, O. (1966) *Grain Size Measurement of Commercial Powders*, Alpine AG, Ausburg.

6   Fruit Juices and Soft Drinks

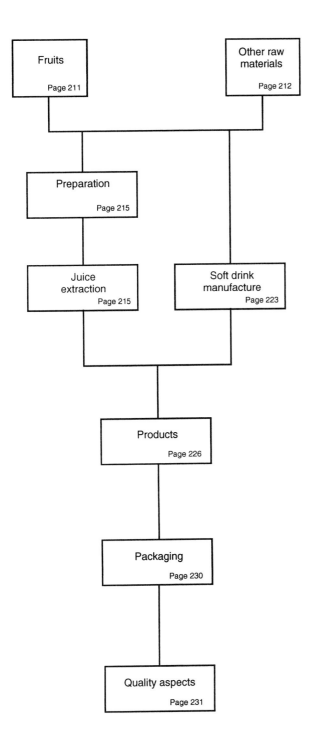

Fruits
Page 211

Other raw materials
Page 212

Preparation
Page 215

Juice extraction
Page 215

Soft drink manufacture
Page 223

Products
Page 226

Packaging
Page 230

Quality aspects
Page 231

INTRODUCTION

Soft drinks are the fastest growing beverage category in the UK. In 1995, 9.6 billion litres of soft drinks, including fruit juices and bottled waters, were sold. This was an 8% increase in volume over the previous year. The industry is aiming to reach a target of 10 billion litres by the year 2000.

Important factors for market growth are changes in lifestyle, particularly an increase in leisure time, an ageing population and a trend towards 'healthier' and more 'natural' products. The next few years will see the first generation brought up on soft drinks become the 20-to-30 year old age group. This is likely to result in a further expansion in the market.

RAW MATERIALS

FRUIT RAW MATERIALS

Fruit juices

Fruit juices, as defined in the UK by the Fruit Juice and Fruit Nectars Regulations 1977 as amended, are 100% pure fruit juices made from fresh fruit or fruit concentrates. Only the flesh of the fruit can be used in juice production and not the pith or peel.

Some fruit juices, such as orange, grapefruit, apple, pineapple, tomato and certain tropical fruits are marketed as pure juice products. They are also often added as ingredients to other soft drinks. Fruit juices that are usually only used as soft drink ingredients include lemon, lime, blackcurrant, strawberry, raspberry and peach.

Citrus juices

The production of citrus juices is now such that in many countries a high proportion of the total fruit production is processed into juice. The major market share is taken by orange but large volumes of grapefruit, lemon and lime are also produced. Most of the citrus juice for use in products manufactured in Europe is concentrated in the country of origin before shipping, but some is exported as single strength juice. The major areas of production of orange for the European market are Brazil, USA (Florida and California) and the Mediterranean area. Grapefruit juice is mainly produced in the USA, Israel and central America, lemon juice in the Mediterranean countries, Argentina and the USA and lime juice in central and South America.

Apple juice

Apple juice is produced and consumed extensively in many countries. The product can be either clear or cloudy. The production of apple juice is essentially a milling and pressing operation. Cloudy apple juice is merely centrifuged after pressing the fruit and clear apple juice is obtained by clarification. The traditional apple growing areas of the world are the temperate zones of North America, Europe, Argentina, South Africa and New Zealand. The most popular cultivar for juice production in the USA is Delicious. In the rest of the world Golden Delicious predominates. However there are many different cultivars with a wide range of taste characteristics that may be used for juice production.

Tropical juices

In terms of volume, pineapple is by far the major tropical fruit juice. The main producing countries are the Philippines, Brazil, Thailand, Kenya, USA and South Africa. The main process varieties are Cayenne and Perola. In most countries, except Brazil, the juice is a by-product of fruit canning. With a wider awareness among consumers, a number of other tropical fruit juices are now popular, particularly in combinations having a 'tropical' flavour. The predominant fruits here are passion fruit and mango, with some banana, kiwi, melon and guava.

Soft fruit juices

The major soft fruit juice is blackcurrant. This is becoming increasingly popular in soft drinks. The main production areas for blackcurrant are Europe and China. Blackcurrant juice is produced by a milling and pressing operation. Pectin degradation is an essential feature of the process and is performed on the milled fruit to enable pressing to take place. The juice is then clarified. Other popular soft fruits mainly used as blends in soft drinks are raspberry, strawberry, redcurrant and cranberry.

Tomato juice

The tomato is not classed as a fruit under the Fruit Juice and Fruit Juice Nectars Regulations 1977 and is included in the EEC Vegetable Juice Directive. Tomato juice is mostly obtained from tomato concentrate. The major concentrate producing areas are Italy, Greece, Turkey, Portugal and USA (California).

Comminuted citrus bases

The comminuted citrus base has been largely developed in the UK for producers of whole fruit concentrated drinks. The bases are prepared in the citrus producing areas from the whole fruit ingredients and exported to the UK. The ingredient combinations are usually producer specific and may contain the following: juice concentrate, essential oil, peel flakes and cloudy concentrate. The main producers of comminuted orange products are Spain, Sicily, Israel and South Africa. Comminuted lemon and grapefruit are produced to a much lesser extent.

VEGETABLE JUICES

Apart from tomato and rhubarb, which should be strictly classified as vegetables, significant quantities of juices are also made from some other vegetables, notably carrot, celery, beetroot and cabbage. These vegetables are all widely grown. The fresh vegetables are coarsely milled and then pressed to extract the juice.

OTHER RAW MATERIALS

Water

Water is the major constituent of soft drinks, including fruit juices. It is important that the water used in production does not affect the products being made in terms of composition, organoleptic and microbiological acceptability.

Waters from different areas have different characteristics. For example, 'hard' waters drawn from limestone sources contain calcium and magnesium salts, 'soft' waters drawn from surface sources usually contain the sulphates, chlorides or nitrates of sodium and potassium. Softer waters with low alkalinity are usually preferred for soft drink production. It is common for soft drink production sites to have their own water treatment facilities to ensure end product uniformity. The types of water treatment will be dependent on the geographical location of the production site and the particular products being manufactured. The water also needs to be sterilized and purified.

There are three main types of water treatment facility, coagulation, ion exchange with dealkalization, and reverse osmosis or nanofiltration. The use of reverse osmosis plants is becoming more widespread as very pure water can be produced with minimal chemical addition.

Where canned beverages are filled into coated steel cans, the level of nitrates in the water needs to be low as they act as a catalyst for corrosion.

Sweeteners

Sugar and sugar syrups

The traditional sweetener for soft drinks was cane or beet sugar, which is 99% or more pure sucrose. The sugar can be either in granulated form, requiring it to be dissolved to make a syrup, or ready dissolved as a syrup. Other syrups such as glucose syrup and high fructose syrups are also used. These are produced by the enzymic treatment of starch, most commonly corn starch.

Artificial sweeteners

These have sweetening power many times that of sucrose. The artificial sweeteners permitted in the EC are:

E950 Acesulfame-K
E951 Aspartame
E952 Cyclamic acid and its sodium and calcium salts
E954 Saccharin and its sodium, potassium and calcium salts
E959 Neohespiridine

Saccharin is still widely used in soft drinks due to its wide availability, low cost and good solubility. It is permitted at a maximum level of 80 mg l^{-1}. Aspartame is not totally stable under the acidic conditions found in soft drinks and

slowly decomposes to the constituent amino acids. Nevertheless, it has achieved wide acceptance in low calorie soft drinks because of its excellent sweetness profile.

Low calorie or 'diet' drinks are sweetened with mixtures of sacharin and acesulfame-K or aspartame. These mixtures have improved flavour profiles compared with those made with saccharin alone, which exhibit a bitter aftertaste.

Colouring materials

Natural and nature-identical colours

These materials have long been available but now due to increased consumer demand they make up a high proportion of the colours used in the industry. Despite problems of lower intensity and brightness than the alternative synthetic colours, and lower stability to light and (in some cases) to acid and SO_2 preservative, a range of satisfactory products has now been formulated using natural or nature-identical colours. It is usually necessary to ensure an efficient retail chain to avoid the possibility of significant fading before the end of commercial shelf life.

The permitted natural and nature-identical colours are :

(i) Carotenoids – carotenes, β-apo-8-carotenal and its ethyl ester, lycopene, lutein, betanin, capsanthin and capsorubin. These give yellow, orange and red colours.
(ii) Chlorophylls – chlorophyll and chlorophyllins and copper complexes of chlorophyll and chlorophyllins.
(iii) Anthocyanins – these give red and purple colours.

Artificial colours

These are still widely used, both to impart an attractive appearance and to overcome the bleaching effect of SO_2 used as a preservative. The colours need to be stable to fruit acids, sulphur dioxide and light. The permitted synthetic colours are:

E102 Tartrazine (yellow)
E104 Quinolene yellow
E110 Sunset yellow
E122 Carmoisine (red)
E124 Ponceau 4R (red)
E129 Allura Red AC
E131 Patent Blue V
E132 Indigo Carmine (reddish blue)

E133 Brilliant Blue FCF
E142 Green S (greenish blue)
E151 Brilliant Black BN
E155 Brown HT

Due to consumer pressure, the use of tartrazine in soft drinks has been greatly reduced.

Others

There are a few other colouring materials that are also permitted. These include:

E150 Caramel
E153 Vegetable carbon
E172 Iron oxides and hydroxides.

Preservatives

The use of preservatives is essential in concentrated drink and squash products where an open-bottle life of several weeks is to be expected. The presence of preservative also eases the production of carbonated ready-to-drink beverages.

Three groups of preservatives are permitted in soft drinks in the EU:

(i) *Sulphur dioxide* group includes sodium, potassium or calcium sulphites, sodium, potassium or calcium hydrogen sulphites, sodium or potassium metabisulphite. Sulphur dioxide has certain advantages as a preservative particularly in lemon, grapefruit and other pale products, in that it also inhibits browning. Its main disadvantages are that it is gradually evolved from the product and lost even from closed bottles, so that in time the preservative effect disappears. It also has the disadvantage of binding to aldehyde or ketone groups, and thus the effective preservative content may be lower than the total SO_2 content.

Sulphur dioxide is the only preservative permitted for addition to fruit juices and juice concentrates. Non-alcoholic flavoured drinks containing fruit juice are permitted to contain a maximum of 20 mg l^{-1} SO_2 as carry over from the concentrate.

(ii) *Benzoic acid* group includes its sodium, potassium or calcium salts. Benzoic acid has no inhibiting effect on browning, but usually has a less deleterious effect than SO_2 on natural or artificial colours. It requires a little care in use since it can precipitate if exposed to a high local concentration of acid such as might occur during the mixing of a batch. Once precipitated, it is very hard to redissolve.

(iii) *Sorbic acid* group includes its potassium or calcium salts. As a result of increasing resistance recently by consumer groups to sulphur dioxide and benzoic acid, there has been an upsurge of interest in sorbic acid. This material is an effective preservative, is not lost with time and does not combine with other ingredients to reduce its potency. It does not appear to break down in the presence of acids, light, etc. or to cause fading of colours. However, sorbic acid cannot prevent the browning of fruit juices and it does have its own particular flavour.

The use of mixed preservative systems can solve many problems.

Clouding agents

Clouding agents are substances added to a naturally clear drink or one whose cloudiness is less than desired, to give it a denser or a more stable cloud. Some clouding agents, also called 'dispersing agents', work in citrus juice preparations by dissolving in the citrus oil and approximating its specific gravity to that of the rest of the drink. Thus the oil does not float to form a coloured ring at the surface but stays uniformly dispersed throughout the drink, so contributing to cloudiness.

Extract of quillaia or various emulsifiers, stabilizers or modified starches may be used to hold natural cloud particles or cloud-forming oils in suspension.

Flavours

Natural and nature-identical flavours

Natural flavours are extracted in water or other solvents from various parts of plants such as fruit, seeds and leaves. The extracts may then be further purified. This type of flavour is common for citrus drinks, an example is the essential oil of the fruit extracted as part of the fruit processing. Nature identical flavours are chemically synthesized to match the natural flavours, for example, benzaldehyde. The natural flavour is extracted from almonds or the nature identical flavour can be produced from the oxidation of benzene carbinol.

Artificial flavours

Synthetic flavours are usually blends of chemicals mixed together to give the required flavour product. The chemical constituents are mainly volatile components such as esters, aldehydes and ketones. Flavour matches can be made in this way for products such as pineapple, banana and strawberry and for non-fruit drinks such as colas and lemonades.

PRIMARY MANUFACTURE

FRUIT AND VEGETABLE PREPARATION

The incoming fruit and vegetables for juice processing should be as free as possible from rotten fruit, leaves, stalks and other debris. The specific preparation depends on the type of fruit or vegetables being processed but all should be washed to remove dirt, debris and any pesticides from the skins. The wash can be a weak detergent solution. For citrus fruits, brushes are used to remove dirt. Fruits such as apples are hand sorted to remove rotten fruit.

JUICE EXTRACTION

The methods used to recover juices from fruits and vegetables vary widely depending on the fruit or vegetable type. Fruits fall into two main classes:

(i) Those which need to be reduced to a pulp before extraction of the juice can take place. The pulp is usually obtained by a milling process and the juice is generally removed by pressure.
(ii) The citrus fruits and others such as pineapple which are normally processed with specialized extraction machinery.

Milling

In fruit processing, four main types of mill may be encountered, roller, grater, hammer and paste mills. The first three are used to obtain a fruit pulp for juice extraction. The fourth is for production of specialized products.

Roller mills

Roller mills are commonly used in grape processing and consist of two rollers each fluted down its length. Frequently the flutes are quite wide and deep, so that the two rollers mesh together like gear wheels. Their effect is to crush the fruit suffi-

ciently for its contents to be released, while leaving large coarse pieces of skin, etc., intact. These large particles make the pulp porous and assist the removal of juice in the later extraction process.

Grater mills

Apples and pears require the use of grater mills which fall into two categories. The first of these has a drum with serrated knives arranged lengthwise at intervals around its circumference; the drum rotates close to the casing which surrounds it. The fruit is retained by the casing until it has been completely shredded by the knives. The second type has similar serrated knives arranged around the cylindrical casing with slots between knives. A rotor with two or more arms on it spins just clear of the knives and as fruit falls into the mill it is pushed over the knives by the rotor. Again the fruit is retained in the mill until it is completely shredded. The pulp escapes through the slots between the knives.

The advantages of the grater mill are that it does not break up the seeds, which would release bitter principles into the pulp, and also that the pulp itself, while fine enough to allow its juice to be extracted, still retains its cellular structure which assists the normal juice pressing process. Such mills are frequently used in the processing of soft fruits because of these factors.

Hammer mills

There are numerous variants of the hammer mill, which consists basically of a shaft carrying a number of projecting blades rotating within a casing, the outlet of which is usually fitted with a screen. The blades themselves may be sharp to deal with fibrous materials, or blunt to smash up cellular products; they may be rigid or hinged. These machines tend to be used for special purposes, such as producing fruit purees for confectionery uses. As a rule hammer mills are not suitable for milling fruit before juice extraction, as their action is usually too drastic.

Paste mills

Paste mills have entered the field of fruit juice processing with the advent of the comminuted type of citrus beverages. In this case, a proportion of the citrus peel has to be blended into the juice and the paste mill is ideal for this purpose. Essentially, the machine comprises a carborundum stone wheel rotating in close proximity to a similar fixed stone, the gap between the stones being as small as 0.001

inch (0.02 mm). Intense shearing forces are set up as material passes through the gap, disintegrating solids and homogenizing liquids.

Extraction – temperate fruits

The temperate fruits give pulps of widely differing physical characteristics. Raspberries and strawberries yield highly fluid pulps, while apples can give pulps that are granular and easy to extract, or they may be viscous and almost impossible to deal with. Blackcurrants give a finely divided pulp, the particles of which tend to deform and compact under pressure, while fermented grape pulps with their content of hard seeds, woody stalks and large pieces of fruit skin have an open structure from which high yields of juice can be obtained by simple drainage. Consideration of these differences between fruits and their pulps, together with practical experience in dealing with the extraction of juice from them, leads to the conclusion that the extraction efficiency (as a function of yield and processing time) depends on four main factors:

(i)   the juice viscosity,
(ii) the resistance to deformation of the solid phase of the pulp,
(iii) the porosity of the pulp,
(iv) the pressure or force applied.

Most of these factors are dependent on physical characteristics of the pulp to be extracted which are subject to change during the course of the extraction. Thus, besides the obvious decrease in volume and the proportional increase in the solid phase that occurs during the pressing operation, it can be shown that pulp adjacent to the perforated surface through which the juice escapes is subject to a greater compression than the pulp elsewhere in the mass. In general, it follows that juice extraction is best carried out using thin layers of pulp, or else the pulp should be mixed at stages during the process.

Batch presses

Modern batch presses are largely based on the early basket press in which a moveable platen is pressed into a perforated cylinder containing the pulp. Typical of new designs are machines in which the cylinder is arranged horizontally and a large number of plastic strands link the face of the movable platen and the closed end of the cylinder. As the platen moves forward, the strings are randomly embedded in the pulp and the juice from the pulp is led along the strands towards the per-

forated wall of the cylinder, from which it escapes. As the platen is withdrawn, the strands tighten to break the compressed pulp and this, together with rotation of the cylinder, has a good mixing action. In a somewhat similar press, the cylinder walls are not perforated and the plastic strands are replaced by heavier plastic rods carrying small grooves which conduct the juice away from the centre of the pulp mass. The rods finish in special manifolds in the end of the cylinder and in the platen from which the juice is led away.

Another version of the basket press has a tough rubber tube fitting inside the perforated cylinder, which is filled with pulp. It rotates to encourage drainage and to spread the pulp evenly around its walls. At the same time the rubber tube is inflated with compressed air, squeezing the pulp between it and the wall of the cylinder.

The conventional pack press, despite its demand on labour, is still widely used. With this system, the pulp is contained in cloths separated by wooden or metal racks so constructed as to have a number of channels on both sides. The pulp is thus split up into thin layers and the resulting 'cheese' is squeezed in a vertical hydraulic press. The method has the advantage of flexibility which may be desirable if more than one type of fruit is to be processed in the same installation.

Continuous presses

Broadly, there are four main types of continuous pressing systems:

(i)   Intermittent or 'continuous batch' presses
(ii) Roller presses
(iii) Belt presses
(iv) Screw presses.

Intermittent presses consist of a perforated conveyor belt which moves intermittently and carries the fruit pulp beneath a press platen which descends when the belt is stationary. Special gates arranged around the sides of the platen prevent the pulp from exuding sideways when pressure is applied. Such presses have been used for grape juice extraction.

Two basic types of roller press have been developed in America and have been used for reclaiming juice from cannery fruit wastes. One type has two large horizontally mounted rollers with perforated faces covered with filter cloth. A special dome is arranged above the point where the drums contact each other, so that the fruit pulp may be fed to this point under a small pressure. Pressing takes place between the two drums, with the liquor flowing to their interior to be removed by a special discharge system, while the residue is removed from their outsides by a scraper or string-discharge system. Another type employs two drums with vertical axes rotating one inside the other and in the same direction. They are pressed together at one point, and again the liquor passes through the perforations in the drums and the solid residue is scraped from them. A third variant uses a single perforated drum, the pressure being applied by a series of solid pressure rolls mounted on it. Difficulties arise mainly in the selection of the size of the perforations in the drum walls and in the cloths used to cover them. The efficiency of these machines tends to be low since the applied pressure and the duration of its application are both small.

With belt presses, the pulp is conveyed between two continuous perforated belts to be squeezed between sets of solid rollers. When the pressing is complete, the belts are separated and the residue removed from them by devices such as rotating brushes. Several variants have been developed, such as a machine which employs a single belt to compress the pulp against a large roller. Two of these have been used in series to give good results with the expression of apple juice.

Screw presses generally consist of a screw conveyor running inside a reinforced perforated cylinder, the discharge end of which is closed by a cone or valve system positioned by various mechanical means. Several modified systems have been developed for special purposes. Unfortunately, fruit pulps are quite fluid and tend to slip back over the screw flights instead of moving forward with them, so that in general, this type of press is only used to extract juice from pulps which have been thickened by previous drainage or similar treatment. Processing aids such as pretreatment with enzymes or the addition of diatomaceous earth may be used to increase yields.

Centrifuges are not widely used in primary fruit juice extraction. Probably, the relatively high cost of centrifugal separators, together with mechanical limitations, have prevented the development of this technique. They may however be used for the preliminary clarification, before final filtration of extracted juices.

Extraction – citrus fruits and pineapple

The extraction of juice from citrus fruits is achieved by passing the unpeeled fruits through special extractors. The basis of the majority of designs is the rotating reamer which tears the juice cells from the previously halved fruit. Machines of this type may be semi-automatic, single head units or fully automatic machines of high output. Older Sicilian extractors generally halve the fruit and present the halves to a row of reamers so that the machines have an intermittent action. A variant of this system halves each fruit vertically as it rolls down a conveyor, then turns the halves to lie with their peel sides uppermost and directs them into two rotating turrets on either side of the feed conveyor. Here fingers grip the halves and hold them over reaming heads which are lifted into them while rotating; fruit and reamers move together in a circular path. In the FMC extractor, which is the most efficient and most popular in use worldwide, the fruit drops into a lower cup fitted with a cutting tube which removes a plug from the base of the fruit; an upper cup comes down, pressure is applied evenly to all surfaces of the fruit and the inner contents are pushed out into a finishing tube where the juice and cells are separated from the seeds and membranes.

Pineapple juice can be obtained by a whole fruit extraction. The 'Pine-O-Mat' extractor halves the topped fruit which is then passed between an extractor drum and perforated grid. The flesh is sheared and pressed from the fruit half shells to give a pulpy slurry which then passes to a finisher.

Extraction – tomato

The juice is usually extracted by multistage pulping equipment using one of three techniques depending on the properties desired for the final product.

Cold-break. Fruit is chopped cold and some time elapses before the pulp is heated, allowing pectin to be destroyed by natural enzymes. This gives a less viscous final product.

Thermal-break. Fruit is chopped cold but is immediately heated prior to extraction of the juice. This gives less time for the natural enzymes to act on the pectin.

Hot-break. Fruit is chopped hot, thereby inactivating the enzymes before they destroy the natural pectin.

Pulp wash

With non-citrus fruits, where it is intended to make a juice concentrate, it is not unusual to wash the solids from the first extraction with fresh water and press again in order to extract as much of the remaining juice as possible; the diluted juice from the second pressing goes forward for concentration, which removes all of the wash water.

With citrus fruits, however, the pulp wash is of inferior quality to the juice from the first pressing because it contains relatively large amounts of extractives from the peel (e.g. pectins and bitter-tasting substances) in addition to any recoverable fruit juice. The presence of pulp wash in a citrus juice is considered to be adulteration.

CLARIFICATION

Many fruit and vegetable juices are cloudy when first prepared and some are definitely preferred by the consumer to be cloudy. This is particularly true of orange and lemon juices. Blackcurrant and other soft fruit products are invariably presented in a clear form whilst apple juice may be either clear or cloudy. The cloud consists of fragments and colloidal suspensions of components from the cells, the composition varying greatly from product to product. The rate at which the natural cloud settles to leave a clear supernatant juice is related to its composition, especially the particle size and the amount and types of any pectins present. Hence the settling rate also varies considerably between different products.

Clarification is practised particularly in the processing of blackcurrants, other soft fruits and apples. Clarified lemon, orange and lime juices are also produced. The practice may be subdivided into chemical processes, in which cloud retaining elements are destroyed, and physical processes, in which minute particles of tissue, etc., are removed.

Chemical treatments

Pectolytic enzymes

Commercial preparations contain a large number of separate enzymes, among them polymethylgalacturonases, pectinmethylesterases and polygalacturonases. These act on the pectin to give degradation products that include pectinic acid (partially demethylated pectins), pectic acids (completely demethylated pectins) and oligogalacturonides of two or more galacturonic acid units.

The use of pectolytic enzymes is essential in the production of blackcurrant and other soft fruit juices to enable the juice to be expressed from the fruit; this is due to the quantities of pectin present. The milled fruit pulp is passed through a tubular heat exchanger and the temperature raised to 40–50°C. Pectolytic enzyme is added and the pulp held for 2–8 h (depending on the activity and concentration of the enzyme) until all pectin has been hydrolysed. In the clarification of lemon and lime juices, the natural pectolytic enzymes of the fruit are utilized. The cloud retaining pectin of the juice is destroyed and the pulp acts as a floc former, clarifying the juice and leaving a clear upper layer on standing.

Gelatin precipitation

This technique is employed to remove colloidal particles which continue to form a haze, particularly in apple juice after enzyme treatment. The addition of gelatin in solution results in the formation of a flocculent precipitate. The settlement of the flocculent deposit is greatly helped by providing efficient enzyme action so that the viscosity of the juice at the time of adding gelatin is very low. It will be found that the quantity of gelatin required will vary throughout the processing season. The deposit of colloidal material and its enveloping tissue from the juice should be formed in about one hour. The rapidity of this settlement is dependent on the right amounts of clarifying agents being used and thus forming the right size and density of flocs. It is preferable to allow the flocs to settle naturally, but the thick deposit which contains much juice may be passed through a centrifuge to recover the greater part of the liquid phase.

Physical treatments

Centrifugation

Centrifuges are normally used for removing relatively large amounts of coarse particles from fruit juices. They are frequently capable of giving a high degree of clarification but do not usually give the brilliance and 'polish' achieved by fine filtration. There are several types of centrifuges the simplest being the tubular bowl clarifier which has a small solids space and limited rate of throughput. The machines most frequently used are of the disc-bowl type, where the bowl contains a number of closely spaced thin metal cones. These divide the liquid flow into thin layers as it moves from the centre outwards, effectively giving a longer path to the liquid and increasing the efficiency of solid/liquid separation. The solids are collected beyond the conical discs on the inner wall of the bowl proper and the clarified liquid is led upwards to a pumping device which discharges it from the machine. Variants of this general type include nozzle designs in which solids are continuously discharged through nozzles arranged on the periphery of the bowl casing.

Intermittent discharge types in which the outer bowl opens at preset intervals to release accumulated discharges, and basket centrifuges, do not usually find application in fruit juice processing. However, the former have been applied to the extraction of juice from pulps, and the latter to separate ice crystals from fruit concentrates prepared by freeze concentration.

Filtration

All methods of filtration involve the use of filter aid, typically a fine, treated diatomaceous earth. This permits the filter to be coarse enough merely to retain peel/pulp particles. A deposit of filter aid is built up on the filter and this is responsible for the main filtration effect, removing the fine and colloidal particles in the juice. The filter aid is mixed with the juice prior to filtration, and a slurry-in-water or juice may be passed through the filter in order to provide a precoating. Since the filtering surface is constantly being renewed as a result of the deposition of more filter aid, it does not become coated with gelatinous pectinous particles and continuous filtration can be carried out. The principal forms of filter are as follows:

(i)   Woven cloths on a plate and frame.
      Woven cloths are useful if large amounts of juice are to be filtered, since this type of filter allows the build-up of a considerable depth of filter aid.
(ii) Asbestos or cellulose pads on a plate and frame.
      The asbestos pad can produce the finest filtration, and if a suitable grade of pad is employed a sterile juice can be obtained. It is customary to pass citric acid solutions through the pads prior to use in order to remove materials which would otherwise taint the product. For small scale operations such as experimental work, it is advisable to wash the filter aid with citric acid also. The citric acid wash is followed by passing water through the pads until all traces of the acid have been removed. Suppliers of asbestos pads advise that in normal use they present no safety hazard. However, because of general concern regarding the use of this material, alternatives are being developed and gradually being introduced.
      If sterile filtration is to be performed, the filter must first be sterilized. This can be achieved by connecting the output nozzle to a clean supply of 'live' steam, allowing this to flow until it is jetting from the input nozzle and then continuing for a further 20 min.

Sterile traps should then be placed on both nozzles while the filter cools. There is a little application for sterile filtration in the UK but it is quite common elsewhere in Europe and in the USA. For normal filtration purposes, the life of the pads can be prolonged by interleaving with sheets of filter paper; these will allow the deposits of filter aid to be easily removed and the pads to be re-used.

(iii) 'Candles' composed of metal washers.
      The 'candle' filter consists of a large number of metal washers, correctly placed on a fluted rod and retained with a nut. Each washer has a flat side and an indented side. These candles hang in a chamber of unfiltered material. Their upper ends are arranged so that the filtered juice is pumped through the candles from their outside inwards and discharged into a separate chamber. The number of candles is variable and depends upon the design throughput of the plant. During precoating, usually carried out with a slurry consisting of filter aid and juice, the filtrate is cloudy until a coat has been formed. It must therefore be passed back into the bulk for refiltration.
      With this type of filter, the maintenance of a continuous pressure on the juice feed is essential, since if it is allowed to fall, the coating of filter aid tends to come off the candle and unfiltered material will pass through the machine.
(iv) Rotary vacuum filter.
      The rotary vacuum filter provides a convenient continuous method of clarifying juices and is commonly used for apple juice. A filter cloth is attached to the curved surface of a perforated drum and a vacuum applied to the inside. The drum is rotated and a precoat of filter aid is applied. A slurry of juice and filter aid is sprayed onto the outside of the drum. The juice passes through the layer of filter-aid and the cloth to the inside of the drum and then to store through a convenient discharge system. The layer of filter aid is prevented from building up beyond a given thickness by use of a 'doctor' knife set at an appropriate distance from the surface of the drum.

CONCENTRATION

Evaporation

The most common and convenient form of concentration employed on fruit juices is vacuum evaporation, usually preceded by a preheating treatment that is adequate to ensure destruction of enzymes and micro-organisms. Some processes for making frozen concentrate utilize evaporation at low temperature throughout and omit the pasteurization step. The juice usually needs screening to remove solid material.

The process of evaporation also removes volatile flavours and various techniques are employed either to remove these before evaporation or to minimize the effect upon the quality of the juice. Practically all evaporators in use are of multiple effect design in which the juice passes from one effect to another at increasing concentrations and decreasing temperature. The vapours from the earlier hotter effects are used to provide the heating in the later, cooler effects, thus producing high efficiencies in terms of steam usage. Thermocompression (the injection of high pressure steam into the vapours, thus increasing the pressure and therefore the temperature) is also widely used. The principal forms of evaporators are outlined below:

Tubular evaporators

In these evaporators, the juice is made to boil inside vertical tubes. A thin film of liquid is formed on the surface of the tube and vapour rises up the centre. The juice may be introduced into either the top or the bottom of the tubes, depending upon whether the evaporator is of the 'falling-film' or 'climbing-film' type. The tubes connect to a cyclone separator from which the vapour passes either to a condenser or to a further evaporator effect. Similarly, the concentrate is led from the separator, either out of the machine or to another effect for further evaporation.

Tubular evaporators also include designs which employ inclined or horizontal tubes and where viscous products are involved, circulation through the tubes may be increased by the use of pumps.

Wiped-film evaporators

These machines consist essentially of a wide vertical steam jacketed tube fitted with a rotating central shaft carrying a number of vanes which almost touch the interior of the tube. The juice enters at the top of the tube and evaporates as a thick film on its surface. This film is agitated by the rotating vanes and their action results in a high evaporation efficiency. Such plants are particularly useful with viscous products and are sometimes employed as finishing evaporators when a high density final concentrate is desired.

Plate-type evaporators

One of the more recent developments, the plate-type evaporator, has the advantage of requiring less headroom than most other types and its surfaces are readily accessible for cleaning. It consists of a number of specially designed plates compressed together in a frame and arranged so that steam is applied to one side of the plate while the product to be evaporated is in contact with the other. The arrangement gives a climbing film effect alternating with a falling film effect as the product passes through the plant. The design lends itself admirably to multieffect operation, and with its rapid evaporation and low hold-up of product, high density concentrates of good quality are obtained.

Centrifugal evaporators

The centrifugal evaporator is essentially a single effect vacuum evaporator, consisting of an outer cylindrical shell inside which is a stack of hollow cone-shaped discs carried upon a vertical shaft. An ingenious arrangement of parts admits steam to the underside of each disc while the product to be evaporated is fed to the centre of their upper sides. Thus, as the stack rotates at high speed, the product is spread over the disc surface in a thin layer and centrifugal force causes it to pass swiftly over the heated surfaces to their edges. A very rapid evaporation takes place and the concentrate is discharged from the stack of discs upwards and out of the machine. The vapour produced is collected in the outer shell and led to a condenser system.

This type of evaporator, with its rapid action, gives excellent concentrates which suffer little heat damage. However, due to mechanical considerations, the size of the machine is limited.

Heat-pump evaporators

Certain evaporators of the tubular type have been designed to operate on the heat-pump principle. This involves the use of a compressor acting on ammonia, or other refrigerant gas, thereby raising its temperature. The hot gas is used to boil the juice in the evaporator tubes; as a result, the gas itself is cooled. The cool gas, still at high pressure, passes through an expansion valve, as in a refrigerator, where it is cooled further and liquefied. In this state, it is used to condense the vapours boiled from the juice and during the process the condensed refrigerant is revaporized and passes back to the suction side of the compressor.

Plants operating on this principle are useful in areas where steam and cooling water are in short supply. They are claimed to have lower running costs than conventional plants. However, because of the high pressures involved on the refrigerant side, their specialized construction involves a high capital cost.

Reverse osmosis

Osmosis occurs when a relatively concentrated solution is separated from a pure solvent by a semi-permeable membrane. It results in the passage of solvent through the membrane and into the concentrated solution which thereby becomes progressively diluted. The migration of the solvent can be halted or reversed by applying a pressure to the concentrated solution which is greater than or equal to the osmotic pressure developed by the solvent. Recent developments have applied this principle to the desalination of sea water and latterly to the concentration of fruit juices. There are still some difficulties to overcome, mainly in respect of the preparation of the membranes, which have to be permeable yet withstand pressure of up to 500 lb in^{-2} (3.45 MPa).

Membrane permeation

This is a technique in which a semi-permeable membrane is interposed between the liquid and the vapour phases in a single effect vacuum evaporator. Depending upon the characteristic of the membrane, this can be arranged so that water vapour will permeate through it but organic vapours are impeded. The process is, therefore, claimed to produce juice concentrates in which the normally volatile aromas and flavours are retained without the use of a conventional volatile recovery step beforehand.

Freeze concentration

Freeze concentration relies on the fact that on freezing solutions initially form crystals of pure solvent. This continues until the solute concentration reaches the so-called eutectic point, at which the whole solution freezes. The degree of concentration of juices is therefore limited by the eutectic point of the juice concerned and is usually not greater than 50° Brix. The process consists of freezing water out of the juice in the form of ice crystals and separating them by centrifugation or

filtration. A high quality product is obtained since all volatile materials are retained.

Cutting back

As previously mentioned, the juices concentrated by evaporation may be devoid of volatile flavour; this effect can be minimized by the process of 'cutting back' the concentrate with a quantity of fresh juice. For example, a sixfold evaporated juice may be blended with fresh single strength juice to give a fourfold concentrate which has some of its volatile flavours restored.

A concentrate is usually described as four-, five- or sixfold, indicating that one volume of concentrate can be diluted with water to give four, five or six volumes of juice.

ESTER RECOVERY

The process of evaporation of juices leads to the loss of volatile flavours which steam distil from the juice and in most cases are lost in the condensate from the evaporator. In the case of apple, blackcurrant and other soft fruits, the volatile flavours can be removed by a controlled partial concentration prior to full concentration, but ester recovery from citrus materials has met with little success until recently.

Around 1940, the Kestner Evaporator and Engineering Company Ltd fitted their fruit juice evaporators with a system which attempted to condense the vapours removed initially from the juice as it entered the plant to give an 'ester' fraction. However, as this fraction was removed under vacuum, many of the more volatile components were not condensed out. The concentration of volatile constituents was very low in any case, so that the stability of the ester fraction itself in storage was poor. If the fraction was added directly to its associated juice concentrate, the latter usually became diluted to an undesirable extent. Nevertheless, the system was probably the first practical attempt to recover volatile flavours, even though its value was limited.

Modern ester recovery techniques consist of heating the fruit juice to boiling at atmospheric pressure and evaporating some 10–25% by volume. The vapours are passed through a fractionating column with reflux and then condensed to give an 'essence' containing the bulk of the volatile constituents of the juice. The volume of the 'essence' is typically 0.5–1.0% that of the original juice, and the material usually has good storage properties which are enhanced if refrigerated storage is available. The so-called 'stripped' juice is rapidly cooled as it leaves the equipment and can be concentrated normally in a suitable evaporator.

The amount of evaporation required in the volatile recovery plant depends upon the type of juice being treated. The major portion of the volatile constituents of apple juice are recovered with about 10% evaporation; with blackcurrant juices, 15% evaporation is required, and up to 25% may be needed with strawberry juices. When the final juice concentrate is reconstituted with water and an aliquot of 'essence', the product closely resembles the original juice. The process stimulated great interest in the examination of the flavour-influencing components of fruit juices, particularly by the use of gas-liquid chromatography (GLC).

Other forms of volatile recovery equipment have also been devised. Some work under vacuum conditions. In others, the volatile constituents are adsorbed on activated carbon, from which they are removed by solvents and later separated. More recent developments have combined volatile recovery systems with evaporators and these systems are now very widely used.

ESSENTIAL OILS

The production of essential oils is a noticeable features of the citrus industry, these materials being used for general perfumery and flavour applications. The citrus essential oils are contained in the flavedo (outer layers of the peel) of the fruit and are extracted by a variety of methods. The traditional process followed in Sicily involves taking the halved peels from the juice extractor and soaking them in calcium hydroxide solution. This precipitates the pectins of the peel as their calcium salt and renders the peel less absorbent to the oil when it is released from the oil sacs. The treated peels are usually allowed to stand overnight for hardening to occur and are then rasped under water sprays. This yields a water–oil emulsion which is centrifuged and the water recycled to the sprays. By this means, the water becomes saturated with water-soluble components of the oil and their loss is minimized.

Other techniques of oil extraction include simply rasping the peels and pressing the oil from the raspings, or performing a water washing and centrifugation of the oil-rich peel flakes produced by the extractors. These may be pressed in a screw press and the resulting emulsion centrifuged to separate the oil phase.

Natural citrus oils contain dissolved waxes from the skin of the fruit and these tend to precipitate in cold conditions. The winterization process involves storing the oil at a low temperature, allowing this precipitation to occur and then clarifying by centrifuging. The oils themselves consist of water-soluble and water-insoluble fractions. Deterpenization of the oils results in a water-soluble material of high flavour intensity. This process is carried out by fractional distillation or countercurrent solvent extraction.

PACKAGING AND STORAGE

Juices which are to remain at single strength must be pasteurized to inactivate enzymes and control microbial growth. Juices that are concentrated by evaporation are pasteurized during the evaporation stage. After cooling, the juice products may be packed in a number of ways.

Freezing

Most concentrated citrus juices are frozen and maintained at a temperature of $-18°C$ for storage and transport. The frozen concentrates may be packed in 200 l (nominal) steel drums with double high density polyethylene liners or stored in bulk storage tanks. Use of the latter is becoming increasingly popular. The concentrate can be filled from the tanks into road tankers and unloaded at customers' sites, again into bulk storage tanks. This making handling very efficient.

Chilling

Temperate fruit juice concentrates such as blackcurrant and apple may be stored satisfactorily under chill conditions. These concentrates may be packed into 200 l (nominal) steel drums with double high density polyethylene liners or into polyethylene drums. Juice concentrates such as peach or plum that are only required in small quantities may be packaged in 40, 25 or even 5 l polyethylene containers.

Aseptic packaging

Aseptically filled fruit juices and concentrates are increasingly becoming available from processors and may be supplied in 208 l bags transported in steel drums, 23 l bags packed in fibreboard boxes and most recently in 1000 l bag-in-box containers.

Aseptically filled products, if unopened, will remain free from microbiological spoilage and can have the same shelf life as their preserved equivalents. They can be stored satisfactorily under ambient or chilled conditions. This method has been in use for some years for packing fruit juices. Cooled pasteurized juice, either single strength or concentrated, is passed to a carefully sterilized filler and filled under aseptic conditions into a sterile polyethylene and metal foil laminated bag, which is either contained in a steel drum or a fibreboard box. Stringent precautions to exclude any spoilage organisms must be taken when using this packaging technique. The filler is often enclosed in an environment of sterile air, although at least one process uses in-line sterilization with hydrogen peroxide and continuous filling.

This method avoids the costly high energy requirements of frozen storage and distribution and is widely used.

Use of preservatives

This technique is not used for concentrates that are to be diluted to yield pure fruit juices. However, it is widely employed for fruit materials that are used as ingredients in soft drinks, for example, squashes, cordials and fruit drinks.

Sulphur dioxide is the preservative of choice; it ensures microbial stability and also acts as an antioxidant, preventing browning. Benzoic acid can also be used and is preferred if the product is to be used as an ingredient in a canned drink, because sulphur dioxide reacts with the can material.

Preserved products are packed into 200 l plastic drums or 1400 l rotoplas containers. More recently, bulk tankers have also started to become available for transporting preserved products, which can be stored under ambient conditions.

Blending of juices

Blending of fruit juices generally occurs at the production site and can be used to overcome the limitations of maturity, varietal characteristics, etc., and thereby extend the production of acceptable fruit juices. For instance, high acidity, early season orange juice can be blended with low acidity, late season juice to produce a product that is much closer to the mid-season optimum. High acidity juice from culinary apples may be blended with low acidity dessert apples to produce a desirable intermediate acidity. Blackcurrants from the Côte d'Or region of France, which have intense colour

and flavour characteristics, are incorporated with benefit into blackcurrant juice blends in other parts of Europe.

SOFT DRINK MANUFACTURE

INGREDIENT MIXING

Fruit juices

It is not uncommon for a manufacturer of fruit juice products or soft drinks to blend juices from different geographical sources in order to achieve a desired quality, or even to blend different fruits to produce new or novel products. This means it is possible to duplicate a blend from season to season, even though the flavour of the individual component juices will vary.

Compounds

A popular method of manufacture of ready-to-drink beverages other than fruit juices consists of dilution of a prepared compound with water after addition of sugar and acid, as appropriate. The compound can contain some or all of the following ingredients:

(i)   The fruit material
(ii) Essences, oils and other flavouring
(iii) Preservative
(iv) Colouring
(v)   Artificial sweetener.

Compounds enable a soft drink bottling operation to be carried out with a minimum of technical control or even consciousness of the composition of the product manufactured. They also fulfil the need of small manufacturers who wish to provide a wide range of products. Compounds can also be obtained that are suitable for the manufacture of squashes or comminuted drinks to be consumed after dilution.

The addition of ingredients to a batch often needs to be carried out in a specific order to ensure that all the ingredients are dissolved properly before the batch is used for production. For example aspartame is sparingly soluble in water and is usually dissolved in a dilute citric acid solution before addition to the batch.

BOTTLING

Soft drinks intended to be consumed after dilution and containing adequate preservatives may be bottled by the simplest means. Soft drinks intended to be consumed without dilution contain less preservative (though this may be supplemented by carbonation), and these products require more sophisticated methods of bottling. Fruit juices are bottled without preservative and therefore should be hot filled, pasteurized in bottle or aseptically filled. The principal steps in the bottling process are outlined below.

Heat treatment

The heat treatments applied to fruit juice and fruit juice products are almost invariably either flash pasteurization, hot filling or in-bottle pasteurization. Pasteurization in this connection does not have the precise legal definition that it has in the dairy industry and implies generally a heat treatment at below 100°C for some period of time.

In flash pasteurization the temperature of the product is raised by means of a plate-type or tubular heat exchanger. The heated product is maintained at a high temperature for the short period necessary to flow through a holding tube, after which the product may be cooled or partly cooled prior to filling.

Flash pasteurization of product and cooling before filling are generally applied to products containing preservative, in which a slight post-pasteurization recontamination may be tolerable. The low pH of practically all fruit juice products, particularly those containing preservative, means that virtual sterility can be achieved by flash pasteurization treatment. These may be at temperatures as high as 95°C and for as long as 60 s in order to ensure adequate bacteriological kill and also the destruction of enzymes.

Hot filling and in-bottle pasteurization are generally employed with pure fruit juices or products which do not contain preservative. Hot filling is achieved by heating the material by means of a heat exchanger, possibly passing it though a holding tube and possibly cooling it slightly. The bottle is filled with hot product, usually at about 70°C. The closure is applied and the bottle may be inverted. The advantage of this procedure is that microbiological contaminants on the inner surfaces of the bottle and closure can be destroyed by the hot liquid. The process thus gives adequate sterility without the expenditure of energy to heat up the container as well as its contents.

In-bottle pasteurization is performed by taking the filled, closed bottles, raising their internal temperature to the region of 70°C and holding at this temperature for periods of about 2 min. This operation can be performed either as a batch process in a steam chest or as a continuous operation in which the bottles are subjected to sprays of water at controlled temperatures. In-bottle pasteurization results in expansion of the product and the consequent production of high pressures in the bottle. The bottle and closure must, therefore, be adequate for this treatment and the minimum headspace usually advised is 4% of the brim-full capacity of the bottles. It is usual for in-bottle pasteurizers to include a cooling section. Provided the temperature and holding time are carefully chosen, the flavour changes resulting from use of the process can be small. Forced cooling is advocated after hot filling to prevent the development of cooked flavours.

Filling

Still beverages

The filling of still beverages into bottles is most simply achieved by exhausting the bottles; air is removed from the bottles and the beverage flows in to release the vacuum. High bottling speeds are obtainable using multihead fillers and the accuracy of the fill is limited by the degree of variation of the bottles themselves. Nozzle design and level of vacuum applied are important, since excessive turbulence at the nozzle can cause aeration of the product which may later give rise to oxidative changes. Similarly, in beverages containing fruit particles, such aeration may cause them to separate upwards. Vacuum fillers are not recommended for use with sterile products or those containing low levels of preservative, since air is being continuously sucked into the nozzles and may carry large numbers of micro-organisms with it.

Fillers which give a positively metered fill include piston fillers, cup-type fillers and time-cycle fillers. The piston filler is basically a pump which delivers the desired volume at each stroke. It is necessarily fitted with valves and this limits its use to clear products or to those containing finely divided particles. Cup-type fillers consist of a number of cups which are first lowered below the liquid level in the tank until they are full and then raised above the liquid level. They are then emptied into bottles or other containers through valves and tubes fitted at the bottom of the cups and protruding through the base of the tank. Fillers of this type lend themselves to sterilization and are suitable for medium speed operation. The time-cycle filler employs a tank in which the beverage is kept at a constant level and valves which open for an accurately controlled time interval to discharge the required volume of product into the bottles or other containers beneath. These fillers are usually of the in-line type and have a relatively low operating speed.

Carbonated beverages

In producing a carbonated beverage, the product has to be held under pressure in order to retain carbon dioxide gas in solution. To fill such beverages, a counterpressure filler is required in which the bottles are first pressurized with gas from the headspace of the filler bowl, after which the product flows by gravity through nozzles into the bottle. The product flow is controlled by mechanically actuated valves and the pressure remaining in the headspace of the filled bottles is released by 'snift-valves' before the bottles are discharged from the filling nozzles. The design of the nozzles is crucial; they must admit the liquid with the minimum of turbulence so that 'fobbing' (the sudden effervescence of the carbonated drink) is prevented.

There have been several developments of the counterpressure filler, including filler nozzles which move upward within the bottles as they are filled with the object of preventing fobbing. Among other refinements are nozzles that enable the bottles to be completely evacuated before they are pressurized with carbon dioxide. This is also said to assist in the prevention of fobbing and to give a better shelf life to the product since oxygen is eliminated from the final pack. With such sophisticated fillers it is now possible to hot-fill carbonated beverages at about 70°C. For lightweight cans and fragile polyethylene terephthalate (PET) bottles, prepurging with carbon dioxide or nitrogen has been used to replace the air instead of evacuation. The can or bottle makes a seal against the filling valve, which is then opened to release the purge gas into the container. The air and gas are released to the atmosphere and the container counter pressurized and filled in the normal way.

On can-filling lines, under-cover gassing is often used to reduce the air content of the package. The can headspace is covered with carbon dioxide or nitrogen as the can end is placed on the can. Similar systems are also available for bottled products.

Aseptic filling

Aseptic filling is a method whereby products may be packed without the use of any preservatives. Such products, if unopened, will remain free from microbiological spoilage and can have the same shelf life as their preserved equivalents. This method has been in use for some years for the packing of milk and fruit juices and more recently for the manufacture of syrups or ready-to-drink beverages which usually contain a high proportion of unpreserved fruit juice.

The syrup is flash pasteurized (15–60 s at 85–95°C) and cooled to near ambient temperature. It is then passed to a carefully sterilized filler and filled into sterilized containers. These may be bottles, cans, laminated cartons or plastic/aluminium foil bags in boxes. Stringent precautions to exclude any spoilage organisms must be taken when using this packaging technique. The filler is often enclosed in an environment of sterile air, although at least one process uses in-line sterilization with hydrogen peroxide and continuous filling.

The process is ideal for packing ready-to-drink products containing ingredients which may be spoiled by the prolonged heating involved with in-pack pasteurization. Typically, they may be uncarbonated or only slightly carbonated and contain no preservative. Even if preserved, however, the levels of preservative permitted in ready-to-drink beverages are very low and this, combined with a lack of carbonation, makes such products particularly susceptible to microbiological spoilage. Products of this type were impossible or very difficult to make before aseptic packing became available.

CARBONATION

The degree of effervescence is probably the most important property of a carbonated soft drink. The quantity of dissolved carbon dioxide gives the beverage its characteristic sparkle and complements the flavour of the drink. The level of carbonation will vary from product to product and for each there is a degree that is just 'right'. Too high a carbonation in a crush is as bad as too low a carbonation in ginger ale or other mixers.

Carbonation is achieved by passing the liquid which is to be carbonated through a vessel containing carbon dioxide under pressure. The volume of carbon dioxide that will dissolve in a liquid is governed by the following factors:

(i) The surface area of contact between gas and liquid: a large interfacial area allows more gas to enter the liquid.

(ii) The time of contact: a longer time allows more gas to enter the liquid.

(iii) The 'absolute' pressure of the gas–liquid mixture: a higher pressure forces more gas into the liquid.

(iv) The temperature of the liquid: the solubility of a gas in a liquid increases as the temperature of the liquid decreases.

(v) The receptivity of the liquid to carbon dioxide: some liquids are more easily carbonated than others, e.g. water is more receptive to carbon dioxide than sugar or salt solutions.

(vi) The purity of the carbon dioxide: the presence of any other gas causes it to be dissolved to the detriment of the amount of carbon dioxide that is eventually taken into solution. This is particularly true of air, where 1 volume will exclude 50 volumes of carbon dioxide from solution. It is for this reason that most carbonation equipment includes the facility for deaeration.

With most carbonators, the amount of carbon dioxide dissolved in the liquid is kept below the maximum that the liquid could absorb. If the liquid is saturated with carbon dioxide at a given pressure, any change in pressure perturbs the equilibrium conditions that apply and the carbon dioxide may come out of solution. This situation is undesirable in the filling operation which follows carbonation. Carbonators therefore produce only partially saturated solutions at pressures that are greater than the equilibrium pressure required to keep the carbon dioxide in solution. This differential is known as 'over-pressure'. The design of carbonators varies greatly, but all embody most of the principles mentioned above.

PRODUCTS

FRUIT JUICES

The principal pure fruit juices marketed today are orange, grapefruit, apple and pineapple. The market for fruit juice has continued to grow as juice is seen as a healthy option in drinks. Fruit juice consumption per capita in the UK still lags behind America and other European countries so the potential for growth still exists. The fruit juice market is becoming increasingly commodity driven and manufacturers are looking into adding value back into the market by packaging and product innovation. Orange dominates the fruit juice market, with apple in second place. Pineapple and grapefruit are sold in much smaller quantities. The major pack types for pure juice products are bottles or cartons.

Orange juice

Orange juice accounts for over two-thirds of the total UK juice market. Recently there has been significant growth in the freshly squeezed market, these products being packed in cartons or plastic bottles. Specialized juices coming from particular geographical areas (e.g. Florida orange) have also entered the market. Blood orange is another specialized product that has also been introduced.

Apple juice

Apple juice is the second most popular pure juice in the UK. It can be either clear or cloudy. Cloudy juice has only recently become popular, being sold as a special product in glass or plastic bottles.

Grapefruit juice

Grapefruit juice has a much smaller market than orange or apple juice. White grapefruit juice predominates but recently specialized products using red or pink grapefruit have been introduced.

Tropical juices

In terms of volume, pineapple is the major tropical fruit juice. With a growing awareness among consumers, a number of other tropical fruit juices are now popular, particularly in combinations of 'tropical' flavour. The predominant fruits here are passion fruit and mango, with some banana, kiwi, melon and guava.

Blackcurrant juice

Blackcurrant is the most popular product in the juice drinks sector, in which it accounts for 25% of sales. The juice of blackcurrants is produced and consumed extensively because of its high vitamin C content. The most popular products are blackcurrant health syrups containing 15–60% by volume of blackcurrant juice. More recently, ready-to-drink products have begun to appear on the market.

VEGETABLE JUICES

The great majority of juices and juice-containing beverages are made from fruits but significant quantities of juices are also made from vegetables. The most important vegetable juice in volume terms is tomato juice but carrot, celery, beetroot and cabbage also provide juices. Apart from tomato, which is widely available, most of these products may be found in specialized retail outlets, particularly in health food stores.

The majority of vegetables yield juices of low acidity, which must be preserved either in cans which are given a full 'botulinum cook' heat process, or by freezing. Tomato and rhubarb juices are sufficiently acid to be stable after pasteurizing. Low acid juices may also be made into mixtures with high acid juices (tomato, rhubarb but also citrus, apple or pineapple) so that the final acidity falls below pH 4.5 and the mixture may be preserved by pasteurizing.

An interesting product in this connection is sauerkraut juice, made in central Europe and the USA. Before sauerkraut (fermented cabbage) is removed from the vat in which it is prepared, most of juice is pumped out. Some of this may be added back to the sauerkraut when it is packed for sale but there is usually a large surplus. This may be filled into cans or bottles and flash pasteurized. The juice contains typically 1.5 to 1.6% lactic acid and about 2% salt.

NECTARS

Fruit nectars are a category of fruit drink very popular on the continent of Europe. Nectars are basically extended fruit juices containing added sugar (up to 20%) and added fruit acid. They may be prepared from almost any fruit. The minimum quantity that must be present in the final product in the UK is specified in the Fruit Juices and Fruit Nectars Regulations 1977 as amended. There are controls on other additives that may be used in nectars and these may also be found by reference to the Regulations.

SQUASHES AND DRINKS

The UK Soft Drinks Regulations 1964 divided compounded products containing fruit into two distinct categories. Squashes are products intended for dilution and must contain a minimum of 25% by volume of juice. Ready-to-drink beverages con- taining a minimum of 5% by volume of juice are normally described as 'crushes'. Products made from comminuted fruit are described as 'drinks', and must contain a minimum of 10% w/v potable fruit if concentrated, or 2% w/v potable fruit if ready-to-drink. Products containing at least the minimum of both materials may therefore be described by either name. Beverages not meeting these minimum requirements must be labelled 'X-flavoured drink' or the name of the predominant flavour followed by the suffix-'ade', e.g. lemonade. There are exceptions to the standards described above, particularly in relation to non-citrus juice products. These may be found detailed in the Regulations.

Barley-containing products are conveniently included here. Beverages such as lemon barley water (containing juice) and lemon barley drink (containing comminuted fruit) contain a boiled barley suspension derived from the addition of barley flour at between 0.5% and 3% in the concentrated product. Clearly, these terms, together with others such as 'Cordial' or 'Juice', may be less than clear to the general public for whom the regulations are intended as a safeguard. Any future revision of the Soft Drinks Regulations may possibly contain some rationalization.

CORDIALS

These are products fairly specific to the UK market. In the case of clear citrus products, other than a clear product made from lime juice, the term 'cordial' is interchangeable with 'squash' and the product is subject to the regulations for 'squashes'. If the product is made from lime juice, the description 'lime juice cordial' is mandatory. All citrus cordials for consumption after dilution must have a minimum juice content of 25% by volume. Non-citrus fruit cordials for consumption after dilution, such as blackcurrant, must have a minimum fruit content of 10% by volume. No minimum fruit content is stipulated for 'fruit flavour' cordials, but such products must be adequately labelled so as not to mislead the consumer. The word 'cordial' may also be substituted for the word 'crush' on the label of a soft drink containing clear fruit juice that is not a comminuted citrus drink and is intended for consumption without dilution. Such products must contain at least 5% by volume of juice, with the exception of lime cordials which need contain only 3% by volume of juice.

CARBONATED BEVERAGES

The term carbonated beverages or carbonates includes a wide variety of drinks. Historically, flavoured carbonated drinks developed from the carbonation of mineral waters and hence they were known as 'minerals'. This term is used less often today, but many small retailers still use it as a general term for soft drinks.

Carbonated drinks dominate the soft drinks market and the sector is one of the fastest growing. About 30% of the carbonated drinks sold are low sugar products. The following serves as a classification of carbonated drinks.

Mineral waters

These may be naturally sparkling or have carbon dioxide added and include the widely used soda water. Natural and 'spa' waters are available from springs in many different places.

Sweetened and flavoured drinks

These consist of an acidified sugar syrup or other sweetener, flavoured with essences that are solutions of synthetic and/or natural substances and containing carbon dioxide under pressure to give the characteristic sparkling drink. This group of products is probably the most varied of the carbonates and includes lemonade, cola and ginger ales.

Sweetened fruit or vegetable drinks

These consist of an acidified sugar syrup, or other sweetener, together with a proportion of fruit or vegetable base. They may also contain other flavouring materials, and are carbonated to give a sparkling drink. The names used for such soft drinks are specified in the UK by the Soft Drinks Regulations 1964 (e.g. if the drink contains at least 5% fruit juice by volume it may be known as a 'crush').

Tonic water

This is more correctly described as 'quinine tonic water' or 'Indian tonic water'. It was first formulated for the treatment of malaria and contains a quinine salt as the bitter principle. It is unique in being allowed to use the word 'tonic' as, generally, a product may not be so described solely because

it contains quinine. The drink is delicately citrus flavoured and according to UK regulations must contain not less than $57 \, \text{mg} \, \text{l}^{-1}$ of quinine sulphate.

Low calorie drinks

These are becoming increasingly popular, although marketed originally for diabetics, they are now consumed by those who wish to control their intake of calories. Low calorie drinks may belong to any of the groups already mentioned. Since aspartame and acesulfame-K have become available, low calorie drinks have become much more popular. This is because products made with these materials taste much better, and do not exhibit the bitter aftertaste characteristic of those sweetened with saccharin. Products using these materials can be made to taste similar to those sweetened with sugar. It is to be expected that they will continue to gain in popularity.

Shandy

This product is, by definition, a mixture of beer and lemonade. Other mixtures of alcoholic and non-alcoholic drinks are included in this group, e.g. ginger beer shandy, lager and lime. These products cannot be described as 'non-alcoholic', but as the alcohol content is less than 1.2% by volume, they are not subject to excise duty and may therefore be included within the soft drinks definition. The maximum alcohol content of the drink described as shandy sold through unlicensed premises is 0.5% by volume.

Preparations from extracts

The most common of these is ginger beer. The product appears to have been so named because it was produced from fermented ginger root. Products today are generally produced from flavourings. Another product in this category is dandelion and burdock, originally a herbal extract, but again now prepared from a flavouring.

Colas

Colas account for the largest portion of the carbonates market. These products really fall into the previous class but because of their popularity are considered separately. They are made from a sugar or artificially sweetened syrup, generally acidified with phosphoric acid and coloured with caramel and are flavoured with extracts of spices, citrus oils and cola nut.

The previous understanding among producers and law enforcement agencies in the UK, that cola drinks should contain between 50 and 200 mg l^{-1} caffeine (unless described as caffeine-free), is now replaced by a requirement to contain not more than 125 mg l^{-1}, on grounds of toxicological safety. If such proportions of caffeine were derived solely from cola nut, the flavour of the product would be unpleasant. Therefore, caffeine is usually added to give the concentration required. There are no regulations on this subject in the USA.

DISPENSE BEVERAGES

Soft drinks syrups have, for a number of years, been supplied to the consumer for dilution with carbonated water. These syrups are similar to those which would be prepared in a manufacturing operation prior to proportioning with carbonated water to give finished beverage. Dispense syrups may be supplied in two distinct package sizes:

(i)   in small containers for retail sale, which the consumer uses for dilution with carbonated water at home;
(ii) in large containers, supplied to catering or similar outlets, for dilution with carbonated water dispensed through a counter-top dispenser.

Dispensing of soft drinks at counters in bars, restaurants and fast food outlets is a rapidly growing sector of the soft drinks market. These dispensed products fall into two categories depending on the form in which the product is delivered to the dispensing outlet: pre-mix and post-mix. Post-mix accounts for the largest volume of sales. In the pre-mix system, the finished carbonated flavoured drink is delivered in pressurized stainless steel tanks and then dispensed under pressure directly into a glass or cup. In the post-mix system the product is supplied as a concentrate, packed in bag-in-box or in stainless steel containers similar to the pre-mix containers. At the dispensing point, the product is mixed at the dispense head with carbonated water, or uncarbonated water in the case of orange or some other fruit drinks. The syrup may be drawn to the dispense head from either a pressurized or an unpressurized container. Water is drawn from the mains supply and carbonated if required in a small cooler/carbonater unit.

NUTRITIONAL DRINKS

Under current UK legislation, drinks may only be called 'nutritional' if it can be shown that they genuinely satisfy a particular nutritional requirement. They may fall into the following categories:

(i)   those which contribute energy and are usually high in glucose;
(ii)   drinks enriched with mixed vitamins, minerals and/or salts, often in combination with (i);
(iii) sugar-free products, essential for diabetics.

SPORTS DRINKS

There is now a range of soft drinks, specially formulated for people engaged in sports or other heavy exercise. The needs to be met include:

(i)   Pre-exercise – drinks to provide rapidly metabolizable fuel for the working muscles;
(ii)   Post-exercise or during prolonged exercise – drinks to replace the water and mineral salts lost through sweating and so diminish dehydration and consequent exhaustion.

The fuel is normally provided from low molecular weight carbohydrates, usually glucose or glucose polymers (maltodextrins). Maltodextrins appear to increase the rate of absorption of the total carbohydrate. This is because absorption is influenced by osmotic pressure, and a given quantity of carbohydrate in solution in the form of glucose polymer has a lower osmotic pressure than the equivalent amount of monomeric glucose. Maltodextrins are commonly included in the formulation for this reason, and also because they are less sweet than glucose.

Whether for pre- or post-exercise consumption, it is necessary to formulate the drink to be approximately isotonic with the human body fluids, so as not to disturb the electrolyte balance when large quantities of water are drunk. Hence the terms 'isotonic' or 'electrolyte' are often used for drinks of these types. One suitable composition is that of Ringer's Fluid, long known in medicine and surgery to have a salt composition similar to that of the body fluids:

Sodium chloride	0.86% w/v
Potassium chloride	0.30% w/v
Calcium chloride (hexahydrate)	0.48% w/v

The composition of sweat is similar but not identical to this, and varies somewhat between people and with different forms of exercise. Some variation around these values in the formulation of the drink is therefore tolerable.

BOTTLED WATER

Water has been the fastest growing sector in the soft drinks market for several years as consumers are looking for healthy, sugar-free products. Both still and carbonated waters are growing, with still waters accounting for about 60% of the market. Bottled waters can be divided into five categories:

(i) Natural mineral waters
(ii) Table water and purified water
(iii) Spring water and natural water
(iv) Mineral water
(v) Flavoured water

All bottled waters, except for natural mineral waters, which have separate guidelines for their production, must comply with the Drinking Water in Containers Regulations 1994 and the Private Water Supply Regulations 1991. Flavoured waters are classed as soft drinks and, as such, are governed by the Soft Drinks Regulations 1964 as amended and not by water directives.

PACKAGING

GLASS

Glass has been used for many years as a packaging format for fruit juices and soft drinks, primarily as a returnable container. However, this pack has declined rapidly over the last few years. Returnable glass bottles are mainly used for fruit juices and carbonated mixers sold to the licensed on-premise trade in pack sizes of 113 ml and 180 ml. A small market still exists for 1 l returnable glass bottles for some carbonated products and cordials.

The use of non-returnable glass bottles is increasing for carbonated drinks and waters, generally in a 330 ml pack size. Non-returnable glass containers are also used for some fruit juices and dilutables.

CANS

Cans are mostly used for carbonated beverages and have been marketed in significant quantities since the 1960s. Two piece cans without side seams, produced by the drawing and wall ironing process, are now used almost exclusively for soft drinks. The ends have easy-open tabs, mostly now of the stay-on variety. Aluminium was increasingly replacing tinned or coated steel as the can material, because of lower cost, improved quality of

printing and greater ease of recycling. However, with recent increases in the cost of aluminium, a number of companies are returning to coated steel.

The majority of soft drinks cans are 330 ml, with smaller 150 ml cans being used for mixers and non-carbonated fruit juices and 440 ml cans for shandy.

PET

Polyethylene terephthalate (PET) bottles are a very popular pack type for carbonated soft drinks. The bottle is manufactured in a two stage process. PET granules are melted and injected at high pressure into preform moulds. The preforms can then be warmed and stretched and blown with compressed air into a mould to give a finished single piece bottle of the desired size.

PET bottles have good gas barrier properties and are lightweight. Once filled with a carbonated drink the bottles become very strong and rigid.

PET bottles are most popular in 1, 1.5 and 2 l sizes, although products are available in sizes from 250 ml to 3 l. In some European countries, notably Holland and Germany, returnable PET bottles are used for carbonated products. These bottles are twice as heavy as one trip bottles, to allow for greater handling abuse.

PVC

Poly(vinyl chloride) (PVC) bottles are used mainly for dilutable products such as squashes and cordials. PVC is unsuitable for carbonated products as a result of its poor carbon dioxide retention properties.

PVC bottles are normally produced by extrusion stretch blow moulding of the molten PVC. The bottles can be round, square or rectangular. The square and rectangular bottles have been introduced to give improved shelf space utilization. The bottle sizes are typically 1–3 l.

CARTONS

Cartons in a variety of formats are used for packing fruit juices and non-carbonated soft drinks. The material used is a paper board coated with a thermoplastic material. Various laminates, including aluminium foil, are used to produce barrier properties for the pack.

The main commercial carton systems are Tetrapak and Combibloc. Pasteurized product is filled aseptically into sterile packs giving a product with a long shelf life at ambient or chill temperatures. Other pack types such as cups and pouches made from multilayer films are also used and are produced on form-fill-seal machines.

BAG-IN-BOX

Bag-in-box (BIB) is the major pack format for dispense beverage syrups, such as cola and lemonade. The bags are multilayer laminates with a nozzle through which they are filled and sealed. The filled bags are then packed in fibreboard boxes. The most popular pack sizes are 10 or 25 l but some 1000 l BIB products are produced.

QUALITY ASPECTS

RAW MATERIALS

Fruit juice

With relatively high value products such as fruit juices, in a large international market, it is not surprising that attempts to produce diluted, altered and cheaper products may be widespread. Adulteration may be considered as the addition of foreign non-fruit substances such as water or various sugars or the addition of other fruit materials such as the juice of different fruits or (with citrus juices) of pulp wash. Pulp wash is the liquid obtained by washing the fruit peels and pulp after the juice has been expressed. Checking authenticity and guarding the consumer against product deception are targets of the European Quality Control System (EQCS), which was founded in 1994 by six countries (Austria, France, Germany, Denmark, Netherlands, UK).

The European fruit juice industry works to the EC Fruit Juice Directive as the legal standard and also to the AIJN Code of Practice (Code of Practice for the Evaluation of Fruit and Vegetable Juices, 1994, Brussels, Association of the Industry of Juices and Nectars from Fruits and Vegetables of the EEC). Previously, individual countries had national standards, for example, the German RSK system (Richtwerte und Schwankungsbreiten bestimmter Kennzahlen or *Standard Values and Ranges of Certain Analytical Characteristics*), and the French AFNOR system (published by the Association Française de Normalisation). The AIJN established the Code of

Practice to transfer national standards into those of the EU.

The AIJN Code of Practice stipulates compositional requirements for fruit juices and products derived from them and provides tables of chemically defined parameters and fruit juices. These constituents are wide ranging and specific and for any meaningful interpretation of results an overall picture of the juice must be obtained.

Typical analytical parameters cited in the AIJN Code of Practice for orange juice are: relative density, corresponding Brix, *l*-ascorbic acid, volatile acids, ethanol, lactic acid, sulphurous acid, sodium, nitrate, heavy metals, titratable acid, *d*-isocitric acid, citric acid, *l*-malic acid, ash, potassium, magnesium, calcium, total phosphorus, sulphate, formol index, hesperidin, water-soluble pectin, total pectin, glucose, fructose, sucrose, reduction-free extract and amino acids.

A number of modern specialized analytical techniques are being applied to specific authenticity problems.

Chromatographic techniques

High performance liquid chromatography (HPLC) or capillary gas chromatography (CGC) may be used to measure oligosaccharides, which may indicate the presence of sugar syrups manufactured by the hydrolysis of starch. HPLC can also be used to obtain the flavanoid pattern giving information about the fruit types present.

Atomic isotope analysis

This may be applied in several ways:

(i) The ^{13}C content of any plant material is related to its botanical class and the metabolic pathway it employs to fix atmospheric carbon dioxide. The commonest group, to which the citrus fruits belong, is the so-called C3 pathway. Sugar cane and maize, on the other hand, follow the C4 pathway and have a lower natural ^{13}C content. Sugars derived from these sources may thus be detected. However, beet is a C4 plant and so beet sugar is not detectable by this test.

(ii) The ratio of $^{18}O/^{16}O$ can give some information about the origins of the water in a fruit or a processed juice; the ratio is infuenced by the geographical region where the fruit was grown, by evaporation (e.g. if the juice was concentrated), and by the source of any water used in processing or subsequent dilution. Interpretation is difficult but the test has been used to distinguish fresh juice from diluted concentrate, or the presence of beet sugar grown in a different region from the juice to which it was added.

(iii) Deuterium (^{2}H) may be measured by mass spectrometry or by NMR. The proportion relative to hydrogen (^{1}H) varies with the origins of any water present and also with a number of biochemical factors related to the metabolism of the fruit. The deuterium content may therefore be used in a similar way to the ^{18}O/^{16}O ratio and can indicate, for instance, the presence of beet sugar in a citrus juice.

Statistical methods

These, including multivariate analysis and computer simulation, have been used as tools for the recognition and validation of relatively complex patterns in the prolific analytical data which all of the above methods can provide. This in-depth testing cannot be carried out on every batch of material received and is usually carried out to a statistical sampling plan. Intake testing for fruit juices normally involves checking key parameters against a specification. For an orange juice or comminute, these key parameters are typically Brix, acidity, enzymic activity, organoleptic characteristics and microbiological assessment.

Brix. The expression 'Brix' is used rather loosely throughout the soft drink industry. Primarily, it is a measure of sucrose concentration as determined by a Brix hydrometer which is calibrated to read directly in terms of percent w/w of sucrose and, as such, should only be used for pure sucrose solutions. It is, however, common practice to measure the refractive index of sugar-containing concentrates, soft drinks or syrups on a refractometer calibrated to the International Sugar Scale, which itself has an imperfect correlation with the Brix hydrometer, and express the result as 'degrees Brix'.

The Brix reading is affected by temperature and unless the instrument is maintained at, or compensated to 20°C, corrections must be applied. The presence of acids also affects the Brix reading and further corrections must be applied; these can be found in standard tables.

Acidity. It is usually sufficient to determine the titratable or free acidity, calculated as the predominant acid; these are tabulated below. An aliquot of the juice or concentrate is titrated with standar-

dized sodium hydroxide (usually 0.1 N) to pH 8.1 using a pH meter or phenolphthalein indicator to detect the end point.

1 ml of 0.1 N NaOH represents:

0.0070 g Citric acid monohydrate
0.0064 g Citric acid (anhydrous)
0.0067 g Malic acid
0.0075 g Tartaric acid

Fruit type	Predominant acid
Orange	Citric
Grapefruit	Citric
Lemon	Citric
Lime	Citric
Apple	Malic
Pineapple	Citric
Blackcurrant	Citric
Tomato	Citric
Grape	Tartaric
Pears	Citric
Cherry	Malic
Apricot	Malic.

Enzymic activity (pectinesterase). Inactivation of pectinesterase in citrus juices is important, since the enzyme will cause separation of the natural cloud in the juice and clarification of product. This is not considered acceptable and will give rise to customer complaints. The determination works on the principle that if juices with active pectinesterase are mixed with a pectin solution, the pectin will be de-esterified. If calcium ions are present they will react with the low ester pectin formed to give a gel. The reaction proceeds more rapidly at neutral pH.

Other methods

For juices such as clear apple or lime, the colour and clarity of the product are important. These can be determined by measuring the absorbance and transmission of known strength solutions at set wavelengths. Measurement of these parameters would be included as acceptance tests.

Details of methodologies for most of the parameters as mentioned in the AIJN Code of Practice can be found in the IFU (International Fruit Juice Federation) Handbook of Analyses 1996 (International Fruchtsaft-Union, Zug, Switzerland).

Organoleptic characteristics. This covers colour, appearance, aroma and flavour of the product and will be specific for the material type and detailed in the specification.

Microbiological assessment. It is important that the fruit raw material is microbiologically sound before use in production. Usually the number of yeasts, bacteria and mould present are determined.

Water

It is important that checks are carried out to ensure that the water treatment plants are operating satisfactorily and that a uniform and consistent supply is obtained for the process plant. Key parameters important for the quality of soft drinks are checked regularly, usually on a daily basis at a minimum. Parameters checked include:

Appearance/taste. This would include the colour and turbidity of the water and taste for any taints such as chlorine.

pH. The pH value determined directly by a pH meter is used to give an indication of any major problems with the treated water.

Conductivity. Conductivity is determined directly by means of a conductivity meter. Its value gives an indication of the total ion content and can be used as a general assessment for the efficiency of the water treatment.

Total alkalinity. This is determined by simple titration against 0.02 N sulphuric acid to a set end point where methyl orange indicator changes colour. The total alkalinity is usually expressed as calcium carbonate.

Total hardness. This is routinely determined by titration of a buffered sample with EDTA (ethylenediamine tetraacetic acid) after addition of an organic dye, calmagite. The end point of the titration is a colour change from red to blue. The normality and volume of EDTA titrant can then be related to the hardness content, usually expressed in mg l^{-1} as calcium carbonate.

Free, residual and total chlorine. This test is performed where chlorine has been used as the sterilizing agent. It is usually determined with the aid of a test kit.

Nitrate. This test is undertaken to check on nitrate removal, where a low level is important, for example, in canning. The nitrate content is usually determined routinely by test kit but anion exchange high performance liquid chromatography can also be used.

Aluminium. Normally, aluminium is only determined if aluminium sulphate is used in the water treatment process. It is routinely determined by test kit.

Microbiological assessment. The water is tested to ensure that it is microbiologically satisfactory. Normally a total count and a total coliform count are carried out.

Other analyses. A number of test kits are commercially available for measurement of the key parameters of water quality. These enable the tests to be carried out easily on a routine basis. They all give rise to coloured solutions; the amount of colour present can be determined by comparison with a comparator disc, or photometrically, measuring the percent transmission of the solution at a set wavelength. These readings are then converted to a mg l^{-1} content of the analyte of interest. More in-depth testing to check levels of trace metals, anions and cations can be carried out on a non-routine basis to monitor the overall quality of the water supply and treatment.

Dry ingredients, sugar and sugar syrups

Intake testing for dry ingredients, such as artificial sweeteners, preservatives and some acids, normally involves checking key parameters to specification. These key parameters include typically, the appearance, chemical purity, Brix value, moisture content and a microbiological assessment; impurities such as trace elements may be determined on a non-routine basis.

Flavours

Intake testing for flavours normally involves checking key parameters to specification. These key parameters are typically, appearance, organoleptic characteristics, density or refractive index. A gas chromatographic (GC) profile may be obtained; key peaks can be identified by reference to internal standards or by comparison with a library of product profiles.

FINISHED PRODUCTS

The requirements for product testing are dependent on the type of product being manufactured. Typically, finished product testing would be carried out to ensure conformance to specification of specific product parameters. These usually include Brix, acidity, artificial sweeteners, preservatives, carbonation (for carbonated products), organoleptic properties and microbiological assessment. A brief overview of some of the methodology is given below:

Artificial sweeteners

For the artificial sweeteners acesulfame-K, aspartame and saccharin, the most convenient method of analysis, now widely used wherever the equipment is available, is reverse phase high performance liquid chromatography (HPLC). Solvent extraction methods used for the determination of saccharin have largely been superseded.

Benzoic acid and sorbic acid

Both preservatives are most conveniently determined by reverse phase HPLC. Solvent extraction methods used for the determination of benzoic acid have largely been superseded.

Sulphur dioxide

Most methods for the determination of sulphur dioxide depend upon its distillation from an acidified sample. The displaced gas may be determined either by absorption into water containing iodine and starch indicator, or into neutralized hydrogen peroxide. The standard distillation method for sulphur dioxide determination is that of Monier Williams.

Alternatively, sulphur dioxide may be determined by direct titration of the acidified product with iodine. This method is subject to interference by colours, ascorbic acid and other reducing substances. However, these may be allowed for and this method is commonly used for routine analysis.

Sulphur dioxide is found in two different forms in soft drinks, either free or bound (loosely combined with sugars or flavouring materials). The distillation method determines total sulphur dioxide, whereas direct titration gives only the free form unless the sample has been adjusted with alkali to pH 13 before acidification. Only free sulphur dioxide is effective as a preservative, but preservatives regulations usually refer to total (free and bound) sulphur dioxide. It is therefore important to determine the proportion of sulphur dioxide bound up in any particular formulation in order to assess its preservative effect.

Carbonation

Determination of the carbon dioxide content of a carbonated beverage may be made by measuring the temperature of the beverage and the equilibrium pressure of carbon dioxide in the vapour phase. The pressure can be determined using a puncture-type pressure gauge fitted with a rubber seal to prevent loss of gas from the can or bottle. The presence of air inside the container, gaseous or dissolved, may cause variable results; this is minimized by 'snifting' the sample (venting off the gas in the headspace) to sweep out the contained air. The method of testing depends on whether the sample has been in storage and has attained equilibrium, or whether it has come straight from the production line, in which case there is little or no carbon dioxide pressure in the headspace. It is recommended that the temperature of the sample should be between 13°C (55°F) and 18°C (65°F). The use of charts enables carbonation to be determined as 'volumes Bunsen'. This is defined as the volumes of carbon dioxide at atmospheric pressure and 0°C (32°F) which are dissolved in unit volume of liquid. With some samples it is not possible to use a pressure gauge, owing to the type of closure and carbonation must be determined by some other method.

The Lunge Nitrometer is suited to the routine determination of carbonation. In this method an aliquot of the sample is stirred under partial vacuum to remove all gas from the liquid, and the volumes of gas and liquid phases are then measured at ambient temperature and atmospheric pressure. The volume of gas divided by the volume of liquid determines the carbonation as 'volumes Ostwald', which unlike volumes Bunsen, does not require a correction for temperature.

In a number of European countries, carbonation is quantified in gl^{-1}, 1 volume Bunsen being equivalent to 1.96 g of carbon dioxide per litre.

Microbiological assessment

An assessment of the finished product is made to ensure that it has not been contaminated during production and that procedures such as pasteurization have been satisfactorily carried out. This should ensure that the product will be microbiologically satisfactory throughout its shelf life. Normally, total count, yeasts, mould and bacteria are determined.

Quinine

This may be determined by extraction with chloroform after making the drink alkaline with ammonia. The absorbance of the chloroform extract is measured at 335 nm. With a clear tonic water, the absorbance at 347.5 nm may be measured directly without extraction and compared with that of a blank containing the same concentrations of sugar, acid and any other ingredients.

HPLC may also be used where the equipment is available.

Caffeine

Caffeine can conveniently be determined by HPLC and this is now the method of choice. It may also be determined by extracting with chloroform from the soft drink made alkaline with ammonia and measuring the absorbance at 273 nm.

Ascorbic acid

The classical method for determining ascorbic acid is by direct titration with 2,6-dichlorophenolindophenol. Where products are preserved with SO_2, it is necessary to remove this before titration. In deeply coloured solutions, direct visual titration with the dye solution is not possible. As the titration is based on an oxidation/reduction reaction, it is possible, however, to use an autotitrator with a redox electrode. Chromatographic techniques are also used.

Alcohol

There are two recognized methods of analysis. These are:

(i) Distillation followed by specific gravity determination and reference to official tables;
(ii) Gas–liquid chromatography (GLC). This method is based on the separation of the volatile components of a product by distribution between a stationary and a mobile phase. Specific methods may be obtained from suppliers of GLC equipment. The method is extremely rapid and is therefore suitable for routine control of production

Colours

Artificial colours may be extracted from the sample by a dyeing procedure using defatted white wool or polyamide powder, followed by thin layer or paper chromatography. There is no single method of analysis for natural or nature-identical colours; various methods may be found in standard analytical textbooks. Suppliers of the materials will also provide specific methods for their products.

Fruit content or juice content

Three parameters are commonly used to determine the fruit or fruit juice content of a soft drink.

These are potassium, phosphorus and nitrogen content. These three analyses may be used individually or in combination to estimate the fruit content. The three values used in combination will give a more accurate estimation than each individual value. However, by virtue of ease and accuracy of determination, potassium is the single parameter most often used. Tables of composition may be found in McCance and Widdowson (Holland *et al.*, 1991).

Sugars

The sugars most frequently found in soft drinks are sucrose, glucose syrup, fructose and invert sugar. The total sugar content is most commonly estimated in terms of degrees Brix using a refractometer. However when sugars other than sucrose are present the Brix value will not refer exactly to the sugar content by weight. If more precise compositional details are required, the individual sugars may all be determined by HPLC or by the use of specific enzyme assay kits which are available for all the common sugars.

REFERENCES

AIJN (1994) *Code of Practice*, Association of the Industry of Juices and Nectars from Fruits and Vegetables of the European Economic Community, 1st edn.
Holland, B. *et al.* (1991) *McCance and Widdowson's The Composition of Foods*, 5th edn and 7 supplements (1988–1995), Royal Society of Chemistry, Cambridge.
IFU (1996) *Analysis Methods Manual*, International Federation of Fruit Juice Producers.

FURTHER READING

Association of the German Fruit Juice Industry (1987) *RSK Values: the Complete Manual*, Flüssiges Obst GmbH, Bonn.
British Pharmocopoeia (1993), HM Stationery Office, London.
Hicks, D. (ed.) (1990) *Production and Packaging of Non-Carbonated Fruit Juices and Fruit Beverages*, Blackie, Glasgow.
Kimball, D. (ed.) (1991) *Citrus Processing: Quality Control and Technology*, Van Nostrand Reinhold, New York.
Kirk, R.S. and Sawyer, R. (1991) *(Pearson's) Chemical Analysis of Foods*, 9th edn, Longman, London.
Mitchell, A.J. (ed.) (1990) *Formulation and Production of Carbonated Soft Drinks*, Blackie, Glasgow.
Nelson, P.E. and Tressler, D.K. (eds) (1980) *Fruit and Vegetable Juice Processing Technology*, 3rd edn, AVI, Connecticut.
Nagy, S., Attaway, J.A. and Rhodes, M.E. (eds) (1988) *Adulteration of Fruit Juice Beverages*, Marcel Dekker, New York.
Paul, A.A. and Southgate, D.A.T. (1991) *(McCance and Widdowson's) The Composition of Foods*, 5th edn, HM Stationery Office, London.
Ting, S.V. and Roussef, R.L. (1986) *Citrus Fruits and their Products: Analysis and Technology*, Marcel Dekker, New York.

7 Alcoholic Beverages

WINE

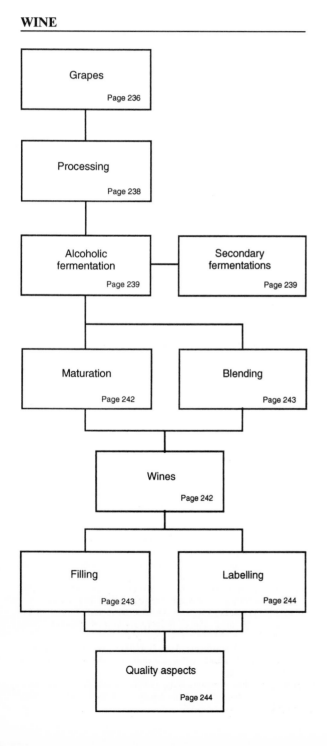

Grapes	Page 236
Processing	Page 238
Alcoholic fermentation	Page 239
Secondary fermentations	Page 239
Maturation	Page 242
Blending	Page 243
Wines	Page 242
Filling	Page 243
Labelling	Page 244
Quality aspects	Page 244

GRAPES

Wines are traditionally made from the fruit of *Vitis vinifera* of which there are a multitude of varieties growing in many parts of the world. Wine can be made from other species. American species are used as phylloxera-resistant rootstocks (see below), while in the eastern states of the USA the indigenous species *Vitis labrusca* is commonly grown. The *labrusca* flavour is different from *V. vinifera* and though quite acceptable in America it is not preferred and therefore not grown in European countries.

Table 7.1 lists some of the best known wine-making grape varieties.

Wine grapes are normally different from table grapes and the skin colours, ranging from dark red or purple to pink or from deep golden to greenish-white, allow a wide range of different wines to be made. Most varieties have white juice, so that white wines can be made from red, as well as white, grapes.

Ripeness is normally achieved about 100 days after the flowering of the vine and is judged by the thickness and colour of the skins, but principally by the accumulation of sugars and diminution of acids (see Figures 7.1 and 7.2). Some of the finest wines need the grapes to ripen beyond normal maturity and for the juices to be concentrated. This concentration can be done in a number of ways: leaving the bunches on the vine; letting 'noble rot' concentrate the juice; spreading bunches on straw mats in the sun; leaving bunches on trays under cover in an airy place; letting the grapes freeze on the vine and pressing the frozen grapes; or freezing the juice after pressing.

A preharvest cull can focus the vine's energies on a smaller number of bunches and special qualities can be made by selective picking of the ripest grapes. These demand labour-intensive hand harvesting. Mechanical harvesting, which may shake the grapes off the stalks onto a moving band, can reduce costs for ordinary wines, but can also allow the makers of fine wines to seize 'windows

Table 7.1 Some of the most widely grown winemaking grape varieties

Variety	Wine			Remarks
Airèn	White			Spanish origin, large quantities
Cabernet Franc		Rosé	Red	
Cabernet Sauvignon			Red	
Carignan (Cariñena)			Red	
Chardonnay	White			especially for Burgundy
Chénin Blanc	White			
Gamay		Rosé	Red	especially for Beaujolais
Gewürztraminer	White			means 'spicy Traminer'
Grenache (Garnacha)	White	Rosé	Red	
Malvasia (Malvoisie)	White			best known for Malmsey Madeira
Merlot			Red	
Müller-Thurgau	White			Swiss origin, large quantities
Muscat	White	Rosé	(Red)	several sub-varieties
Nebbiolo			Red	especially northwest Italy
Palomino	White			especially for sherry
Pinot Blanc (Pinot Bianco)	White			
Pinot Gris (Pinot Grigio)	White	Rosé		
Pinot Noir	(White)	(Rosé)	Red	especially for Burgundy (and Champagne)
Riesling	White			
Sangiovese			Red	especially for Chianti
Sauvignon Blanc (Blanc Fumé)	White			
Sémillon	White			
Sercial	White			for dry Madeira
Silvaner (Sylvaner)	White			
Syrah (Shiraz)			Red	
Tempranillo			Red	especially in Spain
Torrontes			Red	principally in Argentina
Traminer	White			originally from north Italy
Trebbiano	White			principally in Italy
Ugni Blanc	White			Italian origin, large quantities
Verdelho	White			

of opportunity' during changeable weather. Rain during the harvest can produce dilute wines, either by swelling the grapes or by incorporating the rain on the outside of the bunches during crush.

Care and vigilance are needed to protect the vines and grapes against pests and diseases. Some will respond to curative measures, but others like mildew and similar cryptogamic diseases demand preventive spraying. Phylloxera, caused by a root louse, requires the fine grape-bearing scions to be grafted onto resistant root-stock varieties. The grey mould *Botrytis cinerea* can ruin the crop when it attacks unripe grapes, yet paradoxically, on fully-ripe grapes it is called the noble rot and helps to produce the top wines of Sauternes and Tokay, as well as the German Trocken-beeren-auslesen.

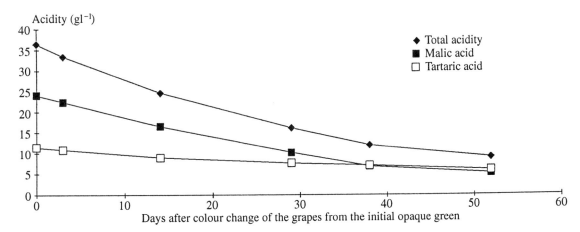

Figure 7.1 Acidity of ripening grapes. ◆, Total acidity; ■, malic acid; □, tartaric acid.

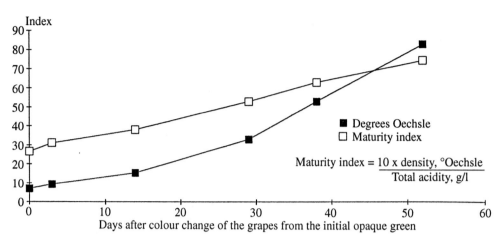

Figure 7.2 Maturity of ripening grapes. ■, degrees Oechsle; □, maturity index. Maturity index = $10 \times \dfrac{\text{density}}{\text{total acidity}}$ where density is measured in degrees Oechsle and acidity is in g l^{-1}.

OTHER RAW MATERIALS

Brandies and similar spirits have long been distilled from wines and compulsory distillation has for a number of years been a principal means of reducing the 'wine lake' in Europe. Such brandies and rectified grape spirits are used in the making of fortified wines, and when added to unfermented grape juice, making wines such as Pineau des Charentes, vins de liqueur or Mistelas.

There is also significant production of unfermented grape juice in Europe and in America, partly for the production of grape juice soft drinks and partly as another means of reducing the 'wine lake'. In Europe the juice is usually concentrated to give concentrated grape must.

In the European Union, wines made from concentrated grape juice or from dried grapes are denominated 'made wines', while 'fruit wines' are those made from fruits other than grapes. Fruit flavours, such as peach juice, are added to some sparkling wines. Aromatized and flavoured wines, such as Vermouths, as well as so-called 'tonic wines', make use of a wide range of traditional herbs and spices, with their essential qualities obtained in a variety of ways according to the desired style of product.

Cane or beet sugars are commonly added to the grape juice in more northerly vineyards, and also in unsatisfactory vintages, to make up for a deficiency in grape sugars. The acidity of the grape juice inverts the sucrose of cane or beet sugars into the glucose (dextrose) of grape sugar. In France, sucrose is added in dry form (dry sugaring), also known as 'chaptalization', whereas in Germany it has been traditional to dissolve the sugar in water. This wet sugaring has the further advantage of reducing the acidity slightly. The

excessive production of grapes in the European Union, as noted above, has led also to a campaign to replace chaptalization with the use of rectified concentrated grape must. Many German wines are sweetened after fermentation, using sterile-filtered grape must (Süssreserve) which may have been kept back from a previous vintage.

PROCESSING OF GRAPES

For the finest types of wine the bunches may be sorted to eliminate faulty or unripe grapes. Normally the next process is to separate the grapes from the stalks and crush the berries. This used to be done by treading with the foot but is now effected with a crusher-destemmer.

For nearly all white wines it is important to separate the juice from the skins, pulp and pips without delay, especially when white wine is being made from red-skinned grapes as may be the case with champagne. This is done in one of many types of wine press. It can be a batch process (for champagne, fine white wines or sherry) or in a continuous press for the speedy and economic handling of quantities of lower priced wines.

For red and rosé wines the colour, which is under the skin of the grape, needs to be extracted into the juice. This skin contact (maceration) may last for only one night or 24 h (for pale rosé wines) or continue for three weeks or more, for deep red wines. Colour extraction is related to temperature (see Thermovinification, under TEMPERATURE CONTROL below) as well as time. Some wines are made by putting whole bunches without destemming or crushing into closed vats, where the start of fermentation from the ripest grapes crushed under the weight of the mass bathes the remainder of the bunches in carbon dioxide gas (carbonic maceration) giving special flavour and colour extraction.

During maceration, bubbles of carbon dioxide gas tend to lift the skins, pips and pulp to the surface to form a 'cap'. Various techniques are used to keep the skins in contact with the juice, such as punching down the cap, pumping the juice over the cap, keeping the cap submerged with a grid part of the way down the vat or by having mechanical paddles inside the vat. When sufficient colour has been achieved or the fermentation finished, the free-run wine can be drawn off, after which the pulp, pips and skins can be put in a wine press and the press wine extracted from the pomace. If desired, pips can be removed from the pomace with a special depipping machine and cold pressed to yield grape seed oil as a by-product. The yield is small (6 to 10% of the seeds) compared with 20 to 50% from other oil-producing seeds and nuts.

Vinification

Grape juice may be processed as such (see Chapter 6) but for it to be considered as wine the presence of alcohol is essential. This may simply be added to the unfermented grape juice, as in Mistelas, Pineau des Charentes or Tokay Essenzia, or added after fermentation in the case of wines like port and sherry. However, most winemaking requires the action of fermentation micro-organisms on the sugars and other constituents of the grape juice.

$$C_6H_{12}O_6 \rightarrow 2C_2H_6O + 2CO_2 + 23.5 \text{ cals}$$

This is called the primary fermentation and is carried out by the enzyme zymase, produced by yeasts, usually *Saccharomyces ellipsoideus*. As a rule-of-thumb one can reckon that each 18 g sugar per litre of juice has the potential to produce 1% alcohol; hence a grape must with 216 g l^{-1} sugar (about 12° Baumé) would be needed for a dry wine with 12% alcohol by volume (ABV).

Yeast cells absorb nutrients (sugars, etc.) through osmosis and dispose of the products (alcohol and carbon dioxide) similarly through the cell wall. They are therefore influenced by the medium in which they find themselves. If the sugar concentration in the grape must is too great the yeast cells are apt to dehydrate and collapse, and similarly if the concentration of alcohol in the wine becomes too high. Thus fermentation may be difficult to start with very rich grape musts, while special alcohol-resistant strains of yeast may be needed for the vinification of high strength wines.

A second fermentation by adding more sugar and yeast is used to make sparkling wine, where the production of carbon dioxide gas is the principal objective. A different second fermentation, called the 'governo', is used in some Italian wines, adding unfermented grape juice and yeast to the wine to give it extra strength and body. Fino sherries have a 'flor' second fermentation by film yeasts growing on the wine surface which strip the sherry of oxygen creating a reductive state and keeping the wine fresh and crisp. Proliferation of yeast cells during fermentation results in 'wine lees', mainly dead yeast cells, tartrates and grape fragments, so that young wine requires clarification or racking, or disgorging in the case of sparkling wines.

Lactobacilli are the ferments in a different secondary fermentation in which the sharp tasting malic acid of some young wines is converted into the milder lactic acid.

$$\underset{\text{Malic acid}}{HO_2CCH(OH)CO_2H} + H_2O \rightarrow \underset{\text{Lactic acid}}{HO_2CCH(OH)CH_3} + CO_2$$

This malo–lactic fermentation also produces CO_2 gas, giving the characteristic 'pétillant' style of certain wines, like the Vinho Verdes of northern Portugal. Lactobacilli seem to have a symbiotic relationship with the yeasts of the primary fermentation, relying on yeast by-products for their activation.

Fermentation stops naturally when all the fermentable sugars have been converted to alcohol. It can also end if the alcoholic strength reaches the limit of tolerance of the strain of yeast involved, causing the cells to dehydrate and collapse. It can therefore be stopped artificially by adding alcohol (fortification) either to the fermenting wine (in the case of port) or even to unfermented grape juice, as in Mistelas or Pineau des Charentes. Fermentation can be stopped mechanically by sterile filtration or centrifuging which remove the yeast from the wine. Temperature control, by cooling to inactivate the yeasts or heating to pasteurize them, is another method available. Sulphur dioxide or sorbic acid preservatives may also be used.

Acids

The predominant acid during the early stages of development of the grape is the 'green-apple' malic acid (see Figure 7.1). As the grape changes colour and matures, aided by warmth and

Table 7.2 Dosage of sulphur dioxide for the treatment of grape musts and wines

SO$_2$ content required (mg l^{-1})	SO$_2$ (applied as)			
	Liquid SO$_2$ (g hl^{-1})	SO$_2$, solution (ml hl^{-1})	5% Potassium metabisulphite (g hl^{-1})	Sulphur candles (no. of thin sheets)
5	5	100	10	1–1½
10	10	200	20	2–3
15	15	300	30	3–4½
20	20	400	40	4–6
25	25	500	50	5–7½
30	30	600	60	6–9
35	35	700	70	7–10½
40	40	800	80	8–12
45	45	900	90	9–13½
50	50	1000	100	10–15
55	55	1100	110	
60	60	1200	120	
65	65	1300	130	
70	70	1400	140	
75	75	1500	150	
100	100	2000	200	
125	125	2500	250	
150	150	3000	300	

moisture, the malic acidity diminishes while the typical grape acid, tartaric, begins to predominate. During the alcoholic fermentation some of the tartaric acid comes out of solution in the form of tartrates (known in the past as 'argols' or wine stones) which can be a useful by-product, manufactured, for example, into baking powder.

During vinification and maturing, the winemaker has to guard against or control volatile acidity. This is caused by the action of *Acetobacter*, which convert alcohol into acetic acid, and if uncontrolled can cause spoilage and turn the wine into vinegar. In very small quantities, however, *Acetobacter* can help to develop esters, giving finesse to the aroma.

A number of wines, such as the Moselles, owe their crisp and refreshing character to a quite high content of malic acid, but for other styles of wines, expected to be mild and smooth, the winemaker may wish to encourage the action of lactobacilli to convert the malic acid into lactic acid in the malo–lactic fermentation. The total acidity of a must or a wine is a useful indication of the maturity of the grapes and the development of the wine.

Other acids

Citric acid has been used in the past to complex and control metal contamination in wines. Traces of succinic, propionic and some other acids may result as by-products of fermentation.

Sulphur dioxide

Sulphur dioxide has a distinctive role in winemaking, as a preservative, a sterilizing agent and as an antioxidant. It has a tendency to combine with sugars which reduce its effectiveness both as an antioxidant and against micro-organisms. As sulphurous acid it can impart a harsh or 'hot' end flavour to a wine. It can also react to produce mercaptans or bottle stink in an old wine. Table 7.2 shows quantitative relationships among various practical sources of SO$_2$ when added to wine.

Carbon dioxide

Carbon dioxide or carbonic acid is a by-product of the alcoholic fermentation. From still wines it is allowed to disperse into the atmosphere but for pétillant and sparkling wines it is retained in solution in the bottle. Its solubility is greater at lower temperatures. It may also be lost from solution by formation of complexes with sugars. The size and frequency of the bubbles in a sparkling wine are referred to as the mousse.

Sparkling wines

The classic 'méthode champenoise' consisted first of making a still wine from the champagne grapes, then adding sugar and a special yeast, bottling the wine in strong bottles and corking it securely, often using crown corks. Here the primary objective is not the conversion of sugar into alcohol but the production of carbon dioxide gas for the sparkle.

The subsequent removal of the dead yeast cells and tartrates resulting from this process requires their manipulation on to the corks. Traditionally this is done with the bottles placed with their necks in racks slanted at about 30–40° to the vertical. Each bottle is first shaken to loosen the sediment, rotated fractionally on its axis and oscillated and gradually titled to become neck downwards. These operations are repeated every few days, over a period ranging from a few weeks to several months. The necks of the bottles can then be immersed in a freezing bath, so that the lees are secured in a lump of ice attached to the cork. Thus the bottle can safely be stood upright, and the cork with its attached ice and sediment removed. The dryness of the wine may be adjusted to give Brut (very dry), dry, half-dry or sweet styles by the addition of an appropriate dosage of wine and sugar solution, then the bottle is recorked and wired to prevent the cork from blowing off later.

The 'Charmat' or 'cuve-close' method was invented to avoid such labour-intensive processes for lower priced sparkling wines. In this system the second fermentation is carried out in an insulated pressure vessel; cooling allows more CO_2 to be dissolved. The sparkling wine is then forced out under counterpressure of the CO_2 gas through a filter to remove the yeast sediment, into the bottles and corked.

An intermediate procedure, known as the 'transfer method', is used for some medium quality sparkling wines. The secondary fermentation takes place in bottle and in due course the bottles are uncorked, placed in a star wheel, emptied under CO_2 counterpressure which drives the wine through a filter to remove the yeast, thence into clean bottles which are recorked.

Some of the labour of the traditional methods may be reduced, for instance if the shaking of the sediment onto the cork is assisted by mechanical means, after which the necks can be frozen and disgorging and recorking proceeds normally.

Partial fermentation is used in making a number of special types of wine. Low alcohol wines like some Lambrusco, are sterile-filtered under CO_2 counterpressure, leaving semi-sparkling wine with residual sweetness. Port has its fermentation stopped by the addition of brandy, producing a rich sweet high strength wine.

Table 7.3 Solubility of potassium hydrogen tartrate (potassium bitartrate) in water and alcohol

Temperature (°C)	Solubility (g l^{-1}) in water containing the stated percentages of alcohol			
	0%	10%	20%	30%
0	0.30	0.17	0.11	0.07
5	0.32	0.19	0.13	0.07
10	0.41	0.21	0.16	0.09
15	0.44	0.24	0.16	0.09
20	0.49	0.29	0.17	0.11
25	0.54	0.36	0.21	0.12
30	0.69	0.40	0.25	0.13
35	0.84	0.49	0.29	0.19
40	0.96	0.54	0.38	0.23
45	1.13	0.73	0.44	0.26
50	1.25	0.87	0.54	0.30

Tartrates

As the alcohol content of the fermenting wine increases, the saturation point of tartrates in solution falls and they may crystallize out into the wine lees or onto the walls of the wine vessels. A fall in temperature will also increase tartrate precipitation (Table 7.3).

To avoid deposition of tartrates in the wine after bottling it is usual, particularly with white wines, to hold the bulk wine just above its freezing point for 7–14 days, then centrifuge or filter (or both) to remove the crystalline deposit. Clarification may be accelerated by 'seeding' the cold wine with a small quantity of potassium bitartrate.

Tartrates can be extracted from wine lees, pomace or the deposits in storage vats, by treating them with boiling water and then cooling the resultant solution. One litre of water at 90°C can dissolve 57 g potassium bitartrate (cream of tartar) but when cooled to 10°C it will retain only 4 g, the rest crystallizing out. The potassium bitartrate is usually accompanied by a small quantity of calcium tartrate. Precipitation of tartrate from a wine also reduces the acidity of the flavour.

Temperature control

Temperature control is important in the art of winemaking. Micro-organisms have different metabolic rates at higher or lower temperatures and chemical reactions such as oxidation are similarly influenced. In the simplest way, temperature may be controlled by judging the best size for the wine vessels and encouraging air currents (in above ground wineries), or by storing the wine in underground cellars.

Technology adds thermovinification to extract colour and flavour from the skins, autovinification to reduce the labour of breaking the 'cap' and pumping-over, while cooling equipment can provide low temperature fermentation.

MATURATION

Although some wines are made for early drinking, as 'Primeur' wines, most wines benefit from careful maturing, during which sharpness and roughness diminish while aromas and flavours can develop and gain in complexity.

The winemaker's skill is required to judge the oxidoreductive state of the intended wine. This process may best be envisaged through the different styles of sherry. After the primary 'tumul-tuoso' fermentation is ended, the various casks are tasted and graded. Some are neutral rayas suitable for blending. Others are destined for maturing in a reductive state state as finos, in which a flor yeast strips the wine of oxygen. Others again are left with an airspace in the cask to develop in an oxidative state as olorosos. Combinations of the two maturing methods yield other styles and subtleties. Once the style has been established in the nursery cellar ('criadera'), it is maintained by transferring the young wine little by little, cask by cask and row by row, through the stack of casks making up the 'solera' of that particular style and quality. Sherry is also different from most other wines by being left on the lees so that the yeast residues remain at the bottom of the casks and are slowly autolysed producing aldehydes.

$$CH_3CH_2OH + \tfrac{1}{2}O_2 \rightarrow CH_3CHO + H_2O$$
$$\text{alcohol} \qquad\qquad\qquad \text{acetaldehyde}$$

Such aldehydes contribute significantly to the typical aromas and flavours of sherries. However if oxidation is unchecked acetic acid can be produced.

$$CH_3CH_2OH + O_2 \rightarrow CH_3COOH + H_2O$$
$$\text{acetic acid}$$

This imparts vinegar tastes and 'volatile acidity aromas'. Other than the sherries, most wines in cask need to be clarified by taking them off the lees ('racking' them) into clean casks. Carbon dioxide or other inert gas such as nitrogen may be used during racking if the wine needs to be protected from the oxygen in the air so as to be kept in a reductive state. A relatively high acidity helps to keep a wine in a reductive state, as does the presence of dissolved carbon dioxide under pressure.

Clarification may need to be aided by 'fining'. Traditionally, high quality red wines were fined with egg white and white wines with gelatin or isinglass. Other products such as bentonite are now used. Most finings act by flocculation and adsorption and they need to be thoroughly dispersed throughout the wine. Centrifuging and filtering are other ways of clarifying grape must and wine; filter aids such as diatomaeceous or infusorial earth may be used.

Wine vessels

Winemaking and storage vessels, originally of stone, pottery or wood, may now be of stainless steel, fibreglass or ceramic-lined concrete. Self-emptying vats reduce the labour of handling pomace from red wine production. Insulated pressure vessels are now used for the 'Charmat' system of sparkling wine production.

TYPES OF WINE

Wine is a long term product. Clearing the land and planting may require several years before the first wine is marketed and there is a return on the investment. Therefore wineries may start by buying in grapes from existing growers and marketing a range of styles: red, white and rosé dry (fully fermented) wines. Wineries use noble grapes for classic styles needing maturation and more ordinary varieties for making 'quaffing wines' for cash-flow. They may produce medium-dry or sweeter table wines, either by starting with more sugar-rich grape juice or by stopping fermentation and leaving residual sugar and they may make sparkling wines by secondary fermentation (and even sometimes by carbon dioxide gas injection!).

Fortified wines were originally developed to withstand longer sea journeys and traditionally come from warmer vineyard regions; madeira owes its particular keeping qualities to the 'estufa' technique, whereby the casks of wine are 'baked' in hot cellars, which stabilizes and oxidizes the wine giving the characteristic 'maderized' style.

Mediterranean countries such as Greece and Cyprus have marketed their surplus grape production as raisins and by making vacuum-concentrated grape must. This has been shipped to countries like the UK to be reconstituted by adding water and fermented with cultured yeast to make 'British Wine' (in EU parlance - 'made wine'). This should not be confused with 'English wine', made from fresh grapes grown in English vineyards.

Ginger wine is made from grape concentrate and ginger, while tonic wines may include phosphates.

Vermouths are normally made with base wine from fresh grapes and their characteristic ingredient is wormwood, *Artemisia absinthium*; in Italian vermouth it is usually Roman Wormwood.

Japanese Saké, although it is sometimes called 'rice wine', is in fact brewed, using the symbiotic relationship of a starch digesting mould and a sugar converting saccharomyces.

Blending

A grower will find that some of his vineyard sites have better exposure than others, and that the grapes ripen at different times; it can be the same with different grape varieties. So a winemaker, using grapes of several varieties from a number of different sites, may have to make a selection from many vats of that vintage. Furthermore, from each vinification of red wine he may have two vats, one of the free-run wine and another of press wine. If the vintage is such that the free-run is lacking a bit in body, an admixture of the more robust press wine may well produce a more balanced wine.

A proportion of the crop may be vinified or matured in new oak in order to complex the aromas and flavours, while some may be in older casks where differences in the timber and its porosity may give greater or lesser oxidative effects. With the element of chance and so many variables at play, the winemaker cannot expect every cask or vat to be exactly the same. Although in the past, top German winemakers may have sold individual casks separately and put the cask number on the label, present-day marketing demands a degree of standardization. In order to iron out minor variations, winemakers put wines of similar quality into a vat for equalization before filling them into container, cask or bottle.

The ultimate in equalization is achieved in Spain with the solera system (see under MATURATION above), where wine for despatch is drawn from every cask in the bottom row of the stack, each containing a blend of all the vintages going back to the time when the solera was first established. It is thus the ultimate in non-vintage wine.

Producing a non-vintage wine has two main advantages. If for example one vintage is lacking in acidity and the next has high acidity, blending the two together may well produce a more balanced wine, superior to either of the ingredients. Moreover, customers will get a more standard product, irrespective of weather variations from vintage to vintage and will not have to keep changing the vintage year shown on their lists.

Vintage wine is assumed to be exclusively of the named vintage. In the case of vintage port it is traditionally bottled two years after the year when it was harvested. The exceptions to this are denominated 'late bottled 19** port' or 'port of the 19** vintage' or 'Colheita 19**'. Old vintage port will normally need to be decanted to avoid getting sediment in the glass, as will 'crusted port'.

It should be remembered that a vintage wine from a southern hemisphere vineyard will be 6 months older than the same vintage from a northern hemisphere vineyard.

FILLING

Until fairly recently most wines were delivered to point of sale or marketing area in casks. Transport casks were of thick-staved chestnut wood, whereas the cellar casks for maturing were more usually of oak. (Tonnage of ships originated in the number of 'tuns' of Bordeaux wine they could carry: 1 tonneau = 4 Hogsheads, each holding about 225 l or 50 Imperial gallons). These transport casks are now largely replaced by containerization: regular stainless steel wine tanks for road, rail and sea transport; road tankers (bowsers) for 'roll-on-roll-off' (Ro-Ro) ferry transport; multicompartmented tanker ships which ply the Mediterranean with vermouths, sherries and other bulk wines.

Bottling at source, in the past restricted to the finest wines, is now an economic possibility even for keenly priced wines, thanks to containerization which allows cartons of bottled wines to travel direct from producer to point of sale without the handling costs of breaking bulk.

Bottling

Bottles come in a variety of colours, shapes and sizes, many dubbed with esoteric names. The European Union has agreed some standardized types and sizes; correctness of fill according to the EC Quantities Directive may be indicated with a lower case letter 'e' on the label.

Corks, from the inner bark of the cork oak, *Quercus suber*, have been standard closures for bottles for many years since they replaced oil-soaked rags. 'Full-long' corks are used for the finest wines which may need to remain in bottle for many years.

The shortage of top quality cork has led to the

development of composition stoppers, combining compressed granulated cork with layers of ordinary cork. Champagne and sparkling wines need over-sized corks, which are squeezed and driven forcibly into the bottle neck, then mushroom out into the typical champagne cork shape and wired down to resist the CO_2 gas pressure.

The corks of vintage port were for many years covered with sealing wax to prevent the entry of cork weevils. It is now more usual to protect the cork and decorate the bottle with a capsule, originally of lead but now of aluminium alloy or plastic. Sparkling wines have foil over the cork and neck, concealing the ullage resulting from the space needed for the gas; 80 cl bottles for sparkling wine have the normal fill of 75 cl. Plastic stoppers are used for some lower priced sparkling wines and crown corks are an alternative, also for some semi-sparkling (pétillant) wines. Stopper corks can be used on wines like sherry which do not need to be laid down and are convenient for the consumer as no corkscrew is required. Closures for inexpensive wines include 'roll-on, pilfer-proof' (ROPP) closures which do not require capsules.

Other retailing containers

Pottery containers like the classic amphora, which were the norm in ancient times, are occasionally used in place of glass bottles.

Plastic bottles (polyethylene terephthalate – PET) are mainly for airline use to save weight as wine packed in them has limited shelf life. Blown plastic PVC containers had the disadvantage that traces of solvents and plasticizers used in their manufacture could leach into the wine and be detectable on nose and palate. They have effectively been replaced by the now ubiquitous 'bag-in-box'; these cartons with multilayer collapsible linings come in various sizes, usually 3 or 5 l, with a variety of taps. Smaller lined cartons (e.g. Tetra-pak) with various pouring devices can also be found and appear to be growing in popularity, particularly in the New World.

Cans are sometimes used for lower priced wines but seem to have limited appeal and little or no price advantage.

STORAGE

Wine needs to be stored at an even temperature. Wide variations can cause premature ageing. Wine is a liquid which expands with a rise in temperature when it may force its way into and through the cork, while a drop in temperature will cause contraction, drawing in air, and causing oxidation. Wine also needs to be protected against light, which can cause unwanted ageing. Both these requirements were met in the past by storage in underground cellars and this still has many advantages. However, mechanical handling and palletization can make temperature-controlled warehouses more advantageous, with greater headroom allowing pallet racking.

With high rates of excise on alcoholic beverages, Customs and Excise bonded warehouses may be used for long term storage, since excise duty and value added tax (VAT) only become payable on removal from bond and duty free re-export can be arranged if required.

LABELLING AND DISTRIBUTION

Labelling laws lay down the information that must be given for wines offered for sale, together with the size of lettering to be used for much of the essential information. In recent years, labelling has become more 'user friendly', with back labels providing data about how and where the wine was made, information about relative fullness or sweetness and suggestions how and with what foods it may best be enjoyed.

Wine trade education is now well-organized and widely available and is beneficial at all levels of the distribution chain. It is arguable whether the wine trade is supply-led or demand-led. Vineyards are planted for an economic lifespan of 20 to 30 years. Public likes and dislikes may change several times during that period. The winemaker and the merchant need all their skills to be responsive to changing tastes.

QUALITY ASPECTS

Quality is easier to recognize than to define. In many wine growing areas of the world certain wines have been recognized for their quality and have set benchmarks. These wines may be recognizably different from one another but the common factor of them all is the harmony of the various characteristics which make up the whole.

In red wines the winemaker is balancing the extraction of colour and flavour components of the anthocyanins (oenotannins) lying under the grape skins, judging their ripeness, deciding how far they should be oxidized and the possibilities of complexing them with wood tannins (gallotannins) from casks or vats.

White wines are pressed so as to extract the juice quickly, which results in limited extraction of the grape tannins, which in turn makes the decision whether or not to use oak aging more important. In white wines the balance between fruit flavours (or soil characteristics) and acidity becomes important; the fruit flavours and aromas will differ according to the grape variety and the ripeness of the vintage, as will the acidity.

Acidities, alcohol and any residual sugars need to be in harmony, to avoid either 'empty', 'flabby' wines, or 'sharp', 'harsh' wines at the other extreme. The extract content (see later under COMPOSITION) plays a part in the body of a wine while the persistence of flavours on the palate creates the impression of 'length' and 'aftertaste'.

A number of the flavour components of a wine are aroma linked. Wine aromas may be divided into primary aromas, such as the flower or fruit notes related to the grape variety, and secondary aromas, related to esterification and the complexing of the primary notes.

Deposits

The behaviour of tannins in red wines is related to the pH of the wine. When the isoelectric point of tannin is reached, tannins can change their charge, link to other substances (often in colloidal form) in the wine, flocculate and fall out. So tannin precipitation may be related to the malic acid content and the state of the malo–lactic fermentation, or to the tartaric acid content and the degree of tartrate precipitation.

Tannin precipitation also depends on whether trace iron associated with the tannins is in the bivalent or the trivalent state. For instance, if a red wine is fined with a colloidal fining under vigorous agitation, oxygen will be introduced, iron converted to the ferric state and tannin precipitation will be encouraged.

Deposits may also relate to the quality of the vintage: in less ripe years, the nitrogenous matter may be mainly amino acids and short chain peptides and fallout may be powdery; in ripe vintages, longer chain peptides may be formed and granular fallout can occur; in super ripe years, polypeptide chains may be formed with the result that old bottles of fine port, for instance, may have 'beeswing' formation on the sides of the bottles.

Bottles of white wine may have angular crystals at the bottom or on the cork. They look rather like sugar crystals and are sometimes mistaken for fragments of glass but they are in fact deposits of TARTRATES (see above). When they form on a cork this may have been triggered by some treatment such as cork bleaching by the cork merchants, or seeded by cork dust coming out of the stomata of the cork.

Composition

Typical laboratory figures for the composition of grape juice are given in Table 7.4.

Tables 7.5 and 7.6 show relationships between several measures of specific gravity and sugar content. These are most useful when the degree of maturity of grapes is to be estimated (see also Figures 7.1 and 7.2). Table 7.7 shows relationships between different measures and expressions of alcohol content. The classic analysis of wines includes measuring the dry extract by evaporating a measured quantity of wine to drive off the liquid components. The extract may also be estimated from the specific gravity (SG). For this one needs to remember that increasing quantities of alcohol lower the density or specific gravity of a mixture of alcohol and water (Table 7.7) whereas sugars, etc. increase the specific gravity (Tables 7.5 and 7.6). The rule-of-thumb formula for the calculation of extract is: $d_2 = (d + 1) - d_1$ where

$$d_2 = \text{SG of the extract}$$
$$d = \text{SG of the wine}$$
$$d_1 = \text{SG of the alcohol content.}$$

Results of this calculation are shown in Table 7.8. With sweet wines one may wish to calculate the balance of extract or 'extract without sugar' by subtracting the residual sugars in $g\,l^{-1}$ from the total extract in $g\,l^{-1}$.

Typical analyses of wines, as quoted in modern buying specifications, are given in Table 7.9. Table 7.10 shows relationships between alcohol content, acidity and extract to be expected in wines with a reasonable balance.

Table 7.4 Composition of ripe grape juice (%)

Water	70–85
Extract (solids)	15–30
Sugars	12–27
Pectins	0.1–1
Pentosans and	
pentoses	0.1–0.5
Acids:	
malic	0.1–0.5
citric	trace
tartaric	0.2–0.8*
Tannins	0.0–0.2
Proteins	0.5–1
Ash	0.2–0.6

*Mainly as potassium bitartrate.

Table 7.5　Measures of density

Specific gravity	Degrees Oechsele	Degrees Brix	Degrees Baumé
1.000	0	0.00	0.00
1.005	5	1.25	0.71
1.010	10	2.50	1.43
1.015	15	3.75	2.12
1.020	20	5.00	2.75
1.025	25	6.25	3.43
1.030	30	7.50	4.14
1.035	35	8.75	4.86
1.040	40	10.00	5.71
1.042	42	10.50	6.00
1.045	45	11.00	6.11
1.049	49	12.20	7.00
1.053	53	13.00	7.22
1.057	57	14.00	7.77
1.061	61	15.00	8.32
1.063	63	15.70	9.00
1.070	70	17.50	10.00

This table is an approximation based on French and German tables, using different instruments.

Table 7.7　Measures of alcohol in mixtures of alcohol and water at 15°C

Alcoholic strength (ABV)	Alcohol (grams per 100 g of mixture) (% by weight)	Alcohol (grams per litre of mixture)	Specific gravity
0	0.000	0.000	1.00000
1	0.795	7.936	0.99844
2	1.593	15.873	0.99695
3	2.394	23.809	0.99592
4	3.196	31.745	0.99413
5	4.001	39.682	0.99277
6	4.807	47.618	0.99145
7	5.616	55.554	0.99016
8	6.426	63.491	0.98891
9	7.238	71.427	0.98770
10	8.050	79.364	0.98652
11	8.867	87.300	0.98537
12	9.685	95.236	0.98424
13	10.503	103.173	0.98314
14	11.324	111.109	0.98206
15	12.146	119.045	0.98100
16	12.969	126.982	0.97995
17	13.794	134.918	0.97892
18	14.621	142.854	0.97790
19	15.499	150.791	0.97688
20	16.279	158.727	0.97587
21	17.111	166.663	0.97487
22	17.944	174.600	0.97387
23	18.779	182.736	0.97286
24	19.616	190.472	0.97185
25	20.455	198.409	0.97084
26	21.495	206.345	0.96981
27	22.138	214.281	0.96876
28	22.984	222.218	0.96769
29	23.832	230.154	0.96759
30	24.683	238.091	0.96545

Table 7.6　Measures of sugar content[*]

Specific gravity	Degrees Oechsele	Sugar content from Brix tables (g l^{-1})	Adjusted sugar content (g l^{-1})
1.060	60	155.7	130.3
1.070	70	181.9	156.7
1.080	80	208.1	183.1
1.090	90	234.2	209.5
1.100	100	260.6	235.9
1.110	110	286.8	262.3

[*] The presence of non-sugars in grape must means that sugar measurements made by specific gravity or refractometer need to be adjusted.

Table 7.8　Calculation of wine extract

Specific gravity at 15°C	Extract (g l^{-1})	Specific gravity at 15°C	Extract (g l^{-1})	Specific gravity at 15°C	Extract (g l^{-1})
1.0040	9.6	1.0105	25.2	1.0170	41.4
1.0045	10.8	1.0110	26.4	1.0175	42.9
1.0050	12.0	1.0115	27.6	1.0180	44.3
1.0055	13.2	1.0120	28.8	1.0185	45.8
1.0060	14.4	1.0125	30.0	1.0190	47.2
1.0065	15.6	1.0130	31.2	1.0195	48.7
1.0070	16.8	1.0135	32.4	1.0200	50.1
1.0075	18.0	1.0140	33.6	1.0205	51.6
1.0080	19.2	1.0145	34.8	1.0210	53.0
1.0085	20.4	1.0150	36.0	1.0215	54.4
1.0090	21.6	1.0155	37.2	1.0220	55.9
1.0095	22.8	1.0160	38.5	1.0225	57.3
1.0100	24.0	1.0165	40.0	1.0230	58.8

Table 7.9　Typical range of analysis of table wines, as bottled

Alcohol	7–14	% v/v	Variable, depending on type etc.; (normally declared on the label, with a tolerance of ±0.5%)
Acidity, total	4.5–10	g l^{-1}[*]	Variable, depending on type and quality
volatile	0.2–0.5	g l^{-1}[*]	Excess is considered a defect; up to 1.0 mg l^{-1} may be tolerated in some wines
pH	3–3.5		
Iron	5	mg l^{-1}	Sometimes as high as 20 mg l^{-1}
SO$_2$, total	50–250	mg l^{-1}	Legal maxima vary according to wine
free	20–50	mg l^{-1}	
Sorbic acid	200	mg l^{-1}	Maximum

[*] Acidity is usually expressed in grams of sulphuric or tartaric acid per litre: sulphuric acid, H_2SO_4, 1 M = 49 g l^{-1}; tartaric acid, $C_2H_4O_2(COOH)_2$, 1 M = 53 g l^{-1}.

Table 7.10 Alcohol, fixed acidity and extract of wines

Alcohol	Fixed acidity[a] (minimum) (g l^{-1} as tartaric)	Balance of extract[b] (g l^{-1})	
(% by volume)	Red & white wines	Red wine	White wine
7	6.6	11.0	9.0
8	5.7	11.5	9.5
9	5.0	12.0	10.0
10	4.5	12.5	10.5
11	4.1	13.0	11.0
12	3.8	13.5	11.5
13	3.6	14.0	12.0

[a] Fixed acidity is calculated by subtracting the volatile acidity from the total acidity.
[b] Balance of extract represents the weight of glycerine and similar substances (glycols), mineral salts, nitrogenous matter, tannins and colouring matters and other substances of secondary importance. The figure for balance of extract is obtained by subtracting any sugar content from the total extract and then subtracting the fixed acidity.

Source: Benvegnin *et al.* (1951).

Sensory profiles

Quality is produced by a combination of grape varieties, site and soil (terroir), weather. and human effort. It has to be judged through the senses of the winemaker and eventual consumer. The winemaker may rely to a considerable extent on laboratory analyses to assess the development of wines, but the ultimate assessment by the consumer is on appearance, aroma and taste. These can be put together as a sensory profile, as in the following example of some of the elements which can be considered for table wines:

Tasting aide-mémoire	
Appearance	
Colour:	purplish, ruby red, brick red, brownish, rosé, salmon, pale pink, blush; greeny-white, water-white, pale yellow, golden yellow, deep gold, brownish
Richness:	depth of colour, fullness of flow
Brilliance:	limpid, hazy, cloudy, piecy
Nose	
Flowery:	spring flowers, summer flowers, potpourri
Fruity:	soft fruits, stone fruits, apples and pears
Spicy:	vanilla, cinnamon, peppery, etc.
Herbs:	lavender, thyme, rosemary, bay, etc.
Soils:	flinty, slatey, smoky, earthy, etc.
Yeasty: or	winey (vinosity) or volatile (acetic)
Palate	
Taste buds:	sweet, acid, (bitter), (salty)
Mouth-feel:	rough/smooth, round/sharp, thin/full, warmth/coolness, flow, length
Remarks	

too young, young and fresh, ready to drink; better in '*x*' years, mature, will keep for '*y*' years; over the top, worn out

Wine jargon relies on terms culled from previous experience or adopted from pundits. As people's tastes and experiences are so varied, a person's wine vocabulary needs to be constantly checked and revised.

If wines can be tasted at different stages of their development, a timescale image may be built up and used, by analogy, to prognosticate the future of a wine. With such a gamut of styles and qualities available in the world of wine, it is good to taste widely to establish a personal range of preferences and wherever possible to sample the best in order to have benchmarks of quality.

BEERS

```
┌─────────────────┐     ┌─────────────────┐
│     Barley      │     │  Other cereals  │
│        Page 248 │     │                 │
└─────────────────┘     └─────────────────┘
         │                       │
┌─────────────────┐              │
│   Germination   │              │
│        Page 249 │              │
└─────────────────┘              │
         │                       │
┌─────────────────┐              │
│      Malt       │              │
│        Page 249 │              │
└─────────────────┘              │
         │                       │
┌─────────────────┐              │
│     Milling     │──────────────┘
│        Page 250 │
└─────────────────┘
         │
┌─────────────────┐
│     Mashing     │
│        Page 250 │
└─────────────────┘
         │
┌─────────────────┐
│  Fermentation   │
│        Page 252 │
└─────────────────┘
         │
┌─────────────────┐     ┌─────────────────┐
│  Conditioning   │     │ Alcohol reduction│
│        Page 253 │     │        Page 253 │
└─────────────────┘     └─────────────────┘
         │
┌─────────────────┐
│     Filling     │
│        Page 254 │
└─────────────────┘
         │
┌──────────────────────┐
│ Transport and storage│
│            Page 255   │
└──────────────────────┘
         │
┌─────────────────┐
│ Quality aspects │
│        Page 255 │
└─────────────────┘
```

INTRODUCTION

Beer is a beverage produced by a fermentation of a hopped water extract of germinated barley. This means that the four main raw materials to become beer are malt, hops, water and yeast.

Barley cannot directly be used for the production of beer but must be malted. The endosperm of barley contains a large amount of starch but only very small amounts of sugars, whereas the brewing yeast can convert only fermentable sugars to alcohol and carbon dioxide. The starch has to be degraded by enzymes to fermentable sugars, of which the most important for the brewing industry are maltose, maltotriose and glucose. The process that makes barley suitable for brewing is called malting. During malting, enzymes are produced or activated.

The malting process involves the collection of stocks of suitable barley, steeping the grain in water, germinating the grain and finally drying and curing it on the kiln. The brewing process involves the enzymatic extraction of malt with water, filtration of the wort, boiling of the wort with hops, clarification and cooling of the wort.

After the brewing process the fermentation of the cooled wort starts by adding yeast, followed by the maturation. Finally the beer can be filtered and bottled, canned or kegged. The world production of beer is estimated at 900 million hectolitres per annum.

MALTING

The germination of the grain is a prerequisite for the production of malt. The growth of the germ or embryo is incidental to the making of malt and leads to depletion of the endosperm material through respiration of the embryo and its growth. The maltster is especially concerned with the degradation of the endosperm, the mobilization of the enzymes of the grains and the yield of the production. Thus the maltster's requirements deviate from those of the farmer whose primary interest is the growth of the embryo into a mature fruiting plant.

In the past only two rowed barley was used for the brewing industry, now two and six rowed barley are used.

Selection of the barley

The selection is based on many criteria including brewing capacity, rapid and synchronous germination of the grains, uniform enzymatic degradation of the endosperm, an adequate complement of enzymes remaining after kilning, low levels of fibrous materials and total nitrogen, purity of variety, suitability for mechanical harvesting, large grain size (measured as thousand corn weight), disease resistance and an acceptable dormancy period. It is also very important that the grains do not start to germinate on the field before the harvest in damp weather conditions. For the farmer it is necessary that the barley varieties suitable to the maltster also have desirable agricultural properties, especially good yields. Samples of barley may be assessed before purchase by submitting them to a pilot- or microscale malting process.

Cleaning up the barley

When harvested, barley cannot be immediately used for malting. The crop is polluted with dust, straw, stones and possibly small fragments of iron. These contaminants are taken out by air cleaning, sieving and magnetic cleaning.

A batch of barley will also contain foreign grains such as wheat, oats or maize, and broken corns of barley which are a great risk for mould growth. These impurities may be removed in a machine where the corns are sorted out by length. Furthermore barley is not homogeneous, so when it is steeped in water it may not give synchronous germination. For that reason the grains are selected for size using sieves. The small grains are sent for cattle food.

Storing the barley

Many maltsters prefer to receive grain as soon as possible after harvest, then to dry, clean and store it themselves to assure that its quality is maintained. The maximum content of water in the barley should be 16%. Large amounts of barley are purchased from the same area to avoid large differences in germination.

It is necessary to store the barley because immediately after harvesting the corns are very water sensitive, that is, they are not strong enough to germinate and during steeping the germ may easily die. Normally, after a period of six weeks 'dormancy' the corns can be used for germination. During storing the barley is still alive, using oxygen and producing heat, water and carbon dioxide, so during storage the temperature and moisture must be strictly controlled.

Steeping the barley

Barley starts to germinate when its water content is high enough. On steeping in water, the moisture in the grain will be increased from about 12 to 45%. Steeping the barley is also a washing step.

The water steeping periods alternate with dry steeping periods. In the dry steep the barley is aerated to activate germination and carbon dioxide is removed, e.g. by suction. The steeping process takes about two days at a temperature below 20°C.

Germination

The main purpose of germination is the formation or activation of the enzymes that will be needed in the brewhouse. This must be done without too much loss of reserve substances.

Germination (usually 5 to 7 days) is stopped when the germ or acrospire, as it is termed by the maltster, has the length of about 80% of the corn itself. The maltster lets the barley germinate under very strict conditions. The temperature is normally kept between 14 and 18°C and the air humidity is kept high. The barley is aerated and briefly turned over several times.

Kilning the malt

During kilning the moisture from the green malt is brought down from 45% to less than 5% by blowing hot air through it. This takes about one day. Kilning produces a dry product that is stable during storage. It also adds character to the malt, altering its colour and flavour, but it reduces its enzyme potential because the enzymes are heat sensitive, especially at high humidity. Therefore the maltster keeps the temperature of the drying air at 40–45°C at the beginning of the kilning. When the moisture content of the malt is below 10% the maltster can use drying air from 80–85°C. For dark malt higher final kilning temperatures are used. The enzymatic activity of dark malt is therefore less than that of pale malt.

The two main systems of germination are floor malting, in which the steeped barley is spread out in a bed which is about 15 cm high, and pneumatic malting with a bed height of about 1.2 m.

Cleaning the malt

The rootlets must be removed from freshly kilned malt. These rootlets are very hydroscopic and taste bitter making them undesirable for brewing. The rootlets are taken from the malt in a special machine. These rootlets represent a rich source of protein and are used as cattle food.

Storing the malt

Malt has to be stored for at least four weeks before being used in the brewhouse. Malt which is not stored long enough presents difficulties during the brewing process, in filtration, fermentation and clarification. Before it is stored the malt must be cold, dry and free from rootlets. It is important that during storage the uptake of water is avoided.

Different malt types

The brewer sets specifications for the types of malt that are needed; the maltster buys barley and processes it to malt which will meet these specifications. Maltsters usually blend malts from different batches to meet particular specifications. The main types of malt are pale malts, caramel malts, dark malts, roasted malts, amber malts and acid malts.

BREWING

The use of raw materials of high quality, modern equipment and the right technology is prerequisite for making good beer. As the market extends more and more, the physical and flavour stability of the beer is of great importance.

The brewer uses malted barley, water and hops as primary raw materials. In most countries unmalted cereals such as maize, rice or corn sugar are commonly used as partial substitutes for malt (adjuncts). In some countries sorghum is also used.

Water

About 92–95% of the weight of beer is water. It is clear that the water quality has to be suitable for human consumption. The presence of various inorganic ions influences the taste and flavour of the beer. During the past millennium, various centres have become renowned for the quality of their beers, the type of beer in each location being distinctive. These distinctions can be attributed, at least in part, to the water composition in each area. Note that in the mashing process, enzyme activity and enzyme stability are influenced by the inorganic ions in solution and therefore the yield of extract is also under this influence.

Brewing requires considerable volumes of water. At the present time breweries use on average between 6 and 8 times as much water as the beer that they produce.

Milling or preparation of the grist

To allow degradation by the malt enzymes of the food store of the grain, the grain has to be mechanically broken down and water has to be added. Malted barley is carefully ground in a mill in such a way that the husk of the grain is left substantially intact (to serve as a filter material later in the brewing process) while the rest becomes a coarse powder. This powder is particularly rich in starch and in the enzymes capable of degrading it rapidly when water is added.

The object of the milling is also to produce a mixture of particle sizes which will secure the highest yield of extract as quickly and efficiently as possible. Milling procedures are determined by the size distribution of the malt grains, their moisture content and their 'modification' (the degree to which enzymes have broken down the polymers of the endosperm cell walls). They are also determined by the intended mashing method and the wort separation method.

The size reduction of the malt corns in the brewing industries is normally achieved by roller or hammer mills, depending on the techniques to be used in the brewhouse. Any adjuncts to be used are finely ground or precooked and added at this stage.

Mashing

By thoroughly mixing of the ground material, called 'grist', with water, compounds such as sugars, amino acids and minerals are dissolved. The enzymes of the malt, which were formed or activated during malting, attack starch and its degradation products (amylolysis). Breakdown of nitrogenous material (proteolysis) also occurs. By these means a part of the insoluble material of the malt is dissolved. The process is called 'conversion' and the total dissolved material is called the 'extract'.

Temperature, pH, time and the ratio of malt to water are all significant. The pH of the mash is kept at the optimum level for the desired enzyme activity. Each malt enzyme has its own optimum temperature. For this reason the mash may be warmed up to different temperatures and kept there for enough time to let the enzymes do their work. Relevant temperatures are:

(i) 37 and 45°C – are important temperatures for the breakdown of hemicellulose and gums;

(ii) 45 and 52°C – peptases and peptidases are active, by which polypeptides, peptides and free amino acids are brought into solution. The first two of these components are important for the head retention and the fullness in taste of the beer. The last is necessary for the yeast growth. Proteins are also important for the stability of the beer;

(iii) 63 and 72°C – are the most important temperatures of the brewing process. 63°C is optimum for α-amylase activity, 72°C is optimum for β-amylase; the two enzymes degrade starch to soluble sugars and dextrins.

The temperature of the mash must not exceed 78°C or most of the enzymes will be destroyed. Components of the husks are also detrimental to the final beer quality so it is important not to allow them to go into solution.

There are several methods of mashing to obtain a satisfactory aqueous extract or 'wort' (see Table 7.3).

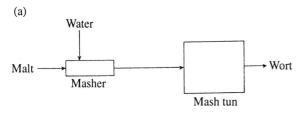

(a)

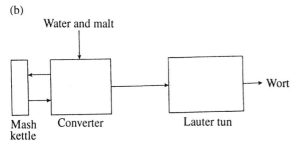

(b)

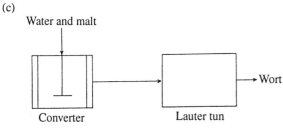

(c)

Figure 7.3 Mashing procedures. (a) Infusion mashing, (b) decoction mashing, (c) temperature profile mashing.

(i) *Infusion mashing.* This is the simplest. It involves adding hot water to the ground material and pumping the mixture to the mash 'tun' or vessel. The mash produced has a final temperature of 65–72°C and has a thick porridge-like consistency.

(ii) *Decoction mashing.* Decoction mashing differs from infusion mashing in several respects. The endosperm of the malt that is used may be less completely enzymatically degraded (less 'modified') and requires more extensive enzymatic action during mashing. To aid this, the malt is both ground more finely than with the infusion process and is mixed with water at a lower temperature (35°C). Extensive proteolysis and solubilization occur. During the conversion a portion of the mash (often one-third) is boiled in a mash kettle and returned to the rest of the mash in the mash vessel for mixing. The temperature rises to about 50°C. Later, another one-third of the mash is withdrawn and boiled. On return to the main mash, the temperature increases to about 65°C. After a last decoction with one-third of the mash, all enzyme activity ceases in the consolidated mash at about 75°C.

(iii) *Temperature-profile mashing.* The mashing temperature is raised by stages through the process, so as to allow the best combinations of time and temperature for each of the desired enzymatic changes. In modern breweries the whole process may be under computer control.

Selection of a mashing method involves consideration of the grist composition, the equipment available, the amount and the type of adjuncts used, the brewing liquor available, the type of beer to be brewed and the number brews required each day.

Wort filtration

The liquor in the mash tun may be left for a period to allow maximum solubilization of extractable material. The extract or 'sweet wort' has then to be separated from the spent grains and any other undissolved matter. It is important for the beer quality to produce bright sweet worts. Any leftover grains or other unsuspended matter remaining in the later processes will give inferior flavour and poor flavour stability in the final beer.

After infusion mashing, any necessary delay together with the final separation take place in the mash tun. After decoction or temperature-profile mashing the mash is transferred to a 'lauter' tun (lauter comes from the German word for purification) for the delay period and final separation. Above the perforated bottom of the lauter tun the layer of husks and spent grains in the mash forms an effective filter to clarify the wort draining through it. (The wort may be recirculated once or twice at first, to consolidate the layer.) Instead of lauter tuns some breweries use more efficient mash filters in which the wort is drawn off through perforated pipes set below the level of the spent grains in the tank.

The spent grains are worthless for brewing but they represent a rich source of proteins and are sold as cattle food.

Wort boiling

After filtration the wort is boiled for one hour or more in a steam-heated copper. The boiling process is associated with the addition of hops or hop extracts; sugars or syrups may also be added.

Many complex reactions take place during wort boiling. As the wort is heated the residual enzymes are inactivated and the wort is sterilized. Various other reactions occur, e.g. colour production, coagulation of proteins and tannins. The wort is concentrated, volatile materials are evaporated and the added hop is extracted and isomerized. Boiling may take place under ambient pressure or at higher pressure and temperature. Normally 10% of the volume is evaporated.

Wort clarifying

Boiled wort is very hazy and has to be clarified. The suspended particles are called 'hot trub' or 'hot break'. The hot trub is made up of proteins, protein–tannin material, insoluble salts, some hop resin material and some of the lipid material that was present in the sweet wort and the hops. If whole hops were used in the copper the spent hops have to be removed. The hot trub is taken out of the wort by the whirlpool tank or centrifuges.

Wort cooling

In order to bring the wort to fermentation temperature the wort is cooled, to 6–10°C for lager beer or 18–25°C for top fermented beer. Most breweries now use plate heat exchangers to cool the wort. The plate heat exchangers may comprise two or more stages so that wort may run counter-current to water in the first section while a second stage may reduce the wort temperature still further by using glycol as refrigerant.

An important consideration of wort coolers is their ability to function as heat economizers; they generate considerable volumes of hot water which may be used for mashing or for cleaning equipment.

FERMENTATION

Primary fermentation

The most important conversion during the fermentation of wort with yeast is that of sugars to alcohol and carbon dioxide. The complete fermentation process is very complex.

The process starts in aerobic conditions. After the hopped clear wort is cooled it is aerated and yeast is added. The yeast consumes the dissolved oxygen, sugars, amino acids and other wort components during its own propagation. In the next step, under anaerobic conditions, fermentation starts. During the fermentation about 70% of the flavour active components of the beer are produced as esters, sulphur compounds, carbonyls, higher alcohols and ketones. The yeast performance is influenced by the yeast strain, yeast condition, the amount of yeast added to the wort, wort composition, the fermenter, the degree of aeration and the fermentation temperature and pressure. The yeast cannot convert all the extract of the wort. High molecular sugars and proteins which are not converted by yeast are important for the fullness of the taste and the head retention.

The equipment may consist of open or enclosed tanks. In order to minimize the risk of microbiological contamination all tanks and rooms must be scrupulously clean. When closed tanks are used, the carbon dioxide produced can be recovered and sold as by-product.

After the fermentation the yeast is harvested in order to avoid contamination with products of autolysis. In the brewing industry two types of yeast are used, bottom and top yeast.

(i) *Bottom yeast* ferments the wort at low temperature (6–12°C) and sediments to the bottom of the fermentation vessel towards the end of the fermentation. The yeast concentration at the start of the fermentation is about $15–25 \times 10^6$ and the multiplication factor is about 3. The fermentation time is about one week.

(ii) *Top yeast* ferments the wort at higher temperature (18–25°C) and the yeast rises to the top of the fermenter towards the end of the fermentation. The yeast concentration at the start of the fermentation is about $10–12 \times 10^6$ and the multiplication factor is about 5. The fermentation time is about four days.

Conditioning

After the primary fermentation the beer is 'green', it contains little carbon dioxide and its taste and aroma are inferior to those of mature beer. The conditioning process, also called 'maturation' or 'lagering', is carried out in closed vessels at low temperature (-1 to 5°C). At this low temperature the clarification of the beer is promoted. Yeast, proteins, tannins and hop resins precipitate.

The small amount of yeast remaining after the primary fermentation ferments the rest of the fermentable carbohydrates. The carbon dioxide produced largely dissolves in the beer. Complex biochemical reactions take place, such as esterification and certain reductions which modify and improve the flavour of the beer.

Nowadays, before the cold maturation, the breweries use a 'diacetyl rest' which is a period of warm conditioning (about 7°C) to encourage the oxidative decarboxylation of α-acetohydroxyacids to the vicinal diketones followed by reduction to the corresponding diols, e.g.

$$
\begin{array}{ccccc}
CH_3C(OH)CO_2H & & COCH_3 & & CH(OH)CH_3 \\
| & \rightarrow & | & \rightarrow & | \\
COCH_3 & & COCH_3 & & CH(OH)CH_3 \\
\text{α-acetolactic acid} & & \text{diacetyl} & & \text{2:3 butylene glycol}
\end{array}
$$

The concentration of the total vicinal diketones is used as a marker of the duration of warm conditioning (about 7°C). When the concentration of the vicinal diketones is less than 0.1 ppm the cold maturation can start.

Reduced-alcohol beers

To meet the increasing consumer demand for beers of low or negligible alcoholic strength, there are basically two options: the brewer may arrange the brewing processes to restrict the production of alcohol or he may remove alcohol from otherwise normally produced beer. The difficulty with the former is the retention of flavours from the wort, e.g. excessive sweetness, which otherwise would have been removed during the alcoholic fermentation. The difficulty with the latter option is to arrange the removal of alcohol without removing other desirable qualities. Restricted production of alcohol may be achieved by

(i) starting with a low gravity wort containing little fermentable sugar. This gives beer of low alcoholic strength but also little flavour and the process appears to be little used;

(ii) mashing at high temperature (about 80°C) restricts the activity of β-amylase but not that of α-amylase and gives a wort high in dextrins and relatively low in fermentable sugars. Cold mashing restricts the activity of all the enzymes;

(iii) the yeast *Saccharomyces ludwigii* may be used, which ferments only glucose, fructose and sucrose or about 15% of the normally available fermentable sugars, consequently the beer has a high content of maltose;

(iv) checking fermentation or cold fermentation - the initial fermentation may be checked by rapidly reducing the temperature and removing the yeast, or fermentation may be carried out entirely at low temperature; the beer may have a high maltose content; with cold fermentation an immobilized yeast may be used, this allows the yeast concentration to be high but reduces the risk of autolysis of the yeast.

Alcohol reduction of a beer may be achieved by:

(i) brewing a concentrated wort to give a high strength beer, then diluting with water to the desired lower alcoholic strength. With suitable adjustment of the fermentation conditions as described above, the development of flavouring substances may be made to predominate over that of alcohol so that reasonably well-balanced flavour is produced on dilution;

(ii) distillation at atmospheric temperature which can reduce the alcohol content to below 0.5% although a burnt flavour is unavoidable;

(iii) vacuum distillation (at 50–60°C) which produces less off-flavour and thin film evaporation (at 30–40°C) which gives virtually none, however in both cases some of the volatile flavours are lost. This may be countered by ester recovery at the beginning of the process and by adding other flavour substances as desired;

(iv) reverse osmosis or dialysis which give low alcohol beers with otherwise good flavour but both processes are expensive. To reduce the alcohol content below 0.5% is said to be either not possible or prohibitively expensive.

Filtration and clarification

It is not possible to remove all the haze particles during the cold maturation, therefore, a fine filtration is needed in order to get bright beer by removing yeast, hop resins and the chill haze. The chill haze or colloidal haze, which consists of tannin protein complexes can make the beer hazy after bottling, canning or kegging.

A prerequisite for good filtration is a cold maturation and a beer temperature below 0°C during the filtration. Some breweries flash cool the beer immediately prior the filtration. The major principles of clarification used in the brewery are sedimentation by centrifugation, filtration and absorption by the use of filter aids.

Centrifugation

The settling rate of particles is increased by increasing the gravitational forces and by reduction of the settling distance. After centrifugation the beer usually is still not bright enough, so centrifugation is often used in combination with another filtration technique.

Sheet filters

Here the beer is pressed through filters of paper or paper-like material. Normal sheet filters not only sieve out the haze particles, they also absorb some other materials. Thus, yeast and bacteria fail to penetrate sheets whose mean pore size should in theory permit them to pass through. Many breweries rely on sterilized fine sheet filters to remove all micro-organisms from the beer prior to bottling.

Powder filters

Since World War II, powder filters have become popular as complete filtration systems or in combination with the systems mentioned above. The most popular powders for this purpose are diatomaeceous earth and perlite (a filter aid of volcanic origin). Some breweries use silica hydrogel as a substitute for diatomaeceous earth. The principle on which powder filters operate is similar to that of sheet filtration, but with the powder filters it is necessary to build up a sufficient thickness of powder on a sheet or on a perforated septum to permit efficient filtration of particles from the beer.

Pasteurization

To ensure that there are no unwanted micro-organisms in the beer at the end, the beers are pasteurized. For ales and lager beers 20 pasteurization units and for alcohol-free beers 40 pasteurization units are enough to ensure that all beer spoiling micro-organisms are killed. One pasteurization unit is equivalent to one minute at a constant 60°C.

Bottled and canned beers are pasteurized after filling by passing the sealed containers continuously through steam- or water-heated pasteurizing tunnels. These normally have an initial heating up zone, a central zone of constant temperature and a final cooling zone. The cans or bottles may exit at about 60°C to allow the surfaces to dry by evaporation before labelling and packing. Beer for filling in bulk to kegs or tankers is pasteurized before filling, usually through plate heat exchangers.

Packing of beer

The main methods of packing beer today are in bottles, cans, kegs and transportable tanks or tankers. Wooden casks may still be used to a small extent in smaller breweries. The casks are washed and steam cleaned before re-use. Sulphur dioxide preservative may be added to the beer.

Preliminary cleaning of containers

In bottling, returnable or non-returnable bottles may be used. The cleaning of non-returnable bottles is usually done with sterile water. Returned bottles must be inspected to remove damaged or very badly soiled bottles, also any remaining old closures. They are then cleaned in bottle washers where they are washed both internally and externally and old labels are removed. A typical sequence of events in the bottle washers is:

(i)   prerinsing with cold water,
(ii) prerinsing with hot water at about 50°C,
(iii) caustic soaking at 70°C, with caustic soda of about 2% and sometimes other additives,
(iv) caustic spraying at 70°C,
(v)   hot water rinsing at 50°C and
(vi) first and second cold water rinsing.

The maintenance of the bottle washer, the use of the right operation conditions and chemicals are very important for the quality of the cleaning. Generally after washing the bottles are inspected by a bottle inspector.

Cans are normally inverted and rinsed with jets of clean air. Kegs and tanks are washed internally and externally with care to ensure no excess water remains afterwards. In modern equipment there is completely automated input of kegs, washing, deterging and steam sterilizing. The selection of detergents to be used depends on the material of the kegs which nowadays are normally aluminium or stainless steel.

Filling

The 'bright' (filtered) beer is put into the container by an appropriate filling machine, which may be quite complex. It is important to avoid oxygen contact with the beer since dissolved oxygen is detrimental to the flavour and physical stability. Modern bottle filling machines therefore evacuate the bottles of air; with cans the air is washed out by carbon dioxide; with kegs also, after steam sterilizing the residual water and air are removed with carbon dioxide.

All the containers are filled under counterpressure with carbon dioxide. Before closing a bottle with a crown cork or a can with a lid, the headspace air is removed mechanically by overfoaming.

Most bottled or canned beer is pasteurized after bottling or canning. With kegs and tankers the beer is prepasteurized or sterile filtered.

Tanked beer is delivered by the tanker directly to the sales outlet. The beer is pumped through a hose from the tanker into receiving tanks in the cellar of the inn.

QUALITY ASPECTS

Types of beer

Lager beers

These are also known as bottom-fermented beers, since the yeast used sinks to the bottom of the tank at the end of fermentation.

Different types of lager beer include 'Pilsener', which is characterized by a medium hop flavour, an alcohol content of 3–4% by weight, its brightness and pale colour. The brewing water for this type of beer is soft. 'Dortmund' beer has less hops then pilsener and is made with water which contains large quantities of carbonates, sulphates and chlorides. 'Munich' beer is a dark aromatic beer with a somewhat sweet flavour. This beer has a very mild hop flavour and an alcohol content of 2.5–5% by weight. The brewing water contains high amounts of carbonates but small amounts of other salts. 'Bock' beer is a seasonal beverage made with caramel malt or heavily roasted malt.

Top-fermented beers

The most important types of beers brewed with top fermenting yeast include ale, porter, stout and weiss beers.

'Ale' is a British type of beer brewed with water of high calcium sulphate content. The hop flavour is pronounced and the alcohol content is 2.5–4% by weight. 'Porter' is dark coloured, less hopped and sweeter than ale. 'Stout' is a very dark with a somewhat burnt flavour and a strong malty aroma. This beer is heavily hopped and contains 4–6.5% alcohol by weight. 'Weiss' beers are less hopped, unfiltered beers brewed with a percentage of malted wheat.

Microbiological control

There has always been a measure of disagreement among brewers on the degree of microbiological control needed in breweries. Most modern breweries however, wish to eliminate all micro-organisms from the brewery except for pure culture yeast and thereby help to achieve a consistently satisfactory quality of their products.

Fortunately for brewers, microbiological control needs to be exercised only over a limited range of bacteria and yeasts. Pathogenic micro-organisms and also many other bacterial strains fail to grow, or even to survive for extended periods, in beer. This is because the micro-organisms are inhibited to different degrees by the low pH, high concentration of alcohol, high content of hop resins and low concentration of fermentable sugars.

The bacteria that may be found in wort or beer are classified in the usual way according to their shape, flagellation, Gram-staining, other structural features and biochemical characteristics. The first group of bacteria that can be found in a brewery are the lactic acid bacteria; the group is divided into *Lactobacillus* (Gram-positive, catalase-negative rod-shaped bacteria) and *Pediococcus* (Gram-positive, catalase-negative cocci, mainly found in pairs and tetrads). Lactic acid bacteria grow in anaerobic conditions in the presence of carbon dioxide. They can ferment a wide range of sugars to lactic acid and some strains produce acetic acid, ethanol and carbon dioxide. Strains of *Pediococcus* are rarely found in top-fermentation breweries but are more common in bottom-fermentation breweries. Spoilage caused by lactic acid bacteria gives rise to acidity, turbidity and off-flavours.

A second group of beer spoiling bacteria are the acetic acid bacteria, Gram-negative, rod-shaped bacteria of the *Acetomonas* and *Acetobacter* species. The latter is usually motile. Most strains oxidize glucose and other sugars and some oxidize ethanol to acetic acid. The growth of the acetic acid bacteria is not restricted by low pH or hop resins, but most strains require oxygen, therefore these bacteria develop best in wort and beer when the liquids are exposed to air. Spoilage by them causes acidity, off-flavours and ropiness.

A third group of beer spoiling bacteria are the Gram-negative, rod-shaped, strictly anaerobic strains of *Pectinatus* and *Megasphera*. They may arise from a secondary contamination while the beer is being bottled.

Yeast strains that are not part of the culture yeast used by the brewery are called 'wild yeast' and can cause spoilage because their metabolic products are different from those of the fermentation yeasts. They can cause off-flavours and turbidity.

Chemical control

Alcohol

For brewery control purposes, alcoholic strength is usually nowadays measured by rapid methods, gas-liquid chromatography (GLC), densitometry or rapid distillation in the laboratory or even on-line. For taxation purposes as required in most countries, official reference methods must be used; these usually require distillation under prescribed conditions in prescribed apparatus.

In the UK until recently the excise tax on beers was based not directly on the alcoholic strength but on the original gravity or OG, the specific gravity of the wort from which the beer was brewed. To find the OG, a sample of beer is distilled, the distillate is made up to a standard volume, likewise the residue in the distillation flask (the 'extract') and the specific gravities of both are measured. The values of the specific gravities are manipulated according to an official formula and tables, whereby the OG can be calculated (Kirk and Sawyer 1991). This use of the OG for taxation purposes was presumably because in the earliest days of the tax more than two centuries ago, it was easier to measure the OG before fermentation than to measure the alcohol content afterwards. Nowadays however, when a value of the OG calculated as just described from an analysis of the finished beer may be meaningless for a low-alcohol beer or other modern product produced by some large modification of the traditional brewing processes, the use of OG for taxation purposes has been abandoned. The alcoholic strength of the final product is now, much more sensibly, used instead.

Other chemical analyses

Significant components which may be measured using normal laboratory methods include total solids (extract), sugars, acidity, pH, CO_2 and SO_2.

Organoleptic and physical controls

Flavour (aroma, taste and after-taste), mouthfeel and appearance are the most important quality factors in any beer. Systematic, regular controlled tasting tests should be a feature of every brewery operation. Colour and turbidity or clarity may be measured and controlled by standard laboratory methods. Foam height (the 'head') and foam stability can be measured empirically by pouring from the bottle or can into a graduated vessel under standardized conditions.

SPIRITS

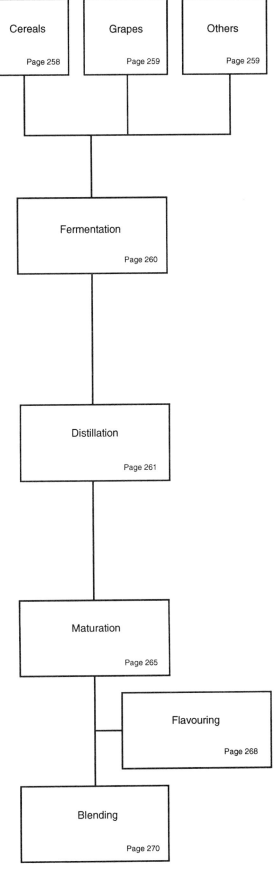

Cereals	Grapes	Others
Page 258	Page 259	Page 259

Fermentation

Page 260

Distillation

Page 261

Maturation

Page 265

Flavouring

Page 268

Blending

Page 270

STARTING MATERIAL

Ethanol is the product of the metabolism by yeasts of low molecular weight sugars derived from a variety of agricultural produce. Sugars from fruits such as grapes and apples can readily be extracted in the fermentable form of the pressed juice. In cereals most sugars are in the form of storage polymers and require an extra stage, known as mashing, to extract the sugars and convert the polysaccharide to a readily fermentable form. Which particular raw material is used for the production of spirits depends on the geography and agriculture of the country or region from which the spirits originate. Thus whiskies have come from grain producing countries such as the UK, United States and Canada, brandies and eaux-de-vie from France and other wine producing countries, and rums from sugar producing areas such as the islands of the West Indies and Guyana.

Differentiation of spirits arises from the imperfect nature of the distillation, so that the final product contains not just ethanol and water but components of raw material and fermentation. Historically raw materials were limited by local availability but with the arrival of modern transport, the individual nature of some spirits has been protected by legislation. This defines which raw materials may be used for production and frequently the geographical area within which it may be produced. Flavoured spirits, including vodka, are based on neutral alcohol of agricultural origin. This may be fermented from any suitable substrate, but is distilled to a high purity and any flavour from the original raw material is lost.

Whatever raw material is used, it is critical that it is of high quality and free from defects that may affect the flavour of the product distillate. It is possible only in the case of rectified spirit to distil out all undesired flavours; in other spirits, where flavours derived from the materials and the process are required, it is not often possible to remove off-flavours by distillation. Many grape-based products are controlled by regulation, and the grape varieties allowed (*Vitis vinifera* and hybrids) are those which have traditionally been used in the production area concerned. The same general criteria apply as in the case of wine grapes and choice of harvesting time is equally important.

For whiskies, the main criterion in choice of material is starch content which has a major influence on ethanol yield. The major cereals are used for whisky production: maize (*Zea mays*), wheat (*Triticum vulgare*), rye (*Secale montanum*), barley (*Hordeum polystichum*) and occasionally oats

(*Avena sativa* and related species). Rice (*Oryza sativa*) is used for many south-east Asian spirits and other cereals are used in other parts of the world. The varieties selected are normally not controlled by regulation but are those favoured for agronomic reasons and providing satisfactory performance in use. In most cases this means a high starch content and ease of processing. In cereals to be malted, good malting performance is important, producing predictable and uniform modification of the endosperm and thus a high fermentable extract. However, the primary function of malt for Scotch grain whisky is to hydrolyse the starch of the main cereal grain, and so it is selected largely on the basis of enzyme activity. Malting of cereals for whisky production generally follows the same practice as for beer brewing, except that Scotch whisky malts are traditionally flavoured with peat smoke. Sufficient peat smoke is added to the drying air to provide the desired flavour, normally assessed on the basis of the total phenols content of the malt (up to 50 ppm). Whiskies distilled in the Scottish islands are typically characterized by higher levels of peating of the malt. Nitrosamine contamination of malt has recently been of concern in Scotch whisky, arising from the use of natural gas in directly fired kilns. This can be controlled by the use of burners designed to reduce nitrogen oxides, by burning sulphur to add sulphur dioxide to the air stream or by indirect firing.

PREPARATION OF FERMENTABLE EXTRACT

Cereals

Mashing is the process of producing a fermentable extract from ground cereal. Two major routes may be followed depending on whether the wash is to be distilled in batch or continuous stills. In the former case, the process is essentially similar to the production of wort for beer brewing, a clear or filtered extract being required to prevent 'burning' in batch pot stills. In the latter case the separation stage has become redundant and the fermentation (and distillation) are carried out with the total solids present.

Mashing for batch distillation

The initial step is a coarse milling of the grain to obtain the maximum extraction of fermentable sugars. The degree of milling is important. If it is too fine then filtration problems can occur and if it is too coarse there will be a loss of extract. A typical mashing process involves a multistage extraction of the malt. A batch of malt is loaded into the mash tun, followed by the first water (4–4.5 tonne/tonne malt) at 64–68°C. This is drained and a second water at a slightly higher temperature is added (1.5–2 tonne/tonne malt). This is followed by a third and possibly fourth water at a higher temperature, up to 95°C, the quantity being adjusted to give the intended starting gravity for the fermentation. Alternatively, later extracts may be stored and recycled into the next mash.

For certain whiskies legislation does not permit the use of exogenous enzymes so the malt must be well modified and the temperature of the first water must be controlled to minimize enzyme damage. Conversion of some starch to fermentable sugars occurs during and after mashing. This 'secondary conversion' is obviously inhibited by the destruction of malt enzymes, particularly α-amylase. The time required for this process is typically 8–12 h, and in an effort to reduce these cycle times many distilleries use lauter tuns. In order to achieve faster drainage rates, a lauter tun uses a shallower bed and a careful design of rakes. Other systems may also be used but all are aimed at producing a clear wort with maximum recovery in the minimum time.

Mashing for continuous distillation

A substantially different process is required for preparation of a fermentable wort for continuous distillation. Mashing may either be a batch or continuous process. Batch mashing, traditionally used in Scotch grain distilleries, is straightforward and can be applied equally to maize or wheat. The grain may be milled or whole; the cost of milling must be set against the energy saving achieved by quicker cooking of the milled grain. The grain is loaded into batch pressure cookers with water (2.5 tonne/tonne grain) and steam injected to raise the pressure to typically 200 kPa for 2 h. The cooked cereal is then transferred to the mash tun and cooled to 62.5°C and ground malt added to achieve the necessary conversion of starch. The malt may be dried or may be used as green malt, but the most important characteristic of the malt is that it has a high enzymic activity in order to minimize the quantity required, typically 10–15% of the dry weight.

In a continuous process, the milled cereal slurry is mixed with a premalt (a proportion of the malt used for starch conversion) and preheated. At this stage some conversion occurs and the viscosity of the mixture is eased for future processing. The main cooking stage follows at about 165°C, followed by cooling to 60°C before the addition of malt for conversion. The wort is finally cooled for transfer to the fermentation stage. The Scotch whisky industry developed continuous mashing using maize as the main non-barley cereal, and the conversion to the use of wheat introduced some problems requiring the removal of pentosans, small starch granules and proteins. The total economics of the process may be affected by the income obtained from the disposal of gluten. The energy demand of the high temperature cooking stage has prompted the investigation of 'cold cook' processes, but such methods require fine grinding of the cereal grain and therefore some of the energy saved must be set against that required for grinding.

Grapes

There are two main methods of producing fermentable extracts from grapes. The first method produces distillates from wines fermented from free-run grape juice, the premier examples of which are Cognacs and Armagnacs. Alternatively distillates may also be produced from crushed fruits or marc. Two types of grape marc spirits are known. Pisco from Chile and Lozovatcha from Montenegro are produced from fermented crushed whole grapes. More commonly the crushed fruit is the residue left after wine making and the resulting spirit is called eau-de-vie de marc (France), grappa (Italy), aguardente (Spain), bagaceiras (Portugal) and tsipouro (Greece).

Wines for distillation

Grape distillates are produced in most wine growing regions of the world. Production encompasses the full spectrum of quality. Highest quality spirits such as Armagnac and Cognac are produced from selected wine varieties which are vinified and distilled with great care. At the opposite end of the spectrum, distillates are produced as a way of salvaging defective wines or production surpluses.

Cognac and Armagnac

Wines for the distillation of Cognac must originate from Ugni Blanc, Colombard and Folle Blanche vine varieties supplemented up to a maximum proportion of 10% by Semillon, Blanc ramé, Jurançon Blanc, Montils and Select grown in the designated geographical area which includes nearly

all of the Charente Maritime, a large part of the Charente and some neighbouring communes. Wines for the distillation of Armagnac are produced from Ugni Blanc, Baco 22 A, Folle Blanche and Colombard vine varieties grown in the designated geographical area. Most vineyards for Cognac and Armagnac use mechanical harvesters to harvest the grapes and a continuous wine press to extract the juice.

Vinification presents few problems provided that over-crushing or pressing of the marc is avoided and that there is rapid processing of the grape stock. Over-crushing or pressing of the marc produces higher levels of solids which increases the concentration of higher alcohols in the spirit. Storage time should be as short as possible before fermentation as certain enzymic reactions have negative effects on spirit quality. Examples of these are the enzymic degradation of carotenoids that results in a kerosene-like defect caused by 1,1,6-trimethyl-1,2-dihydronaphthalene (TDN), the degradation of fatty acids produces green character in the spirit caused by C6 aldehydes and alcohols and the degradation of pectins increases the level of methanol in the spirit.

Marcs for distillation

Marcs may be prepared in three ways. For the first method the juice and any washing liquid are filtered before fermentation and distillation. For the second and third methods, the marcs are fermented with the fruit solids present. The solids may or may not be separated prior to distillation. Depending on storage and fermentation conditions the enzymic degradation of pectin can raise methanol concentrations to levels that require the inclusion of a demethylating column in the distillation. In large fermenting vessels spoilage by anaerobic bacteria in the deeper layers can result in sensorially important concentrations of butanoic acid and ethyl butanoate in the final spirit. The aroma of the final distillate rarely shows the same varietal characteristics as the grape variety with the exception of Muscat distillate (Chilean Pisco and grappa from White Muscat of Piemonte).

Other fruits

After grapes, apples are the next most abundant fruit used for the production of eaux-de-vie. The most famous apple brandy is Calvados produced in Normandy. Apples are harvested, milled then pulped followed by a resting time of at least 1 h. Juice extraction may either be a continuous or batch process. The addition of cold water to the marc after the initial press for a subsequent extraction is allowed depending on the initial concentration of sugar in the fruit. However addition of either sugar or apple juice concentrates is prohibited.

Eaux-de-vie and brandies are also produced from other stone fruits, most frequently in Eastern Europe. Examples are Sliwowitz (from plums), Kirsch (from sweet black cherries) and Barazk (from apricots). Production of the fermentation substrate follows the same general procedure as for grapes and apples. A major problem of most fruits other than grapes is that the distillates produced are frequently rather bland and lack any of the character of the original fruit. This can be overcome by using soft, nearly over-ripe fruits. However due to the high pH of the mash produced, it is susceptible to bacterial infections and spoilage of the distillate due to high concentrations of acrolein or butyric acid.

The high concentration of methanol in the spirit is also a problem associated with the production of this type of eau-de-vie. Methanol is produced by esterases from pectin and in extremes can exceed toxicological limits. The esterases may be inactivated by heat treatment of the fruit immediately after mashing and this is most desirable. The addition of a demethylation column to the still, used for some marc spirits, results in a significant loss of flavour volatiles.

Another possible toxicological problem concerns the level of ethyl carbamate in some stone fruit spirits. During preparation of the mash, amygdalin in the stones of fruits is either released from broken stones or extracted during storage and subsequently broken down to hydrogen cyanide and benzaldehyde. The cyanide reacts with ethanol in the still to yield high concentrations of ethyl carbamate in the final spirit. Destoning fruit greatly reduces the levels of carbamate produced but also results in greatly reduced levels of benzaldehyde which is an important flavour volatile. Carbamate may be effectively controlled by gentle mashing techniques (aided by the use of soft, nearly over-ripe fruits and pectinases to aid fruit disintegration) and rapid processing of mashes to reduce time for the extraction and breakdown of amygdalin.

Sugar

The raw materials for rums are the by-products of the refinement of cane sugar. This may include sugar syrups and spent sugar syrups (molasses) which may be mixed with sugar cane juice. A special category 'rhum agricole' is produced by the fermentation of sugar cane juice only. The molasses, syrups or juices are diluted to around 12% sugar for fermentation using local water supplies.

Spoilage of spirit and ethanol losses may be caused by poor sanitation during harvesting and mashing. Bacterial contamination can produce undesirable acidity in the distillate and can result in the production of acrolein (a potent lachrymator). Bacterial spoilage of the canes is also possible, so rapid processing of the canes is necessary. Burning of canes increases Streptococcaceae which can reduce alcohol yields. Finally the water used in the preparation of fermentation washes is rarely treated and has its own microflora including sulphur-reducing bacteria, which are thought to be responsible for the production of some sulphur compounds during fermentation.

Neutral alcohol of agricultural origin

For the production of flavoured spirits most regulations only permit the use of ethyl alcohol of agricultural origin, i.e. alcohol fermented from a carbohydrate source. This precludes the use of alcohol produced from fossil fuels. The production of alcohol for the preparation of flavoured spirits is a separate operation from any subsequent botanical distilling or compounding operation and is normally undertaken at a different location by a different company. The alcohol may be distilled from a number of carbohydrate sources including grain (wheat or maize), molasses, grapes, potatoes and lactose (from whey). The choice of carbohydrate source depends upon quality, availability, price and any local tariff restrictions in the country of manufacture. Alcohol distilled from grain is often presented as premium quality, but with appropriate rectification during distillation neutral alcohols can be obtained from other substrates. Where cereals are used there is no requirement to use whole grains (as is the case for whisky); the distiller can recover bran and protein from the grain before mashing. The conversion of starch into fermentable sugars and dextrins can be achieved using either natural malt enzymes or manufactured enzymes allowing the use of cold cook processes.

FERMENTATION

The fermentation stage is essentially the same as for beer or wine production, but since the aim is a product which can be distilled rather than consumed directly, the objectives of the process are slightly different. Unless the final product is to be a highly purified rectified or neutral spirit, it is important that no defects arise in the fermentation and that no off-flavours are allowed to form.

The fermentation may be spontaneous, as is largely the case with cognac, or a known pure strain of a cultured yeast may be used. The basic reactions are the same as those in beer and wine production, that is the conversion of simple sugars to ethanol and carbon dioxide. In addition there are many secondary reactions which lead to the production of the minor flavouring compounds such as esters, higher alcohols, glycerol, organic acids, etc. Different yeasts have different abilities to produce the secondary compounds and the expression of these abilities depends on the fermentation conditions (temperature, pH, acidity, degree of oxygenation). Thus it is to be expected that there will be considerable variation between batches and producers especially when spontaneous fermentations are used, whereas a pure cultured yeast is likely to produce less variation.

Fermentation for grape spirits is essentially the same, though some minor variations occur in different areas, such as whether the use of sulphur dioxide is permitted to stabilize the fermented must before distillation. In the Charente winemaking process for Cognac, the clarified grape juice is fermented in the vats in which it will remain until distillation. The yeasts are naturally present in the must, though a starter may be used at the beginning of the season. Although 650 yeasts have been recorded, the balance between species varies between the vine, the warehouse and the fermenter and only one, *Saccharomyces cerevisiae*, is dominant in the fermentation. The product flavour, the sum of the varietal and fermentation contributions, is the result of the interaction of the vine variety, local climatic conditions which affect both the grape composition and the wild yeasts, and the grape treatment and fermentation conditions. Commercial active dried yeast has not been successful and has been shown to produce inferior distillates compared with those obtained under natural conditions. However, it is likely that a selection of suitable strains will be found, representative of and adapted to the local conditions. The malo–lactic fermentation occurs naturally after alcoholic fermentation and the resulting wine varies between about 8 and 12% v/v ethanol, depending on the grapes used and local conditions.

Cereals are invariably fermented with a cultured yeast, and in most regulations *S. cerevisiae* is specified as the only organism. Malt and other cereals may be contaminated with a wide range of organisms, but the cooking stage required when unmalted cereals are used (e.g. in grain whisky) should eliminate them all. Fermentations are started by pitching with a known yeast culture, normally a specific strain of high performance distilling yeast. In some cases, typically Scotch malt whisky production, a supplementary brewer's yeast may also be used, because it is believed to produce a distillate of more desirable flavour. In small scale production, fermenters are closed vessels of traditionally wooden construction with no means of temperature control. In larger scale production, the fermenters are more commonly stainless steel with cooling facilities. In this case carbon dioxide may be collected. A typical fermentation will produe a wash at around 8% v/v ethanol after 40–48 h, a much shorter time than has traditionally been allowed and much shorter than the time allowed for fermentation of grape or fruit musts. In the case of Scotch whisky at least, shorter fermentation times are thought to be detrimental to spirit quality. In longer fermentations there is the danger of substantial bacterial growth, which can both reduce ethanol yields and cause flavour defects. Bacteria found in Scotch whisky fermentations have included *Lactobacillus, Leuconostoc* and *Pediococcus*.

DISTILLATION

The distillation is the core of the production process for beverage spirits. The origins of distillation are not clear, but a grain spirit is said to have been produced in Ireland in the twelfth century and there is good evidence that the process was well understood in thirteenth century Europe. However it seems that knowledge of distilling was transferred into Europe by the mediaeval alchemists, who had learned it from Arab chemists. The word alcohol comes from Arabic, though the process may in fact have originated further east. Whatever its origins, distillation presumably evolved as a means of preserving the relatively low ethanol beers and wines from deterioration in storage, though there have been more fanciful explanations, and early uses seem to have been for the production of plant extracts as perfumes and medicines.

Without modern science and technology it is very difficult to produce beers or wines that will remain stable and palatable for long, and until the work of Pasteur and the other pioneer microbiologists in the mid-nineteeth century even the causes of spoilage were a mystery. The central function of distillation can therefore be regarded as a means of increasing the ethanol concentration in the beverage. Within this aim, a number of different systems have evolved, but the permitted variations in the process are circumscribed by regulation. During a distillation process, the ethanol, water and other components of the beer or wine are separated primarily by volatility. Since the ethanol is the major component required in the distillate, the other compounds which will remain present are those relatively close in volatility to ethanol. Thus as the ethanol concentration of the distillate increases towards the limit (the azeotropic point, 95.57% w/w), the level of other compounds (flavouring compounds, congeners) falls. As a result the strength of flavour remaining in the distillate is a function of the ethanol concentration. The primary tool of regulation of the distillation stage in the spirit processes is therefore the ethanol concentration of the distillate, though some regulations, for example for specific designations such as Cognac, may be much more restrictive. Regulations commonly specify a maximum, which limits the distillers' freedom to vary the flavour strength of the product and ensures that

traditional products maintain a characteristic flavour.

In practice, the stills used can be separated into three principal types. These are the pot still, characteristic of Scotch malt whisky, Cognac and 'rhum agricole'; the 'simple' column, characteristic of many other French spirits such as Armagnac and of Scotch grain whisky; and the 'complex' column, used for neutral spirit production for British and North American gins and vodkas.

Pot stills

The pot still is the simplest form of still used and in its traditional form can be traced back to the earliest days of distillation. Slightly different variants of the still have evolved in different areas and have become associated with different products, but the basic principles are similar. The still consists of a copper pot with a neck and a near-horizontal connecting link to a condenser (Figure 7.4). Copper is the material traditionally used, being easy to work and shape, a good heat conductor, corrosion resistant and relatively long

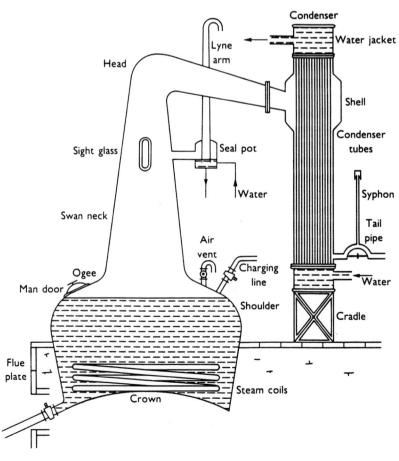

Figure 7.4 Plain wash (first) still for Scotch malt whisky distillation. From Nicol (1989).

lasting, though stills may require to be replaced after 20 years or so. A secondary effect of copper, which has led to its retention even though modern materials such as stainless steel might appear to offer advantages in corrosion resistance and longevity, is that it can be very reactive and both catalyses reactions during the distillation and reacts directly with some compounds of the distillate. In this way the content of sulphur compounds in the distillate is thought to be reduced, particularly in whisky.

A pot still is not a very efficient distillation apparatus, and therefore in order to achieve both an acceptable yield of ethanol and an acceptable flavour, double distillation is almost universally used. The two aims of distillation, recovery of and concentration of ethanol, can be separated and attention given to optimizing both. In the first distillation, the aim is maximum recovery of ethanol from water and other congeners; in the second the aim is to collect a spirit fraction with the required ethanol concentration and flavour level.

The process is fundamentally the same, whether applied to Scotch malt whisky, Cognac or other spirits. In the first distillation, the wine or beer (wash in whisky terminology) is heated and a distillate collected at 20–30% ethanol. The actual ethanol concentration depends on the product and within a product type, on practice within the distillery concerned. Heating of the still is similarly varied, with Scotch whisky distillers using direct heating by coal or gas or indirect by steam coils or jacket, though Cognac distillation is restricted to direct heating. In the case of directly heated stills, a stirrer or rummager may be fitted to prevent burning of the still charge on the heated surface. Distillation is continued until all recoverable ethanol has distilled over, a process which may typically take 8 h for a whisky still. In the second distillation, three fractions are collected. These are:

(i) a heads fraction containing the most volatile components of the distillate;
(ii) a heart, middle or spirit cut which forms the desired product, at typically 65–70% ethanol;
(iii) a final fraction (seconds, feints) which contains the less volatile compounds that are not required in spirit.

The first and third fractions are then recycled for redistillation. The product characteristics are set by the choice of cut points onto and off spirit collection, and by the manner in which the recycling and balancing of foreshots and feints is controlled. A typical analysis of a Scotch malt whisky is shown in Table 7.11.

Table 7.11 Chromatographic analysis of a malt whisky new make[a]

Period	4	4	4
Charge	12	14	16
Strength (% v/v)	63.5	71.5	69.4
Acetaldehyde	3.2	3.8	6.8
Ethyl acetate	23.7	25.5	27.0
Diethyl acetal	1.7	1.2	2.2
Methanol	5.1	4.6	5.3
Propanol	40.8	42.7	41.9
iso-Butanol	79.8	80.8	80.5
o.a. Amyl alcohol	47.7	44.7	49.5
iso-Amyl alcohol	142.5	145.5	142.5
Total higher alcohols	331.1	313.7	314.4
Ethyl lactate	4.7	2.5	4.1
Ethyl octanoate	1.6	1.9	1.7
Furfural	3.3	3.9	4.2
Ethyl decanoate	5.7	5.6	4.5
β-Phenethyl acetate	5.7	7.5	5.9
Ethyl laurate	2.1	2.6	2.1
β-Phenylethanol	3.8	0.6	0.6
Ethyl myristate	0.6	1.1	0.6
Ethyl palmitate	2.7	3.3	2.6
Ethyl palmitoleate	1.5	1.9	1.4

o.a. = optically active.
[a] Concentrations are g per 100 l alcohol.

Source: Nicol (1989).

Column stills

A simple form of small single column continuous still was developed in the mid-eighteenth century and was rapidly taken up in France. It is of the type predominantly used for distillation of Armagnac and many other French spirits. However it seems to have made no impact in Scotland and column distillation of whisky did not start on any scale until the introduction of the Coffey or Patent still in 1830. Variants of these two forms of still are now used for the production of many other spirits around the world.

The boiler, distillation column, wine heater and cooler form the main parts of the apparatus, normally constructed in copper (Figure 7.5). The wine heater acts as a preheater for the wine and a first stage condenser of distillate, the cooler subsequently ensuring complete condensation and cooling. The preheated wine is then introduced to the column, which typically contains 12 plates equipped with bubble-caps or similar devices to restrict the free downward flow of liquid and ensure good mixing and heat transfer from rising vapour to falling liquid. As the wine flows down the column, volatiles are stripped and passed to the condenser system in the wine heater and cooler. The first condensing products (less volatile, tails, compounds) may be removed from the first few

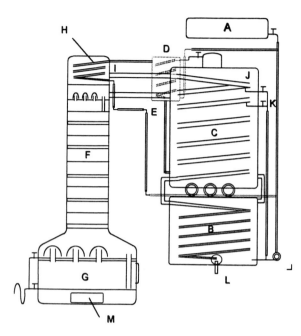

Figure 7.5 Armagnac still. A: head of wine; B: cooler; C: wine heater; D: head condenser; E: wine arrival; F: column; G: boilers; H: head column coil; I: swan neck; J: coil; K: drawing and recycling of tailings; L: alcohol-meter holder; M: furnace. From Bertrand (1995).

turns of the coil and recycled into the wine inlet. A head condenser may also be added to control the most volatile products and a head column coil to help to condense the least volatile products. A typical analysis is shown in Table 7.12.

The Coffey still (Figure 7.6) used for Scotch grain and much other whisky, though normally physically consisting of two columns, can logically better be regarded as a single column, cut in halves and the two placed side-by-side. The column may be square, to facilitate the traditional

Table 7.12 Distallation of volatile substances (averages from 50 wines and their corresponding spirits)

	Wine (mg l^{-1})	Spirit (mg l^{-1})	Recovery (%)
Ethanol (% vol.)	11.1	59.7	
Higher alcohols	373	2043	102
Methanol	41.4	194	87
2-Phenyl ethanol	53	32	10
Higher alcohol acetates	2.61	12.7	90
Volatile acid ethyl esters	1.46	14	177
Ethyl acetate	41	207	94
Diacetyl	0.65	2.91	83
Volatile acids (C$_3$–iC$_5$)	4.12	10.79	38
Volatile acids (C$_6$–C$_{12}$)	11.2	34.2	56
Ethyl lactate	340	248	14
Acetic acid	400	118	5.5
2,3-Butanediol	549	14.9	0.5

Source: Bertrand (1995).

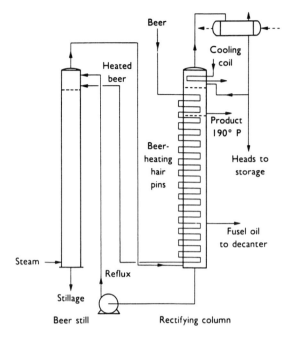

Figure 7.6 Coffey still. From Panek and Boucher (1989).

wood and copper construction, or more likely circular and stainless steel. In this case copper must still be present in the system to assist in the removal of undesirable flavour compounds, and may be added as a 'demister', a pad of copper mesh at the top of the analyser. Internally the column consists of a series of plates (originally simple sieve plates, but may also be bubble cap plates) with holes to allow upward flow of vapour and linked by downcomers to allow a restricted downward flow of liquid. The fermented liquor is preheated by passing through a tube winding through the second column (rectifier) and is then fed into the first column (analyser) towards the top. Steam is sparged into the base of the column and volatiles are stripped from the descending liquid and removed from the top of the column. The vapour then passes to the base of the rectifier and spirit product is removed from a level towards the top of the column. Fusel oil is taken out near the base of the column, and foreshots (from the top) and feints (from the base) are recycled into the top of the analyser. North American practice for producing relatively strongly flavoured grain spirits is to use a single distillation column, augmented by a doubler (a continuous pot still).

Multiple column stills

The Coffey still produces spirit at about 94.5% v/v ethanol, with a relatively strong flavour. The

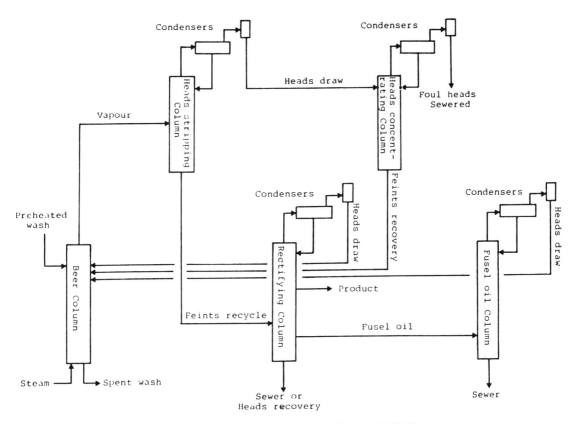

Figure 7.7 Five-column still for neutral spirit production. From Wilkin *et al.* (1983).

addition of further columns to the system (Figure 7.7) facilitates the production of higher purity spirit (rectified or neutral spirit, at higher ethanol concentration) in an economic way from almost any fermentation feed (e.g. grain, molasses, wine, whey). These are the spirits commonly used for gin, vodka and other flavoured products where regulations simply call for the use of agricultural alcohol. It is possible to produce high purity spirit from a Coffey still, but it would be very inefficient. The addition of further columns allows the recovery and recycling of ethanol from other fractions.

MATURATION AND OTHER FLAVOURING

Maturation

The tradition of maturation probably arose because of irregular crops of the raw materials. Excess production from one year would be stored for subsequent years when crops may be poor. Historically wooden barrels were the only suitable container for bulk liquid storage. The improvement in the organoleptic properties of the spirits during storage became obvious and in time this storage or maturation became necessary to produce the required aroma and taste. Carried through to modern times it has become a legal requirement to turn eaux-de-vie into brandies and grain spirits into whiskies.

Legal requirements

The legal minimum maturation time for most whiskies is three years in wooden casks not exceeding 700 l capacity. For American 'straight' whiskies a minimum of two years is specified. For brandy the legal minimum is one year in oak receptacles or at least 6 months in oak casks with a capacity of less than 1000 l. The organoleptic characteristics of a spirit however continue to improve well beyond the legal minimum age so most spirits are matured for longer. Where an age is specified on the bottle this is the minimum for all spirits used in production. AOC wine spirits may also follow the commerical designations 'three stars', VSOP and XO or Napoleon which require maturation periods of at least two, four and five years, respectively.

Cooperage

Though many species of wood have been used historically, the physical and chemical requirements of a maturation container have reduced this to a few species of oak. Physically the wood should be suitable for holding liquids (tight cooperage) while chemically the extraction of wood components that occurs during maturation should enhance the sensory properties of the distillate.

Only the best grown oak is suitable for the manufacture of spirit cooperage. The grain of the wood should run from top to bottom of the log. If the grain corkscrews (i.e. deviates in both the vertical and horizontal planes), then porosity is increased and there is a risk of leakage. Other desired features are the blocking of heartwood vessels by tyloses, again to reduce porosity, and the absence of knots. The principle sources are the eastern part of the USA and continental Europe. In the USA the most widely used species is *Quercus alba* (white oak) though other related species of oak, e.g. *Quercus muehlenbergii* and *Q. macrocarpa* (chinkapin and bur oaks, respectively) may be used. The European species used are *Quercus petraea* and *Q. robur*, the sessile and pedunculate oaks. Sessile oaks most commonly come from the forests of the Troncais and Allier regions of France, and pedunculate oaks most commonly from the forests of Limousin and Gascony in France and Galicia in north west Spain. Both species hybridize freely so differentiation is most frequently done on the region of production.

Trees should be close grown and straight and only parts of the trunk free of branches and defects can be used. At the mill the selected logs are cut to stave length then split or sawn in line with the heart. These cleft pieces are known as bolts and from these staves are split or sawn. Staves and headings require seasoning to reduce their moisture content from greater than 50% to between 12 and 18%. To achieve this, stave blanks are stacked in the open air for 24–36 months in France and Spain or for 12–18 months in the drier atmosphere of the USA. At the end of prohibition in the USA the quantity of air seasoned wood was unobtainable for the production of Bourbon and the process was accelerated by kiln drying. This allowed partially dried wood (22–30% moisture) to be reduced to a working moisture content within 21 days. Further innovation led to the development of predriers that produce workable wood within one month of tree felling. Research suggests, however, that there is a substantial decrease in the quality of the matured spirit with kiln dried wood. Some of the benefits of air seasoning are attributed to the leaching effect of rain and the growth of fungi on the surface of the wood.

New oak casks are used only for the maturation of Bourbon, Cognac and Armagnac, though for cognac and armagnac new wood ageing is limited to months rather than years before racking into used casks. In virtually every other spirits industry, however, previously used barrels are the containers of choice. In the production of Cognac and Armagnac, French oak is used, either from the forests of Allier or Troncais, which have a fine grain, or from the forests of Limousin or Gascony, which have a coarser grain. The new distillate is put initially into a new cask, generally 400 l capacity, for a period of about 6 to 12 months before being transferred to an older barrel known as a Roux, to avoid the appearance of astringency and marked bitter flavours.

In the United States there are strict regulations regarding cask type and new charred casks are required by law for the maturation of Bourbon, rye, wheat, malt and rye malt whiskies. Other whiskies may be matured in new or used oak cooperage, but the new cooperage must not be charred. After one fill with Bourbon (at least 4 years) casks are then used for the maturation of other whiskies in America, Scotland, Ireland and Canada. The casks may be reused unaltered but more commonly capacity would be increased by the introduction of additional staves and new oak ends.

Prior to 1965 a significant proportion of Scotch whisky was matured in oak casks used for the import of sherry from Spain. These sherry casks were 500 l butts, manufactured in Spain from both American and Spanish oak, well-seasoned by use in the grape fermentation process and then used to ship various types of fully matured sherries to the United Kingdom. Total contact time with sherry could be of the order of 6–9 months. After 1965 most sherry was shipped by bulk tankers and alternative measures were introduced to simulate the seasoning and sherry treatment processes. These failed to reproduce the sensory characteristics of the home emptied sherry cask and now new butts (500 l capacity) and puncheons (460–500 l) are built for Scotch, Irish and Japanese whisky distillers in Spain but are used for the fermentation and maturation of sherry prior to their use for spirits.

Casks used for the maturation of whisky other than Bourbon are used, repaired and reused indefinitely until the cask is no longer a sound and viable container or until it has lost the ability to effect any sensory improvement over an economical maturation period. Therefore in a whisky warehouse the vast majority of casks will be refill casks, used an unknown number of times, with various degrees of repair work carried out on them and consequently with varying abilities to mature whisky. Normal practice is to average out these variables in the make up of the final product by blending various ages and wood types.

Warehousing

Distilled spirits were traditionally matured in stone-built, single- or multistorey warehouses which were located beside the distillery. Casks were stored in 'stows' usually two or three high, sitting on top of one another with wooden runners between each layer. Damp conditions (very damp in Scotland) were often created on the ground floor of sloping sites and were reputed to produce the best quality spirits. Expansion in production meant warehouse accommodation was insufficient and so large centralized units were built. Basic construction was brick walls and insulated aluminium roofs. Steel racking with wooden runners was installed to allow casks to be stored up to 12 high depending on size. Sites closest to the roof are the driest and have the least stable temperature. Those on the ground were the most stable in temperature and the wettest. Traditional single storey sites tended to be relatively wet with unstable temperatures.

Climatic conditions vary between various countries. In Scotland the cool humid climate encourages preferential loss of ethanol relative to water and consequently strength decreases during maturation. In the USA, the relatively hot and dry climate encourages preferential loss of water vapour relative to ethanol and consequently strength increases during maturation. Under controlled conditions, the non-volatile content is significantly influenced by temperature, cask type and to a lesser extent humidity. In multitiered warehouses in the United States significant temperature differences between top, middle and bottom tiers result in differences in the content of both volatile and non-volatile components. In the warmest (top) tier, the physical and chemical reactions typical of maturation proceed at a greater rate but there is no accepted optimal position for maturation.

Sensory changes during maturation

Maturation should produce a significant improvement in flavour quality to justify the cost of storage. This improvement results from the development of mellow or mature characteristics from the wood and a loss of the harsh or immature characteristics of the new distillate. Mature flavours that develop during maturation include vanilla, spicy, floral, woody and smooth. The harsh or immature characteristics have been described as sour, grassy, oily and sulphury, though there is considerable variation in the vocabularies used. Both the magnitude and the rate of change during maturation are dependent on the type of cask used. Charring has been shown to increase the intensity of mature characteristics (such as smooth, vanilla and sweet) and to decrease the intensity of immature characteristics (pungent, sour and oily). Cask reuse conversely decreases the intensity of mature characteristics and increases that of immature characteristics.

Chemical changes during maturation

During the maturation period the distilled spirit becomes highly modified as a result of its contact with the cask wood. Though many reactions have been identified as occurring during maturation, there is no reliable chemical or physical index available to indicate the progress of maturation. Consequently, the surest means of following the progress of maturation is by sensory assessment.

Chemically, the distillery and its operating practices are the dominant factors in determining the concentrations of volatile compounds, while cask type is the dominant factor in determining the amounts of non-volatile compounds present in the matured whisky. It is the combination of these compounds that produces the flavour and aroma of the final product. The concentrations of distillate components generally either increase or do not change significantly during maturation and where this occurs such changes are frequently related to cask type. The increase in concentration is due to the evaporation of ethanol and water. The rate of evaporation is greatest for ethanol, water and acetaldehyde with much smaller losses for iso-amyl alcohols and ethyl hexanoate. Evaporation however is thought to be the main route for the loss of dimethylsulphide and dihydro-2-methyl-3(2H)-thiophene. The rate of evaporation may be affected by cask stave thickness, the air flow round the cask, humidity and temperature.

Chemical reactions can also occur that alter distillate components resulting in less volatile compounds. Examples of these are oxidation and acetal formation. Oxidation results in the formation of acetaldehyde and acetic acid from ethanol and the formation of dimethyl sulphoxide from dimethylsulphide. Oxidation in the maturing spirit is enhanced by the presence of wood extractives, particularly vicinal hydroxyphenols which together with traces of copper from the still are thought to act as a catalyst. Acetal/aldehyde equilibria are established for most aldehydes and are important for aroma as aldehydes frequently have sour and pungent odours, while acetals are pleasant and fruity. Acetal formation is also important for removing acrolein, a potent lachrymator, from distilled spirits. The equilibrium between free aldehyde, hemiacetal and acetal is affected by spirit pH and ethanol concentration.

During maturation extraction of components of the cask wood takes place. Many wood components are not present in a free state in the barrel wood and are the result of the decomposition of macromolecules forming the framework of the wood, such as lignin, cellulose and hemicellulose. Polymeric material is also extracted from the wood and this in turn may be broken down by the spirit solution. Differences in the composition of wood extract occur between the different varieties of oak used for maturation. In general French and Spanish oaks yield higher concentrations of tannins and lower concentrations of oak lactones, scopoletin and vanillin than American oaks. This may partly be explained by different cooperage practices. In Europe stave blanks are commonly seasoned by air drying and casks produced with only light charring. In the United States stave blanks are commonly kiln-dried and casks produced with a heavy char.

Cask charring is an important contributor to the flavour of Bourbon whisky. Thermal degradation of the inner face of casks produces a layer of 'active' carbon, greatly increases the yield of oak lactones and coloured and phenolic extracts, and results in the formation of maltol and 2-hydroxy-3-methyl-2-cyclopentenone and the destruction of resinous wood flavours. Central to the increase in phenolic extract is the degradation of lignin to aromatic compounds such as vanillin, syringaldehyde and coniferaldehyde. During maturation these compounds are extracted by the spirit and further breakdown of lignin occurs through hydrolysis and ethanolysis during maturation.

When repeatedly re-used for maturation the yield of wood compounds extracted during maturation decreases. In tandem with this decrease in extract is a decrease in the development of mature characteristics and less suppression of immature characteristics. Eventually a point is reached where the cask fails to produce any sensory improvement and is termed 'exhausted'.

Another cask parameter that affects the course of maturation is the surface-to-volume ratio. Cask sizes range from the 558 l puncheon down to the 190 l American standard barrel. Smaller barrels have a higher surface to volume ratio, resulting in quicker extraction of wood components but also a higher rate of evaporation of ethanol and water. Given the same wood type and history, the smaller casks would be expected to produce a higher extract and to mature the whisky in a shorter period of time.

There is also interaction between components of the original distillate and wood extracts. An example of this is esterification, though theoretically oxidation and acetal formation should also be included. During maturation the concentration of esters increases, due to the esterification of free acids by ethanol. A large part of this is due to the formation of ethyl acetate from acetic acid, either extracted from the cask wood or the product of ethanol oxidation. Transesterification reactions are also thought to occur, which in the presence of the large excess of ethanol favours the formation of ethyl esters. Aromatic acids extracted from the cask wood such as syringic and vanillic acids are also known to form ethyl esters during maturation.

Changes in pH during maturation, which may be cask dependent, affect the ionization state of weak bases and consequently their volatility. Decreases in pH have the greatest effect on pyridines due to their pK_a values and greatly reduce their perception in the aroma of whisky.

Wood components have been shown to change the solubility of many compounds present in malt distillates and have a stabilizing influence on solute agglomerates formed when samples are diluted for consumption or sensory analysis. The formation of these stable solute agglomerates suppresses the volatility of hydrophobic aroma compounds, such as alcohols, aldehydes and esters and may be important in reducing the sensory impact of some constituents of the new distillate.

Flavoured spirits

Flavoured spirits are based on high purity alcohol and are subject to further processing to give flavour characteristics. This category, which includes gin and aniseed flavoured and related products, are normally colourless and are not subject to maturation processes, as are whisky, brandy and rum. Their alcohol base is normally neutral alcohol of agricultural origin. The alcohol is flavoured either by redistilling in the presence of botanical material, or by the use of natural essences. Vodka is simply based on pure alcohol, though some may be flavoured to give special organoleptic characteristics such as a mellow taste. The water used for the production of these spirits must be clear, pure and without odour or taste.

Water obtained from local supplies requires demineralization followed by carbon filtration and UV irradiation to eliminate any microbiological activity in the water. Demineralized water is normally subjected to both sensory and chemical checks.

Gin

The three key ingredients in gin manufacture are botanical materials, neutral alcohol and water. The essential botanical ingredient is the juniper berry (*Juniperus communis*) though a number of other herbs and spices are also included in the recipes. Gin may be prepared in two ways. Distilled gin is prepared by distilling neutral alcohol and water in a traditional gin still in the presence of juniper berries and other botanical ingredients. Compounded gin (which may not be labelled as distilled) is simply a mixture of juniper-based flavours or essences in neutral alcohol and water. For both, the exact recipes are closely guarded trade secrets but most include juniper berries, coriander seed, angelica root, orange and lemon peels, cinnamon bark and cardamom seeds.

Gin distilling is a traditional batch process using steam heated copper or stainless steel pot stills similar in design to those used for Scotch malt whisky. Like Scotch stills, designs tend to be constant over generations and specific to a particular brand. The process begins with the still being charged with water and then alcohol to the desired alcoholic strength for distillation. The botanical ingredients are then added to the still, either loose or in a bag, which may be suspended above the pot (containing the liquid). The still is then heated and three fractions collected from the condenser. The initial distillate, the heads, and final fraction, the tails, contain the more and less volatile constituents from the botanicals and are combined as feints and purified in a separate distillation to recover the alcohol. The middle fraction is collected as gin product with a strength of 80% v/v and is subject to rigorous sensory checks before blending, reduction and bottling. Dutch-type gin or genever is made by the redistillation of a triple pot distilled malt spirit (moutwijn) in the presence of juniper berries.

Anis flavoured spirits

This category of spirit was developed mainly in countries of southern Europe, each producing its own regional variation, hence ouzo in Greece, Pastis in southern France, sambuca in Italy and raki in Turkey. The main aromatic component is

trans-anethole which may be obtained from aniseed, star anise oil or fennel. Neutral alcohol of agricultural origin may be flavoured by maceration and/or distillation, redistillation of the alcohol in the presence of botanicals, the addition of natural distilled extracts or any combination of these.

For a spirit to be called anis, its characteristic flavour must be derived exclusively from aniseed, star anise oil or fennel and to be termed distilled anis, alcohol flavoured by distillation in the presence of seeds must constitute at least 20% of the drink's alcoholic strength. For an aniseed flavoured spirit to be called pastis it must also contain natural extracts of liquorice roots giving a glycyrrhizic acid content of between 0.05 and 0.5 g l^{-1} and should have an anethole content between 1.5 and 2 g l^{-1}. Pastis may also contain sugar up to 100 g l^{-1}. For an aniseed-flavoured spirit to be called ouzo it must be produced exclusively in Greece and have been produced by blending alcohols flavoured by means of distillation or maceration using aniseed and possibly fennel seed, mastic from a shrub (*Pistacia lentiscus* var. *chia*) indigenous to the island of Chios and other aromatic seeds, plants and fruits. Alcohol flavoured by distillation must represent at least 20% of the alcoholic strength of the ouzo and must have been produced in traditional copper stills with a capacity of 1000 l or less. Ouzo must be colourless and have an alcoholic strength between 55 and 80% v/v and have less than 50 g l^{-1} added sugar.

Vodka

Vodka production simply requires high purity alcohol so that the character comes from ethanol. The alcohol is normally distilled from grain fermentation, although potatoes may be used as a carbohydrate in Poland and Russia. Vodka spirit is further purified with activated carbon in order to remove trace congeneric material which may impart a sensory character. This may be achieved by either dispersing and agitating powdered charcoal in a large volume of spirit followed by its removal by filtration, or by passing the spirit through one or more columns containing granular carbon. Some vodkas are reduced to packaging strength simply with pure water, filtered and bottled. Others are mixed with trace amounts of additives such as sugar and glycerol to impart smooth mouthfeel and also to give a residue on analysis in markets where simple alcohol–water mixtures are not permitted.

BLENDING AND FINISHING

Most bottled whiskies are in fact blends of spirits matured in various types and ages of casks. Before bottling the contents of the different casks are blended to average out the different wood ages and types to produce a consistent product. True blended spirits, which are typically whiskies, consist of light bodied spirit mixed with a number of heavier bodied spirits in a wide range of proportions. The aim of this blending is to produce a consistent product that has a distinctive flavour. 'Light-bodied' spirits are those distilled to high ethanol concentrations using continuous column stills and include Scotch grain whisky and American light whiskies, grain spirits and grain neutral spirits. 'Heavier bodied' whiskies are either batch still products or column still products distilled to lower ethanol concentrations.

The nature and components of blends are determined by the traditions and regulations of the country of origin. The actual process of blending is, however, very similar. Approved whiskies are delivered to the blending house and drained from the casks in correct proportions into passivated steel troughs. The troughs convey the whiskies to a blending vat where they are thoroughly mixed using mechanical agitators and compressed air. When the blend is correct, deproofing water is added to the blend to reduce the strength for bottling. Minor variations do occur. In Scotland, blending may be followed by a further period of maturation and in Canada, distillates may be mixed prior to any maturation (preblending).

For Scotch whisky the light bodied spirits are the products of up to 10 grain distilleries, situated mainly in the central belt of the country. The heavier bodied spirits are the products of up to 100 malt distilleries, mostly in the Highlands and Islands. Blends tend to be 60 to 70% grain whisky with as many as 50 malt whiskies. Recipes are often complex to guard against variation in the quality or unavailability of a single whisky having a noticeable effect on the quality of the blend. The complexity of blends is maintained by purchasing or exchanging new whiskies which are matured in the producer's warehouses and delivered when they have been matured to the level required by the blender.

In the United States where there are fewer distilleries and trading between competitors is uncommon, the components of a blend tend to be produced at only a limited number of distilleries. To increase the variety of components available to blenders, different cereals, fermentation conditions, distillation parameters and maturation periods and cooperage may be used. Heavier bodied spirits include Bourbon, rye, wheat, malt, rye malt and corn whiskies while the lighter bodied spirits are light whiskies or grain and grain neutral spirits. In addition, the addition of blenders up to 2.5% by volume is allowed and these may include sherries and blending wines.

Irish, Japanese and Canadian blenders have the same problems as blenders in the United States, in that they have a limited number of distilleries to chose from. Again variations in mash cereals, fermentation conditions, distillation parameters and maturation time and cooperage are used to increase the variety of flavours available to the blenders. In Canada blended whisky may contain as much as 9.09% flavouring on a litres of absolute alcohol basis, although this level is not usually achieved in practice. In Japan blends frequently include imported malt whiskies to give more flexibility in their formulation.

Filtration

Many spirits are filtered prior to bottling to reduce the risk of haze formation. Spirits are traditionally matured at 50 to 70% alcohol by volume but are bottled at 40 to 45% alcohol by volume. For heavier bodied older spirits and spirits matured at higher strengths this can result in haze formation when reduced due to high molecular weight lipids and esters and ethanol soluble lignins. These compounds are less soluble in water than ethanol. Haze formation is controlled by chill filtration, where the whisky is cooled to between $-10°C$ and $10°C$ and held for a specified period of time before the problem compounds are removed by physical separation and adsorption by a filter.

The conventional filter used is a plate and frame variety using preformed pads made of cellulose, or cellulose impregnated or precoated with diatomaceous earth. Typical particle retentions are of the order of 5 to 7 μm. Operational parameters depend on the batch size, the nature of the product and the filtration rate required. In general higher filling strengths and new wood require more filter area per unit volume.

REFERENCES

Benvegnin, E., Capt, E. and Piguet, G. (1951) *Traite de Vinification*, Payot, Lausanne.
Bertrand, A. (1995) Armagnac and wine spirits, in *Fermented Beverage Production*, (eds A.G.H. Lea and J.R. Piggott), Blackie Academic and Professional, Glasgow.

Kirk, R.S. and Sawyer, R. (1991) *(Pearson's) Chemical Analysis of Foods*, 9th edn, Longman, London.

Nicol, D. (1989) Batch distillation, in *The Science and Technology of Whiskies*, (eds J.R. Piggott, R. Sharp and R.E.B. Duncan), Longman, Harlow.

Panek, R.J. and Boucher, A.R. (1989) Continuous distillation, in *The Science and Technology of Whiskies*, (eds J.R. Piggott, R. Sharp and R.E.B. Duncan), Longman, Harlow.

Wilkin, G.D., Webber, M.A. and Lafferty, E.A. (1983) Appraisal of industrial continuous still products, in *Flavour of Distilled Beverages*, (ed. J.R. Piggott), Ellis Horwood, Chichester.

FURTHER READING

Barker, A. and Guy, S. (1996) *Process Engineering for Brewing*, Chapman & Hall, London.

Bertrand, A. (ed.) (1991) *Les Eaux-de-Vie Traditionelle d'Origine Viticole*, Lavoisier Tec & Doc, Paris.

Birch, G.G. and Lindley, M.G. (eds) (1985) *Alcoholic Beverages*, Elsevier Applied Science, London.

Briggs, D.E., Hough, J.S. and Young, T.W. (1981) *Malting and Brewing Science, Volume 1, Malt and Sweet Wort*, Chapman & Hall, London.

Campbell, I. (ed.) (1990) *Proceedings of 3rd Aviemore Conference on Malting, Brewing and Distilling*, Institute of Brewing, London.

Campbell, I. and Priest, F.G. (eds) (1986) *Proceedings of 2nd Aviemore Conference on Malting, Brewing and Distilling*, Institute of Brewing, London.

Campbell, I. and Priest, F.G. (eds) (1995) *Proceedings of 4th Aviemore Conference on Malting, Brewing and Distilling*, Institute of Brewing, London.

Cantagrel, R. (ed.) (1992) *Élaboration et Connaissance des Spiritueux*, BNIC/Lavoisier Tec & Doc, Paris.

Charalambous, G. (ed.) (1988) *Frontiers of Flavor*, Elsevier, Amsterdam.

Daiches, D. (1976) *Scotch Whisky*, Fontana/Collins, Glasgow.

EC (1989) Council Regulation (EEC) No. 1576/89 of 29 May 1989. *Offic. J. European Communities* 12.6.89 No. L160/1–17.

Gunn, N. (1995) *Whisky and Scotland*, Souvenir Press, Edinburgh.

Hough, J.S., Briggs, D.E., Stevens, R. and Young, T.W. (1982) *Malting and Brewing Science, Volume 2, Hopped Wort and Beer*, Chapman & Hall, London.

Jackson, R.S. (1994) *Wine Science*, Academic Press, London.

Lea, A.G.H. and Piggott, J.R. (eds) (1995) *Fermented Beverage Production*, Blackie Academic and Professional, Glasgow.

Lewis, M.J. and Young, T.W. (1995) *Brewing*, Chapman & Hall, London.

Marrison, L.W. (1973) *Wines and Spirits*, Penguin, Harmondsworth.

Neve, R.A. (1990) *Hops*, Chapman & Hall, London.

Piggott, J.R. (ed.) (1983) *Flavour of Distilled Beverages*, Ellis Horwood, Chichester.

Piggott, J.R., Sharp, R. and Duncan, R.E.B. (eds) (1989) *The Science and Technology of Whiskies*, Longman, Harlow.

Piggott, J.R. and Paterson, A. (eds) (1989) *Distilled Beverage Flavour*, Ellis Horwood, Chichester.

Priest, F.G. and Campbell, I. (eds) (1983) *Current Developments in Malting, Brewing and Distilling*, Institute of Brewing, London.

Priest, F.G. and Campbell, I. (1996) *Brewing Microbiology*, 2nd ed., Chapman & Hall, London.

Storm, D. (1995) *Winery Utilities*, Chapman & Hall, London.

Vine, R.P. (1981) *Commercial Winemaking*, Chapman & Hall, London.

Zoecklein, B.W., Fugelsang, K.C., Gump, B.H. and Nury, F.S. (1995) *Wine Analysis and Production*, Chapman & Hall, London.

8 Fats and Fatty Foods

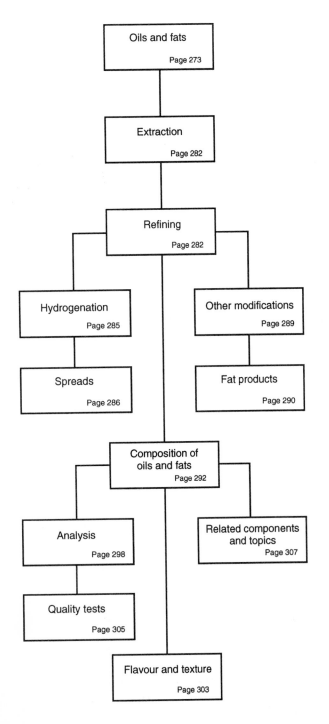

Oils and fats	Page 273
Extraction	Page 282
Refining	Page 282
Hydrogenation	Page 285
Other modifications	Page 289
Spreads	Page 286
Fat products	Page 290
Composition of oils and fats	Page 292
Analysis	Page 298
Related components and topics	Page 307
Quality tests	Page 305
Flavour and texture	Page 303

INTRODUCTION

Oils and fats occur naturally in a wide range of sources. There are several hundreds of vegetable oil-bearing seeds which grow in different parts of the world, but only about 22 vegetable oils are commercially developed on a large scale today and 12 of them constitute more than 95% of the world vegetable oil production. Animal fats are rendered from the slaughtered carcasses of animals considered to be fit for human consumption, as well as from marine sources such as sardine, menhaden and herring.

World production of fats and oils has continued to increase in recent years (Table 8.1). Soya bean oil remains the leading oil but sunflower seed, palm and rapeseed oils have increased markedly. Increased acreages planted with soya beans, sunflowers and oilseed rape have contributed to the increased production of oils from these plants, but the increase in production of palm oil is largely due to improved yield resulting from modern plant breeding techniques. The development and propagation of higher yielding oil palms has played a major role in increased oil production by Malaysia. Rapeseed oil, low in antinutritional components has gained acceptance as an edible oil and there have been big increases in the acreage of oilseed rape in Europe and Canada. Genetic modification of other plants has led to modifications in oil composition including the development of high oleic sunflower oil and high oleic linseed oil.

Traditionally, plant breeding has been used to affect yields, oil content, adaptation to varying climate and resistance to pests and/or pesticides. In recent years, attempts have been made to produce high lauric acid rapeseed oil via a biotechnological route. Colgene scientists at Davis, California, have isolated an enzyme, lauryl ACP (acyl chain promoter) thioesterase, important in the formation of lauric acid in rapeseed crop. Such development and commercialization of the

Table 8.1 World production of major vegetable and animal oils

	Million metric tonnes (of oil or fat)		
	1992/93	1993/94	1994/95
Soya bean oil	17.1	18.1	19.5
Palm oil	12.9	13.8	15.2
Rapeseed oil	9.0	9.6	10.1
Sunflower seed oil	7.9	7.7	8.4
Groundnut oil	4.0	4.1	4.2
Cottonseed oil	3.9	3.6	3.8
Coconut	3.0	2.8	3.4
Olive oil	1.8	1.9	1.9
Palm kernel oil	1.6	1.8	1.9
Corn oil	1.6	1.6	1.7
Sesame seed oil	0.6	0.6	0.7
Tallow and greases	6.9	7.1	7.6
Butter	5.8	5.7	5.9
Lard	5.4	5.6	5.7
Fish oil	1.1	1.2	1.2

Source: ITP (1996).

first genetically modified plant oil have pioneered a new route to plant oils with specific fatty acid compositions.

Dietary advice from many quarters continues to recommend that the total fat content of Western diets should be reduced to 30% of total energy intake. This has led to large increases in the production and sales of low fat foods including low fat spreads and low fat cheese. Moreover, the health benefits of incorporating long chain polyunsaturated fatty acids into the diet have been realized for many years. Since 1990 eggs containing docosahexaenoic fatty acid (DHA) have been sold in Japan. Recently DHA containing eggs are appearing in large supermarkets of European countries. Also, new fat spreads based on refined fish oil are being marketed.

OILS AND FATS

The production of edible oils and fats has increased progressively in recent times. This has been mainly effected by rapid increases in the production of soya bean oil, palm oil and sunflower seed oil, whilst in the EU and Canada the production of rapeseed oil has increased considerably since about the mid-1980s. Table 8.1 shows the world production of edible oils since 1992, while composition and properties of many oils and fats are given in Tables 8.2 to 8.4 and Table 8.7, respectively. Properties and features of the most important oils are reviewed below.

Babassu oil

Babassu palms (*Orbignya matiana* or *O. oleiferae*) grow only in Brazil, mainly in the remoter parts, as a wild crop. The output of oil varies considerably from year to year. Potentially, they are a tremendous source of edible babassu oil, but expansion of production is limited by the technical difficulty of cracking the very hard thick shell of the nut. The oil features to only a very minor extent in international trade. Its composition and properties are similar to those of coconut and palm kernel oils.

Blackcurrant oil

This oil, extracted from the seeds of the blackcurrant plant, has generated interest recently because it is a good source of γ-linolenic acid.

Borage oil

The oil extracted from borage is higher in γ-linolenic acid than evening primrose oil and consequently is of interest as a nutritional supplement.

Cocoa butter

Cocoa butter is obtained by pressing ground, roasted, decorticated cocoa beans and is derived almost entirely from the nib or kernel of the cocoa bean. Cocoa butter is used extensively in making chocolate and other confectionery products, and to a limited extent in the pharmaceutical industry. It is particularly suitable for these purposes because it has a low, but sharp, melting point; it is brittle and fractures readily, and whilst it is not greasy to the touch it melts completely in the mouth. These properties are a reflection of its triglyceride composition. The principal triglycerides present are monounsaturated, disaturated glycerides, predominantly 2-oleopalmitostearin. In all of the triglycerides present in cocoa butter, the 2- or central hydroxyl group on the glycerol backbone of the triglyceride molecule is esterified with an unsaturated acid.

The price of fats has varied considerably over the years, but cocoa butter is normally one of the most expensive. The high and variable cost has stimulated an extensive search for cocoa butter replacement fats.

Table 8.2 Ranges of fatty acid composition of commercially important vegetable oils (% m/m)

Fatty acid	Oils analysed										
	Palm kernel	Coconut	Cottonseed[a]	Soya bean	Maize	Groundnut	Palm	Sunflower seed	High erucic rapeseed	Low erucic rapeseed	Safflower-seed
C6	ND–0.8	0.4–0.6	–	–	–	–	–	–	–	–	–
C8	2.5–4.7	6.9–9.4	–	–	–	–	–	–	–	–	–
C10	2.8–4.5	6.2–7.8	–	–	–	–	–	–	–	–	–
C12	43.6–51.4	45.9–50.3	tr–0.1	tr	tr–0.3	–	ND–0.2	–	–	–	–
C14	15.3–17.2	16.8–19.0	0.7–1.0	tr–0.2	tr–0.3	tr–0.1	0.8–1.3	tr–0.1	0.1	tr–0.2	tr–0.2
C16	7.2–10.0	7.7–9.7	21.4–26.4	9.9–12.2	10.7–13.6	9.2–13.9	43.1–46.3	5.6–7.4	2.8–5.1	3.4–6.0	5.3–8.0
C16:1	–	–	0.3–1.1	tr–0.2	tr–0.4	tr–0.1	tr–0.3	tr–0.1	0.2–0.5	0.2–0.6	tr–0.2
C18	1.9–3.0	2.3–3.2	2.3–3.2	3.6–5.4	1.8–3.3	2.2–4.4	4.0–5.5	3.0–6.3	0.7–1.3	1.1–2.5	2.1–2.9
C18:1	11.9–18.5	5.4–7.4	14.7–21.4	17.7–25.5	24.6–42.2	36.6–65.3	36.7–40.8	14.0–34.0	9.8–49.8	52.0–65.7	8.4–21.3
C18:2	1.4–3.3	1.3–2.1	46.7–57.7	50.5–56.8	39.4–60.4	15.6–40.7	9.4–11.9	55.5–73.9	13.0–22.9	16.9–24.8	67.7–83.2
C18:3	tr–0.7	tr–0.2	0.1–0.2	5.5–9.5	0.7–1.3	tr–0.1	0.1–0.4	tr–0.1	7.0–10.3	6.5–14.1	tr–0.1
C20	0.1–0.3	tr–0.2	0.2–0.4	0.2–0.6	0.3–0.6	1.1–1.7	0.1–0.4	0.2–0.3	0.2–1.0	0.2–0.8	0.2–0.4
C20:1	ND–0.5	tr–0.2	0.1	0.2–0.3	0.2–0.4	0.8–1.7	ND–0.3	0.1–0.2	2.6–9.4	1.2–3.4	0.1–0.2
C22	–	–	0.2	0.3–0.7	0.1–0.5	2.1–4.4	–	0.6–1.0	0.3–0.9	0.1–0.5	0.2–0.8
C22:1	–	–	–	ND–0.3	–	tr–0.3	–	ND–0.2	5.0–51.6	tr–5.0	tr–1.0
C24	–	–	0.1	ND–0.4	0.1–0.4	1.2–2.2	–	0.2–0.3	0.1–0.3	0.1–0.2	0.1
C24:1	–	–	–	–	–	–	–	–	0.3–1.1	0.1–0.4	0.1–0.2

ND = not detected

tr = trace

Where single values are shown, all samples had same concentration within experimental error.

[a] Cottonseed oil also contains small amounts of cyclopropenoid fatty acids. These normally decompose during conventional GLC of the methyl esters.

Table 8.3 Fatty acid compositions of miscellaneous animal fats (ranges and typical values, % m/m)

Fatty acid	Beef fats				Mutton (lamb) fats			Pig fats				Chicken fat
	Brisket	Cod	Flank	Suet	Breast	Body	Shoulder	Back	Belly	Flare	Head	
C12	0.2	0.2	0.2	0.2	0.5–1	0.5–1	0.5–1	0.2	0.2	0.2	0.2	0.2
C14	2–4		3–5	3–4	3.5–5	2.5–4.5	3–4	1.3	1.6	1.7	1.5	1.2
C14:1	2–2.5	1–1.5	1–2	0.5–1	0.5–1	tr–0.5	tr–1	tr	tr	tr	tr	0.2
C16	21–24	25.5–29	25–27	26–28	20–21	20–21	19–21	23.8	25–28.5	28–31	23–26	23.2
C16:1	8–9	3.5–4	4–5	2–3	1–2	1–1.5	1–1.5	3.2	2.5–3.5	1.5–2.5	3–3.5	6.5
C18	7–9	17–19.5	14–18	23–27	16–20	22–26	23–27	11.7	14–17	18–24	11–15	6.4
C18:1	45–48	35–38	38–43	30–35	36–38	33–37	33–37	45.3	37–42	30–38	41–45	41.6
C18:2	1–2	0.5–1.5	1–2	1–1.5	tr–2.5	2–2.5	2–2.5	9.1	tr–13	7–12	8–13	18.9
C18:3	0.5	0.5	0.5	0–1	tr–2	tr–1.5	tr–1.5	0.8	0.5–1.5	0.5–1	0.5–1.5	1.3
C20	tr–0.1	0.1	tr–0.5	tr–0.5	tr–1.5	tr–2	tr–2	0.2	0.2	0.3	0.2	
C20:1	1–2	0.5–1	0.5–1	tr–1	2–3	1.5–2.5	1.5–2	1.0	1.0	1	0.5–1.5	
C22	tr–0.1	tr–0.1	tr	tr–0.1	tr–0.1	tr–0.1	tr	0.3	0.2	0.2	0.3	
C22:1	tr							0.1	0.1	0.1	0.1	
C24	tr				tr–0.1	tr–0.1	tr	0.4	0.3	0.3	0.5	
C24:1								0.3	0.3	0.3	0.5	
Others	5–8	6–8	4–6	4–8	1–2	1–2	1–2	1–3	1–4	1–2	1–2	0.4

tr = trace

Table 8.4 Approximate content of principal fatty acids typical of fish oils (%)

	Capelin	Herring	Sprat	Norway pout	Mackerel	Sandeel	Menhaden	Sardine/ pilchard	Horse mackerel	Anchovy
C14:0	7	7	ND	6	8	7	9	8	6	9
C16:0	10	16	16	13	14	15	20	17	24	19
C16:1	10	6	7	5	7	8	12	9	7	9
C18:1	14	13	16	14	13	9	11	12	13	13[a]
C20:1	17	13	10	11	12	15	1	3	2	5[b]
C22:1	14	20	14	12	15	16	0.2	3	2	2
C20:5	8	5	6	8	7	9	14	17	11	17
C22:6	6	6	9	13	8	9	8	9	16	9
Iodine value	100/140	110/140	125/150	140	135/160	140/175	150/175	140/200	180/195	170/200

[a] contains element of C16:4 content
[b] C20:1 and C18:4 combined
ND = no data

Coconut oil

Coconut oil is extracted from copra, which is dried pieces of coconut kernel. Coconuts are fruit of the coconut palm (*Cocos nucifera*), which is cultivated in tropical coastal areas. Dried copra contains 60–65% oil and in this respect it is the oiliest of the commercial oilseed crops. World production is about 3.4 million metric tonnes (Table 8.1), about half of which is traded internationally. The main producing area is the Philippines; other producing countries are Indonesia, India, West Malaysia, Sri Lanka, Papua New Guinea, Thailand and Solomon Islands.

Coconut oil, formerly used in margarine and cooking fats, is now less common in view of its high price. However, it is still used in the manufacture of cream fillings for biscuits and in other confectionery applications.

Coconut oil is a lauric oil similar in composition to palm kernel and babassu oils. The high content of lauric acid imparts a plastic consistency to the fat at room temperature but this melts rapidly at body temperature, explaining the popularity of the fat in confectionery products.

Corn oil (maize-germ oil)

The main producer of corn oil is the USA, the annual production being almost twice that of South Africa its nearest competitor. In Europe, France is a significant producer with about 12 000 tons per annum. World production is rising slowly. Corn oil is used mainly as a salad or frying oil, but health margarines based on corn oil have been marketed. The low cloud point makes it a useful edible oil.

The kernel of the corn plant, *Zea mays*, contains only 3–7% oil, the oil being concentrated in the germ or embryo. The germ is separated from the kernel in the milling process and is crushed to obtain the oil. The composition and properties of the oil are given in Tables 8.2 and 8.7. Corn oil contains a higher level of sterols (up to 2%) than most other liquid vegetable oils.

Cottonseed oil

The production of cottonseed oil varies with the cotton fibre industry. In 1994/95, 3.8 million tonnes of cottonseed oil were produced (Table 8.1), the bulk being used in the main producing countries of the USA, Brazil, mainland China, the former USSR, India and Pakistan. Egypt and countries in the Middle East formerly featured as main cottonseed producers and exporters but are not so prominent now in this industry.

Cottonseed has an oil content of 15–25%, but the extraction rate is often only about 16%. The kernel contains 30–38% of oil, but industrially processed seeds invariably have adherent linters. Cottonseed oil has a strong flavour and has a dark red colour due to the presence of gums and gossypol. Cottonseed oil can be refined to give a pale yellow oil, however, but it requires thorough refining. Oil produced in the USA is generally considered to be of the best quality. Cottonseed oil is used in production of margarine, shortenings and as a cooking fat. Winterized cottonseed oil is used as a salad oil. Cottonseed oil may be detected by the HALPHEN TEST. The fatty acid composition is shown in Table 8.2, other properties being given in Table 8.7.

Evening primrose oil

This expensive oil is a rich source of γ-linolenic acid (GLA) (see ESSENTIAL FATTY ACIDS), an EFA said by some to have potent properties.

Fish oils

Fish oils are obtained by the extraction of oil from the whole fish. They are highly unsaturated, containing a proportion of penta- and hexaenoic acids, as shown in Table 8.4. This renders them particularly susceptible to oxidation. They must, therefore, be carefully hydrogenated and refined before they can be used in foods.

Since the 1960s there has been a great upsurge in the amount of fish oil available throughout the world. In the early 1960s an exceptional growth in production of fish oil from anchoveta took place in Peru, reaching 151 000 tons in 1962. However, production of Peruvian fish oil has since diminished due to a combination of overfishing and a change in the Humboldt ocean currents off the coast of Peru. In the mid-1960s there were also substantial increases in the amounts of Norwegian and Icelandic herring and American menhaden oils. Consequently, the amount of fish oil used in the western world for margarines and shortenings has increased. In the UK, fish oils have been a major edible oil, accounting for 20–30% of UK consumption, ranking second to rapeseed oil. In recent years, due to controversy surrounding *trans*-fatty acids, the use of hydrogenated fish oils in margarine and other food products has declined sharply.

The demand for fish oils has stimulated production elsewhere, and sardine and oils are now produced from mackerel in Japan, from menhaden in the USA, capelin in Norway and Iceland, sand eel and Norway pout in Denmark, and pilchard and horse mackerel in Chile.

Grapeseed oil

Grapeseed oil is derived from the seeds of grapes (*Vitis vinifera* Linnaeus) as a by-product of the wine industry, mainly in Argentina and Italy. The oil content varies from about 6% for black grapes to 20% for white grapes. The oil is high in polyunsaturated fatty acids, mainly linoleic acid. It is used as a substitute for linseed oil in the manufacture of paints but may also be used for edible purposes after refining.

Groundnut oil (arachis or peanut oil)

Although production of groundnut oil has varied over the years (Table 8.1) the amount traded internationally has stayed fairly static. In 1938, 800 000 tons of groundnut oil were exported by the producing countries. This fell to 700 000 tons in 1954, rose to 900 000 tons in 1977/78, but fell again to 650 000 tons in 1980/81. The fact that much of the oil is consumed in the country of origin is illustrated by the fact that about 3 195 000 tons of oil are produced throughout the world at the present time. India is currently the largest producer of groundnuts, having produced 5 million tons in 1980/81 and 6 million tons in 1981/82. However, none of this is exported. Mainland China both produces and exports groundnuts for oil production, while the USA produces groundnuts mainly for the edible groundnut trade.

The groundnut consists of a shell, a thin red skin and a kernel. Shelling is usually carried out at the point of harvesting. The kernels contain 40–50% oil, which is obtained by a combination of pressing and solvent extraction. The residual meal is a valuable animal feedingstuff because of its high protein content. The oil is pale yellow and has the characteristic flavour of groundnuts when fresh and produced from good quality kernels. This aromatic oil is prized in some parts of the world, but in general the oil is refined and deodorized. The oil is widely used on the continent as a frying and salad oil.

Groundnuts infected with *Aspergillus flavus* mould are likely to contain aflatoxin, and while this is removed from the oil during normal refining, it remains in the oilseed meal or cake.

This has resulted in dramatic fall in the use of groundnut meal in animal feedingstuffs and must in time rebound on the price and availability of groundnut oil.

Hard butters

Hard butters are vegetable fats which melt over a narrow temperature range. They are often used in the manufacture of cocoa butter replacement fats. The most common are COCOA BUTTER and illipe, sal and shea butters.

(i) *Illipe butter (Borneo tallow)*. This fat is derived from the seeds of jungle trees such as *Shorea stenoptera* which grow wild in Sarawak and other parts of northern Borneo. The composition of the fat resembles that of cocoa butter and its main use is as a cocoa butter equivalent *per se*, or as a component of cocoa butter equivalent (CBE) blends. The trees crop erratically and the nuts are harvested by the local population on a sporadic basis. This has in the past led to rumours that a crop is available once every seven years.

An unfortunate confusion has arisen in botanical circles with nuts from the *Bassia* species which grows in India and is sometimes referred to as 'true illipe'. Fat from the seeds of *Bassia* species is of little value in chocolate or CBE production and is seldom harvested or traded.

(ii) *Sal fat*. Salt fat is one of the newest additions to the range of hard butters available for CBE formulation. Sal trees grow in remote areas of India, bearing nuts about the size of European acorns. The nuts fall to the ground shortly before the onset of the seasonal monsoon, leading to considerable difficulties in harvesting and collection. Nevertheless, the Indian authorities have made considerable advances in the collection of sal seeds. The nut contains only about 14% oil, which must therefore be removed by solvent extraction. The oil often has a very intense green colour, but this can be alleviated by a fractionation process, most of the colour remaining with the softer or olein fraction. The harder stearin fraction may be slightly modified by a mild hydrogenation, after which it is bleached and deodorized for use in chocolate and CBE manufacture. Salt fat is now exported to many parts of the world and is becoming an important source of foreign currency for the Indian government.

(iii) *Shea butter.* Shea nuts are obtained from a tree (*Butyrospermum parkii*) which grows wild in West Africa, Upper Volta and Uganda. The nut contains 40–55% of a hard edible fat, which has up to 10% of unsaponifiable matter. Shea butter therefore has one of the highest unsaponifiable matter contents of any natural vegetable oil. The unsaponifiable matter comprises a mixture of terpenoid hydrocarbons. Shea butter can be processed to remove the unsaponifiable matter and solvent fractionated to give hard and soft fats. The hard stearin fraction can be used as a constituent in chocolate or as a component in a CBE formulation. Trade in shea nuts and shea butter varies sporadically with varying climatic and harvesting conditions in West Africa, where prices offered by Government agencies sometimes lead to smuggling.

Hazelnut oil

Hazelnut oil is a liquid oil, often sold in health food shops as a salad oil. It is rich in monounsaturated fatty acids and has an attractive nutty flavour. The kernels contain 50–68% oil, usually extracted by cold pressing.

Jojoba oil

Jojoba oil is derived from the seeds of the jojoba shrub (*Simmondsia chinensis*) and is the subject of growing interest as a replacement for sperm whale oil. As the shrub grows in arid regions such as Arizona, Mexico, the Negev desert, northern Australia, the Sahel region of Nigeria, etc., it is seen as a means of bringing agricultural industry to these areas. Chemically the oil is a liquid wax, comprising esters of the long chain eicosenoic and decosenoic (erucic) acids with eicosanol and docosanol alcohols. Its main applications are in the industrial sector and in cosmetics, but several food applications, including use as a coating agent for dried fruits, are foreseen.

Lard

The most useful form of lard is obtained by wet rendering of pig fat, and this is known as prime steam lard. Refined lard of commerce is usually prime steam lard which has been dried, clarified and solidified. Lard is also obtained by a dry rendering process and, when made this way, tends to be darker and to be more strongly flavoured. A further type is continuously rendered lard, which is continuously comminuted, heated and then centrifuged.

The composition of lard varies considerably (Table 8.3) with the animal from which it is obtained, the part of the animal and the diet on which the animal has been fed. The fat from an individual animal becomes increasingly unsaturated as one goes from the internal organs outwards toward the back. Lard from the USA tends to have a higher iodine value than European lard. Lard has a relatively low resistance to oxidation, due to a deficiency of natural antioxidants. Because of this, lard is often treated with synthetic antioxidants before sale. Lard is also important as a shortening and has been used for many years for this purpose. However, its creaming properties are not good and it possesses a decided flavour, which may be a disadvantage. The properties of lard can, however, be modified by interesterification and the consistency and performance in cake making are thereby improved. World production of lard is shown in Table 8.1.

Linseed oil

Linseed oil is extracted from the seed of the flax plant (*Linum usitatissimum* Linnaeus) which contains 35–45% oil. The main producers include Argentina, India, Canada and states of the former Soviet Union. Linseed oil is mainly used as a drying oil for the oleochemicals industry particularly for paints, varnish and linoleum. The high α-linolenic acid content (57%) makes it suitable for these applications. However, the oil is also used for edible purposes. In India about 35–40% of the oil is consumed as cooking and frying oil. Hydrogenation of the oil or blending with other oils are often needed to give the oil sufficient stability for use as a cooking oil. However, recent nutritional interest in *n*-3 fatty acids which include α-linolenic acid has led to cold pressed linseed oil and ground flaxseed products being sold in health food shops. The world production of the oil in 1994–95 was 0.7 MMT (mega million tonnes).

Olive oil

Olive oil is derived from the fruit of *Olea europaea*, the olive tree, which grows extensively in countries with a Mediterranean climate. The yield of olive oil varies considerably from year to year according to the conditions, leading to price variations. Because of this, most olive oil producing countries fix a guaranteed minimum price for the oil. They may reserve stocks of the oil from a good year to tide them over a bad year.

Olive oil is considered to be the best of the liquid edible oils for salad and culinary purposes and the finest quality virgin olive oil is used without refining. It has a fine flavour and a very good shelf life. Refined olive oils are also obtained from crude olive oils of inferior organoleptic properties, or oils obtained by solvent extraction. The olive also contains a kernel, the oil from which has a similar composition. Refined olive oil may at times contain small amounts of olive kernel oil.

The high esteem in which olive oil is held in some Mediterranean countries can lead to political and other problems. The SPANISH OLIVE OIL SYNDROME was caused by unscrupulous dealers selling contaminated rapeseed oil as a substitute for olive oil in blends with it. In the EU countries, olive oil has a guaranteed price. World production of olive oil is shown in Table 8.1.

Palm kernel oil

Palm kernel oil (PKO) is extracted from the kernel of the oil palm (see PALM OIL, below). It is a lauric oil, very similar in appearance and constitution to coconut oil, containing 83% saturated fatty acids, mainly lauric (Table 8.2). PKO was formerly extensively used in margarine manufacture but it is seldom used now as technological advances have enabled the use of other cheaper oils. PKO is also used as a confectioners' hard fat in the production of chocolate-type coatings for baked products. The oil may be fractionated to produce a hard fat stearin which is very useful as a cocoa butter substitute. However, palm kernel stearins are not compatible with cocoa butter and cannot therefore be used in the formation of CBEs. PKO is also widely used as a component in the filling creams in cream biscuits and in coffee whiteners for vending machines and domestic use.

West Malaysia is the main exporter of PKO and Nigeria is also a major producer. World production of PKO is shown in Table 8.1.

Palm oil

The oil palm (*Elaeis guineensis*) is native to West Africa but is also cultivated in southeast Asia, especially Malaysia. The fruit consists of mesocarp or outer pulp, endocarp or shell and the palm kernel itself. The mesocarp contains 25–55% of palm oil, whilst the kernel contains 45–55% PKO. Although both oils are derived from the same fruit, they differ markedly in composition (Table 8.2) and the palm kernels are therefore carefully separated from the pulp during processing. Palm oil must be extracted at the point of harvesting, as enzymes present in the fruit would otherwise cause hydrolysis and deterioration of the oil. The extraction units are therefore located in the centre of plantations. As the trees bear fruit all year round, and as transportation to the processing plant is so short, the palm oil plantation and processing plant is the most productive source of vegetable oil in the world. As the fruit is a tree crop, the harvest cannot be altered on a short term basis by planting schedules.

World exports of palm oil have risen considerably in recent times. In 1938 400 000 tonnes were traded internationally in export markets. This rose to 1 million tonnes in 1971/72, 2 million tonnes in 1976/77, and the oil production was 15.2 million tonnes in 1994/95 (Table 8.1). These dramatic production increases are expected to continue in the future, as selected palm tree strains are now being cultivated by cloning. Pollination of the flowers in Malaysian plantations is now being assisted by the *Elaeidobius* species of weevil, which will probably have far reaching effect on the production of palm and palm kernel oils in Malaysia.

Palm oil is a semi-solid fat and as such it needs no hydrogenation prior to use in food applications. It is often used as a component in margarine, biscuit for blends, etc. Palm oil may be fractionated into several components. In Malaysia, it is often fractionated by dry or detergent fractionation processes into a hard stearin fraction and an almost liquid olein. This olein is much used in warmer countries as a frying oil but has less attraction in Europe for this purpose as it tends to throw down a deposit of intermediate melting triglycerides in temperate climates. Palm oil may also be solvent fractionated to give a more liquid olein, a middle melting point fraction and a very hard stearin or upper fraction. The middle melting fraction contains large quantities of the triglyceride POP (palmitic–oleic–palmitic), a major component of cocoa butter. Palm mid-fraction is for this reason an important CBE component. As it is usually the cheapest component in the blend and may be the only component not subject to political and/or climatic pressure on its availability (see HARD BUTTERS, above), much effort has been spent on obtaining palm mid-fractions of the very best quality. The composition and properties of palm oil are given in Tables 8.2 and 8.7.

Premier jus

Premier jus (first juice) is an old established name applied to carefully rendered fresh beef suet (see TALLOW, below).

Rapeseed oil

Rapeseed is grown in temperate and warm temperate zones, both summer and winter varieties being sown. Several varieties of *Brassica* give rise to rapeseed, including *Brassica napus* and *B. campestris*. The oil is known as colza oil in some parts of Europe.

In the early 1950s rapeseed oil contained a high level of erucic acid and this was implicated in certain dietary and health problems. Canadian researchers were able to breed a new variety of rapeseed in which the oil was free of erucic acid. This oil is now known as low erucic acid rapeseed (LEAR). As it is a temperate climate crop, its cultivation has been encouraged in Canada and in the EU, where it now forms a major agricultural item. There are now many cultivated strains of rapeseed available in Europe, each having particular features related to agricultural yield, value of the meal as a feedingstuff or oil quality. The strain most often grown in the UK is Jet Neuf, the seed of which contains 40–45% oil. World production in 1994/5 reached 10.1 million tonnes of oil, most of which was used in the producing countries, notably India and China, the main exporting countries being Canada and France.

Rapeseed oil is a liquid oil with a dark amber or green colour. This can usually be removed on bleaching or deodorization, but difficulties can arise if the seed is harvested in an unripe condition or allowed to sprout during storage. Unripe seeds contain more chlorophyll, which leads to dark coloured green oils and this can give rise to problems during refining. In recent years, extraction of oil from decorticated seed has been attempted.

In view of its high level of unsaturation, the oil is frequently hydrogenated prior to use in food products. In the hydrogenated form, it is used in the production of margarine, in fats used for the production of milk powders and in dough fats. Oil which is lightly hydrogenated and then fractionated to remove harder components may also be used as an industrial frying oil.

Oil high in erucic acid is still grown in China, in India, in Eastern Europe and in parts of Scandinavia. The use of this oil in the UK falls under the 'Erucic Acid in Food' Regulations 1977, as amended, which limit the amount of erucic acid in

human foods to 5%. The high erucic acid varieties of rapeseed oil can, however, be used if suitably blended with other oils of low erucic content. High erucic acid rapeseed oils are also used for specific industrial purposes, particularly in the production of lubricant additives for jet engines. The composition and properties of rapeseed oil are listed in Tables 8.2 and 8.7.

Rice bran oil

Rice bran is a product of rice milling and although little oil is produced from this source at present, it has enormous potential. The main difficulty in the production of rice bran oil is that enzymes present in the bran degrade the oil if it is not quickly extracted.

As rice bran is produced predominantly in the Third World, much effort is being spent in overcoming problems of rice bran collection and oil extraction. In recent years, rice bran oil has gained worldwide attraction due to the presence of high valued minor components namely oryzanol and tocotrienols.

The mixture of esters of ferulic acid with normal sterols predominantly campesterol and β-sitosterol and triterpene alcohols, mainly cycloartenol and 24-methylene-cycloartanol has been named oryzanol. The amount of oryzanol in crude oil is about 2%. Oryzanol has been reported to be very effective in lowering total and LDL (low density lipoprotein) serum cholesterol level.

Safflowerseed oil

Safflower is an annual plant, a member of the Compositae. The major producer is the USA, but safflower is also grown in India, North Africa, Mexico and Canada. The oil content of the seed is 25–45%. The oil is semi drying, but it is widely used for edible purposes having recently gained popularity for dietary use in view of its high EFA content. It has the highest linoleic acid content of any commercially available oil. Its composition is listed in Table 8.2. In the late 1960s a high oleic acid variety of safflowerseed oil was introduced. They may be used for dietary purposes, such as baby milk.

Sesame seed oil

The sesame crop is grown mainly in China, India, Africa, Japan, Indonesia, Thailand, Egypt and

Mexico. The seeds are white and dark coloured, commercially known as white and black. The white varieties yield oil superior to that from the black. In some countries, e.g. India, sesame seed oil at a level of 5% is added (as a marker) to various fat products such as vegetable GHEE/ vanaspati. An important characteristic of sesame seeds is that they contain, in addition to tocopherols, various other antioxidant agents such as antioxidant precursors, predominantly sesamin and sesamolin, trace amounts of sesamol, an active antioxidant, and other minor lignan phenolic analogues, including sesamolinol and sesaminol.

Soya bean oil

Soya bean oil is the most important single edible oil, world production having reached 19.5 million tonnes in 1994/5 (Table 8.1). The soya bean is the seed of a leguminous plant. Although it is native of Asia, it is extensively cultivated in other areas, notably in North and South America, where the climate is suitably warm and damp.

Soya beans provide a useful source of protein as well as of edible oil, the residual cake from soya bean extraction being a valuable feedingstuff in considerable demand. It is only comparatively recently that soya bean has been primarily produced as an oilseed crop. In most eastern countries most of the crop is still eaten directly by farm animals.

Crude soya bean oil is usually obtained by solvent extraction of the beans, which contain from 17–20% oil. The crude oil contains a high proportion of phospholipids, which are removed during processing. This is a valuable source of LECITHIN, an emulsifying agent widely used in food products. Crude soya bean oil has an objectionable 'beany' flavour, which is removed by the refining process. However, similar flavours can return during storage of refined or hydrogenated and refined soya bean oil. These new off-flavours are called reversion or hydrogenation flavours.

Soya bean oil is used in the manufacture of margarine, compound cooking fat and as a salad oil. For the latter application it is frequently given a mild hydrogenation, after which the harder components are removed by fractionation.

The major producing areas are the USA, Brazil, Argentina and China. The composition and properties of soya bean oil are listed in Tables 8.2 and 8.7.

Sunflowerseed oil

Sunflowerseed oil is an important oil seed crop, production of which reached a record 8.4 million tonnes in 1994/5. The major production areas are the former USSR, the USA, Argentina and China. In Europe, France is a significant producer.

The popularity of sunflowerseed oil in recent years is attributable to its good flavour stability, coupled with its high linoleic acid content (Table 8.2). This has made it useful for the manufacture of margarines, spreads and other foods high in EFA.

Tallow

The better quality edible tallow is derived from beef. The composition of tallow varies with the animal's diet and, as with the pig, the more unsaturated fat is found nearest the skin (Table 8.3). The degree of unsaturation therefore depends to some extent on the technique used in cutting and rendering. Tallow is likely to go rancid quickly, as with lard, due to the absence of natural antioxidants. Premier jus, or oleostock, is a high grade of tallow made by wet rendering of fresh beef fat at low temperatures. This material is light yellow and of good flavour. It can be fractionated to give a hard material, oleo stearin, which contains all the fully saturated glycerides. It may be used as a hard fat in pastry margarines. The soft fraction known as oleo oil is used in newer varieties of margarine, said to have texture and flavour properties closely related to those of butter. Beef oleo oil is also used in the preparation of baby foods. Tallow intended for non-edible use is graded according to the scheme in British Standard 3919. World production of tallow in 1994/5 was 7.6 million tonnes (Table 8.1).

Whale oil

Before 1960 whale oil was a common component of margarine and other fatty foods. However, whale oil declined in importance as a result of its decreased availability. The whale oil industry passed completely out of British hands and became concentrated mainly between Japan and the USSR. Continued hunting of whales led to a fall in their numbers and as a result restrictions were imposed on the whaling industry. This failed to halt the decline and many species of whales are now classed as endangered.

The sperm whale was formerly one of the most

important. Its oil has an unusual composition in that it comprises a high proportion of fatty acids esterified with long chain fatty alcohols. These compounds are commonly known as wax esters. The high proportion of wax esters gives sperm whale oil attractive properties in some fields, especially mechanical lubrication. Efforts have been made to find vegetable oil alternatives for sperm whale oil in these specialist applications. One of the most attractive of these vegetable oil alternatives is jojoba oil (see above).

EXTRACTION

The process of oil extraction involves obtaining crude oil from the oil source. When the oil source is animal the extraction process is termed rendering which involves melting the fat out of fatty tissue by heat. From vegetable sources, the oil is extracted by pressing, solvent or combination of both in many cases.

RENDERING

Nearly all animal fats, tallow or beef fat and lard, are obtained by rendering. There are several methods of rendering animal tissues.

In batch steam rendering, the fat cell walls are destroyed by steam under pressure. The process involves pressure cooking fatty issues for 1–3 h with steam at 120–145°C corresponding to pressure of 300 to 525 kPa (45–75 psi). This dissolves the tissue, releasing the fat from proteinaceous material. The liquid fat floats on the top water, which can be separated thereby producing edible fat products where colour, flavour and keeping quality are of great importance.

In batch dry rendering, fatty tissues are heated under pressure or under vacuum, at atmospheric pressure or under pressure in a horizontal steam-jacketed tank. In order to improve heat transfer, the rendering tissues are mechanically agitated during heating for 4 h at 110–120°C. The melted fat is separated from the other material for further processing. Generally, the dry method is preferred for inedible fat products where the colour and flavour are secondary and the prime objective is to produce larger quantities of high quality residue. Final recovery is completed by pressing either in a hydraulic or continuous screw press. Screw pressing, after boiling, is used to extract the oil from fish such as sardine, herring and menhaden.

Several continuous systems are also available to render animal fats. The fatty tissues are shredded in mechanical disintegrators. The shredded tissues are then heated to liberate free fat. Centrifugal separation clarifies the oil phase by precipitating the sludge and water allowing a purified oil layer to be removed continuously.

PRESSING AND SOLVENT EXTRACTION

The oil from a seed is either squeezed out by sheer physical pressure using either a hydraulic press or screw press or the combination of the two, or it can be dissolved out by using solvent hexane. In many cases the two methods are combined depending upon the nature of the seed (e.g. rapeseed and sunflowerseed) and the cost considerations. As the oil within the seed or nut is trapped in tiny cell walls, there is a need to soften up the walls of each cell to obtain the maximum amount of an unaltered oil. The seeds are effectively cleaned prior to processing in the extraction equipment. The cleaning not only removes sand, stones and dirt, but also removes other foreign matter such as leaves, stems and pods. Depending upon the seed type, the hull is removed usually through cracking and air classification or aspiration. The hull-free material (meats) is then conditioned or dried to the moisture level required (9–10%) for flaking. In the case of soya beans, the recovered meats are flaked to a thickness ranging from 0.25–0.45 mm in order to cause sufficient cell breakage to release the oil for ease of extraction with solvent hexane in the ratio of 1:1 to 1:3. Extraction conditions vary: times of 15–60 min and temperatures of 40–65°C are used. The recovered solvent–oil mixture is called 'MISCELLA' and the extracted flakes are termed as 'spent flake' or 'cake'. The crude oil is freed of the extraction solvent which is recovered and re-used.

Solvent extracted crude vegetable oils contain about 95% neutral lipids (triglycerides) and a broad spectrum of non-triglycerides. These co-extracted compounds may include phosphatides (commercially called 'gums'), sugars, resins, proteinaceous material, flavour components, trace metals, pigments, partial glycerides, free fatty acids, sterols, hydrocarbons, tocopherols (vitamin E) and many breakdown products.

REFINING

Crude oils and fats, especially vegetable oils, contain various kinds of extraneous matter such as dirt, moisture, gums, waxes, carbohydrates, proteinaceous material, pigments, flavouring substances, trace metals, antioxidants (commonly called tocopherols or vitamin E), free fatty acids and other breakdown components. To remove most of these undesirable components and to achieve a pleasing colour and bland flavour, oils are refined. Conventional (caustic) refining involves the processing steps of degumming, neutralization, bleaching and deodorization under high vacuum.

In some countries, the term 'refining' refers to the degumming, neutralization, bleaching and deodorization under high vacuum.

In some countries, the term 'refining' refers to the degumming, neutralization, washing and drying stages only. Therefore, in the oil and fat industry the terms 'fully refined', 'deodorized' or 'refined, bleached and deodorized (RBD)' are used.

PHYSICAL (STEAM) REFINING

Physical or steam refining is a term often used for deacidification of degummed/phosphoric acid treated and/or light bleached oils by high temperature steam distillation (see below). The prime objective of refining of oils and fats is to remove various undesirable components, while keeping triglycerides intact and with minimum losses of antioxidant tocopherols present in vegetable oils. The various stages of refining, described below, should not have any significant effect upon the fatty acid composition of the triglycerides of oils and fats.

CRUDE OIL

The term 'crude oil' is applied to the unprocessed oil directly after it has been extracted from the vegetable or animal raw material. Crude oils normally need refining to render them fit for human consumption. Separate stages of degumming, neutralization, bleaching and deodorization/or deacidification (steam refining) are normally applied to crude oils before they reach the consumer. In some cases a crude oil may be of adequate quality and does not need any further processing, for instance, olive oil produced by the first pressing of olives. Such oils are often termed 'virgin oils'.

DEGUMMING (DESLIMING)

Crude oils often contain non-glyceride substances, either dissolved or in suspension. These impurities consist of carbohydrates, proteins, phospholipids, resins, etc. Oils such as groundnut and soya bean are rich in these impurities, whilst others such as coconut or palm kernel oils are relatively free of them.

The simplest method of removing these impurities is to heat the oil to just below 100°C and add 3–5% hot water or a weak salt or alkali solution. The oil is then stirred and the impurities become hydrated and flocculate to form an oil-insoluble gum or mucilage. The gum may be allowed to settle and run off, or may be separated from the oil by centrifuging.

Various organic and inorganic acids may be used to improve the efficiency of the process. Aqueous citric and phosphoric acids are often used. In some cases, the phospholipids may react in the seed or in the oil to form calcium salts, or possibly other substances which do not readily hydrate and which remain dissolved in the oil. If left behind, they interfere with subsequent processes such as high temperature deacidification. One of the advantages of phosphoric or citric acid degumming is that these non-hydratable salts are rendered hydratable and can be removed with other phospholipids. The gum separated from phosphatide-rich oils such as soya bean and rapeseed oil is often used as a commercial source of vegetable LECITHIN.

NEUTRALIZATION

The neutralization of crude oils to remove free fatty acids (FFAs), as distinct from other methods of deacidification, consists of treating the crude oil with aqueous alkali, usually caustic soda or sodium carbonate (soda ash). The oil is heated to between 75–95°C and aqueous alkali added as a fine spray from above (sparging). The fine droplets of aqueous alkali fall through the oil, reacting with the FFAs to form soaps which are soluble in the hot water. The aqueous soap solution is then separated from the oil either by settling or by centrifuging. In the former case, the soap solution or soap stock is run off and the oil is then washed with several doses of hot water which remove any residual soap from the oil. The wash waters are then separated from the bulk oil by settling or centrifuging as before. The oil is dried under vacuum before treatment with earth in a bleaching process. Variations of the techniques are employed, particularly with regard to the quantity and strength of alkali, stirring speeds, use of salt solution, etc. These variations are designed primarily to prevent emulsification of the water and oil and to minimize occlusion of neutral oil in the soap stock. Variations may also be introduced in order to assist removal of non-triglyceride impurities present in the original oil, depending on the crude oil quality.

The traditional plant for neutralization consists of mild steel vessels holding up to 25 tons of crude oil. The vessels have conical bottoms, mechanical stirrers, heating coils and means for spraying alkali into the oil. Often, but not always, the bleaching process is carried out in the same vessel, which is therefore capable of being closed and evacuated for drying and bleaching. However, continuous plant is also employed, based usually on continuous alkali and wash water addition, in conjunction with centrifugal separation. In continuous neutralization the time of contact of the oil with alkali is considerably shorter than is the case with batch processing. Among companies providing continuous processing plants are Alfa-Laval Co., Sharples Corp., Clayton and Refining Inc. and De smet Rosedowns Ltd.

Other neutralizing methods include the use of ammonium hydroxide instead of conventional alkali and neutralization of oil in a solvent or in the miscella, methods designed further to decrease the loss of neutral oil.

BLEACHING

Bleaching is one of the operations in the refining of oils and fats. It normally follows the neutralizing–washing–drying procedures. Bleaching removes colouring matter such as carotenoids and chlorophyll and other minor constituents such as oxidative degradation products or traces of transition metals from the oils or fats. Some of these minor constituents are hydrogenation catalyst poisons and it is therefore essential to bleach an oil thoroughly before hydrogenation. Transition metals, especially iron and copper, are pro-oxidants and their removal therefore improves the oxidative stability of the oil.

Bleaching is achieved by stirring the oil under vacuum with 0.1–2% activated fuller's earth. Temperatures of treatment range from 90–130°C, whilst times range from 10–60 min. The bleaching takes place in vertical vessels made of mild steel which hold up to 25 tons of oil. Each vessel is fitted with a mechanical stirrer, a means of heating and cooling, valves, sight glass, a vacuum gauge and a thermometer. The earth is sucked into the oil under vacuum at 60–80°C. After treatment, the earth is removed by filtration, usually in a plate-and-frame filter press. Considerable interest has also developed in continuous bleaching. Other methods include the use of mixed adsorbents (e.g.

mixtures of fuller's earth and carbon), the possibility of bleaching oil miscella (i.e. treating the oil plus solvent mixture obtained in oil extraction plants before recovery of the oil), use of high temperatures during deodorization to effect thermal breakdown of colour material and hydrogenation. Chemical bleaches are not suitable for use with food fats, as these usually oxidize the oil, causing an increase in peroxide value and subsequent off-flavour development.

After bleaching, oils and fats may either be deodorized for immediate use or they may be hydrogenated. If the oil or fat is hydrogenated, a second bleaching treatment is usually carried out on the hydrogenated product before it is deodorized, mainly in order to remove traces of hydrogenation catalyst.

DEODORIZATION

Deodorization is the final processing step in the preparation of an oil for edible purposes, such as a salad oil or cooking fat, or in the preparation of an oil stock for margarine. Deodorization removes from the oil volatile impurities, among which are many of the foreign odours and off-flavouring which would render the oil unsatisfactory. Hardened fats are always deodorized after hydrogenation to remove characteristic off-flavours formed during the hydrogenation process.

Deodorization is essentially a steam distillation under low pressure. The oil is heated at a temperature of between 180 and 270°C, at pressures ranging from 1–5 Torr, and live steam is injected. The time of treatment varies according to the design of plant and the temperature used, from 5 h in low temperature batch processes to as little as 15 min in some continuous systems.

Batch deodorization vessels consist of vertical steel tanks holding up to 25 tons of oil. A large vapour space is maintained above the liquid level to contain the oil during the violent agitation caused by steam injection. Provision for reducing the entrainment of oil in the escaping steam is made at the top of the deodorizer, where it connects with the vacuum line. The heating of the oil to deodorization temperature and its subsequent cooling are carried out under vacuum since it is important that the hot oil should not come into contact with atmospheric oxygen, which would lead to the production of new off-flavours. The deodorized oil is often filtered after deodorization, a process known as 'polishing'.

DEACIDIFICATION

Deacidification, as distinct from neutralization, consists of the removal of free fatty acids from an oil by steam distillation. The oil is first treated with hot water and optionally phosphoric or citric acid to remove phosphatides in a process known as DEGUMMING. The oil is then subjected to a light bleach and charged into the deodorization vessel. The oil is heated to a temperature of up to 270°C under vacuum and live steam is injected. Fatty acids are thereby volatilized and drawn off. This process avoids the loss of some of the neutral oil, as occurs in alkali neutralization, and a higher yield of final product is thereby obtained. There is also a possibility that some of the free fatty acids originally present combine with partial glycerides to form triglycerides, thus increasing the yield further. Deacidification may be carried out in batch, semi-continuous or fully continuous plant. Some heat bleaching also takes place.

HYDROGENATION

Hydrogenation, or hardening, causes an increase in the melting point of fats through classical addition of hydrogen to some, if not all, of the double bonds present in the fatty acids of the triglycerides. It also stabilizes the fat toward oxidation and often has a bleaching effect. Hydrogenation has great commercial significance in the edible fat industry as many of the raw materials are liquid oils at room temperature and are liable to oxidative deterioration. Many of the cooking fats and margarines available today contain a proportion of hydrogenated fat blended with natural oils. Fish oil is an exception among food oils in that it is very seldom used in the unhardened condition as off-flavours very quickly develop in the unhydrogenated oil.

The increase in melting point (mp) with decreasing unsaturation and double bond geometry can be demonstrated in the series of eighteen carbon fatty acids as follows: linolenic acid, 3 cis double bonds, mp −11.5°C; linoleic acid, 2 cis double bonds, mp −6°C; oleic acid, 1 cis double bond, mp 16.2°C; elaidic acid, 1 trans double bond, mp 42°C; stearic acid, 0 double bond, mp 72°C.

The difference between cis and trans double bonds lies in the positions of the two hydrogen atoms attached to the carbons linked by the double bonds. In the cis form these both lie on the same side of the double bond, whereas in

trans isomers they lie on opposite sides. The cis form occurs almost exclusively in natural fats. This has led to some criticism of hydrogenated fats containing high levels of trans double bonds. In the preparation of edible fats, the majority of oils are only partially hydrogenated, the double bonds which remain may be in the cis or the trans form, and there may be either one or two double bonds per fatty acid molecule. Hydrogenation conditions and catalysts are carefully selected in order to regulate the ratio of cis to trans double bonds and the proportions of saturated, mono- and polyunsaturated acids. It is the balance of these different factors which governs the physical properties of the product and suits it for a frying oil, a soft tub margarine, a harder block margarine, a fat for cream fillings in biscuits, a dough fat or a coating.

Hydrogenation is generally carried out in special pressure vessels, with an operating pressure of from 1–7 bar, and a temperature of 100–180°C. After the catalyst is added, hydrogen is pumped into the oil which is vigorously stirred. Hydrogenation is an exothermic reaction with fresh nickel catalysts, but not so with poisoned catalysts, and the heating requirements must therefore be balanced depending on the catalyst. Normally, hydrogenation is a batch process, but efforts are being made to devise continuous processes in which the catalyst is supported in heated columns.

The hydrogenation process can be controlled by measuring the refractive index, iodine value or melting point of samples withdrawn at intervals, Hydrogenated fats are filtered to remove the hydrogenation catalyst, subjected to a light earth bleach and deodorized before they can be used for edible purposes. In some countries such as the USA and Canada, the earth bleach and catalyst removal are sometimes combined into a single operation, but in these cases the nickel catalyst cannot be recovered. Patterson (1994) gives an excellent review of hydrogenation.

Hydrogenation catalysts

A catalyst is a material which when added in minute amounts to a mixture of substances capable of reacting with each other greatly increases the rate of the reaction. In the oils and fats industry, hydrogenation catalysts are the most important. Nickel is the usual hydrogenation catalyst, although patents have been issued for the use of many other noble metal catalysts. Nickel catalyst was at one time prepared by the dry reduction of nickel carbonate, the wet reduction of nickel formate or from Raney alloy. With the increased costs of nickel, supported nickel catalysts are now preferred, however.

In the production of a supported nickel catalyst, the support material is stirred into an aqueous solution of a nickel salt such as sulphate. The solution is gradually rendered more alkaline, e.g. by the addition of sodium carbonate, whereupon basic nickel carbonate is precipitated onto the surface of the support. The mixture is filtered, washed free of sulphate ions, dried and reduced with hydrogen. The size of the nickel crystallites on the surface of the support and the size of the pores left in the surface of the nickel determine the activity and selectivity of the resulting catalyst.

Catalysts prepared on this basis normally contain between 15–25% nickel. As they are pyrophoric, the fresh catalyst is normally suspended in a fully hydrogenated oil which is allowed to crystallize and is converted into flakes. A review of hydrogenation with non-poisoned nickel catalysts is given in a 1970 symposium of the American Oil Chemists Society (*J. Am. Oil Chem. Soc.*, **47**, 463–500).

In some cases special attributes are imparted to the catalyst by the addition of promoters such as zirconium, or of poisons such as sulphur. Sulphured nickel catalysts are especially useful in converting *cis* double bonds in a fat to the *trans* form, a process which gives a steep melting fat suitable for uses in the confectionery industry.

The amount of catalyst used in hydrogenation is usually about 0.1% of nickel on the weight of the oil for a fresh catalyst but rising to 1–2% for a poisoned or spent catalyst. Catalysts can normally be re-used for a number of batches, artificially poisoned or sulphured catalysts lasting much longer than fresh catalysts, as in the latter case impurities in the oil gradually poison the catalyst and lower its activity. After hydrogenation, the catalyst must be removed from the hardened fat by filtration to give a clear oil. In the USA and Canada this is often accomplished by adding bleaching earth to the oil and filtering off the combined catalyst and bleaching earth slurry. In such an operation it is not economic to recover the nickel from the filter cake. In Europe, however, it is more common to filter off the bulk of the catalyst separately and then subject the oil to an additional earth filtration to remove the last traces of nickel catalyst. Incomplete removal of the nickel catalyst may be criticized from a nutritional point of view and it can also give the oil a greenish or greyish appearance.

Selectivity

Selectivity is a term applied to catalytic hydrogenation and refers to the ability of a catalyst to promote the preferential hydrogenation of polyunsaturated acids. Various forms of selectivity are recognized, depending on whether the manufacturer wishes to hydrogenate the linolenic acid but not the linoleic acid, or wishes to hydrogenate both linolenic and linoleic, but not the oleic. These selectivities are called selectivity 2 and selectivity 1, respectively. Selective catalysts are usually less active than non-selective catalysts and are more likely to promote a higher degree of *trans* bond formation.

SPREADS

In recent years there has been considerable activity in new product development in the area of margarine and spreads. In fact the growth of these spreads in the UK has benefited from the NACNE (National Advisory Council on Nutrition Education) and the COMA reports (DHSS, 1984). The NACNE report (1983) recommended that the fat content of the diet should be reduced from 42% of energy to 34% in the short term and 30% in the long term. Saturated fat intake should be reduced from 20% to 15% in the short term and 10% in the long term. The main types of product manufactured can be divided as follows:

(i) *Table margarine*. It must contain a minimum fat content of 80% and maximum of 16% water. There are a number of different types, e.g. soft/tub, health/diet or high polyunsaturated fatty acids (PUFA), foil wrapped, whipped and liquid/shortening. Table margarine can be used for spreading, frying and baking.

(ii) *Reduced fat spreads*. These spreads contain 60–70% fat and can be used for spreading and in most cases frying and baking.

(iii) *Low fat spreads*. They contain maximum 40% fat and can be used for spreading only.

(iv) *Very low fat spreads*. There are a number of products available, these days, with fat content 3 to 25%.

MARGARINE

Margarine is a fatty food closely resembling butter. The fat of margarine is not derived from milk fat however, or at most only to a minor extent. Margarine was invented in the nineteenth century by a Frenchman, Mége Mouriés, who patented his original process in 1869. Mouriés won a competition organized by the French Government for a butter substitute, cheaper and less perishable than butter, needed to alleviate the shortage of fats among poorer parts of the population and the armed forces. The process was taken up in Holland by Jurgens of Oss and in a number of other countries.

The term 'margarine' is derived from the Greek word margarites, meaning pearl, and refers to the pearly lustre of the fat then known as margaric acid and subsequently shown to be a mixture of stearic and palmitic acids. It should be pronounced with a hard 'g', but it is often mispronounced.

The basic method of margarine production consists, as it did in Mége Mouriés' day, of emulsifying a purified oil blend with skim milk, chilling the mixture to solidify it and working it to improve the texture.

Margarine is subject to statutory regulations, which vary throughout the world. In the UK it must not contain more than 16% moisture, nor less than 80% fat, the balance being salt, protein, colouring matter, emulsifiers, antioxidants and vitamins, ingredients which are also subject to regulations.

Raw materials

Oil

The selection of oils for margarine is made by the manufacturer with regard to cost, quality and desired properties in the margarine. The stability, consistency, plastic range and EFA level must all be considered. In selecting a blend, information about the ratio of solid and liquid glycerides present at various temperatures is important.

The pattern of usage of oils for margarine manufacture has changed considerably in recent years. Thus, in 1960, 56% of the oils used for margarine manufacture in the UK were vegetable oils, whilst in 1965 this figure had dropped to 37% due to the increased availability of fish oils, which almost completely replaced whale oil. Lard has also been used when it has been available cheaply, notably from 1962 to 1965. In recent years, rapeseed, soya bean, palm and sunflowerseed have been the most important vegetable oils used, partly in view of their availability, but also in the case of soya bean and sunflowerseed oils as a result of their high EFA contents. Of course the pattern of raw mate-

rial usage will vary in different parts of the world and the increasing quantities of rapeseed oil currently grown within the EU have been reflected in the increasing use of rapeseed oil in European margarines. Speciality margarines may of course contain a preponderance of a particular component, and in this respect it should be noted that the dietary margarines high in essential fatty acids are normally based on sunflowerseed oil, while margarines said to have a very close resemblance to butter are often based on fractions derived from beef tallow.

The aqueous phase

About 16–18% of margarine consists of an aqueous milk preparation, except in certain special products such as pastry and kosher margarines, when just water may be used. In the production of the aqueous phase pasteurized fresh milk or reconstituted dry milk is subjected to a ripening process. During this process diacetyl and other aroma giving substances are developed. To induce the ripening process, carefully selected strains of *Streptococcus cremoris* and *S. citrovorus* are added to milk held in ripening vats at 20 to 22°C and at a pH of 5.3 to 6.3

Production methods

The first step in the production of margarine is the preparation of the fat blend. A weighing tank is often used for measuring out the fat ingredients; this may be done automatically or metering pumps may be used. The fat blend may contain natural crude fats or processed fats such as fractionated, hydrogenated or interesterified mixtures. It is first refined and deodorized and then emulsified with the aqueous phase, emulsifiers being generally added at this point. Other ingredients such as vitamins, dyes, flavours, etc., are usually incorporated just before emulsification. If emulsification is to take place in a Votator, the fat and aqueous phases can be fed separately by proportioning pumps and emulsification carried out in the chilling tubes. 'Votator' is the proprietary name of the Girdler Corporation of the USA for their continuous, totally enclosed, scraped-surface heat exchanger used in the manufacture of shortenings, cooking fats, margarine and other fat products. There is a family of related scraped-surface heat exchangers available on the market, differing slightly from one another in design but effectively functioning in the same manner. Minor modifications may be built into the machine and its ancillary equipment, depending upon customers' individual requirements.

Packing

Margarine is moulded and packed directly from the production unit. Rectangular half-pound or 250 g blocks are a common unit while the softer margarines are packed directly into plastic tubs. Modern automatic packaging machines, such as those manufactured by Forgrove Machinery Co., Benz and Hilgers and Kunster Frères, operate at very high speeds of over 150 packets per minute. It is now common to use a mechanical case packer.

Whipped margarines

Margarines containing a relatively large volume of finely dispersed gas (air or nitrogen) and hence of high volume in relation to weight are sold in the UK. These products are claimed to spread easily, to mix more readily in cake making and to be better as cooking fats.

Refrigerator (table) margarine

Margarines usually have a fat blend which is so designed that the product will spread well at normal room temperatures. Use of refrigerators for storing margarines has led to the introduction of margarines that will spread well immediately after removal from the refrigerator as well as at normal room temperature. These margarines normally contain a high proportion of liquid oil and of solid fats which melt quickly in the mouth, although they must give sufficient body to the margarine over the desired temperature range.

Cake margarine

Although ordinary table margarine is suitable for cake making, industrial bakeries use special cake margarines with a wide plastic range, high creaming power and an ability to impart shortness to certain classes of baked goods. To achieve these aims, cake margarines are made by blending high melting point fats with a fair amount of liquid oil. A typical fat blend might consist therefore of a mixture of palm oil with hydrogenated fish oil. The moisture content of cake margarines is lower than that of domestic table margarines.

The most desirable quality of margarine in cake making is its creaming power, i.e. its ability to take up air in the form of finely divided bubbles. The volume and cellular structure of a cake, its lightness and to a large extent its palatability depend on this creaming power. During baking, the entrapped air, saturated with water vapour, expands because of the increased pressure of the water vapour. Creaming power and the ability to take up a high percentage of moisture are improved by incorporating into the margarine emulsifiers such as glycerol monostearate. Work carried out at the Leatherhead Food Research Association and since substantiated by others showed that fats which crystallize in a β-polymorphic modification comprise microcrystals and have improved creaming properties.

Margarines intended for use in short pastry do not contain emulsifiers (see SHORTENINGS). The main difference between a cake and a short pastry is that the latter is essentially dry. It contains only about 4% of moisture and there is no need for volume and cellular structure. The moisture content of cakes is about 20%. The amount of sugar that can be incorporated into a cake depends on the amount of moisture and eggs present. As the percentage of water can be increased in a cake when using a margarine containing an emulsifier, the percentage of sugar can also be increased.

The manufacture of cake margarine is similar to that of ordinary REFRIGERATOR MARGARINE. The books by Schwitzer (1956) and Stuyenberg (1969) are useful references on this topic.

Pastry margarine

Pastry margarine is used for making flaky or puff pastry. The margarine must have a smooth texture and a tough consistency because it should form thin coherent layers in the pastry. It must stand up to heavy mechanical working and rolling which occur when it is rolled into thin layers.

The margarine has to be relatively dry and free from occluded air, because the puff effect of the pastry is obtained by expansion of the dough between the fat layers. The composition of pastry margarine is usually made up of a high melting fat and a proportion of liquid oil.

Dairy spreads

There are also products in which cows' butter or cream is blended with polyunsaturated vegetable oils to give a dietary product with a high EFA content. These may have a little less than the 80% fat content of margarines and butters and may not therefore be described as such. Dairy spread has the taste of butter but can be spread straight from the fridge.

Reduced fat and low fat spreads

Since the mid-1980s a wide variety of reduced fat and low fat spreads have appeared on the market. As these contain less than the statutory amount of fat, it is not legally possible to call them margarines, although they are often referred to as such by the public. Some of these spreads are labelled to contain zero or little *trans* fatty acids. The plasticity in oils being used for making *trans*-free spreads is, sometimes, created through mechanical bonding by using new, isoelectric dispersion technology. The 'Super' spread sold in UK by Whole Earth Foods contains 70% oil as emulsion in water, 10% saturated fat and 60% other fats. It should be mentioned that these spreads are notable for the mouthfeel they exhibit which is different from soft/tub margarines containing hydrogenated oils.

As low fat spreads contain a higher proportion of water, a blend of stabilizers and emulsifiers is necessary to retain the plasticity range of these products. Low fat spreads do not offer performance in frying, baking and may even make hot toasted bread soggy. Also, they may not have the creamy and pleasant taste provided by the higher fat table spreads. Nevertheless, the future opportunities for low fat spreads are bound to expand. As the controversy about *trans* fatty acid will not go away, a number of zero-*trans*, low fat spreads will appear on the market. Moreover, the antioxidant group of chemicals such as vitamins A, C and E are receiving favourable media attention for their health effects. Low fat spreads easily could incorporate a high content of the antioxidant group and other nutritional substances such as phospholipids and conjugated linoleic acid. It is anticipated that low fat spread products containing such health beneficial components will start appearing in the near future.

OTHER FAT MODIFICATIONS

FRACTIONATION

Fractionation is a process applied to semi-solid fats such as palm oil. It involves a physical separation of higher and lower melting triglycerides and results in a semi-solid fat being split into a low melting oil (olein fraction) and a solid fat (stearin fraction). There are three types of fractionation, namely dry, solvent and detergent fractionation.

In dry fractionation, the fat is melted completely and then cooled with agitation until the higher melting fraction crystallizes. The crystals of the stearin fraction are then removed by filtration.

In solvent fractionation, the fat is dissolved in a solvent such as acetone or hexane. The warm solution is cooled with agitation until the stearin fraction precipitates and can be removed by filtration. Solvent fractionation gives a more efficient separation than dry fractionation but the costs involved in using organic solvents limit the application of the process. Solvent fractionation is commonly applied in the preparation of palm mid-fraction which is commonly included in cocoa butter equivalent fats for use in chocolate.

Detergent or Lanza fractionation involves the use of an aqueous detergent solution to reduce entrainment of the liquid fraction in the solid crystals in a dry fractionation process. Although the separation efficiency of the dry fractionation is improved, the problems of the effluent have limited the application of this process.

WINTERIZATION

Winterization is a process of fractional crystallization of oils in which the higher melting glycerides are removed, giving the oil a clear, bright appearance even after refrigerator storage. The name is derived from the original method of carrying out the process, by allowing the oil to stand in outdoor tanks during the winter. Winterization is now carried out in chilled brine tanks under carefully controlled conditions. Alternatively, other forms of fat fractionation may be applied; the resulting clear oil is nevertheless termed 'winterized' oil. Winterization is most often applied to cottonseed oil which is grown extensively in the USA and tends to throw down a deposit during refrigerator storage. On the other hand it has a better flavour stability than soya bean oil which does not need winterization. In the American market, winterized cottonseed oil is therefore often considered optimum for a salad or table oil.

Dewaxing of sunflower and corn oils for the removal of waxes to prevent clouding is usually carried out by the winterization process. These oils must be winterized if they are intended for use as salad oils.

INTERESTERIFICATION

Triglycerides consist of three fatty acids esterified with the three free hydroxyl groups of the original glycerol molecule. The physical and chemical properties of the resulting triglycerides vary depending upon the nature and positions of the fatty acid residues in the triglyceride molecule. In many natural oils and fats the positions of the individual fatty acids are controlled during biosynthesis. The physical properties of an oil can be modified by rearrangement of the fatty acids in the triglyceride molecules. To achieve this, the oil or fat is heated to a temperature of about 100°C together with an interesterification catalyst under a vacuum. Under these conditions the fatty acids are momentarily liberated but almost immediately recombined with different hydroxyl groups in a random manner. Interesterified fats are therefore often referred to as 'randomized'. This feature can be especially attractive when two different oils are first blended and then interesterified.

Interesterified oils have different properties from those of the original oil and are frequently used in tailor-making fatty products for specific uses. However, it is almost impossible to predict the properties of an interesterified oil or oil blend without recourse to experimental information, as the number of triglycerides generated becomes extremely high, leading to complex eutectic and intersolubility interactions between the different triglyceride groups.

FAT PRODUCTS

COCOA BUTTER REPLACEMENT FATS

Cocoa butter replacement fats are used in the confectionery trade in chocolate coatings or for the partial replacement of cocoa butter in the manufacture of chocolate. The earliest cocoa butter substitutes were prepared by fractionation of palm kernel oil, the higher melting components of the fat being separated. These palm kernel stearins melt at approximately the same temperature as cocoa butter, but as they contain different triglycerides they are not compatible with cocoa butter and form eutectics with it. As a result, the melting point of a blend of cocoa butter and palm kernel stearin is lower than that of either of the components, which makes it unsuitable for use in chocolate. Another class of cocoa butter substitute is manufactured by a combination of hydrogenation and fractionation of cheaper vegetable oils such as soya bean oil or palm oil. These fats have high *trans* values, and have moderate compatibility with cocoa butter. Nevertheless, they are far from ideal and coatings made with such fats often have a waxy palate response. The most sophisticated cocoa butter replacement fats, commonly known as cocoa butter equivalents (CBEs), are prepared by blending carefully selected natural fat fractions. Commercially successful CBEs are manufactured on a large scale by blending a middle melting fraction from palm oil with naturally occurring (Borneo) illipe butter, the upper melting fraction (i.e. stearin) of shea oil and sal stearin. Each of these triglyceride fractions provides a concentrate of triglycerides normally present in cocoa butter, giving a final blend with almost identical physical characteristics to those of cocoa butter itself.

The use of cocoa butter replacement fats in products labelled 'chocolate' is strictly controlled, the legislation varying from country to country. Various analytical methods must be brought to bear in order to detect the presence. Fats such as palm kernel stearin may be detected by their high lauric acid content, whilst products based on hydrogenation and fractionation may be detected by the increased *trans* value imparted to the mixture. Cocoa butter equivalents are more difficult to detect and sophisticated methods must be brought to bear.

SHORTENINGS

The term shortening arises from the use of this type of fat to impart shortness in the preparation of flour confectionery, such as shortbread, short pastry or cakes. The main domestic use is for pastry making, frying and cake making. The term 'shortening' is of US origin and the product was originally developed as an outlet for cottonseed oil available as a by-product of the cotton growing industry.

Shortenings contain no moisture and consist entirely of fat. They frequently have glyceride compositions resembling those of margarine fat blends. The traditional shortening in Europe is lard and this is still used extensively.

The consistency and plasticity of shortenings are important; too soft a shortening will mix poorly with flour and cause problems in baking, whilst too hard a product gives a tough shortbread which does not aerate properly. Shortenings are improved by aeration and domestic shortenings generally fall into one of two categories, moulded products containing up to 10% air and liquid filled products containing 10–30% air. Moulded products are usually parchment wrapped, and aim to be competitive with lard and to have better flavour and better cake making properties. Liquid filled, highly aerated products are more expensive, but offer advantages in ease of use.

High ratio shortenings allow a greater proportion of sugar to fat in cake recipes, due to a higher dough strength which is dependent upon the emulsifying properties of mono- and diglycerides in the shortenings (see EMULSIFIERS). Because such shortenings contain a higher proportion of combined glycerol in the form of mono- or diglycerides, they are referred to as 'superglycerinated' shortenings in the USA.

Pumpable shortenings

Pumpable shortening was developed in response to a trend in the flour confectionery industry for bulk delivery of raw materials. Flour, sugar and liquid were all delivered in bulk and a desire therefore developed for a pumpable shortening which could also be delivered in bulk. Pumpable shortenings are delivered in bulk road tankers and pumped to holding vessels from which they can be easily pumped to any point of a factory. Pumpable shortenings are tempered in bulk, usually in 6 to 12 tonne vessels into which they are fed directly from a Votator processing unit. They are delivered to customers at temperatures of 23–25°C, a temperature which must be maintained in the tanker.

Procter and Gamble hold a number of patents covering the preparation of pumpable shortenings. Liquid shortenings have also been developed and these offer the convenience of being pourable from a bottle. The base material is a liquid oil, such as cottonseed oil, containing a finely dispersed emulsifier.

FRYING OILS

Frying oils can be considered under two headings, (i) for industrial consumption and (ii) for domestic consumption. The larger market is in the industrial area where enormous quantities are used for frying fish fingers, precooked chips, crisps and various snack foods. In the domestic market the appearance on the supermarket shelf is of considerable importance and oils which will remain clear and bright without the deposition of solid fat crystals are in greatest demand. Vegetable oils rich in polyunsaturated acids are satisfactory in this respect, but can oxidize quickly during use in the kitchen, developing off-flavours. This problem is particularly severe with soya bean and rapeseed oils, as these contain up to 10% of linolenic acid (C18:3), the triple unsaturation of which renders the oil easily oxidized. This problem can be alleviated by selective hydrogenation of the oil to reduce the linolenic acid to a low level. Hydrogenation also produces high melting triglycerides, but these may be removed in a dry fractionation process. In a well-controlled plant, a yield of 70% of clear liquid oil with a very low linolenic acid content can be obtained from soya or rape oils by this means.

Commercial frying oils must also meet strict quality requirements and here again fluidity at low temperatures can be important. This is especially the case if, for instance, the oil is stored outside in bulk storage tanks from where it is piped to the frying installation. It has been known for frying oils to set up in these pipes in cold weather or overnight, leading to considerable production difficulties. It is also important with snack foods which are eaten cold that the frying medium should give no waxy palate response. Industrial frying oils are also used to produce fried foods which must have a long shelf life, a problem seldom encountered with domestic frying oils. Measures are therefore taken to preserve the quality of the frying oil and thus ensure adequate shelf life of the fried food. The addition of suitable antioxidants and of methyl polysiloxane antifoam agent are often recommended as is the continuous filtration of the frying oil in a recirculatory system. The fried food normally absorbs a proportion of oil, which is made good by 'topping up' with fresh oil. In a well-organized system the amount of 'top up' is sufficient to maintain the oil in adequate condition, so that there is no need to reject frying oil on the basis of deterioration in quality.

In both domestic and industrial frying operations, smoke haze can be a problem. Smoke arises from the volatilization of breakdown products in the oil, in particular free fatty acids. Lauric oils are unsatisfactory frying oils since the constituent fatty acids are more volatile than with the other oils.

In the frying process, the oil acts as a medium for heat transfer from the fryer to the product. The industrial fryer can be heated directly, often with gas jets under the base of the fryer or can be heated by internal coils or by external heat exchangers. Many modern industrial frying operations now have external heat exchangers through which the oil is pumped, in conjunction with a filtration system.

SALAD OIL

Salad oils are vegetable oils which remain liquid when kept in a refrigerator at about 5°C, and are used for salad dressings, for the preparation of MAYONNAISE, etc. The total consumption in the UK is small compared with that in Mediterranean countries where large quantities are consumed for cooking purposes. Salad oils may be naturally flavoured oils such as virgin olive oil, or refined deodorized oils which are preferred in the UK and the USA. Winterized cottonseed oil is often used in North America as it does not become turbid on keeping.

GHEE, VANASPATI

See Chapter 3

COMPOSITION OF OILS AND FATS

Oils and fats comprise mainly triglycerides, a small portion of free fatty acids, mono- and diglycerides and a wide range of minor components. These minor components include sterols, tocopherols, phospholipids, hydrocarbons, waxes, pigments, etc. The fatty acids esterified with glycerol comprise about 95% of most oils and the fatty acid compostion is the most important chemical characteristic of an oil. The fatty acid compositional ranges for a number of oils are shown in Tables 8.2, 8.3 and 8.4.

TRIGLYCERIDES

Structurally most oils and fats are esters of glycerol, which has three alcoholic hydroxyl groups, with three fatty acids. These fatty acids may be the same as one another or different. Figure 8.1 illustrates the structure of POS (palmitic–oleic–stearic), the major triglyceride in cocoa butter. In

natural fats, mixed triglycerides strongly predominate, but it is now accepted that in the vast majority of vegetable fats the central or 2- position is esterified with an unsaturated fatty acid, as shown in Figure 8.1. The physical properties of a fat or oil are dependent upon the physical properties of the constituent triglycerides, and the interactions between these triglycerides. A redistribution of fatty acids amongst the triglycerides, for instance by INTERESTERIFICATION, can therefore change the physical properties of the fat. In some cases, for instance with cocoa butter, this can lead to a serious deterioration in performance.

FATTY ACIDS

The types and proportions of the fatty acids present in the triglycerides of an oil have a major influence on the physical, chemical and nutritional properties of the oil. The fatty acid composition of an oil is therefore its most important chemical characteristic. In view of the recent dietary concern over the composition of oils and fats, it has become customary to group various categories of fatty acids according to their degree and type of unsaturation. The following groups are commonly used.

Saturated acids (saturates)

These are acids such as lauric, myristic, palmitic, stearic, etc., the hydrocarbon chain of which has no double bonds (but note that in the COMA report (DHSS, 1984) the description 'saturated' includes the *trans* unsaturated acids). Some experts recommend that only C12, C14 and C16 saturated acids should be counted for dietary purposes. The individual members of this group are described below.

(i) Butyric acid (C4) occurs in butter and is responsible for the characteristic butyric note of rancid butter.

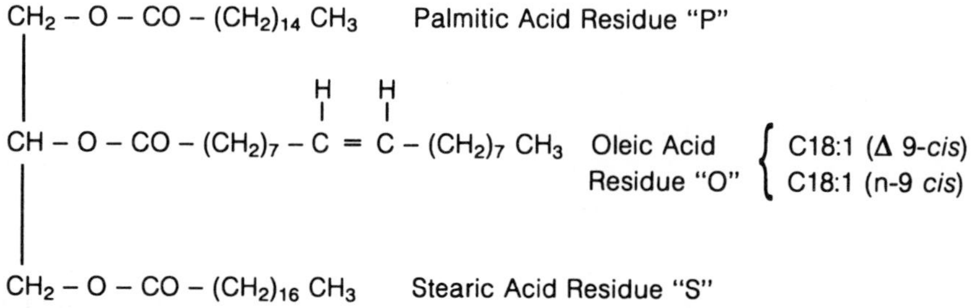

Figure 8.1 Triglyceride POS, a major constituent of cocoa butter.

(ii) Caproic acid (C6) occurs to a small extent in butter and coconut oils.

(iii) Caprylic acid (C8) and capric acid (C10) are present in babassu, coconut and palm kernel oils and to a slight extent in butter.

(iv) Lauric acid (C12) is the major fatty acid constituent in babassu, coconut and palm kernel oils; these oils are, therefore, collectively known as the lauric oils. The acid also occurs to a small extent in butter.

(v) Myristic acid (C14) occurs in babassu, coconut and palm kernel oils and to a trace extent in cottonseed and palm oils, as well as in a number of animal fats such as butter, lard, tallow and fish oils.

(vi) Palmitic acid (C16) is an acid of wide occurrence found in practically all oils and fats. It is the major constituent of palm oil, from which it derives its name.

(vii) Stearic acid (C18) is found in almost all naturally occurring animal, marine and vegetable oils.

(viii) Arachidic acid (C20) is present to a small extent in groundnut (Arachis) oil from which it derives its name. It also occurs in trace amounts in other oils, e.g. soya bean oil, cocoa butter and sal fat.

(ix) Behenic (C22) and lignoceric (C24) acids occur to a small extent in groundnut oil.

Unsaturated acids (unsaturates)

Monounsaturated acids

These have a single double bond in the hydrocarbon chain. *Trans* monounsaturates such as elaidic acid are classed as saturated in the COMA report (DHSS, 1984). The individual members of this group are described bleow.

(i) Myristoleic acid (C14:1, 9-*cis*) occurs to a small extent in butter, tallow and fish oils.

(ii) Palmitoleic acid (C16:1, 9-*cis*) is present in relatively large amounts in many fish oils, menhaden oil, for instance, containing up to 15%. It is also present in small amounts in palm oil, cottonseed oil, butter and lard.

(iii) Oleic acid (C18:1, 9-*cis*) is the most commonly occurring fatty acid. It is present in almost all vegetable and animal fats, in particular olive, palm, low erucic rapeseed, groundnut, tea-seed, almond and fish oils.

(iv) Elaidic acid (C18:1, 9-*trans*), the isomer of oleic acid, is present to a small extent in animal fats and mammalian butters, and also in partially hydrogenated oils and fats.

(v) Ricinoleic acid (C18:1, 9-*cis*) contains one hydroxy group at the 12 carbon position and occurs as the principal acid in castor oil.

(vi) Erucic acid (C22:1, 13-*cis*) is present in many of the oils from brassica seed, such as mustard and *Sinapis arvensis*. It was formerly present in most rapeseed oil, and rapeseed oil grown in the Third World is still mostly of the high erucic acid form. In view of evidence that erucic acid may cause cardiac lipidosis, new varieties of rapeseed have been introduced which are free of erucic acid. The use of high erucic acid rapeseed oil in foods is subject to legislation (Statutory Instrument No 691, 1977—The Erucic Acid in Foods Regulations 1977, as amended (UK)). Where necessary, the erucic acid content of the oil must be diluted before use to a level of less than 5%, by blending.

(vii) Cetoleic acid (C22:1, 9-*cis*) is an isomer of erucic acid and occurs in fish oils; it has not been implicated as a health hazard and does not come within the scope of the legislation on erucic acid in foods.

The *trans* isomers of some of the above acids have trivial names, e.g. myristoelaidic (C14:1, 9-*trans*), palmitoelaidic (C16:1, 9-*trans*) and brassidic (C22:1, 13-*trans*).

Polyunsaturated acids (polyunsaturates or PUFA)

These have more than one double bond in the hydrocarbon chain. In natural vegetable oils and unhydrogenated fish oils these are in the *cis-cis*-1,4-diene structure (all *cis*-) and these are the essential fatty acids (EFAs), see also below). Hydrogenated oils may contain polyunsaturates in which double bond migration or isomerization has taken place, so that not all the bonds are *cis-cis* or 1,4-diene. Identification of the proportion of polyunsaturates which are still *cis-cis*-1,4- may be carried out by the lipoxidase technique (IUPAC method 2.312), but this is of low accuracy with less than about 15% EFA. Capillary column gas chromatography (GC) may be used but here there may be a problem in separating the peaks of positional isomers. The individual members of this group are described below.

(i) Linoleic acid (C18:2, all *cis* -9,12) is the principal acid in safflower, soya bean, sunflower and corn oils. It also occurs in smaller amounts in other vegetable, animal and marine oils. It is regarded as the most significant EFA in the diet.

(ii) Linolenic acid (C18:3, all *cis* -9,12,15) occurs as the principal acid in linseed oil, from which it derives its name. It is also present in smaller amounts in soya bean and rapeseed oils. The common variety is the α-isomer. The less common variety, γ-linolenic acid (GLA) (C18:3, all *cis* -6,9,12) is a mammalian metabolite of linoleic acid which also occurs in a few vegetable oils, the best known being evening primrose seed, blackcurrant pip and borage seed oils. Oils containing GLA are said to have enhanced dietary value. Eleostearic acid is the *cis*-9, *trans*-11, *trans*-13 isomer of linolenic acid found in tung oil, useful in the paint and varnish industry.

(iii) Arachidonic acid (C20:4, all *cis* -5,8,11,14) is a major constituent of unhydrogenated fish oils.

(iv) Eicosapentaenoic acid (C20:5, all *cis* -5,8,11, 14,17) (EPA) and

(v) Docosahexaenoic acid (C22:6, all *cis* -4,7,10, 13,16,19) (DHA) are constituents of unhydrogenated fish oils. For example, menhaden oil contains EPA 13% and DHA 8%.

Other PUFAs with chain lengths of up to 26 carbons and up to 6 double bonds occur in fish oils. PUFAs with *trans* double bonds are sometimes given names such as 'linolelaidic', but these names lack specificity and are not often used.

Trans *acids*

The *trans* acids content is most easily measured by infrared, using IUPAC Method 2.207. However, this may give inaccurate results when *cis-trans* or *trans-trans* dienes are present. The *cis-trans* diene absorbs infrared at 85% of the absorption of the *trans*-mono-diene and the *trans-trans* diene absorbs at 166% (instead of the 200% expected). These influences compensate to some extent in oils containing both *cis-trans* and *trans-trans* dienes. At present, a good, reproducible and accurate method for the sum of C18:1*t*, C18:2*ct*, C18:2*tt* trans fatty acid levels above 5% is the combined GLC–IR method, which is AOCS Method Cd 14b-93.

In nutritional calculations a *trans*-mono-diene, a *cis-trans* diene and a *trans-trans* diene each contribute one fatty acid to those classed as *trans* and are grouped with the saturated acids in the COMA report (DHSS, 1984).

Some individual trans acids, such as elaidic, have been discussed above.

See also Chapter 19.

Essential fatty acids

Certain polyunsaturated fatty acids (PUFA) are essential in the diet of animals if proper growth is to be maintained and these acids are known as essential fatty acids (EFA). Three fatty acids which are effective in curing EFA deficiency syndromes are linoleic, γ-linolenic and arachidonic. Arachidonic acid is the most effective and can be synthesized from linoleic acid in the body by mammals. Considerable importance has been attached to this topic since the mid-1980s, as it has been maintained that essential fatty acids may be effective in preventing coronary heart disease and other vascular problems (FAO, 1980: Holman, 1981). Mammalian biosynthesis converts arachidonic acid into prostaglandin metabolites which have far reaching effects in the body.

The richest common source of linoleic acid is safflowerseed oil, but as this is expensive, sunflowerseed oil is far more widely used. Soyabean oil, corn oil and cottonseed oil are also rich sources of linoleic acid. In recent years margarines rich in linoleic acid have been marketed. The immediate metabolite of linoleic acid is γ-linolenic acid (GLA). This differs from the normally occurring α-linolenic acid in the position of its double bonds. The position of the double bonds in unsaturated fatty acids has considerable dietary importance, and in order to emphasize the relationships of classes of compounds metabolized from dietary fatty acids, the position of the double bond is often counted from the methyl end of the fatty acid chain. This is called the '*n*'-nomenclature.

α-Linolenic acid is in the *n*-3 series, whilst GLA is in the *n*-6 series, as is naturally occurring linoleic acid. GLA is found in very few natural oils, the best known being evening primrose seed, blackcurrant pip and borage seed oils. Eicosapentaenoic acid (EPA), a natural component of many fish oils, is a mammalian metabolite of α-linolenic acid.

It has been claimed that GLA is effective in curing essential fatty acid deficiency and other related medical problems. EPA is also said to have beneficial properties, especially for the nervous system and brain formation. According to the latest recommendations of FAO/WHO, essential fatty acid intake should include 4–10% of total calories supplied by linoleic acid, with a ratio of linoleic to α-linolenic acid of 5:1 to 10:1. Also, long chain PUFAs namely EPA and DHA are essential in infant diets.

Conjugated linoleic acids (CLA)

Conjugated linoleic acid is a collective term for positional and geometric isomers of linoleic acid. While linoleic acid contains methylene interrupted double bonds between C9 and C10 and C12 and C13, CLA has conjugated double bonds at carbon atoms 10 and 12 or 9 and 11 with possible *cis* and *trans* configurations. The 9-*cis*, 11-*trans* isomer is thought to be the most biologically active form of CLA. Predominantly, CLA occurs in meat and dairy products. For instance, milk contains $5.5 \, \text{mg g}^{-1}$ of fat, while lamb fat has a CLA content of $6 \, \text{mg g}^{-1}$ fat. From studies on several animal species, many researchers have shown that CLA has a significant protective effect with respect to carcinogenic and cardiovascular diseases. CLA in a triglyceride form was as effective as free conjugated linoleic acid. In one study, the addition of either linoleic acid or CLA as 2% of calories resulted in 20% reduction in LDL cholesterol levels. However, reduction in systems of early atherogensis were approximately three times greater in the CLA diet (45%) versus the linoleic acid diet (15%). The present findings indicate that CLA when oxidized produces furan fatty acids which may have antioxidant properties. It is worth emphasizing here that CLA is a newly recognized nutrient that functions to regulate energy retention and metabolism. The feed efficiency and growth rate markedly improved when chicken, rats, mice and rabbits were fed CLA. Their body fat dropped, but body protein increased. Certainly, further research concerning beneficial health attributes of CLA would throw valuable light on the exact biochemical mechanisms about the working of CLA.

P/S ratio

The ratio of polyunsaturated fatty acids to saturated fatty acids is called the P/S ratio. In the COMA Report (DHSS, 1984) and elsewhere, P, the polyunsaturated acids, are taken to include only those with a *cis*, *cis*-1,4-diene structure, i.e. the essential fatty acids, whereas S, the saturated acids may or may not include the mono- or polyunsaturated acdis with one or more *trans* double bond. It is to be recommended that the terms used should be clarified in any case where confusion might occur.

MONOGLYCERIDES

Monoglycerides are partial esters of fatty acids with glycerol in which only one hydroxyl group is esterified while the other two remain free. Commercial monoglycerides are made by reacting triglycerides of fatty acids with glycerol at an elevated temperature. The product is an equilibrium mixture of glycerol, free fatty acid and mono-, di- and triglycerides. The economic equilibrium mixture contains 35–40% of monoglyceride. A second type of product contains over 90% monoglyceride and is made by molecular distillation of the 35–40% product. Monoglycerides are excellent emulsifiers and have been widely used in the food industry (see EMULSIFIERS) in bread, cakes, margarines, ice-creams, etc.

LECITHIN

Lecithin is a phospholipid, known chemically also as phosphatidyl choline. The lecithin molecule contains glycerol and fatty acids, as do the simple glycerides, but also contains phosphoric acid and choline residues. Structural formulae of various phospholipids are presented in Figure 8.2. An α-lecithin is represented by the formula shown in Figure 8.2 and β-lecithin contains the phosphoric acid and choline moieties on the centre carbon atom of the glycerol. Lecithins are widely distributed in body cells and are found particularly in egg yolk and liver.

The most important source of commercial lecithin is soya bean oil. The commercial product contains other phospholipids (cephalin, phosphatidyl serine and inositol). Soya bean lecithin is separated by heating the crude oil with water and centrifuging. The water is evaporated and the crude residual lecithin may be bleached with peroxide. However, where the product is needed for food, unbleached crude soya lecithin should be obtained as the bleached lecithin contains fatty acid peroxides which can lead to off-flavours in the finished food.

Soya lecithin contains 60–75% phospholipid and up to 40% oil. Lecithin has marked surface

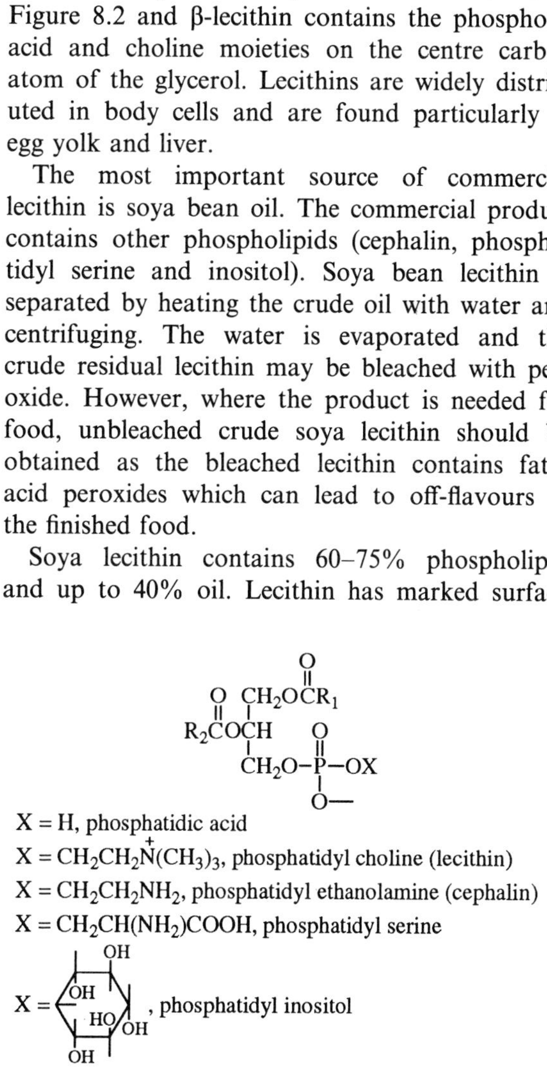

$$
\begin{array}{l}
\qquad\qquad\quad\ \text{O} \\
\qquad\qquad\quad\ \| \\
\quad\ \text{O}\ \ \text{CH}_2\text{OCR}_1 \\
\quad\ \| \ \ | \\
\text{R}_2\text{COCH} \quad\ \text{O} \\
\qquad\quad | \quad\ \| \\
\qquad\quad \text{CH}_2\text{O–P–OX} \\
\qquad\qquad\qquad | \\
\qquad\qquad\qquad \text{O–}
\end{array}
$$

X = H, phosphatidic acid

X = $CH_2CH_2\overset{+}{N}(CH_3)_3$, phosphatidyl choline (lecithin)

X = $CH_2CH_2NH_2$, phosphatidyl ethanolamine (cephalin)

X = $CH_2CH(NH_2)COOH$, phosphatidyl serine

X = (inositol ring structure with OH, OH, HO, OH, OH), phosphatidyl inositol

Figure 8.2 Structure of phosphatides.

activity properties and is widely used in the food industry, for example as an antispattering agent in margarine and for viscosity reduction in chocolate. A review of lecithin production and properties is given in symposium papers (American Oil Chemists Society, 1980).

COLOUR OF OILS AND FATS

Most oils when extracted from oilseeds are green or orange in colour due to the presence of chlorophyll or carotene. Only virgin olive oil is sold as a green coloured oil; the colour of most oils is removed by the BLEACHING stage of the refining process. This leaves the oil as a pale yellow colour.

The chemical identity of the colour components in oils and fats is not fully understood, but the characteristic yellow-red of many natural fats is due to the presence of polyene carotenoid pigments. These pigments derive their colour from the presence in their structure of a sequence of multiple conjugated double bonds which absorb light in the region of 440–450 nm. Of the vegetable oils widely used, palm oil has the highest concentration of carotenoids, usually about 0.03–0.15% (m/m) in the crude, unbleached oil. Figure 8.3 shows the relationship of β-carotene to vitamin A.

Colour may be added to a fat to improve the appearance of a food product. For example, colour is normally added to margarine to give it an attractive yellow appearance. The tendency to avoid the use of synthetic dyestuffs in food has concentrated attention on the use of natural colours, such as carotene and annatto. β-Carotene is the natural colouring matter of butter and is thus an obvious choice for addition to margarine. Often it is used in the form of a dilute solution in palm oil. Crude palm oil contains about 0.1% of a mixture of carotenes and if refined carefully can be used to colour margarine. Annatto seed is the seed of *Bixa orellana*, the main compound respon-

sible for the colour being the carotenoid bixin. The colouring matter is found in the seed coat and may be extracted into a vegetable oil such as soyabean oil. Other food colours may be used in the preparation of pastel icing, ice-creams and cream fillings of biscuits. Local legislation varies with regard to colouring materials permitted.

HYDROCARBONS

Hydrocarbons, both saturated and unsaturated, are present in a very small amount in oils and fats. Among these hydrocarbons, squalene, $C_{30}H_{50}$, containing six double bonds is widely distributed. With the exception of rice bran oil (containing about 0.3%) and olive oil (0.08–1.2%), the squalene content of other oils and fats is lower than 0.05%.

Certain oils like coconut oil may contain trace quantities of polycyclic aromatic hydrocarbons (PAHs). These PAHs probably result from the traditional/local method of drying coconut. It should be pointed out here that most of these PAHs are removed by the deodorization stage of oil processing.

STEROLS

Plant sterols (phytosterols) occur in the unsaponifiable matter of vegetable oils. Cholesterol (a zoosterol) occurs in animal fats and fish oils, but in only trace quantities in vegetable oils and was once considered to be absent from these. Figure 8.4 gives structural formulae of normal, 4-desmethylsterols. The sterols may be characterized and identified by a combination of thin layer chromatography and gas-liquid chromatography according to the International Standard method (ISO 6799). Sterol compositions have been used to prove purity, contamination or adulteration of fats and oils, as each oil type has its own sterol composition. The level of sterols in an oil can be

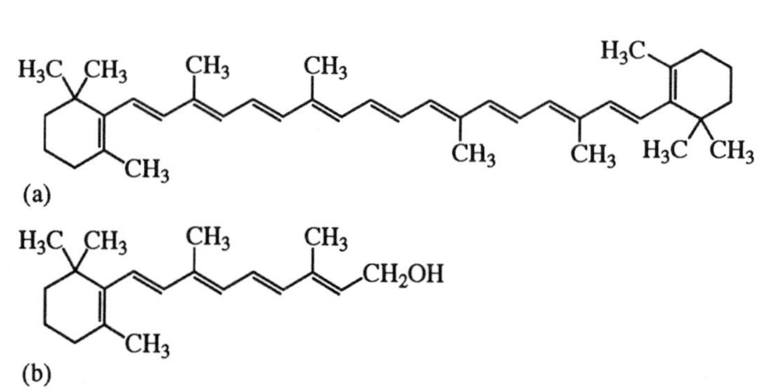

Figure 8.3 Relationship of (a) β-carotene to (b) vitamin A.

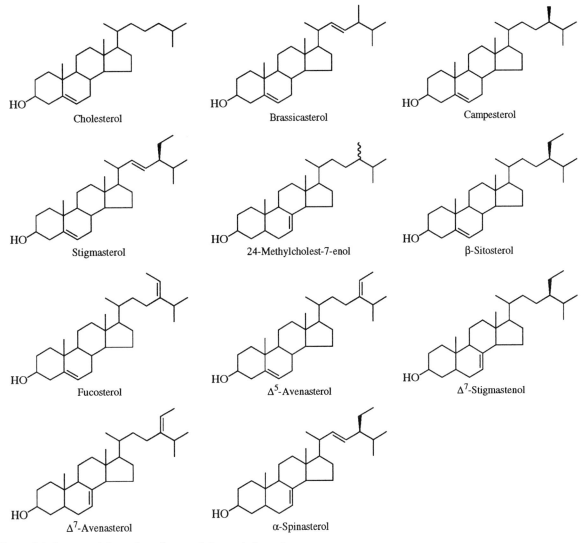

Figure 8.4 Structural formulae of some 4-desmethylsterols.

reduced or modified by processing, however, and skilful interpretation of the results is therefore needed. A comprehensive review on the influence of different stages of processing on the contents and compositions of sterols has been published by Kochhar (1983).

TOCOPHEROLS

Tocopherols are naturally occurring antioxidants (vitamin E). They occur in nearly all vegetable oils, there being several related compounds (Figure 8.5). Tocotrienols also occur, especially in

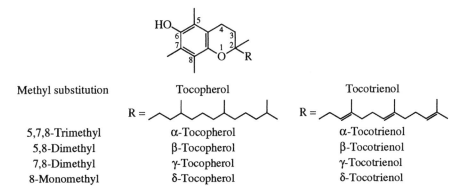

Methyl substitution	Tocopherol	Tocotrienol
5,7,8-Trimethyl	α-Tocopherol	α-Tocotrienol
5,8-Dimethyl	β-Tocopherol	β-Tocotrienol
7,8-Dimethyl	γ-Tocopherol	γ-Tocotrienol
8-Monomethyl	δ-Tocopherol	δ-Tocotrienol

Figure 8.5 Structural formulae of naturally occurring tocopherols and tocotrienols.

Table 8.5 Ranges and mean values[a] of tocopherol and tocotrienol levels in selected vegetable oils (mg kg^{-1})

Tocol	Palm kernel	Coconut	Cottonseed	Soya bean	Maize	Groundnut	Palm	Sunflowerseed	High erucic rapeseed	Low erucic rapeseed
αT	– 13[b]	ND–17 5[b]	136–543 (388)	9–252 (99.5)	23–573 (282)	49–304 (178)	4–185 (89)	403–855 (670)	39–305 244[b]	100–320 (202)
βT	–	ND–11	ND–29 (16.9)	ND–36 (7.7)	ND–356 (54)	0–41 (8.8)	–	9–45 (27)	24–158	16–140 (65)
γT	–	ND–14	158–594 (429)	409–2397 (1021)	268–2468 (1034)	99–389 (213)	6–36 (18)	ND–34 (11)	230–500 430	287–753 (490)
δT	–	ND–2 6	ND–17 (3.3)	154–932 (421)	23–75 (54)	3–22 (7.6)	–	ND–7 (0.6)	5–14 12	4–22 (9)
αT3	– 21	ND–5 5	–	–	ND–239 (49)	–	4–336 (128)	–	–	–
βT3	–	– 1	–	–	ND–52 (8)	–	–	–	–	–
γT3	–	ND–1 19	–	–	ND–450 (161)	–	42–710 (323)	–	–	–
δT3	–	–	–	–	ND–20 (6)	–	tr–148 (72)	–	–	–
Total	– 34[b]	tr–31 36[b]	410–1169 (788)	575–3320 (1549)	331–3402 (1647)	176–696 (407)	98–1327 (630)	447–900 (709)	312–928 686[b]	424–1054 (766)

[a] Mean values are given in brackets.
[b] Typical values.
T = tocopherol, T3 = tocotrienol, ND = not detected.

palm oil, while soya bean oil contains δ-tocopherol. The determination of tocopherol content in vegetable oil used to be difficult as the tocopherols were liable to oxidation during the analytical procedure. In recent years, however, the introduction of HPLC methods, coupled with fluorescence detection, has enabled much more accurate determinations. Table 8.5 lists ranges and mean values of tocopherols and tocotrienol levels in ten commercially important crude vegetable oils.

ANALYSIS OF OILS AND FATS

COMPOSITION AND IDENTITY

Sampling

When the composition or properties of a bulk consignment of oil or fat are in question, a sample is taken and submitted to the laboratory for examination. However, the value of the analytical results is only as good as the sample; a perfect analysis on a non-representative sample is of less value than an imperfect analysis on a fully representative sample. The importance of starting with a fully representative sample cannot be overemphasized and International Standard ISO 5555 therefore lays down procedures for taking samples from bulk lots in a variety of circumstances. One of the most important factors is that the oil or fat

should be sufficiently warm to be fully fluid, otherwise higher melting components will crystallize and settle to the bottom of the container or adhere to its sides. On the other hand, it is important not to damage the cargo by overheating. Table 8.6 gives the recommended temperature limits for sampling.

Solid fats are best melted and rendered completely homogeneous, e.g. by stirring, before any sample is taken. Sometimes this is not feasible, however, and in these cases a metal fat trier with a C-shaped cross-section is used. It is pushed into the fat, rotated and withdrawn with a plug or core of the fat.

Table 8.6 Temperature limits for sampling

Fat or oil	Temperature (°C) min.	max.
Olive, maize, rape, safflower, sesame, soya, sunflower, linseed	15	25
Groundnut (arachis), tung	20	25
Cottonseed	20	30
Castor	25	35
Whale, fish oils	30	35
Sperm	35	40
Coconut	35	45
Palm kernel	40	45
Shea nut butter	45	50
Greases	45	55
Lard, palm oil	50	55
Tallow	50	60
Palm stearin	55	70

Acetyl value

The acetyl value is defined as the number of milligrams of potassium hydroxide required to neutralize the acetic acid formed when one gram of acetylated fat is saponified. The hydroxyl value is defined as the number of milligrams of potassium hydroxide required to neutralize the acetic acid capable of combining by acetylation with one gram of an oil or fat. Both values are thus measures of the hydroxyl groups present in a fat; they are effectively the same for values of less than 2.0. Hydroxyl groups may occur in a fat as a result of the presence of mono- or diglyceride emulsifiers, or they may be present in a fatty acid combined with glycerol in a triglyceride, a typical example being ricinoleic acid the main fatty acid of castor oil.

The preferred forms of the test are specified in British Standard 684, Section 2.9: 1977 (as amended by amendment slip 2653, July 1978). In this, the fat is acetylated with a measured quantity of acetic anhydride in pyridine; the excess acetic acid is decomposed by boiling water and the acetic acid formed is titrated with sodium hydroxide solution in ethanol. A control test with acetic acid in pyridine, but without the fat, is carried out to determine the amount of acetic anhydride available for acetylation and a similar test is carried out with fat but omitting the acetic acid, to determine the free fatty acids present. Because of the lengthy nature of this test the determination is not often carried out as a routine.

Ash

As oils and fats are organic substances, they should burn completely and leave no residues. Measurement of the inorganic residue after ignition is therefore a useful test for the determination of inorganic impurities. In the test a weighed amount of about 10 g of fat is heated in a crucible to the ignition of the fat and left to burn. When the burning ceases, the crucible is heated to a dull red heat until no more change is observed. The ash content is calculated from the initial weight of sample and the difference in weights of the empty crucible and the crucible together with the ash. The method in British Standard 684: Section 2.2 may be contrasted with the IUPAC procedure, which is claimed to avoid loss of relatively volatile alkaline compounds.

Density

The relative density or specific gravity of an oil (at $t/20°C$) is defined as the ratio of the apparent mass, determined by weight in air, of a given volume of the oil at $t°C$ to that of a same volume of water at 20°C; whereas the apparent density (at $t°C$) is the apparent mass in grams, determined by weighing 1 ml of fat at $t°C$ in air. The method commonly used in the UK is that in British Standard 684: Section 1.1. The density of an oil should always be determined at a temperature of at least 10°C above its melting point but preferably below 65°C.

The main value of a density determination is in the calculation of the weights of oil in large tanks. The volume of the oil in the tank can be readily determined by calibration and its temperature can be measured. Careful measurement of the density of the oil at that temperature can then be used to calculate the weight of oil. A term normally used for oil densities at the dock side is 'litre weight in air', a form in which density may be expressed in commercial contracts.

Flash, fire and smoke points

These three tests are a guide to the content of volatile inflammable organic materials in an oil. They all involve observation of the surface of a fat as it is heated. The smoke point is defined as the temperature at which the sample begins to smoke when tested under the specified conditions of the test; the flash point is the temperature at which combustible products are volatilized in sufficient quantity to allow instantaneous ignition, and the fire point is the temperature at which evolution of volatile proceeds with sufficient speed to support continuous combustion.

The smoke point is of most benefit in assessing the quality of a frying oil, when it is mainly the FFAs produced in the frying operation which contribute to the smoke haze. The flash point is of most value with regard to the safety during storage or transport of an oil; the contracts issued by Federation of Oils, Fats and Seeds Association (FOSFA International) for oils traded in bulk specify that the flash point should be above 121°C (250°F). There are two methods for the determination of flash point, namely those of the open or closed flash tests. The two methods give different results.

Iodine value

The iodine value (IV) of a fat is a measure of its unsaturation and is a useful criterion of purity of identity. The measurement is based on the fact

Table 8.7 Analytical properties of some oil and fats

Oil or fat	Refractive index n_D^{40}	Titre (°C)	Melting point (°C)	Saponification value	Unsaponifiable matter	Iodine value
Babassu	1.448–1.455	22–23	24–26	245–256	0.2–0.9	14–18
Borneo tallow (Illipe)	1.456–1.457	51–53	37–39	190–205	0.7–2.0	
Castor-bean	1.466–1.473		−12 to −10	176–187		
Cocoa butter	1.456–1.459	45–50	31–35	188–189	0.1–0.2	33–42
Coconut	1.448–1.449	20–24	23–26	248–265	0–0.5	7.5–10
Corn (maize)	1.465–1.468	14–20	−12 to −10	187–193	0.5–2.8	103–128
Cottonseed	1.464–1.468	30–37	−2 to 2	189–198	0.2–1.5	99–115
Groundnut	1.460–1.465	26–32	−2	187–196	0.2–0.8	84–105
Kapok	1.460–1.466	27–32	30	189–195	0.5–1.0	
Kokum	1.456	60	40–42	192	ca. 2.3	
Linseed	1.474–1.475	19–21	−20	188–196	0.1–1.7	155–205
Mustard-seed	1.461–1.469	6–8	−16	170–184	0.7–1.5	106–113
Olive	1.460–1.463	17–26	−3 to 0	182–196	0.7–1.1[b]	80–88
Palm	1.449–1.456[a]	40–47	33–40	190–207	0.3–1.2	50–54
Palm-kernel	1.449–1.452	20–28	24–26	230–254	0.2–0.8	16–19
Rapeseed (high erucic)	1.465–1.469	11–15	−9	168–181	0.2–2.0	97–110
Rapeseed (low erucic)	1.465–1.467	20–22	−20	188–193	0.2–1.8	106–126
Rice bran	1.466–1.471	25	0–8	179–195		
Safflower	1.467–1.469	15–18	−18 to −13	186–198		135–150
Sal fat	1.456–1.457	51	33–39	186–194	0.3–1.3	
Sesame-seed	1.465–1.469	20–25	−4–0	187–195	0.9–2.0	104–120
Shea nut	1.463–1.467	49–54	37–42	178–190	4–8	
Soya-bean	1.467–1.470	20–21	−23 to −20	188–195	0.5–1.6	125–136
Sunflower-seed	1.466–1.469	16–20	−18 to −16	188–194	0.3–1.3	85
Walnut	1.469–1.471	14–16	−16 to −12	189–198	0.5–1.0	
Butter (cow's)	1.452–1.457	33–38	28–35	233–240		25–42
Beef tallow	1.457–1.459	40–47	40–48	190–202	0–1.0	45–57
Lard	1.458–1.461	32–43	33–46	192–203	0–0.8	59–70
Mutton tallow	1.455–1.458	43–48	44–51	192–197	0–1.0	35–46
Cod liver	1.470–1.475			180–190	0–1.5	
Herring	1.465–1.467	23–27		179–194		
Menhaden	1.472–1.475	31–33		189–193		
Whale	1.465–1.472	22–24		185–194	1.2–2.0	110–135

[a] At 50°C
[b] Pressed oil; extracted oil up to 2.5

that halogen addition occurs to unsaturated bonds until these are completely saturated. Because substitution as well as addition reactions can occur it is important to carry out the determination under carefully controlled and standardized conditions. Not all unsaturated bonds are alike in reactivity and those near a carbonyl group hardly absorb iodine. These acids are, however, rare. When the double bonds are conjugated, they react more slowly than non-conjugated double bonds.

Several methods are available, those in common use being the tests of Wijs, Hanus and Rosenmund–Kuhnhenn. The original Wijs method is now adopted almost universally and is described in British Standard 684: Section 2.13: 1981 (ISO 3961-1979).

The main differences between the methods cited are in the halogenating agents, Wijs' method uses iodine monochloride, the Hanus method uses iodine monobromide and the Rosenmund–Kuhnhenn method employs the milder pyridine–sul-phate–bromide reagent. All of the methods give similar results when applied to the majority of ordinary oils and fats, but differences can arise when large amounts of sterols or other unsaponifiable matter are present. Iodine values of common oils are given in Table 8.7.

Moisture and volatile matter

The amount of water in a fat may be determined by several methods. The titrimetric Karl Fischer method described in BS 684: Section 2.1 is very useful for the determination of water in fats or fatty foods having a water content in the range of 0.5–1%. Three rapid and more frequently applied methods to determine moisture and other volatile matter by drying are described in BS 684: Section 1.10. These are a method using a sand bath or hot plate which is applicable to all oils and fats; a method using a drying oven, which is applicable only to non-drying fats and to oils with an acid value of less than 4; and a method using a stream of nitrogen to protect the oil from oxidation during the test, which is particularly applicable to drying oils such as linseed oil. The determination of the moisture content can be particularly important in commercial transactions of large quantities of oils and fats.

Refractive index

The refractive index (RI) is a property closely related to those of density and molecular weight and to the constitution of oils. It is helpful in the identification of oils. It can be determined either by means of the Zeiss refractometer, which reads on an arbitrary scale, or by means of the Abbé refractometer, which gives the true refractive index. Monochromatic light is used in the measurement of RI, and when this is the sodium D line the RI measured in the refractometer is given the abbreviation n^t_D, where t is the temperature of measurement. As RI is temperature dependent, the instrument and the oil sample must be thermostatted during the measurement. This is normally achieved by circulating water from a thermostatically controlled bath.

RI has been used to follow the course of hydrogenation, but in this case the temperature of measurement must be sufficiently high to ensure fluidity throughout the whole course of the hydrogenation. The RIs of a number of oils and fats are given in Table 8.7.

Slip/melting point

Natural fats are complex mixtures of glycerides and have no sharp melting point. Fats also exhibit polymorphism (i.e. solidification in more than one crystalline form, each one possessing a different melting point). Melting points of the constituent fatty acids can, however, be accurately determined. This property of the liberated fatty acids of a fat is called the titre and is described in British Standard 684: Section 1.6. Melting points and titres of several fats are given in Table 8.7.

To obtain reliable results for the melting point of whole fats, it is necessary to follow carefully a standardized procedure. The melting point method most commonly used in the UK is the slip melting point, described in British Standard 684: Section 1.3. This is the temperature at which a column of fat of specified length rises in an open capillary tube suspended in a water bath of gradually increasing temperature. Stabilization of the fat prior to measurement of the slip melting point is extremely important, different procedures of stabilization being necessary for fats with pronounced polymorphic behaviour such as cocoa butter. It is also important to ensure that the capillary tubes used in the melting point determination have parallel sides with no tendency to taper toward one end.

Titre

The titre is the value of the maximum of a temporary temperature rise during the crystallization of the constituent acids of a fat. If the latent heat is not sufficient to cause a rise in temperature, the temporary interrruption of the cooling process is considered as the titre. The titre is determined according to BS 684: Section 1.6. Titres of several fats are given in Table 8.7.

Unsaponifiable matter

'Non-saponifiable' and 'unsaponifiable' are synonymous terms and are used to refer to that material present in oils and fats which, after reaction of the oil or fat with caustic alkali, remains insoluble in the aqueous alkali and non-volatile during drying. The amount of unsaponifiable matter in a fat is determined according to BS 684: Section 2.7.

The unsaponifiable matter of a fat contains sterols, higher alcohols, hydrocarbons, pigments and in some cases (e.g. fish liver oils) vitamin A. Crude fats contain varying amounts, up to 8% in the case of shea oil, while refined oils contain lower amounts of unsaponifiable matter. Halibut liver oil contains a high level (7–20%) of unsaponifiable matter, most of which is vitamin A. Determination of the unsaponifiable matter content can be useful if contamination of the oil with a mineral oil or other non-triglyceride contaminant is suspected. Unsaponifiable matter contents of several oils are given in Table 8.7.

Identification agents

In some countries the addition of identification agents to most food fats is compulsory by law. The process is intended to make adulteration of noble fats such as olive oils easily detected. Development of modern analytical techniques should make such additions unnecessary, however. Up to 5% sesame oil may be added to a fat blend, after which it can be detected by the Baudouin test (below). An alternative is 0.2% starch, which can be detected by the blue colour developed on addition of iodine.

In the EU, butter is sometimes sold at a subsidized price in an effort to reduce the butter 'mountain'. This cheap butter, known as 'intervention butter', may only be used for specified purposes. In order to control its use and detect fraudulent use in non-permitted outlets, various identification agents are added. These are stigmasterol, ethyl butyrate, vanillin, enanthic acid (C7:0) and carotene, in various combinations.

Tests for identity or adulteration

Baudouin test

Crude sesame oil gives a deep pink colour when it is reacted with ethanol and ammonia and then with acidified sucrose solution. The test is known as the Baudouin test and is described in Section 2.30 of British Standard 684. The test enables detection of small amounts of sesame oil in any other product. In Italy, all oils other than olive oil are treated with crude sesame oil to enable Government officials to detect adulteration of olive oil. In India the test may be used to detect adulteration of ghee.

Halphen test

The presence of cottonseed oil in vegetable or animal fats or oils can be detected by the Halphen test. The fat is heated with a solution of sulphur in carbon disulphide in the presence of amyl alcohol. The presence of cottonseed oil is shown by the appearance of a red colour due to reaction of the constituent cyclopropenoid acids. Some oils, such as kapokseed oil, also contain these cyclopropenoid acids and can therefore confuse the interpretation of the test. Hydrogenation removes the cyclopropenoid acids from cottonseed oil and hydrogenated cottonseed oil therfore gives a negative result.

Reichert-Meissl, Polenske and Kirschner values

All three values are a measure of the shorter chain fatty acids which are present in some oils, especially dairy butter, coconut and palm kernel oils. The Reichert-Meissl value is a measure of the water-soluble volatile acids, while the Polenske value is a measure of the water-insoluble volatile fatty acids. The Kirschner value is a measure of the water-soluble fatty acids not heavier than butyric acid. The determinations are carried out according to the procedure in BS 684: Section 2.11, but they have fallen into disuse in recent years in favour of the accurate determination of fatty acid composition by GC.

Saponification value

The saponification value (SV) is defined as the number of milligrams of potassium hydroxide required to neutralize the fatty acids liberated on the complete hydrolysis or saponification of one gram of the fat or oil. It is determined by refluxing 2 g of the oil with 25 ml of 0.1 M alcoholic potash for 1 h, on a water bath, then back titrating the excess alkali. The value obtained is subtracted from the value for a blank carried out simultaneously.

The esters of low molecular weight fatty acids require the most alkali for saponification, so that the saponification value is inversely proportional to the mean molecular weight of the fatty acids in the glycerides present. Because many oils have similar saponification values, the test is not universally useful in establishing identity or indicating adulteration and should always be considered along with the iodine value for these purposes. Fats with relatively high proportions of the lower fatty acids include coconut oil (SV = 255), palm kernel oil (SV = 247) and butter fat (SV = 225) and their presence may frequently be detected in this way. Paraffin has a negligible SV and can therefore be detected and estimated. The SV of some oils and fats are given in Table 8.7.

Fatty acid profile

With the widespread application of gas chromatography (GC) has come the possibility of rapid and convenient estimation of all the component fatty acids in the glycerides present in an oil or a mixture, and this is now a powerful tool for establishing the identity or purity of any oil or fat.

In the International Standard method (ISO 5508: 1990) the glycerides are first hydrolysed to the component fatty acids with sodium hydroxide and neutralized, then the fatty acids converted to their methyl esters with methanol under suitable conditions and extracted into heptane or hexane for GC analysis. A portion of the solution is passed through the gas chromatograph with a suitable detector. Successive peaks corresponding to the individual fatty acids are noted by the chart recorder and from their relative areas the proportions of the fatty acids are calculated.

In addition to the chemical diagnostic value, a simplified version of the fatty acid profile, showing the totals of the saturated, monounsaturated and polyunsaturated fatty acids, gives useful nutritional information (for fatty acid profiles of a number of fats and oils see Tables 8.2, 8.3 and 8.4).

FLAVOUR AND TEXTURE OF OILS AND FATS

The flavour of a properly refined and deodorized oil is mild and bland. Off-flavours can develop on storage and these are variously defined as rancidity, hydrogenation off-flavours or reversion flavours. The food chemist is often concerned with preventing the development of these off-flavours in the processed oils.

Flavours are often added to an oil or to the food recipe when these are not provided by the other ingredients. In the case of margarine, the flavours are added to give the margarine a flavour approaching that of butter. Early work on butter showed that the lower fatty acids and diacetyl were important flavour constituents. Addition of these compounds to margarine gives an approximation of the flavour of the butter. A comprehensive examination of the trace components of butterfat has shown the C8–C14 lactones of hydroxy fatty acids, particularly γ- and δ-lactones, to be important. Other flavour components of butter were identified as acetylmethylcarbinol and the lower chain fatty acids. Over 80 flavouring materials were identified. Much of this information is available in the patent literature and clearly many flavour blends are possible, each contributing its own specific taste to a fatty product.

Mono- and diglycerides and the nature of the free and combined fatty acids also play a role in flavour, and the addition of specific manufactured mixtures of such compounds is also practised. Heated fat generally has its own flavour and this is very marked when butter is used for frying or other similar purposes such as baking. Margarines or shortenings which develop the appropriate flavour attributes during cooking are said to have 'carry-through flavour'.

The texture of a fat can also influence the flavour, in that it influences the release of fat-soluble flavour components. Liquid oils clearly release the components very quickly, whilst solid fats which do not melt on the palate are unlikely to release any flavour at all. Hard fats which melt quickly on the palate, such as cocoa butter, release their flavours quickly giving a sudden flavour impact, whilst fats such as suet or some types of cocoa butter substitute melt progressively and give a progressive release of fat-soluble flavours. In this case, the senses become dulled by the initial moderate impact of the flavour and do not respond to the slow release of additional flavour as the fat slowly melts. It is partly as a result of this that some chocolate coatings appear to have inferior flavours.

Solid/liquid ratio

The ratio of solid fat to liquid oil is the most important physical property of an oil/fat mixture, as it will govern the physical properties of any fatty food into which it is incorporated. The proportion of solid fat changes with temperature and at ambient can vary from zero in a liquid frying oil, to 20–30% in margarines, around 50% in shortenings and 80% or more in fats intended for chocolate or coatings. The ratio of solid fat to liquid oil was formerly measured by dilatometry, but more recently Nuclear Magnetic Resonance (NMR) has been used.

As the ratio of liquid oil to solid fat varies with temperature it is important to control the temperature carefully during the determination and to report the temperature along with the results. This is normally carried out by reporting the dilatation at 20°C as D20, or the actual solid content as determined by NMR at 20°C as N20.

Fats have complicated polymorphic and crystallization behaviour and it is therefore necessary to temper and stabilize the fats carefully at the measuring temperature prior to the solid/liquid ratio measurement. This is particularly the case with fats of the cocoa butter type, which crystallize in the β-modification. Standard procedures for tempering and stabilization of fats prior to solid/liquid fat determinations have therefore been established. The procedure used in British Standard 684: Section 1.12 (Dilatation) is also used in the IUPAC methods for dilatometry and solid fat content by NMR. The largest source of error with both techniques is the time and temperature of stabilization, an aspect often overlooked, as duplicate determinations are often conducted in the same thermostatic bath.

Cooling curve

The cooling curve of a fat is a measure of its crystallization properties. It is particularly useful in the assessment of cocoa butter and cocoa butter replacement fats, as it can form a guide to the tempering behaviour of the chocolate. Two different methods are used, namely the Schukov cooling curve and the Jensen cooling curve. The latter is described in British Standard 684: Section 1.13.

Dilatometry

Dilatometry is a term applied to the measurement of the temperature/volume relationships of a fat over the range of temperature during which it changes from solid to liquid. The isothermal expansion on melting of the fat is measured, which provides information about the ratio of solid fat to liquid oil in the fat over the temperature range used. These measurements have great practical value in providing an assessment of the suitability of a fat or fat blend for use in any particular food product. Dilatations were widely used in manufacturing industry for selecting fat blends for margarines, cocoa butter replacement fats, etc.

The method used in Europe is described in British Standard 684: Section 1.12, and gives dilatation units measured in mm^3 per 25 g of fat. In the BSI system, fats may have dilatation values of up to 2500. The American Oil Chemists Society (AOCS) method (Cd 10–57) gives the dilatation values in ml kg^{-1} of fat. These values are, therefore, about one twenty-fifth of those obtained in the BSI method. This conversion factor is not accurate, however, as the tempering procedures in the two methods differ. The fact that the AOCS method gives values ranging from 0 to about 100 has led to the mistaken impression that the AOCS dilatation, also known as the solid fat index (SFI), is a true measure of the proportion of solid fat present.

Dilatometry is at present rapidly being replaced, however, by wide line, continuous wave and pulse NMR measurements of the solid fat-to-liquid oil ratio.

Hardness measurements

It has for a long time been customary to describe the hardness of fats by reference to their melting points or solid fat content. This method gives a rough and ready idea of hardness, but in more recent years direct measurement of hardness by use of a penetrometer has become established.

Nuclear magnetic resonance

Nuclear Magnetic Resonance (NMR) spectroscopy can distinguish between the protons of liquid oil and solid fat. It can therefore be used to measure the liquid/solid ratio in fats and oils and also to measure the oil content of oilseeds or other products. As the protons in water also give a signal, samples should be dried before evaluation.

Two forms of NMR instrument are on the market, namely continuous wave wide line NMR and pulsed NMR. In the former, the instrument is adjusted to respond only to the signal from the proton in the liquid oil phase. The intensity of this signal is measured and related to the amount of sample used. As the signal varies with temperature, a correction factor must be derived from measurements on an oil which is fully liquid at all temperatures of measurement. Olive and soya bean oils are frequently used for this purpose.

A similar measurement enables the determination of oil content in fatty foods or oilseeds. This technique has now become established as the standard IUPAC method 2.323. The method has been shown by collaborative test to be highly reproducible between different laboratories. However, in this author's view it has not been adequately shown that the method gives exactly the same answer as solvent extraction methods for the determination of oil content of oilseeds. Pulse NMR is used primarily for the determination of solid fat content. Instruments based on two techniques have been developed. In one of these the liquid signal only is measured, the procedure therefore relating closely to that of the wide line instrument. In a second version of this technique, the one most favoured in Europe, solid and liquid signals are measured and distinguished by the rapidity with which the excited protons relax in the liquid and solid phases. Separate signals from the liquid and solid phases are obtained and converted into a digital readout showing the actual percentage of solid fat on an illuminated display. A general review of the subject is given by Waddington (1986).

Plasticity

Margarine, cooking fat and other semi-solid food fats consist of discrete crystalline particles embedded in a considerable proportion of liquid oil. There is some loose adhesion between the crystals which breaks down rapidly when the fat is subjected to working and a shearing stress is applied; this characteristic is known as plasticity.

The plasticity of a food fat is of great practical importance. It determines, for instance, the spreadability of margarine, the shortening properties of baking fats and the creaming power of cake margarine. It can be controlled during manufacture, important factors being:

(i) Content of solids
(ii) Size and shape of crystals
(iii) Persistence of crystal nuclei during heat treatment
(iv) Mechanical working of the fat etc.

Plasticity per se is difficult to measure and instead the consistency of a fat at different temperatures is measured. Penetrometers are often used for this purpose.

Thermal analysis

Thermal analysis of fats may be by differential thermal analysis (DTA) or differential scanning calorimetery (DSC). In both techniques a small quantity of fat is slowly heated and compared with a reference sample. In DTA the temperature of the fat is compared with a reference, whilst in DSC the temperature of the fat is kept the same as that of the control, the heat necessary to do this being measured. Thermal transitions such as melting are displayed by a fall in temperature of the sample relative to the control, or a necessity for the supply of extra heat. Cooling runs can also be carried out, in which case the crystallization phenomena are followed. It has been claimed that the solid-to-liquid ratio of fats may be determined by these techniques but this author is sceptical about this approach as it is impossible to obtain continuous baselines before and after melting or crystallization, because the specific heat of a solid is always different from that of its liquid. Thermal analysis can also be used to study polymorphism, but as the thermal history of the fat has a pronounced influence on the shape of the thermograms, the results of thermal analysis can be very difficult to interpret. Some heating rates can give rise to spurious peaks, which again hinder interpretation.

OTHER QUALITY TESTS

Colour measurement

The colour of an oil is of considerable commercial importance. Pale, bright colours are desired in refined oils as this is regarded as the criterion of quality and purity and also facilitates production of finished foods with a desired colour shade.

Colour removal is generally effected by BLEACHING. Oil colours are measured in the Lovi-bond system by comparing the colour of the cell containing the oil with that of coloured glasses, the glasses being adjusted until a good match is obtained. Normally, only red and yellow colour units are quoted, although blue and neutral glasses may also be provided. There are a number of different Lovibond systems and care must be taken to specify which system is being used and the size of cell used when comparing the colour ratings of different oils. In the UK, British Standard 684: Section 1.14 is normally used for colour measurement. Staff making such colour measurements should, of course, be tested for colour blindness, an aspect sometimes overlooked.

Automatic versions of the Lovibond Colorimeter are now available, but it is generally accepted that these may not always be suitable for measurement of colours in crude oils.

Free fatty acids

The presence of free fatty acids (FFAs) in an oil or fat is an indication of previous lipase activity, other hydrolytic action or oxidation. The FFA content as commonly determined is a measure of the uptake of caustic alkali during the titration. The values may be quoted as acidity or as a percentage by weight of a specified fatty acid in the oil. In the UK acidity and FFA content are determined by the method in British Standard 684: Section 2.10.

Peroxide value

Oxidation of an unsaturated oil or fat takes place via the formation of hydroperoxides. The hydroperoxides subsequently decompose into secondary oxidation products, the majority of which have unpleasant odours or flavours. Although hydroperoxides themselves have no off-flavours, they are an important aspect of rancidity development.

The method used for the determination of peroxide value in the UK is that described in British Standard 684: Section 2.14: 1976. In this test a solution of the fat in a mixture of acetic acid and chloroform is treated with a solution of potassium iodide. The peroxide oxygen liberates iodine from the potassium iodide and the liberated iodine is then titrated with a solution of sodium thiosulphate.

Specific extinction in ultraviolet light

Oxidation of an unsaturated fat leads to the formation of conjugated dienes which have characteristic ultraviolet absorption spectra. Different absorption spectra are associated with conjugated trienes which may be generated during earth bleaching of unsaturated oils at high temperatures. The specific extinction in UV light is determined by British Standard 684: Section 1.15.

Accelerated stability tests

(i) The Swift test is a method of measuring the resistance of a fat to oxidation. It is sometimes called the active oxygen method or AOM. In this test the oil is heated to 100°C and air is then bubbled through it. Samples of the oil are periodically taken and the peroxide value determined. The increase in peroxide value is plotted against time, and a sharp break in the curve is noticed when the natural resistance to oxidation is exhausted and oxidation proceeds at a progressively increasing speed. The time during which the oil exhibited resistance to oxidation is called the induction period (IP), or Swift test life. There are various methods of interpreting the results from this test, that used in the UK being described in BS 684: Section 2.25. The AOM procedure is also standardized by the AOCS (method CD 12–57).

(ii) In recent years the automated Rancimat apparatus has appeared in which the effluent gases are led into distilled water and the electrical conductivity of the water automatically measured (Allen and Hamilton, 1995). The time at which the conductivity rapidly increases is taken as the end of the induction period.

(iii) The oxidative stability instrument (OSI), fully automated equipment, is available commercially to measure oxidative stability of 24 samples of oils and fats. The OSI is also based upon the measurement of conductivity of the water in a conductivity cell through which the resulting volatiles are carried continuously. Software algorithms compute the induction period (time in hours) for the oil or sample based upon the mathematical calculations of the second derivatives of the data curve. The processor stores the data for recall and manipulation.

(iv) The Oxidograph, employing newly developed pressure gauges and specially designed sample vessels, is based on the Sylvester principle for measuring oxidative stability of oils and fats. The oxygen in the air above the sample reacts with the sample and the resulting pressure drop is measured at intervals and plotted against time. A sharp decrease in pressure indicates the end of the induction period.

The induction period data, from the above tests, are then employed to predict the shelf life of oils and fats.

Thiobarbituric acid (TBA) value

The TBA value is a guide to the oxidation or deterioration of fats. TBA, it is believed, reacts with malondialdehyde formed during the oxidative decomposition of polyunsaturated fats to form a coloured product. One attraction of the test is that it can be carried out on whole foods and may pick up oxidation damage to materials other than the triglyceride fats themselves.

Polar compounds in heated oils

Used frying oils deteriorate through oxidative polymerization of the unsaturated fatty acids. This leads to increases in viscosity, darkening of colour and decrease in the nutritional quality of the oil. The polar compounds formed during frying can be determined by a column chromatographic separation followed by weighing of the separated polar fractions. It is recommended that the oil be discarded if the level of polar compounds reaches 27%.

RANCIDITY

Freshly deodorized oils and fats have bland flavours, but off-flavours develop on storage, some oils being more resistant to the development of off-flavours than others. When an oil has developed off-flavours it is known as a rancid oil. The phenomenon is called rancidity. There are several forms of rancidity, the two most common being hydrolytic rancidity and oxidative rancidity. In hydrolytic rancidity a catalyst such as a residue of alkaline soap, an enzyme, or an active mould or yeast together with moisture cause hydrolysis of the constituent triglycerides. This leads to the liberation of FFAs. In most oils a low level of FFAs is no problem, but in the lauric oils (palm kernel and coconut) the liberated acids have a distinctive soapy taste and the phenomenon is therefore also known as 'soapy rancidity'. Oxidative rancidity is, as the name implies, caused by oxidation of the fat (Kochhar, 1993). This type of rancidity may be alleviated by adequate provision of antioxidants.

ANTIOXIDANTS

When fats are exposed to atmospheric oxygen they gradually become oxidized. Some fats oxidize more readily than others and develop off-flavours. This is known as oxidative rancidity and it may also give rise to deleterious nutritional effects, for example, destruction of essential fatty acids and vitamins. Any substance which retards the oxidative deterioration of fats may be referred to as an antioxidant.

Tocopherols which occur naturally in vegetable oils have good antioxidant properties and may provide effective protection. Animal fats are deficient in natural protection and may need added antioxidants, so many fats required to have long shelf life, such as certain baking fats, or oils used for frying. In the latter case, oxidation can occur on the fried product which may have a high fat content and a large surface area.

The addition of artificial antioxidants is controlled by law and only those substances which have been rigorously tested toxicologically are permitted. In EU countries permitted synthetic antioxidants include n-propyl, n-octyl and n-dodecyl gallate, butylated hydroxyanisole (BHA) and butylated hydroxytoluene (BHT). Adverse reports were made about the physiological effects of BHT in the early 1960s but these have not been substantiated.

Combinations of the antioxidants, with one another or with citric acid, may be beneficial. Although citric acid would not normally be classed as an antioxidant it complexes with any trace transition metal catalyst present, neutralizing its catalytic effects on oxidation.

Tocopherols, particularly γ-tocopherol, possess good 'carry-through' properties. 'Carry-through' refers to the ability of certain components to escape destruction on baking or frying. BHA and BHT also possess good carry-through properties during baking, but their volatility gives them reduced effectiveness during frying operations, BHA being less volatile than BHT.

Antioxidants function in two ways. They may act as oxygen scavengers, removing oxygen from the oil, or by virtue of their ability to inhibit autoxidation by capturing free radicals (Hudson, 1990). General reviews of oil rancidity and antioxidant effect are given by Allen and Hamilton (1995).

RELATED COMPONENTS AND TOPICS

ACETOGLYCERIDES

Acetoglycerides are fatty acid glycerides in which at least one of the hydroxyl groups of the glyceride molecule is esterified with acetic acid. Thus, acetostearin, acetopalmitin and aceto-olein are names used to decribe acetoglyceride mixtures in which the predominant acid present in addition to aceitic acid is stearic, palmitic or oleic acid, respectively. Acetoglycerides are manufactured either by direct acetylation of partially esterified glycerides with acetic acid or by intersterification of appropriate oils and fats with triacetin.

Acetoglycerides possess unusual physical properties which are valuable in food products. They are non-greasy, waxy, translucent solids which are highly flexible and can be formed into films that will bend without cracking. This is due to the existence of acetoglycerides in a stable α-polymorphic form, which contrasts with that of most other fats. Acetoglycerides are low melting and are generally stable to heat and oxidation. Because of their physical nature they are suitable as edible coatings for food products. They can be applied by dipping and are preferred to paraffin wax for two reasons. On one hand, they are less brittle and, on the other hand, they have an enhanced consumer appeal over that of paraffin wax, which is criticized by some customers who doubt the wisdom of using mineral hydrocarbons in foods. Acetoglycerides can be blended with hard fats and thus they have been used to coat cheese, nuts, chocolate pieces, meat and poultry. When blended with hard fats, acetoglycerides form eutectics with other glycerides in much the same way as oleic acid glycerides. In fact, the acetic acid chain appears to influence the interaction of acetoglycerides in determining eutectic or solid solution behaviour in much the same way as an oleic chain in the corresponding oleic glyceride.

Acetoglycerides are useful constituents of shortenings and increase the plastic range of the shortening; that is, they reduce the tendency of the shortening to be too hard at low temperatures and too soft at high temperatures. They are also said to confer improved spreadability in margarine, perhaps due to their eutectic forming behaviour. Their use as salad oils and mayonnaise substitutes has also been suggested. Confectionary uses of acetoglycerides include use as an enrobing fat for chocolate, as slab dressings and mould release agents and as a substitute for chicle. In baking they can be used as pan release agents.

The physiological properties of acetoglycerides

appear to be similar to those of the conventional long chain triglycerides, while digestibility compares favourably with that of a commercial shortening. Acetoglycerides are permitted food additives in the EU, USA, Canada and New Zealand.

ANTISPATTERING AGENTS

When margarine is used for frying (a frequent practice in continental Europe, rare in the UK), bubbles of water vapour escape and burst, causing hot fat to fly. This unpleasant effect is known as spattering. Some emulsifiers reduce the effect if incorporated in margarine and these materials are called antispattering agents. Egg yolk was formerly widely used, but this has been replaced by citric acid esters of mono- and diglycerides and blends of these with lecithin. Polyglycerol esters of fatty acids, polydimethylsiloxane and monostearin sodium sulphoacetate are also effective. Salted margarines spatter less than salt-free margarines.

COMA REPORT

See Chapter 19

DIGESTIBILITY

The digestibility of a fat is the percentage of total ingested fat which is absorbed. Vegetable and animal fats melting below 50°C are almost completely absorbed by humans, but those having higher melting points are less easily digested. There has been some suggestion that digestibility may depend on chain length and patients suffering from tropical sprue are advised to eat medium-chain-length fats. These are fats in which the fatty acids are predominantly those with 6, 8 and 10 carbon atoms.

EMULSIFICATION

Emulsification is the dispersion of one liquid phase, in the form of fine globules, in another liquid phase with which it is incompletely miscible. In the food industry the two phases most commonly emulsified are oil (O) and water (W), thus forming either oil-in-water (O/W) or water-in-oil (W/O) emulsions. Emulsifiers are usually added to one of the phases to promote ease of formation or stability of the emulsion. The emulsifier reduces the interfacial tension between the two phases and makes possible the formation of the greatly enlarged interfacial area with a much reduced energy input via mechanical agitation. In addition, the emulsifier can stabilize the formed emulsion, e.g. by formation of a semi-rigid interfacial film, by its relative partial solubility in the two phases or by its contribution to the formation at the interface of an electrical double layer of charge which inhibits coagulation.

The particle size of an emulsion is generally decreased by more vigorous agitation during its preparation, by arranging for the viscosities of the two phases to be more nearly equal and by use of the correct emulsifier. The main factors in the choice of emulsifying equipment are the apparent viscosity in all stages of manufacture, the amount of mechanical energy input required, the type of agitation and the heat exchange requirements. Thus the planetary stirrer, in which the axis about which the paddle rotates itself follows a circular orbit, is used at fairly slow speeds for the intimate mixing of very viscous emulsions, e.g. heavy batters, and at higher speed for the aeration of low viscosity emulsions. A mixer consisting of one or more propellers mounted on a common shaft is widely used for preparing low and medium viscosity emulsions.

In recent years, there has been an almost universal adoption of the continuous Votator system for the manufacture of margarine, which is a W/O emulsion produced by blending about 16% water and 4% of other substances with 80% of fat. In this system the oil phase (containing 0.1–0.5% of an emulsifier such as lecithin or a monoglyceride) and the aqueous phase are fed continuously by separate proportioning pumps through externally refrigerated cylinders equipped with rapidly revolving coaxial shafts bearing scraper blades. Centrifugal force and the resistance to rotation offered by the mixture cause the scraper blades to bear lightly against the cylinder walls, thus giving a very high rate of heat transfer. The emulsification process is considerably promoted by this efficient cooling, since a proportion of the globules are solidified at the lowest temperatures. The whole system is enclosed, thus giving complete protection from atmospheric contamination. The margarine emerging from the cylinders is ready for sizing, packing, etc. Whereas a fine distribution of water globules in margarine is essential, the water droplets must not be too small as the product would then give a sticky, fatty sensation to the palate. In a good margarine, some 95% of the globules have a diameter of 1–5 mm, 4% of 5–10 mm and 1%

of 10–20 mm. There are 10 000 to 20 000 million water globules per gram of margarine.

Emulsification has many other applications in the food industry apart from margarine, for example in the manufacture of ice-cream, cakes and mayonnaise. Other means of agitation used include turbine agitators, colloid mills, homogenizers and ultrasonic oscillators. A useful review of emulsification is given by Becher (1965).

EMULSIFIERS

Emulsifiers play an important role, as the manufacture of many foodstuffs involves the formation and stability of emulsions of two basically immiscible phases, usually oil and water. The use of an emulsifier increases the ease of formation and promotes the stability of an emulsion (see EMULSIFICATION). The concept of hydrophilic–lipophilic balance (HLB) (Griffin, 1949) is now widely used in the selection of emulsifiers for a particular purpose. Each emulsifier is assigned an HLB value, which is a number expressing the balance of the number and strength of its polar as compared to its non-polar groups, and this serves as an indication of the emulsifying action it will perform. Whereas the HLB system has proved quite adequate in applications in which only water and oil are involved, the selection of a proper emulsifier for food applications is often considerably complicated by the fact that the emulsions concerned are so complex that the determination of their 'required HLB' is extremely difficult. Nevertheless, the HLB system can be useful in narrowing the selection from the many emulsifiers available.

In addition to the HLB of an emulsifier, attention may have to be given to its physical nature, which may be dictated by the form of the complete product, and to its chemical composition, minor differences in which can cause vast differences in performance. The method of preparation of an emulsion can also be important; for example, a homogenizer or a heat exchanger may give either an O/W or a W/O emulsion, from the same two phases. In some cases, such as in ice-cream, bread and cake, emulsifiers exert important influences on the product by means of their effect on other ingredients like starch and protein, their unique crystallization characteristics or their indirect induction of interactions lowering the interfacial tension between the two phases. Blends of emulsifiers usually give better results than a single compound, and techniques such as response surface methodology have been used to determine the optimum levels and proportions of several

emulsifiers for a blend designed for a specific application, for example in cake mix shortenings. Some of the most common applications of emulsifiers in foodstuffs are:

(i) *Bread.* At 0.1–0.5% (of the flour) emulsifiers improve the loaf volume, crumb structure and tenderness of the bread and markedly decrease its rate of staling. Diacetyl tartaric acid (DATA) esters are used together with a small proportion of hard fat in bread improvers.

(ii) *Cakes.* At 1–4% level (of the fat) emulsifiers improve the cake volume, crumb softness, texture, moisture retention and keeping quality of the cake.

(iii) *Ice-cream.* At 0.1–0.5% level (of the mix) emulsifiers improve the aerating properties (overrun), dryness, texture, body and apparent richness of the ice-cream.

(iv) *Margarine.* At 0.5% level (of the fat) emulsifiers improve the stability, texture and spreadability of the margarine and some reduce its spattering when used for frying (see ANTI-SPATTERING AGENTS).

(v) *Whipped fillings.* At 2% level (of the fat) emulsifiers improve the whipping characteristics, appearance, volume and stability of the fillings, toppings and icings.

(vi) *Chocolate.* The main advantage of adding emulsifiers to chocolate is to reduce the chocolate viscosity and thus reduce the fat requirement. Lecithin is the most widely used emulsifier, the optimum level of utilization being about 0.4% of the total mix. Other emulsifiers, such as phosphorylated monoglycerides (both ammonium and sodium salts), polyglycerol polyricinoleate and sucrose dioleate, are useful in the reduction of chocolate viscosity. Sucrose dioleate is not at present permitted in the UK for use in chocolate. Sorbitol tristearate and other sorbitol esters are sometimes used to reduce bloom tendency and enhance texture, gloss, keeping qualities and meltdown on the palate. However, these emulsifiers are found to have more effect in coatings than in real chocolate.

Emulsifiers used in foods may be present in some of the basic ingredients, for example lecithin in egg yolks and casein in milk, or they may be added at a given stage in the process. The added emulsifiers may have been obtained from natural sources, for example, lecithin extracted from soya bean oil is widely used as an emulsifier in bakery fats, margarine, confectionary and pan release agents. However, while these natural emulsifiers have proved very useful, they are somewhat limited in their application to the great variety of technological problems currently facing the food industry. Thus in recent years a whole range of synthetic products has been developed, some of which have been specifically designed for particular applications. These synthetic compounds have all been subjected to rigorous toxicological examination before they are permitted for use as food additives. The following classes of synthetic emulsifiers are currently permitted in the UK.

Stearyl tartrate

This is used as a bread improver, sometimes in combination with tartaric or diacetyl tartaric acid esters (see below).

Partial glycerol esters

The partial esters include any compound formed by incomplete esterification of the hydroxyl groups of glycerol with:

(i) Any single fatty acid or mixture of fatty acids. This class comprises the well-known monoglycerides, diglycerides and their mixtures which were the first synthetic emulsifiers to be used in the food industry. Because of their great diversity of application, stability and relative cheapness, they are still by far the most widely used emulsifiers in foodstuffs. They are used in the preparation of bread, cakes, margarine, ice-cream, whipped toppings, confectionery, cereal products, starch products and frozen and dehydrated foods. Saturated monoglycerides inhibit oiling out of peanut butter.

(ii) Any mixture of fatty acids with one of a series of organic acids. These are modified mono- and diglycerides and are generally more hydrophilic than the parent compounds.

The permitted organic acids are:

(i) *Acetic acid.* The acetoglycerides, which are exceptionally stable in the α-polymorphic form, are particularly effective emulsifiers in high ratio cakes and in cake mixes designed for single stage preparation. They are also used in edible protective coatings, enrobing fats, spreads, salad oils, pan release agents and lubricants for machinery used to process foods.

(ii) *Lactic acid.* The lactoglycerides, such as glycerolactopalmitate, tend to be stable in the α-

polymorphic form and are widely used as emulsifiers in shortenings, both for conventional and high ratio cakes, and in cake mixes. They are particularly effective aerating agents when incorporated in liquid shortenings for cakes and are also used as whipping agents in topping creams. Their other applications include ice-cream mixes, baked goods, bread and confectionery coatings.

(iii) *Citric acid.* These esters are not widely used, their main application being as antispattering agents in margarine. They exert a mildly antioxidative effect when incorporated into oils and fats.

(iv) *Tartaric or diacetyltartaric acid.* These esters improve bread, cakes and other baked goods. They also improve confectionery, margarine and whipped toppings. The diacetyl tartaric (DATA) esters are permitted in the UK and are widely used, but the simple tartaric acid esters are not allowed.

(v) *Phosphoric acid.* These esters are not widely used, their main application being as viscosity reducers in molten chocolate (see Chapter 11).

Partial esters of polyglycerols

Polyglycerol mixtures, which are made by the alkali-catalysed thermal dehydration of glycerol, consist of chains of glycerol molecules joined together by ether linkages. The remaining free hydroxyl groups are incompletely esterified with:

(i) Any single fatty acid or mixture of fatty acids,
(ii) Dimerized fatty acids of soya bean oil,
(iii) Interesterified fatty acids of castor oil.

Since the length of the polyglycerol chain, the degree of esterification and the nature of the fatty acids can all be varied independently of each other, a wide range of esters, covering the HLB range, can be prepared. This versatile class of emulsifiers has found application in margarine, peanut butter, ice-cream, confectionery, chocolate, cakes, baked goods, low calorie spreads, toppings and bakery release agents. They retard crystal formation in salad oils. The esters tend to be hydrophilic in character and are therefore particularly effective in synthetic creams and high ratio shortenings.

Propylene glycol esters

Propylene glycol may be completely or partially esterified with:

(i) Any single fatty acid or mixture of fatty acids. These esters tend to crystallize in the α-polymorphic form and are very effective emulsifiers in cake mixes and in plastic and liquid shortenings for cakes. They impart excellent whip stability to dessert toppings, ice-cream and icings made from prepared dry mixes. They are also used in confectionery products.

When a molten equimolar mixture of glycerol monostearate and propylene glycol monostearate is cooled, mixed crystals are formed in which the majority of glycerol ester is in the normally unstable α-polymorphic form. These crystals disperse readily in water and this property is retained for over one year. The aqueous dispersions foam when shaken, and due to this unique property, the mixed crystals exhibit enhanced effectiveness in bread, cake mixes, sponges, whipped sauces, purees, toppings and confectionery.

(ii) Any mixture of fatty acids and lactic acid. Mixtures obtained by the reaction of lactic acid with mixed fatty acid esters by propylene glycol and glycerol are particularly effective as emulsifiers in cake mixes.

(iii) Alginic acid. Propylene glycol alginate is used as an emulsifier in baked goods, brewery products, canned goods, soft drinks, sauces and margarines containing highly unsaturated oils. It is also used as a thickening agent in a wide varietry of foods.

Fatty acid esters of sorbitan and their polyoxyethylene derivatives

The sorbitan esters, prepared by the reaction of sorbitol with fatty acids, are more hydrophilic than monoglycerides and their hydrophilicity can be still further increased by condensing them with ethylene oxide. The esters cover a wide HLB range and have found applications in baked goods, whipped toppings, ice-cream, confectionery coatings and beverages. A blend of sorbitan ester and a polyoxyethylene derivative usually gives better results than a single ester.

Monostearin sodium acetate

This is an effective antispattering agent for margarine, but it is not permitted in the UK or the USA.

Miscellaneous emulsifiers

Other compounds, such as cellulose ethers and sodium carboxymethyl cellulose, are stabilizers rather than emulsifiers and are used in baked goods, frozen foods, sauces, soft drinks, soups, jellies, etc. The fatty acid esters of sucrose comprise a versatile class of emulsifiers which when pure are entirely non-toxic. Until recently, however, great difficulty was experienced in completely removing the toxic solvents, usually dimethylformamide, used in their manufacture and for this reason they are not permitted in some countries. However, methods of manufacture which do not involve toxic solvents have now been found and the esters are gaining wider acceptance. They have proved to be effective emulsifiers in cakes, biscuits, confectionery, chocolate, bread, margarine, shortenings and salad oil. The compounds calcium stearoyl lactylic acid and sodium stearyl fumarate, which have been gaining wider acceptance, are effective improvers for bread and other baked goods.

Table 8.8 shows the acceptable daily intake (ADI) figures recommended by the FAO/WHO Expert Codex Committee on Food Additives, together with the current status in the EU and the USA. The UK position is equivalent to the EU status and is controlled by the Miscellaneous Food Additives Regulations 1995, S.I. 3187.

FAT SUBSTITUTES

'Fat substitutes' have been defined as synthetic or natural substances whose molecules closely resemble fat molecules and can replace fats in all applications, even frying. 'Fat mimetics' (or fat imitators) is a name given to substances which imitate the food textural effects of fat, including mouthfeel, but not all of the other properties. The term 'fat replacers' is also used to cover either type of material.

Fat has a number of functions in food which need to be considered when reducing fat levels. It is a high energy nutrient, which in many modern cases is a major reason for wishing to reduce its amount. It contributes characteristically, depending on the extent to which it is emulsified, to the texture, mouthfeel and 'richness' of food. It is a carrier of flavours, whose perception in the mouth may also be influenced by the degree of emulsification. It may influence the appearance of the product before or after cooking. With such a complex of properties and effects it is not surprising that there are few true fat-free substitutes.

Table 8.8 Food emulsifier status (1986)

Chemical name	Abbreviation	ADI value	EU no.[1]	USA FDA 21 CFR[2]
Lecithin	–	Not limited	E 322	SS 184.1400[7]
Mono- and diglycerides	MG	Not limited	E 471	SS 182.4505[7]
Acetic acid esters of mono- and diglycerides	AMG	Not limited	E 472a	SS 172.828[8]
Lactic acid esters of mono- and diglycerides	LMG	Not limited	E 472b	SS 172.852
Citric acid esters of mono- and diglycerides	CMG	Not limited	E 472c	SS 172.832[9]
Diacetyl tartaric acid esters of mono- and diglycerides	DATA	0–50	E 472e	SS 182.4101[7]
Succinic acid esters of mono- and diglycerides	SMG	–	–	SS 172.830
Salts of fatty acids (Na, K, Ca)	–	Not limited	E 470	SS 172.863[10]
Polyglycerol esters of fatty acids	PGE	0–25[3]	E 475	SS 172.854
Propylene glycol esters of fatty acids	PGMS	0–25[4]	E 477	SS 172.856
Sodium stearoyl-lactylate	SSL	0–20	E 481	SS 172.846
Calcium stearoyl-lactylate	CSL	0–20	E 482	SS 172.844
Sucrose esters of fatty acids		0–10	E 473	SS 172.859
Sorbitan monostearate	SMS	0–25[5]	E 491	SS 172.842
Polysorbate 60	PS 60	0–25[6]	E 491	SS 172.836
Polysorbate 65	PS 65	0–25[6]	E 436	SS 172.838
Polysorbate 80	PS 80	0–25[6]	E 433	SS 172.840

(1) EU Council Directive (74/329/EEC)
(2) US Code of Federal Regulations
(3) Calculated as polyglycerol esters of palmitic acid
(4) Calculated as propylene glycol
(5) Calculated as total sorbitan esters

(6) Calculated as total polyoxyethlene (20) sorbitan esters
(7) Generally Regarded as Safe (GRAS)
(8) Relates to monoglycerides only
(9) Relates to mono-oleate citric acid ester
(10) Specifies salts of Na, K, Ca, Mg and Al.

Jojoba oil, used for centuries for frying by some American Indian tribes, is coming into increasing use. It is not a triglyceride oil but a wax, therefore is likely to be only poorly metabolized by humans. However it contains a significant proportion (3%) of erucic acid which, if it were liberated in the gut, might have detrimental side effects; there is therefore some resistance to its widespread use without further testing.

Other substances with roughly similar molecular structures to the fats include medium chain and very long chain triglycerides (e.g. caprenin), glyceryl ethers, glycoside fatty acid esters and other polyesters (e.g. sucrose polyester, 'Olestra'). Most attempts at fat substitution tend to avoid applications where all or most of the characteristic properties of fat are essential (such as frying), concentrating instead on products where only some of the effects are required and might be copied using different ingredients. A wide range of such products, carbohydrate-based, protein-based and other, is now available for the enterprising product development department to test. A good list of currently available materials in all the above classes is given in Leatherhead Food Research Association (1991).

In January 1996, the US Food Drug Administration approved the use of 'Olestra' in selected snack foods. Olestra, prepared by esterifying six to eight of the -OH groups of sucrose with long chain fatty acids is a truly fat-based fat substitute. The large Olestra molecule contributes no calories because it is indigestible, i.e. not metabolized as it passes through the digestive tract. Theoretically, Olestra could be used in any product where normal fats and oils are used, including baking and frying applications. It should, however, be mentioned here that the FDA approval of Olestra's use in snack products requires specific labelling. Because Olestra absorbs Vitamins A, D, E and K from foods eaten at the same time, these vitamins will be added to Olestra containing snack to help compensate for the potential loss.

ISOACIDS

During hydrogenation, double bond isomerization and migration take place. The resulting positional and geometric fatty acid isomers are sometimes termed isoacids. This term is less often used nowadays, *trans*-isomers being specifically referred to as such.

LIPASES

Lipases, or glycerol ester hydrolases, are enzymes which catalyse the breakdown of oils and fats. They are concerned in the digestion of fat, splitting triglycerides into fatty acids, diglycerides, monoglycerides and glycerol. Resynthesis of triglycerides may take place partly in the intestinal lumen, where pancreatic lipase has synthetic as well as hydrolytic functions. Lipases may originate from plant, animal or microbiological sources. Controlled action of these enzymes is useful in the development of desirable flavours in cheeses and other fermented dairy products. Qualitatively, lipases release fatty acids from fat-based products, but variations exist in the kinds or quantities of specific fatty acids released, depending upon the nature and source of the lipase. Lipases are usually destroyed by common pasteurization treatment, but those of microbiological origin may be heat stable and may be destroyed only by autoclaving treatment. Lipase is generally assayed by the indoxyl acetate test.

The selectivity of pancreatic lipase in partial hydrolysis of triglycerides has been used as a biochemical method of analysis in establishing the configuration of some natural fats. Acyl groups attached to the 1- and 3-positions of the glycerol moiety are hydrolysed preferentially, allowing the residual 2-monoglycerides to be separated and subjected to independent fatty acid analysis. This has enabled theories of fatty acid distribution in triglycerides to be developed, the application of which often allows the calculation of triglyceride composition from simple fatty acid data. However, confusion can arise through misapplication of the theory. The technique may be used as a criterion of purity.

More recently, specific lipases have been used to carry out microbiological modification of oils and fats by incubation of triglycerides with carefully chosen mixtures of fatty acids, or fatty acid methyl esters, and lipase enzymes. These techniques may be particularly useful in manufacturing fats with properties similar to those of cocoa butter.

MAYONNAISE

Mayonnaise is an emulsion of vegetable oil, egg yolk or whole egg, vinegar, lemon juice and salt, mustard or other seasoning. Usually, mayonnaise contains 70–85% of oil, and it is difficult to produce a product having a sufficiently stiff body with a proportion of less than 70% oil. A typical formulation is: oil 75%, vinegar (4.5% acetic acid) 10.8%, egg yolk 9%, sugar 2.5%, salt 1.5%, mustard 1% and white pepper 0.2%. The oil forms an internal discontinuous phase dispersed in an external aqueous phase of vinegar, egg yolk and other ingredients. In a good mayonnaise the largest droplets are not more than 8 μm in size and many particles are in the size range of 2–4 μm.

The temperature of mixing is important; too thin a product may result if mixing is carried out above 16–21°C. Thicker emulsions can be made by mixing at 4°C, but it is usually inconvenient to work at this temperature.

In making mayonnaise the egg yolks, sugar, seasonings and part of the vinegar are mixed, the oil is gradually beaten in and the emulsion is thinned by addition of the remainder of the vinegar. During the mixing, about 10–20% air is usually incorporated. A coarsely emulsified product can be further treated in a colloid mill. It is essential to use best quality oils, corn oil, sunflowerseed oil or winterized cottonseed oil being suitable. Oils which throw down a deposit or crystallize at refrigerator temperatures are generally unsatisfactory, as crystallization of the oil breaks the emulsion and can cause separation of the phases when the mayonnaise warms to room temperature (see also Chapter 9).

MEDIUM CHAIN TRIGLYCERIDES

Medium chain triglycerides (MCT) are triglycerides based on fatty acids with 8–10 carbon atoms. Fats containing high levels of these triglycerides are beneficial to patients suffering from digestive illnesses such as tropical sprue and idiopathic steatorrhoea, who are unable to assimilate normal triglycerides because they lack the necessary enzymes.

MISCELLA

The solution of oil in solvent obtained during solvent extraction of oilseeds.

PROSTAGLANDINS

See ESSENTIAL FATTY ACIDS

SINGLE CELL OIL

Single cell oil is the term applied to triglyceride fats generated by microbial means on non-fat substrate. Mould fermentations were first used to generate proteinaceous material for animal feedingstuffs, and it was subsequently realized that a similar technology could be used to produce oils or fats of specified properties. At the time of writing no single cell oil has yet been successfully produced on a fully commercial basis, but the most likely candidate is an alternative to evening primrose oil, a rich source of the EFA γ-linolenic acid (see ESSENTIAL FATTY ACIDS).

SOLID FAT INDEX (SFI)

See FLAVOUR AND TEXTURE OF OILS AND FATS

SPANISH OLIVE OIL SYNDROME

In early 1981 some industrial rapeseed oil, denatured with aniline, and which therefore attracted a low rate of duty, was imported into Spain. This oil was apparently processed to remove the aniline. It was then blended with olive oil and sold in street markets in the working class areas of Madrid. Unfortunately, the treatment with aniline and the subsequent processing had apparently generated highly toxic components in the oil and people using the oil subsequently became seriously ill. Over 500 people died, while 10 000 or more required hospital treatment. The exact cause of the illness is still not known, as the reported symptoms do not correspond to those known for aniline or anilide poisoning. A paper by Noriega (1982) describes the symptoms of this sickness, while a World Health Organization monograph reviews all aspects of the catastrophe (WHO, 1984).

VITAMINS

Vitamins are accessory food factors which cannot be synthesized in the human body and have to be supplied in the diet, albeit in small quantities. The lipid-soluble vitamins are coded A, D, E and K. Vitamins A and D are present in butter and it is compulsory to add them to margarine in the UK. In the USA it is compulslory to add vitamin A and optional to add vitamin D. Carotene (Figure 8.3) is a natural source of vitamin A and occurs in high quantities in palm oil. Tocopherols (Figure 8.5), a naturally occurring source of vitamin E, occur in most vegetable oils. Vitamin D is a sterol and is essential for bone formation.

WEEVIL

Palm trees growing in Malaysia were found to be suffering from incomplete flower fertilization, leading to a sub-optimum fruit yield. Considerable research with insects pollinating flowers in Cameroun and other parts of West Africa led to the identification of the weevil *Elaidobius kamarunicus*. This species was therefore introduced to the oil palm plantations of Malaysia, where it has bred extensively. Much better flower fertilization resulted and this led to a dramatic increase in fruit production. A consequence of this appears to be a smaller yield of palm oil in each individual fruit, whilst the yield of palm kernel oil remains more constant. The overall yield plantation-wide of palm kernel has therefore increased more dramatically than that of palm oil following the introduction of the weevil.

REFERENCES

Allen, J.C. and Hamilton, R.J. (eds) (1995) *Rancidity in Foods*, 3rd edn, Elsevier Applied Science, London.

American Oil Chemists Society (1980) Lecithin—papers from a symposium at the AOCS Annual Meeting in New York City, April 28 1980. *J. Am. Oil Chem. Soc.*, **58**, 885–916.

Becher, P. (ed) (1965) *Emulsions, Theory and Practice*, 2nd edn, Reinhold, New York.

Department of Health and Social Security (DHSS) (1984), *Diet and Cardiovascular Disease*, DHSS Report on Health and Social Subjects No 28, HMSO, London (so-called COMA report by the Committee on Medical Aspects of Food Policy – Report of the Panel on Diet in Relation to Cardiovascular Disease).

Food and Agriculture Organization (1980) *Dietary Fats and Oils in Human Nutrition – Report of an Expert Commission*, Food and Agriculture Organization of the United Nations, Rome.

Griffin, W.C. (1949) Classification of surface-active agents by HLB. *J. Soc. Cosmetic Chem.*, **1**, 311–326.

Holman, R.T. (1981) Essential fatty acids and prostaglandins. *Prog. Lipid Res.*, **20**, 907.

Hudson, B.J.F. (ed) (1990) *Food Antioxidants*, Elsevier Applied Science, London.

ITP (1996) Oils and Fats International Directory, International Trade Publications, Redhill, UK.

Kochhar, S.P. (1983) Influence of processing on sterols of edible vegetable oils. *Prog. Lipid Res.*, **22**, 161–188.

Kochhar, S.P (1993) in *Atmospheric Oxidation and Antioxidation* (ed. G. Scott), Vol II, Elsevier, London, pp. 71–139.

National Advisory Council on Nutrition Education (NACNE) (1983), *Proposals for Nutritional Guidelines for Health Education in Britain*, Health Education Council.

Noriega, A.R. (1982) Toxic epidemic syndrome, Spain 1981. *Lancet*, 697–702.

Patterson, H.B.W. (1994) *Hydrogenation of Fats and Oils: Theory and Practice*, AOCS Press, Champaign, USA.

Schwitzer, M.K. (ed) (1956) *Margarine and Other Food Fats*, Leonard Hill (Blackie), London.

Stuyenberg, J.H. van (1969) *Margarine*, Liverpool University Press.

Waddington, D. (1986) Applications of wide-line NMR in the oils and fats industry, in *Analysis of Oils and Fats*, (eds R.J. Hamilton and J.B. Rossell), Elsevier Applied Science, London, pp. 341–400.

World Health Organization (1984) *Toxic Oil Syndrome – Mass Food Poisoning in Spain*. WHO Regional Office for Europe, Copenhagen (92 pp.).

FURTHER READING

British Nutrition Foundation (1995) *Trans Fatty Acids*, Report of the BNF Task Force, London.

Hamilton, R.J. and Rossell, J.B. (eds) (1986) *Analysis of Oils and Fats*, Elsevier Applied Science, London.

Hoffmann, G. (1989) *The Chemistry of Edible Oils and Fats and their High Fat Products*, Elsevier Applied Science, London.

Laning, S.J. (1991) Fats, oils, fatty acids and oilseed crops, in *Biotechnology and Food Ingredients*, (eds I. Goldberg and R. Williams), Van Nostrand Reinhold, New York.

Leatherhead Food Research Association (1991) *Fat Substitutes – an Update*. Food Focus Series, November, 1991, LFRA, Leatherhead, UK.

Nestel, P.J. (1995) Comment on *trans* fatty acids and coronary heat disease risk. *Am. J. Clin. Nutr.*, **62**, 522.

Pomeranz, Y. (1985) Additives II – food emulsifiers, in *Functional Properties of Food Components*, Academic Press, London, pp. 329–464.

Rossell, J.B. and Pritchard, J.L.R. (eds) (1991) *Analysis of Oilseeds, Fats and Fatty Foods*, Elsevier Applied Science, London.

Salunkhe, D.K. Chavan, J.K., Adsule, R.N.and Kadam, S.S. (1992) *World Oilseeds – Chemistry, Technology and Utilisation*, Van Nostrand Reinhold, New York.

9 Salt, Acid and Sugar Preserves

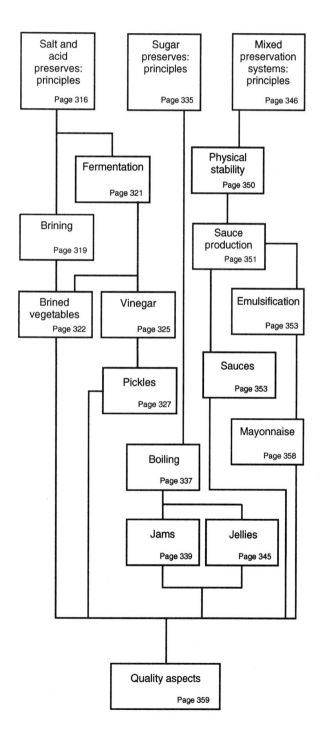

INTRODUCTION

This chapter is concerned with the manufacture of food products whose stability and preservation from spoilage are ensured entirely or mainly through the beneficial effects of components of the food itself – salt, sugar and acid. Of course, some of these processes have been known and used for thousands of years and many of the products can certainly be called 'traditional', but it is also true that the understanding and finer applications of the principles behind their effectiveness have come about only with the growth of understanding of food science and technology in the twentieth century.

The application of the same principles in meat product manufacture was dealt with earlier. In this Chapter we deal with fruit and vegetable products.

PART 1: SALT AND ACID PRESERVES

PRINCIPLES

Water activity or equilibrium relative humidity

One of the most important food preservation processes is the control of water activity (a_w) or equilibrium relative humidity (ERH). Water activity is a measure of the availability of the water to assist in the metabolic processes of organisms living in it. At high values of water activity (the a_w of pure water is 1.0) bacteria and other micro-organisms can grow in food, other factors being favourable. Some substances dissolved in the water reduce the water activity, the greater the concentration the greater the reduction in a_w. Sufficiently high concentrations of salt or sugar therefore will reduce the a_w to levels where microbial growth is prevented. The relevant quantitative relationships are set out in Table 9.1.

Table 9.1 Water activity relationships in food (approximate)

Water activity a_w	Organisms whose growth is inhibited at a_w below the value shown	Concentration of solute with the a_w shown (%)	
		Salt	Sugar
1.0		0	0
0.95	most food poisoning bacteria	7	40
0.9	most other bacteria	12	55
0.85		15	68 (saturated)
0.8	yeasts	26 (saturated)	
0.7	moulds		
0.6			
0.5			

Preservation by salt or sugar

It is well known that micro-organisms may grow well on media including foodstuffs containing low or moderate proportions of salt or sugar. However Table 9.1 shows clearly that a food product whose water content is saturated with salt or sugar will be preserved from spoilage by bacterial organisms with the possible exceptions of yeasts and moulds. This of course is the basis of most of the old-fashioned curing and preserving processes.

Note incidentally that as can also be seen from Table 9.1, when the old processes were properly carried out, they were effective not only in preserving the food against spoilage but they also avoided the growth of food poisoning organisms and so ensured food safety.

In modern times when public taste and fashion call for ever milder flavours and smaller amounts of salt and sugar, it has become necessary for preservation to seek additional aids to those provided by higher concentrations of the ancient preservatives.

Acidity and pH

Increasing acidity in food media, as indicated by lower pH values, also has inhibitory effects on most bacteria. In detail, these effects are quite complicated. For our purposes it is probably sufficient to note

(i) the general principle that lower pH values are beneficial to preservation against spoilage and

(ii) the specific point that pH 4.5 is considered the upper limit for protection from the growth of food poisoning organisms.

There also are specific influences of acetic and lactic acids which will now be considered.

Acetic acid preservation

The acetic acid present in pickles and sauces, whether deriving from vinegar or from acetic acid incorporated as such, is the main factor responsible for the self-preservation of these products, coupled with, to a lesser extent, other preservative influences which may be present such as salt, low water activity and permitted preservatives. Unpasteurized pickles containing no added preservative rely on the acetic acid for self-preservation. Pasteurized pickles rely on the combined effect of acetic acid and heat processing prior to the pack being opened and on acetic acid after opening.

In pickles and sauces the preservative action appears to be bacteriostatic rather than bactericidal. This action is not directly dependent on pH (although the low pH of pickles and sauces is responsible for their freedom from the risk of growth of pathogenic and many other micro-organisms); the inhibitory effect of acetic acid is due to the undissociated acetic acid molecules (possibly due to the greater ease with which the uncharged molecule can penetrate the cell membranes of the micro-organisms). pH is, however, indirectly involved, for the degree of dissociation of the acetic acid depends on the pH of the product in which it is incorporated; the lower the pH, the lower the degree of dissociation and the greater the proportion of acetic acid present that can exercise an inhibitory effect. Fortunately, at the pH generally obtaining in pickles and sauces, nearly all of the acetic acid is in the undissociated form.

In evaluating the microbiological keeping properties of a pickle or sauce in relation to its composition and where the action of heat processing or preservative is not additionally involved, the acetic acid acidity calculated as a percentage of the whole product is not a reliable criterion, but rather the acetic acid acidity calculated as a percentage of the total volatile constituents, i.e. a value which may be called the preservation index PI, equal to:

$$PI = \frac{\text{total acetic acid acidity} \times 100}{(100 - \text{total solids})}$$

Experience has shown that if the preservation index is not less than 3.6%, a reasonable degree of freedom from microbiological spoilage can be expected. However, it must be emphasized that this is a relationship arrived at empirically which is subject to some important considerations.

Certain lactobacilli and yeasts may occasionally

be encountered which are capable of tolerating acetic acid at a preservation index of up to 3.6%, and the mould *Moniliella acetobutans* is capable not only of tolerating acetic acid but of assimilating it as a carbohydrate source.

Secondly, the relationship will be observed to take care of the influence of soluble and insoluble solids, but probably only within a limited range so far as sugar and salt are concerned. For flavour reasons, high salt contents (3–4%) are not normally desirable. High sugar contents may, however, be utilized in certain sweet pickles and especially mango chutney. Beyond a certain point, sugar and salt probably exercise an inhibitory influence over and above their contribution as part of the total solids, i.e. where the sugar content is high there begins to be an element of sugar preservation due to significantly reduced water activity as well as acetic acid preservation of the pickle, and where this occurs a lower preservation index can provide adequate safety.

Thirdly, it will be seen that the preservation index can be increased by increasing the total acetic acid content or the total solids content or both. This provides a useful means of attaining a safe preservation index without simultaneously causing an excessively acid flavour, through a suitable balance of acetic acid content and sugar content, both contributing to the preservation index, and the sweetness of the sugar partly offsetting the sharp flavour of the acetic acid.

For mayonnaise and similar products a more complicated formula for determining microbial stability has been proposed by the Comité des Industries des Mayonnaises et Sauces Condimentaires de la Communauté Economique Européene (CIMSCEE, 1985), namely

$$\frac{\text{\% undissociated acetic acid in the water phase}}{4} + \frac{\text{moles salt + moles sugar per kg water phase}}{3.5} \geq 1$$

or 15.75 $(1 - \alpha)$ (total acetic acid %) + 3.07 (salt %) + (hexose %) + 0.5 (disaccharide %) $\geq$ 63, where $(1 - \alpha)$ is the undissociated acetic acid fraction, from the classic equation:

$$pH = pK + \log \frac{\alpha}{1 - \alpha}$$

The pK for acetic acid is 4.757. The percentages in the above formula are expressed on the weight of aqueous solution, i.e. water plus acetic acid plus salt plus sugars. CIMSCEE also provides a

table of typical $(1 - \alpha)$ values at the pH range found in acetic acid-containing sauces. Although this formula may guarantee an unacceptable medium for pathogenic micro-organisms, it does assume that 'no organisms such as *Moniliella acetobutans* will be present.'

The equilibrium concentration of acetic acid within the tissue will take some time to achieve, but fortunately the spoilage organisms are not found in the deep tissues. However, for quality control purposes it is usually necessary to take into account the degree of equilibrium which the product may have reached as it is normally more practical to apply control by the analysis of covering liquor.

The preservation index rule or the CIMSCEE standard may also be invalidated by the use of raw materials abnormally heavily contaminated with micro-organisms or by unhygienic conditions of manufacture.

Acetic acid preserves with acidities below 3.6% can usually be safeguarded by adequate pasteurization or other heat treatment.

Lactic acid

Whether produced in the product by fermentation or when added as an ingredient, lactic acid plays an important role, along with salt, in the brining preservation of vegetables, both for sale as such (e.g. olives, sauerkraut) and for subsequent use in manufacture of acetic acid pickles (e.g. silverskin onions, gherkins, cauliflower, walnuts).

In unpasteurized pickles relying for preservation on an acetic acid preservation index of 3.6%, lactic acid cannot be used to replace any significant part of the acetic acid, for it does not possess a comparable inhibitory effect. It might be thought that lactic acid, by slightly lowering the pH of the pickle, would increase the proportion of undissociated acetic acid and thus enable less acetic acid to produce the same inhibitory effect. But when it is appreciated that in most pickles the pH is such that about 97% of the acetic acid is undissociated, it will be seen that the scope for increasing the proportion of undissociated acetic acid by lowering the pH is negligible. In pasteurized packs, lactic acid can be used in conjunction with acetic acid to produce a milder flavour and odour in the product.

It has been suggested that the effect of lactic acid in lowering the pH is advantageous in improving the colour of pickled beetroot and pickled red cabbage, the natural colours being, in effect, acid–base indicators changing colour at low pH. For this purpose 0.5–1.0% lactic acid in the packing liquor has been proposed. In the case of red cabbage, which is not pasteurized, any lactic acid used must be in addition to sufficient acetic acid (whether derived from vinegar or added as such) to give a preservation index of 3.6%.

PROCESSES

Almost any vegetable or fruit can be successfully preserved by pickling with acid or salt or both. In Europe cabbage, red cabbage, gherkin, cucumber, marrow, several varieties of onion, olives and walnuts are commonly used; mangoes, apples, pears, maize, water chestnuts and citrus fruits are used in other parts of the world. Several styles of preservation, with associated preservation processes, may be distinguished:

(i) *Fully preserved in salt only*: treated with saturated brine or dry salting; suitable for further processing after longer times but excess salt will need to be removed;

(ii) *Semi-preserved*: treated with dilute salt, acid or (sometimes) chemical preservative, to give short-time preservation of materials for further processing;

(iii) *Fermented – preserved with salt and lactic acid*: lower salt content than (ii) so may be consumed without desalting; good keeping quality;

(iv) *Preserved with salt and acetic acid*: preserved with vinegar (usually) and relatively low salt content; (i), (ii) or (iii), or fresh produce, may be used as raw material.

(i), (ii) and (iii) are dealt with below, (iv) which concerns vinegar pickles is considered in detail under the relevant products later.

Brining

Fully brined vegetables

Fully brined vegetables may be regarded as stable subject to reasonable safeguards during storage. They may be either fully fermented products or in a brine so treated as to inhibit the onset of fermentation completely. They may be stored for periods of months or years before use and may have been subjected to considerable preparatory processing, e.g. walnuts and olives.

The main commercially available supplies of brined vegetables are also fermented. Fermentation results in physical and chemical changes

Table 9.2 Typical state of vegetables on receipt by pickle manufacturers in the UK

Vegetable	Fresh/chilled	Semi-preserved	Fully brined
Onions and silverskin onions (*Allium cepa* L.)	+	+	+
Cauliflower (*Brassica oleracea botrytis*)			+
Gherkins and cucumbers (*Cucumis sativus* L.)	+	+	+
Walnuts (*Juglans nigra* or *J. regia*)			+
Olives (*Olea europea*)			+
Capers (*Capparis spinosa* L.)			+
Red cabbage (*Brassica oleracea capitata*)	+		
Beetroot (*Beta vulgaris* L.)	+	+	
Marrow (*Cucurbita pepo*)	+	+	+
Carrot (*Daucus carota* L.)	+	+	+
Rutabaga or swede (*Brassica rutabaga*)	+	+	+

which affect the appearance, texture and flavour. When a manufacturer is brining vegetables for his own subsequent use, it is a matter of choice whether a fermentation process is adopted.

Semi-preserved vegetables

These may be received in a covering liquor (or may have been drained of one) which is not of itself sufficient to ensure long term stability during storage, but will suppress or inhibit short term deterioration, e.g. gherkins in weak acid solution and onions in weak acidified brine; see Table 9.2.

Fermentation brining

Vegetables received at the brining station have an extensive flora of micro-organisms, the majority of which are inhibited when the vegetables are placed in a brine giving between 8–11% equilibrium salt content overall. Certain types, however, are capable of tolerating this salt concentration. These bacteria carry out desirable fermentation changes. Some (homofermentative) convert sugars almost entirely to lactic acid, others (heterofermentative) produce lactic acid, carbon dioxide and traces of alcohol and acetic acid.

In an historical context with some vegetables such as runner beans, dry salt was applied and mixed so that all microbial growth was inhibited. These days dry salting is mainly used for sauerkraut production, but the salt addition is limited to allow tissue fluid extracted by osmotic action to form a brine in which fermentation can proceed.

The essential requirements for traditional fermentation brining are:

(i) An initial brine strength which, with the chosen vegetable:liquor ratio, gives an equilibrium salt content of 8% or more overall.

(ii) Sound vegetables at the correct stage of maturity, undamaged and suitably size-graded.

(iii) Suitable vessels, which may range from brining tanks to the drums in which the brined vegetables are subsequently stored.

(iv) The prompt brining of vegetables on arrival at the brining station.

(v) Careful weighing of vegetables and brine for each vessel, thus ensuring adherence to the predetermined ratio.

(vi) The placing of part of the brine in the vessel before adding the vegetables (to assist in ensuring that there are no clumps of vegetable unpenetrated by brine).

(vii) Suitable intermittent mixing, especially during the early stages of brining, to prevent stratification.

(viii) Careful checks and topping up with brine where necessary during the fermentation period (which may vary, depending on temperature).

(ix) Possible final draining off of the brine and its replacement by a fresh brine containing up to 1% lactic acid and provision of an equilibrium salt content of about 15% overall. This ensures the incidental removal of cloudy exudate, dirt, enzyme activity and surface yeasts and the provision of a sufficiently high salt content and lactic acid content to prevent further microbiological activity.

If a very low strength brine were used, the vegetables could develop undesirable organisms and experience has shown that a 40° salinometer brine (10% salt) is about as low a salt concentration as can safely be used. Of course it is necessary to use a stronger brine in the first place to allow for brine and vegetables to reach equilibrium. Alternatively, and more usually, a careful check may be kept on the brine strength during processing, and this is adjusted by the addition of salt or strong brine if the strength should fall below 40°.

In the brining and fermentation process, the bacteria use up the natural sugar present in the vegetables to produce lactic acid, also releasing carbon dioxide gas which bubbles to the surface of the brine and having an agitating effect. The fermentation stops when there is no further available sugar. Very strong salt solutions, say over 17% salt, inhibit fermenting bacterial growth until the brine has been diluted by some means. Unless carefully controlled this can cause problems during freshening, as the sugars are still available for fermentation.

When the fermentation is over, as long as the container is airtight, the brine strength is at least 10% and there is at least 0.3% lactic acid in the brine, then the vegetables can be stored and will stay in good condition for many months.

The overall microbial picture in conventional brining is similar to that occurring in sauerkraut fermentation, the main organisms involved being members of the Lactobacteriaceae (producing lactic acid), *Acetobacter* (producing CO_2 and H_2) and yeasts (producing CO_2 and alcohol).

Various other organisms, both desirable and undesirable, may develop according to the temperature at which the fermentation takes place. Ideally this is between 15–20°C; at lower temperatures the growth of organisms such as *Leuconostoc mesenteroides* is encouraged. The general procedure followed in brining individual commodities varies and by way of example the brining of cauliflower is given later.

Freshening or debrining

Before brined vegetables can be used to make pickles, their salt content normally has to be reduced to an acceptable level for the eventual consumer, usually about 5% in the freshened vegetables. This is done by a reverse of the brining process, washing the salt out of the water. Stratification can be a problem, because salt coming out of the vegetables will tend to sink to the bottom of the container. This can be avoided by some means of agitation, e.g. rolling drums between changes of water or using a continuous flow of water so that all the salt is washed away quickly. The use of freshening water that contains a high level of calcium can improve the texture of the vegetable in the end product.

It is, of course, important to use potable water, to avoid contamination with iron and copper and to ensure that freshened vegetables are used in a reasonable period of time. It is not unusual for the freshened vegetables to be acetified pending use, so as to ensure the rapid availability of stable desalted stock.

Many factors affect the rate of removal of salt from vegetables, including the type and size of vegetable, the temperature at which the extraction is carried out and the concentration gradient of salt between the outer part of the vegetable and the liquid immediately in contact with it. The latter factor is of course affected by several other factors such as the relative quantities of vegetables and water, flow rate or frequency of changing of the water and agitation.

The process can be speeded up by the use of warm water and this has been advocated by some, but it adversely affects the quality of the resulting pack as well as increasing the risk of microbiological spoilage of the vegetables during debrining and in the writer's opinion cold water should be used. Methods of freshening fall into three main categories: continuous, multi batchwise and single batchwise.

Continuous methods

The crudest continuous method is simply to drain the brine off, put a hose into the cask or drum and leave the water running, allowing it to overflow. This is unlikely to give satisfactory or uniform results. Because the denser brine tends to fall, the correct technique for continuous debrining is to introduce water at the top and run it to waste at the bottom.

Continuous methods are the most rapid and require the least labour of handling, but are normally rather wasteful of water. The optimum results may, however, be obtained by combining a slow water flow with agitation. The ideal system consists of a large stainless steel or plastic tank, not more than a metre deep and capable of holding the vegetables from several drums of vegetables. It has a water inlet valve at the top and an outlet valve at the bottom, protected by a stainless steel wire mesh false bottom a few centmetres above the base of the tank. The tank is fitted with a means of agitating or circulating the water and the rates of outflow and inflow of water are balanced to maintain a constant level.

The rate of debrining will obviously depend on the precise factors of size and shape of tank, stirring conditions, type and quantity of vegetables (onions lose salt more slowly than gherkins or cauliflower) and flow rates. Initial experiments with the equipment available will establish a standard procedure. Once this has been done for each type of vegetable, one can operate under the standard conditions for the standard debrining times thus established, merely checking the salt content at or near the end of the debrining period. After debrining, acidification of the vegetables may be carried out in the same tank, with agitation but with valves closed.

Multi batchwise methods

These involve repeated changes of water, and the speed of debrining depends on the frequency of the changes and the extent of any agitation. Without agitation, this method inevitably gives rise to stratification. Thus, the multi batchwise system without agitation must be regarded as unsatisfactory. A similar setup to that described under CONTINUOUS METHODS could be operated, with several complete changes of water instead of a continuous inflow and outflow, but this requires extra activity without compensating benefits.

Single batchwise method

This involves the suspension of the vegetables in a stainless steel wire mesh basket in the top part of a vessel of water, and turns to advantage the otherwise disadvantageous occurrence of stratification. Thus, the brine formed by extracted salt falls into the lower part of the vessel and at no stage is there a significant variation of salt content between the vegetables at the top and bottom of the basket. In this method, agitation must be avoided. Again the precise vegetable to water ratios and times to be used will need to be determined by experiment and analysis for each size and type of vegetable.

Fermentation

Acetic fermentation

Vinegars are made by the fermentation of alcohol to acetic acid, the alcohol itself having normally been made by fermentation of carbohydrate materials into beer, wine, cider, etc.

Lactic fermentation

This is the normal process occurring when vegetables or fruits are prepared in salt brines of low or intermediate concentration (up to about 10%). Natural ferments on the vegetable surface convert available sugars to lactic acid, up to a final concentration which in favourable cases may be as high as 2.5%. This lactic acid makes a significant contribution to the preservation of products with relatively low salt concentrations.

If a saturated (24%) or near saturated brine is used initially, then the salt concentration in the whole product is likely to fall to the region of 10 or 12% by dilution with water from the vegetable tissues and lactic fermentation is unlikely to take place. The preservation of such products will depend almost entirely on the salt concentration.

PRODUCTS

Brined vegetables

Brined cauliflower

Cauliflower is harvested in the UK during the summer and autumn until the hard frosts come, while broccoli is more hardy and withstands frost to become available in spring and early summer.

The pickle manufacturer will require a head of brined cauliflower which is hard in texture, as white as possible in colour and is to use a trade expression 'good and tight'. The terms 'tight' and 'loose' are used to describe what is really the relative proportion of stalk to curd. Ideally, the stalk should be as short as possible, thus producing a head which is hard to break apart into the smaller segments. In the case of a 'loose' head, excessive stalk growth will have occurred and the smaller segments will not hold together well. It is also, of course, extremely important that the curd has not begun to run to seed.

In order to achieve these points in the brined product, it is necessary to start off with the same characteristics in the cut flower. Harvesting at the right times is essential and such factors as frost, rain and exposure to sunlight can all cause curd discoloration. It is common practice to break a leaf over an unharvested plant to protect the curd from the light and sometimes from frost. It is also important to keep the leaves on the plant to afford protection from bruising and soiling during transit to the brinery.

Leaves and surplus stalk are removed from the whole heads, the minimum amount of stalk which will allow the heads to hold together being left. Grading may be carried out at this stage or after the initial brining process. Alternatively, a rough grading can be done at the time of trimming, to be followed by a more selective grading during subsequent processing.

The grading standards required are virtually identical to those required in the UK for the fresh vegetables offered for sale through wholesale channels, except that the sizing requirements are not particularly relevant and the progressive defects allowed should be cut out as a precaution against enzymic softening.

Apart from the whole heads there is some demand for specially cut or diced cauliflower and even for stalk and stump. This can provide some

outlet for heads which have been partly damaged or have suffered partial discoloration or bruising.

The brining of cauliflower may conveniently be divided into an initial treatment known as 'shrinking' and the 'brining proper' where full fermentation occurs. The shrinking process consists of putting the cauliflower into brine for a short period, during which time although the volume change is negligible, it does become more pliable so that it is then possible to pack 50–90% more into the same container.

Shrinkage is normally carried out in suitable vats or in drums, the vegetables being held in the shrinking brine for 24–48 h, after which they are packed into their final containers, covered with brine and allowed to ferment naturally. During shrinkage, a certain amount of extraneous matter which may not have come to light during the trimming and grading, e.g. leaves, straw, insects, etc., may float to the brine surface. This should be carefully skimmed off.

It is generally recommended that a 14% salt brine should be quite satisfactory for both shrinking and storage. However, it is important to remember that the container will contain somewhere between 60–75% cauliflower and the brine will naturally come into equilibrium with this during processing. It is therefore, necessary to check brine strengths fairly regularly during processing and to adjust them if necessary by the use of a stronger topping-up brine, or by the direct addition of salt with subsequent mixing. With polydrums, less evaporation of water occurs, there is no leakage of brine and the use of a 16% salt brine for the second brining and topping up has proved more satisfactory. The brine strength should be maintained between 10% minimum and 16% maximum during fermentation and subsequent storage.

Good agitation is vital during the early stages of brining to avoid stratification. Agitation is usually achieved by rolling the drums at least once a day for the first week to ten days. If other containers are used it is necessary to employ some form of mechanical agitation to ensure that the brine is thoroughly mixed.

Gaseous fermentation will proceed for around 6–8 weeks, depending on the temperature and during this time some provision for the venting of the container and topping up with brine at regular intervals must be made. It is most important that the venting provision should restrict the re-entry of air to a minimum. The entry of oxygen will allow the growth of scum yeasts on the brine surface, with a resultant possibility of low acidity and texture deterioration.

Lactic acid acidity will gradually increase during the process and at the end of fermentation the brine may be expected to contain around 0.5% lactic acid, but can range from 0.3% to as high as 1.0%. In the event of the brine acidity being less than 0.3% it should be adjusted to this figure by the addition of edible lactic acid, or increasing the sugar level to allow fermentation to progress further.

On completion of the fermentation process the vegetable is repacked if applicable, sulphur dioxide added if required and the containers are finally topped up, sealed and are ready for storage or use. A completely fresh brine for the final packing may be used if desired, containing appropriate levels of salt and lactic acid.

Brined cucumbers and gherkins

Pickled cucumbers represent the main pickle trade in the USA and continental Europe and an extensive industry and technology has developed in the handling of various types, including brine-fermented cucumbers, cucumbers fermented with dill and spices in a mildly acid brine and pasteurized fresh cucumbers.

In the UK the main interest is in small cucumbers, known as gherkins. Brine-fermented gherkins are still very largely imported from sourthern Europe and Holland, in graded sizes. Size is normally expressed in terms of number per kilogram and each grade is quoted as a range of sizes, e.g. 30/40; 80/100; 180/200. The sizes larger than 50 are normally used for cutting into rings, dice or spears, for use in mixed pickle, sweet pickle and piccalilli, as are 'crooks' (crooked or misshapen gherkins) of any size. The number of gherkins and pickling cucumbers being grown in the UK is very small, although efforts are being made to increase it.

Gherkins should be brined promptly after harvesting, as delays result in progressive quality deterioration. Where delays between harvesting and brining are unavoidable, it is possible to minimize deterioration by controlled atmosphere refrigerated storage.

The fermentation brining of cucumbers is subject to the general considerations listed previously, but also involves the peculiar problem of 'bloaters' and 'floaters', caused by the presence of cavities in some fruits. Apart from the undesirability of this defect in itself, bloaters are liable to emerge from the brine and develop mould. Various factors have been said to influence the incidence of bloaters, including variety,

size, the use of nitrogenous fertilizers, salt concentration, the addition of sugar or lactic acid to the brine, excessive yeast contamination of brine and high brining temperature. Gas production is due to gas forming micro-organisms occurring within the tissue. Pricking the gherkins with spiked rollers or needles is sometimes practised, primarily to assist in securing rapid brine penetration, but it is also beneficial in allowing release of internally formed gas and thus in minimizing incidence of bloaters. The addition of 500 ppm sodium benzoate to the brine also minimizes the incidence. In Europe there is an AIFLV (the European trade association for manufacturers of pickles) Code of Practice for Gherkins in Brine.

Brined olives

As these products are not freshened but are packed in brine and as they are usually considered to be pickles if not actually acetic acid preserves, they are conveniently considered at this point. To produce the end products, the vegetables are repacked from fermented brined stock, either in original brine, or repacked in a brine containing 8–10% salt and 0.5% lactic acid. They are normally hot filled or pasteurized to prevent scum yeast formation, but spoilage may still occur after opening. Oil is sometimes added to avoid this and preservatives may be considered. Other spoilage to occur may be 'yeast spots', which are really clumps of *Lactobacillus plantarum* on the fruit surface.

The brining of olives is interesting in that the glucoside oleuropein, which makes them very bitter when fresh, is hydrolysed by a lye treatment before brining, which thus reduces the bitterness. The processes used can be summarized as follows:

Green olives:

(i)   2% sodium hydroxide for 48 h to penetrate to about two-thirds into the fruit
(ii) Water washed thoroughly
(iii) Brine covered to allow a normal lactic fermentation.

Black olives:

(i)   Pickled red and put into lye, or
(ii) Exposed to air over five days (the colour changes involve catechol-like compounds which oxidize to a very dark colour in alkaline conditions)
(iii) Brine covered and fermented. Olives may be pitted and/or stuffed after brining.

Brined onions

The two main types of onions for pickling are silverskin onions and what are usually referred to as 'pickling onions' or 'brown onions'. Silverskin onions are virtually all imported into the UK from Holland; Israel is a smaller supplier. They are fully fermented in brine, fresh chilled or delivered in acid liquor. When brined, they are graded by size (e.g. 18–21 mm), colour and shape, but only for size and major defects if fresh or in acid liquor. Sulphur dioxide is usually present in the brine or the acid liquor. Fermented silverskin onions should be of white, translucent appearance, crisp texture and characteristic flavour.

Brown pickling onions used to be fully fermented in brine but changing tastes and economies have meant that now virtually all are quick brined, i.e. semi-preserved in weakish brine. Onions used to be mainly supplied as outgrades from ware crops, but now specific varieties are available to produce small onions suitable for pickling. Most supplies are grown in the UK and Holland and may be obtained ready peeled from specialist vegetable preparers. In the quick brined process, the onions do not ferment significantly and are only in brine for a short time, thus retaining much of the fresh onion flavour.

The onions are brined for at least 24 h to prevent cloud formation in the jar, but with a low salt content to save freshening, and a high lactic acid content in the brine to retard fermentation by the repression of microbial growth. The most probable causes of trouble in this process are excessive delay between peeling and getting the onions into the brine, and failure to ensure that the onions are properly submerged below the brine surface. Ideally, there should at least two clear inches between the uppermost onions and the brine surface.

Similar techniques can be used with other vegetables, e.g. silverskin onions and gherkins, and may involve the use of acids other than lactic, with or without the additional use of permitted preservatives and processing techniques such as blanching.

Red cabbage is most frequently used in the fresh state, possibly being subjected to a brief brining after shredding. It must be handled on the same day as it is received at the factory. Walnuts are harvested in late summer while underripe and before the shell was hardened. Once the woody shell has formed, it cannot be softened. In order to facilitate penetration of brine through the tough outer tissue, the nuts can be punctured. Brine fermentation follows the normal pattern.

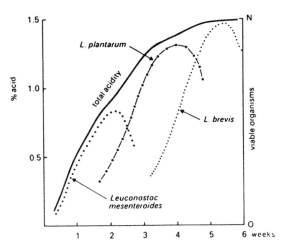

Figure 9.1 Typical sauerkraut fermentation showing cumulative acidity and microbial population.

Sauerkraut

Dry salting is the usual process for sauerkraut production, with the salt addition limited to allow tissue fluid extracted by osmotic action to form a brine in which a lactic fermentation can proceed.

Shredded cabbage with 2–2.5% dry salt is evenly mixed and pressed down under weighted covers. Fermentation takes 3–6 weeks, giving a lactic acidity of 1.8–2.2% and pH 3.5–3.7. The typical course of fermentation is illustrated in Figure 9.1. Sauerkraut can be stored and will stay in good condition for many months.

Other fermented foods

In many parts of the world the diet includes fermented products made from cereals or pulses. In most cases the fermented product has much better palatability than the starting material, sometimes otherwise inedible or poisonous material is thereby converted into nutritious food. In most cases the fermentation does not function to any significant extent as a preservation process, improving the durability of the food. Furthermore the majority of the products are mostly made on the small scale by traditional methods. These traditional foods therefore lie largely outside the scope of this chapter, many of them outside the scope of this Manual. It may be useful nevertheless to note a few of their features.

Fermented soya products

Tofu (sufu), tempeh and a range of related foods in the form of coagulated cheese-like materials are made from mashed or slurried soya beans, fermented by various moulds, yeasts or bacteria. They may be dried or salted for preservation. Natto is made in China by fermentation with *Bacillus subtilis* strains and some of the output is canned.

Beers

In most parts of the world germinated cereals of different kinds are converted into beers by yeast fermentation. Beers are covered in Chapter 7.

Fermented starchy roots and fruits

Cassava (tapioca), taro and banana form the bases of a number of fermented cakes or cheeses in Africa, Indonesia and elsewhere.

Vinegar

Vinegars are the products of two successive fermentations (without intermediate distillation, except in the case of spirit vinegar); a yeast fermentation converting sugars to alcohol and an acetic fermentation by micro-organisms of the *Acetobacter* group converting the alcohol to acetic acid. Many intermediate and minor side reactions also occur, which make an important contribution to the character of the finished product.

Almost any material containing sugars (or starches which can be converted enzymically to sugars) and small amounts of nitrogenous substances can be made into a vinegar. Among starting materials which have been so used are apples, oranges, bananas, dates, prunes and honey. In Europe, wines (themselves the result of a yeast fermentation) are widely used, but in the UK malt vinegar, prepared from malted barley, predominates, followed by spirit vinegar (mainly prepared from molasses).

Malt vinegar

The following is a brief outline of the manufacture of malt vinegar, which involves three main progressive conversions, viz. starch to solution of sugars and dextrins ('wort') to alcoholic liquor ('gyle') to vinegar.

The first stage in the 'mashing' process, in which ground malted barley (often supplemented with other cereals) is steeped in water for 2 h at carefully controlled pH and temperature (normally in the region 60–66°C) and the enzymes present in the malt convert the insoluble starch to soluble sugars and dextrins. Simultaneously, proteolytic enzymes convert some of the protein to amino acids.

The 'wort' is then drained or filtered from the spent grain (which can provide cattle food), cooled by passing through a head exchanger, and inoculated or 'pitched' with yeast, with controlled (but not excessive) aeration in a vessel equipped with internal or external cooling facilities to prevent the temperature rising above about 21°C. The yeast in reproducing itself converts the sugars to alcohol and carbon dioxide, and in conjunction with enzymes also converts dextrins to alcohol. (In this instance maximum alcohol production is required, whereas in beer brewing, the wort is boiled before pitching, to destroy enzymes and thus ensure some dextrin retention to provide 'body'.) The resulting 'gyle', which may contain up to 6–9% alcohol, is stored and filtered, when a secondary fermentation of residual carbohydrate can occur.

The next stage is an acetic oxidative fermentation (acetification) in which the matured 'gyle' is brought into contact with acetifying bacteria under controlled conditions with abundant air supply. The traditional acetifier is a tall vat containing birch twigs, wood shavings or other neutral materials holding the *Acetobacter* culture, in which the gyle is circulated by sparging or spraying from the top, falling through a false bottom and continuously recirculated for several days, until most of the alcohol is oxidized to acetic acid. The modern alternative method, typified by Fring's acetator, involves submerged culture fermentation. The culture is dispersed in the gyle itself and forced aeration is supplied from the base of the vessel providing considerable upward turbulence. This type of equipment requires less space and gives far more rapid conversion of alcohol to acetic acid, but power consumption is higher, and continuity of air supply is critical.

The resulting raw vinegar is filtered and pumped to storage vats (often containing beech chips) where it matures for several months, with minor but important chemical changes, the deposition of colloidal matter and the development of the characteristic flavour and bouquet. After maturing it undergoes a final clarifying filtration to a crystal clear, pale gold coloured liquid.

The vinegar thus prepared may be required for bottling (with or without the addition of caramel and suitably adjusted with water to the desired acidity which must not be less than 4% w/v), or for bulk storage for pickle and sauce manufacture. In either case, the vinegar requires protection from spoilage due to growth of *Acetobacter* micro-organisms, which can lead to cloudiness and sometimes the production of the mass of extracellular cellulosic growth commonly referred to as 'mothers'. These micro-organisms are very sensitive to salt and the addition of 1–2% salt is adequate to prevent spoilage. Unfortunately the caramels suitable for vinegar tend to be precipitated by salt and if caramel addition is required this method may be unsuitable.

The two main alternative methods used, therefore, are either continuous pasteurization or further filtration through a sterilizing filter. For bottling, the resulting vinegar may be filled into hot presterilized bottles under semi-aseptic conditions (e.g. the use of ultraviolet irradiation of the filling line and of the closure feed); alternatively the filled capped bottles may be subsequently pasteurized, the sterilized vinegar may be protected from reinfection by airborne *Acetobacter* by ultraviolet lamps installed in the headspace of the vessel.

Malt vinegar used to be sold in the UK by 'grain', and reference to this occurs in the literature. Sixteen grain contains 4% acetic acid, 20 grain, 5%, and 24 grain, 6%. This should not be confused with US nomenclature in which 'grain' is ten times acetic acid percentage.

Distilled vinegar

Distilled vinegar (in the UK, usually distilled malt) is prepared by distilling off the acetic acid and other volatile constituents, the colour and soluble solids remaining behind as a residue in the distillation vessel. Distilled vinegar is water white and retains much of the distinctive flavour and aroma of the original vinegar. It is used in products where the absence of contributed colour and refinement of flavour is of importance, for example in the preparation of pickled cocktail onions, salad dressing and tomato ketchup. It is also bottled for sale, and is particularly popular in Scotland.

Spirit vinegar

Spirit vinegar is a brewed vinegar containing not less than 4% m/v and not more than 15% m/v acetic acid. It is normally prepared by yeast fermentation of molasses, distillation and acetification of the distilled alcoholic liquid thus formed. In practice, the distilled alcohol is available as an article of commerce and merely requires acetification.

Spirit vinegar normally contains between 10–12% acetic acid, is water white and has a strong acid flavour, but lacks the characteristic 'malt' flavour. It is extremely useful in pickle and sauce manufacture, for uses where absence of colour is an important factor and especially where its higher acidity can be conveniently utilized.

Concentrated vinegar

Concentrated vinegar of up to 40% acetic acid content can be made by freeze concentration. Vinegar is continuously passed through a Votator scraped-surface heat exchanger in which it is cooled to −8°C, whence it emerges as a slush containing 20–25% ice which is centrifuged to remove the ice.

Evaporation of vinegar results in an increase in the acetic acid concentration in the residue but most of the volatile flavour constituents are lost, together with some acetic acid, so this method is not really a practical one.

Spiced vinegar

Spiced vinegars are vinegar infusions of spice and/or herbs. Their main use in the past was as a means of incorporating the spicing in pickles or sauces in which the visible presence either of whole spices or of specks of ground spice in the product was undesirable.

Widely differing spice and herb mixtures have been quoted (and there are certainly many more which have not been quoted), with levels of total spice addition ranging from 2–50 g l^{-1} of vinegar. The simplest method is to steep the spices and/or herbs (possibly in a muslin bag) in cold vinegar for several days with occasional stirring. The process can be shortened to several hours by heating the vinegar. An alternative procedure is to place the spice in the perforated upper chamber of a percolator and heat the vinegar in the lower part. Spiced vinegar may also be prepared by dispersing spice oils or oleoresins in the vinegar.

Spiced vinegars, however, are outdated as a means of introducing spices or herbs into pickle and sauce products. This is due to the availability of standardized and true flavour spice and herb extracts concentrated on salt or dextrose, or prepared as essences which can be incorporated direct or as concentrated solutions into pickles and sauces with no trace of specking and with considerable saving of labour and trouble. Similarly, a filtered 10% solution in vinegar can be used as desired to dose batches of packing liquor for pickled vegetables, but some manufacturers prefer to incorporate whole spice in the jars.

A very small trade exists in bottled spiced vinegar, mainly for home pickling, though it is probable that most people bottling their own pickles prefer to add their own pickling spice.

In some countries, herbs and spices are put into vinegar storage vats to steep in the vinegar for considerable periods of time. The resultant flavoured vinegars are used both in pickle manufacture, e.g. dill, and for retail sale as such, e.g. tarragon and garlic.

'Non-brewed condiment'

Solutions of acetic acid of 4 to 8% acidity were formerly sold in the UK under the name of 'non-brewed vinegar' or 'artificial vinegar'. However it is now established that the name 'non-brewed vinegar' is a false trade description and 'non-brewed condiment' is now the proper name to use. A product suitable for bottling under this description may be prepared by diluting 5 parts of 80% acetic acid to 100 with water, with addition of salt, caramel or flavour as required.

Salt and vinegar pickles

'Vinegar pickles' or 'clear pickles' are the generic terms used in the UK to describe the group of products in which a single type of vegetable or mixture of vegetables is preserved in a clear vinegar or acid liquor, with or without sugar or spices. The main varieties are onions, silverskin onions (including 'pearl' and 'cocktail' onions), gherkins, mixed pickles (usually involving onions, cauliflower and gherkins), red cabbage, beetroot and walnuts. Many features of these products are dealt with in detail elsewhere in this chapter under the appropriate headings, but a number are discussed here.

Pickled beetroot

Pickled beetroot is the largest volume pickle sold in the UK and consists of cooked and skinned, sliced beetroot packed with vinegar with or without a little salt, sweetener and spice, in glass jars with self-venting or twist-off closures, and pasteurized. A lesser trade exists in a pack prepared in a similar manner but containing whole 'baby' beets (about 25–35 mm in diameter) instead of sliced beetroot.

It is also feasible to pack beetroot in plastic laminate pouches, while an intermediate, short shelf-life product now exists which is halfway between fresh cooked beetroot and the pickled product. For this, small whole cooked and peeled beetroot are given a very strong acetic acid dip; although most of the acid drains off before packing in trays or pouches, sufficient is retained to defer the onset of spoilage for several days and hence give a reasonable shelf life 'fresh cooked' product.

The UK packing season normally runs from October to February, but it is advantageous to pack as much as possible of the required quantity early in the season as the longer stored beetroot yields inferior results in respect of colour, texture and flavour. Baby or whole beetroot may be obtained as a specific early crop from July onwards, while if the additional costs can be justified, the use of chill and controlled atmosphere storage can extend the packing season well beyond February and the end of the traditional supply from clamp.

Beetroot preparation: Deep red spherical varieties of beetroot (*Beta vulgaris* L.), such as Detroit, Libero and Bolthardy are normally used and should be purchased against an agreed specification for grade and quality. If necessary the beetroots are first washed to remove adhering earth. This may be done batchwise in tanks of water or continuously through a spray washer. The washed beetroots are then cooked in boiling water or steam until tender, the time required depending on the age and quality of the material, the size grade, cooking temperature and the equipment in use. Continuous and batch cookers are commercially available for both steam and water cooking. If the boiling water method is used, it can be advantageous to incorporate salt at this stage at the rate of 2% of the weight of beetroot to elevate the boiling point and speed cooking. Keeping the size grade range narrow also minimizes texture variation of the cooked beet.

After cooking, the skins are removed, usually in an abrasive peeling machine and the beets are normally cooled before slicing. Cooling is desirable as it reduces the proportion of broken slices, fragments and debris liable to occur if the beetroot is machine-sliced while still warm. On the other hand, the longer the delay between cooking and packing, the greater the deterioration of colour due to oxidation, and the greater the risk of microbial spoilage. Both of these problems can be minimized by water or air cooling the cooked beets prior to peeling, or by immediately immersing the peeled beets in bulk in the liquor which is subsequently to be used for packing and allowing them to cool. Other systems have included steam peeling followed by a cooking operation and cooling.

After cooling, the liquor is drained off, filtered and used for packing. The beetroots are sliced by machine and the slices are promptly filled into glass jars, which are topped up with liquor to 3 to 5 mm in headspace.

Packing liquor: The liquor is usually prepared from spirit vinegar, but acetic acid solution may be used provided this is appropriately shown in the ingredients declaration on the label. Salt and sweetener may be incorporated into the liquor as desired, as may herbs and spices.

With young beetroots, it is a matter of opinion whether added sweetening is necessary, but it is generally agreed that the older the beetroots, the more need there is for added sweetening to compensate for loss of sugars during storage. Sweetening may be added in the form of sugar, glucose syrup or aritificial sweetener, incorporated in the liquor to an extent judged to give a satisfactory flavour in the finished product and appropriately declared in the ingredients list. Although there is nothing to prevent the use of added colour if desired, this should be unnecessary. The natural beetroot pigment is available as a food colour (betanin).

Packing ratio: The packing ratio of beetroot slices to liquor may vary somewhat from one manufacturer to another, but a reasonable ratio to aim for is 60% slices : 40% liquor after pasteurizing. If the acidity of the packing liquor is 4%, a vegetable : liquor ratio of 60:40 will result in a final overall acidity of about 1.6%, a fairly usual level for this product.

Closures: After the jars are filled with beetroot slices and topped up with liquor, they are immediately capped with acid resistant lacquered metal closures, either of the self-venting type such as Omnia® which give a vacuum on cooling after pasteurization, or of the Twist-off® type applied with steam injection to create vacuum.

Pasteurizing: The capped jars are pasteurized either with steam or hot water. The heat process required will vary according to the method and equipment, and according to the size and shape of the container, but must be such that a P_{70} value of not less than 10 at the slowest heating point in the pack is obtained. This is the bare minimum requirement and should preferably be exceeded, commonly used processes being 20 min or longer at 80°C or higher. For example, if a steam cabinet is used for batchwise pasteurization, the temperature of the cabinet should be slowly raised to 85°C over a period and held at that temperature for suggested periods ranging from 30 min for 340 g jars to 60 min for 2500 g jars. (These suggested conditions are of course subject to the aforementioned provisos.)

After pasteurizing, the jars may be allowed to

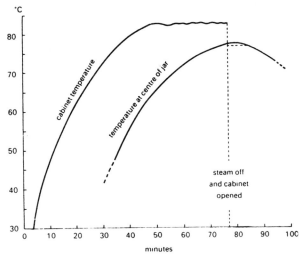

Figure 9.2 Typical time/temperature curve for batch pasteurization of beetroot.

cool slowly in the air. Beetroot differs from most other pickles in that further 'cooking' does not adversely affect the quality. During the early stages of cooling, the temperature at the centre of the jar remains at or above 74°C for a time and thus the first part of the cooling process actually contributes towards the effective pasteurization process. This effect is clearly shown in Figure 9.2.

After cooling, the jars are washed externally to remove any traces of liquor spillage occurring during pasteurizing, allowed to dry, inspected for removal of any jars not holding a vacuum (as evidenced by lack of concavity of the closure) and labelled. Batch water pasteurizing will incorporate a cooling and washing operation in the cycle, while most continuous pasteurizers will also present clean jars at the take-off belt.

A white deposit of calcium oxalate may sometimes be formed in jars of pickled beetroot, arising from reaction of the natural oxalic acid in the beetroot with calcium ions in the packing liquor. It is harmless, but detracts considerably from the appearance. Precautions against this occurrence include the following.

(i) The use of beetroot grown in well-limed soil. This does not reduce the oxalate content, but means that a lower proportion of it is in the form of free oxalic acid.

(ii) The lining compound in the closures should be free from calcium extractable by hot packing liquor.

(iii) The packing liquor should be virtually free from calcium. If the vinegar (and/or acetic acid), the salt and any other ingredients used are very low in calcium content, and any water incorporated in the liquor is soft (or softened), this will suffice. Alternatively, the packing liquor, after preparation, could be passed through a column of cation-exchange resin in the sodium form, and this would have the added advantage that not only would calcium ions be removed, but any trace of iron contamination also.

Pickled gherkins

Gherkins and cucumbers, *Cucumis sativus* L., which for our purpose may be regarded as synonymous, are pickled both fresh and from brined stock. The latter may be fully fermented or may have received only a short treatment allowing much of the natural sugar to be retained. The so-called 'fresh-pack' gherkins are often flavoured with dill, particularly in Germany, Eastern Europe and North America. In recent years some further trade has developed in supplying gherkins in a weak acetic acid solution in cans or drums for repacking.

In the UK, although the market for fresh-pack products is increasing, the majority of production is from brined stock, imported mainly from the Mediterranean countries and Holland. Within the EU, trade codes of practice have been developed under the auspices of the AIFLV covering gherkins in brine for subsequent processing and for some final products such as fresh pasteurized gherkins.

Brined gherkins should be checked on receipt to ensure that they meet the required specification for salt, pH, and lactic acidity. Should the total as lactic be less than 0.3%, then this should be immediately adjusted to 0.5% by the direct addition of 80% lactic acid. Salt is normally checked by measuring the specific gravity of the brine.

Gherkin sizes are commonly designated in terms of count per kilogram, e.g. 80/100 means that the fruits so described will be of weights equivalent to 100 per kilogram at the smallest and 80 per kilogram at the largest for this grade. Grades are selected appropriate to pack size, consumer demand and traditional practice, and are freshened (debrined) to a salt level of about 5%. After washing and inspection they are packed into jars and covered with vinegar, spiced and coloured with caramel as and if desired, usually at an acidity to give 1.0–2.5% acetic acid in the final product for pasteurized packs or 3.6% or more for unpasteurized products. Sugar is sometimes also used in the covering liquor. Capping, pasteurizing, labelling and packing proceed in the normal way.

Gherkins should be protected from light at all stages of manufacture and distribution to avoid fading or bleaching, and particular care should be taken to avoid the ingress of copper and iron during processing as these are prime causes of discoloration.

Pickled onions

Pickled onions (*Allium cepa* L.) in their various forms are very popular in the UK and may be divided into two main groups depending on the raw material used, and into further subdivisions depending upon the preacetification processes employed. Improved technology and raw material supplies, coupled with a reluctance to incur the high costs of financing large stocks of brined vegetables, have mitigated against the traditional fully fermented brined raw material and have favoured more rapid processing techniques.

Silverskin onions: A quantity of packs continue to be produced from raw material fully fermented in brine and subsequently sorted by the briner into high standard size and quality groupings. These are marketed under a variety of descriptions, including 'silverskin', 'pearl' and 'cocktail' onions, the latter usually involving the smallest sizes (10–14 mm diameter). Other retail packs are usually based on 18–21 mm or 21–23 mm diameter (sometimes 23–25 mm or 25–28 mm in the larger jars). The onions are received in polydrums in 10–15% salt brine. The lactic acid content should be checked on receipt and if necessary adjusted by the addition of lactic acid to bring the level to 0.5% of the total contents. They should be inspected from time to time and where necessary topped up with brine containing 0.3% lactic acid.

The onions are debrined to about 5% salt content and packed into jars with a vinegar giving a preservation index of 3.6 and a salt content of around 2%. If desired, sugar and/or spicing may be incorporated in the liquor, which may be uncoloured or coloured with caramel or other colours for some exotic cocktail packs.

Although the pasteurization of brined silverskin onions has been investigated, it does not appear to have found favour as a commercial process, possibly because the fermented silverskin onions do not stand up well to pasteurization but tend to soften and develop a 'cooked' flavour.

Fresh-pack silverskin onions are prepared either directly from the fresh peeled vegetable, normally imported chilled and ready peeled from Holland, or from slightly acidified stock in a weak covering liquor held in chill storage. These onions are normally washed and packed into jars, covered with vinegar suitably spiced and coloured as preferred, capped and pasteurized. The vinegar used should contain sulphur dioxide to the permitted level, some of which may already have been incorporated in the carrying liquor and hence be present as carryover. The preservative is essential as its antioxidant properties help to maintain a pristine white colour and prevent 'pinking'.

Silverskin onions have been grown, harvested and peeled successfully in the UK, but unfortunately no viable commercial operation has emerged. Over recent years an increasing supply from Israel has developed. An AIFLV Code of Practice for some silverskin packs has been proposed.

Brown onions: These are the traditional British pickled onions and were originally produced from small outgrades from the ware crop, supplemented by Dutch and Egyptian imports. With breeding and growing developments, specific varieties, e.g. New Brown Pickling, have been selected to give a neat spherical shape and white end product in the desired size ranges of 28–35 mm and 35–45 mm (unpeeled). Peeling is usually carried out mechanically, or by flame-peeling through a furnace with hand finishing and may be carried out by the pickle manufacturer or by a supplier of ready prepared vegetables. No mechanical systems are entirely satisfactory as yet and manual completion is inevitable to a greater or lesser degree.

Ideally, onions should be in a senescent state when peeled and processed and prior to peeling may be stored for considerable periods under controlled atmosphere storage conditions, especially if treated with sprout suppressant and mould inhibitors. Recent development include the use of overwintered varieties of Japanese origin. Problems commonly found with raw material are sprouting, *Botrytis* and *Fusarium* infections and onion fly.

Peeled onions are subjected to a brining treatment. This precipitates soluble protein within the tissues which would otherwise cause proteinaceous cloud and precipitation to arise in the covering vinegar. Alternatively, a blanching treatment may be effective. It is interesting to note that this problem rarely affects silverskin onions which may be fresh packed, because the peeling system used for them does not give rise to such a significant area of cut surface from which tissue fluid may be lost, the protective effect of the underlying semipermeable membranes between the sheaths being retained.

In most cases a 'quick-brining' treatment is used, with brine of around 10% salt containing 1% or more lactic acid as a fermentation suppressant and this is followed 24–96 h later by washing

and packing. It is, of course, quite possible to carry out a full lactic fermentation, producing a stable brined product which can be stored for lengthy time periods before freshening and packing. Such production is now a rarity due to the high labour and financing costs involved.

Onions are particularly susceptible to the adverse effects of iron contamination and heed should be given to this risk from peeling, through the brining process and vinegar preparation right up to the final filling and capping equipment. After the brining process the onions are washed and filled into jars, probably using vibratory table fillers or horizontal drum tumbler fillers. The vinegar is prepared to give a final acidity in-pack at equilibrium of around 2% for pasteurized packs to obtain the requisite preservation index for unpasteurized products. Spirit vinegar is preferred as the vinegar base, although cheaper products are sometimes packed in non-brewed condiment. Caramel may be used to give the desired colour and the appropriate spicing also added to the vinegar, with sugar and preservative if required. Sulphur dioxide is useful in protecting the white colour of the onions and the bright appearance of the vinegar. Where sulphur dioxide cannot be used for some reason, added ascorbic acid can be of value for its antioxidant properties.

After capping, for pasteurized packs, a pasteurization process is given to achieve a centre pack temperature of 80°C for 10 min ($P_{70} = 200$). Lower processes may be used where there is sufficient confidence in the reproducibility of the process characteristics, to optimize enzyme inactivation and minimize cooking effects. The smaller onion grade of 28–35 mm unpeeled (No. 1's) is used in the smaller retail packs while the 35–45 mm grade (No. 2's) is used in large retail and catering packs.

'Pinking'. Both the onions and the liquor may develop a pink discoloration, eventually darkening to brown. This probably results from a reaction between trace amounts of aldehydes in the acetic acid and an enzymic reaction product in the onions. Certain grades of acetic acid manufactured catalytically from acetylene have given rise to this pink coloration, but grades of acetic acid now generally available are sufficiently free from aldehydes to avoid pinking. If acetic acid is to be used for unfermented onions, this point should always be investigated before deciding on a supply of acetic acid. The discoloration is eliminated by SO_2 and reduced by adequate salt penetration prior to acetification.

Quercetin deposition. This occurs as spots of material like flowers of sulphur immediately below the tissue surface. (It also occurs in capers, and more rarely in gherkins.) The pigment is pentahydroxyflavone, precipitated at low pH by enzyme action from the naturally present colourless and soluble glycoside. Some varieties seem more susceptible than others and bacterial infection of raw material may be involved. Attention to processing conditions can reduce its incidence.

Pickled red cabbage

Red cabbage (*Brassica oleracea capitata*) should be packed as soon as possible after receipt from the field or chill store. Good coloured varieties with tight heads are used. It is trimmed and shredded either by hand or machine. Stainless steel knives or cutters must be used and the cabbage may be packed directly or be brined with a brine of about 10% salt. Dry salting as an alternative is now little used. The brine may be slightly acidified with lactic acid, left overnight and then drained. If the resulting shreds are packed directly into the acid liquor, the salt content of the finished pack may be excessive for most tastes and it can be advantageous at this stage to give the shreds a brief rinse in water and drain thoroughly.

The equilibrium acidity of the final pack will depend on the acidity of the packing liquor and on the ratio of cabbage to liquor. The latter can vary widely, depending on the tightness of packing of the cabbage, and unless care is taken some jars may show excessive acidity and others insufficient acidity to prevent microbiological spoilage. Whatever the packing ratio chosen, close control must be exercised in maintaining it and variability be taken into account in setting the target acetic acidity. Some manufacturers prefer to add the liquor to the shreds in bulk and allow it to stand for a few days before repacking into jars. This minimizes the effect of minor variations in cabbage-to-liquor ratio on the equilibrium acidity of the finished pack.

The liquor may be made from natural or distilled malt vinegar, spirit vinegar, acetic acid or any desired combination, provided the ingredients are properly declared on the label and the required acetic acid content is attained. Spices may be incorporated conveniently in concentrated extract form and may include ginger, pimento, black pepper, chillies, cloves and coriander. If desired, 0.5 to 1.0% of lactic acid may be incorporated, if not carried over from the brining process.

If sulphur dioxide is to be used in the UK it should also be incorporated in the liquor to achieve as near 100 mg kg^{-1} as practicable

without exceeding this current legal maximum. The use of sulphur dioxide greatly minimizes the rapid deterioration of colour and texture that otherwise occurs. This deterioration is primarily oxygen dependent and may be observed to occur most rapidly at the top of the jars. Although some softening may result from the action of polygalacturonase enzymes, the main cause has been shown to be a non-enzymic oxidative degradation of cellulose by an as yet unidentified constituent present in red cabbage.

Pasteurization is not advantageous because of the non-enzymic nature of the degradation, and in any event tends to soften the cabbage by cooking. The main counter measures are the use of sulphur dioxide, close topping-up with liquor so as to minimize the volume of the headspace, an effectively hermetically sealed closure and optional vacuumization during capping.

In some other countries firm texture is not as important and low acid pasteurized packs, sometimes incorporating apple or onion, are produced and pasteurized. Scandinavian packs tend to be very sweet, containing 20% or more sucrose in some brands.

Pickled walnuts

The standard pickled walnut pack is based on cooking brine-fermented walnuts. Brined walnuts are thoroughly drained of brine, placed in perforated containers and steamed either in a steam cabinet for 2½ h or in a retort at 0.035 bar for 20–30 min. After cooking, the nuts are pricked with a needle at one of the ends, preferably near the shoulder, to check that shell formation has not commenced. Before or after this, the nuts are soaked for 24–72 h in vinegar or a weak acetic acid solution. Some manufacturers use spiced vinegar at this stage but this is unnecessary as all the desired spicing can be incorporated in the final liquor. The vinegar or acid is drained off, and may be used in brown sauce manufacture. The acidified nuts are packed into jars which are topped up with liquor, capped, washed, dried and labelled. Care must be taken not to crush the cooked nuts during filling.

The liquor is normally prepared from vinegar, sugar, soya sauce or hydrolysed protein, spices and caramel. Many varying recipes have been quoted, but up to 10% of sugar in the vinegar appears to be widely accepted. Spicing generally includes ginger, pimento, cloves, chillis and black pepper, and additionally mace, tarragon and garlic have been used, while in some cases spicing is achieved by using Worcestershire sauce.

The ingredients are brought to the boil and simmered in a boiling pan for sufficient time to dissolve the sugar. If concentrated spice extracts are used, no further heating is necessary, otherwise simmering may continue for 30–45 min. The liquor may be filtered and may be filled either hot or cold. If a suitable hermetic closure is used, a liquor of lower acidity may be used and the filled capped jars pasteurized.

To prepare packs retaining most of the vitamin C content of the walnuts, fresh unripe green walnuts may be used when these are available. These are peeled thinly, to give a green-coloured product, or completely, when the colour is white, and brined overnight in a 15% salt brine. The brine is drained off, the nuts are rinsed with water and packed in a liquor similar to that described above, but with at least 7.5% acetic acid acidity. These packs should be left for several weeks to mature.

In all cases there is a bitter flavour initially present with walnuts which is said to diminish during storage while other flavours improve. There is said to be merit in using brined stock at least six months old and keeping the pickled product for a similar period before sale.

Mixed pickle

Mixed pickle may contain any types of vegetable in any desired proportions, but most often consists of cauliflower, onions and gherkins, usually in a 60 : 20 : 20 to 40 : 30 : 30 ratio. Most of the ingredients are from fully fermented brined stock but onions may be used from quick brined sources or from acid carrying liquors. Silverskin and brown onion varieties are used. Cauliflower may be obtained as cut florets or be cut from whole heads, while gherkins or cucumber may be cut to form rings or slices. Brined vegetables are freshened before use.

Packing the vegetables into the jar may be done by hand or by machine. With the advent of machine packing, the elaborate hand packed 'pattern-packs' have largely disappeared and the alternatives are a 'throw-in' pack (i.e. a mixture of vegetables without any discernible pattern), or a simple pattern of layers of vegetables (e.g. a layer of onions at the bottom of the jar, then a layer of cauliflower, then a layer of gherkin rings, another layer of cauliflower and finally a layer of onions at the top).

The filled jars are covered with a spiced vinegar (traditionally chilli based), formulated to give the desired end product acidity and flavour and capped. If required, the pack may be pasteurized,

in which case a low acidity may be used and the liquor may be filled hot.

Sweet pickle

Sweet pickle consists of a mixture of acidified chopped vegetables and sometimes fruits also, with a thick, sweet, brown fruit sauce. Among products of this type wide variations exist, especially in the range of vegetables used and in the extent to which the vegetables are softened by cooking. Some manufacturers aim at maximum crispness, others at a substantial degree of softening.

General considerations: The vegetables, and fruit if used, and the sauce should be separately prepared, both in accordance with the need to ensure the spoilage inhibition effect of acetic acid. Due attention must be given to this because attainment of equilibrium of acid and sugars between the vegetables and the sauce is slow, particularly if the vegetables are not cooked in the sauce, and localized fermentation of vegetables could occur. Moreover, significant differences of acidity, salt content, and particularly sugar content between the vegetables and the sauce can result in gradual development of syneresis in the liquor surrounding each piece of vegetable.

The proportion of prepared vegetables to sauce is according to choice, but in most sweet pickles is found to lie between 4 : 6 and 6 : 4. The refractometric solids usually are in the range 15–50%. A refractometric solids content of 35% and an acetic acid content of 2.4% provide both a satisfactory sugar/acid flavour combination and a safe preservation index.

Vegetables: The chopped or diced vegetables normally consist of onions, gherkins, cauliflower curd and/or stump, marrow and cucumber. Some manufacturers use carrots, swedes or turnips, which can be precooked to soften them.

Other vegetable ingredients include by-products of other pickle manufacture, such as beetroot (oversize and damaged slices), mis-shapen and broken gherkins, cauliflower stalk, stripy or blemished onions and materials resulting from production trials, etc. Much of this material may be collected, chopped or diced and stored in an acetic acid liquor or strong brine until required for use.

Fruit, such as chopped dried fruits, vine fruits, mangoes, apple, etc. may sometimes be included with the vegetables.

Brined vegetables (e.g. onions, gherkins, cauliflower) are debrined to 4–6% salt content. They may then be acidified and syruped if required.

Acidifying and syruping may be carried out in a single operation. For this, the debrined vegetables are placed in a suitable stainless steel or plastic vessel and covered with a syrup acidified with acetic acid to a level of about 5%. Some mixing is advisable to avoid stratification effects as sugar enters the vegetables and water leaves them. When required for use, usually after 24 h, residual syrup is drained off and can conveniently be used as an ingredient for a subsequent batch of sauce, or rectified for re-use. The prepared vegetables will then need to be thoroughly drained prior to mixing with the sauce.

Sauce preparation: The sauce is prepared along one or other of the general lines described under THICK SAUCE, possibly utilizing the drained-off acid syrup from a previous batch of vegetables.

If it is desired the prepared vegetables may be cooked for a short time in the sauce. The subsequent procedure depends on whether a gelling or non-gelling stabilizer has been used. In the former case, the pickle is cooled and stored for at least 5 days before stirring vigorously to break up the gel structure. In the latter event, the pickle is either cooled and filled, or filled and capped at over 80°C.

An alternative procedure is to cool, store and sieve the sauce, mix the prepared vegetables and the sauce cold in the desired proportions and then fill. Of course, it is quite possible to add some vegetables to the sauce to be cooked and to add others later. Their addition may even be used to speed cooling.

Filling: The resulting pickle is machine filled into jars, which are capped, washed externally, dried, inspected and labelled. Careful manufacture, along the foregoing lines, with strict hygiene, will give a product which is satisfactorily safeguarded against microbiological spoilage, not liable to separation 'cracking' or syneresis and with a shelf life of 2 years or more. If a low acid product is required, hot filling is to be recommended for products of this type due to the relatively poor heat transfer in the jar in pasteurization systems.

Piccalilli

Piccalilli consists of a mixture of vegetables (traditionally onions, cauliflower and gherkins) with a thick spiced mustard sauce. Unlike sweet pickle (vegetables in a brown fruit sauce), where opinion is divided on the degree to which the vegetables should be heat softened, in piccalilli the vegetables are expected to be firm and crisp, so they are not incorporated at the time of

boiling of the sauce but are mixed with it later before filling.

General considerations: The vegetables should be prepared separately from the sauce so that some acetification of the vegetables occurs before mixing. It is dangerous to neglect to do this and to proceed with lower vegetable acidity and higher liquor acidity, even though the final calculated overall acidity of the mixture may be correct. Equilibrium of acid between the thick sauce and the vegetables is relatively slow and under these circumstances fermentation of the vegetables is possible. Moreover, significant differences in acidity, salt content or sugar content between vegetables and liquor can result in syneresis in the liquor surrounding each piece of vegetable. Normally piccalilli sauce is prepared with 4–7% sugar content, and this is of the same order as the residual natural sugars of the vegetables. There are sweetened piccalillis, however, in which the sauce contains up to 20% sugar, where it can be necessary to syrup the vegetables to a similar sugar content. The proportion of vegetables to liquor is according to choice, but in most piccalillis is found to lie between 4 : 6 and 6 : 4.

Vegetables: The vegetables normally consist of small silverskin onions, cut gherkin rings and/or dice, and cut cauliflower curd. The onions may include the 'stripey' silverskin onions which most pickle manufacturers prefer to exclude from their silverskin onion packs on grounds of appearance. Similarly, while large gherkins may be cut to provide gherkin rings and/or dice, sound but misshapen gherkins are also used for chopping. A proportion of chopped onion may also be included in the mix.

The proportions of the different vegetables are a matter of choice, but the ingredients must be declared on the label in accordance with the labelling regulations and the mix must be relatively uniform between jars.

The vegetables, ex brine, are debrined to about 5% salt content, and then acidified with an acid liquor of up to 7% acetic acid content to give vegetables containing 2–3% salt and up to about 3% acidity. (In the case of sweetened piccalilli, an acid syrup may be used to provide, say, 20% sugar content in the vegetable, and 1–3% acidity.)

Piccalilli liquor: As with other thick sauces, piccalilli liquor may be made in two basic ways; by use of a thickener giving viscosity without any gel structure (e.g. gum tragacanth or certain modified starches); or by use of thickener (e.g. cornflour) giving viscosity and a gel structure which is allowed to develop fully and is then physically destroyed by sieving; or by a combination of the two.

The stabilizer is dispersed in part of vinegar by vigorous whisking (in the case of starches, when required for use; in the case of gum tragacanth, preferably a few days in advance). The remainder of the ingredients are heated to boiling with continuous stirring, and the sauce is prepared in accordance with one or another of the methods referred to under Sauces, later.

Apart from the stabilizer, the basic ingredients of a piccalilli liquor are vinegar, possibly fortified with acetic acid; sugar (4–7% in piccalilli, 15–25% in a sweetened piccalilli); salt (around 2%); pickling mustard. The remaining ingredients may vary very widely, but usually include finely minced onions, garlic and spices. Ginger is usually the predominant spice; others that have been quoted include white pepper, capsicum, cayenne, nutmeg, cinnamon, fenugreek and cumin. The spices can be incorporated as a single concentrate on dextrose or salt base.

Traditionally, turmeric is added to piccalilli liquor to provide flavour and a yellow colour, although its contribution to the former may be small. The colour is not stable in light, fading quickly in direct sunlight or strong artificial lighting. Alternative natural or synthetic colours are sometimes used in addition to the turmeric itself to help mask bleaching.

Filling and capping: After the sauce and the well-drained pre-treated vegetables have been well mixed in the desired ratio, the resulting piccalilli is machine-filled into jars, which are capped using injection of super-heated steam if desired. The capped jars are washed externally, dried, inspected and labelled.

Careful manufacture along the foregoing lines should give a product which is satisfactorily safe-guarded against microbiological spoilage, not liable to separation, 'cracking', or syneresis, and with a shelf life of at least a year. Low-acid picca-lillis may be produced and are preferably hot-filled rather than pasteurized.

Piccalilli sauce: Piccalilli sauce consists basically of piccalilli liquor as described above, filled into sauce bottles. If it is desired to reproduce the spice effect of a piccalilli, the level of spicing used should be about two-thirds of that incorporated in the piccalilli liquor.

Here also a lower-acidity product may be made and satisfactorily hot-filled.

Chutney

Chutney derives from a Hindi word meaning a strong, sweet relish. It is a type of pickle usually with a lower acidity and higher sugar content than other pickles, often using fruits as a base.

Mango chutney

Mango chutney is made from sliced mango fruit, sugar, vinegar, tamarinds, salt and spices. It is largely imported in bulk from India and Pakistan in a wide variety of types, although mango slices in brine may be purchased for manufacture of mango chutney.

Purchased bulk mango chutneys vary widely in flavour, from mild to very hot, and in consistency. Generally the refractometric solids content lies in the range 55–65% and the volatile acidity in the range 0.5–0.8%. Thus, the preservation index of these chutneys, calculated in the usual way is far below the normally accepted safe level of 3.6% and yet microbiological spoilage does not ordinarily occur. It is evident that, apart from the influence of sugar on the total solids and hence on the preservation index, at this level of sugar content a substantial element of preservation by virtue of reduced water activity is also involved.

Some manufacturers modify the bulk-purchased mango chutney before repacking, sometimes by altering the mango slices/syrup ratio, and/or modifying the spicing or reducing the stiffness of the consistency to suit their customers' tastes. These modifications may include extending volume by syrup or glucose syrup addition, keeping an overall soluble solids of about 60% or adjusting the acetic content to a preferred level.

In preparing mango chutney from brined sliced mango, the liquor portion is prepared by boiling together sugar, tamarinds, vinegar, spices (usually ginger, pimento and mace) and apple pulp, if desired, to 70% refractometric solids and an acetic acid acidity of 1–2%. Debrined mango slices are added (2 parts slices to 7 parts liquor by weight), and the mixture simmered for a further 30 min. If a higher slices/liquor ratio is required, it is desirable to presyrup the debrined mango slices, using for example a hot acid syrup of 65% refractometric solids and 2.5% acetic acid content. The excess syrup drained off after syruping can be incorporated as an ingredient in a subsequent batch of liquor. The formula of the liquor is so adjusted that when the chutney has been boiled to 60% refractometric solids, a total acidity of about 1.5% is achieved.

Pickled eggs

Small size hens' eggs are normally used and are simply hard boiled, shelled, packed into jars and covered with a spiced vinegar (clear or coloured with caramel) with added salt to taste and the jars capped. Quails' eggs may be packed as a speciality product in the same way. Retail and some catering size packs are of low acidity and are pasteurized but some pub trade still exists for eggs in strong vinegar which do not need to be pasteurized.

PART 2: SUGAR PRESERVES

PRINCIPLES

Jams, marmalades and their associated jellies are basically sugar/acid/pectin gels containing fruit or fruit juice. The basic principles underlying their manufacture are to control the composition so that:

(i) the sugar concentration is high enough for satisfactory preservation of the product; from Table 9.1 and the earlier discussion of a_w it can be seen that this requires the sugar concentration to be saturated (about 68%) or near saturated; if this sugar concentration is considered too high for other reasons, e.g. for reduced-sugar jams, then additional means of preservation must be included;

(ii) a satisfactory stable sugar/pectin gel is formed with the following properties: the texture should be such that it can be easily spread, yet before use it should retain its shape without syneresis (separation of free syrup), crystallization of sugar or becoming rubbery; the fruit pieces in the product should be readily recognizable and neither tough nor distintegrated.

Pectin gel formation

The formation of pectin gels is an extremely complex subject, not least because pectin is not a single specific chemical compound. In simple terms, pectin is able to form gels with sugar solutions which have soluble solids in the range 60–70% and pH values of 3.0–3.4.

The controlling factors – (i) pectin type and quantity, (ii) sugar concentration and (iii) pH must be balanced to obtain optimum gel conditions. A reduction in sugar level gives a weaker structure as does a pH level in excess of 3.5, whereas a pH below 2.9 gives an increase in gel strength but also a tendency to syneresis. In between the limits above, it will be found that pH value has an effect on the rate of setting. Optimum conditions depend to some extent on the manufacturing process used and hence the time available between boiling and filling. Control is achieved by using sodium citrate or sodium carbonate to raise the pH, or citric acid to lower it.

Effect of sugars

The sugars present in jam comprise the natural sugars originating from the fruit together with the added sugars. These together constitute almost two-thirds of the product.

The bulk of the added sugar is sucrose which may be from cane or beet sources. During the boiling process some of the sucrose is converted to invert sugar, a mixture of dextrose and fructose. This conversion is accelerated by increase in temperature and by decrease in pH. It should be noted that the change gives an increase in sugar solids, since 19 parts of sucrose plus 1 part of water together yield 20 parts of invert sugar. Inversion is advantageous since a solution of sucrose is saturated at about 66% at 20°C and may crystallize at higher concentrations. The solubility of a mixture of sucrose and invert sugar is higher, although an excess of dextrose will produce dextrose crystallization. In general, an invert level of 20–35% of the sugars will avoid either type of crystallization in products up to 72% total soluble solids.

Sugar may be used in dry granulated form, or is often purchased as a 67% aqueous solution of sucrose, a 'sugar solution'. Glucose syrup may be incorporated into the recipe (it is often cheaper on a dry weight basis than sugar) up to that level of reducing sugars in the finished product mentioned above, but the use of glucose syrup influences the setting characteristics of the jam by raising the setting temperature.

pH control

The fruit used in the manufacture of jam contributes acid to the product. It is usually necessary, however, to add additional acid, usually citric although others such as malic or tartaric can be used. In addition, a quantity of sodium citrate or carbonate or bicarbonate is added as a buffering salt in order to achieve better pH control. Measurement of pH is carried out on a 50% aqueous solution of the jam since it is not always easy to obtain accurate readings from the gel.

PROCESSES

Jam manufacture

Ingredients

Glucose syrup: Glucose syrup is now in general use in preserves manufacture. As a generalization, 63 DE (dextrose equivalent) is the most often used. A small amount of residual sulphur dioxide is often present, together with salts which result from neutralization of the acids used for the original hydrolysis.

Pectin: The quantity and quality of useful pectin occurring in fruit is dependent upon the amount naturally present, the state of maturity when the fruit is harvested and the level of post-harvest enzymic activity. When making jam, this naturally occurring pectin is of considerable importance. The useful quantity will vary depending on the type of fruit used and upon the conditions stated above. Some fruits have a fairly high level of natural pectin, e.g. apples, citrus fruits, plums, whereas others have low levels, e.g. blackberry, cherry. It is because of variation in natural pectin levels that it is necessary to incorporate commercial pectins in order to obtain consistent products.

Naturally occurring pectin has a high degree of methylation and falls into the rapid-set classification (see Chapter 14). Selection of the type of added pectin to be used depends on the product. If fruit pieces are present, a fast set is required to ensure that setting takes place before fruit flotation or sinking occurs. If a clear jelly is being manufactured, then a slow set is advantageous so that air bubbles can be completely removed to give a clear product. In practice, a compromise is usually achieved, with the rate of setting being the result of a combination of pectins, pH and sugars solids control.

Formulation

When formulating recipes, it is convenient to specify a standard theoretical output and relate quantities to this. It is necessary to know the actual fruit content of fruit pulps used and also the refractometric solids. Pectin should be used as a solution, the solids content of which must be known. Some of the sucrose will be inverted during processing and this also needs to be taken account of. A measure of 1 kg of sucrose yields 1.052 kg of invert sugar, so if it is assumed that 25% of the sucrose is inverted, then 0.987 kg of sucrose yields 1 kg of mixed sucrose + invert sugar.

As an example, a strawberry jam is desired 'prepared with 35 g of fruit per 100 g' and with 'total sugar content of 66 g per 100 g', to quote the label declarations. The product is to be made using sulphited pulp at 87.5% fruit content. Thus, for an output of 100 kg, $35 \times 87.5\% = 40$ kg of fruit pulp are needed. The refractometer solids on this pulp are 8% and so the pulp contains $40 \times 8\% = 3.2$ kg of sugar solids. By experiment, 6 kg of pectin solution containing a measured 10% soluble solids are needed, so the pectin contributes 0.6 kg of solids. The final refractometric solids are to be 66% and so the sugars to be added are $66 - 3.2 - 0.6 = 62.2\%$. This could be added as sucrose, in which case $62.2 \times 0.987 = 61.39$ kg of dry sugar would need to be added. But 20% of the added sugars must be replaced with glucose syrup supplied at 80% dry solids: $62.2 \times 20\% \times 80\% = 15.55$ kg of glucose syrup.

This leaves $62.2 \times 80\% \times 0.987$ as the amount of dry sugar which is needed to make up the recipe $= 49.11$ kg, which if used as a 67% sugar solution (syrup) means that 73.30 kg must be added. The recipe thus becomes:

	kg
Strawberry pulp	40.00
Pectin (solution)	6.00
Glucose syrup	15.55
Sugar solution	73.30
(plus acid, colour and buffer) *to give*	
Output	100.00

The water to be removed becomes 29.63 kg less the weight of acid, etc. There will be small errors in this calculation, since no account has been taken of the effect on refractometer reading due to acid and buffer, the variation in degree of inversion and the variations in the raw materials used. Another variable will occur since it is not possible to boil to precisely the same end point on each occasion. While a basic formulation may be obtained in this manner, it is necessary with any new product or change in material supplied to carry out trial boils in order to assess the actual requirements of pectin, acid and buffer.

Jam boiling

The boiling process, in addition to removing excess water, also has other effects, in particular partial inversion of sucrose, development of characteristic flavour, textural changes in the fruit and destruction of yeasts and moulds. Boiling may be carried out at atmospheric pressure or at reduced pressure (vacuum boiling) and may be in batches or continuous.

When boiling at atmospheric pressure the

product reaches about 105°C at completion and this process gives the characteristic flavour of traditional jam. With vacuum boiling where the bulk of the excess water is removed at 50–60°C, the flavour changes are less marked, with a reduced level of caramelization. Which flavour characteristics are preferred is a matter of opinion. Vacuum boiling also restricts the degree of sugar inversion and removal of sulphur dioxide and these factors must be considered when formulating the product. Continuous operation has obvious production advantages if the required output is sufficient to warrant it.

If reduced sugar jams are being produced, or those with relatively low fruit content, evaporation may not be necessary. In these cases it is only necessary to heat the product to achieve flavour development, sterilization and correct conditions for setting.

Atmospheric pressure boiling: The traditional method of jam boiling (Figure 9.3) is to use steam heated open pans normally capable of holding batches of 75–100 kg. Steam at 4–5 bar is supplied to a jacket or to internal coils for faster boiling. Pans are used in sets of between four and eight so that a continuous supply of product can be obtained. Pans which are discharged through bottom outlet valves are more convenient (and

safer) in use than tipping pans. Each pan is supplied with a hood connected to an exhaust system for removal of water vapour. The pans are loaded by gravity or by blowing or pumping from a premixing vessel. Once loaded, the steam supply is turned on and the mix boiled to the required solids level.

This may be determined by refractometer or by temperature which will need to be corrected for changes in atmospheric pressure. If controlling by temperature, then it is quite practical to use electrical thermometers which can be connected to a central control point which can be preset to show when the required temperature is reached. The system can be expanded into a total control system controlling valves, pumps, etc.

Batch vacuum boiling: Vacuum boiling vessels are available to take charges of 500–2000 kg. The vessel, made of stainless steel, is provided with heating coils, a vacuum pump or ejector system, a means of sampling or a built-in refractometer, entry points and an inspection window.

The mode of operation (Figure 9.4) is to prepare a premix in a stirred heated vessel. The premix is drawn into the process vessel by vacuum. When charged, the pressure in the vessel is controlled such that evaporation takes place at 50–60°C until the required solids level is reached. When this is achieved, the vacuum is released and

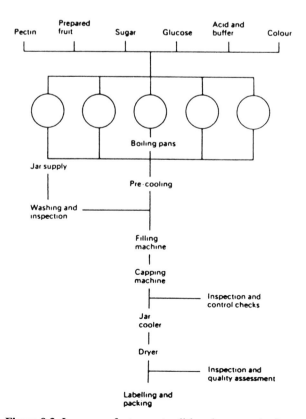

Figure 9.3 Jam manufacture – traditional pan method.

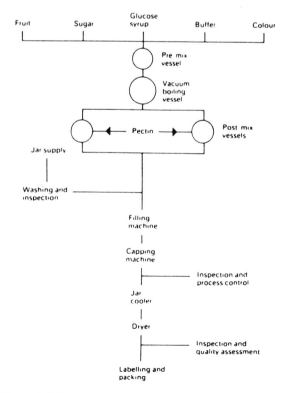

Figure 9.4 Jam manufacture – batch vacuum boiling.

the temperature is allowed to rise to about 90°C. This ensures that sterilization occurs and that conditions for formation of the pectin gel are achieved. The vessel is emptied by gravity, pump or air pressure.

Since large batches of product are obtained, problems with presetting may occur and it is necessary therefore to add the acid and/or the pectin after the boiling is complete, making appropriate allowance in the solids level of the boil. This later addition can be by metering or by the use of two receiving vessels which are used alternately.

Continuous boiling: Continuous vacuum boiling (Figure 9.5) may be carried out by two methods. The APV system utilizes a plate evaporator whereas the Alfa-Laval system makes use of scraped-surface heat exchangers. The former system is limited to handling jellies or sieved products whilst the latter can handle larger fruit pieces.

The APV system, manufactured by APV Baker Ltd of Crawley, England, requires a premix system using either tanks or continuous metering. Since the contact time in the evaporator is short, it will be necessary to install a preheater and flash tank to remove sulphur dioxide if this is present in the raw material. The evaporator is of the rising and falling film type, in which the feedstock passes through a system of steam-heated stainless steel plates where it is raised to boiling point. The mixture of liquid and vapour passes to a flash chamber in which the vapour is removed by vacuum and the concentrated product removed by a pump from the base of the flash chamber. The efficiency of the system can be further improved by using multistage evaporation whereby vapour from the first stage is used to provide the heat for the second stage. This system would normally be used where the fruit ingredient is particulate and fresh, frozen or canned. If sulphited ingredients were to be used, and a low SO_2 content in the finished product were required, an SO_2 stripper as described above would be needed.

This system, whilst very efficient in terms of heat transfer, will not handle fruit pieces due to the small distances between the plates. It is, however, very useful in the production of sieved bakery jams and for jelly jams and marmalades.

The Alfa-Laval system, supplied by the Alfa-Laval Company, is based on vertical scraped-surface heat exchangers. As with the previous system, a means of providing a premix is required to give a continuous supply of feedstock which is fed to the heating units which raise the temperature to boiling point. The product then passes to the evaporator which is a modified scraped-surface heat exchanger operated under reduced pressure. A vortex is formed to increase the surface area for increased evaporation. The resulting liquid and vapour mixture is carried to a separator from which the vapour is drawn off to a condenser and the product removed using a positive pump for transfer to the filling system.

When using continuous systems, it is necessary to control refractometric solids continuously, and this is done either using a simple in-line unit or, preferably, an automatic unit with electrical feedback to control the evaporator. When starting up continuous systems or when re-circulating (for instance, because of filling problems), water should be added to ensure that the feedstock is of constant water content. A simpler version of the Alfa-Laval system is available for non-evaporative processing.

Filling and capping

Filling is carried out using rotary multiple-piston displacement machines. Jars which have been washed and preheated travel round a carousel at constant speed. Each jar is filled and the filler

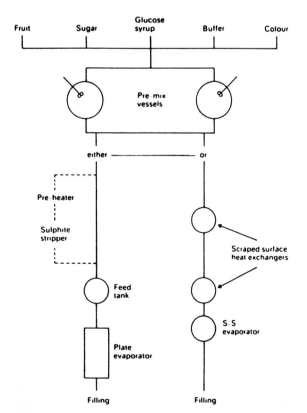

Figure 9.5 Jam manufacture – continuous evaporators.

cylinders refilled during each circuit. Rotary fillers are capable of filling between 100 and 600 jars per min, depending on size and the number of filling heads. Capping is carried out immediately afterwards using a steam-flow machine which will ensure that a vacuum is obtained in the finished jar.

Filling temperature: Whichever method is used to prepare the jam, the temperature at filling should be controlled to between 85–95°C. This is to ensure optimum conditions for setting, distribution of fruit, minimization of weight variation through density changes and to obtain a sterile product. When atmospheric boiling has been used some cooling will be necessary to achieve correct filling temperature, whereas with vacuum boiling the exit temperature may be adjusted directly. Cooling may be carried out using a horizontally stirred mixer fitted with a water jacket. This may be preceded by a sloping stainless steel inspection tray, where some cooling may also take place.

Cooling and finishing

In most cases when the product has been filled at not less than 85°C and capped using steam flow closure, the pack will be sterile with respect to yeasts and moulds and at a pH which precludes the development of bacteria. If this is not the case it will be necessary to use a steam sterilizing unit prior to cooling.

Jars are cooled using a continuous band cooler fitted with water sprays. The initial sprays should be at about 60°C to avoid thermal shock and also to ensure that initial cooling is not too rapid, since if a high vacuum is induced too early during the cooling process, after-boiling may occur in the centre of the product. Later stages of cooling are carried out with water at 20°C.

Jars should emerge from the cooler with a surface temperature above ambient and are passed through an air knife to remove residual moisture from the surface. To ensure that an hermetic seal has been obtained, a non-vacuum detector is used which automatically checks for concavity in the cap. Any jars which fail the check are rejected automatically.

Jars should next pass a visual inspection point provided with suitable illumination so that any which show undesirable characteristics such as foreign material, fruit flotation, bubble formation, etc., can be removed.

Jars then pass to labelling machines and are finally packed, either in cases or on trays which are shrink wrapped.

PRODUCTS

Jams and marmalades

Extra jam

'Extra jam' is a term introduced via the EC Preserves Directive 79/693/EEC. All extra jams are jams, but not all jams are extra jams, as the latter require a higher fruit content – basically 45 g per 100 g compared with 35 g per 100 g for jams (although there are some exceptions to these general levels).

Extra jams have to be made from 'fruit pulp' which is defined as the sliced or crushed edible part of the fruit, with or without peel, skin, seeds or pips but which has not been reduced to a purée. Extra jams cannot be made from purée and so, legally, it is not possible to make a 'seedless raspberry extra jam' as the fruit will have to be sieved before use. It is, of course, quite permissible to make a 'seedless raspberry jam' with a fruit content of 45 g per 100 g or more.

The definition of 'fruit' has been increased recently by amending legislation and now includes, for example, carrots, carrot jam being a delicacy of the Portugese. In addition, there are differing limits for residual sulphur dioxide content for the two classes of product.

Sorbates or benzoates are permitted in extra jams, added colours are not permitted.

Under the present legislation, there are no such products as 'extra reduced sugar jam' or 'extra marmalade', as such products are not defined.

Reduced sugar jam

Although still pectin/sugar gels, the low sugar contents permitted for reduced sugar jams means that normal high methoxyl (HM) pectins are ineffective, and so it is necessary to use the calcium/low methoxyl (LM) (often amidated) pectin gelling system. The texture of the low sugar content gel may be modified by the incorporation of additional stabilizers and/or gelling agents, many of which are listed in the UK Regulations. The low sugar content also has two other effects: first, there is no chance of crystallization and second, the sugar content is insufficient to provide a preservative effect (hence the reason for the permitted use of preservatives).

Formulation and processing: The basic principles outlined above apply, and it is possible to adjust the amount of added water so that a non-evaporative process can be used. Gelling agents and pH control are assessed using test batches. If no preservative is used in the formulation, the processing temperatures must be at least 85°C to provide sterility at the time of filling.

The low sugar contents permissible give different flavour profiles to traditional jams while allowing the strength of flavour of some of the more unusual and exotic tropical fruits to be used to advantage. Low sugar contents also give products with a lower density than normal jams and so special consideration must be given to the selection of an appropriate size of jar.

Jam for diabetics: Diabetics are intolerant to the sugars normally present in jams and marmalades, but they are unaffected by sorbitol, a polyhydric alcohol derived from glucose. Specially prepared jams and marmalades are therefore manufactured for diabetics in which sorbitol is used as a direct replacement for sucrose and glucose. The same manufacturing methods are used but it is necessary to increase the quantity of pectin as pectin/sorbitol gels are weaker than comparable sugar-based gels.

Sorbitol-based products do not have low energy value since the sorbitol is metabolized. The daily intake of sorbitol should be limited since it has a laxative effect.

Diabetic preserves must clearly be labelled as such and in the UK must bear the statutorily required nutrition information demanded by the Food Labelling Regulations. The minimum soluble solids and prescribed quantities requirements for other jams do not apply to diabetic products.

'No-added-sugar' fruit spreads

In these products, the source of added sugar is fruit juice and not one of the sweetening agents listed in the Jam and Similar Products Regulations 1981. Because of this, they do not have to conform to the legal requirements for reduced sugar jam, but in the UK must comply with the Food Labelling Regulations 1996.

Marmalades

Marmalades, as understood in the UK, are effectively jams or jellies made from citrus fruits. The procedures for boiling, filling and finishing are as previously given under JAM MANUFACTURE.

Orange marmalade is by far the largest sector of the market. Bitter oranges are used, the bulk of which are grown in southern Spain, particularly in the Seville province. Small quantities are also available from Sicily, Cyprus, Israel and South Africa. A particular advantage of South African fruit is its availability in September/October,

whereas northern hemisphere fruit is harvested in January/February. Grapefruit, lemons, limes, mardarins and sweet oranges (as well as some of the hybrid types) may also be used in marmalade manufacture. In many cases, their availability is almost year round.

The peel which has been separated as quarters is cut into shreds of the desired size by means of hand-fed or automatic shredding machines. It is also practical to use machines in which quartering and shredding are carried out in one unit, although production rates tend to be slower due to the more complex nature of the machines. Peel is stored in sulphur dioxide solution, or canned in water, or frozen.

The cut peel should be cooked prior to use in order to ensure adequate sugar penetration and absence of toughness in the finished marmalade. Cooking is carried out in steam or water. Cooked shreds may than be mixed with sieved 'dummy' (see below) in the proportions required for the marmalade prior to sulphiting or canning. Alternatively, the shreds may be cooked and either used immediately or held in syrup for a short time prior to use.

Various styles of marmalade are manufactured, the differences being in the size and manner of cutting the peel and the proportions of peel to centre which are used. The usual styles of marmalade are 'coarse cut' with the peel in strips about 5 mm wide, and 'medium' and 'fine' cut with strips of about 3 mm and 1.5 mm wide, respectively. In each of these cases the peel includes the albedo, the white portion lying beneath the skin. In the finished marmalade the albedo becomes translucent when bitter oranges are used. If sweet oranges are used, however, the albedo appears white and opaque.

Fruit is first washed and then sorted to remove defective fruit and 'buttons' which are the points at which the stalks were attached. Size grading may also be carried out at this point. The fruit is then fed to quartering machines in which it is cut into quarters and the peel is separated from the centre. The thickness of the albedo can be controlled at this point by the setting of the machine. The centres, together with blemished but otherwise sound fruit, are minced and removed for further processing. The pulped centres are known as 'dummy' and this must be cooked to prevent enzymic activity. Cooking may be carried out either by continuous methods involving thermal screws and direct steam injection, or by batch methods (usually after dilution to avoid burning) using stirred steam-heated vessels or autoclaves. The cooked dummy is then sieved to remove pips and unwanted fibrous material. It is either used immediately or preserved by sulphiting or canning. As with all processing of fruit materials, it is essential to maintain an accurate record of yields so that the actual fruit content of the product is known.

Marmalades are formulated using the principles described for JAM MANUFACTURE, the fruit content being derived from both peel and dummy. Where the albedo is retained with the peel the natural ratio of peel to centre is approximately 1 to 2, whereas in the case of chips without albedo a ratio of 1 to 10 is more realistic. Whilst a marmalade may be formulated with a higher or lower proportion of peel, it is advisable to balance the use of derivatives overall to avoid imbalances in raw materials. Legislation requires marmalade to contain a minimum of 20 g of citrus fruit per 100 g, of which 7.5 g (at least) shall be from the epicarp (centre).

Jelly jams and marmalade

Jelly jams and marmalades are prepared from the juice of the selected fruit and do not contain any of the insoluble fibrous material, except that jelly marmalade may contain small pieces of peel. The juice is obtained from the fruit either by mechanical pressing or by aqueous extraction. The latter method gives a better yield and is often easier.

When calculating the fruit content of the product, legislation distinguishes between jelly and extra jelly, where the fruit content is specified as the quantity of juice used in the formulation, and UK standard jelly, which is based on the original fruit from which the juice has been obtained.

When using aqueous extraction to obtain the juice, it is essential to know the actual fruit content of the extract; this is achieved by weighing batches of in-going fruit and then weighing or measuring the quantity of extract obtained from that quantity of fruit. If sulphited fruit pulp was used as the initial material, this must be accounted for in the calculation. In practical terms, it is best to use a known quantity of fruit for each batch and then to make to a constant final quantity.

Jelly is formulated and manufactured in the same way as jam. There may be a tendency for excessive frothing to occur on boiling, and this can be controlled by the addition of a silicone antifoam (max. 10 mg/kg) to the juice. When aqueous extracts are used, the quantity of water to be removed may be considerable if sugar syrups are also used. This will be reduced if the sugar is dissolved in the extract and supplied to the boiling system as a ready prepared blend. Since the components can all be supplied in liquid form, they may be metered into the boiling system.

In the production of jelly marmalade, clarified juice or aqueous extract from the dummy is used in the preparation of the jelly. Where peel is added, this is in the form of finely cut 'chips' without albedo. As with the thicker cuts of peel, it is necessary to precook the chips to ensure adequate sugar penetration. The chips are syruped before use in order to obtain a density which is similar to the finished jelly to ensure suspension of the peel through the jelly. Essential oils, which may be obtained from the peel during processing, are added to jelly marmalade to enhance the flavour.

Concentrated juices (the exact strength of which, in relation to normal strength juice and in relation to the fruit from which it is obtained, must be known) can be obtained through many fruit suppliers and such concentrates make the manufacture of jelly jam a much easier proposition.

The product requirements for a good jelly are good colour, clarity and flavour. To ensure freedom from bubbles, the use of slow set pectin will delay setting sufficiently for bubbles to disperse. When pieces of fine peel are to be included in jelly marmalade, these should be precooked in syrup to ensure that the solids level, and hence the density of the peel are the same as the jelly. This will help to ensure that they can be evenly distributed through the jelly. Addition of the peel should be made at the completion of the boiling process.

Fruit curds

Fruit curds, of which lemon curd is the most popular, are gelled emulsions consisting of sugar, fat or oil, egg, pectin, fruit or fruit juice and colour. Fruit flavour curds are basically similar, with the fruit being replaced by flavour.

The additional requirements specified in the UK Jam and Similar Products Regulations 1981 are that one kg of fruit curd shall contain not less than 40 g of oil or fat, 6.5 g of egg yolk solids from whole egg or egg yolk, and that the quantity of fruit, fruit pulp, fruit purée, aqueous extract of fruit and essential oil of fruit is sufficient to characterize the product. The soluble solids of the product must be not less than 65% by refractometer at 20°C.

The traditional British 'lemon cheese' has to comply with the requirements for lemon curd and normally uses butter as the fat ingredient. Commercially made curds are based on starch gels and usually use pectin.

Where starch alone is used as the gelling agent, some oil separation may take place before complete gelatinization has occurred, whereas if pectin is also used, the formation of an initial pectin gel will prevent this separation. The combination of starch and pectin enables the nature of the final gel characteristics to be better controlled in terms of firmness, consistency and mouthfeel. The choice of starch is important in obtaining the desired characteristics, and the specialist starch manufactuer will usually assist in the selection of the correct material.

Formulation: A typical recipe to produce 100 kg of lemon curd is as follows:

Sugar	45 kg
Glucose syrup	18 kg
Unsalted margarine	6 kg
Liquid pectin (5 grade)	7 kg
Modified starch	5 kg
Liquid egg	6 kg
4/1 Concentrated lemon juice	1 kg
Trisodium citrate	250 g
Citric acid	200 g
Glyceryl monostearate	200 g
Lemon oil	200 ml
Water	25 kg
Colour	as required

In the above formulation, concentrated lemon juice has been suggested, although an appropriate quantity of sieved lemon dummy could be used. Glyceryl monostearate (an emulsifier) is used to improve the smoothness of the product. Citrus pectin should be used in preference to apple because of its much lighter colour.

Processing: It is necessary to prepare premixes of various ingredients before bringing them together for final processing. Starch would be solubilized according to the specification of the supplier, quite possibly using hot water. Self-emulsifying glyceryl monostearate is premixed in hot water and then cooled whilst stirring to produce an aqueous gel. Margarine or other fat is melted before use and a premix of this together with egg and emulsifier is prepared.

This type of formulation enables fruit curds to be made in a similar manner to jam. A jam boiling pan, preferably of stainless steel, is used, although copper can be used unless butter is specified in the formulation. A 'slow' pan should be used in preference to one which has internal heating coils, since this reduces the risk of burning. If this occurs, there will be discoloration and the possibility of charred particles giving an unsatisfactory and unsaleable product.

All the materials except the lemon oil are mixed together in the pan and boiled to the desired refractometric solids. After the boiling, the curd is sieved, cooled slightly and the oil mixed in. Filling in carried out at 85–90°C into prepared jars which are immediately capped and then cooled. Since the specific gravity of fruit curd is less than that of jam, it is normal practice to fill 14½ oz (411 g) into a nominal 1 lb (454 g) jam jar, with *pro rata* weights for other sizes.

Stiffening of the gel will continue for some time after manufacture, hence the final texture should not be assessed for at least two weeks, although experience will enable earlier judgements to be made. As with other products, the recipe should be adjusted in terms of flavour, consistency and colour to achieve the characteristics which are deemed to be most desirable.

Bakery preserves

Appreciable quantities of jam, mincemeat, fruit curds and fruit fillings are used by the baking industry for such commodities as sponge cakes, Swiss rolls, tarts, biscuits and pies. The products must be capable of meeting specific applications and as a result a wide range is manufactured. Since most bakery jams are deposited through small orifices or spread in thin layers, they are usually prepared from sieved fruits and purées.

The refractometric solids levels used are higher, generally, than those in similar products packed for retail sale, of the order 72–74%, or up to 80% when used for biscuits. The high levels of solids reduce the risk of fermentation and also contribute to higher viscosity levels. In addition, the low water availability which corresponds to high sugar concentration reduces the risk of seepage into baked products. Products with much lower solids are used where a short shelf life is a desirable and is also a practical proposition.

The performance requirements vary widely. In the case of tarts, the jam or curd must be capable of being deposited without 'tailing' and then submitted to baking temperatures of around 200°C without 'boiling out'. After baking, the product should reset to a firm, non-rubbery texture. An alternative method of tart production is to bake the tart base first in the manner of a biscuit and to add the filling after cooling the base. When using this method, quick setting may be achieved by preparing a jam at high pH (about 4.5) to prevent setting, then adding citric acid solution to the reheated jam at the time of depositing in order to reduce the pH to the setting value. When used in Swiss rolls, the jam must be capable of spreading continuously in a thin layer without breaking or tearing the surface of the sponge, nor should there be excessive seeping of the jam into the sponge.

These various requirements inevitably mean a

compromise between solids content and gel strength to obtain the optimum product. The formulation of the baked product itself may also influence the performance of the jam due to moisture content, pH or surface texture so that the same jam may show variation in performance in different bakeries.

The manufacture of jams for bakery use follows the basic principles described above. The type of pectin used will depend on the product use. Rapid-set pectin should predominate where the product is to be baked or hot deposited, whereas slow set may be used where the product is to be used at lower temperatures. The use of amidated pectins is appropriate to pie fillings.

Fruit curds for bakery use are usually made stiffer than the domestic products, with higher soluble solids. This is achieved by increasing the added sugar at the expense of the same quantity of the water present in the formulation, thus maintaining the levels of fat, egg, etc., as in the standard formulation given previously.

Bakery jams and curds are usually filled into 12.5 kg polythene-lined cardboard cartons, which if they are hot filled should not be sealed or close stacked until the contents have cooled. These precautions are necessary to prevent caramelization and over-inversion of the sugars, as well as condensation which may cause surface dilution with the consequent risk of fermentation and mould growth. It is possible to use heat exchangers just prior to filling to cool the product, when larger (up to 750 kg) containers can be used.

The legislative requirement for the fruit content of bakery jam is the same as that for ordinary jam, but the restriction on the use of additional ingredients as laid down in Schedule 2 of the UK Jam and Similar Products Regulations for ordinary jam do not apply. In fact, the use of additional ingredients other than those shown in that schedule is permitted in bakery jam manufacture.

Fruit fillings

Fruit fillings consist of fruit, sugars and stabilizers, together with colouring and flavouring where appropriate. Though mostly known in the context of pie fillings, they are also used in yoghurts, desserts, etc. The main feature of fruit fillings is their consistency, which is that of a viscous suspension of fruit rather than a 'set' as obtained in jam.

Preparation is based on fruit and/or fruit purée. Fruit purées may be reformed into appropriate shape using alginate and/or pectin to maintain the structure. Where reformed fruit is used, the labelling of the product should, of course, indicate this.

The consistency of pie fillings should not vary greatly whether hot or cold, since the pies may be consumed at either temperature and may also be stored as frozen commodities. In the UK all the common soft fruits are used, also some imported fruits such as apricots and cherries. There are no specific compositional requirements at present. As a general guide it is suggested that a fruit content of 50% and a soluble solids level of 25–35% should be used.

Fruit: Although fresh fruit may be used in season, it is necessary to use preserved fruit for year round production. Preservation should be by freezing or canning, since if sulphited fruit is used, the lack of extended boiling time in the process presents problems in the removal of sulphur dioxide, which may cause corrosion if the product is subsequently canned. On the other hand, there are advantages in the treatment of apples with sulphur dioxide solution to prevent browning.

Sugar solids: The level of sugar solids in fruit fillings is normally in the range 25–35%. A proportion of these solids is contributed by the fruit, the remainder being added as sugar or glucose. There is no problem with crystallization at these solids levels, and the ratio of different sugars present is freely variable.

Stabilizers: The stabilizers used for fruit fillings are modified starches, pectins and gums. Starches which are specially produced to provide the desired characteristics of high viscosity without production of a 'set' are available from the specialist starch manufacturers. Pectins, if used, must be of the low methoxyl type and amidated pectins are to be preferred, since the level of sugars is insufficient to provide gelation with high methoxyl pectins. Gums are used in some cases to provide increased viscosity.

The exact requirements for appearance, texture, mouthfeel, etc., are a matter of choice and the use of combinations of stabilizers allows a wide range of these characteristics.

Formulation: A wide variation in formulation is possible. The following formulation to make 100 kg is an example:

	kg
Fruit	50
Sugar	15
Glucose syrup	10
Pectin solution	6
(as 3% solution)	
Modified starch	4
Citric acid	0.2
Water	15

The water is used to dissolve the sugars and to provide a starch premix. Colours and flavouring may be added if required.

Processing: Processing may be by batch or continuous methods. There is no requirement for evaporation, but heating is required to gelatinize the starch fully and to sterilize the product. The method used for heating may be a steam-jacketed jam boiling pan or a scraped-surface heat exchanger. The starch should be presolubilized in hot water, as for FRUIT CURDS, and sieved before use.

If the process is to be carried out in a boiling pan, all ingredients except fruit, starch and flavouring (if used) are added to the pan and brought to boiling point. The fruit and other items are then added and the mixture brought back to boiling point. A stirred pan is to be preferred, since the viscosity of the product is such that there is a risk of burning against the pan wall. Care must be taken to avoid continuous boiling, since evaporation of water will cause deviation from the intended formulation.

If a steam heated scraped-surface heat exchanger is to be used, a premix of all the ingredients is prepared in a stirred feed tank, ensuring that there is consistent mixing and no fruit flotation. The use of a heat exchanger ensures that no evaporation occurs and burning is prevented.

Filling may be into jars, cans or plastic containers. Temperatures of fill should be 85–90°C in order to achieve a sterile pack. Containers should be inverted immediately after seaming in order to sterilize the headspace. An alternative process may be adopted for cans, in which filling is carried out with a premix at a much lower temperature and the contents then processed by rotating the can. Fruit fillings for yoghurt are generally packed in larger containers such as 25 kg polythene bags which are filled at 85–90°C, as above, to achieve sterility but which after sealing are rapidly cooled by immersion in cold water.

Another alternative processing technique which may be used for the 25 kg packs, or for larger tanks of up to 1 tonne, is aseptic filling into the sterile container. Here, the recipe mixture is heated to achieve sterility and then is cooled, for instance using scraped-surface heat exchangers, and discharged into a sterile holding tank from where it is aseptically filled into the chosen container. Obviously, the strict maintenance of sterile conditions is essential to prevent subsequent spoilage of the product.

Mincemeat

Mincemeat is a traditional product composed basically of apples, sugar, vine fruits and citrus peel together with smaller quantities of suet, acetic acid and spices. The colour most often used is caramel. Most mincemeat is manufactured during September to November in preparation for the Christmas period, although the flavour of the mincemeat matures if it is kept for six months or more.

Although derived historically from a meat-containing product, the name 'mincemeat' is nowadays a misnomer. In some countries the name is unacceptable and in these cases the term 'fruit mince' may be used.

The UK Jam and Similar Products Regulations 1981 specify that each kilogram of mincemeat shall contain:

(i)   A minimum of 300 g of vine fruits and citrus peel of which at least 200 g must be vine fruits
(ii) A minimum of 25 g of suet or an equivalent fat
(iii) A maximum of 5 g of acetic acid
(iv) Minimum refractometric solids of 65% at 20°C.

Mincemeat is not heat processed and it is necessary, therefore, to ensure preservation by means of the formulation, hence the need for a minimum sugar level and the incorporation of acetic acid. The maximum permitted level of acetic acid should be used if the soluble solids are close to the minimum permitted level. If, however, the solids are 70% or more, then an acid level of down to 0.3% may be used satisfactorily. The minimum suet level is specified at 2.5%. Shredded suet is often used which is about 85% fat, the remainder being rice flour or wheat flour which is added to assist in maintaining free-flow characteristics.

Formulation: Whilst the legal requirements provide some limitations, there is considerable scope for variation in the recipe used, such as the addition of nuts, glacé cherries, spirits, etc., in the manufacture of more 'luxury' grade mincemeats. A typical basic formulation to make 100 kg is as follows:

	kg
Apple	24
Sugar	40
Currants and sultanas	25
Mixed citrus peel	5
Shredded suet	4
Mixed spices	0.5
Caramel solution	0.2

Culinary apples should be used; the Bramley Seedling is the preferred variety. Since these are not available until late September, the varieties Grenadier or Lord Derby are used as alternatives in the early part of the season. The apples should be washed, but peeling is optional. Coring is preferred but not always carried out, since the apple is finely chopped prior to use. Inspection must be carried out to remove unsuitable fruit.

Vine fruits may include raisins, sultanas, currants and muscatels, although usually only currants and sultanas are used, the currants helping to produce the dark colour traditionally expected of a mincemeat. Vine fruits always present a foreign-body hazard, particularly with respect to stones and stalks, and so thorough cleaning of the fruit before use is essential. This is carried out by washing, spin drying and visual inspection.

Citrus peel may include a mixture of orange, lemon and citron. It is normally added as cut mixed peel but may be added in the form of a marmalade which may also act as a vehicle for the incorporation of glucose syrup into the recipe.

Suet may be added as broken or grated block material but is usually added as the shredded suet mentioned above. Vegetable fat of similar characteristics may be used to replace the suet and legislation in some countries makes this essential. Acetic acid is added as a concentrated solution. Vinegar should not be used, because an excessive quantity of water would be included. Most mincemeat has spice added, although in some geographical areas the unspiced product is preferred. The spices are normally clove, nutmeg and cinnamon.

Mixing

Mixing is carried out in a batch-type mixer. The action required is to mix thoroughly without causing undue breakdown of the structure, particularly avoiding smearing of the fat. The problem of smearing is reduced if the mixing is carried out below 15°C, and this can be achieved by holding the fruit in chill prior to use. If it is necessary to pump the mincemeat to the filling point, this should be done using a low-shear pump to reduce structural breakdown.

Packing

Freshly mixed mincemeat is quite fluid and is filled into jars using piston filling machines in a manner similar to that used for jam. When filled into standard jam jars, the filled weight is lower due to the density difference, so that mincemeat is filled to 14 oz (400 g) in a nominal 1 lb (454 g) jam jar. The jar is sealed using steam flow closure. No cooler is required, but a rinser should be used to remove any adhering material from the surface

of the jar. The jars are then labelled and packed. When packed, the product is quite liquid and there is some tendency for liquid separation to occur. The mincemeat will gradually set to a much firmer texture due to gelling of the apple pectin and liquid absorption by the dried fruit. It is essential to leave the product undisturbed until this 'setting' has taken place, and therefore freshly made mincemeat should be stored for 2–3 weeks after manufacture before distribution. If kept for maturation for 6 months or more, the textural changes are considerable (as are the changes in flavour).

Although the solids level is generally above the saturation point of sucrose, a rapid inversion of sugar takes place in the product and sugar which is initially present in the crystalline form is gradually taken into solution. With higher solids levels, there is a possibility of dextrose crystallization, although this can be overcome by the incorporation of glucose syrup into the formulation.

Bakery mincemeat

Large quantities of mincemeat are prepared for bakery use. In order to meet the particular requirements of pie filling depositors and to reduce the available liquid, it may be necessary to reduce the size of the particulate material and to increase the viscosity by the addition of rusk or a thickener such as modified starch or sodium carboxymethylcellulose, which should be thoroughly mixed with the sugar prior to the addition of the other ingredients. Mincemeat for bakery use is packed in polythene-lined containers of between 12.5 kg and 200 kg, or in stainless steel tanks of up to 1 tonne.

Table jellies

Table jelly tablets are produced as firm jellies designed to be dissolved in a specified quantity of hot water so that the resultant solution will set on cooling to give a jelly of satisfactory quality. Alternatively, jelly crystals may be provided which, when dissolved in hot water, produce similar results.

Jelly tablets are composed of sugars (sucrose, invert sugar and glucose syrup) together with gelatin, water, citric acid, flavour and colour. Occasionally fruit juice is incorporated. It is necessary to ensure that the tablets are self-preserving and also to prevent syneresis, for which a refractometric solids content of at least 72% is required. Glucose syrup is incorporated to eliminate the possibility of sucrose or dextrose crystallization. The desired characteristics of jelly tablets are satisfactory flavour, colour and gel strength coupled with freedom from crystallization or mould formation.

Formulation: Jelly tablets are normally manufactured such that a 5 oz (283 g) tablet makes 1 pint (568 ml) of finished jelly. Since the strength of gelatin may vary, the quantity used in formulation will vary slightly in order to maintain a standard set in the finished product. Normally, gelatin will constitute 9–11% of the tablet. The following formulation is based on a 100 kg output at 75% refractometric solids (RS).

Syrup (80% RS)	80 kg
Gelatin	11 kg
Water	8 kg
Acid, colour, flavour	1 kg

This formulation includes 8 kg of water which is used to presoak the gelatin and presumes no evaporation. If fruit juice is incorporated, evaporation may be required.

Gelatin

Acid-extract gelatin is preferable to lime gelatin for table jellies since it avoids the possible production of calcium salts which may cause turbidity. Sulphur dioxide may be present as a preservative in the gelatin, up to 100 mg kg^{-1}, and there may well be a carryover into jelly tablets. Preservatives are not otherwise permitted in table jellies in the UK.

It is important to assess the gel strength of gelatin supplies, and samples should be prepared based on the product formulation to be used. Gel strength is measured using the Bloom gelometer or the FIRA-Stevens jelly tester. Earlier UK Regulations, now rescinded, included a requirement that a 3% solution of the jelly tablet should set to a jelly.

The gelatin is presoaked before use so that swelling occurs, which makes dissolution in the syrup much easier. Presoaking may take place in the vessel into which the other ingredients are then added to prepare the mix.

Syrup

The syrup used may consist of sucrose and glucose syrups in the ratio of 55 : 45 based on solids together with invert sugar at, say, 55:25: 20. As has been indicated previously, the presence of glucose will greatly reduce the risk of subsequent crystallization.

Acids

Jelly tablets are acidified with fruit acid, usually to about 1% as citric acid. This has the dual benefit of flavour contribution and lowering the pH to a level which is more satisfactory microbiologically. If lime-extracted gelatin is being used, a problem could arise due to the low solubility of calcium citrate. The problem can be overcome if malic acid is used since its calcium salt is more soluble.

Colouring and flavouring

Any of the permitted colours which is suitable for jam is suitable for jelly tablets. Colour is added to the mix as an aqueous solution. A wide variety of suitable flavours and juice concentrates, if required, is available from the many flavour suppliers. Flavours must be assessed by the user to determine which is preferred. If fruit juice is to be used, it must be completely clarified to avoid turbidity in the jelly.

Manufacture

Syrup at 70–80°C is added to the presoaked gelatin in a stirred vessel and gently mixed until all the gelatin is dissolved. Acid, colour and flavour are then added and blended into the solution. Extended heating should be avoided since this will cause degradation of the gelatin. This solution is then filtered and taken to the filling machine. Filling may be either into trays which produce slabs of jelly which are later cut into cubes, or into moulds so that a ready-to-pack jelly tablet is produced. A release agent (liquid paraffin applied as a mist) used to be used to assist the removal of the tablets from the mould, but PTFE-coated moulds make this unnecessary.

Packaging

Jelly tablets are individually wrapped, since the unit of sale is a single jelly tablet. The tablets are machine wrapped in semi-moisture proof cellulose film and may then be packed into individual cartons. Alternatively, printed film may be used so the tablet may be sold uncartoned.

PART 3: MIXED PRESERVATION SYSTEMS

PRINCIPLES

Some food products are not preserved wholly or mainly by one or two of the three factors so far considered – salt, acid or sugar – but by combinations of factors which may include in addition pasteurization or the use of chemical or microbiological preservatives.

Pasteurization

Pasteurization of pickles involves the controlled and limited heat processing of the final product within its container. Pasteurization has three possible advantageous effects:

(i) The heat destruction of spoilage micro-organisms, thus preserving the pack during its shelf life, the acid present continuing to preserve the product after. This permits a lower acetic acid content than would be necessary for preservation in the absence of pasteurization.

(ii) Total or partial heat inactivation of enzymes of vegetable or of microbiological origin, thus preventing or inhibiting deterioration reactions caused by enzymes. Deterioration includes darkening, softening, quercetin formation in onions, clouding of liquor, sediment formation and flavour deterioration.

(iii) In the case where self-venting closures are used, removal of air during pasteurizing and retention of the headspace vacuum on cooling, which minimizes oxygen-dependent and oxidative deterioration reactions, including darkening and softening.

Pasteurization of pickles in the USA was practised from the late 1930s and process definition in terms of in-pack temperature and time requirements started in the 1940s. In the 1950s, process requirements were related to microbial thermal death times, and more recently, aspects of thermal inactivation of enzymes implicated in softening have been considered. The thermal destruction characteristics for some enzymes and an acid-tolerant yeast are shown in Figure 9.6.

In general terms, the process required for the inactivation of potential softening enzymes is more severe than that needed for destruction of acetic acid-tolerant micro-organisms, and hence it may be beneficial to pasteurize even high acid products for this reason alone. Also, where high pasteurization temperatures are used, this can contribute further cooking effect to some products if so desired. Thus, pasteurization of lower acid pickles is ideally suited to the production of crisp, mild vinegar pickles of good appearance and stability and may also benefit other types.

Pasteurization of filled capped jars may be effected in a variety of ways, including batch methods, in which a batch of jars is heated in a

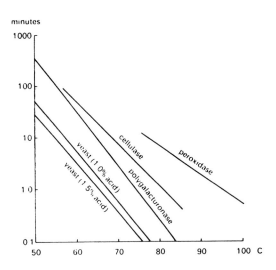

Figure 9.6 Thermal destruction curves for enzymes and an acid-tolerant yeast.

steam cabinet or a tank of hot water, and continuous methods, in which jars are conveyed through a steam chamber, a hot-water spray chamber or a hot-water bath.

Continuous pasteurizers have considerable advantages and economy in labour of handling, but tend to be costly as they need to be of large capacity (especially for beetroot pasteurization) and of acid-resistant construction. A substantial volume of production is therefore needed to justify a continuous pasteurizer, while competent technical skill in its operation is essential if full benefit of the advantages is to be obtained. Whichever pasteurizing method is adopted, the pasteurizing conditions (temperature, time) depend on:

(i) the temperature–time requirement at the point of slowest heating in the jar (normally referred to as 'the centre of the jar', but in most packs two-thirds to four-fifths of the way down the central axis);

(ii) the time needed to raise the contents at that point to the required temperature, which in turn depends on the rate of heat penetration through the glass and contents; and

(iii) the initial temperature. In this respect it can be advantageous to fill the liquor at as high a temperature as possible, but it must be noted that the vegetables themselves (representing about twice the weight of the liquor) are likely to be initially at room temperature.

Reference should be made to the literature for a detailed account and discussion of the work that has been carried out on pasteurization. As Figure 9.6 shows, the heat process required for complete inactivation of peroxidase is greater than that required for polygalacturonase inactivation, which

Table 9.3 Pasteurization conditions for various pickle products (after Dakin, 1957)

Product	Temperature		Time (min)
	°C	°F	
Onions, quick-brined (1–3) days)	79	175	10
Onions, medium-brined (4–14 days)	77	170	10
Onions, long-brined (28 days or more)	74	165	20
Silverskin onions	71	160	15
Gherkins	74	165	20–25
Mixed pickle	71	160	15
Walnuts	71	160	15
Beetroots	82	180	20
Piccalilli, sweet chutney Tomato ketchup Brown fruit sauce	71	160	15

in turn is greater than that required for heat destruction of spoilage organisms. For effective pasteurization, the temperature–time combinations shown in Table 9.3 are recommended; the temperatures are those in the liquid at the point of slowest heating in the jar. The recommendations have stood the test of time and are relatively easily measured in batchwise processing equipment.

The processing conditions which will result in the achievement of these temperatures and times depend on the nature of the pasteurizing equipment and method of operation, the size and shape of the jars or bottles, the precise composition of the sauce or liquor and the vegetable/liquor ratio. It is therefore impossible to generalize, and for a given set of circumstances processing conditions must be established experimentally.

Process determination in continuous pasteurizers is more difficult, but the greater uniformity of processing attainable has led to wider use of the pasteurization P value calculation for process evaluation and definition. Using the principles described by Shapton et al. (1971) and Stumbo (1973), and assuming $Z = 7$°C, for most purposes a minimum process value of pasteurization $P_{70} = 10$ can be considered. Where this has been related to Dakin's original recommendations, it is seen that he recommended generally significantly higher processes, e.g. 20 min at 74°C roughly equals a P_{70} value of 75. However, it should be remembered that at the time of Dakin's recommendations it is likely that considerable variation occurred in processes actually achieved commercially and perhaps demanded a larger safety margin than some of those found today. Also, the effect of thermal process itself is logarithmic.

Measurement of in-pack temperatures is difficult in continuous pasteurizers and it is essential that measurements are made that include significant parts of the warming up and cooling down zones. A travelling recorder able to withstand conditions in the pasteurizer may be used for this purpose. For products likely to suffer from overcooking, it is essential to achieve the shortest warming and cooling times consistent with good glass-handling practice. In some cases, particularly for very large packs, a compromise will need to be made between enzyme inactivation effects and adverse cooking effects.

Excessive headspaces in jars of pasteurized pickled may result not only in poor appearance but in failure to comply with the declared average weight of contents. This may arise from trapped air (especially in beetroot), interstitial air in the vegetables, or the presence of dissolved carbon dioxide, driven off and expanding during the pasteurizing and either rising to the headspace or actually driving out liquor (with possible continuance of liquor loss by capillary action). For self-venting closures this effect may not be too serious, but where Twist-Off or Roll-On caps are in use it can be disastrous. The effective debrining of brine-fermented vegetables, the avoidance of fermentation in short-brined fresh vegetables and vacuum brining and/or vacuum liquoring, will all help to minimize this occurrence. Another possibility is a two-stage pasteurization. In the first stage, jars filled with vegetables and liquor pass uncapped through a waterbath in which the contents are raised to pasteurizing temperature and at the same time exhausted of air and dissolved gases. On emerging from the first stage, the jars pass through a spray filler and are topped up with hot liquor, capped and then pass through the second stage of pasteurization. Filling with hot vinegar and blanching the vegetables before filling also help.

Another problem encountered is the trapping of small particles of vegetable (e.g. beetroot fragments, loose fragments of onion skin, cauliflower crumb) between the jar rim and the cap, either at the time of filling, or in the course of liquor loss during pasteurization. This can prevent satisfactory seating of the closure and results in the pack being susceptible to microbiological reinfection and consequent spoilage and to oxidative deterioration. When the pasteurized jars are cool, they should therefore be inspected for satisfactory hermetic seals as evidenced by the headspace vacuum. With many types of closures, this is visually evident from the concavity of the closure. There are types, however, where this does not apply, but where the presence or absence of vacuum can be detected by the sound when the closure is tapped.

Piccalilli, sweet pickle, thick fruit sauce and tomato ketchup may be pasteurized, provided that a non-gelling stabilizer thickener has beed used. Otherwise, because of the slow rate of heat penetration, excessive heating times are necessary, which can result in softening of silverskin onions in piccalilli, and caramelization of tomato ketchup. If a low acid product is desired (as is normally the case with tomato ketchup) hot filling at not less than 80°C with suitable precautions provides a good alternative, while for fluid products flash pasteurization and aseptic filling may be efficacious. In all cases the same principles of process calulation described above may be applied.

Pasteurized low acid pickles have a preservation index of less than the recommended minimum of 3.6%. When these pickles are opened in the home, however, they do not usually undergo microbiological spoilage, the probable reason being that as acid tolerant pickle spoilage organisms occur largely in pickle factories, they are not generally distributed and are unlikely to be found in the home. Those micro-organisms which have been found in the home can be inhibited by the lower level of acetic acid present in the pickles.

Preservatives

The presence of preservatives in pickles and sauces is governed in the UK by the Miscellaneous Food Additives Regulations 1995, a complicated set of regulations arising from EC Council Directive 95/2/C. These permit pickles and sauces to contain up to 100 mg kg^{-1} of sulphur dioxide and up to 2000 mg kg^{-1} of a mixture of sorbic and benzoic acids (under the category 'vegetables in vinegar').

Vinegar (the category in the Regulations is 'fermentation vinegar') may contain up to 170 mg kg^{-1} of sulphur dioxide. As with all food ingredients, the current legislative position in respect of composition and labelling should be checked by the intending user prior to final commitment, whatever the destination of the finished article.

The US equivalent to these regulations is Title 21 Code of Federal Regulations Section 172. Under these regulations sulphur dioxide is generally regarded as safe (GRAS) and is permitted in pickles and jams according to good manufacturing practice (GMP). Similarly benzoic acid and sorbic acid are GRAS. There is a residual limit for all three substances in jams of 0.1%.

Sulphur dioxide

The UK Miscellaneous Food Additives Regulations do include a proviso whereby ingredients may include preservatives (and other additives covered by the Regulations) even if these are not listed in the Regulations if they are to be used in a food which is permitted to include that substance or substances.

Sulphur dioxide at 100 mg kg^{-1} inhibits the growth of pickle spoilage lactobacilli, yeasts and moulds in media in the presence of acetic acid, but some of the yeasts and moulds grow readily in the presence of 50 ppm SO_2. However, only uncombined sulphur dioxide is inhibitory, and since substantial proportions of the sulphur dioxide in a pickle or sauce rapidly combine with components of the product, 100 mg kg^{-1} sulphur dioxide is not an effective preservative from a microbiological viewpoint, except in olives where it has been shown that it effectively prevents the formation of scum yeasts.

Apart from this aspect, the benefits deriving from the use of 100 mg kg^{-1} sulphur dioxide spring not from its antimicrobial properties but rather from its reducing or antioxidant properties and its property of inhibiting the Maillard nonenzymic browning reaction. Thus, it minimizes storage deterioration of colour, odour and flavour in gherkins; retards darkening of onions, with improved retention of satisfactory odour and flavour; confers whiter colour, minimizes storage deterioration of odour and flavour in silverskin onions; gives whiter cauliflower; reduces discoloration to turmeric coloured cauliflower in mixed pickle; greatly minimizes loss of crispness and dulling of colour in red cabbage; minimizes storage deterioration of flavour in walnuts; prevents blackening by trace iron contamination; prevents oxidative 'blackneck' in tomato ketchup; minimizes surface darkening in non-mayonnaisebase horseradish relish; and prevents (but will not reverse) 'pinking' of onions.

The main adverse effects are in salad cream and products based on it (e.g. sandwich spread and horseradish cream), where sulphur dioxide addition results in emulsion breakdown and development of off-flavours and in pickled beetroot, where it results in marked colour loss.

In brined vegetables, sulphur dioxide is of value in preserving the white colour of cauliflower and silverskin onions, although the time of addition is important and it is this writers' view that it should be added only after the fermentation is complete, as earlier addition may retard fermentation to an undesirable extent. It may also mask any defects, such as iron contamination, which will come to light during the pickle processing and may create bad relationships between briner and user. This can occur where cauliflower is coloured with tumeric for use in clear mixed pickle, when a dirty brown colour can result as the SO_2 is lost, rather than the bright yellow colour desired.

As mentioned before, the presence of sulphur dioxide in jams is limited (and more so in extra jam). The permitted maxima are 10 mg kg^{-1} in extra jam, 50 mg kg^{-1} in jelly and 100 mg kg^{-1} in other products. Its presence in these products is residual and is due to carryover from the ingredients used. Residual sulphur dioxide in excess of 30 mg kg^{-1} must be declared as such in the list of ingredients. The amounts of sulphur dioxide carried over into sugar preserves are insufficient to have any significant preservative effect. Sulphur dioxide preserved fruit must not be used in 'extra' jams or the permitted maximum in the product may be exceeded.

Combinations of effects

In general terms, the quantities of salt, sugar, acid (or pH value) or the degree of pasteurization (e.g. P_{70} value) necessary to preserve a food product, when any of these is used by itself, are known. They have been discussed in previous paragraphs.

The question may be asked, is it possible to calculate the necessary amounts when several of these are used together and the final preservation of the product does not depend solely on any one of them? That is, can one design a product containing, say, X% salt, Y% sugar, with a pH of H and pasteurized for M minutes at D degrees, where X, Y, H, M and D are computed in advance to give a stable product?

The problem has been discussed for some time among microbiologists and food processors. The idea of 'hurdles' has been put forward – the idea that for a microbe to grow it must be able to surmount a number of 'hurdles' such as a_w, pH or heating effect, and each 'hurdle' might therefore be raised high enough to ensure that the microbe cannot grow. The idea is interesting and gives a helpful qualitative description of the various individual effects, but so far it has been difficult to derive reliable, simple relationships for the effects of two or more 'hurdles' together. Some progress has been made through Food Micromodel, a computer database compiled in the UK, which plots the growth curves of particular spoilage and food

poisoning organisms under defined conditions of temperature, a_w and pH, but these data require careful interpretation before they can be applied directly in practical situations. So for the time being it appears best to rely on traditional experience where it is available and careful experiment otherwise.

Physical stability of sauces

Gravitational stability

Thick sauce, whether prepared for bottling as such, or for use in pickles such as piccalilli or thick sweet pickle, may be considered as a suspension of finely divided insoluble fibres of fruit, vegetables and spices if ground spices are used, in an aqueous medium containing acetic acid, salt, natural fruit acids and salts, sugars, other soluble extractives of fruit, vegetables and spices, and various hydrophilic colloids including added stabilizers and pectic substances deriving from the fruit and vegetables.

A physically stable sauce is one which shows no tendency to gravitational separation of solid and liquid phases either by rising or by sedimentation of the suspended particles, no tendency to gelling or syneresis and negligible change of consistency during the shelf life of the sauce or the pickle in which it is incorporated.

Separation results from the tendency of solid particles to rise or fall owing to the difference between their density and that of the fluid in which they are suspended. The velocity of falling V of an individual sphere in an infinite liquid medium is theoretically expressed by Stokes' equation:

$$V = \frac{2}{9} \times \frac{(D-d)\,gr^2}{\eta}$$

where D is the density of the particle, d is the density of the suspending fluid, r is the radius of the particle, η is the viscosity of the suspending fluid and g is the gravitational constant.

Although this equation is not strictly valid when applied to the movement of large numbers of irregularly shaped particles in a finite liquid medium, the significance of the factors $(D-d)$, r and η is not thereby diminished, and clearly the rise or fall of particles can be prevented:

(i) By making the density of the suspending fluid the same as that of the solid particles, so that $D-d = 0$ and therefore $V = 0$;

(ii) By reducing the particle size so that r is extremely small and r^2 negligible;

(iii) By increasing the viscosity of the suspending fluid to the greatest possible extent consistent with the requirement that the sauce should be dispensable from the bottle.

By combining these three factors, a sauce can be prepared with, for practical purposes, no tendency to gravitational separation. In practice it is impossible to guarantee that there is no difference between D and d, but if $(D-d)$ is very small, r is very small and η has as high a value as possible, V becomes so small as to be negligible within the expected shelf life of the product.

Normally the difference in densities of solid and liquid is very small, but one factor which may increase it, to the point where the solid particles rise leaving clear liquid at the bottom, is aeration of the sauce; thus decreasing the value of D so that $D-d$ becomes a substantial negative value. The probable mechanism is occlusion of air in the fibre particles. Hence the importance of avoiding aeration.

Colloid-milling or homogenization not only reduces the fibre particle size but also increases the apparent viscosity, thus doubly contributing to stability. It is, however, not always necessary; given satisfactory provisions in relation to the other factors, fine machine sieving may be found adequate.

It should be noted that gum- and starch-based sauces are non-Newtonian liquids, i.e. they do not have absolute viscosities, but exhibit 'apparent viscosity' which is variable and inversely related to the rate of shear. Thus, for a given fluid, η is not a constant. Any factors tending to decrease V, e.g. reducing the value of $(D-d)$, or reducing particle size and thus tending to reduce the rate of shear, simultaneously tend to increase the effective viscosity, i.e. the resistance to gravitational movement encountered by particles, and thus further decrease V.

A sauce which is initially stable may eventually lose its stability through gradual acid hydrolysis of some types of thickener, with consequent loss of apparent viscosity. This may not only result in the sauce becoming too 'runny' from the consumer viewpoint, but may also permit gravitational separation to occur.

Gelation and syneresis

Gravitational separation should be distinguished from syneresis in which there is also a separation of fluid. Syneresis in a sauce on its own is usually the consequence of a change in gel structure, which as it strengthens and shrinks exudes liquid.

Syneresis may also occur in mixtures of sauce and vegetables (e.g. sweetened piccalilli and sweet pickle) where the vegetables have not been pre-syruped; the process of equilibration between each piece of vegetable and the surrounding layer of sauce results in the formation of 'pockets' of clear liquid around the pieces of vegetable. This can be avoided by pre-syruping the vegetables.

Syneresis from a gel is seen in its most extreme form when, for example, a cornflour-based sauce is boiled and filled into bottles while still warm. The gel structure soon becomes evident then becomes increasingly rigid and shrinkage occurs, so that after a few weeks or months, dependent on the exact formulation and processing conditions, the result is a rigid, shrunken cylinder of gel in the centre of the bottle, surrounded by a substantial volume of exuded liquid. No amount of shaking of the bottle will remedy this type of separation.

In a sauce, however, gelling is itself a serious defect to be avoided, and if steps are taken to avoid gelling occurring in the finished product, the type of syneresis which is a consequence of gelling will also be obviated.

PROCESSES

Manufacture of sauces

Thick sauces

A wide variety of techniques have been described for the manufacture of thick sauces (and there are probably others practised which have not been described). Some of them, however, are conducive to subsequent gelling or syneresis. Three basic techniques are described below, which both on theoretical grounds and in the light of extensive experiment and experience give satisfactory all-round results.

The cornflour techniques: The traditional technique here is to allow the maximum development both of viscosity and of gel structure and then to destroy the gel structure thoroughly by mechanical means. The required amount of cornflour, expressed as a percentage of the finished product, will vary according to the amount and pectinous nature of any fruit present and according to the precise consistency desired. Thus, a typical brown fruit sauce may require about 3.5% cornflour, whereas a piccalilli liquor may require as much as 5%. The exact requirement for a particular formulation can only be arrived at by experiment.

Sufficient vinegar (or acid liquor) is held back for making a 15% slurry of the cornflour. This is prepared on a high speed mixer and fine sieved before use. The remaining sauce ingredients are simmered in the boiling pan, with stirring, to the predetermined refractometric solids. At this point, the steam is turned off and the cornflour slurry immediately added with continued stirring. The sauce is then passed, via a mechanical sieve, to a cooling tank where it is continually stirred while cooling. It is then transferred to suitable storage containers (e.g. 200 l high density polythene drums), where it is stored until required for bottling or for combining with vegetables, e.g. in piccalilli. The storage time must be not less than 5 days, but with a preservation index of not less than 3.6 and good hygiene, may be extended safely to several months. This system also permits the subsequent blending of batches to even out minor batch-to-batch variation in apparent viscosity. When required for use, the stored material, which will have a distinct but not rigid gel structure, is vigorously agitated by suitable high speed stirring and/or mechanical sieving, so that the gel structure is completely destroyed.

The non-gelling modified starch technique: Examination of a wide range of commercially available modified starches has brought to light some which have proved to give thickening without any trace of gel structure, to be satisfactorily resistant to heat and acetic acid and to yield an apparent viscosity remarkably tolerant of minor variations in processing conditions. The required amount for a given formulation and consistency appears to be about three-quarters of the amount of cornflour that would be required.

The method of preparation and addition to the sauce is exactly as described in the cornflour technique. The sauce is then machine sieved and may be cooled and stored if desired, but may be filled hot without any risk of gelling or syneresis.

This starch is more costly than cornflour, and although there is some economy in the amount required to produce the required consistency and some saving in labour of handling, it remains a markedly more expensive technique than the cornflour method in terms of ingredient costs. On the other hand, it is the most easy and convenient of the three methods to operate, is cheaper than the gum tragacanth technique and yields a satisfactory product with no significant variation. It is hence probably now the most commonly used system.

The gum tragacanth technique: Various techniques have been described for the addition of powdered gum tragacanth directly to the hot sauce, either premixed with dry sugar or prewetted with vinegar, water or sugar syrup (to prevent 'balling-up' of the gum). Although this direct addition appears superficially attractive in terms of convenience, it does not allow the full potential viscosity contribution of the gum to be realized, leads to erratic results and may result in syneresis of the sauce on storage. The most effective way of using gum tragacanth is to prepare (in advance) a cold dispersion of gum in vinegar or acid liquor containing not more than 2% gum. This is vigorously whisked, allowed to stand for at least 4 days and whisked again vigorously before use. The gum dispersion is best added to the simmering sauce, with efficient stirring, about 10 min before the end of the heating period. It is impossible to specify the required level of addition of gum, as this depends on the viscosity characteristics of the particular gum supply, the amount and pectinaceous nature of the fruit in the sauce and the pressure at which may subsequent homogenization is carried out, the must therefore be established by experiment in relation to any given gum specification, formulation and processing procedure. When using the best quality white ribbon, a 0.5% level of usage may provide a useful starting point for experiment. This is by far the most expensive method.

Combination techniques: With ever-increasing pressure to improve process efficiency, minimize holdings of work-in-progress and meet marketing requirements for subtle changes in product character, many manufacturers have found advantages in using mixed stabilizer systems to give them the rheological properties required. For example, the conversion of a manufacturing formula from boiling pan production to plate or scraped-surface heat exchanger will almost certainly require a revision of stabilizing or thickening components both to achieve the best physical properties to optimize the new production method and to achieve acceptable character in the end product. Popular systems include mixes of flour and modified starches, starches and alginates, gum mixtures and modified starch mixtures.

Stabilizers and thickeners

The main technical requirements of a stabilizer or thickener for use in sauces and thick pickle liquors are:

(i)   the provision of the required viscosity;
(ii) stability to heating in the presence of acetic acid;
(iii) preferably some tolerance of minor variations in processing conditions;

(iv) preferably yielding opaque rather than clear dispersions (except in the case of tomato ketchup and salad cream, when this is of no significance);

(v) stability in the presence of acetic acid on prolonged storage;

(vi) either no formation of a gel structure, or the formation of a gel structure which can be mechanically destroyed and does not reform.

The combination of these requirements directs attention largely to certain natural and modified starches and gums.

Providing the method of processing is suitably designed to achieve satisfactory results with the stabilizer used, the other important consideration is the cost of stabilizer in relation to its rate of usage. For general purpose sauces and thick liquors (excluding salad cream and tomato ketchup), cornflour and rice flour or other flours can be made to give excellent results when used alone, following the cornflour technique described above or in combination with gums and/or modified starches.

Manufacture of emulsion products

Mayonnaise and salad creams

Mayonnaise is an oil-in-water emulsion of vegetable oil, egg yolk or whole egg, vinegar, lemon juice and salt, mustard or other seasoning. Normally mayonnaise contains 70–85% of oil; the egg and mustard act as emulsifying agent, and the concentrations of salt and acid in the water phase are sufficient to give satisfactory preservation of the final product.

At the usual recipe concentrations emulsification is relatively easily achieved using high speed mixers or similar equipment, to give oil droplet sizes in the range 4–8 μm. The temperature of mixing is important and should not be higher than 16–21°C. Thicker emulsions can be made by mixing at temperatures down to 4°C.

Salad creams are made as cheaper products than mayonnaise, with lower proportions of oil sometimes as little as 30%. The proportion of emulsifying agent (egg, mustard or other) may be correspondingly lower than in a mayonnaise but additional thickening agent is required to augment the viscosity of the water phase. Emulsification may be carried out in high speed mixers but it is usual to finish with a colloid mill, votator or similar emulsifier.

More practical details of these manufacturing processes are given later.

PRODUCTS

Sauces

Thick sauce

The term 'thick sauce' refers here primarily to the brown fruit sauce bottled for sale as such, but also covers thick sauces for incorporation in sweet pickle.

Thick brown fruit sauce consists essentially of a suspension of very finely divided fruit and vegetable particles in a thickened, spiced acid syrup. The consistency should be neither too runny nor too stiff, such that after shaking the closed bottle and removing the cap, the desired amount can be dispensed without undue shaking or spattering and will remain as a coherent mass on the plate, without running out over the plate, and also such that separation of solid and liquid does not occur in the bottle.

Ingredients: There is scope for a great number of variations in formulation, virtually the only limitations being the need to provide a preservation index of not less than 3.6%, unless hot filled, and a stable non-separating product of the required consistency.

The acetic acid may be provided by malt vinegar, spirit vinegar, acetic acid or any convenient combination, bearing in mind that, with the presence of acids deriving from the fruit, it is the volatile acidity that must be taken into account in the preservation index.

A wide variety of fruits may be used. Apples, dates and tomatoes (in the form of tomato purée) are almost invariably used, and other fruits may also be incorporated. Total fruit contents ranging from 15–50% have been quoted. If any significant quantity of sulphited fruit pulp is used in the UK, care must be taken to ensure that the sulphur dioxide residual in the finished sauce does not exceed 100 mg kg^{-1}, if necessary precooking the pulp to drive off sulphur dioxide. Prepulping, or at least mincing, is advantageous in permitting rapid softening of tissues during sauce boiling.

Vegetables may be incorporated, but are frequently limited to onions and garlic, added primarily for flavour. Other vegetables may be used, to supplement the insoluble particles provided by the fruit and may be extensively incorporated in the sauce for manufacture of sweet pickle. Some of the ingredients will arise as by-products of the manufacture of other products, e.g. minced onion, apple pulp. The refractometric solids value of the finished product is normally about 30–40%. The added sugars may be incorporated in the form of refined or unrefined dry sugar, glucose syrup, golden syrup or any desired combination, with or without the addition of artificial sweetener. These supplement the natural sugars of the fruits and vegetables used.

The amount of added sugars used may be varied as desired, limited by the important requirements of:

(i) the flavour as affected by sugar/acid ratio; and

(ii) the preservation index as influenced by the contribution of sugar content to the total solids content.

Among spices frequently quoted are ginger, nutmeg, cayenne, cloves, coriander, cardamom, cinnamon, cassia and pepper. Frequently additional flavouring materials may be used, including soy sauce, yeast extract and protein hydrolysate. Salt may be added if insufficient derives from the other ingredients, and caramel to give the desired depth of brown colour.

The other main ingredient is the thickener/stabilizer. Gum tragacanth and certain modified starches can be used to give viscosity without gel structure. Alternatively, a thickener such as corn-flour may be used to give viscosity and the gel structure, which is allowed to develop fully and is then physically destroyed by sieving. Details are given above. In either case, the stabilizer is dispersed in part of the vinegar by vigorous whisking (in the case of starches, when required for use; in the case of gum tragacanth, preferably a few days in advance).

Manufacture: All the ingredients, except for the stabilizer dispersion and the spice concentrate, are placed in a boiling pan and simmered with continuous stirring for the shortest time necessary to soften the fruit and vegetable tissues so that they can subsequently be broken down to fine particles by mechanical sieving. The exact finishing point of this stage is best determined by a refractometer check (to a predetermined refractometer solids value, such that after mixing in the stabilizer dispersion and the spice, the desired final refractometric solids value is attained). The stabilizer dispersion and spice are then added, the heating discontinued, but stirring continued for at least 10 min and the hot sauce passed through a mechanical sieve.

At this point it is desirable to check the acetic acid content and if necessary adjust by addition of a calculated amount of acetic acid. When speed is essential, as a volatile acidity determination does not yield a quick result, an approximate rapid

check can be obtained by determining total acidity and subtracting an average value corresponding to the acidity contributed by the fruit ingredients in the given recipe from the result obtained.

The sauce can only be filled hot into bottles if it is based on a non-gelling stabilizer. In any event, provided the preservation index is not less than 3.6, there is no advantage in hot filling. Normally, the sauce is partially cooled, with continued stirring, and transferred to bulk storage in closed tanks or high density polythene drums. If a gelling stabilizer has been used, the sauce should be stored for at least a week before use, and should be subjected to sieving or high speed mixing to destroy the gel structure before vacuum filling into bottles. If filling is done too soon, a gel structure may subsequently develop in the bottles.

For continuous or larger scale manufacture, components may be prepared in premixes, suitably milled, homogenized, sieved or slurried. They are maintained as required and then brought together and metered into the continuous system or final batch mix. Providing that the appropriate stabilizer system has been chosen and process controls are adequate, this type of production lends itself well to microprocessor or computer control.

Tomato ketchup

Because of the heat sensitivity of tomato, ketchup is usually manufactured by vacuum boiling. The desirable characteristics are a bright red tomato colour; a smooth appearance, free from specks unless deliberately introduced; a consistency which is neither too runny nor too stiff, such that after shaking the closed bottle if necessary and removing the cap, the desired amount can be dispensed without undue shaking or spattering and will remain as a coherent mass on the plate without running out over the plate; that separation of solid and liquid does not occur in the bottle; and a mild, 'clean' characteristic flavour.

Ingredients: Fresh tomatoes are not used commercially for the manufacture of tomato ketchup, but rather it is prepared from canned tomato purée. As the quality of the tomato purée has a great bearing on the quality of the product, only the best quality should be used. If onion is to be used, it may be incorporated, at about 2%, as minced onion or in the form of good quality onion powder free from dark specks. Ground spices or spice extracts concentrated on salt or dextrose may be used. Most recipes include cloves and cinnamon, but beyond these widely varied additional spices are quoted, including pepper, paprika, mace, pimento, nutmeg, coriander and cardamom.

The strength of vinegar should be such as to give a volatile acidity of about 0.8% or more, as required in the finished ketchup. Added salt may be required to the extent of 2–3% of the finished product, and sugar up to 25%. As the conditions in vacuum boiling may favour multiplication of thermophilic anaerobes, it is advisable to use sugar which is free from these organisms. The quantity of water used should be such as to allow not more than about 12% of the original weight of ingredients to be boiled off in attaining a refractometric solids of about 38%. If gum tragacanth is used as a stabilizer, best quality white gum should be used, preferably predispersed in part of the vinegar and water.

Manufacture: Many different variants exist, but the basic method involves premixing and preheating of the ingredients, sieving, homogenizing, vacuum concentration, deaeration, heating to about 88°C and filling into presterilized bottles and capping under aseptic conditions.

The ingredients other than the gum dispersion are heated in a premix vessel, with continuous stirring, to 50°C. The gum dispersion, if used, is added and heating and stirring continued for a further 10 min. This premix is transferred via a centrifugal sieve and a homogenizer operating at 70–175 bar to a vacuum concentrator, where it is concentrated under vacuum to 38% refractometric solids. By this arrangement, the vacuum concentration can be arranged to give simultaneous deaeration. After vacuum concentration in which the temperature of the ketchup is about 66°C, the ketchup is transferred to a holding tank in which the temperature is raised to 88°C (preferably under nitrogen), then the hot ketchup fed to a vacuum filler where it is filled into preheated sterilized bottles which are immediately capped with steam injection. If the filling and capping operations are carried out under aseptic conditions (e.g. ultraviolet irradiation of the filler, conveyor and capchute) inversion of the bottles is not necessary; otherwise, the bottles should be inverted immediately after capping to sterilize the headspace and interior of the cap (which should be an effective vacuum closure). At one time it was common practice to fill the ketchup at lower temperature and pasteurize the filled capped bottles, but that tended to give an inferior result.

After capping and inverting, if necessary, the bottles are cooled as rapidly as possible, e.g. by passing through a water spray mist cooler so arranged that the temperature difference between the water and the glass is minimized, to avoid the risk of breakage due to thermal shock. The bottles are then dried and labelled.

Procedures basically similar to that outlined above overcome the main problems of tomato ketchup quality:

(i)   it may be a low acid product, with a preservation index of only 1.3 or thereabouts and therefore requires heat preservation;

(ii) it is sensitive to heat, with regard to colour and flavour;

(iii) it is sensitive to oxygen and the presence of retained air leads to darkening, the formation of a thick plug at the surface and separation or lyophoresis.

Other things being equal, the viscosity of the final product depends on the homogenizing pressure, which should be chosen accordingly. Although a ketchup made carefully in this way will keep well, slight browning will gradually occur under normal conditions, so that a deterioration of the colour may become noticeable after some time.

Other manufacturing systems involve the use of more concentrated ingredients obviating the need for lengthy evaporation procedures. Such formulations may be coupled with closed processing systems incorporating flash pasteurization and subsequent aseptic filling.

Mushroom ketchup

Mushroom ketchup is prepared from brined mushrooms, soy sauce (a fermentation product derived from soya beans), salt, spices and caramel, and depends for its preservation on its salt content of not less than 12%. It may be bottled for retail sale (preferably filling at about 80°C into hot, presterilized bottles). The retail outlet is, however, limited (mainly because of the high salt content) and mushroom ketchup is mainly used as an ingredient for other products, e.g. sauces, and in liquor for pickling walnuts. If prepared for use in other products, a higher salt content is no disadvantage and provides extra safety in storage.

Recipes have been quoted indicating a fresh mushroom equivalent of 35–50% in the finished product. The mushrooms are prepared by intimately mixing 5 parts mushrooms to 1 part salt and standing for several days, when water is extracted from the mushrooms and a brine formed. The brined mushrooms, with or without the brine, and necessary amounts of salt and water (to give the required mushroom content and salt content) are simmered for 1–2 h and the soy sauce, spices and caramel added. If ground spices are used, a further simmering period is necessary, but if concentrated spice extracts are used, only sufficient time is needed to obtain satisfactory dispersion and solution. The batch is sieved and filtered free of sediment. The spices used generally include ginger, cloves, pimento and black pepper.

Walnut ketchup

Walnut ketchup or sauce is primarily prepared for use as an ingredient in sauces and as an outlet for broken, soft or woody walnuts unsuitable for pickling as such. Other ingredients vary, but generally include onion, vinegar, soy sauce, anchovies and spices, e.g. ginger, clove, pepper, mace and sometimes also hydrolysed protein and sugar.

In the case of brined walnuts, the nuts are steamed or pressure cooked to soften them and simmered in a boiling pan for 1–2 h with the other ingredients. The resulting batch is finely sieved. The formulation should be such that the salt content is about 4% and the acetic acid content 3.5%. Recipes have been quoted with walnut contents of anything from 20–45% of the finished ketchup.

Mustard and mustard products

Mustard powders are derived from the milling of black mustard, *Brassica nigra*, white mustard, *Brassica (Sinapis) alba* or Indian mustard, *Brassica juncea* (brown mustard). Black mustard contains the glycoside sinigrin which produces the pungent allyl isothiocyanate or volatile oil of mustard by enzyme action when wetted. Similarly, white mustard yields *p*-hydroxybenzyl-isothiocyanate (sinalbin mustard oil) as a result of myrosin action when wetted. Legislation has traditionally allowed the blending of mustards with flour and spices up to 20% by weight so long as the condiment yields not less than 0.35% allyl isothiocyanate after maceration with water for 2 h at 37°C.

Prepared mustards may contain mixtures of powder from the various sources, together with flour, vinegar and optionally salt and other spices. 'French mustard' is typically mild and is based on brown mustard powder, while 'English mustard' is based on a more pungent mix of white and brown powders. 'Dijon mustard' is pungent when fresh and is made from a higher brown seed content. Within the EU various compositional standards have been debated by the European Mustard Trade Association (CIMCEE).

Horseradish

Horseradish products are prepared from the root of the plant *Cochlearia armoracia*. The wild roots are particularly pungent. Grated horseradish is virtually self-preserving except for browning due to oxidation of polyphenols.

Whole roots are peeled and trimmed prior to shredding, mincing or milling. Most horseradish cream comprises minced or milled horseradish with vinegar, skimmed milk powder, vegetable oil and appropriate starch or gum stabilizer and spices. Sulphur dioxide, used as an antioxidant, is also a useful preservative in these and in grated horseradish products.

Thin sauces

Thin sauces

Thin sauces are, in effect, hot highly spiced and flavoured vinegars or vinegar/acid liquors, with no added stabilizing agent present, and either no insoluble particles (as in clear relish) or a limited amount of finely divided insoluble matter appearing as a sediment in the bottle (as in Worcester sauce). They are used both as sauces (e.g. with meat, meat pies, fish, etc.) and as concentrated flavouring agents for incorporation in small quantities in soups, stews, meat puddings, tomato juice, etc. In the former use, the quantity used tends to be much smaller than that of a thick sauce.

The choice of ingredients and recipes is very wide indeed. In addition to vinegar or a vinegar/acid liquor (comprising 70–80% of the total), most recipes include soy, tamarinds, anchovies, onions, sugar (and/or molasses), salt, garlic, caramel and spices. Other ingredients sometimes used include fruit pulps, fruit juices, lemon skins, lemon oil, meat extract and protein hydrolysate.

The variations on the spice theme are virtually limitless, but generally speaking, hot spices predominate. Capsicum, ginger and pepper appear in most recipes and form the major proportion of the spice mix. Other spices frequently used additionally are cloves, coriander, cardamom, mace and pimento. In thin sauces and with adequate quality control, the use of natural ground spices is generally recommended, particularly for the Worcester type sauce, where a sediment is required.

Many varying methods of manufacture have been described. The simplest and most straightforward involves precooking the tamarinds with water or part of the vinegar and fine sieving. This sieved material, plus preminced onions and garlic and the other ingredients are placed in a boiling pan or autoclave and simmered; times quoted vary from 20 min to several hours. The sauce is then passed through a mechanical sieve and preferably a colloid mill, and is stored to mature, possibly for several months.

During maturing, an oily film may form on the surface and if bottled too soon, may form an unsightly oily ring inside the bottle neck. The normal remedy is to skim the oil off the surface of the bulk matured sauce before bottling. This, however, involves the removal of spice essential oils and the material removed thus makes no contribution to the flavour. A similar product could therefore be obtained more economically and with less trouble by reducing the level of spicing. It is open to argument whether the use of emulsifying agents to obtain increased solubilization of spice oils is necessary or desirable.

After maturing, skimming if required and filtration in the case of a clear relish, the sauce is vacuum filled into bottles. In the case of Worcester-type sauce, the sauce is mechanically stirred continuously during filling to ensure an even distribution of solid particles among the bottles. Before capping, in the case of Worcester Sauce and Yorkshire Relish, it is usual to insert into the neck of the bottle a press-fit polyethylene dispenser, which enables the thin sauce to be dispensed cleanly and without subsequently running down the outside of the bottle.

If the vinegar or acid liquor used is of 5% acidity, the finished product, containing upwards of 70% liquor, will have a preservation index in excess of 3.6% and the product will not be subject to microbiological spoilage.

Relishes

There is something of a problem of definition for products in this category, as the name may be used to describe thin sauces, e.g. Yorkshire Relish, thick sauces, e.g. tomato (where the word 'relish' is included in UK legislation as synonymous with 'ketchup' and 'sauce'), traditional, hot spiced sweet pickle and a whole range of North American-type products including corn, mustard, cucumber, tomato, barbecue and hamburger relishes.

This last category is briefly dealt with here, as the other types of products are covered elsewhere of this chapter. These relishes have provided a growth area for the industry in recent times, largely stimulated by uptake in the catering area and particularly fast food outlets.

Corn relish in the UK is produced from frozen or canned corn, dispersed in a sweetened, starch thickened liquor and normally containing onion and red pepper in small particles to enhance visual appearance and flavour. Typical acidity is 1.0–1.5% as acetic acid.

The other relishes are characterized by chopped vegetables dispersed in sweet viscous vinegars or sauces. There are no positive guidelines for the formulation of these products other than those indicated by their designations, which may refer either to their content or to their intended use. Barbecue relishes usually contain a smoke flavouring. Most relishes have an overall acidity between 1.0–2.0% as acetic acid, and hence require a thermal process, normally by hot filling.

Soy sauce

Soy sauce is a centuries-old part of Chinese cuisine. It is unlike other thin sauces in that it does not have a vinegar base. Rather it is essentially water with extract of soya bean, salt and colour, usually caramel. Other ingredients include wheat flour, flavour enhancer (usually monosodium glutamate) and a preservative, e.g. potassium sorbate. It is available in dark and light forms.

Anchovy sauce

Anchovy sauce or essence is traditionally prepared from fermented, salted anchovies which are washed and sieved before milling and pressure cooking to produce the base for a salty sauce containing 20–40% anchovies, usually stabilized with gum tragacanth and flavoured with cayenne pepper. Thick sauce variants may also be produced containing less anchovies and less salt.

Mint products

Mint sauce consists of chopped leaves of mint (*Mentha viridis*) in vinegar, with or without added sugar. Concentrated mint sauce is also manufactured; this may be complete in itself, requiring only dilution with vinegar for use, or may also require the addition of sugar.

Fresh mint or rubbed dried mint may be used. Where fresh mint is used, the leaves are stripped from the stalks and chopped. Where dried mint is used, it should be free from stalk and dark brown particles, and not contain excessive leaf stalk. Microscopical examination should reveal no adulteration with foreign leaves. Not more than 20% should be retained on an 8-mesh British Standard Sieve (BSS), and not less than 85% should be retained on a 30-mesh BSS. In use, dried mint should be gently sieved on a 30-mesh sieve, the dust and fine particles passing though being discarded.

A complete concentrated mint sauce could be based on 10% dried mint or 55% fresh mint, 15% sugar, 0.5% salt and green colour as required, the remainder in each case being made up with vinegar and/or acetic acid and/or water providing about 3% acetic acid acidity overall and thus a safe preservation index.

All the ingredients except the mint are placed in a boiling pan and heated with mixing to boiling point. This hot liquor is poured on to the prepared mint, mixed and allowed to cool. Essential oil or flavouring may be added to fortify the natural mint flavour. The mixture is filled into jars which are capped with suitable acid-resistant closures.

Mint jelly is a sugar–pectin jelly containing chopped mint leaves or dried mint, vinegar, green colour if required and mint flavour if desired. If dried mint is used, it is presoaked in the vinegar and colour.

For 100 kg jelly, a typical formulation would require 63 kg sugar or glucose syrup solids, 16 kg pale 5-grade liquid pectin and 15 l vinegar of 5% acetic acid content. The quantities of mint, colour and flavour (if any) are a matter of choice depending on the type of finished product required. Sufficient water is used to bring the total up to 100 kg. The ingredients other than mint and flavour are heated together, with mixing, in a stainless steel pan and brought to about 88°C. The mint and flavour are stirred in and the jelly filled hot into glass jars, which are capped with acid-resistant closures prior to cooling, labelling and packing.

Oil emulsion products

Mayonnaise

A typical formulation is: oil 75%, vinegar (4.5% acetic acid) 10.8%, egg yolk 9%, sugar 2.5%, salt 1.5%, mustard 1% and white pepper 0.2%. It is essential to use best quality oils, corn oil, sunflowerseed oil or winterized cottonseed oil being suitable. Oils which throw down a deposit or crystallize at refrigerator temperatures are generally unsatisfactory, as crystallization of the oil breaks the emulsion and can cause separation of the phases when the mayonnaise warms to room temperature.

In making mayonnaise the egg yolks, sugar, seasonings and part of the vinegar are mixed, the oil is gradually beaten in and the emulsion is thinned by addition of the remainder of the vinegar. During the mixing, about 10–20% of air is usually incorporated. A coarsely emulsified product can be further treated in a colloid mill. The mustard mix may be added straight into an aqueous phase mix containing all the non-oil ingredients, including egg. Oil may be run into this material as it is fed into a high speed mixer or colloid mill. The vinegar may have been omitted to be included in a final mixer with the emulsion just prior to filling.

The main factors influencing mayonnaise quality are the egg yolk content, the relative phase volumes, mixing method, water hardness, viscosity of the mix and the emulsifying effect of the mustard. In a good mayonnaise the largest droplets are not more than 8 µm in size and many particles are in the size range of 2–4 µm.

Salad cream

Salad cream is a viscous oil-in-water emulsion with relatively low oil content, in which egg yolk is the primary or only emulsifying agent and which contains an acidifying agent which is normally vinegar. Other significant ingredients may include sugar, mustard, salt, thickener, spices, flavouring and colouring. In most salad creams the oil globules are about 5 µm in diameter dispersed throughout the aqueous phase. Typical formulations are: 30–40% oil, 8–16% sugar, about 2% salt, 55–60% total solids, 1.5–1.6% acidity (as acetic acid), not less than 1.35% egg yolk solids, 1–2% mustard, plus a stabilizer to assist in providing the required consistency and emulsion stability and other ingredients as required.

The oil may be winterized cottonseed oil, groundnut, sunflowerseed, soya, rape or even the more traditional but now costly olive oil. It may contain permitted antioxidants and proportionate carryover of antioxidant into the salad cream is permissible.

Preservation requires that sufficient acetic acid be present to provide a preservation index of at least 3.6. Because of colour considerations, distilled vinegar or spirit vinegar is normally used, sometimes with a small addition of lactic acid. The oil used should be low in free fatty acids, not rancid, not have a high incorporated air content and be clean and bright in colour.

Although other emulsifiers can be used, with emulsions in the compositional range mentioned above, no additional emulsifiers will be necessary if the lecithin present in the recipe amount of egg yolk provides sufficient emulsifying power. Egg may be added as whole egg or egg yolk, in the form of fresh eggs or frozen, dried or bulk liquid pasteurized egg.

Thickeners or stabilizers should meet the general requirements previously outlined and have good resistance to shear. Suitable materials are gum tragacanth, arabic and carob bean, certain carrageenans and alginates and certain modified starches.

Manufacture: Manufacturing methods vary, but are based in principle on the following procedures.

A premix is prepared as appropriate to the chosen thickener, and this is mixed with a water/mustard mix already held at 40–50°C for a prescribed time period (activation) to optimize the enzymic production of pungent mustard flavour. The remaining ingredients, except for oil and egg, are added and the temperature is brought up to about 85°C, which may serve both to gel the thickener and to deactivate enzymes. This mix is then cooled to below 60°C.

Meanwhile, the egg is dispersed in a little water and the oil is slowly added in a fine stream with good mixing to produce a coarse emulsion free from pockets of oil. The mixer needs to be a high speed non-aerating type. This emulsion is then blended with the premix with good mixing to provide feedstock to a high pressure (175 bar) homogenizer from which the product may be filled anaerobically into the final container.

Variations on the above theme can include feeding the oil into the premix after the egg has already been added. The production facility can be arranged to handle batch or continuous manufacture.

Acetic acid content at the right level safeguards against microbial spoilage, while lipase activity is prevented by heating the premix. Oxidative rancidity is minimized by the use of good quality oil and the presence of antioxidant therein, the avoidance of trace metal contamination (especially copper and iron), the minimizing of entrained or headspace oxygen and the avoidance of warm or sunlit storage conditions.

Salad cream-based sauces and spreads, dressings and dips

There has been a proliferation in recent years of dressings products in the UK, due in part to the all year round availability of salad crops. Some of these are based on high quality mayonnaise recipes with new ranges of flavours and this in turn has led to a market for dips. The latter have recently found a market in association with snack foods and are sometimes packaged along with a snack food.

The general market has also widened to include low fat and low calorie products. These may use fat substitutes or replacers. There are also blends of mayonnaise with yoghurt (particularly popular in Continental Europe) and of egg-free products (which may use egg replacers).

PART 4: QUALITY ASPECTS

SPOILAGE

The main kinds of spoilage of pickles and sauces are:

(i) Multiplication of yeasts, moulds or bacteria.
(ii) Deterioration of appearance, texture or flavour through the action of enzymes of vegetable or microbiological origin.
(iii) Deterioration of appearance, texture or flavour through oxidative reactions.
(iv) Deterioration due to the action of trace metals.
(v) Deterioration (mainly of appearance) due to the physical or chemical interaction of component substances.
(vi) Physical deterioration of sauces due to physical or chemical changes.
(vii) The presence of foreign matter or contaminants of any kind.
(viii) Deterioration due to unsatisfactory packaging or adverse storage conditions.

In many real instances of spoilage, of course, two or more of these basic types may be involved simultaneously, e.g. oxidative rancidity in salad cream accelerated by trace metal contamination; or discoloration due to the reaction of a component with a substance resulting from an enzymic reaction; or the influence of storage temperature on any of items (i) to (vi). Many of these kinds of spoilage are referred to elsewhere in this chapter, under appropriate headings, but a brief survey is given below.

Multiplication of yeasts, moulds or bacteria

Because of their low pH and the presence of undissociated acetic acid, correctly formulated pickles and sauces present a very low risk of microbiological food poisoning. There is, however, a possible risk of the formation of staphylococcal toxin in incorrectly handled ingredients, e.g. egg might be mishandled prior to use in a salad cream, with subsequent heating being inadequate to destroy any toxin which might be formed. This emphasizes the special importance of using microbiologically sound egg, handling it with strict hygiene and once dispensed, incorporating it without delay; but the principles also apply to other ingredients.

Leaving aside the question of chemical preservatives, the problem of microbiological spoilage is largely one of micro-organisms capable of tolerating acetic acid and can be considered for three ranges of this acid.

(i)   Up to 1.0% acetic acid in the volatile constituents: many organisms can grow at low levels of acetic acid, depending upon the medium, its pH and the presence of other preservatives.

(ii) From 1.0 to 3.6% acetic acid in the volatile constituents: at these levels only acetic acid-tolerant organisms can grow. These are various strains of lactobacilli (*Lactobacillus brevis*, *L. buchneri*, *L. fructivorans*), yeasts (*Saccharomyces bailii*, formerly *acidificiens*) and *Pichia membranaefaciens*.

(iii) Above 3.6% acetic acid in the volatile constituents: apart from some acetous fermenters, etc. which lack salt tolerance, the only significant organism is a mould *Moniliella acetobutans*. This organism is unique in its ability to tolerate and metabolize acetic acid. It is only a problem in aerobic situations, e.g. opened packs, sauce and vinegar storage vats, etc. Other moulds have been reported from time to time, but on investigation have invariably been found to be growing on an intermediate substrate of acid-tolerant yeast.

Apart from pasteurization and preservatives, the only real safeguard against this type of spoilage is strict hygiene.

With the advent of salad products with low acetic acid levels and higher pH values, new microbial spoilage hazards must be assessed. This is a less well-known area. As well as organisms of the type described above, other *Lactobacillus*, *Candida* and *Saccharomyces* species have been identified as potential spoilage organisms.

It should not be overlooked that spoilage by a wider range of micro-organisms may occur at immediate stages by a wider range of processing ingredients and subsequent incorporation may spoil the finished product. For example, prolonged debrining of vegetables, with poor hygiene and inadequate acidification, may result in fermentation by any of a number of common yeasts. If these vegetables are then mixed with, say, a piccalilli liquor, the fermentation may continue unabated even though the overall acidity of the mixture provides a theoretical preservation index of 3.6% or more, since the establishment of equilibrium acidity between vegetables and a thick liquor is very slow and tends to be confined to the thin layer of liquor surrounding each piece of vegetable.

Another instance arises from inadequate lactic acid in fermented brined vegetables. This may allow the growth of otherwise inhibited proteolytic bacteria, elaborating enzymes capable of breaking down proteins. This may in turn provide amino acids which can react with reducing sugars present to give non-enzymic browning reactions of the Maillard type. It may also provide sulphur compounds which can result both in faecal odours and black discoloration by reaction with trace amounts of iron.

Enzymatic spoilage

Enzymes naturally present in vegetables, or elaborated by micro-organisms, are responsible for reactions causing various forms of deterioration in pickles, including darkening and softening of pickled vegetables and vegetables stored in brine, flavour deterioration and the formation of quercetin in pickled onions.

The enzymes responsible for softening are mainly those causing degradation of pectin. The pectinesterase group of enzymes attack and split off the methoxyl side groups, while the polygalacturonases attack the links between the galacturonic acid units making up the chain. Mould is a primary source of these enzymes.

The rate at which enzyme reactions proceed is temperature dependent and there is much to be said, therefore, for storing stocks of vegetables in brine in the coolest possible conditions. Enzymes can be inactivated by heat and the pasteurization of pickles serves not only to prevent microbiological spoilage, but dependent on pasteurizing conditions, to prevent or minimize enzymic deterioration. Similarly, judicious use of blanching techniques on fresh vegetables prior to processing can help to minimize subsequent deterioration.

Oxidative deterioration

Examples of oxidative deterioration are the softening of pickled red cabbage, oxidative rancidity development in salad cream and the browning of tomato ketchup.

Trace metals

Apart from the role of trace metals, especially copper, in accelerating oxidative rancidity in salad cream and the role of trace iron in the sulphide blackening of vegetables (see above), the main form of trace metal deterioration is discoloration due to reaction of trace amounts of iron with tannins deriving from ingredients or wood.

Deterioration due to interaction of component substances

In this category is the physical interaction between positively charged caramel and negatively charged onion cells which results in rapid staining of onions in a vinegar coloured with an unsuitable caramel. Even with the most suitable caramels, slow absorption of caramel occurs, with gradual browning of the onions, accelerated if they have been abrasively peeled.

Chemical interaction of components present is exemplified by Maillard browning reactions and the 'pinking' of onions. Separation in sauces or thick liquors may take the form of separation of a layer of clear liquid at the top or at the bottom. In sauces these originate from the actual rising or sedimentation of solid particles through the liquid phase. In the special case of salad cream, the separation of an oil layer at the top, or more frequently, of an aqueous layer at the bottom, is the consequence of emulsion breakdown. This type of separation should be distinguished from syneresis, in which there is also a separation of fluid. This is often a side effect of gelling, but may also result from mixing vegetables with a thick liquor of markedly higher sugar content. Gelling, usually accompanied by syneresis and often by 'cracking', may result from the use of an unsuitable thickener, or from incorrect processing of liquor, or both. Finally, decreasing viscosity on storage may result from the gradual acid hydrolysis of some types of thickener.

Adverse storage conditions

As the rates of the progressive deterioration reactions are temperatures dependent, it follows that optimum shelf life will be obtained under conditions of the coolest possible storage. From purely quality and shelf life considerations, there is much to be gained by providing chill storage at about 5°C. Conversely, storage at higher temperatures in warehouse or shop will accelerate deterioration

and shorten effective shelf life. Display in sunlit shop windows is particularly to be deplored, not only because of the effects of temperature, but also in relation to the effects of sunlight (colour fading, acceleration of oxidative rancidity). Where products are hot filled or warm labelled, the risks of stackburn should be considered.

CONTAMINATION AND FOREIGN MATTER

The avoidance of carryovers and contaminants which might cause taints or off-flavours may be difficult, and there is a great advantage in using known and reliable sources of supply in this context. In general, the measures required are those applicable to all food manufacture and may be summarized under five headings:

(i) measures to inspect and decontaminate raw materials;
(ii) measures to prevent the entry of foreign matter during processing;
(iii) measures to treat the finished product before filling;
(iv) visual and instrumental inspection of the filled containers;
(v) organoleptic checking of end product and work-in-progress against off-flavours and taints;
(vi) vigilance on the part of all operatives and strict supervision.

Glass breakage is a particular hazard; personnel should be adequately trained in its handling, precautions and breakage procedures, and warning notices and clear instructions for the procedure to follow in case of breakage should be prominently displayed.

UNSATISFACTORY PACKAGING

Imperfect sealing of closures may be due to imperfect neck-ring finish on glass containers, defective closures, incorrect capping machine adjustment or the trapping of particles (e.g. vegetable) between the glass rim and the closure. It may result in leakage of liquor, oxygen-dependent deterioration reactions and, in the case of pasteurized low acid packs, post-process leaker contamination and consequent microbiological spoilage. Information on closure integrity assessment may be scant and the producer may have to develop his own protocol in this area.

The interior of closures must be suitably resistant to acetic acid (under pasteurization or hot filling conditions where appropriate), in order to safeguard against corrosion. Breakage of filled glass containers may occur in handling and transport if the outer packaging (cartons, shrink-wraps, etc.) fails to provide adequate protection.

PARTICULAR PROBLEMS IN BRINED VEGETABLES

The processes described under brining show the basis of the brining process, and in practice a few problems regularly crop up. It is not proposed to discuss such things as leaky drums or casks, use of wrong strength brine and so on, but only to deal with more fundamental problems.

Softening

This can occur for several obvious reasons, such as excessive exposure to air, excessively long storage (all vegetables gradually get softer anyway during storage in brine), or through being subjected to extremes of temperature, causing freezing or cooking effects. Some brined vegetables may be soft simply because poor quality raw materials were used to start with. Softening for other reasons than these is almost invariably due to enzyme action. This can rise from the inclusion of mouldy vegetables in the container or from other microbial spoilage.

The occasional drum or cask of cauliflower or other vegetables which has softened to a remarkable degree is rare, but casks which have softened to a small extent are, unfortunately, more common. This is often due to the enzyme polygalacturonase (PG), originating from mould growing in empty barrels which have not been thoroughly cleaned out. The PG can be absorbed by the wood, subsequently to be slowly released into the contents, even after a superficial cleaning.

Polydrums have been found to be very good containers for brined vegetables and have almost completely superseded wooden barrels in the UK. One of the problems with polydrums, compared with wooden casks, is that of heat absorption in sunlight, inducing cooking and softening effects. Wood was a good insulating material and also allowed some degree of evaporation, which gave a cooling effect. Also, wooden casks sprayed with water will stay damp for some time, allowing more evaporation and cooling effect, whereas water sprayed on to plastic containers simply runs straight off again. With white polydrums, even stored in direct sunlight, it has been found that

their contents increased only marginally in temperature as compared with others of black and grey colours. Thus dark coloured polydrums should be shaded in some way from direct sunlight; white polydrums have much less need for shading.

Scum yeast growth

Some yeasts can grow in strong salt solutions and form a scum across the surface of the brine. Indeed, Dutch briners have effectively allowed the development of scum yeast at the beginning of brining for its oxygen scavenging effect. Apart from producing enzymes as described above, they can also metabolize the lactic acid present, rendering the brined goods more liable to other sorts of spoilage. Hence it is of primary importance to maintain the anaerobic state.

The use of polydrums without gaskets in the lids has proved a particular danger in this respect. Whenever the lid of a polydrum is replaced, care should be taken to see that the gasket is present and in good condition and that an airtight seal is obtained. Expansion pots used by some briners also present risks unless especially closely inspected throughout fermentation and storage.

Bacterial spoilage

Bacterial spoilage sometimes takes place during the storage of brined vegetables, causing objectionable off-odours and off-flavours. It may be due to lack of salt or lactic acid or can happen because of yeast growth, or because of topping-up with water or the wrong strength brine.

Spoilage during the early stages of brining can happen because of uneven salt concentrations in the container (stratification). If not regularly mixed, water which tends to come out of the vegetables while the salt goes into them, floats towards the top as it is lighter than the brine. Thus if the contents are not mixed up for some time the brine at the top can become very weak and spoilage can occur in that area. It usually takes about 10 days for the salt to have balanced out between the brine and the vegetables: if fermentation starts sooner than this, the gas bubbles will keep the brine mixed up.

Discoloration

Various forms of discoloration can arise, usually due to excessive exposure to air or light, the use of a container previously having contained a coloured substance such as red wine or to contamination with iron and copper. Discoloration is one of the most frequent causes of complaint in brined vegetables and several investigations have been reported. Pink discoloration has been shown to be caused by the use of brine acidified with acids other than lactic. Some apparent discoloration does, of course, result from the use of raw material with initially unsatisfactory colour.

Probably the most usual form of discoloration is the darkening attributable to iron contamination. The necessity to take every possible precaution to avert this cannot be over emphasized and the pickle factory quality controller needs to be constantly alert and monitoring for its possible incursion.

REFERENCES

CIMSCEE (1985) *Code for the Production of Microbiologically Safe and Stable Emulsified and Non-emulsified Sauces Containing Acetic Acid*, Comité des Industries des Mayonnaises et Sauces Condimentaires de la Communauté Economique Européene.

Dakin, J.C. (1957) *Microbiological Keeping Quality of Pickles and Sauces*, Research Report 79, Leatherhead Food Research Association.

Shapton, D.A., Lovelock, D.W. and Laurita-Longo, R. (1971) The evaluation of sterilization and pasteurization processes from temperature measurements in degrees Celsius. *J. Appl. Bacteriol.*, **34**, 1.

Stumbo, C.E. (1973) *Thermobacteriology of Food Processing*, 2nd edn, Academic Press, London.

FURTHER READING

Anderson, K.G. (1981) *The Technology of Clear Pickles*, Mastership in Food Control Dissertation, Serial No. 34, Institute of Food Science & Technology (UK), London.

Arthey, D. and Ashurst, P.R. (1996) *Fruit Processing*, Chapman & Hall, London.

Arthey, D. and Dennis, C. (1991) *Vegetable Processing*, Chapman & Hall, London.

Binstead, R., Devey, J.D. and Dakin, J.C. (1971) *Pickle and Sauce Making*, 3rd edn, Food Trade Press, Orpington.

Garrido Fernandez, A., Diez, F. and Adams, M. (1996) *Table Olives*, Chapman & Hall, London.

Harrison, L.J. and Cunningham, F.E. (1985) Factors influencing the quality of mayonnaise: a review *J. Food Quality*, **8** (1), 1–20.

Rausch, G.H. (1950) *Jam Manufacture*, Leonard Hill (Blackie), London.

10 Hot Beverages

COFFEE

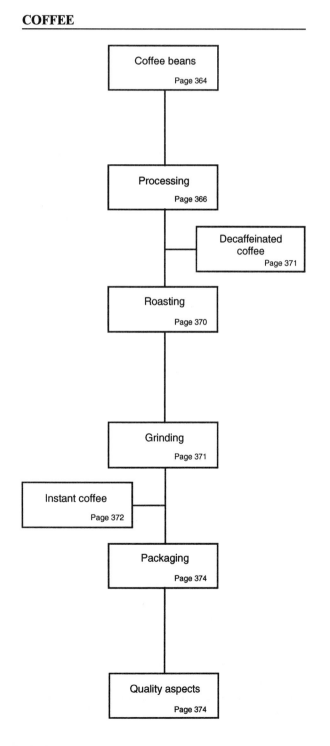

Coffee beans
Page 364

Processing
Page 366

Decaffeinated
coffee
Page 371

Roasting
Page 370

Grinding
Page 371

Instant coffee
Page 372

Packaging
Page 374

Quality aspects
Page 374

PRODUCTION AND MARKETING

Total world production of coffee is some 5.5 million tonnes. Coffee is a massive international industry, second only to oil in its economic importance. It is the major world agricultural commodity, earning more than wheat, rice, sugar, tea or rubber, and is therefore of vital concern to the economy of the Third World countries where it is grown.

COFFEE BEANS

Varieties

There are two commonly grown distinct species, *Coffea arabica* and *C. canephora* var. *robusta*, generally known as arabica and robusta, respectively. A third species (*C. liberica*) represents only a small proportion of commercial planting and is largely confined to West Africa and Malaysia. It is, however, highly disease-resistant and therefore useful for development of disease-resistant hybrids. Arabica is considered to have an economically useful life of around 30 years. Robusta is generally grown in wet, low lying areas unsuitable for arabica. It is more vigorous and tends to have larger, coarser leaves. It is longer lived and less prone to disease. Fruit bearing laterals are usually in the form of a single stem in robusta, whereas arabica develops sublaterals which also bear fruit.

A number of well-defined varieties of arabica have been widely planted in the different coffee growing areas and these may be successfully propagated by seed or cuttings. These varieties give rise to types of coffee recognized in the trade, e.g. Blue Mountain, Mocha, Bourbon, etc. Since robusta relies on cross-pollination, no traditional varieties have been developed, and the only reliable way of propagating selected trees is by vegetative propagation using cuttings or graftings. Interestingly, Angolan robusta is distinctive chemically and organoleptically, but neither the histor-

Table 10.1 Major producing countries with average annual production levels (1989–1994)

Producer	Production (tonnes × 10³)
Brazil	1549
Colombia	814
Indonesia	332
Mexico	236
Ivory Coast	202
Ethiopia	169
India	164
Uganda	152
Costa Rica	152
El Salvador	125
Ecuador	113
Honduras	110
Kenya	94
Zaire	85
Cameroon	82
Venezuela	70
Peru	68
Thailand	66
Papua New Guinea	57
Madagascar	52

Table 10.2 Average annual consumption rates (1989–1994) for major consuming countries

Importer	Imports (tonnes × 10³)
USA	1290
Germany	820
France	380
Japan	340
Italy	300
Netherlands	178
United Kingdom	180
Spain	179
Belgium and Luxembourg	115
Sweden	105
Austria	110
Switzerland	67
Finland	68
Denmark	59
Norway	44

ical nor the botanical origin of this material is known with certainty. With recent breeding programmes and the introduction of hybrids there is increasing interest in vegetative propagation. There are many wild species of coffee and a few, e.g. *C. racemosa* and *C. congensis*, have been exploited locally as a crop in Africa and the latter as a parent in hybridizations.

Harvesting

Coffee beans are the seeds of the coffee tree encased within the fruit or cherry. There are nor-

mally two beans within each cherry. Depending on local climatic conditions and the maturity of the tree, flowering may occur several times a year and the subsequent fruit crop ripens some 6–9 months after flowering.

Ideally, the cherries should be uniformly ripe at harvesting. This requires selective picking from a given tree several times throughout the season, which may be from 2–3 months (as in Brazil) to around nine months (in Colombia), or show several distinct cropping periods during the year, as in parts of east Africa.

In practice, harvesting procedures vary considerably in different areas according to the variety, the cropping pattern and the processing methods to be used. For wet processing of arabica and a long cropping period, selective picking of ripe fruit every 10–14 days is practised. Where the season is short and natural processing is the general rule, as in most areas of Brazil, it becomes necessary to strip trees of fruit completely. Inevitably this results in a harvest containing variable proportions of unripe, overripe and partially sun-dried cherries, some of which may require collecting from the ground beneath the trees. Robusta fruit tends to remain on the tree and in some areas is allowed to ripen and dry completely before stripping altogether.

Buni is the name given to coffee which is left in the field until dry. It is collected by picking or from the ground beneath the tree, and is generally hard and poor in quality.

Storage

Coffee is principally traded in the form of green beans. Provided these have been properly processed and are stored under suitable conditions, they remain in good condition for several years. After drying and hulling, the moisture content should be close to 12%. Any rise in moisture content is dangerous and at 15% or above mould growth is highly likely and quality deteriorates rapidly.

High temperatures also exert an adverse influence on quality. It must be appreciated that green beans are still living seeds. At high temperatures and with sufficient moisture present they respire quite rapidly, undergoing changes in composition as a result of the metabolic activity which, in natural situations, precedes germination. Thus for storage of green beans cool, dry conditions are desirable and for prolonged storage air conditioning is essential if quality is to be preserved.

Coffee is commonly shipped in 60 kg bags but

bag sizes vary considerably according to country of origin. It is usually stored on pallets or, in the case of some large purchasers, may be held in silos prior to use.

PROCESSING

The objective of processing the coffee cherry, after preliminary cleaning and classification, is to remove the outer skin and pulp layers, leaving the beans in a clean and undamaged condition.

Classification and cleaning

It is best to sort the berries by size at an early stage, to permit efficient operation of the pulping and other machines. This may be done by flotation, which also achieves the removal of dirt, stones and other foreign matter, or by sieving. Underripe and overripe berries and hard partially dried berries may be removed by hand.

Natural processing

In its simplest form as applied to much robusta coffee, the cherries are left on the trees until virtually dry. They are then harvested all at once, the drying is completed as necessary in the sun and the dried fruit passed through a hulling machine which removes the outer skin, dry fruit and parchment, collectively known as coffee husk, in a single step.

The fruit of arabica tends to fall when overripe or partially dry, thus where natural processing of arabica is practised, e.g. for much of the Brazilian crop, it becomes necessary to dry the fruit after harvesting.

The harvested cherries are usually cleaned by sieving to remove dirt, stones, twigs, leaves, and so on and often passed through a flotation tank. Partially dried fruit floats and is treated separately. The ripe fruit which sinks is drained and dried from about 70% moisture down to about 12%. This may be carried out on terraces or trays in the sun over a period of several weeks with frequent turning and raking. Hot air driers may also be employed (see DRYING).

Wet processing (or washed coffee)

Cherries are usually classified by flotation (as above) and the ripe fruit or 'sinkers' are pulped

and the skins and pulp removed. The beans at this stage are still encased in a mucilaginous parchment layer. The latter is removed by natural fermentation in steeping tanks and after final washing the beans are dried. The moisture content prior to drying is about 53% and the moisture content of the outer pulp is much higher than that of the bean. Thus, despite the use of so much water, this process which selectively removes the pulp results in a product with a lower moisture content for final drying. Drying of the coffee produced by this method therefore requires less energy than does natural processing.

Wet processing is generally under rather better control than natural processing and is therefore considered more reliable in producing the best quality product. The best natural coffee, while of a recognizably different character, referred to as 'Santos' in the trade, is probably about as good as the best washed or 'mild' coffee. On the other hand, wet processing requires considerably more equipment and a reliable supply of clean fresh water, something not always easily available.

Pulping

After cleaning and sorting, cherries to be wet processed are pulped. Pulping machines are of two main types, drum and disc.

A drum pulper consists of a rotating cylinder with a roughened surface having a stationary cover or 'breast' over about one-third of its surface. The gap between the breast and the drum can be adjusted and decreases in the direction of rotation of the drum. Cherries are fed from a hopper or by siphon directly from the classifying tanks to the pulper and squeezed through the gap. This ruptures the fruit and expels the beans still encased in the parchment layer. The capacity of a drum pulper depends on the dimensions of the drum but a typical medium sized drum will process about $1200-1500$ kg h^{-1} of cherries. A variation of the drum pulper has a breast in the form of a slotted plate, the slots being just wide enough to allow the beans to pass through the slots while the skin and pulp pass down through the gap.

In the disc pulper, a vertically mounted disc with a roughened surface rotates against a near vertical flat plate, the gap between the two being adjustable and narrowing in the direction of rotation of the discs. Disc pulpers may be fitted with several discs (commonly up to four) and each disc can process about 900 kg h^{-1}. Whatever design of pulper is employed, a steady stream of clean water is passed through during the pulping operation.

The pulped cherries are discharged from the pulper on to a sieve which may be in the form of an oscillating flat tray or a hollow rotating cylinder. The sieve is inclined at a slight angle to encourage the passage of material along its length. The holes allow the beans to pass through while retaining pulp, skins and any unbroken fruit. This is a wet sieving operating assisted by water sprays throughout. The skins and unpulped cherries are sent to a further small pulper, known as a repasser. This is a single disc or small drum set with a smaller gap than the main pulper so that it effectively pulps the smaller fruit which passed through.

The beans are then washed and graded by passing along channels with a strong flow of water and weirs to retain the beans. This requires manual stirring and wooden paddles and is fairly labour intensive.

The functions of sieving, washing and grading may be usefully combined in the so-called pregrader which saves labour, water and space. The pregrader is a tank filled with water and fitted with a submerged shaking sieve. The discharge from the pulper passes directly to the pregrader together with a strong flow of water. As before, the material retained on the sieve is passed back to the repasser while the best beans are taken from the bottom of the tank. Lighter floating beans are taken off near the top of the tank. Washed and graded beans are then passed to the fermentation tanks.

Throughout the whole operation of pulping, sieving, washing, grading, and so on, it is desirable to avoid any delays. However use of a pregrader so reduces contact time with water that it makes a two-stage fermentation stage desirable (see FERMENTATION).

While wet processing operations are usually carried out in a specifically designed factory, small hand operated pulpers are used in various smallholder schemes. These allow the individual farmer to process his own crop on a small scale.

Fermentation

The objective of the fermentation stage in the wet processing of coffee is to remove the layer of mucilage surrounding the beans before the final drying stage. The mucilage layer is only around 0.8 mm thick and is slippery, translucent and colourless when freshly pulped, but turns brown on exposure to air. It is an amorphous gel quite insoluble in water and clings tightly to the parchment-like skin surrounding the bean itself. It is only present in this form in ripe fruit and since it greatly aids expulsion of the bean from the cherry during pulping, the importance of harvesting at the correct stage can be appreciated. However if it is not removed after the pulping stage drying will be impeded and the surface of the beans becomes sticky, causing handling problems later.

Chemically the mucilage consists principally of pectins, which must be digested to become soluble and easily removed. Traditionally this is achieved by transporting the beans by water into concrete tanks, draining off the water and holding for 2–3 days. Each day the tank is partly refilled with water and the beans turned over thoroughly to ensure an even process. Fermentation tanks should be protected from direct sunlight. Depending on the scale of the crop, tanks may be from about 0.5–30 m^3 and may have sloping bottoms to assist washing after fermentation is complete.

The mucilage is digested initially by the action of naturally occurring enzymes and subsequently as a result of microbiological attack. The proliferation of suitable micro-organisms depends on the extent to which the tanks are cleaned and the source of the water used, hence the process is not easily controlled. However, the simple test of feeling a sample of beans after washing in clean water is an adequate means of judging the complete removal of the mucilage. The beans must then be washed rapidly and thoroughly to avoid microbiological damage.

Where a pregrader has been used the beans have experienced inadequate contact time with water for removal of substances which cause the bean to turn brown. In this situation a so-called second fermentation stage is instituted where the washed beans, now free from mucilage, are simply steeped in clean fresh water overnight. The final washing may be carried out in channels similar to those used for washing and grading immediately after pulping.

The speed of the fermentation stage is markedly dependent on temperature and therefore may at times be very slow, especially at high altitude. Many process modifications have been investigated in attempting to speed the process and to keep it under better control. Warm water, dilute acid or alkali, or pectinolytic enzymes all greatly accelerate the dispersal of the mucilage. Likewise, a number of machines have been devised which are able to remove most of the mucilage by physical attrition in the presence of water.

Apart from saving time, rapid processing

reduces losses of several percent which otherwise occur during natural fermentation. This factor alone should make it economically sensible to use acclerated processing. However, there are indications of adverse influences on quality when some of these procedures are used, and the trade is extremely conservative in its attitude to innovation. Consequently these accelerated procedures are not very widely used. Of those that are used, the addition of pectinolytic enzymes is perhaps the least open to criticism. It can have quite dramatic effects, allowing fermentation to be completed within one working day instead of three or more.

Another simple procedure which may be employed is to recirculate water used in pulping or earlier washing stages. This contains some sugars, is slightly warm and therefore promotes the growth of micro-organisms. After the initial stages of pectin degradation by enzymes, natural or added yeasts take over with an alcoholic fermentation of the simple sugars. Some of the organisms present at this stage also produce pectinolytic enzymes and directly assist the digestion of the mucilage. Subsequently bacteria become dominant, producing lactic and acetic acids with rapid fall in pH.

It is generally considered that the yeast stage is beneficial, imparting 'body' to the final cup quality. This may be one of the important benefits of natural fermentation. However, while lactic and acetic acid-producing bacteria are quite acceptable there is always the possibility under conditions of low sugar content and relatively high pH, that propionic acid-producing bacteria will become important and impart undesirable off-flavours. Clearly close control over the fermentation stage is highly desirable and not particularly easy to exercise.

Drying

Sun drying

In Latin America, sun drying of coffee is traditionally carried out on large concrete terraces. Washed beans, still with their parchment coat intact, are spread in layers 5–10 mm deep and allowed to dry in the sun. Frequent stirring or turning of the beans is carried out during the day. At night the beans may be heaped up and covered with waterproof sheeting. This conserves heat and aids moisture equilibration between the beans. Alternatively, some terraces are fitted with mobile covers which are used at night and can be rapidly deployed in the event of rain. Drying of washed coffee from about 53% moisture to the required

10–12% takes approximately one week on a concrete terrace, given continuous good weather.

In Africa, cheaper and more versatile wire trays are usually used for sun drying of washed coffee. Commonly two stages are employed. Freshly washed beans are first spread on portable wire mesh trays at only 2–3 mm depth. The beans are frequently stirred and sorted by hand, defective beans and undesirable debris being removed. When the moisture content has fallen to about 45%, the outer parchment is substantially dry. At this point the beans are transferred to the drying tables, which are much larger wire mesh surfaces stretched over wooden frames. They are dried for a further five days or more to the required moisture level, being regularly stirred during this period. To prevent uneven drying and cracking of parchment, some shading may be employed when the sun is at its height. The tables are, of course, also covered at night and in the event of rain. During drying the appearance of the beans after removal of the parchment is a rough indication of moisture content. When the moisture content falls below 30% the beans turn from whitish to nearly black. Once 12% is reached they become grey-green and at 10% are grey-blue. Overdrying introduces a yellowish colour. While these colour changes are a useful indication, regular measurement of moisture by standard procedures or instruments is highly desirable to check the operation.

After drying, parchment coffee is usually conditioned by storing in heaps with regular stirring, or (better) in conditioning bins which have perforated floors through which currents of air may be passed. This serves to equilibrate the moisture throughout the mass and eliminates small differences between individual beans.

Hot air driers

Many different machine driers have been designed and used in the coffee industry, mostly in Latin America. The main criterion is to avoid raising the temperature of the beans themselves too high. Actual bean temperatures should not be higher than about 50°C. However, because evaporative cooling effects it is permissible for the air surrounding the beans to be as warm as 90°C during the early stages while the bean is still relatively moist.

The simplest form of hot air drier is, in effect, a screen table similar in principle to those used in sun drying, but fitted with a fan to blow warm air through the coffee even, in some cases, while continuing sun drying. The hot air is supplied usually via a heat exchanger from a furnace burning wood, dry coffee waste, and so on, or oil.

The Wilken drier is a simple development of this concept with automatic loading and unloading and power driven rakes to stir the coffee and keep the drying even.

One of the most widely used machines is the Guardiola drier. This is essentially a horizontal revolving cylinder made of perforated mesh in which the coffee is placed. Warm air passes into the centre of the rotating cylinder, through the mass of tumbling coffee and out through the walls of the cylinder. These machines are essentially batch driers, one charge being dried in 30–36 h with a carefully controlled temperature programme. The largest of this type, about 5 m long by 1.8 m in diameter, will produce about 2 tonnes of dry parchment coffee per batch.

The Torres drier is used primarily for natural (dry) processed coffee in Brazil. It consists of two concentric cylinders of perforated mesh, the coffee being charged into the annular space between the cylinders, with warm air passing from the centre outwards as in the Guardiola machine. Using a complicated system of elevators, conveyors and chutes, this drier can be charged and discharged while running and operated semi-continuously. When drying natural coffee, the beans are discharged to conditioning or equalizing bins for a rest period and returned to the drier several times. This allows a final moisture of 12% to be achieved without overdrying a proportion of the charge.

The Moreira drier, designed and manufactured in Brazil, consists of a tower with two hollow legs with vertical screens forming the inside walls of the legs. A wood-burning furnace designed for complete combustion of all gases permits the use of direct firing without a heat exchanger. Hot air from this furnace passes into the chamber between the legs of the tower, hence through the screen and the coffee which is charged into the tower.

The coffee is slowly withdrawn from the bottom of the legs and returned to the top of the tower using a bucket elevator. Thus the coffee is continuously recycled until the required moisture content is obtained. The upper part of the tower effectively acts as a large equalizing bin.

More sophisticated driers based on designs used for cereals drying have also been adapted for coffee drying. They have features in common with the Moreira drier in that recirculation of coffee by elevator and gravity are employed. In one version the coffee passes down louvred columns. Hot air is forced into the chamber between the columns and exits through the louvres by passing through the coffee. The flow of coffee can be regulated by controlling the exit of the beans at the bottom. These may be recirculated directly to the top of the columns or discharged to equalizing bins as required. Recirculation of the drying air may also be employed to improve the fuel efficiency. After drying, parchment coffee may be bagged and traded direct, but is more usually hulled before bagging, storage or shipping.

Hulling

The purpose of hulling is to remove the parchment from washed coffee after drying, or the dry skin, pulp and parchment from natural dry processed coffee. Hullers for the two operations differ somewhat in detailed design but have major features in common.

A horizontal screw with its pitch increasing towards the discharge end rotates within matching covers. Part of the lower cover is perforated and connects to a lower chamber to which is applied a suction by means of a fan. As the beans pass along the screw, pressure and friction are applied. Fragments of parchment, dust, and so on are drawn into the lower chamber and collect there in a cyclone interposed before the fan inlet. An adjustable discharge gate regulates pressure and residence time of the beans within the huller. As the beans fall from the discharge end they meet the air flow exiting from the fan which serves to carry away fragments of parchment, etc. not previously removed by suction. A typical small hulling machine for parchment coffee will have a three-foot screw and a capacity of around 500 kg h^{-1} of clean beans. For the hulling of naturally processed coffee the screw is longer and the housing is fitted with projections to increase the tearing action to ensure effective removal of the whole husk. In both types of hulling operation the pressure may be adjusted so that the beans are polished in the final stages to remove the fine outer membrane or 'silverskin'.

Grading

Green coffee is sorted by size and density. Sorting by size using screens or reel graders allows the larger beans and peaberries, which might attract a premium, to be segregated if required. Consistency of bean size facilitates even heat transfer and more uniform changes during roasting thus giving better control of the roasting process and product character. Broken and misshapen beans and some extraneous debris can be removed. Pneumatic and gravimetirc sorting removes dust and other light particles as well as beans of low density, or which are deformed, discoloured or insect atacked.

Colour sorting may be performed either by hand or more commonly now by machine. Black beans in particular need to be removed. Monochromatic and bichromatic sorters are available, computer controlled and capable of handling 100–150 kg h^{-1}. The bichromatic sorters, e.g. the British-made Sortex, have greater discrimination and can even be instructed to select for certain characters rather than just to reject certain characters. This capability has facilitated the accumulation of certain 'atypical' beans in sufficient quantity for test roasting thus allowing certain physical characters to be linked causally with certain defects or even particular desirable sensory properties.

There are no internationally accepted grade sizes or names. Each producing area has its own grade names or numbers with different specified screen analyses. The specification of the various grades also usually includes maximum permitted numbers of various recognizable types of defective beans.

ROASTING

Roasting of coffee beans involves rapid heating, first to drive off the retained moisture, then to raise the bean temperature to about 180°C. Pyrolysis takes place and exothermic chemical reactions occur which raise the bean temperature a further 20–30°C. Including the moisture loss, the weight loss during roasting is 15–22%. Dry matter losses may reach 9–10% in very dark roasts. Obviously the longer the heating time, the greater the loss and the darker the colour. The degree of roast also affects the flavour and is a matter of customer preference.

Typically, heat for roasting is supplied by burning oil or natural gas. There are two basic designs of coffee roasters, European and American. American systems use a rotating perforated cylinder containing the coffee beans. Hot gases from the burner pass directly through the perforated drum and the coffee beans are then partially recycled through the burner.

European equipment generally employs a cylinder without perforations and relies partly on heat transfer through the walls of the cylinder. After circulation round the outside of the cylinder, the hot gases enter one end and pass along the length of the cylinder through the tumbling beans to be discharged at the opposite end.

In both designs, a suitable arrangement of baffles keeps the beans tumbling through the hot gases. When these are arranged spirally they transfer the beans through the cylinder hence allowing the possibility of continuous operation.

A large installation may use continuous roasters with a capacity of up to 5½ tonne h^{-1} and a residence time of 3½–5 min. Other installations may operate with batch roasters taking 200–250 kg per batch with a 15 min roasting time. Smaller installations may utilize roasters taking 30–50 kg per batch, while still smaller installations may consist of very simple batch roasters, which, particularly in developing countries, may use solid fuel and even be rotated by hand.

The degree of roast is controlled primarily by visual appearance and the timing is quite critical. Colour reflectance measurements afford the best method of checking roasting instrumentally. At the end of the roasting period the beans must be cooled rapidly. This is usually achieved by the passage of a current of cold air through the beans.

In the 1980s much effort went in to developing rapid roasting equipment. Interestingly, there is now some interest in developing slower roasting procedures.

Chemical changes during roasting

During roasting, significant chemical changes occur which are responsible for the development of the desirable flavour. Most important is the formation of an extremely complex mixture of volatile aroma constituents. While they constitute only about 0.04% of the weight of the roasted bean, they are responsible for most aspects of flavour and aroma. Over 700 compounds have been identified. They arise as a result of partial decomposition and interactions involving most of the constituents of green beans. Interestingly, most of these reactions take place at a high pressure (5–8 atm) at elevated temperatures and essentially in the absence of oxygen. Decomposition of proteins gives rise to various amines and from the sulphur-containing amino acids, various mercaptans and dimethyl sulphide considered to be organoleptically significant components. Decomposition of carbohydrates gives rise to volatile aldehydes, organic acids and carbon dioxide. Partial decomposition of pentosans leads to the formation of furfural. Most of the sucrose disappears. It is either caramelized, or after hydrolysis or fragmentation, participates in Maillard-type reactions with protein and amino acids. Most of the chlorogenic acids decompose yielding a range of non-volatile lactones and a range of volatile simple phenols (e.g. catechols, guaiacols, etc.) which contribute to smoky notes especially in dark roasts. Approximately one-tenth of the chlorogenic acid initially present in the green bean is lost for every 1% dry matter that is destroyed, i.e. little chlorogenic acid remains in roasts where the dry mass loss is 10%.

Caffeine is chemically stable under roasting conditions, but a small proportion may sublime. The exhaust stacks of coffee roasting plants have been used as a source of caffeine. The relatively linear loss of chlorogenic acids relative to the stability of caffeine has led to the chlorogenic acid:caffeine ratio of a roasted bean being used as a chemical index of the severity of roasting. It has been suggested that the cafestol:dehydrocafestol ratio might serve the same purpose. A proportion (about 10%) of the trigonelline decomposes, forming nicotinic acid and pyridine. There is relatively little chemical change in the coffee oil during roasting. Some minor components are volatilized, while others are degraded. The oil is physically important, however. During roasting it melts and is forced to the surface due to the development of considerable pressure as carbon dioxide, water vapour, and so on are generated. It thus traps many volatile substances that might otherwise be lost, and serves as a hydraulic seal maintaining the internal pressure to a greater extent than would otherwise occur. This high pressure undoubtedly influences the chemical transformations occurring within the bean and which *inter alia* are responsible for the unique odour and flavour of roasted coffee.

While some of the water-soluble substances in green beans are either decomposed or rendered insoluble, certain of the polysaccharides, especially arabinogalactan but also mannan are broken down and therefore made more soluble. As a result around 25–30% of the roasted bean is still readily soluble in hot water and extractable by domestic brewing. Commercial extraction at higher temperature hydrolyses some of the otherwise insoluble mannan and increases the solubles yield substantially. Although essentially tasteless this solubilized polysaccharide is important in binding coffee volatiles in the soluble powder during the drying process.

GRINDING

Before extraction with hot water, it is necessary to grind the roasted beans in order to obtain a satisfactory yield and rate of extraction. The particle size of the ground coffee influences not only yield and rate of extraction but also the actual substances extracted with hot water. Very fine grinds allow large quantities of colloidal substances, mostly carbohydrates and oils in nature, to enter the brew. Thus taste is also influenced by the fineness of the grind. Grinding releases quite large amounts of carbon dioxide (generated during roasting and partly retained within the structure of the bean) and also some of the volatile aromatic substances. After grinding the flavour becomes rather unstable. Some of the volatile substances are lost and others are oxidized, leading to a recognizably stale flavour after several days.

Roasted beans are, of course, commonly ground in the domestic situation. There is, however, a large non-domestic market for roast and ground coffee. Large scale commercial operations usually use a series of contrarotating toothed rollers with adjustable gaps.

DECAFFEINATED COFFEE

Caffeine is perceived negatively by many members of the public and some scientists and physicians. It could well be the most studied, or most unnecessarily studied, substance in coffee, cocoa, tea, and so on. The negative perceptions are largely without foundation and quite irrelevant to the vast majority of the population. Caffeine is rapidly and totally absorbed by man, being detected in the bloodstream about 5 min after consumption and reaching peak values after some 20–30 min. It readily crosses the blood–brain barrier and enters all body fluids including serum, saliva, milk and semen. The rate of absorption depends on the vehicle, being slower from low pH cola beverages compared with tea and coffee. In contrast to some early reports the tendency for caffeine to complex strongly with certain phenols in tea does not seem to influence the rate of absorption.

The half-life of caffeine in man is typically 2.5–4.5 h but during the last trimester of pregnancy it may rise to about 15 h. Age *per se* has little effect, although the newborn and especially the premature infant may have a greatly extended half-life for caffeine (more than 100 h) due to incomplete development of the metabolizing enzymes situated in the liver. Exercise may increase the rate of elimination whereas alcohol may retard it. Excretion is slower also in the grossly obese and those suffering from liver disease. Oral contraceptives delay clearance by about one-third. Smoking enhances clearance by inducing the caffeine-metabolizing enzymes in the liver, and the more rapid reduction in serum caffeine levels with consequent need for more frequent replenishment, may explain the strong statistical association between smoking and consumption of caffeinated beverages. It is particularly important to recognize that clearance and metabolism vary with the species. Data from rodent studies must not be extrapolated to humans without due regard to certain well-known differences. Failure so to do can lead to erroneous projections.

Coffee may delay the onset of sleep and result in lighter sleep, especially if consumed immediately before sleep. Inability to sleep could just as accurately be thought of as increased alertness. However, many people adapt very quickly and sleep quite soundly after a caffeine-containing nightcap. Subjects who are sensitive to caffeine may become more anxious. Children are not more sensitive than young/middle aged adults, but sensitivity does increase with age thereafter. Caffeine has a vasoconstricting effect on the brain and a vasodilating effect on the peripheral blood flow explaining the use of caffeine-containing medications for the relief of migraine.

The available animal studies do not point to any clear permanent and adverse neuro-behavioural effects in rodents up to dose levels causing severe toxicity. Massive doses of caffeine, e.g. 5–10 g equivalent to say 75 cups of coffee in a very short period of time, can be dangerous and may be fatal but these cases are irrelevant to the majority of the population.

Habituation to caffeine has been established and withdrawal symptoms can be expected within 12–24 h, usually peaking within 48 h and occasionally lasting a week or even longer. However, caffeine is not considered to be addictive in the strict definition of the term (American Psychiatric Association), but may be termed habituating. Newborn infants may exhibit symptoms if their mothers were heavy consumers of caffeine during pregnancy.

The negative perceptions have led to a significant trade in decaffeinated coffee. Currently more than 15% of US coffee consumption is decaffeinated and a similar proportion is being approached in some European countries. There is increasing interest in developing caffeine-free beans by genetic engineering.

Decaffeinated coffee is usually manufactured by the extraction of green beans with a suitable solvent prior to roasting. The most commonly used solvent is dichloromethane (methylene chloride), which is quite selective and does not remove other components important to quality to any significant extent. Legislation in the USA and Europe requires that the residue of this solvent should not exceed a few parts per million in the product for sale. The maximum residual caffeine content is also controlled.

While all the evidence now points to dichloromethane being perfectly safe and posing no health hazards in this context, considerable efforts have gone into developing alternative procedures. Ethyl acetate and coffee oils have both been used commercially but result in inferior products. Although a naturally occurring substance, ethyl acetate is still an organic solvent and unlike dichloromethane is highly inflammable. Consequently a flame-proof extraction plant is required resulting in higher investment costs.

Caffeine, together with many other components, can be extracted from coffee beans with water. If the caffeine is then selectively removed from the water (e.g. by extraction with dichloromethane) the water may be returned to the coffee beans and used for further extraction of caffeine. This forms the basis of the 'water decaffeination' operation which is used commercially and claimed by some to be a 'natural process'. It would seem likely that traces of dichloromethane would quite easily still end up on the coffee beans. Alternative methods for selective removal of the caffeine from the water, e.g. by use of a specific selective absorbent, have also been patented.

Decaffeination by extraction with supercritical carbon dioxide at high pressure (up to 350 bar (5000 psi)) is also now being used commercially in USA and Europe. Although there are many attractions, e.g. CO_2 is quite selective and leaves no residues, the very high pressures utilized make the capital cost of the processing plant very high. Consequently the minimum economic size of plant is considered to be around 6000 tonnes per annum which is much larger than plants using alternative solvents.

INSTANT COFFEE

Extraction

As much as 90% of the coffee consumed in the UK is in the form of instant or soluble powder. The proportion is very much lower in the other major consuming countries with a long established tradition of coffee consumption. There are several hundred instant coffee factories around the world, operating both in the consuming and growing countries.

Instant coffee plants generally take in green beans and have their own blending, roasting and grinding operations. Extraction of the coffee solubles is usually carried out in a series of percolating columns. Depending on the output of the plant the columns usually number five to eight and vary in height between about 2 m and 6 m with a height to diameter ratio of five or six to one.

The system is operated at pressures of up to 20 bar (300 psi), and temperatures of 180°C. These conditions bring about hydrolysis of some of the polysaccharide components of the beans and result in very much higher yields than obtained in a domestic situation.

The columns are operated in a batch counter-current fashion. Fresh water enters the bottom of the columns containing the most completely extracted grounds and the very dilute extract from this column passes to the base of the next most completely extracted column and so on. On entering the bottom of the final column charged with freshly roast and ground coffee the extract passes up the column, thoroughly wetting the grounds and displacing air and CO_2 upwards.

The final extract is withdrawn from the last column at a concentration of up to about 35% solids and is quickly cooled. At any one time one column is out of the circuit being emptied and refilled. It then becomes the last in the sequence, and the first, containing completely spent grounds, is taken out.

Alternatively extraction systems operating on a continuous basis have been developed and used but the percolation column system remains the most popular.

Coffee oil

Coffee beans have significant oil content. Arabicas contain some 16% and robustas some 10%. Compared with typical vegetable oils, coffee oil has a low proportion of triglycerides and a high proportion of unsaponifiable matter (25%). Linoleic (C18:2) is the major fatty acid present in the triglycerides. The unsaponifiables consist primarily (some 75%) of sterols accompanied by significant quantities of waxes (serotonins) and a group of peculiar diterpenes (cafestol and kahweol derivatives, see COMPOSITION). Cafestol, but not kahweol, is now known to be responsible for the reversible hypercholesterolaemic (blood cholesterol raising) and hypertriglyceridaemic (blood triglycerol raising) effects of boiled coffee as commonly consumed in Italy and Scandinavia, but not the UK. During roasting there is a progressive but incomplete conversion of cafestol and kahweol to dehydrocafestol and dehydrokahweol, respectively.

While the oil is itself unimportant for quality it

may contain valuable volatile flavouring depending on the way it is isolated. Indeed, it may be utilized as a carrier for volatile constituents for aromatizing instant coffee. The oil is effectively insoluble in water so remains in the grounds after brewing. The large quantities of coffee grounds associated with instant coffee manufacture have, in some places, been used as a commercial source of oil.

Drying

This is usually carried out by spray drying. The extract is atomized at the top of a tower through which is passing a current of hot air. Each droplet dries individually, resulting in a hollow sphere. These spheres then form a free-flowing dry powder which is readily soluble in hot water. Adjustment of the drying conditions allows control over the density of the powder so that one 5 ml spoonful corresponds to one cup.

While much instant coffee is retailed in the form of a spray dried powder, the product may be further treated by agglomerating. The powder is slightly rewetted and redried in such a way as to cause the particles to stick together in the form of granules. The size and density of the grounds can be closely controlled. This results in a product which superficially resembles coarse ground coffee in appearance while retaining its solubility. It is also possible to introduce extra aroma components at the agglomeration stage to replace those which are inevitably lost during spray drying; consequently an improved quality is presented to the consumer.

There is no doubt, however, that the best quality products are obtained by freeze drying and this product occupies the upper segment of the retail market. The concentrated extract is frozen, comminuted and sieved to give uniform granules, these operations being necessarily carried out in a cold room. The frozen granules are loaded on to trays and dried under high vacuum. This may be achieved using a series of batch-operated vacuum chambers or by passage through a vacuum tunnel fitted with air locks. In both systems the trays are placed on a heated surface to accelerate the drying. The vacuum must be sufficiently good to prevent melting. The gransules then retain their shape and do not fuse together. A much higher retention of aroma constituents is achieved by freeze drying, so that the final product more readily represents freshly roast and ground coffee in the cup. The operation is, however, expensive in respect of both capital investment and running costs and this is reflected in a higher retail price.

Decaffeinated versions of each of these types of retail instant products are now available. These are generally manufactured by roasting and extracting previously decaffeinated green beans.

PACKAGING

Roast and ground coffee

To overcome the problem of staling, ground coffee is usually sold in vacuum or gas-packed cans or flexible packs offering satisfactory barriers to oxygen and moisture. Account must be taken of the evolution of carbon dioxide after grinding. A delay between grinding and packing to allow dispersal is necessary unless a sufficiently high vacuum can be applied to draw off all the CO_2 prior to sealing. The packing of single-cup portions in bags in the same manner as tea-bags is used to a limited extent in the catering market but has not been successful in the retail trade.

Instant coffee

Instant coffee powder is extremely hygroscopic, and the universal container for retail purposes is a glass jar with a sealed foil cover and screw cap to avoid any possibility of moisture uptake. Ingress of oxygen must also be minimized.

QUALITY ASPECTS

Composition

The composition of green coffee beans is summarized in Table 10.3; that of roasted coffee beans and a representative soluble coffee powder are summarized in Table 10.4. Some 25% of the dry matter is water soluble. This includes all the

Table 10.3 Approximate composition of green coffee beans (% dry material)

Component	Arabica	Robusta
Carbohydrates		
sucrose and reducing sugars	5.3–9.3	3.7–7.1
araban (polymeric arabinose)	4	4
mannan (polymeric mannose)	22	22
galactan (polymeric galactose)	10–12	12–14
glucan (polymeric glucose, i.e. cellulose)	7–8	7–8
Proteins	12	12
amino acids	0.4–1.0	0.4–1.0
Alkaloids		
caffeine	0.6–1.5	2.2–2.7
trigonelline	1	1
Chlorogenic acids		
caffeoylquinic acids	5.2–6.4	5.5–10.0
feruloylquinic acids	0.3–0.5	0.7–1.5
dicaffeoylquinic acids	0.7–1.0	1.4–2.5
other chlorogenic acids	0.1	0.2–0.3
Other acids (malic, citric, tartaric, etc.)	2.5	2.5
Lipids		
triglycerides	10–14	8–10
sterols	2–3	2–3
diterpenes	1–2	0.3–0.6
Minerals		
ash	4	4

simple sugars, organic acids and alkaloids and a small proportion of the proteins and polysaccharides. Chlorogenic acids have the general structure shown in Figure 10.1 and occur as a series of isomers with slightly different properties including particularly their ability to complex with caffeine.

The polysaccharides have only been fully characterized since 1987 and most earlier publications are incorrect. Starch and typical pectins (polygalacturonic acids) are present in the bean at no more than a trace, if at all. They are replaced by a Type II arabinogalactan and a mannan. The mannan has a 1β4 backbone with very few branches and a mass probably in the region of 5000–6000. It resembles Ivory nut mannan and is not, as once thought, a *galacto*mannan.

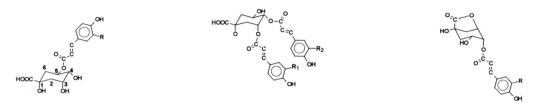

Figure 10.1 Structures of chlorogenic acids and lactones. Left: 5-*p*-coumaroylquinic acid, R = H; 5-caffeoylquinic acid, R = OH; 5-feruloylquinic acid, R=OCH₃; the 3 and 4 isomers also occur. Middle: 3,4-dicaffeoylquinic acid, R₁=R₂ = OH; 3-feruloyl, 4-caffeoylquinic acid, R₁ = OCH₃, R₂ = OH; 3-caffeoyl, 4-feruloylquinic acid, R₁ = OH, R₂ = OCH₃; the 3,5 and 4,5 isomers also occur. Right: 4-caffeoylquinide, R = OH; 4-feruloylquinide, R = OCH₃; since quinides are 1,5 lactones, the 3 isomers exist but the 5 isomers cannot.

Table 10.4 Approximate composition of medium roasted coffee beans and a representative soluble coffee powder (% dry material)

Component	Medium roasted	Soluble powder
Carbohydrates		
sucrose and reducing sugars	0.1	1.8–2.5[a]
araban (polymeric arabinose)	0.7	1.4–2.3
mannan (polymeric mannose)	14	13–18
galactan (polymeric galactose)	6	16–18
glucan (polymeric glucose, i.e. cellulose)	5	0.5–0.7
Proteins (denatured and transformed)	12	12
amino acids	0.4–2.4	0.8–0.9
Alkaloids		
caffeine	0.6–1.5	2–5
trigonelline	0.4	1.0–1.7
Chlorogenic acids		
caffeoylquinic acids	1.4–2.5	2–8
feruloylquinic acids	0.1–0.2	0.7–1.9
dicaffeoylquinic acids	0.1–0.5	0.2–1.2
other chlorogenic derivatives e.g. lactones	0.25–0.35	
Other acids		
(malic, citric, tartaric, etc.)	2.9–3.8	
Lipids		
triglycerides		
sterols	11–16	trace
diterpenes		
Minerals		
ash	4	7–13

[a] Soluble sugars of authentic (i.e. unadulterated) soluble coffee powder include arabinose (about 1%), galactose (about 0.5%) and mannose (about 0.5%) arising from hydrolysis of polysaccharide and thus accounting for the much higher level of total sugars when compared with roast coffee.

The arabinogalactan has a more complex structure (see Figure 10.2) with a galactose-1β-galactose backbone and side chains of galactose (linked 1β6) and arabinose (linked 1α3 or 1α5). This glycan occurs at a higher level in robusta beans compared with arabica beans.

The lipids are of some interest due to their very high content of unsaponifiable constituents which form some 25% of the crude lipid. Among the unsaponifiables are a group of diterpenes (Figure 10.3) which seem to be unique to coffee. These include cafestol, kahweol (found only in arabicas) and 16-*O*-methylcafestol (found only in robustas). Cafestol has been associated in humans with the reversible elevation of serum cholesterol and triglycerides.

The chlorogenic acids (Figure 10.1) are major and characteristic components of green coffees with significant differences in profiles between arabicas and robustas. In addition coffees of certain origins, in particular Angolan robustas, are quite distinct due to the presence of caffeoyl tyrosine and a second similar but unidentified constituent. The chromatographic profiles of typical (Indian) arabica and robusta are shown in Figure 10.4.

Adulteration

There has been much concern over the contamination, accidental or deliberate, of instant coffee powders with extracts of coffee husk. Considerable effort at an international level has led to the development of analytical methodology for carbohydrates characteristic of husk (xylose and mannitol) but absent, or virtually absent, from beans. Standards have been established amidst some controversy arising primarily from the burden placed on suppliers because of the difficulty and expense of the analyses required.

It has been proposed that robustas can be detected in arabicas, and the relative proportions estimated by measurement of the headspace volatiles (particularly sulphur-containing volatiles) which are higher in robustas. The measurement of methylisoborneol (MIB) has also been suggested for this purpose since it has been suggested to be

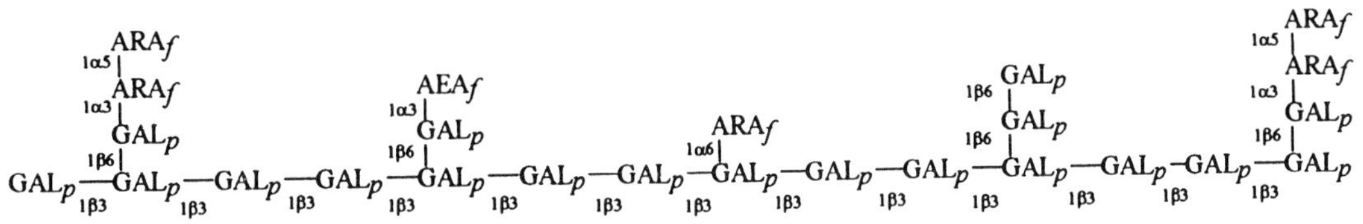

Figure 10.2 Probable structure of robusta arabinogalactan.

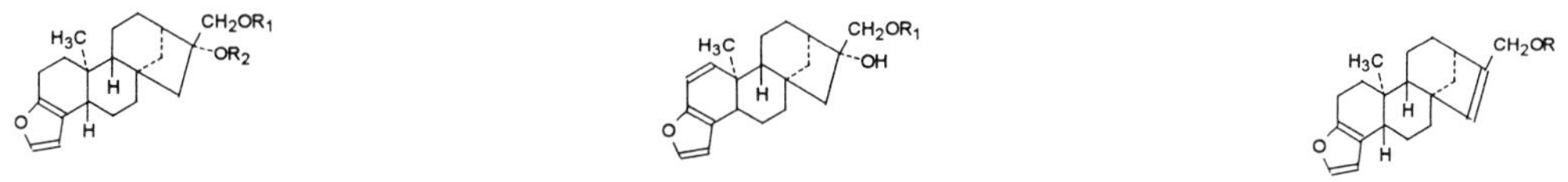

Figure 10.3 Coffee oil diterpenes. Left: cafestol esters, R_1 = fatty acyl, R_2 = H; 16-*O*-methylcafestol esters, R_1 = fatty acyl, R_2 = CH$_3$. Middle: kahweol esters, R_1 = fatty acyl. Right: dehydrocafestol esters, R_1 = fatty acyl; dehydrokahweol also occurs.

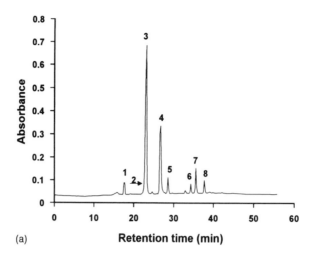

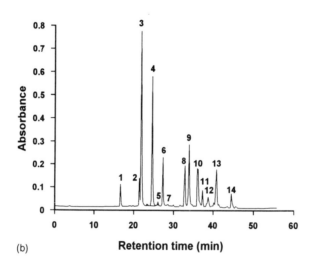

Figure 10.4 Chlorogenic acids profiles of typical (Indian) (a) arabica and (b) robusta coffees. (a) 1 = 3-CQA; 2 = 4-CQA; 3 = 5-CQA; 4 = caffeine; 5 = 5-FQA; 6 = 3,4-diCQA; 7 = 3,5-diCQA; 8 = 4,5-diCQA. (b) 1 = 3-CQA; 2 = 4-CQA; 3 = 5-CQA; 4 = caffeine; 5 = 4-FQA; 6 = 5-FQA; 7 = unknown; 8 = 3,4-diCQA; 9 = 3,5-diCQA; 10 = 4,5-diCQA; 11 = 3C,4FQA + 3F,4CQA; 12 = 3C,5FQA + 3F,5CQA; 13 = caffeoyl tryptophan, 4C,5FQA + 4F,5CQA; 14 = unknown. (NOTE: CQA = caffeoylquinic acid, FQA = ferulolylquinic acid, C = caffeolyl, F = feruloyl, etc.)

the 'robusta' odour impact compound but this view is not universally accepted. Alternatively, 16-O-methylcafestol, which is found only in robustas and is stable to roasting, may be measured by high performance liquid chromatography (HPLC). The related compound, kahweol, is found only in arabicas. Robustas from Angola, even after roasting and brewing, are detectable by the presence of caffeoyl tyrosine and a similar but un-identified constituent.

typical arabicas and at about 2.5–3% in typical robustas where it is strongly associated with chlorogenic acids. After roasting, which destroys most of the latter, the alkaloid is readily extractable with hot water. This leads to a caffeine concentration in the beverage of 40–100 mg/100 ml. Trigonelline is present to the extent of about 1% in green arabica and robusta beans but is partly decomposed during roasting, yielding niacin and a range of volatile substances, including pyridines.

Alkaloids

The major alkaloids of coffee are caffeine and tri-gonelline. Caffeine occurs at just over 1% in

Mycotoxins

Ochratoxin A (OTA) is a carcinogenic mycotoxin which is produced by moulds of the ubiquitous

Penicillium and *Aspergillus* spp. In many species, including humans, kidney is the primary target organ and in animals 30% tumour incidence has been observed over two years' exposure to doses as low as 70 μg kg^{-1} body weight. OTA survives roasting and enters the coffee brew and contaminated coffee will therefore contribute to the human burden. This has led to considerable concern and programmes of monitoring have been introduced. Many interested parties are proposing a properly designed investigation to identify the critical points in coffee production storage, processing, and so on at which contamination occurs and thus identify the necessary practical control measures.

The doses of aflatoxin B1 and sterigmatocystin derived from coffee are generally considered to represent a smaller risk to health. Aflatoxin production by *Aspergillus parasiticus* is inhibited in the presence of caffeine, and in at least one strain studied the reduced specific production was maintained when grown subsequently in the absence of caffeine.

QUALITY ASSESSMENT

The quality of coffee is assessed at three stages.

Raw beans

Ideally beans should be a uniform blue, bluish grey or greyish green. Brown beans result in poor liquor and may be caused by poor processing or nutritional defects. Black beans arise from disease or insect damage. Reddish or 'foxy' beans arise from overripe fruit or inadequate processing. Yellowish or 'amber' beans may be caused by picking underripe cherries or by iron deficiency. 'Coated' beans have silverskins still largely intact. This may arise by hulling at too high a moisture content.

In addition to colour, the shape of the beans is also critically assessed. Large well-rounded beans are considered desirable, while poorly shaped or 'ragged' beans may be thin, pointed or even boat shaped and are poorly regarded. Other defects of shape include triangular beans, formed when three beans instead of the usual two grow within a single cherry, 'elephant beans' formed when the two beans grow partially fused together and 'peaberry' when only a single bean develops within the cherry; peaberries are fairly common in some areas and if separated may fetch a good price. They are said to produce a high quality coffee. Variations in bean size and shape may have significant effects on

heat transfer and hence the rate of roasting and might lead to uneven roasting and impaired quality.

Roast beans

After roasting a sample under carefully controlled and standardized conditions, the beans should ideally be a shiny uniform dark brown. The centre cut should open to reveal a whitish chaff. Brown centre cuts are a sign of defective processing and are undesirable, likewise a dull surface tends to be indicative of a mediocre liquor. Non-uniformity of colour is often associated with non-uniformity of the raw beans. Immature yellow beans give pale roasts and associated poor flavours, while certain pale beans, referred to as 'stinkers', release highly unpleasant odours when ground and can contaminate a whole batch. Stinkers are now known to be associated with the growth of certain micro-organisms on the developing bean.

The liquor

Freshly roasted and ground beans are liquored under standard conditions for assessment (see COFFEE TASTING TERMS).

Coffee tasting terms

Acidy

A desirable flavour character associated with high grown coffee.

Astringency

A generally undesirable flavour associated with high and perhaps irregular chlorogenic acids contents. Such levels may arise from the use of immature beans (which may have high dicaffeoylquinic acid contents). Chlorogenic acids are known to bind to salivary proteins, precipitating them, causing dryness, i.e. astringency, in the mouth.

Body

See 'Winey'

Earthy or musty

Undesirable taste associated with beans contaminated by moulds, particularly *Eurotium* spp.

Fermented or fruity

Associated with high level of aldehydes and may be attributed to overripe cherries or the influence of certain aldehyde-forming organisms during fermentation.

Hard

Harsh and undesirable but distinct from Rioy. Associated with harvesting when dry.

Hay-like

Common flavour note attributed to the presence of furfural.

Nutty

Nut-like taste regarded as undesirable and associated with immature beans from underripe fruit.

Peasy

An unpleasant odour reminiscent of fresh green peas and often, but not exclusively, associated with Central African coffee. Associated with the presence of 2-methoxy-3-isopropyl pyrazine and to a lesser extent the isobutyl analogue. These are produced by moulds possibly as a result of β-ketonic rancidity followed by the β-ketone participating in a Strecker degradation.

Rich or mellow

Well-balanced properly developed flavours and aroma.

Rioy (Rio)

A harsh unpleasant medicinal taste associated with mould growth, particularly *Aspergillus fumigatus*, and the production of 2,4,6-trichloroanisole. This well-known tainting agent has been reported in green coffee at concentrations in the range 12–106 ng g^{-1} well above its odour threshold value which is 10^{-12} ng g^{-1} or less depending on matrix.

Sour

An undesirable taste often associated with 'foxy' beans.

Sweet, soft or neutral

Free from any undesirable character, therefore useful as a base for blending.

Thin

Lacking in body or flavour.

Winey

Having a taste sensation of texture or fullness. Very desirable character associated with high grown coffee.

Physiological effects

Circulating metabolites of caffeine and the chlorogenic acids have been identified in man. Caffeine is discussed more fully under DECAFFEINATED COFFEE. There is growing interest in phenols such as chlorogenic acids and their metabolites as possible beneficial components of the diet and results from *in vitro* and animal studies suggest that this might indeed be the case. Among the possible benefits suggested is protection from the effects of the ubiquitous dietary pyrolysis mutagens and possibly a reduced risk of developing colon cancer. A weak hypocholesterolaemic (blood cholesterol lowering) effect has been demonstrated for some chlorogenic acids (particularly 1,3-dicaffeoylquinic acid, characteristic of artichoke and absent from coffee) but it is unlikely that coffee as commonly consumed has much if any effect. The lactone derivatives of chlorogenic acids formed on roasting have attracted some attention as possible agonists (stimulators) or antagonists (blockers) of opiate receptors. The biological data so far published are somewhat conflicting, but if one or more of these lactones do stimulate opiate receptors then such compounds might explain the mild euphoric and habituating effects of coffee. While the lactones might account for some 10% of the total residual chlorogenic derivatives, these compounds if active are of low potency and are not considered to be of pharmacological significance.

Cafestol (1 mg kg^{-1} body weight per day for four weeks) but not kahweol causes a reversible elevation of serum cholesterol and triglycerides in humans. The slope of the dose–response curve is not known. Cafestol only enters the brew when the grounds are boiled extensively thus facilitating the extraction and emulsification of this lipid-soluble substance. This is not a feature of instant coffee manufacture, nor of the brewing procedures as commonly practised in the UK. It may however be a feature of French, Italian and Scandinavian practice and convincing epidemiological links have been demonstrated between coffee consumption and elevated serum cholesterol in such populations.

TEA

Tea leaf
Page 379

Processing
Page 381

Blending
Page 387

Decaffeinated tea
Page 388

Instant tea
Page 388

Packaging
Page 389

Quality aspects
Page 389

PRODUCTION AND MARKETING

The total world production in 1994 was 2502 million kg (some 2.5 million tonnes) and approximately 41% of this was exported from the producing countries.

TEA LEAF

Varieties

All tea is now considered to be derived from one *Camellia* species having two major and numerous minor varieties and clones. The major varieties are *C. sinensis* var. *sinensis* and *C. sinensis* var. *assamica*, originating from the tropical rain forests of China and Assam, respectively. The pure China type has very small leaves, only a few centimetres in length even when mature, whereas the typical Assam variety has large leaves 10–15 cm or more in length. There is however a large leafed var. *sinensis* found in Yunnan and known to have an unusual and distinctive profile of flavanols. Virtually all tea grown outside China and Japan is of mixed origin, plantations having been established

Table 10.5 World production of tea

Producer	Average production 1991–94 (tonnes × 10³)	Production 1994 (tonnes × 10³)
India	742	744
China		588
Sri Lanka	224	242
Kenya	203	209
Turkey		134
Indonesia	138	130
Japan		86
Bangladesh		52
Iran		45
Argentina		42
Vietnam		36
Malawi	36	35
Tanzania	21	24
Taiwan		22
USSR/CIS		18
Uganda		13
Zimbabwe	13	13
S. Africa		12
Brazil		10
Burundi		7
Malaysia		6
Rwanda		6
Papua New Guinea		6
Mauritius		5
Other African origins		9
Other origins		6

Table 10.6 Imports for consumption (1994)

Importer	Net imports for consumption 1994 (tonnes × 10³)
United Kingdom	148
CIS	127
Pakistan	107
USA	96
Egypt	55
Japan	41
Morocco	34
Poland	31
Iran	31
Jordan	30
Syria	24
Australia	17
Germany	16
Dubai	16
Saudia Arabia	15
Chile	14
Canada	14
France	12
Tunisia	12
Ireland	11
Afghanistan	11
South Africa	10
Taiwan	10
Sudan	10
Libya	9
Netherlands	8
Hong Kong	8
Algeria	6
Others	95

by naturally pollinated seeds. It is possible that hybridization with other closely related *Camellia* species has also occurred. Most newly planted tea is now derived by vegetative propagation of selected clones or from seed derived therefrom by controlled pollination.

Speciality teas

Almost all tea is blended before retailing. Blends from a single country, e.g. Kenya, Sri Lanka, China or regions, e.g. Assam, Darjeeling, may be marketed as specialities. Single estate teas are rarely available except from small specialist retailers.

High grown teas

Wherever tea is grown, it is a common experience that high elevation is beneficial to product quality. While Darjeeling may be regarded as an extreme example, the situation is nowhere better illustrated than in Sri Lanka. Here the nature of the terrain is such that estates within a few miles of each other can be at widely different elevations with marked effects on quality, especially at certain times of the year.

The effects are probably largely associated with differing growth rates which tend to alter the composition of the leaf. A further contributory factor may be the relative ease with which fermentation temperatures can be controlled at higher altitudes.

Eamples of districts associated with particular qualities include:

(i) Sri Lanka: High grown Sri Lankan teas are all manufactured by the orthodox procedure and are generally light in colour with excellent flavours. The best known high grown districts are Nuwara Eliya, Dimbula and Uva.

(ii) Darjeeling: Produced from a very restricted area of northern Bengal high in the foothills of the Himalayas, genuine Darjeeling is valued highly for its characteristic flavour and aroma. Unfortunately, in recent years, there has been a trend to blend Darjeeling with other poorer quality products and to label the result as Darjeeling, thus downgrading the image and benefiting the packer at the expense of the producer. Some Darjeeling blends may contain as little as 60% of true Darjeeling tea.

(iii) Nilgiri: Most of the tea produced in southern India is consumed locally. Small quantities of the best products from the highest tea areas in the Nilgiri hills are exported. It is similar in character to good to medium Sri Lankan.

(iv) Assam: In Assam (the major tea growing state in northeast India) it is seasons and climate rather than altitude that cause striking changes in quality. The so-called 'second flush' teas (i.e. teas produced from the second plucking round after the winter dormant period) have particularly strong bright liquors with good flavour characteristics. The 'later rains' teas are generally duller and plainer in character. While Assam teas are major components of most popular blends, blends of 'pure' Assams are marketed on a smaller scale.

(v) Keemum: The best known quality product from north China with a good flavour and clear coloured liquor.

Oolong

This term is used to describe a partly fermented tea. It is intermediate in character between black and green teas. Oolong tea is produced in China and other parts of the Far East and is very popular in that region. Like green tea it is often flavoured with jasmine flowers.

Flavoured teas

(i) Earl Grey: This popular product is flavoured by the addition of oil of bergamot. It is interesting to note that the major constituent of this oil is linalool which is also quantitatively one of the more important compounds of the tea volatile fraction. Therefore one is simply enhancing one particular aspect of the natural tea flavour.

(ii) Lapsang Souchong: Originally produced in a few factories in China by burning tarry ropes in the vicinity of the tea during processing. The smoky character is now reproduced more or less successfully by absorption from an appropriate source on to the finished product before packaging.

(iii) Others: These include lemon, apple, blackcurrant, mandarin and wild cherry.

Herbal teas

There is a small but growing market for herbal teas. Some of those available are camomile, rose hip, peppermint, fennel, lemon verbena and mixed fruit.

Alongside the European herbals one or two exotic teas are also now available. These include the South American Yerbe Maté or Matte (from the leaves of *Ilex paraguayensis*) and the South African Rooibos (Bushman's Tea). Maté is an important crop and effectively replaces tea and coffee in much of South America. Unlike the true 'herbal' it contains caffeine and theobromine. It is also rich in chlorogenic acid (containing more even than coffee q.v.) but flavanols and particularly condensed tannins are absent despite older and even some contemporary claims for their presence.

In the past some herbal teas have borne exaggerated claims for their beneficial effects. Improved labelling requirements make this more difficult; however, concern has been expressed about the lack of current knowledge on the effects of some herbs (e.g. preparations of comfrey) on metabolism and that some may indeed be harmful, but it is conceded that most are safe if taken in moderation.

Plucking

Young tea bushes begin to yield some crop 3–4 years after planting out in the field. Maximum yields are generally achieved after about 10 years and the economic life of a bush is said to be at least 40 years. Some of the original Assam plantings were still producing healthy crops when more than 100 years old.

Harvesting of tea is still largely a labour intensive operation in which young shoots are plucked by hand. The best tea is produced by plucking only the two terminal leaves together with the included bud, known as 'fine plucking', but it is difficult to maintain this standard in many of the tea growing areas. When tea is growing rapidly, fine plucking necessitates plucking each bush every 7–10 days.

Pluckers generally fill baskets carried on their backs and take these to collecting stations where the weight is recorded. From here the leaf is transported to the factory, usually by some form of motor vehicle. Good practice dictates that the leaf should be handled carefully in open weave sacks at this point to avoid damage and overheating. Unfortunately this point is often ignored.

There is increasing interest in the development of mechanical harvesting equipment. The development of machinery capable of manoeuvring through fields of tea and simulating hand plucking to sufficient degree without inflicting damage on the bushes has posed considerable problems. These now appear to have been very largely solved and the general introduction of mechanical harvesting where the terrain is suitable is now dependent on social, political and economic factors rather than severe technical problems. All Russian tea has been harvested in this way for many years and a few estates are beginning to employ this procedure in Africa.

Of course it has to be appreciated that turning spaces and tracks along which harvesting machines can run have to be provided. When planting new areas this can readily be taken into account, but the adaptation of existing estates would require considerable cutting out or uprooting of bushes.

Yields of leaf vary widely in different parts of the world. Three to four tonnes of made tea per hectare is commonly achieved in east Africa and such yields may be obtained even when where there are substantial dormant periods (e.g. in Assam, or Malawi). In other areas, climatic and other local factors combine to depress yields to less than half this figure.

PROCESSING

Withering

Withering is the first stage in the processing of the leaf and consists principally of a reduction in moisture content from the 75–80% characteristic of fresh leaf to 55–70% according to the process to be used subsequently. The time taken is generally some 18–20 h.

Leaf arriving at the factory is immediately spread out on withering troughs. Commonly these are steel troughs, 30 m^2 in surface area, fitted with mesh to support the leaf and with a fan at one end to blow air through the leaf. A more complex enclosed version is fitted with additional fans and doors to allow air to be blown either upwards or downwards through the bed of leaf without resorting to reversing the fans. Many tea factories have the withering troughs situated in lofts above the main processing factory. In this situation it is easy to make use of waste heat from the drier to raise the temperature and accelerate the moisture loss. However, the modern trend is to house the withering troughs at ground level and supply heat when necessary by other means. The amount of heat needed during this part of the process is largely determined by the ambient conditions.

Withering of one day's plucking is usually continued through the night in order for the leaf to be ready for processing during the succeeding day. As the ambient temperature falls and the humidity rises during the night it becomes necessary to apply heat in order to continue the evaporation of moisture. Under extreme conditions, e.g. during heavy rains, it becomes necessary to use heat for withering virtually continuously. However, it is generally recognized that excessive heat should be avoided in the interests of maintaining quality of the final product. Ideally, sufficient heat should be applied to raise the air temperature only a few degrees so long as the humidity is also lowered sufficiently for this to be effective.

During withering the leaf becomes flaccid and certain changes take place in its chemical composition. As soon as the shoot is plucked the normal metabolic reactions of the leaf are thrown out of balance. Products manufactured within the leaf can no longer be transported to other parts of the bush and tend to accumulate. Likewise, the supply of other nutrients and metabolites entering the leaf from other parts of the plant is cut off. Whereas photosynthesis ceases quite rapidly, respiration continues for much of the withering period and is typically responsible for a loss of around 4% of dry matter in the form of carbon dioxide. Other metabolic activities continue at decreasing rates. Some breakdown of proteins and polysaccharides occurs and is naturally accompanied by increases in the levels of free amino acids, especially asparagine and simple sugars. Synthesis of caffeine apparently continues and other events include

changes in the characteristic citric acid cycle components and the volatile fraction.

These metabolic activities are time and temperature dependent and not closely related to moisture levels. They result in changes in chemical composition of the leaf which have a bearing on the characteristics of the final product. It is desirable that withering should be continued for a minimum of 10–12 h at the lowest practical temperature which will be consistent with moisture removal. Increasingly sophisticated computer controlled systems are under development.

Leaf disruption

Orthodox manufacture

The traditional rolling procedure is a batch process. The roller consists of a vertical metal cylinder open at each end. This moves, under the control of a revolving crank, in an eccentric fashion over the surface of a somewhat larger circular table which is fitted with a series of ridges or battens. In some designs the table also moves. The cylinder is charged with leaf and the latter is compressed by a cap. The cap pressure can be controlled, as can the speed and throw of the crank. The extent of disruption is controlled by cap pressure, speed and the pattern of the battens. Low pressure gives rise to large twisted fragments while higher pressure causes more extensive breakdown of the leaf.

Traditionally each batch is rolled over a period of 20–30 min then sifted and the coarser fragments returned for further treatment and the procedure repeated several times. Many factories now carry out a simple rolling operation, extract the first 'fines' then pass the coarse 'mal' to an alternative process such as CTC (see below).

Rotorvane

This was introduced in the late 1950s in an attempt to mimic the orthodox process and adapt it to continuous operation. The machine resembles a large meat mincer in principle and consists of a horizontal cylinder within which is an electrically driven rotor assembly. The latter consists of a worm to lead the leaf forward and a series of forward propelling vanes. Each segment of the rotor has two opposed vanes and every successive pair is set at right angles to its neighbour. One or more pairs of reversing vanes designed to hold the leaf within the machine for an increased period are also often present. Also fitted within the barrel are ribs against which the leaf is pressed during passage through the machine. The discharge end of the barrel is fitted with an adjustable plate or iris which permits some control of the pressure being applied to the leaf. As an alternative to the iris end plate, the rotorvane may be fitted with a cone discharge attachment. The cone, which has spiral battens on its outer surface, fits on to the end of the rotor shaft and rotates within a conical sleeve. The gap between cone and sleeve, and thus the pressure applied to the leaf, can be adjusted.

Rotorvanes of various diameters, commonly 15 inch (38 cm) and 8 inch (20 cm), are used. A small number of factories operate with rotorvanes alone, producing 'rotorvane orthodox' tea. Such factories generally use one or more 15 inch machines followed by one or more 8 inch machines. However, the commonest use of the rotorvane today is as a pretreatment for the CTC process.

CTC

The CTC machine consists of two contrarotating metal rollers 8 inches (20 cm) in diameter. The surface of each roller is engraved with a matching pattern of teeth and the roller is mounted so that the teeth mesh together, the gap between the rollers being adjustable. One roller is run at 700–1000 rpm and other other usually at one-tenth of this speed. Leaf passing between the rollers is said to experience a crushing, tearing and curling action, hence the name CTC.

CTC machines were originally introduced in Assam in the 1930s to process the coarse 'mal' from orthodox manufacture. Their use has spread throughout the industry worldwide and currently CTC manufacture is by far the most popular process employed.

The withered leaf is usually passed once through a 15 inch (38 cm) rotorvane followed by several (usually three or four) CTC machines in sequence. The rollers on successive machines are set progressively closer together, resulting in a complete breakdown of the leaf into small particles of relatively uniform size.

The main drawback in CTC manufacture is the high level of skill and cost involved in maintenance. Rollers should be sharpened after about 100 h of use and precise setting of the gap is critical.

Lawrie tea processor (LTP)

This machine was introduced in the late 1960s, much of the pioneering work being carried out in Malawi. The machine is essentially a modified

Table 10.7 Typical throughput (kg h^{-1}) from different leaf disruption processes

Process	Withered leaf	Typical moisture content (%)	Made tea
15″ rotorvane	1250	65	440
8″rotorvane			
as primary step	400	65	140
as secondary step	650	65	230
15″ rotorvane followed			
by 3-cut CTC (30″)	1500	68	480
LTP	1800	71	520

hammer mill. Within the barrel of the machine is a shaft carrying a series of knives and tungsten carbide-tipped beaters, the precise arrangement of which can be altered. The shaft rotates at high speed and leaf is blown into and discharged from the machine by means of a centrifugal fan. The advantages are:

(i) Use of a single machine in a one-step operation and hence lower capital cost (nevertheless some users employ it in combination with rotorvane or CTC);
(ii) High maximum throughput;
(iii) Ease of maintenance. Knives and beaters can easily be replaced and sharpening requires no special tools or skill.

The disadvantages are:
(i) Less uniform product;
(ii) Throughput and character of product are highly sensitive to precise moisture content, 70–72% is necessary for optimum performance;
(iii) Because of the high moisture content required, less moisture can be removed cheaply from the leaf during the preliminary withering process and drying costs are increased significantly.

A comparison of the throughputs from different leaf disruption processes is shown in Table 10.7.

Fermentation

'Dhool' is the name given to the leaf at this stage before, during or after fermentation but before drying. After disruption the leaf is allowed to 'ferment' for 1–3 h before passing to the drier.

The term 'fermentation' is a misnomer since micro-organisms are not involved in the black tea that we know. However, in the production of the Chinese Pu'er teas, there is a microbial stage. The typical process is principally one of enzymic oxidation. In the intact leaf, polyphenols are contained within discrete vacuoles. During maceration, the vacuoles are ruptured, allowing the polyphenols to diffuse throughout the leaf particles and to come into contact with the enzyme polyphenol oxidase. This, in conjunction with atmospheric oxygen, initiates a complex series of reactions which bring about large changes in chemical composition. These reactions are vitally important in determining the characteristic appearance, taste and flavour of the finished product.

The critical factors influencing the fermentation reaction are time, temperature and oxygen supply. Naturally the higher the temperature the faster the reaction, and therefore at higher temperatures control of fermentation time becomes important. However, by far the greatest influence on quality is the temperature of the leaf itself during fermentation. This affects the pattern of the reactions. Higher temperatures bring about a higher level of oxidative polymerization and reduce the proportion of the desirable theaflavins in the reaction products. It is therefore desirable to ferment at leaf temperatures of about 20–25°C. Lower temperatures cause the reactions to proceed too slowly and require impracticably large areas for the fermentation operation. Higher temperatures progressively lower the quality of the product.

Measurements of oxygen consumption rates have shown that an adequate supply of oxygen is usually available. However confusion with the effects of passing air through a bed of leaf and therefore influencing its temperature have mistakenly led to the concept that it is necessary to pass air continuously in order to supply sufficient oxygen. The development of various fermentation systems and equipment illustrates this misconception.

Floor fermentation

The traditional method of fermentation was to spread the dhool out on a concrete floor. Provided that the layers were shallow enough, sufficient oxygen was able to diffuse into the bed and the concrete floor acted as a heat sink, absorbing the heat of reaction and keeping a measure of control over the dhool temperatures. Such a system is both unhygienic and labour intensive and has largely been replaced by improved procedures in the more progressive sectors in most of the producing countries.

Tray fermentation

Dhool is spread on large plastic or aluminium trays and these are stacked in a room which is usually supplied with humidified, temperature controlled air. While this improves the handling and hygienic aspects and reduces the space needed, control over the leaf temperature is, if anything, worse than in floor fermentation. Heat transfer between air and moist tea leaf is extremely poor and the temperature attained by the dhool during the fermentation reactions is largely dependent on its temperature at the time it is spread on the tray. Control of the surrounding air temperature serves only to determine the temperature which the dhool reaches by the time most of the important reactions have finished.

Troughs, tubs and skips

Several fermentation systems are based on the concept of holding dhool in trough or tub whilst passing air through the bed.

The system developed by George Williamson & Co. (the so-called G.W.A. system) is commonly used in east Africa. This consists of a series of small troughs mounted on castors and fitted with a mesh floor to support the bed of dhool. Dhool is loaded into the trough to a depth of some 200–250 mm and the trough is then wheeled and coupled to a duct supplying conditioned air which passes through the bed.

While the system has handling advantages, it is still labour intensive and does not provide adequate and effective temperature control. There will always be a significant temperature differential between the top and bottom of the bed and unless great care is taken to regulate the air flow the differential may easily be as much as 10°C with obviously detrimental effects on quality and consistency. With care, however, the system has proved to be reasonably satisfactory despite relatively high maintenance costs.

An earlier, essentially similar, system is commonly used in northern India. Here smaller tubs or gumlahs with perforated bottoms are filled with dhool and placed into outlets situated on top of ducting supplying conditioned air.

A few factories in east Africa have been fitted with mobile skips mounted on overhead monorails. After filling with dhool these are coupled to an air supply which enters the bed of dhool via a perforated pipe. At the end of fermentation the skip passes to the drier on the monorail. While there are obvious handling advantages, temperature control is inferior to the G.W.A. or gumlah systems.

Continuous fermentation systems

Several continuous fermentation systems have been developed, with obvious advantage in handling and labour requirements. However, the belief that it is necessary to pass air continuously through the bed of dhool has led to the use of perforated moving belts or trays which are both expensive and difficult to keep clean. A simpler device (British Patent No. 1478139), which is capable of exerting a satisfactory level of temperature control, consists essentially of a stack of solid belts. The temperature of the dhool is regulated at the cutting stage and it is then deposited on the top belt of the stack. The belts move at controlled speeds and as the dhool falls from an upper to lower belt it passes through a ball breaker which readily dissipates the heat of fermentation and supplies the necessary oxygen for the fermentation to continue.

Fermentation reactions

Polyphenol oxidase in fresh tea leaf oxidizes flavanols and their gallates first to flavanol semi-quinones and then to flavanol *o*-quinones. The semi-quinones are logically the precursors of colourless dimeric flavanols linked through their B rings (bisflavanols) whereas the *o*-quinones are the known precursors of the orange theaflavins and thought also to be precursors of the brown thearubigins and yellow theacitrins (see structures XVII–XX, Figure 10.5). The theaflavins form when one molecule of a simple (dihydroxy B ring) quinone interacts with one molecule of a gallocatechin (trihydroxy B ring) quinone eliminating carbon dioxide and forming a benztropolone. The theacitrins probably form from the interaction of two gallocatechin quinone precursors without the loss of carbon dioxide.

The formation of orange theaflavic acids and red theaflagallins requires the participation of gallic acid quinone. Gallic acid is not a substrate for tea polyphenol oxidase and this essential intermediate is generated by coupled (chemical) oxidation in which a simple (dihydroxy B ring) quinone oxidizes the gallic acid while itself being reduced to the flavanol which might then be reoxidized by the enzyme. The gallic acid residue of flavanol gallates can also behave in a manner similar to free gallic acid. Two molecules of epicatechin gallate can be transformed to an atypical theaflavin where the gallic acid residue in one molecule substitutes for the trihydroxy B ring quinone and the other in the conventional manner provides the dihydroxy B ring quinone. Similar coupled oxidations might well be responsible for the destruction of theaflavin and the theaflavin gallates which are not direct substrates for polyphenol oxidase but which in model systems are

rapidly destroyed when the enzyme and epicatechin are present.

The flavonol glycosides and theagallin seem comparatively stable during fermentation, but some model system studies indicate that myricetin glycosides might be transformed by peroxidase. Other oxidation reactions which take place during fermentation include:

(i) Oxidative deamination of amino acids by flavanol quinones with the formation of the corresponding odiferous aldehydes.
(ii) Oxidation of lipid materials, notably unsaturated fatty acids, with the formation of odiferous aldehydes and acids of lower molecular weight.
(iii) Oxidation of carotenoid pigments with the formation of odiferous terpenoid aldehydes and ketones.

All of these reactions are thought to be significant in the development of the characteristic flavour and aroma of black tea. Degradation of chlorophylls also takes place during fermentation, forming brown and black pigments. These affect the appearance of the leaf but have little influence on the character of the extracted liquor.

Green tea

Green tea is processed without allowing fermentation. After withering, the leaf is subjected to a heat treatment which completely inactivates the enzymes responsible for fermentation. It is then rolled by a process resembling orthodox manufacture and dried in a manner similar to black tea.

Green tea is especially important in China, Japan and other parts of the Far East and a great many different types are marketed. They are graded according to leaf size, appearance, colour and origin and are often flavoured with jasmine or other flowers. A good green tea produces a very pale straw coloured liquor with a very delicate aroma and it is, of course, drunk without milk.

The quality of green tea depends upon a suitable balance between the polyphenols which have remained more or less unchanged during processing, the amino acids, especially THEANINE and caffeine. The volatile substances are also particularly important.

Drying

When fermentation has reached the desired stage, the dhool is passed to the drying or firing stage, where the moisture content is rapidly reduced to

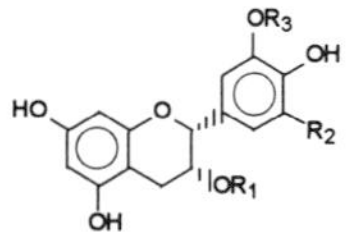

Flavanols

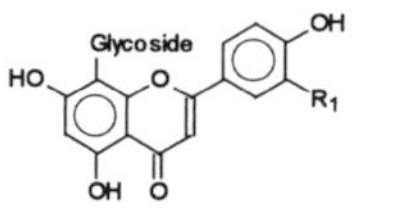

Flavone-*C*-glycosides

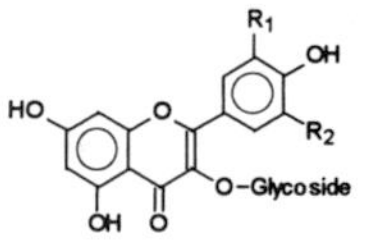

Flavonol glycosides

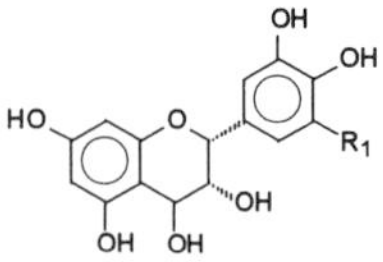

Flavan-3,4-diols

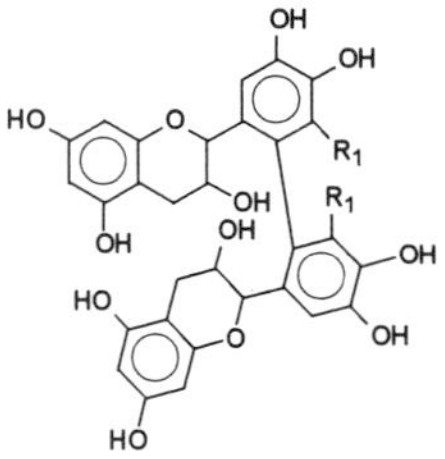

Bis-flavanols

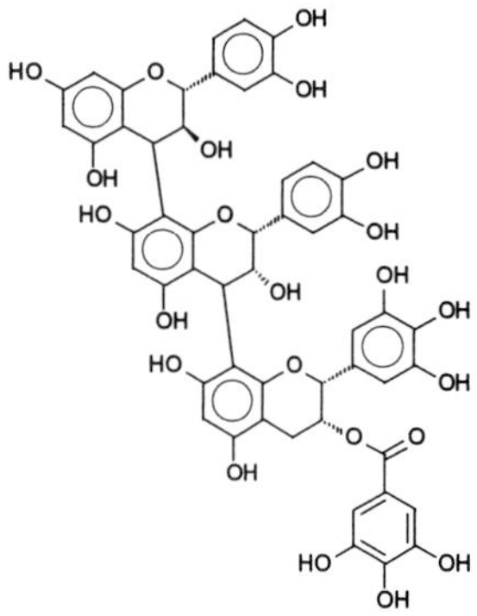

Condensed tannin

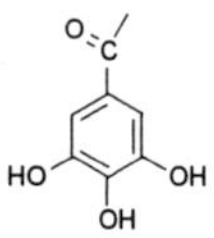

XI   Gallate

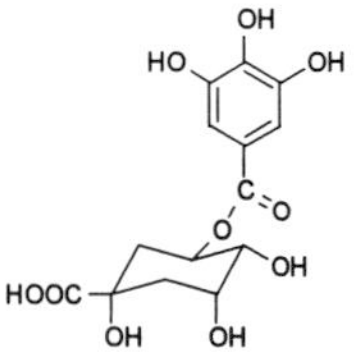

XII   Theogallin (5-galloylquinic Acid)

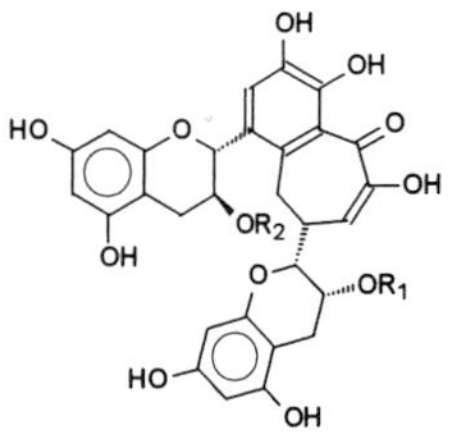

Theaflavins

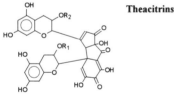 **Theacitrins**

Figure 10.5 Structures of tea polyphenols. Flavanols: I, (−)-epigallocatechin gallate (EGCG), R_1 = gallate, $R_2 = R_3$ = OH; II, (−)-epicatechin gallate (ECG), R_1 = gallate, R_2 = H, R_3 = OH; III, (−)-epigallocatechin (EGC), R_1 = H, $R_2 = R_3$ = OH; IV, (−)-epicatechin (EC), $R_1 = R_2$ = H, R_3 = OH; V, (−)-epiafzelchin, $R_1 = R_2 = R_3$ = H. Flavone-*C*-glycosides: VI, vitexin-8-glycoside, R_1 = H; VII, orientin-8-glycoside, R_1 = OH. Flavonol glycosides: VIII, kaempferol glycoside, $R_1 = R_2$ = H; IX, quercetin glycoside, R_1 = OH, R_2 = H; X, myricetin glycoside, R_1 = R_2 = OH. Flavan-3,4-diols: R_1, R_2 = H or OH. Bisflavanols: R_1, R_2 = H or OH. Condensed tannin, e.g. (+)-catechin-4α8-(−)-epicatechin-4β8-(−)-epigallocatechin-3-*O*-gallate. XI, Gallate. XII, Theogallin (5-galloylquinic acid). Theaflavins: XIII, theaflavin, R_1 = R_2 = H; XIV, theaflavin-3-gallate, R_1 = H, R_2 = gallate; XV, theaflavin-3'-gallate, R_1 = gallate, R_2 = H; XVI, theaflavin-3,3'-digallate R_1 = R_2 = gallate. Theacitrins: XVII, theacitrin, R_1 = R_2 = H; XVIII, theacitrin-3-gallate, R_1 = H, R_2 = gallate; XIX, theacitrin-3'-gallate, R_1 = gallate, R_2 = H; XX, theacitrin-3,3'-digallate, R_1 = R_2 = gallate.

3–4%. The temperature/time regime experienced by the leaf is critical in determining the characteristics of the product both in respect of taste and stability during storage.

In the earlier part of the drying the leaf experiences a period of accelerated fermentation during which important flavour developments take place. Polyphenol oxidase remains active until the moisture content has fallen to around 20%, and at this point the chemical composition of the final product is virtually fixed. Subsequently it is desirable for the leaf to reach a temperature of around 100°C, i.e. the inlet air to the drier should be at 100°C minimum in order to achieve maximum irreversible inactivation of the enzyme, otherwise the product may deteriorate rather quickly. Somewhat lower temperatures are employed in special circumstances where it is desirable to minimize the loss of important volatile flavouring substances (e.g. Darjeeling).

In traditional driers the fermented tea is usually spread in thin layers on perforated stainless steel trays and hot air is directed underneath the tray, passing through the holes and bed of tea. The trays move forward and drop their contents at the other end of the drier on to a similar layer of trays running underneath in the opposite direction. After several passes (usually four to six but sometimes eight) the dry tea emerges at a low level. The deeper the layer of tea or the greater the number of thin layers through which the same air passes, the better is the heat utilization. Nevertheless, with traditional driers there is a considerable exhaust heat loss.

While the traditional drier uses thin layers and several passes, driers are now being installed with deep layers and one pass (so-called fluidized bed driers). Since wet leaf is very difficult to fluidize, such driers are not truly 'fluid bed' in their operation until the later stages. Although their efficiency is good at the feed end, shrinkage of the tea as it dries causes a rise in the exhaust temperature as the tea passes through the drier. The exhaust gases at the exit end of the drier tend to be very hot, hence the efficiency at this point is very poor.

The source of the hot air for the drier is normally a stove operating via an air/air heat exchanger or a steam boiler/radiator system. The stove or boiler may be operated by oil, wood, coal or natural gas. In certain areas where the fuel oil or natural gas is sufficiently clean and free from sulphur it is possible to utilize the heat from the burner directly without employing a heat exchanger.

Whatever the system employed the efficiency of the stove will, in part, be determined by the heat loss from the exhaust chimney. The higher the temperature the higher this loss. Hence for maximum efficiency, the operating temperature should be as low as possible, consistent with the processing requirements.

Sorting

Tea leaving the drier is generally a heterogeneous mixture. Depending on the plucking standard and the method of processing, the ex-drier product will have more or less stalk or fibre and the particles of leaf will be variable in size. The objective of the sorting room is to separate the various fractions and produce a series of even sized clean black teas without visible pale coloured stalk or fibre.

This can be achieved most easily by passing the

Table 10.8 Common grades of tea leaf in descending size order

Whole leaf grades	GFOP	Orthodox manufacture only
	FOP	
	OP	
Brokens	FBOP	Major products from rotorvane manufacture only
	BOP	
	BP	
Fannings	BOPF	Fannings and dusts are the major products from CTC manufacture
	OF	
	PF	
Dusts	PD	
	RD	

tea through a stack of continuously vibrating screens whose mesh size decreases from top to bottom. Each fraction which is retained on a particular mesh is led off to a bulking bin for that particular grade. At various stages during its passage along the various conveyors the leaf passes close beneath or between electrostatically charged rollers which remove the bulk of the stalk or fibre. In some factories traces of fibre are removed in a current of air in a winnower.

The traditional orthodox process produces a wide range of particle sizes and the product is generally sorted into as many as eight to ten grades. These grades are traded under their traditional names: Golden Flowery Orange Pekoe (GFOP), Flower Orange Pekoe (FOP), Orange Pekoe (OP), Broken Orange Pekoe (BOP), Orange Fannings (OF), Pekoe Fannings (PF), Pekoe Dust (PD), Red Dust (RD), and so on, generally referred to by their initials (Table 10.8). Each of these grades may be subdivided in an arbitrary fashion according to the nature of the particular factory's production and marketing requirements.

While the particle size range in each grade is clearly determined by the mesh sizes employed in the various screens, there is unfortunately no standardization even within one producing country and certainly not internationally. Thus the grade names are little more than an approximate indication of method of manufacture and leaf size.

Blending

The character and quality of tea are highly variable depending on origin, method of manufacture and season. Consequently, with the exception of a few speciality products, virtually all tea is retailed in the form of blends. The blender attempts to standardize a given brand as far as possible by careful testing. This blend is then matched as closely as possible from the raw materials available to the blender within the required cost structures, that is to a given cost and quality.

Blending is usually carried out immediately before packaging. In larger blending and packaging plants the tea chests or bags are opened automatically and tipped into hoppers. From these the leaf passes on conveyors under magnets to the blender. Where large leaf grades are being incorporated the tea may be passed through a tea cutter or crushing rollers. This is desirable to adjust the bulk density to ensure that packets are properly filled and to avoid any difficulties with the actual blending operation.

The blender itself consists of either a simple rotating drum, or particularly for very large installations, a double cone blender. This consists of two cones connected by a cylindrical section mounted vertically. Blending is achieved by rotating in the vertical plane. Large scale installations may be fitted with blenders able to handle batches of 5 tonnes. This allows a blend of 30 or more components to be made up without having to split the contents of a tea chest. From the blender the tea is usually discharged into hoppers which feed directly to the packing lines.

DECAFFEINATED TEA

Currently decaffeinated tea represents less than 1% of the UK tea market. This may be because the availability of good quality products has been limiting the development of the market until relatively recently.

The market is dominated by products based on extraction at a suitable stage during the original processing on the tea estate. These are generally far superior to those prepared from black tea whatever solvent is employed (for a discussion of solvents used see DECAFFEINATED COFFEE).

From a quality point of view methylene chloride is the preferred solvent but, paradoxically, tea decaffeinated by this solvent cannot currently be sold in the USA despite its continued widespread use for that purpose with coffee. There is no difficulty at all in ensuring residual solvent levels far below those permitted for coffee.

It will be clear that extraction with supercritical carbon dioxide is most unlikely to be viable using partly processed leaf on a tea estate because insufficient raw material could be made available locally to carry out the operation on a large enough scale. Its use on black tea results in a med-

iocre product at high cost. Ethyl acetate is also used on black tea but the product quality is not very good due to the preferential solubility of theaflavins in ethyl acetate compared with water and so an undesirable tendency to remove some of this important fraction. Supercritical carbon dioxide has also been tried.

INSTANT TEA

Instant tea is manufactured to meet two distinctly different requirements, namely hot and cold water solubility.

Hot water soluble tea

In countries consuming the traditional hot beverage the instant product has not, so far, enjoyed the same fairly wide acceptance as has instant coffee. Only in the last few years has there been any serious attempt to market instant tea in the UK outside the catering and vending sector. The market is still in an early stage of development with several freeze dried products of reasonable quality consisting of 100% tea solids retailing alongside mixes containing whiteners or sugar and other flavourings. At the end of 1991 the total volume was probably only of the order of 1% of total UK tea market, but growing.

The manufacturing process is essentially similar to that for instant coffee, namely the preparation of an aqueous extract, followed by concentration and drying of that extract to yield a completely soluble powder or granules. In the case of tea, however, control of the processing conditions is much more critical if product degradation is to be avoided. The extract may be prepared from conventional black tea or from undried green leaf which may be partly processed before the extraction. In either event the extraction conditions must be precisely controlled in order to provide both a satisfactory quality and a commercially viable yield. After extraction the liquors require very careful handling to minimize undesirable changes before the final drying step. This may be by spray drying or freeze drying. For a retail product, marketing considerations dictate that the product density should be such that one teaspoonful of powder or granules corresponds to a cup of average drinking strength when reconstituted. A somewhat higher density is usually required for most vending machines. Drying conditions must be adjusted to provide the required density. Where mixes are marketed the density is rather less critical.

Cold water soluble tea

Cold water soluble instant or iced tea is a well-established and successful product in North America. When a tea liquor is cooled it forms a precipitate or 'tea cream' which is completely unacceptable for the iced beverage which is normally consumed without milk and presented in a glass. For this purpose it must be perfectly clear even after the addition of ice.

A great deal of research has been applied in dealing with this problem and each of the major producers has its own particular process or variation. In principle, however, the cream must be removed, and since it may represent as much as 30% of the solids originally extracted, it is essential that it be treated in some way to render a high proportion soluble and then added back at some point before drying.

A further problem is met in hard water districts. A product which is perfectly clear on dissolving in demineralized or soft water may become very hazy after standing for a short time if dissolved in hard water. This is considered to be due to the interaction of calcium ions with the pectins present in tea. Further treatment is necessary to overcome this problem and the best products available today are now satisfactory in this respect.

In recent years the cold soluble product has been increasingly incorporated into a whole variety of tea 'mixes'. These usually consist of dry mixes of sugar, soluble tea solids and flavourings. These products are very popular in the USA.

BOTTLED TEA

The concept of a tea flavoured soft drink is an attractive one and it is therefore somewhat surprising that the development of such a product has been largely left to the tea producing countries. The most successful development of such a product has taken place recently in Indonesia. The product is manufactured by relatively unsophisticated technology and based on the locally popular jasmine flavoured green tea. It now accounts for around 20% of the soft drinks market in that country.

PACKAGING

Tea is very hygroscopic and its ability to remain in good condition is highly dependent on its moisture content. Therefore the main technical requirement for packaging of tea at any stage is the provision of an adequate moisture barrier.

Bulk

The traditional tea chest remains as a common form of ex-factory packing in some producing countries, principally for local distribution.

Strong multi-wall paper sacks with laminated foil and polythene linings have now become the main form of export packaging. These provide a superior moisture barrier compared to the tea chest and any possibility of physical damage to the tea can be minimized by palletizing within the factory and shipping in containers.

Recently one company has begun importing tea which has been vacuum packed in laminated foil packs by the producer. It is claimed that this product arrives fresher at the packing factory than with other methods and the subsequent retail pack is nitrogen flushed.

Retail packets

Loose tea is packed in a multitude of different shapes, sizes and types of materials throughout the world. The best is probably a metal box with snap-on lid, while cartons with foil or laminated linings are satisfactory. Shrink wrapping of individual packs with cellophane offers little increased protection. It hardly needs stating, however, that there is little point in good quality retail packaging unless the product enters the packet in good condition and with a low moisture content. All too often this criterion is not met.

Tea bags

Tea bags have now become the most popular form of retail packaging, at least in the UK, USA and western Europe. A good deal of painstaking development has gone into improving the tissue paper used for this purpose as well as the very high speed packaging machinery now used for filling the bags. Once filled, the same considerations for a satisfactory moisture-proof barrier described above are necessary.

QUALITY ASPECTS

Composition

It has become clear in recent years that the 'typical' composition of tea leaf as commonly quoted is not as representative of the polyphenols

Table 10.9 Approximate composition of green tea shoots (% dry weight)

	var. *assamica*	Small leafed var. *sinensis*	So-called hybrid of small leafed var. *sinensis* & var. *assamica*	Large leafed var. *sinensis*
Substances soluble in hot water				
Total polyphenols	25–30	14–23	11–15	32–33
Flavanols				
(–)-epigallocatechin gallate (EGCG)	9–13	7–13	3–6	7–8
(–)-epicatechin gallate (ECG)	3–6	3–4	4–6	13–14
(–)-epigallocatechin (EGC)	3–6	2–4	2–3	1–2
(–)-epicatechin (EC)	1–3	1–2		2–3
(+)-catechin (C)				4
other flavanols	1–2			2
Flavonols and flavonol glycosides	1.5	1.5–1.7		1
Flavandiols	2–3			
Phenolic acids and esters (depsides)	5			1
Caffeine	3–4	3		
Amino acids	4	4–5	2–3	
Simple carbohydrates (e.g. sugars)	4			
Organic acids	0.5			
Substances partially soluble in hot water				
Polysaccharides:				
starch, pectic substances	1–2			
pentosans, etc.	12			
Proteins	15			
Ash	5			
Substances insoluble in water				
Cellulose	7			
Lignin	6			
Lipids	3			
Pigments (chlorophyll, carotenoids, etc.)	0.5			
Volatile substances	0.01–0.02			

Note: blanks in table indicate data not available.

content as it might be. Table 10.9 gives figures from a variety of sources and includes up-to-date values for the phenols content of different tea varieties and hybrids. Unfortunately, similarly up-to-date data for the other major constituents are not easily come by.

After processing the composition changes significantly, especially in regard to the polyphenols (Table 10.10). The composition depends not only

Table 10.10 Polyphenolic composition of black tea manufactured from different varieties of leaf (% dry weight)

	var. *assamica*	Small leafed var. *sinensis*	Large leafed var. *sinensis*
Total polyphenolic substances	10–18		23–24
Theaflavins	1–2		1–2
Thearubigins	3–6	normally used only for green tea production	5
Phenolic acids and esters (depsides)	3–4		1
Flavanols	1–3		15–16
Flavonols and flavonol glycosides	0.4–1.7		1

on the genotype of the particular bush but also on the environmental conditions, particularly shading, under which it is grown. In many areas seasonal climatic changes result in changes in rates of growth and in composition leading to distinctly recognizable seasonal variations in quality of the final product.

Alkaloids

Caffeine: The principal alkaloid of tea occurs at a level of up to 4% of the dry leaf. About 80% of this is extracted into the liquor when brewing tea and the average liquor concentration at drinking strength is about 0.03%. It has been suggested that as a result of interaction with polyphenols, the physiological effects of caffeine in tea are modified considerably when compared with caffeine administered in isolation or indeed in coffee.

Theanine:

$$HOOCCH(NH_2)CH_2CH_2CONHC_2H_5$$

The major free amino acid of tea is theanine, γ-*N*-ethylglutamine, first discovered in tea and amounting to 1–2% of the dry leaf. While this amino acid appears not to be of great significance for the quality of black tea, it is considered to be a particularly important component of GREEN TEA, where its content is much influenced by leaf maturity and shade during cultivation.

Polyphenols

Quantitatively by far the most important components of green tea leaf are the polyphenols, which may amount to as much as 30% of the dry matter (Table 10.9). They may be classified according to their chemical structures (Figure 10.5) as follows:

Flavanols: This is the major group of polyphenolic substances in green leaf and one component EGCG (structure I, Figure 10.5) may exceed 10% of the dry matter of the leaf and usually dominates. The large leafed var. *sinensis* from Yunnan is an exception (see Table 10.9). These substances are largely (90–95%) consumed during fermentation but in some cases somewhat greater amounts remain (see Table 10.10).

Flavones: Flavones occur as the *C*-glycosides (structures VI and VII, Figure 10.5). Little is known of their behaviour during fermentation. These are relatively water insoluble and few quantitative data are available.

Flavonols: These compounds (structures VIII, IX and X, Figure 10.5) are present in green leaf predominantly as 3-glycosides. Typically the Assam variety contains simple 3-glucosides and 3-rhamnoglucosides while the China variety contains in addition 3-rhamnodiglucosides. The pattern of flavonols present may therefore give some indication of the genotype of a particular bush and to some extent the geographic origin. The flavonols survive fermentation nearly quantitatively.

Flavandiols: Flavandiols are very unstable substances, and if they occur may only be biosynthetic intermediates. It is possible that the reports of such substances in tea and tea products are actually attributable to the dimeric and trimeric condensed tannins which have more recently been unequivocally detected.

Condensed tannins (proanthocyanidins): Small amounts of dimeric and trimeric condensed tannins are present. These incorporate (epi)afzelchin monomers as well as the commoner (epi)catechin and (epi)gallocatechin monomers and some gallates (see structures II, IV and V, Figure 10.5).

Phenolic acids and depsides: Gallic acid (structure XI, Figure 10.5) is present in tea extracts at all stages during processing and is known to be formed during fermentation by hydrolysis of the flavanol-3-gallates. The most important depside is theogallin or 5-galloylquinic acid (structure XII, Figure 10.5), a substance first isolated and characterized from tea and rarely found elsewhere. Also present in smaller quantities are the structurally related and widespread chlorogenic acids (caffeoylquinic acids and *p*-coumarylquinic acids) in which the gallic acid residue is replaced by caffeic and *p*-coumaric acids, respectively. For further information see entry under COFFEE. Theogallin and other related depsides are partially consumed during fermentation.

Theaflavins

Theaflavins (structures XIII, XIV, XV and XVI, Figure 10.5) are important reddish-orange pigments formed during fermentation by oxidative condensation of the *o*-quinone of a simple (dihydroxy B ring) catechin with the *o*-quinone derived from a (trihydroxy B ring) gallocatechin and corresponding reactions of their respective gallate esters.

Theaflavins, together with thearubigins, are responsible for the appearance of the extracted tea liquors. They are also significant in taste having a powerful astringency or mouthfeel (see Figure 10.6). The desirable characters of 'brightness' and 'briskness' (see TEA TASTING TERMS) are positively correlated with theaflavin content and within certain teas from particular localities, correlation between theaflavin contents and prices realized at auction have been demonstrated.

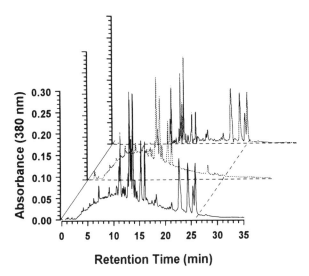

Figure 10.6 Interaction of black tea polyphenols with saliva as studied by HPLC. Rear plot, calculated 'difference chromatogram', material precipitated by saliva; middle plot, supernatant, tea + saliva (5 : 1); front plot, control, tea + water (5 : 1).

Structurally related theaflavic acids, in which gallic acid replaces one of the flavanol moieties giving rise to the benztropolone nucleus, are also present in small quantities.

Thearubigins

In attempting to describe the appearance of tea liquors in chemical terms, the coloured components have been divided into two groups, termed theaflavins and thearubigins. Whereas the theaflavins have been shown to be a structurally distinct group capable of characterization and measurement, the term 'thearubigins', because of the original methods of measurement in fact includes all coloured components of tea extracts except theaflavins. Thus while spectrophotometric analysis of theaflavins as a class has some meaning, similar measurements of thearubigins are of doubtful quantitative significance. This is because thearubigins include a whole range of substances with different colours, molecular weights and extinction coefficients.

Conventional reversed phase HPLC has failed to separate the thearubigins (see Figure 10.7), although cellulose adsorption chromatography has isolated a tawny fraction referred to as theafulvins and a brownish caffeine-precipitable fraction which is also a component of tea cream. Recently (in 1995) there has been success in fractionating the thearubigins by HPLC using new size exclusion column packings. Calibration of the column using purified condensed tannins and ellagitannins suggests that the thearubigins have masses in the range 900–2500 Da (see Figure 10.8).

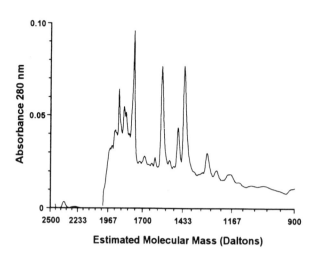

Figure 10.8 Size exclusion chromatogram of caffeine-precipitable thearubigins from a Lattakari Assam black tea.

Thearubigins have been classed as polymeric proanthocyanidins by some workers, but many fractions are negative to this test and the positive response may be due to untransformed condensed tannin passing through from the green leaf. Certain thearubigin fractions yield gallic acid, quinic acid and/or amino acids on hydrolysis, but they can be prepared free from nitrogen and the amino acids should perhaps also be viewed as contaminants rather than integral moieties. It is clear that there is considerable variation in spectral properties and further characterization is under way in the writer's laboratories.

Tannin

Use of the term 'tannin' in connection with tea is misleading. There is no tannic acid present in tea but it is now known that fresh leaf and the derived commercial green and black teas contain small amounts of dimeric and trimeric flavanols known as condensed tannins, which are common components of many leaves, fruits and seeds. It may be these, rather than the flavan-3,4-diols which are responsible for the proanthocyanidin reaction (production of red anthocyanidin pigments when treated with hot acid) of fresh leaf and green teas. A similar response is seen in black tea and with isolated thearubigins. This is probably due to the fresh leaf condensed tannins which have passed through fermentation untransformed, but it is possible that some proanthocyanidins are formed during fermentation. It must be noted however, that the bisflavanols referred to elsewhere, while flavanol dimers, possess a different intermonomer linkage and are not proanthocyanidins. They might well be astringent.

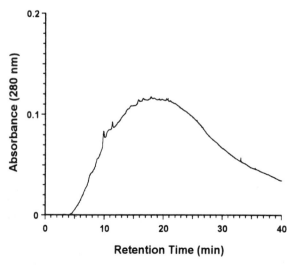

Figure 10.7 Reversed phase chromatogram of caffeine-precipitable thearubigins from Lattakari Assam black tea.

Model system studies have shown that flavanol gallates, some flavonol glycosides, the theaflavins and a fraction of the thearubigins interact strongly *in vitro* with salivary proteins leading to the formation of a precipitate. This is clearly illustrated in Figure 10.6 which compares the chromatogram of a black tea brew with the chromatogram of the supernatant remaining after interaction with saliva and centrifugation and the calculated 'difference' chromatogram of the polyphenols that were in the precipitate. These observations have in general been supported by taste panel studies and it seems that such interactions in the mouth are responsible for the sensation of astringency, which in moderation is described as briskness and viewed favourably. The susceptibility of individuals to this sensation is likely to be a function of the amount of salivary protein secreted and this is known to be variable.

Polyphenol oxidase

Tea polyphenol oxidase has an unusually restricted specificity and should really be classified as a flavanol oxidase (formerly EC 1.14.18.1, now EC 1.10.3.2) since it only attacks the 'B' rings of the tea flavanols. Gallic acid, theogallin, the flavone and flavonol glycosides and the theaflavins are not substrates. Interaction between the semi-quinones and *o*-quinones produced by the enzyme gives rise to structures such as bisflavanols and theaflavins (Figure 10.5) based on two flavonoid residues. However, the primary products of the enzyme-catalysed reaction, *o*-quinones, can bring about coupled oxidation of many substances which are not substrates of the enzyme including gallic acid, theaflavins and amino acids. In the complex environment of the macerated leaf a wide variety of oxidation products is possible.

Volatile substances

As with most foods, the subtleties of flavour in tea are largely determined by the volatile fraction which is quantitatively very small but qualitatively very significant. The essential oil of green leaf is a complex mixture which is further added to and chemically changed during processing, while losses obviously occur during drying. Tea volatiles include a whole series of aliphatic and aromatic alcohols, esters, aldehydes, ketones and acids as well as characteristic components such as *cis*-hexenol, linalool, methyl salicylate and geraniol.

Gas chromatography reveals that highly flavoury teas such as Darjeeling contain relatively large quantities of the terpenes and related compo-nents with high boiling points. Likewise it has been observed that the China variety generally exhibits more of those compounds possessing characteristically floral notes than does the Assam variety. Teas have been classified and to some extent graded by their profile of volatiles.

Tea cream

Tea cream is the precipitate which forms when strong tea liquor is cooled. It consists of a complex between caffeine and certain of the polyphenolic components, particularly the theaflavins. The extent to which thearubigins are removed is more variable but in some teas they may be a significant component of the cream. It is possible that this fraction consists of gallated thearubigins since gallate esters generally have a greater affinity for caffeine.

The formation of this complex is doubtless responsible for modifying the extremely bitter taste of caffeine and is considered to be associated with 'strength' and 'briskness' of tea.

Tea scum

Tea scum forms on black tea in certain specific circumstances and is viewed as undesirable. It occurs when tea is made with temporary hard water. Analysis has showed it to consist of high molecular weight organic material together with some 15% $CaCO_3$ from the hard water as well as small amounts of other metallic salts. Scum will grow continuously for hours on end and the organic part of the scum is formed by oxidation of polyphenolic tea solubles at the brew surface. This requires the presence of both Ca^{2+} (or Mg^{2+} or Mn^{2+}) ions and HCO_3^- ions. The absence of either leads to a scum-free surface. The addition of lemon juice lowers the pH and inhibits the formation. The greater acidity of strong tea brews also diminishes the extent of scumming relative to more dilute brews.

Tea tasting terms

Professional tea tasters apply a large number of descriptive terms to teas which they taste. Tasters examine the leaf before and after infusion and assess both appearance and taste of the tea liquors, generally brewed much stronger than the normal consumer level. Amongst the terms used to describe the liquors are the following:

Bakey (high fired)

The result of drying at too high a temperature without actually acquiring a burnt taste.

Bright

Desirable in regard to cup colour and can indicate a degree of 'live' character.

Brisk

A 'live' characteristic. Can indicate some degree of pungency and maltiness. An indication of correct fermentation and firing. Akin to astringency. Often associated with high theaflavins contents.

Burnt

A taste of burnt organic matter often accompanied by a similar smell on the leaf. Caused by firing temperatures which are excessive.

Coloury

Indicates useful depth of colour.

Cream

The precipiate that appears when a good strong liquor cools. Useful tea creams down brightly.

Dull

Brownish or greenish liquors, which are not clear or bright. Often associated with low theaflavins contents.

Earthy

Mouldy or old tea. Often indicates packing at too high a moisture level and microbial spoilage.

Flavoury

A liquor having a definite character or taste associated with certain seasons and districts. Usually caused by slow growth and comparatively rare in Africa.

Fruity

Suggests a taint. May be due to overfermentation and/or bacterial infection before firing.

Full

A strong coloury liquor with no bitterness. One which will cream down.

Harsh

A liquor which is bitter. Sometimes referred to as 'raw' or 'coarse'.

Light

Lacking body or thickness.

Malty

A desirable characteristic often found in good quality liquoring tea.

Plain

A liquor lacking desirable characteristics, but not necessarily dull.

Pungent

Having astringency without bitterness. This is a very desirable character indicating good leaf and manufacture.

Soft (mushy)

Lacking in 'live' characteristics, often caused by overfermentation or underfiring.

Stewed

A soft liquor lacking brightness and briskness with a marked undesirable character. Usually caused by low firing temperatures and/or insufficient air flow in driers, causing high temperature fermentation.

Strength

A combination of 'body' and fullness without softness. A really strong tea creams down.

Thick

A concentrated full liquor which creams down well. A characteristic of a good CTC tea.

Thin

A dilute characterless liquor of little value. Sometimes described as 'weak' or 'washy'.

Weedy (grassy)

A grass or hay flavour which is undesirable.

Experience has shown that certain of these terms may be correlated fairly well with substances which can be defined in chemical terms. For example, the characteristics of brightness and briskness generally go together and correlate positively with the level of theaflavins present, while dullness, stewiness and softness are associated with a lack of theaflavins and the presence of more than usual quantities of polymeric pigments. Likewise, as might be expected, flavoury teas exhibit relatively high levels of certain characteristic volatile substances whereas plain teas lack these substances.

Physiological effects on the consumer

There is considerable interest in the possible therapeutic and prophylactic properties of both green and black tea beverages. Of particular interest is the recent report that diets rich in complex phenols and related simple phenols have been associated in the Netherlands with a significant inverse effect on mortality from coronary heart disease. Black tea was considered to be an important dietary source of these phenols. If there is a causal relationship between the consumption of such phenols and a reduced risk of disease, the agents responsible and their mechanisms of action have yet to be defined. It has been demonstrated unequivocally that both flavanols and flavonol glycosides are absorbed by man in small quantities and the associated metabolites are found in the circulation. It is tempting to focus on these compounds as the active principles because the results of some *in vitro* studies and some animal studies suggest such compounds protect against oxidative stress and, along with caffeine, possibly protect against the undesirable effects of naturally occurring dietary and environmental toxicants and mutagens. However, not all data agree, and it would be unwise to ignore the possible effect of other components and certainly unwise to make sweeping recommendations about changes to the diet at this stage.

COCOA

```
┌─────────────────┐
│   Cocoa beans   │
│                 │
│      Page 397   │
└────────┬────────┘
         │
┌────────┴────────┐
│  Fermentation   │
│                 │
│      Page 397   │
└────────┬────────┘
         │
┌────────┴────────┐
│   Processing    │
│                 │
│      Page 398   │
└────────┬────────┘         ┌─────────────────┐
         │                  │    Roasting     │
         ├──────────────────┤                 │
         │                  │      Page 398   │
         │                  └─────────────────┘
┌────────┴────────┐         ┌─────────────────┐
│      Nib        │         │     Shell       │
│                 │         │                 │
│      Page 398   │         │      Page 398   │
└────────┬────────┘         └─────────────────┘
         │
┌────────┴────────┐         ┌─────────────────┐
│  Cocoa powder   │         │  Cocoa butter   │
│                 │         │                 │
│      Page 400   │         │      Page 400   │
└────────┬────────┘         └────────┬────────┘
         │       ┌─────────────────┐
         │       │  Alkalization   │
         ├───────┤                 │
         │       │      Page 399   │
         │       └─────────────────┘
         │  ┌─────────────────┐   ┌─────────────────┐
         │  │   Chocolate     │   │   Chocolate     │
         ├──┤ see Chapter 11: │   │ syrups, spreads │
         │  │      Page 435   │   │      Page 402   │
         │  └─────────────────┘   └─────────────────┘
┌────────┴────────┐
│ Cocoa products  │
│                 │
│      Page 401   │
└────────┬────────┘
         │
┌────────┴────────┐
│ Quality aspects │
│                 │
│      Page 403   │
└─────────────────┘
```

PRODUCTION AND MARKETING

Cocoa beans are the seeds of the tree *Theobroma cacao* L. The species can be subdivided into two main groups, Criollo and Forastero, with a third subgroup, Trinatario, produced from a cross between Criollo and Amazonian Forastero. The world cocoa production is predominantly Forastero, with Criollo producing about 1% and Trinatario some 5%. Forastero is more hardy and vigorous than Criollo and additionally has a higher fat content.

Cocoa as a crop has to be grown in tropical areas with adequate annual rainfall and temperatures between 20–35°C. Three main areas can be identified.

(i)   In West Africa, the main producing countries are Ivory Coast, Ghana, Nigeria and Cameroun, with smaller tonnages from countries such as Togo and Sierra Leone. Africa produces some 60% of world cocoa. The Ivory Coast is now by far the largest producer of cocoa, with Ghana and Nigerian tonnages recovering after several years of decline in the 1980s.

(ii) South America is a major producer, the Bahia region of Brazil producing the most. Ecuador, Colombia and Mexico are also producers.

(iii) In the Far East production has increased from 4% of the world total in 1975–1976 to 16% in 1995–1996. At the same time tonnages from Indonesia increased as tonnages from Malaysia declined.

Total production from Central America and the West Indies has been stable since about 1980. Table 10.11 shows world production of cocoa for the second half of this century.

The crop pattern is affected by many factors, such as climate and variety, and is spread over several months. In countries with pronounced wet and dry seasons, the main harvest occurs about 6 months after the start of the wet season. In African countries the majority of the crop is harvested in November. In Malaysia, where the climate is more uniform, the peak month produces as little as 12–13% of the harvest. There is an advantage in a uniform cropping pattern, since the harvesting facilities must be sized to cope with the peak month and wide variation in monthly harvests means that the plant is under-utilized for most of the year.

Approximately 30% of cocoa is processed in the growing country, and 50% is processed in Europe.

Table 10.11 Cocoa production (tonnes $\times 10^3$)

	1955–1956 Actual	1965–1966 Actual	1975–1976 Actual	1985–1986 Actual	1995–1996 Forecast
Cameroun	54	79	96	118	120
Ghana	241	417	397	219	365
Ivory Coast	71	113	231	563	1100
Nigeria	116	185	216	110	140
Other West Africa	43	72	61	77	27
Brazil	171	173	258	385	218
Ecuador	32	36	63	96	100
Other S. American	71	78	89	122	121
Dominican Republic	28	28	30	40	56
Other W. Indies	17	15	11	11	11
Malaysia	0	1	17	131	120
Indonesia	0	0	0	45	290
Papua New Guinea	1	18	32	32	31
Other Asia	10	11	11	17	19
TOTAL	855	1226	1512	1966	2718

This has resulted in a large world trade in cocoa butter, liquor and powders. There are some benefits and disadvantages to buyers of cocoa products. Cocoa products from some origin countries can be variable in terms of quality, but generally are cheaper than those from European processors. The advantages of buying from the more sophisticated processors is their flexibility which allows them to supply a wider range of products. By selection of beans, blending and process adjustments, such factors as powder colour or liquor flavour can be tailored precisely to the customer's requirement. Such products however tend to command a price premium, and thus there is still a preference for large users of cocoa to process their own beans.

The largest amounts of cocoa are used for chocolate manufacture, which causes some imbalance in the trade because chocolate requires relatively high proportions of cocoa butter, relatively lower proportions of cocoa liquor (or cocoa mass) and cocoa powder.

Other significant uses of cocoa are as a beverage and as a flavouring or colouring agent in bakery products, desserts and drinks. These non-chocolate applications use cocoa powder or fat-reduced cocoa powder almost exclusively, which helps to restore the balance and economies of cocoa processing.

Of special interest to the non-confectionery cocoa user is the Dutch cocoa industry, which specializes in the supply of cocoa powders with a wide range of colour and flavours for these applications.

COCOA BEANS

The cocoa tree is relatively low in growth, attaining a height of some 20–30 feet. Trees growing in their native habitat are taller than those grown on cultivated plantations. The tree produces a large number of flowers directly on the trunk or main branches, most of which are not pollinated. The pods take between 150 and 200 days to mature, depending mainly on climatic conditions.

The fruit or pod is elliptical or ovoid in form, from 18–25 cm in length and 8–10 cm in diameter. It has a hard, thick, leathery rind of rich purple-yellow colour. The interior of the fruit is divided into five cells, in each of which is a row of from five to twelve seeds embedded in a soft, pale pink pulp. Each fruit thus contains from 20 to 50 or more seeds and these constitute the 'cocoa beans' of commerce.

Harvesting

This is still an entirely manual process with mature pods being cut from trees and taken to a central area where the pods are split open with knives and the beans removed.

Fermentation

The fermentation process is of major importance in determining cocoa or chocolate flavour. It is also a widely varying process both between countries and between individual operations in the same country. It is important that harvested beans are fermented promptly, otherwise germination will begin.

During the process, changes occur in the bean due to respiration of the cocoa bean and to microbiological activity in the pulp. Together these result in the generation of acidity, heat and carbon dioxide. The production of acetic acid in the fermentation process finally kills the beans.

Fermentation is carried out in many different ways using beans plus pulp removed from freshly harvested mature pods. In Ghana and Nigeria the beans are heaped upon and covered with banana leaves in the plantation. The length of time the beans are left, the number of times they are turned and the size of the heap are decided by the farmer and are thus subject to wide variation. Typically, 200–400 lbs (90–180 kg) of beans are fermented for 2–5 days, with two turnings in this time.

In other countries, baskets, boxes or other containers are used to ferment between 100 and 500 lbs (45–230 kg) of beans. In Malaysia for instance, a semi-industrial fermentation process is practised where beans are transported to a central area and fermentation is carried out in tiers of boxes. Fresh beans start at the top tier and descend into the next box after 24 h. The process takes five days and gives a much more uniformly fermented product than the uncontrolled West African processes. As a result, Malaysian beans have different flavour characteristics. After fermentation the beans are dried.

Drying

It is necessary to dry the beans to a moisture content of less than 7.5% prior to storage. In West African countries, beans can be dried in the sun. They are spread on the ground or on trays which are arranged so that they can be covered when rain falls. Alternatively, they are dried in a building with a sliding roof, which achieves the same effect.

Artificial drying is also practised where large scale fermentation areas are used or in countries where the climate precludes sun drying. Such systems use either blown hot air or natural convection from piped hot water under the floor to dry beans in 24–36 h. It is quite common for these systems to be homemade and poorly constructed, which can result in smoke or fumes from the burner coming into contact with the beans. The resultant smoky taint is very persistent and can carry through to cocoa products or chocolate.

Once adequately dried, the beans can be stored in sacks or silos for extended periods, however the normal method of shipment in jute sacks gives little protection against atmospheric moisture and the moisture content can rise to 10–12%. Mould development may occur on beans whose moisture content has risen above 8–9%. The beans must also be kept free from pests.

PROCESSING OF DRIED BEANS

Several separate processes are required to produce the range of cocoa products, i.e. liquor, cocoa butter, cocoa powder. They are bean cleaning, roasting, winnowing, grinding and alkalization.

Cleaning

The first stage in bean processing is cleaning, where the beans are separated from contaminating materials such as stones, string, twigs and so on. Typically, vibratory mechanisms, sieves, air separators and magnets may be used. At the same time broken and undersized beans are removed.

Roasting

Most modern cocoa factories use continuous roasting ovens of the revolving drum type. Roasting conditions vary according to the flavour required, but the temperature range is typically 100–140°C, for times between 4–6 min. Either whole beans or the separated nib (see WINNOWING, below) can be roasted but separation of the nib is easier if roasting is done first. Continuous roasting requires beans of a uniform size and moisture content to achieve consistent results and small beans may have to be processed separately.

Batch roasters are still used where production volumes are small or when there are special flavour requirements for blending purposes. Roasting reduces the moisture content of the beans and also develops cocoa flavour with the loss of some volatiles.

Winnowing

Cocoa beans consist of an outer shell, unsuitable for human consumption and the inner edible 'nib'. Separation of the shell from nib is required. In most modern processes this stage is carried out after roasting the whole cleaned beans but it may be done before roasting.

Cooled roasted beans are passed between rollers to crack them and then passed through the winnowing machine which is essentially an air classifier. The lighter shell fragments are blown away from the nib by air jets in a four- or five-stage separation process. In practice, it is very difficult to separate fine nib particles from shell completely and a small residual shell content is allowed in the nib. The tolerance is 5% (dry fat-free basis) in the UK and 1.75% (whole alkali-free basis) in the United States. Shell particles are difficult to grind and high shell levels significantly increase mechanical wear of liquor mills.

The separated shell is of low commercial value, since the theobromine content limits its suitability as an animal feedstock. The fat content of the shell is very low (2%) and the fat is chemically different to the cocoa butter in nib.

The nib fraction at this stage contains about 55% fat, contained within the cellular structure of the cotyledon.

Grinding

Grinding is normally a multistage process. The first stage liquefies the solid nib particles by rupturing the cell walls. The resultant cocoa mass or cocoa liquor is a suspension of cell wall fragments in cocoa butter, with a large mean particle size.

Further particle size reduction is carried out by grinding mills which are basically rotating stone or steel discs which reduce particle size by imparting a high shear to the cell wall fragments. The mills are generally water cooled since further heating would undesirably modify the flavour developed during the roasting process.

For use in chocolate manufacture, the particle size in the liquor is not very important at this stage since it will be further reduced during chocolate manufacture. However, for applications such as the production of chocolate beverages finely ground products are required to facilitate dispersion and to minimize rapid sedimentation.

Alkalization

Most cocoa powder produced has been subjected to alkali treatment at some stage in processing. Alkalization (also known as dutching or solubilization) was originally used as a means of increasing the dispersibility in milk or water of cocoa powder when used in beverages. A more important aspect now is that of changing the colour of cocoa for bakery, confectionery or other applications. The technique has advanced to the stage where cocoa powders ranging in colour from an orange/ginger to a very dark, almost black product are possible. The precise details of alkalization treatment are not generally made available, but there is a large number of possible permutations. For instance it is possible to treat either liquor or nib, from roasted or raw bean. Various alkalis may be used, with various acids to neutralize them, as dealt with below under PERMITTED BASES, ACIDS AND EMULSIFIERS. Additionally, the dosage rates, reaction times and temperatures can all be varied to achieve the desired effect.

The physical and chemical reactions involved are complex and poorly understood. They include neutralization of natural acids, hydrolysis of proteins and cell wall polysaccharides and modification of tannins and polyphenolic compounds. It is unlikely, though, that any but the most extreme conditions affect the fat phase.

The term 'pH' is normally used in association with alkalized cocoa powders, but can be misleading. It refers to the pH of water after a quantity, normally 10% by weight, of cocoa powder or liquor has been dispersed in it. It has little value except as an indicator that alkalization has been used. Natural (i.e. non-alkalized) cocoa powder measured in this way has a pH of about 5.5. At pH values much above 7 there may be deleterious effects upon flavour.

Permitted bases, acids and emulsifiers

Cocoa and fat-reduced cocoa may contain basic or acid residues as a result of the alkalizing process. Only permitted acids and bases may be used for this purpose.

In the UK and the EU the permitted bases are the carbonates, bicarbonates and hydroxides of sodium, potassium, ammonium, magnesium or calcium. Permitted acids are citric, tartaric and orthophosphoric, either singly or in any combination. Lecithin and ammonium phosphatides are allowed, either singly or in combination. Table 10.12 shows the permitted levels. Additionally, lecithin and ammonium phosphatides are allowed as emulsifying agents in LECITHINATED COCOA.

Table 10.12 Levels of permitted bases and emulsifiers in the EU

Reserved name	Level of base or emulsifier
1. *Cocoa* Cocoa powder Fat-reduced cocoa powder (In each case in relation to a cocoa product other than in the form of an instant preparation)	Permitted base not exceeding 5%. Lecithins and/or ammonium phosphatides not exceeding 1%
2. *Cocoa* Cocoa powder Fat-reduced cocoa Fat-reduced cocoa powder (In each case in relation to a cocoa product in the form of an instant preparation)	Permitted base not exceeding 5%. Lecithin and/or ammonium phosphatides not exceeding 5%
3. *Drinking chocolate* Fat-reduced drinking chocolate Sweetened cocoa Sweetened cocoa powder Sweetened fat-reduced cocoa Sweetened fat-reduced cocoa powder (In each case in relation to a cocoa product other than in the form of an instant preparation)	Lecithins and/or ammonium phosphatides not exceeding 1%
4. *Drinking chocolate* Fat-reduced drinking chocolate Sweetened cocoa Sweetened cocoa powder Sweetened fat-reduced cocoa Sweetened fat-reduced cocoa powder (In each case in relation to a cocoa product in the form of an instant preparation)	Lecithins and/or ammonium phosphatides not exceeding 5%

COCOA POWDER

Types of cocoa powder

Cocoa powder is produced from liquor by removing some of the fat. This is carried out by hydraulic pressing of the liquor, the quantity of fat removed being carefully controlled. A residual fat content of 22–23% is readily achievable and it is possible to reduce this further to 8%, the legal minimum for low fat cocoa, but the time required is significantly longer. Most low fat cocoas have fat contents of 10–11% which represent a good compromise between pressing time and fat content. The fat content of the powder is controlled by weighing the expressed butter and continuing the pressing until the target fat weight is achieved. The cocoa butter passes forward for chocolate manufacture: the residual press cake, which is a hard disc, must be pulverized before use.

The subsequent pulverizing of the press cake is complicated by the residual cocoa butter. It is important that this is kept solid, and thus the temperature must at all times be kept below 30°C. Liquid fat not only affects the pulverizing process by cushioning the impact of the mill, but also can give problems of colour and flow properties in subsequent handling operations. Various special hammer or impact mills are available for this application, fitted with air temperature control equipment. Generally, the aim of this pulverizing stage is to separate the particles so as to give a similar size distribution to that of the original liquor.

If powder with a fat content below 10% is required (e.g. in some compound coating applications) then a solvent extraction process must be used. Defatted cocoa produced by this means has a poorer flavour than cocoa powders produced by pressing. In the USA three types of cocoa powder are recognized according to their fat content:

USA type	Description
Breakfast cocoa	High fat cocoa prepared by pulverizing the residual material remaining after part of the cocoa butter has been removed from ground cocoa nibs, containing not less than 22% of cocoa butter determined by the method specified.
Cocoa	Medium fat cocoa conforms to the definition of breakfast cocoa except that it contains less than 22% but more than 10% cocoa butter.
Low fat cocoa	Low fat cocoa conforms to the definition of breakfast cocoa except that it contains less than 10% cocoa butter.

In the EU, including the UK, reserved names define the different cocoas and cocoa beverages, with the fat and cocoa contents required:

EU reserved name	Description
Cocoa or cocoa powder	The finely ground particles obtained from the mechanical disintegration of cocoa press cake containing not more than 9% water and not less than 20% cocoa butter calculated on the dry matter.
Fat-reduced cocoa or Fat-reduced cocoa powder	The finely ground particles obtained from the mechanical disintegration of cocoa press cake or fat-reduced cocoa press cake containing not more than 9% water and not less than 8% cocoa butter calculated on the dry matter.
Sweetened cocoa or Sweetened cocoa powder	A mixture of cocoa and sucrose containing not less than 32% cocoa.
Sweetened fat-reduced cocoa or cocoa powder	A mixture of fat-reduced cocoa and sucrose containing not less than 32% fat reduced cocoa.
Drinking chocolate	A mixture of cocoa and sucrose containing not less than 25% cocoa.
Fat-reduced drinking chocolate	A mixture of fat-reduced cocoa and sucrose containing not less than 25% fat-reduced cocoa.

These reserved names are currently under review by Codex. In other countries, legislation is generally more relaxed than in the USA or the EU. For instance India has legislation that defines minimum fat contents as 20% for cocoa powder and 10% for low-fat cocoa powder. Singapore and Malaysia only require that the powder originates from cocoa and fat contents are not specified. Pakistan, China and the Phillipines have no standards defined. Australia also does not define fat contents, but stipulates that the products must be 'powder products prepared from mass whether or not deprived of a portion of its fat'. The Japanese regulations have recently changed and are not currently available.

Cocoa powder storage

Cocoa powder contains a significant proportion of cocoa butter. It is important in storage to keep this fat stable and solid and for this reason a maximum storage temperature of 20°C is recommended. If this is exceeded for any length of time, then fat migration can occur as the softened fat fractionates and recrystallizes. This will give a greyness in the powder due to the growth of fat crystals and can also cause caking of the powder within the container.

The presence of the fat also means that cocoa is particularly sensitive to taints from strong smelling products. Phenolic components are notorious for imparting taints, even at parts per billion level, to fat-based materials.

Packaging for cocoa powder must be moisture proof and sealed to prevent insect or rodent penetration. Recommendations for storage are as follows:

(i)   Temperature less than 20°C
(ii) Relative humidity below 50%
(iii) No direct sunlight
(iv) Avoid areas with strong smelling products
(v)   Avoid stacking bags or sacks more than 20 high.

COCOA PRODUCTS

Mocca

This is a mixture of coffee and cocoa which can be used either as an ingredient in bakery or confectionery products or as the basis of a flavoured milk drink. ('Mocha' is Arabian or similar coffee.)

Lecithinated cocoa

The term 'lecithin' in the context of cocoa and chocolate does not usually refer to the specific phospholipid, phosphatidyl choline, but to a phospholipid mixture extracted from crude soya bean oil at the degumming stage of the refining process. It is a mixture of several phospholipids including phosphatidyl choline, phosphatidyl serine and inositol, together with some residual oil.

Ammonium phosphatides are manufactured from hydrogenated rape seed oil and are the ammonium salts of phospholipids. This material performs a similar function to soya lecithin, but has little or no flavour and is preferred in some applications for this reason. Both of these materials are used widely in the food industry as emulsifiers. They act by lowering the surface tension at the particle/liquid interface.

Legislation permits the use of either lecithin or ammonium phosphatides in cocoa powders and drinking chocolate at a maximum level of 5% to improve the cold wetting performance. Lecithin can be applied as a surface coating to cocoa powder to improve the cold wetting characteristics using a spraying technique in a tumbler mixer.

Care is required in choosing a lecithin of good quality, particularly with respect to flavour.

Drinking chocolate

Drinking chocolate is a mixture of sugar with at least 25% cocoa powder. The cocoa powder must contain a minimum of 20% fat. It is usual for alkalized cocoa powders to be used and most products are lecithinated and instantized. It is increasingly common for other ingredients to be included, e.g. colouring, stabilizers and flow enhancers. These factors greatly facilitate the preparation of the drink and account for the present popularity of drinking chocolate compared with cocoa.

In the UK the compositional standards are governed by the Cocoa and Chocolate Products Regulations (1976).

Fat-reduced drinking chocolate

This is a product showing increased popularity as consumers become more conscious of a need to reduce fat intake. It is similar to drinking chocolate, except that at least 25% of fat-reduced cocoa must be used. It is usually instantized for easy preparation.

Sweetened cocoa

This is a mixture of cocoa and sucrose containing at least 32% cocoa. It is a variant of drinking chocolate.

Fat-reduced sweetened cocoa

As for fat reduced drinking chocolate, but with at least 32% fat reduced cocoa.

INSTANTIZING

Cocoa powder in common with some other powders is difficult to disperse in liquid. The poor wettability of cocoa causes lumps to form, which are externally wet agglomerates of cocoa powder with dry centres. Because some air is trapped, these lumps float on the surface of the liquid. A product such as drinking chocolate can be instantized to overcome this problem. A variety of processes exist, but they all aim to produce porous agglomerates of particles which, by capillary action, are more readily wetted by milk or milk–water mixtures.

Instantizing can be carried out by blowing the powdered mixture into a chamber together with steam or high humidity air. This causes surface wetting of the sugar particles which stick together to form agglomerates. These agglomerates are then dried in a fluidized bed drier to form the familiar granular material. The material from the drier is sieved and any fines are recycled through the process.

The porous structure of instantized granules allows rapid solution because air is no longer trapped within lumps of the powder to prevent ingress of liquid. Instantized granules have better flow properties than their powder equivalents and additionally do not cake on prolonged storage.

CHOCOLATE DRINKS

These products are essentially drinking chocolate plus dried milk, to be made up into drinks with water. Originally, the main application was in vending machines, but there is increasing popularity for catering and home use. Formulations are becoming increasingly sophisticated with a wide variety of products on the market. These include products that contain a proportion of chocolate, branded products and chocolate drinks flavoured with for instance orange, mint or toffee. Additionally the range of packings has increased with single serving pouches becoming available.

Recipe development has seen the inclusion of such ingredients as aspartame, maltodextrin, dried glucose syrup, whey, caseinate and xanthan gum as manufacturers optimize low calorie products formulated to be made up with water.

Chocolate milk

Chocolate flavoured milk is a product that is increasing in popularity. Originally simply a suspension of cocoa powder in milk, the recipes have become increasingly sophisticated by the addition of other flavourings and modified starches or alginates, the latter to provide a 'thicker' drink and to reduce the sedimentation of the cocoa particles.

MALTED COCOA BEVERAGES

There is a wide range of branded cocoa flavoured malted beverages available on the market. They typically contain cocoa powder, malt extract, sucrose, glucose and egg, plus emulsifiers, colourings and flavourings. Sometimes vitamin and mineral additions are made in order to increase nutritional value.

These products are manufactured by mixing the ingredients together to form a 'magma' of around 85% solids content. This material is then dried down to a final product of around 1% moisture. The final drying, usually performed under vacuum to prevent damaging the malt flavour, can be carried out using either batch or continuous ovens. The dried cake is then granulated prior to packaging. The product is hygroscopic and therefore care has to be taken with the choice of packaging materials. As with chocolate drinks, new products have been launched offering low calorie or low fat alternatives to the traditional formulations.

CHOCOLATE SYRUP

For many bakery, ice-cream and cold drink applications, a fluid chocolate flavoured material is required. Chocolate syrup fulfils this need and consists of a sugar syrup with cocoa powder dispersed in it. The sugar syrup is a mixture of invert sugar, glucose syrup and sucrose, formulated to minimize crystallization on storage. Some preparations also contain milk soilds.

Chocolate syrups are prepared by heating and mixing the ingredients together ensuring complete dissolution of the sucrose. Subsequent cooling must be such that the fat globules do not coalesce and the sugar does not recrystallize. Either of these events will lead to a thickening of the syrup with time, which limits its shelf life.

Chocolate syrups can be packaged in sterile cans for large users, in which case syrups of lower soluble solids content, typically around 70%, can safely be used. For retail packs, a high soluble solids syrup, around 80%, must be used to ensure microbiological stability after the pack is opened.

CHOCOLATE SPREADS

There are two distinct types of chocolate spread available on the market. One is similar to chocolate syrups, i.e. a sugar syrup system of sucrose, invert sugar and glucose syrup with fat-reduced cocoa powder suspended in it. Lately, fat-based spreads have appeared on the market, frequently containing hazelnut paste. These spreads are more expensive than the syrup types and particular care needs to be taken with the vegetable fat used to avoid bloom or graining as a result of incompatibility with the cocoa butter.

PACKAGING

Both chocolate and malted drinks need to be protected from taint and moisture and are mainly packed in laminated containers for retail sale, with tubs or single portion pouches being the most popular. These have replaced the traditional glass jars for cost reasons.

QUALITY ASPECTS

Nutrition

Cocoa can make a significant contribution to the nutritional value of foodstuffs. A major factor is the fat content of the cocoa powder used. In the case of drinks, the strength of mix used and whether the drink is made up with water, milk or a mixture, are also important. Tables 10.13 and 10.14 show nutritional data for various cocoas and beverages as purchased and as made up.

Fat content

Cocoa nib and liquor have fat contents of about 55%. Cocoa powder can have a fat content from 0.2% to 22% depending on processing.

Fat-reduced cocoa powder (10–11% fat) is mainly used in bakery and confectionery applications. Since the value of powder lies almost entirely in the fat, high fat cocoa powders are more expensive. Fat-reduced cocoa powder represents a useful compromise therefore between cost and flavour, as well as having useful dietary or nutritional connotations.

Fat-free cocoa powders must be used in compound chocolate formulations in which the principal fat, e.g. hardened palm kernel oil (HPKO), is highly incompatible with cocoa butter. Because they are derived via a solvent extraction process, the flavour of these powders is inferior to powders originating via pressing.

Flavour

Cocoa flavour

Flavour is an almost impossible parameter to specify in any quantitative way. The usual ways of overcoming this difficulty are for a supplier either to match an existing standard or alternatively to supply a range of products for selection by the customer. This would normally be carried out after small scale preparation of the final product.

Table 10.13 Composition of cocoa powders (as purchased)

		Cocoa powder (alkalized)	Fat-reduced cocoa (alkalized)
Total fat	g/100 g	19.7	9.6
Nitrogen	"	3.0	3.4
Protein	"	19.0	21.3
Dietary fibre	"	13.0	15.6
Carbohydrate	"	38.0	41.6
Moisture	"	3.4	4.8
Ash	"	7.1	7.2
Energy	kcal/100 g	395	352
Sodium	mg/100g	1059	863
Potassium	"	1236	1389
Calcium	"	120	136
Iron	"	7	–

Table 10.14 Nutritional data for drinks (as prepared)

	Typical values per serving[a]			
	Protein (g)	Fat (g)	Carbohydrate (g)	Calories
Cocoa serving made with 200 g whole milk	7.4	8.9	9.9	149
Drinking chocolate serving made with 200 g whole milk	7.5	9.1	22.5	202
Malted drink serving made with 200 g whole milk	7.2	8.3	18.9	178
Chocolate drink serving made with water	2.8	3.0	18.3	111
Malted drink (light) serving made with water	5.5	0.9	20.3	107
Chocolate drink (light) serving made with water	1.8	1.5	4.9	40

[a] Manufacturer's serving weight used for calculation

Many factors affect cocoa flavour: cocoa species, country of origin, fermentation conditions, drying conditions, bean roasting conditions, fat content, alkalization and particle size. In the UK, Ghanaian or Nigerian bean is regarded as the flavour standard in both the confectionery and cocoa beverage industries. However, different standards are preferred in other parts of the world, e.g. South American bean in USA and Canada; Malaysian or Papua New Guinea bean in Australia and New Zealand.

Considerable research effort has been put into understanding chocolate flavour, and those parts of cocoa processing that contribute most to it, i.e. fermentation and roasting. There is still much to be learnt.

Flavour defects

The most common flavour defect is caused by mould. Mould damage due to one or several mould species can occur within the pod prior to harvesting or at any later stage where excess moisture is present. Such instances can occur where bean drying is too slow or not completed or when poor storage conditions exist, particularly in ships' holds. Apart from the flavour problem, moulds can also produce mycotoxins (though they have rarely been detected in cocoa) and can also increase the free fatty acid content of the cocoa butter to unacceptable levels. There is no simple correlation between the number of mouldy beans detected by a cut test (see COMMERCIAL QUALITY ASSESSMENT) and the level of off-flavour.

Acidity is a flavour defect associated particularly but not exclusively with box fermentation as practised in Malaysia. Excess lactic and acetic acid remain in the beans and give rise to the defect. Research into how to reduce this acidity is being carried out, both in terms of fermentation techniques and bean processing. Some Malaysian estates offer acid reduced beans, which have been pressed during fermentation to remove more of the acid-containing 'sweatings'.

Underfermented beans have a characteristic slaty colour and are detectable by a cut test. These beans have an astringent, bitter and unpleasant flavour.

Smoky flavours (sometimes referred to as 'hammy') are usually due to smoke contamination during drying. Brazilian cocoa is particularly prone to this defect due to the design of the drier in most common use in that country.

Lack of chocolate flavour is a general defect that can accompany other defects. Underfermenting reduces chocolate flavour as does excess acidity, so reducing or eliminating the immediate cause of such flavour defects does not necessarily upgrade the overall flavour quality of the beans.

Some of these defects are impossible to rectify. The only real remedy with most flavour defective beans is to blend with satisfactory beans. This must be carried out on the basis of tasting tests.

Colour

As outlined under ALKALIZATION, a wide range of colours is available. As with flavour, it is only feasible to assess colour in the finished product, since the colour of the powder itself is modified by several factors such as fat content and particle size. It is also possible after prolonged storage for the colour to change due to recrystallization in the fat phase, a process analogous to bloom in chocolate.

Particle size

The particle size required generally depends on the application. Where cocoa powders are used in drinks, the particle size needs to be very fine in order to reduce the proportion of sedimented particles at the bottom of the drink. A maximum size of around 75 μm is required. In other applications, fineness is less important, though very coarse particles may give a speckled appearance.

The particle size achievable in the powder is related to the size in the original liquor. The mills used for disintegrating press cake only serve to separate particles from one another and it is necessary to grind the liquor to the required particle size at that stage in the processing.

The best method of measuring particle size is dry sieving; wet sieving or instrumental methods involving solvents give distorted results. This is because organic solvents separate clumps of particles as the residual fat is dissolved and the use of water swells the cocoa particles.

Microbiological standards

Dried fermented cocoa beans have a wide microbiological flora due to the nature of the processes used. Mixed farming practices in some countries may cause the beans to come into contact with animals such as goats and chickens and the possibility of contamination with pathogens must be allowed for.

The majority of the micro-organisms are on the outside of the beans, so the roasting and winnowing processes result in significant reductions in the total count. However it remains vital for cocoa products to be sterilized or pasteurized at some stage in the processing. This is especially important where cocoa powders are likely to come into contact with other foodstuffs and where subsequent processing may not be sufficient to inactivate the micro-organisms.

Most cocoa processors now pasteurize all liquor, using for example scraped-surface heat exchangers. Because of the low water content, temperatures between 130 and 140°C for 2–3 min are required to achieve the necessary reduction in microbiological count.

A typical microbiological specification for cocoa powder is:

Aerobic plate count at 30°C	10 000/g maximum
Moulds	50/g maximum
Yeasts	50/g maximum
Escherichia coli	Absent in 1 g
Salmonella and *Shigella*	Absent in 25 g

COMMERCIAL QUALITY ASSESSMENT

Four areas are of interest here: microbiology, flavour, cut tests and financial value.

(i) Microbiological standards are discussed above.

(ii) The flavour of individual parcels of beans is best evaluated by tasting panels. This can either be done by tasting chocolate prepared from a small sample of the beans, or by tasting the liquor. Either method requires the use of a highly trained, expert panel. There is no satisfactory instrumental alternative to this method at present.

(iii) *The cut test* – a standard method exists where 100 beans are cut and examined. Poor fermentation of the beans can be detected: well-fermented beans have cotyledons which are fragmented in appearance and brown in colour, partially fermented beans have a slaty or purple appearance and ideally should be absent. The cut test also allows visual assessment for mould damage, insect damage or signs of germination. The cut test is widely used for quality monitoring but it does not give any indication of flavour quality. Beans that give a good cut test result may have poor flavour.

(iv) In practice, 'value' means fat content, since the value of cocoa beans lies predominantly in the cocoa butter. Other factors which enter into the calculation are moisture content and shell content, which is related to bean size.

FURTHER READING

ASIC (1989) *Treizième Colloque International sur le Café*, Paipa, Association Internationale du Café, Paris.
ASIC (1991) *Quatorzième Colloque International sur le Café*, San Francisco, Association Internationale du Café, Paris.
ASIC (1993) *Quinzième Colloque International sur le Café*, Montpellier, Association Internationale du Café, Paris.
ASIC (1995) *Seizième Colloque International sur le Café*, Kyoto, Association Internationale du Café, Paris.
ASIC (1997) *Dixseptième Colloque International sur le Café*, Nairobi, Association Internationale du Café, Paris.
Beckett, S.T. (ed.) (1988) *Industrial Chocolate Manufacture and Use*, Blackie, Glasgow and London.
Clarke, R.J. and Macrae, R. (eds) (1985) *Coffee, 1. Chemistry*, Elsevier Applied Science, London.
Clarke, R.J. and Macrae, R. (eds) (1987) *Coffee, 2. Technology*, Elsevier Applied Science, London.
Clarke, R.J. and Macrae, R. (eds) (1988) *Coffee, 3. Physiology*, Elsevier Applied Science, London.
Clarke, R.J. and Macrae, R. (eds) (1988) *Coffee, 4. Agronomy*, Elsevier Applied Science, London.
Clarke, R.J. and Macrae, R. (eds) (1988) *Coffee, 5. Related Beverages*, Elsevier Applied Science, London.
Clarke, R.J. and Macrae, R. (eds) (1988) *Coffee, 6. Commercial and Technico-Legal Aspects*, Elsevier Applied Science, London.
Clifford, J. and Clifford, M.N. (eds) (1997) *Chemical and Biological Properties of Tea Infusions*, Deutscher Medizinischer Informationsdienst, Frankfurt am Main, Germany.
Clifford M.N. and Willson K.C. (1985) *Coffee: Botany, Biochemistry and Production of Beans and Beverage*, Chapman & Hall, London.
Debry G. (ed.) *Coffee and Health*, John Libbey Eurotext, Paris.
Graham, H.N. (1992) *The Polyphenols of Tea Biochemistry and Significance – a review*. In XVIe Journées Internationales Groupe Polyphénols and The Royal Society of Chemistry, Lisbon. 2 Anon DTA, Lisbon, Portugal, pp. 32–43.
IARC (1991) *Coffee, Tea, Maté, Methylxanthines and Glyoxal*, Volume 51, IARC monographs on the evaluation of carcinogenic risks to humans, IARC, Lyons, France.
Minifie, B.W. (1980) *Chocolate, Cocoa and Confectionery: Science and Technology*, 2nd edn, AVI, Westport.
Willson, K.C. and Clifford, M.N. (eds) (1992) *Tea: Cultivation to Consumption*, Chapman & Hall, London, UK, 792 pp.
Wood, G.A.R. and Lass, R.A. (1985) *Cocoa*, 4th edn, Longman, Harlow.
Yamanishi T. (ed.) (1995) Tea. *Food Rev. Internat.*, **11**, 371–544.

11   Sugar and Chocolate Confectionery

INTRODUCTION

In the three years which have elapsed since the last edition of the Food Industries Manual was published, little change has occurred in the confectionery industry and there is little evidence that new products are of much importance to the manufacturer or consumer. Sugar replacement and the so-called 'healthier' products have had little impact on the main market in the UK and traditional products, most of which have been made for 50 years, still dominate sales. In the chocolate sector shell moulding seems to be used for many sweets which used to be enrobed and in chocolate assortments many more centres are now truffles or truffle variants. It remains to be seen if the current problem with BSE in the UK will affect sales of products containing gelatin, or if the solution of that problem will result in difficulties with the supply of some milk products. However neither of these seems particularly likely, nor can one predict any other significant factor likely to affect raw material supplies, so we may assume there will be little overall change in that situation.

SUGAR CONFECTIONERY

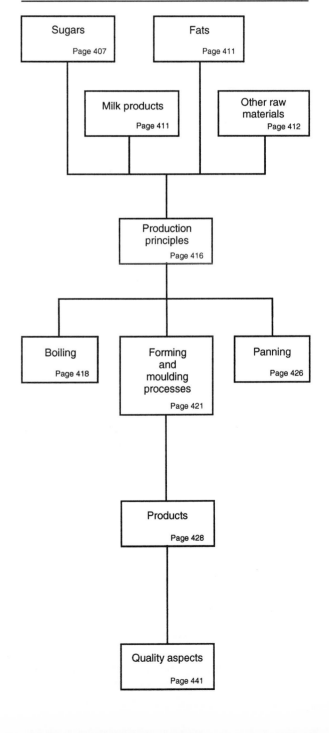

RAW MATERIALS AND INGREDIENTS

Sugars: Sucrose and derivatives

Cane or beet white sugar

White sugar extracted from sugar cane or sugar beet is very nearly pure sucrose with traces of mineral matter and water. It is probably the purest of all ingredients available to the food manufacturer. The occasional differences in behaviour relate to the grain size and to the type of natural mineral matter entrapped in the crystals. These mineral components produce a buffering effect in solution and prevent the rapid inversion of sucrose to invert sugar in the presence of an acidic ingredient. This inhibitory effect caused the early confectioner to classify sugar as 'hard' or 'weak' according to the ease by which it was inverted by cream of tartar.

At 20°C (68°F) sucrose has a solubility in water of 67.1%. The addition of other sugars normally depresses the solubility of sucrose, but increases the overall concentration of sugar in solution.

Liquid sugar

The relative merits of purchasing liquid sugars or granulated sugar in bags or bulk have been discussed for many years, but mainly depend on the scale of operation. Bulk syrup installations and distribution systems within factories are much cheaper to install than bulk granulated sugar systems. The costs of liquid sugars are higher, however, and in most cases more water has to be evaporated during processing. Many liquid sugars are below 75% solids and can ferment, so hygiene standards have to be maintained at a high level. Blends of liquid sugar and glucose syrups are available which reduce this problem and due to the scale of preparation are constant in composition. These can be justified if sufficient demand exists for a particular ratio of sugars. Most large sugar users tend to purchase bulk granulated sugar and process on site.

Brown sugar

Brown sugars used to be produced from lower grade syrups in the cane sugar refineries. Many are now produced by blending granulated sugar with refinery syrups or molasses; although very similar they are not exactly the same as the original product.

Because ordinary brown sugars cannot be handled in bulk, many are now produced in microcrystalline aggregates which are free flowing. Some manufacturers, however, use granulated sugar and add lower grade sugar syrups themselves to simulate the flavour of brown sugar. Some brown sugars are artificially coloured and have no characteristic flavour. These are not used in confectionery processing.

Icing sugar

Icing sugar or milled sugar is difficult to handle in bulk and large users mill granulated sugar on site. Sugar dust can give rise to dust explosions and milling procedures must be rigidly controlled to prevent dust buildup. Mill rooms should have blast venting panels and ideally a sugar milling installation should not require operatives in the mill room during operation.

Ready milled sugars can be purchased with or without anticaking agents. In general, process sugar does not contain these agents, but they are useful when the sugar is to be used as a dusting medium, as for boiled sweets or Turkish delight where a proportion of starch is frequently added.

Golden syrup

Golden syrup is a standardized product with a soluble solids content of 83%, a sucrose content of 31–33% and a reducing sugars content of 47–60%. Its characteristic flavour, although mild, has been utilized to produce subtle modification to the flavour of a number of confectionery products.

Refinery syrups

Refinery syrups are darker than golden syrup and have more flavour. They are frequently used as additions to granulated sugar to simulate brown sugars.

Molasses and dark treacle

Molasses and dark treacle are the lowest grade of refinery products and are used in confectionery specifically for their flavour.

Invert sugar

Any acid and certain enzymes will split sucrose (commercial cane or beet white sugar) into two simpler sugars – dextrose or D-glucose and fructose or D-laevulose. Dextrose can be crystallized from the solution, but fructose is highly soluble and only crystallizes under extreme conditions. The mixture of the two sugars is more soluble than sucrose alone and has a greater tendency to attract moisture from the atmosphere. A solution of the two sugars formed by 'inverting' sucrose is known as invert sugar and contains nearly equal amounts of the two simpler sugars. Natural inversion of sucrose takes place when sucrose solutions are heated in even mildly acidic conditions. Commercially, invert sugar is prepared as a syrup of about 70% soluble solids concentration. Invert sugar can by produced by holding a 65% sucrose solution containing 0.25% hydrochloric acid at 50°C (122°F) for one hour. Sodium bicarbonate should then be added to neutralize the acid. Alternatively, a suitable enzyme solution can be added to the sucrose solution and the mixture held overnight under moderate heat.

Invert sugar is an excellent humectant for sugar confectionery, controlling the loss of water and improving the tenderness of the product. It depresses the solubility of sucrose in a mixed solution, but increases the total amount of sugars that can be held in a syrup. The main problem encountered when using invert sugar is its ability to attract moisture from the air on to the surface of the confection.

Invertase

Invertase is used for the conversion of sucrose into invert sugar, either in a syrup or in the stored confection. Satisfactory dosage rates are 0.5% for a syrup and 0.1% for a flowing cream or liqueur centre. Invertase is heat sensitive and inactivation takes place progressively at temperatures over 60°C (140°F). It is essential to add the enzyme at as low a temperature as practicable. Enzymic activity is at its maximum between pH 4.2 to 4.5. The enzyme can be purchased either as a powder or as a solution. It should be stored in a cool place and stocks should be held at a minimum compatible with production requirements.

Glucose syrups

Glucose syrup

'Confectioners' glucose' is a name still used occasionally to describe glucose syrup, the abbreviation 'Glucose' is in common use in the industry.

The grading of glucose syrups is based on the solids content as indicated by the Baumé reading and on the reducing sugar content of the solid matter calculated as dextrose (termed DE value or dextrose equivalent). Many grades of glucose syrup are available to the sugar confectioner,

including 35 DE (low DE), 42 DE (regular), 55 DE, 63 DE (high DE) and high maltose glucose syrup. The average total soluble solids content for glucose syrup ranges between 78 and 83%. Glucose syrups contain a number of different sugars all built up from dextrose units, with dextrose and maltose predominating. Increasing quantities of the higher sugars create higher viscosity syrups. The optical rotation of a glucose syrup varies with DE value.

Colour formation on boiling is normally caused by the presence of small traces of metals and nitrogenous materials, the latter also being sometimes responsible for foaming during processing. Ion exchange treatment removes the colour forming components but lowers the buffering power of the syrup. The pH of glucose syrup ranges between 4.8 and 5.2. For products which require low viscosity, tender consistency and good whipping qualities, e.g. marshmallow. high DE glucose syrups are recommended. Low DE glucose syrups give a tougher texture and a resistance to moisture gain and for these reasons they are used in rock manufacture. Regular grade glucose syrups (42 DE) are used for general manufacturing purposes.

High maltose glucose syrup

High maltose glucose syrups up to 52 DE were originally made from starch by a combined process using hydrochloric acid followed by a maltogenic enzyme. In the processes now most commonly used bacterial amylases are employed to replace the acid hydrolysis stage and the subsequent treatment with maltogenic enzymes produces syrups containing up to 75% maltose with very low dextrose contents.

High fructose glucose syrups

High fructose glucose syrups (HFGS) are produced by isomerization of dextrose to fructose by enzymes. Confectionery products made with these syrups have a high resistance to crystallization and have improved stability at high temperatures. The lowest and possibly most widely used grade contains 42% fructose calculated on a total solids basis. Other grades contain from 55 to 90% fructose. The lowest grade also contains some 50% dextrose, the dextrose concentration falling with increasing concentration of fructose. All grades contain small proportions of the higher saccharides. The base material is corn (maize) starch which is subjected to a series of enzymic treatments to develop a high concentration of fructose.

Confections made with HFGS are sweeter than those made with conventional glucose syrups; the actual sweetness ratio is affected by pH, presence of other flavourings and concentration.

Hydrogenated glucose syrups

Hydrogenated glucose syrups are prepared by the catalytic hydrogenation of a base material which in turn is produced by the enzymatic hydrolysis of starch. High DE syrups give sorbitol syrups on hydrogenation and from these sorbitol may be crystallized. Similarly, high maltose syrups give maltitol syrups from which maltitol is crystallized. By controlling the enzyme reaction different compositions of hydrogenated syrup can be produced to suit different applications. For this reason syrups from different manufacturers are not necessarily interchangeable.

Hydrogenated syrups can perform a similar function to glucose syrups in conventional products as they can increase viscosity and retard crystallization of other sugar replacers. It is possible to produce a boiled sweet with these syrups alone, but in many cases the sweet is prone to cold flow. To combat this, it is desirable to have a very low moisture content in the sweet, which means boiling at high temperature of the order of 160°C; due to the heat stability of hydrogenated syrups discoloration is not a problem.

In common with other polyols the hydrogenated syrups are considered to provide 2.4 kcal g^{-1} (10.02 kJ g^{-1}) based on dry matter. They have low cariogenicity. They are laxative and daily intake should be restricted to a maximum of 80 g per day for adults and half this for children. The suppliers should be consulted concerning their suitability for consumption by diabetics.

Other sugars

Dextrose

Dextrose is the monosaccharide produced by the complete hydrolysis of starch and is of widespread occurrence in nature. It is commercially produced by acid or enzymic hydrolysis of maize or potato starch. It is available in the anhydrous form, but is usually supplied as the monohydrate.

Dextrose exists in two forms, α and β. The crystalline monohydrate is the α-form but when dissolved in water it gives an equilibrium mixture of α- and β-forms. The saturation solubility of this mixture is 51%. Dextrose monohydrate melts at 85°C.

Due to its negative heat of solution and rapid solubility, dextrose has a cooling effect in the mouth and is sometimes used for this purpose. Other uses in confectionery are to improve fluidity, increase Maillard browning and give a more tender texture to products such as Turkish delight and marshmallows. Dextrose should not be used to excess in these products as it can crystallize, forming large granular crystals.

Fructose

Fructose (fruit sugar) may be added to confectionery recipes to increase the sweetness, inhibit crystallization and to act as a humectant, although invert sugar is much more frequently used. In some countries fructose is regarded as an acceptable sweetener for use by diabetic sufferers. Confections which contain fructose are more likely to pick up water from the atmosphere than those containing other common sugars. A suitable dosage rate for fructose is 2–5%.

Lactose

Lactose is the natural sugar present in milk and milk products, and is sometimes known as milk sugar. It has a low solubility and a marked tendency to grain from solution. The use of lactose in confectionery recipes increases the colour formation, modifies the flavour and develops a softer, more chewy texture.

Lactose hydrolysates are prepared by the enzymic hydrolysis of whey lactose. The resulting products are sweeter than equivalent quantities of cane or beet sugar. The humectant properties of a lactose hydrolysate are superior to those arising from the use of invert sugar and there is a comparable reduction in viscosity. Confections made using lactose hydrolysates show an increased tendency to brown under raised temperature conditions.

Honey

The water content of most honeys ranges between 15–20% depending on the area of origin. When fresh, honey is a clear liquid ranging in colour from pale straw to brown. In time the dextrose will crystallize and the higher the concentration the greater this tendency becomes. Viscosity is not a good guide to the solids content as it is affected by the presence of proteins. Honey contains 30–40% fructose, 34–38% dextrose and 4–5% sucrose. The flavour of Turkish delight and nougat is enhanced by the incorporation of honey.

Other sweeteners

In sugar confectionery, the term 'sugar replacer' is often used for non-carbohydrate bulk sweeteners. For fuller information about them see Chapter 14.

Bulk sweeteners

These are mostly prepared from HYDROGENATED GLUCOSE SYRUPS For fuller detail of their preparation and properties see Chapter 14. All these could be used to make simulated confectionery products of one form or another but their use for this purpose has not been fully developed. Few commercial products have been manufactured in the UK apart from diabetic products. Acceptance has been greater in continental Europe and Scandinavia.

Like the hydrogenated syrups, they have laxative properties to a greater or lesser extent and daily intake needs to be controlled. However, maximum usage is not legally defined but given as *quantum satis*. All are claimed to be less cariogenic than the carbohydrates but not all are non-cariogenic. Xylitol is claimed to help in remineralization of decayed teeth. All are considered to have a calorific value of 2.4 kcal g^{-1} (10.02 kJ g^{-1}). All are more expensive than carbohydrates, some significantly so.

Properties considered to be critical when selecting a bulk sweetener include cariogenicity, sweetness equivalent, solubility and viscosity, hydroscopicity, the cooling effect of polyols, the laxative effect and cost. In most cases the best results are obtained by using mixtures of two or more.

Sorbitol: Sorbitol is used as a sugar replacer in diabetic chocolate. Due to its hygroscopic nature it can cause viscosity problems during processing. Sorbitol is also used as a humectant in conventional confectionery at levels between 5–10%. It is also claimed to modify the sugar crystal structure of fudge.

Mannitol: Mannitol is more expensive than sorbitol. It has about 60% of the sweetening property of sucrose and in confectionery it is usually used with sorbitol.

Isomalt or palatinit: Isomalt can be used to make a boiled sweet without the addition of other sugar replacers. It is however difficult to dissolve and when used for boiled sweets requires a high boiling temperature and produces a very fluid product which requires additional cooling facilities to achieve a consistency suitable for further processing on conventional equipment.

It has approximately half the sweetness intensity of sucrose. It appears to be non-cariogenic and to cause no significant changes in the blood glucose levels in humans. These properties indicate isomalt as a suitable sweetening agent for diabetic foods, low calorie products and 'tooth-friendly' confectionery.

Maltitol: Maltitol is available in crystalline form but is used largely in the form of a hydrogenated high maltose starch conversion syrup. It is sold by a number of manufacturers under a wide variety of trade names. Product compositions differ slightly between manufacturers.

Maltitol will crystallize, so a maltitol syrup of appropriate composition is analogous to a syrup containing sugar and glucose syrup and syrups of this type are commercially available. Because maltitol is relatively sweet, with a sweetening power of 90% that of sucrose, intense sweeteners may not be necessary.

Lactitol: Lactitol is available as a crystalline monohydrate or dihydrate and has a sweetening power of a about 40% that of sucrose.

Xylitol: Xylitol has a pronounced cooling effect when eaten or dissolved in water.

Intense sweeteners

The intense sweeteners, acesulfame K, aspartame, cyclamates, saccharin, thaumatin and neohesperidin, are so-called because they exibit sweetness many times that of sugar. Their use is more widespread in chewing gum than other confectionery. There seems to be little or no other application in conventional confectionery products, but if bulk sweeteners or polydextrose become established as ingredients, since they are mainly less sweet than sugar, it is possible that additional sweetening may be required.

The maximum quantities permitted in cocoa-based confectionery and in chewing gum are different and should be checked before use. Under US legislation only saccharin, acesulfame K, aspartame and thaumatin are permitted; thaumatin is permitted in chewing gum only.

Bulking agents

These are non-calorie or low calorie fillers often used in low calorie confectionery. They may be either derived from cellulose which is insoluble in water and non-calorific or polydextrose which is soluble, non-crystallizing, and has a calorie value one-quarter that of sugar. Functionally polydextrose is similar to glucose syrup. Now that the polyols have been assigned a calorie value of only

2.4 kcal g^{-1} (10.02 kJ g^{-1}) they too may be regarded as partial bulking agents.

Fats

Vegetable fats

Fats for confectionery use must be completely or almost completely melted at blood heat (about 37°C). If they have higher melting points they give an unpleasant greasy sensation in the mouth. On the other hand, low melting fats lack body and tend to make sweets oily, which can also result in bloom on chocolate covered products. Cocoa butter has almost ideal properties in this respect but because of its high cost its use is almost entirely restricted to chocolate products, see later.

Lauric fats, such as hardened palm kernel and coconut oils, used to be very popular due to their short melting range, but economic changes have resulted in the increasing use of fats derived by hydrogenation of soya, groundnut, rapeseed, palm and some less common oils, either alone or as mixtures. These fats have a wider melting range and are usually softer than the lauric fats. To achieve an adequate slip point, some higher melting fractions may be present, which on occasions are detectable in products. For further information see Chapter 8.

Butter

Butter is used in confectionery primarily for its flavour, although it also functions as any other fat. The melting point of butter fat lies between 21–23°C. Most confectionery fats have higher melting points and the mixture with butter gives a softer consistency.

Best results in toffees and caramels are obtained with strong flavoured butters and in the past it was not unknown for rancid butter to be used deliberately. Currently, enzyme-modified butter fats are available which serve the same purpose.

To try and reduce EU butter stocks, subsidized butter or butter oil is available to sugar confectionery manufacturers and usage has tended to increase.

Emulsifiers

The usual emulsifiers used in sugar confectionery are lecithin or glyceryl monostearate. Both are used to assist the emulsification of fat into product but whilst lecithin is generally used for boiled sweets such as butterscotch, glyceryl monostearate is preferred for toffees, nougats and chews as it provides better lubrication for cutting knives during processing.

Typical dosage rates of lecithin in caramel and toffee manufacture are 0.1–0.2% of the total batch weight. Glyceryl monostearate (GMS) is an excellent emulsifier which is highly suitable for use in caramel and toffee manufacture. In addition to the emulsifying properties, GMS has antifoam properties and it is claimed reduces sticking on cutting knives.

Milk products

The usage of milk products in the EU has been dramatically changed by the efforts to dispose of surplus butter and butter fat.

Powdered milk

Available as spray dried or roller dried product, the choice is governed by the characteristics required. Spray dried powder has a considerably larger surface area than roller dried material, which affects the texture of fat-based products. The flavour produced is affected by the degree of caramelization produced during drying which is higher with roller drying than with spray drying. Some milk powders are deliberately caramelized during processing. Milk powders are difficult to reconstitute into solution or condensed milk compositions without some grittiness being noticeable.

Sweetened condensed milk

This is available as the full cream product with 8% butter fat and added sugar, or as the skimmed product with a much lower fat content. Typical analyses are:

	Full cream (%)	Skimmed (%)
Fat	8.5	1.0
Sugar	42.0	46.0
Lactose	11.5	13.0
Protein	9.0	10.0
Water	27.0	28.0

The use of full cream condensed milk is currently uneconomic for products containing 4% or more of butter fat. Mixtures of skimmed condensed milk with subsidized butter or with butter oil are cheaper.

Skimmed condensed milk equivalents

These are made by concentrating milk by ultrafiltration. Because the milk protein is subjected to less heating than in the evaporative processes it retains more of its functional properties. The composition also tends to be more consistent than condensed milk.

Whey powder

Spray dried whey contains 9–13% protein and 60–75% lactose. Its use enhances browning. Delactosed whey has excellent foaming properties.

Sweetened condensed whey

Condensed or powdered whey has a considerable price advantage over the comparable condensed or powdered milks and can be used to advantage in the cheaper ranges of caramels and toffees. Products made with whey tend to be darker and have a noticeably different texture and taste. Ingredients based on whey give lower viscosities than whole milk and softer textured confections which in the case of toffees are subject to cold flow. It is advisable to use whey products as partial substitutes for milk rather than as complete replacement. Blends of this type are available commercially and do give toffees that resemble those from whole milk. There is however still a tendency to cold flow, which makes them more suitable as chocolate centres than plain toffees.

OTHER RAW MATERIALS

Chocolate

Both plain and milk chocolate and compounds are used as ingredients in some filled boiled sweets and enrobed products such as eclairs. Unsweetened chocolate may be used as a flavouring agent. The preparation and properties of chocolate and chocolate products are considered later in this chapter, under CHOCOLATE CONFECTIONERY.

Foaming agents

The most common foam producing agents used in sugar confectionery are egg albumen, hydrolysed milk protein, hydrolysed wheat protein, gelatin and soya protein.

Egg albumen

Once imported from China as crystal albumen, much egg white is now produced in Scandinavia and America. It is available in both crystal and spray dried forms. Egg white solutions are the traditional basis of many food foams and a large amount of research has been carried out on whipping properties and foam stability. Many of the conclusions, however, appear to relate only to specific applications and are not of general benefit. Egg albumen denatures at surfaces, forming a solid film and it is mainly this which imparts stability to the foam. When overwhipped, the films coagulate and the foam collapses. Heat will also coagulate egg albumen and this is said to be a factor in stability of some nougats.

Hydrolysed milk protein and hydrolysed wheat protein ('Hyfoama' products) do not denature and cannot be overwhipped. Their aerating power is thus rather higher than for egg albumen but the foam stability is appreciably lower so gelatin or agar are often added as stabilizers.

For confectionery use, frappés are frequently used. These are foams of egg white or other whipping agents with sugar, glucose syrup or invert sugar. These are much more stable than albumen foams, and if the dissolved solids levels are above 75%, can be kept for one to two days if necessary. High levels of hygiene must be maintained and the containers must be covered, but ultimately separation will occur.

Gelling agents, thickeners and stabilizers

Gelatin

Gelatin is used for producing jellies, as a binder in lozenges and tablets and as a stabilizer in chews and nougats and finds use in small quantities in many other products. The main criteria for gelatin to be used in confectionery manufacture are its colour, odour and gel strength. A wide range of gel strengths find use in different products.

Gel strength is measured in Bloom Units which increase with increasing strength. Low Bloom gelatins, 80–100 Bloom, are used in wine gums, confectionery jellies use 130–150 Bloom gelatin and 180–220 Bloom gelatin is often used for marshmallows. The lower bloom gelatins give a firm texture but are less elastic than high bloom gelatins.

Pectin

Pectin is made from citrus or from apple sources; for confectionery purposes, citrus pectin is normally chosen. Slow set (high methoxyl) pectins require accurate pH control in the range 2.6–3.5.

Pectins are difficult to disperse in water and are best mixed with five times their weight of sugar and added gradually with rapid stirring. The time taken to form a gel depends on the pH of the solution and the presence of other soluble solids. Excessive heat will cause a breakdown in the jelly strength of a pectin. Depositing should take place as soon as possible after the addition of acid to the pectin boil. A broken set will occur if the batch is left for too long after the addition of the acid or if it is deposited at too low a temperature.

Rapid set (low methoxyl or amidated) pectins require a higher range of pH 2.9–3.5 to produce firm gels with sugar but they also give good gels without sugar or acid but with calcium, within the pH range 2.8–7.0.

Gums

The important criteria in purchasing gums are purity, mesh size and viscosity. Colour is not always a reliable guide to quality. Gums are used as binders in lozenges and in products such as fruit gums and pastilles.

Gum arabic: Gum arabic is an excellent emulsifier and inhibits sugar crystallization by increasing the viscosity of the mix. It will produce a solution in water up to concentrations of about 50%. The viscosity of a solution of this gum is inversely proportional to the temperature and is adversely affected by acid conditions, e.g. in solutions of pH 4.0 or less. Excessive heat causes the breakdown of the gum and discoloration of the solution. Gum arabic solutions must be filtered through a medium mesh before use. Gum arabic is now available as a spray dried material which is free from contamination and much easier to handle.

In recent years, the supply of gum arabic has been very variable, leading to enormous price fluctuation. This has led to the development of a number of materials produced by modification of starches which may be used as partial or complete replacements.

Guar gum: Guar gum can be used in conjunction with other gums to improve stability. It is soluble in cold water, has excellent pH stability and is compatible with gelling agents. A solution of this gum will slowly lose its viscous nature when left to stand for some time.

Locust bean or carob gum: Roasted carob or locust bean flour is used as a replacement for cocoa in some dietetic coatings. The main advantage is the absence of the alkaloids caffeine and theobromine present in cocoa. The gum can be used in conjunction with agar and carrageen to improve gel strength and stability. It swells in cold water and forms a highly viscous solution. Mixtures of this gum in water should be used at a raised temperature.

Gum tragacanth: Gum tragacanth will swell, albeit slowly, when added to cold water and must be soaked overnight before it is used in confectionery manufacture. Tragacanth has a wide pH stability, the maximum viscosity being achieved at pH 5.0. Heat degradation takes place under adverse conditions. Gum tragacanth will deteriorate under long storage and stocks should be limited to not more than three months' supply.

Xanthan gum: Solutions of xanthan gum are very stable to variations in pH and to a wide range of temperatures. The gum combines with guar gum and locust bean gum to form high viscosity solutions or gels. Xanthan gum can often be used as a substitute for gum tragacanth.

Agar: Agar strips require soaking in water before use, powdered agar need not be. Normal usage levels of agar in confectionery jellies are from 1–1.5%.

Alginates: The main uses of alginates are as stabilizers and thickeners.

Soya

Soya flour is available as full fat or defatted powder or flakes. Soya concentrates, isolates and textured and spun soy proteins are also available. Good water absorption and binding properties are obtained using soy flour and the soy concentrates. Emulsification and flavour binding require the use of either a concentrate or an isolate, while foaming properties need the addition of either an isolate or a hydrolysate.

Starch

Starches are the most widely occurring of all polysaccharide substances used in the manufacture of food. Natural starch is insoluble in water. On heating a dispersion of starch in water, the granules swell and absorb moisture. A change of state takes place at the gelatinization temperature and the granules lose their characteristic shape, releasing amylopectin and amylose into solution. The viscosity on cooling is related to the concentration and type of starch, the effect of temperature on the starch during cooking and the presence of other ingredients. Chemical modification can be used to develop a range of starch products which are tailormade for the needs of the sugar confectioner.

The suitability of a particular starch depends on its source and the treatment it receives during manufacture. There is a gradual increase in viscosity during cooking which is related to the increase in temperature. This viscosity falls as the starch granules are ruptured. The manufacturer should aim to develop process conditions which will neither under- nor over-gelatinize the starch. The strength of a starch jelly is dependent on the final cook temperature, the degree of agitation and the pH of the mix. Variation in the weight of any component in a recipe invariably alters the final texture.

Acid-modified thin boiling starches reduce the time taken to cook a batch and still form an effective gel on cooling. Temperature readings taken while boiling a starch/sugar solution may be misleading; the cooking process should be monitored by following the change in total soluble solids contents as observed on a refractometer. Oxidised starches produce softer gelled confections and have a reduced tendency to flocculate.

Wheat flour: The main use of wheat flour for the manufacture of sugar confectionery is in the production of liquorice paste. Colour is unimportant for this purpose and a high gluten content is not required. Wheat flour gelatinizes between 58 and 70°C (136 and 158°F) but the presence of sugar raises the gelatinization point. Texture can be controlled by varying the amount of water used to slurry the wheat flour.

Colours

Many foods in the natural state are coloured and in many cases the colour influences the consumer's perception of flavour. Most confectionery products are not naturally coloured and so colours are added to assist in flavour perception and to provide a means of differentiation between sweets in an assortment of flavours.

Natural colours

These may be used in response to consumer demands, but they have many disadvantages. They include:

(i) the range of available colours is limited
(ii) many lack brightness
(iii) dosage rates are therefore generally high
(iv) lakes are not available

(v) some of the colours are derived from the spices and have residual flavour which can be difficult to mask
(vi) they may contain other ill-defined substances;
(vii) in many cases they are not stable to light, heat, acid or reducing agents such as SO_2 or ascorbic acid
(viii) they are usually more expensive than artificial colours.

Caramel

Caramels are available in a wide range of colour values, all with the same 'E' number (E150). Positive and negative caramels should never be mixed as this leads to precipitation of colouring material.

Artificial colours

The advantages of artificial colours are the very low dosages required, the bright colours and the wide range possible; they are mainly fairly stable and they are soluble.

They are also available as 'lakes' where the colour is adsorbed on to calcium or aluminium hydroxide. Lakes are very useful, for example for the dispersion of colours in fats or where moisture migration occurs during processing. Whereas with soluble dyes, colour migration may lead to uneven distribution, the immobilization of the colour on an inert carrier prevents this and the greater opacity of the lake gives a deeper colour. So for instance in an operation such as sugar panning, the same intensity can be achieved in fewer coats using a lake.

For more information see Chapter 14.

Flavours

All flavours are to a greater or lesser extent volatile and in confectionery applications often have to be added at high temperatures.

Natural flavours

Fruit flavours, mainly juice concentrates or distillates, may be used but high levels are usually required. They are therefore frequently reinforced with added components.

Essential oils

The quality of an oil varies according to the country of origin and the season when the fruit was gathered. They may be purchased on the basis of their genuineness, using appropriate standards, e.g. those listed in the national pharmacopoeias. However, the final test is their behaviour in trial batches of confectionery.

The oils become rancid when left open to the air and off-flavours can develop if they are held in contact with metal surfaces such as iron or copper, or on exposure to bright sunlight. Plastic bottles and measures will soften and discolour when left in constant contact with essential oils. In practice, they may perform relatively well at the high temperatures of sugar confectionery manufacture. In jellies and clear sweets some essential oils will produce a cloudy appearance; this problem can usually be overcome by using terpene-less oils.

Vanilla

The original flavouring used in confectionery was vanilla, an alcoholic extract of vanilla beans. Although now very expensive, vanilla extracts could return to favour if the current pressures for natural ingredients continue. Synthetic vanillin has largely replaced vanilla for flavouring chocolate and confectionery products. Suitable dosage rates are between 0.01 and 0.03%. Vanillin has a melting point of 82°C and vaporizes readily.

Artificial flavours

These include nature identical flavours and artificial flavourings; they are dealt with more fully in Chapter 14.

Acids

Acetic acid

Acetic acid at about 0.1% can be used to delay fermentation in coconut ice and other products with syrup phase solids lower than 75%. It is also used for acidification of some wine gums.

Citric acid

Citric acid has a high solubility (59.2% w/w) and is only mildly sour. Dry storage conditions are necessary to prevent caking. Suitable dosage rates are 0.5% for jellies, 1.0% for boiled sweets and 1.5% for acid drops.

Lactic acid and buffered lactic acid

Lactic acid was not used much in confectionery until the introduction of boiled sugar depositing plants. Buffered lactic acid is a solution of sodium lactate and lactic acid. At high concentration this makes possible an acceptable acid taste in a boiled sweet without the inversion which would inevitably develop at lower pH values. This in turn made the very clear bubble-free boiled sweet possible and buffered lactic acid is now virtually standard for this process.

Malic acid

Malic acid is available commercially as anhydrous crystals or powder, with a melting point of 130–132°C, and is readily soluble in water (about 50% w/w at 20°C). The acid taste is similar in character but slightly milder than citric acid at the same concentration; the two acids may be used as alternatives, particularly in boiled sweets.

Tartaric acid

Tartaric acid is highly soluble, more astringent than citric acid and is particularly suitable for use in fruit flavoured confectionery.

Nuts

Information about the various nuts which may be used in confectionery production is given in Chapter 12.

Processing aids

Release agents

Over the years many manufacturers evolved their own release agents, particularly for vacuum cookers. These were usually based on mixtures of beeswax, petroleum jelly and mineral oil, and frequently talc was added as a filler to help retention on vertical surfaces. On slabs, mineral oil (liquid paraffin) was used. Mineral oils and greases are no longer permitted and only release agents based on vegetable products may be used. These have not the resistance to oxidation that the mineral oils had.

Some applications subject the release agent to high temperatures and problems may arise with off-flavours in products with long shelf life. Many companies supply vegetable-based release agents, some containing antioxidants, but performance varies and trial assessment is necessary to select a suitable product.

Release agents tend to be used in excess, frequently because suitable applicators are not provided. This is undesirable, and minimum quantities only should be permitted. Thin coatings such as silicones and PTFE can be applied to plant, but due to the amount of wear and rough usage encountered, their life is usually fairly short. Wherever possible, heavier layers of coating are to be preferred. Self-adhesive PTFE tape is sometimes very useful on guide bars and heat-sealing jaws.

Talc

Talc or French chalk has been used as a dusting powder, particularly on boiled sweets, for many years. Although its use is considered by some to be undesirable it has properties which are virtually unique and suitable replacements do not yet exist. For some applications magnesium stearate is a possible alternative but it is not a complete replacement and is considerably more expensive.

PRODUCTION PRINCIPLES

All confectionery products have a number of common requirements. They must have an extended shelf life under ambient storage conditions and although this may be assisted by the use of protective packaging their inherent properties must provide stability against microbial deterioration and stability of shape. In sugar confectionery, shelf life is achieved by control of water activity or ERH (equilibrium relative humidity) through control of the composition. Stability of shape is aided by the inclusion of ingredients such as gelling agents or fats which create structure and prevent flow during storage, by formation of a glass as in boiled sweets, toffee and nougat, or by allowing crystal growth as in fondant creams. Also, soft or liquid centres may be held firm by being enclosed within a rigid shell.

Composition

The prime ingredients of sugar confectionery are sugar, glucose syrup or invert sugar and water. The purpose of the reducing sugars is to control or prevent crystallization in the product; whether crystallization occurs or not, it is essential to ensure that the level of dissolved solids in the water never falls below 76%, in order to prevent microbial deterioration.

Reference to data on the mutual solubility of the sugars in water shows that in any product the ratio of reducing sugars to water must be at least 1.5:1. The proportions of reducing sugars to sugar depend on the type of product required and particularly on whether and how much crystal sugar is wanted.

Table 11.1 gives the data for the specific gravity of different sugar concentrations and the associated volume changes when sugar is dissolved in water.

Equilibrium relative humidity

In a perfectly dry atmosphere all sweets will lose moisture. As the relative humidity of the atmosphere increases, a point will be reached when the sweet ceases to lose moisture and at higher relative humidities the sweet will absorb moisture from the air. The relative humidity of the atmosphere when a sweet neither gains nor loses moisture is the equilibrium relative humidity (ERH) of the sweet and is a function of its water vapour pressure.

This is mainly dependent on the moisture content of the sweet, or of the liquid portion in the case of sweets containing liquid and solid components. It is affected to a lesser degree by the molecular weight of the dissolved sugars. A number of equations exist for calculating ERH if the composition of the sweet is known. The best known of these equations is probably that by Money and Born:

$$ERH = \frac{(W \times 100)/18}{W/18 + 1.5 \times (S_1/M_1 + S_2/M_2 + \text{etc.})}$$

where W is the percentage of water, S_1, S_2, etc. are the percentages of carbohydrates, and M_1, M_2, etc. are the respective molecular weights of these carbohydrates.

Crystalline solids such as sugar crystals have no effect on the ERH of a syrup in which they are dispersed. Other solids such as cocoa, starch and desiccated coconut absorb moisture and the amount depends on the ERH of the solution in which they are contained. The effect is usually small.

Typical ERH values for confectionery products are:

20%	for high boilings
45%	for toffees and nougat
60–65%	for fudge
70%	for jellies
75–80%	for fondants and coconut ice.

It follows that if sweets of different ERH are packed together in a sealed container moisture will migrate from sweets with a high ERH to those with a lower value. Similarly if a composite sweet

Table 11.1 Comparing Brix and specific gravity scales and showing the relationship between these, pounds of sugar to be added to 1 gal of water, volume of syrup prepared from 1 gal of water and weight of sugar in 1 gal of syrup

Degrees Brix (at 20°C)	Specific gravity (at 20°C)	Weight of sugar to add to each gal of water (lb)	Volume of syrup from 1 gal of water (gal)	Weight of sugar contained in 1 gal of syrup (lb)
10	1.040	1.11	1.067	1.04
11	1.044	1.23	1.076	1.14
12	1.048	1.36	1.085	1.25
13	1.053	1.49	1.093	1.36
14	1.057	1.62	1.101	1.47
15	1.061	1.76	1.111	1.58
16	1.065	1.90	1.119	1.70
17	1.070	2.04	1.127	1.81
18	1.074	2.19	1.137	1.93
19	1.078	2.34	1.146	2.04
20	1.083	2.50	1.157	2.16
21	1.087	2.66	1.167	2.28
22	1.092	2.82	1.176	2.40
23	1.096	3.00	1.187	2.52
24	1.101	3.17	1.198	2.64
25	1.106	3.34	1.208	2.76
26	1.110	3.52	1.220	2.89
27	1.115	3.70	1.231	3.01
28	1.119	3.89	1.243	3.13
29	1.124	4.09	1.256	3.26
30	1.129	4.30	1.269	3.38
31	1.134	4.50	1.281	3.51
32	1.139	4.72	1.294	3.64
33	1.143	4.94	1.309	3.77
34	1.148	5.17	1.323	3.90
35	1.153	5.40	1.338	4.03
36	1.158	5.64	1.353	4.17
37	1.163	5.89	1.369	4.30
38	1.168	6.14	1.384	4.44
39	1.173	6.41	1.401	4.58
40	1.179	6.69	1.419	4.71
41	1.184	6.97	1.437	4.85
42	1.189	7.26	1.454	4.99
43	1.194	7.56	1.474	5.13
44	1.199	7.88	1.494	5.27
45	1.205	8.20	1.514	5.42
46	1.210	8.55	1.536	5.57
47	1.215	8.90	1.558	5.71
48	1.221	9.26	1.580	5.86
49	1.226	9.64	1.604	6.01
50	1.232	10.03	1.628	6.16
51	1.237	10.44	1.654	6.31
52	1.243	10.86	1.681	6.45
53	1.248	11.31	1.710	6.61
54	1.254	11.77	1.739	6.77
55	1.260	12.26	1.770	6.93
56	1.265	12.77	1.803	7.08
57	1.271	13.29	1.837	7.23
58	1.277	13.85	1.871	7.40
59	1.283	14.03	1.907	7.57
60	1.288	15.05	1.948	7.73
61	1.295	15.69	1.988	7.89
62	1.301	16.37	2.032	8.05
63	1.307	17.08	2.077	8.21
64	1.313	17.84	2.124	8.39
65	1.319	18.62	2.174	8.57
66	1.325	19.47	2.229	8.75
67	1.331	20.39	2.287	8.92
68	1.337	21.32	2.344	9.10
69	1.343	22.23	2.411	9.27
70	1.350	23.40	2.480	9.44

contains two components of different ERH, moisture will migrate within the sweet from the component with the higher ERH to the component with the lower value although the moisture in the total sweet remains unchanged.

Chocolate is said to have an ERH of 75–80% but it will absorb some moisture from centres with high ERH values. This will only be of consequence if the ratio of chocolate to centre is high, when moisture moving slowly into the chocolate will allow centres such as fondant creams and jellies to dry out eventually.

Humectants

As far as sugar confectionery is concerned, a humectant is any ingredient which retards moisture loss. Since moisture loss is dependent on ERH, any ingredient which reduces ERH is a humectant. ERH is related to the molar concentration of water in the sugar solution. The molecular weight of sucrose is 342 and the average molecular weight of 42 DE glucose syrup is about 340, so any non-volatile soluble substance with molecular weight lower than 340 will impart some humectant property and the lower the molecular weight the more effective it will be. Thus glycerol is more effective than sorbitol or invert sugar.

Sugar solubility

Sugar dissolves in water to give a saturated solution of 67.1% solids at 20°C. Up to about 74% solids, however, an undisturbed sugar syrup will not crystallize unless seed crystal is present or mechanical or thermal shock is encountered. This metastable state as it is known is utilized in the wet crystallization process. The presence of invert sugar or glucose syrup decreases the solubility of sugar but increases the total solids in solution at saturation. Sugar solubility is affected to the same extent by the presence of glucose syrup or invert sugar solids.

A number of empirical equations have been found which allow the solubilities to be calculated. They are based on the ratios of sugars to water. If g = glucose/water, i = invert sugar/water, $(g + i)$ = (glucose + invert sugar)/water or 'non crystallizing' solids/water, S = sucrose/water and c = total dissolved solids/water, the empirical equation for total dissolved solids is:

$$c = 1.994 + 0.661\,(g + i) + 0.038\,(g + i)^2$$

and the empirical equation for sugar solubility is:

$$S = 1.994/[1 + 0.1775\,(g + i)]$$

Both equations are valid for $(g + i) = 2$ to 6 at 20°C and for up to about 84% dissolved solids; little reliable information exists for higher solids levels.

A fair agreement with the experimental results for total solids up to 84% can also be obtained by assuming that glucose or invert solids dissolve in water in the ratio 6.69 to 1 to give an 87% solution and the remaining water dissolves sugar in the ratio 1 to 2.03 to give a 67.1% solids solution.

This implies, however, that sugar will not dissolve in an 87% glucose syrup which is not consistent with the facts. It is also rather surprising that glucose solids and invert sugar affect the solubility of sugar to the same extent, unless the controlling factor is a property of sucrose and not the reducing sugars. It seems that something as yet unexplained does occur at about 87% dissolved solids. For example, if the solubility of sugar and the boiling point of the solution are plotted on the same chart the curves merge at about 87% solids which suggests that at this level the remaining water is bound by the sugar and in effect becomes an impurity in the melt. The concept of bound water in carbohydrate water systems occurs frequently. In many equations constants are introduced to compensate for the effect of moisture binding.

Sugar boiling

The quantity of water evaporated when a sugar is boiled under constant conditions is related to the boiling temperature. Water boils at 100°C (212°F) under standard atmospheric conditions. The addition of sucrose, or any other sugar, increases the boiling point in a non-linear manner but related to the concentrations of sugars in solution. A number of tables have been produced which relate boiling point to concentration. The agreement between them is not good, particularly at the higher solids levels, and at concentrations up to about 80% it is much more reliable to use a refractometer.

A sugar syrup of constant composition boiled to the same temperature under the same conditions will give syrups of a similar and repeatable content of total soluble solids. Problems arise at higher temperatures because heat transfer through fairly viscous syrups is not good, thermometer response times and accuracy in factories are not always up to laboratory standards and proximity to heating surfaces can distort readings. The relationship between the concentration of sucrose syrups and boiling temperature is given in Tables 11.2 and 11.3.

Table 11.2 Boiling points of sucrose solutions at atmospheric pressure

Sucrose concentration (%)	Boiling point (°C)
40	101.4
50	101.9
60	103.0
70	105.0
75	107.2
80	110.0
85	115.1
90	123.3
95	140.0

Table 11.3 Boiling point of sugar syrups under vacuum (batch boiling only)

Sugar solids (%)	Vacuum (inches mercury)					
	15"	20"	22"	24"	26"	28"
	Boiling point (rounded) (°C)					
90	102.0	91.0	85.0	77.0	68.0	54.0
91	104.0	93.0	87.0	79.5	70.0	55.5
92	107.0	96.0	89.0	82.0	72.0	57.5
93	109.5	97.5	92.0	84.0	74.5	60.0
94	113.0	101.5	94.5	87.0	77.0	62.0
95	117.0	104.5	98.0	91.0	80.0	65.0
96	121.5	109.0	103.0	95.0	84.0	68.5
97	127.0	114.0	107.5	99.5	88.0	72.0
98	132.0	120.5	114.0	105.5	94.5	78.0
99	143.0	129.0	122.2	113.5	101.5	84.5

The presence of sugars other than sucrose will affect the boiling point of the solution in differing ways. Ingleton's method is to relate the effect of the other sugar to that of an equivalent sucrose concentration. On this basis a series of factors is derived and these are listed in Table 11.4. The Ingleton factors are used in the following manner: glucose syrup solids $= S_C \times F_G$ and invert sugar solids $= S_C \times F_I$, where $S_C =$ sucrose concentration, $F_G =$ Ingleton factor for glucose syrup and $F_I =$ Ingleton factor for invert sugar.

Thus at 110°C (232°F), a boiling sugar syrup will contain:

(i) 80% total solids (Table 11.3), when all sucrose solids
(ii) 87.2% total solids (80.0 × 1.09, Table 11.4, low DE syrup), when all glucose syrup solids
(iii) 76.8% total solids (80.0 × 0.96, Table 11.4), when all invert sugar solids.

Mixed syrups can be calculated by proportioning the various total solids contributions of the different sugars. Additionally, the boiling point

Table 11.4 Factors relating the concentration of glucose syrup and invert sugar solutions to that of sucrose solutions boiling at the same temperature

Sucrose (S_C, %)	Invert sugar (F_I)	Glucose syrup (F_G)			
		Low DE	Regular	High DE	Enzyme
60	0.94	1.21	1.18	1.15	1.13
70	0.95	1.17	1.14	1.12	1.09
75	0.96	1.11	1.08	1.06	1.03
80	0.96	1.09	1.07	1.05	1.03
85	0.96	1.06	1.05	1.04	1.02
90	0.97	1.03	1.03	1.02	1.01
95	0.98	1.01	1.03	1.00	1.00

Table 11.5 Variation in the boiling point of water with change in barometric pressure

Barometric pressure (cm Hg)	Boiling point (°C)
72.5	98.7
73.0	98.8
73.5	99.0
74.0	99.2
74.5	99.4
75.0	99.6
75.5	99.8
76.0	100.0
76.5	100.2
77.0	100.3
77.5	100.5
78.0	100.7

of mixed sugar solutions can be calculated using the information given in these tables.

Variations in barometric pressure will affect the boiling point of a sugar syrup. The effect of pressure on the boiling point of water is shown in Table 11.5. Barometric pressure must be taken into account in products whose properties are affected by minor changes in moisture content.

The variation in the height above sea level for the location of a factory can give rise to differences in the boiling point of a standard sugar syrup, see Table 11.6.

It follows that if the atmospheric pressure can be reduced by mechanical means, then a syrup will boil at a considerably lower temperature than that previously observed. Boiling under vacuum is frequently used in sweet manufacture and can make up to 22°C (40°F) difference in the observed boiling point of a sugar syrup. The lower temperature will affect both colour and the amount of moisture that can be removed from the syrup. Although tables have been quoted which relate vacuum to the boiling point of sucrose solutions, these are of little value except in a batch boiling situation. Many modern sugar cookers boil under atmospheric conditions and discharge into vacuum, so the final solids are affected by the final temperature and pump stroke which governs the time the syrup is held under vacuum.

Table 11.6 The effect of geographic situation on the boiling point of water

Height above sea level		Depression in boiling-point of water
(ft)	(m)	(°C)
250	76.2	0.3
500	152.4	0.6
750	228.6	0.8
1000	304.8	1.1
1250	381.0	1.4
1500	457.2	1.7
1750	533.4	1.9
2000	609.6	2.2
3000	914.4	3.3
4000	1219.2	4.4
5000	1524.0	5.6
6000	1828.8	6.4
7000	2133.6	7.2
8000	2438.4	8.3

Crystallization

The structures of products such as fondant creams, coconut ice and fudge are produced with formulations which contain enough sugar to crystallize on cooling. When a supersaturated solution is seeded with existing crystals, usually by the addition of fondant, crystallization occurs throughout the mass. The crystal size may be con-

trolled by varying the degree of supersaturation, the amount of seed crystal and the temperature. By varying these, textures ranging from firm pastes to hard products (where the crystals have grown into each other and formed a rigid structure) may be achieved.

Crystallization rate

Whilst it is possible to calculate if a particular composition will crystallize and how much sugar crystal will be formed, it is not possible to predict how long this will take to occur. It may be affected by many factors such as temperature; and particularly temperature fluctuation, composition of the reducing sugars, the effect of other ingredients and the treatment the product has had during cooling.

Threshold value

This is the moisture content of a boiled sweet below which crystallization will not occur spontaneously and will not progress even if initiated. The threshold value is determined by the glucose solids content and is equivalent to an 87% solution of glucose in water.

Glassy state

In the case of boiled sweets, toffees and nougat, stability of shape is achieved by forming a glass or supercooled liquid where random bonding by hydrogen bonds increases the viscosity as the product cools down until it becomes rigid. If exposed to moisture these sweets will form crystals or grain on the outside, but if the ratio of glucose syrup solids to water is greater than 6.69:1, and the amount of moisture pick-up is small, then crystallization will not progress. This appears to be independent of the sucrose content or the actual moisture content, and implies a moisture binding function of the glucose syrup solids which prevents further crystallization.

Toffees and nougats are much nearer the borderline of stability in this respect and in fact are sometimes below the ratio required. It is not unusual for them to contain a few small sugar crystals but these do not appear to impair shelf life of the product. This once again suggests a moisture binding effect which may be related to the protein or hydrocolloid present in the sweet. It would appear that some other factor in addition to just glass formation is involved and further investigation in this area is highly desirable.

Grain

This is a term used in sugar refining and sugar confectionery which refers to the development and growth of sugar crystal in a product or a syrup. Graining in a product may be induced by subjecting it to high shear forces or by the addition of a relatively small quantity of crystal seed such as by the addition of fondant or milled sugar. It is not used where all the crystal sugar has been added as such.

Reducing sugars

Strictly speaking this is a chemical definition embracing all sugars which have the ability to reduce Fehling's solution. In confectionery terminology it is frequently used to refer to sugars and polyols such as sorbitol which under the condi-

tions of use will not crystallize but will increase the level of dissolved solids at equilibrium. Problems may arise if this terminology is used in relation to lactose which chemically is a reducing sugar but may occur as crystals in products containing significant levels of milk solids.

Tempering

In sugar confectionery, this is the process of cooling a sugar mass uniformly to achieve the correct plasticity for further processing such as plastic forming or extruding.

Aeration

Aeration may be applied for a number of reasons such as increasing volume for a given weight of product and creating opacity, but the main reason is texture modification. Strictly interpreted, aeration refers to the incorporation of air, but other gases such as nitrogen and carbon dioxide may be used. Aeration may be achieved by chemical or mechanical means in batch or continuous processes. It may be produced using an atmospheric process, a pressurized system or under vacuum. The aeration may be achieved by beating where foaming agents and stabilizing agents are necessary, or it may be done by pulling or liberation of gas under vacuum where the setting of the matrix by cooling stabilizes the structure.

PRODUCTION PROCESSES

Some processes are virtually unique to specific products. In these cases they are included with the product entry and are not repeated in this section.

Cooking

Gas fired cooking pans are now rare and are used only by small producers or for specialist products. They consist of cast iron stoves with a circular aperture in the top plate, a firebrick lining and a large capacity gas burner in the base which in the days of coal gas was intensified by an air blast. A copper bowl fitted the top aperture closely. Heating was largely radiant from the incandescent firebrick and allowed high boiling temperatures to be achieved, frequently associated with some degree of caramelization of the sugars. These stoves gave off a lot of heat and since the product was usually boiled sugar, needed care in operation.

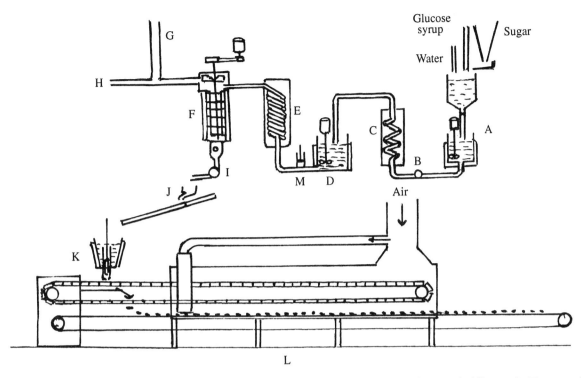

Figure 11.1 Boiled sweets process. A, Ingredient mixing; B, feed pump; C, dissolving; D, holding tank; E, precooker; F, film cooker; G, vapour vent; H, vacuum (if required); I, cooker discharge pump; J, flavour mixer; K, depositing hopper; L, cool box, M, metering pump.

Modern cookers tend to be relatively small. They are usually fitted with stirring gear which allows heat sensitive confectionery products to be produced in small batches. A typical process line for boiled sweets is shown in Figure 11.1. See also CONTINUOUS SWEET PRODUCTION, later.

Drop rolling

The drop roller was one of the earlier sweet forming machines. It consisted of two synchronized brass rolls engraved with matching impressions. The plastic mass was fed into the rolls and forced into the impressions. A thin web remained between the sweets, and after cooling this had to be removed by sieving. Removal of the web left the sweets with rough edges which were often masked by sugar sanding.

Moulding

Boiled sweets

These are deposited into PTFE-coated aluminium moulds fitted with ejector pins. After cooling, the pins are depressed from the underside of the moulds to eject the solid boilings. More recent

developments have made it possible to emboss the top surface of the deposited sweets and also to produce a soft-centred sweet by simultaneous depositing in the same manner as 'one-shot' chocolate depositing.

Toffees

These can be deposited in silicone rubber moulds. After setting, the mould is deformed by plungers which push the toffee out.

Release agents are not required with either of these processes.

Jellies and fondant confections

These can be conveniently and continuously deposited into impressions made in moulding starch. Machines for this purpose, called Moguls, have been used for many years, but it is only recently that radical rethinking has resulted in equipment with vastly improved performance. The sequence of operations for starch moulding is as follows:

(i)   Trays of previously cast, fully set sweets are automatically fed into the plant. These trays, usually 81×40 cm $\times 5$ cm deep, hold 4–5 kg starch. Product weight depends on sweet size and number of deposits.

(ii) The sweets and moulding starch are discharged on to vibratory sieves.

(iii) The discharged starch is circulated into empty trays, levelled and impressions made using prepared moulds of the desired shape.

(iv) Prepared jelly or fondant mix is deposited into the moulds from swinging steam- or water-heated hoppers. The pump bar for controlling depositing is normally interchangeable to permit flexibility in size and number of sweets per tray.

(v) The trays are removed and stacked to permit the product to set, or transferred to drying stoves for further removal of moisture from products such as gums and, in some cases, mallows.

A proportion of fresh dry starch should be incorporated at stage (iii) to maintain a low moisture content, while a similar proportion of 'damp' starch should be removed for drying. Figure 11.2 shows the sequence of operations in a Mogul depositor.

Moulding starch

Maize starch is normally used for forming the impressions for deposited sweets. Many confectionery products will be left in the starch for at least eight hours, and the weight of starch in a tray is at least three times the product weight. With machine outputs of about 12 tonnes/day about 36 tonnes of starch is constantly in use. This has to be kept in condition.

(i) It must take a good impression. New starch does not print well and to improve this some suppliers add about 0.5% of vegetable oil to improve bonding.

(ii) The starch needs to be aerated. The mould board which prints the impressions compresses the starch in the tray. If this is too dense, it compacts badly and prevents deep mould impressions. Aeration is normally achieved by passing through fine sieves.

(iii) The moisture content must be controlled

(iv) Tailings, fragments of deposited products, must be removed. This is done by circulating the starch through sieves, but ultimately fine sugar dust can build up in starch and it becomes difficult to handle. In these cases, some starch is usually removed from the system and new starch added.

(v) Starch moulding plants are dusty and hence in some circumstances may be involved in dust explosions.

(vi) The starch can become microbially infected and it must be kept in condition.

These hazards have led to many attempts to find alternatives and starchless moulding plants are now available.

Starchless moulding

These are large expensive plants and are only really suited to mass production of limited ranges of sweets. The sequence is similar to starch moulding, but the moulds are plastic. Ejection is by compressed air through very small holes in the moulds and a release agent is essential. As cooling and setting occur in a cool box, the sweets must set rapidly for demoulding and no loss of moisture can occur, so high solids or slow setting products are not really possible yet.

While two or three centres of different types and shapes can be deposited simultaneously and the sweets are discharged in ordered rows ready for feeding to enrobers, the cost of changing moulds is very high, so use has been restricted to well-established products with assured markets. Some products such as fudge and fondant creams may be deposited in silicone rubber moulds and smaller plants such as those used for toffees may be adapted for these processes.

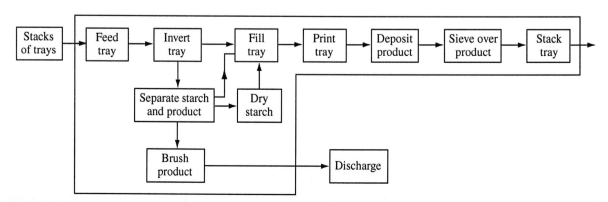

Figure 11.2 Starch depositing process

Extrusion

Extruders of various types are used for forming bars and sheets from pastes and plastic confectionery bases. Roller extruders are made in many configurations, some of which are designed to extrude two or more sheets which are layered together simultaneously. Lozenges are formed by passing the paste through a screw extruder and die to form a sheet which is then reduced in thickness by subsequent sizing rollers. Extruded sheets are formed into sweets by stamping and the remaining web is removed continuously and returned to the extruder. In the case of bars, the sheets are slit by rotary knives and guillotined to length. Extruders of this type are frequently linked directly to chocolate enrobers.

Recently much interest has been shown in the possibility of using cooker extruders for confectionery production. Theoretically, due to the high pressures and temperatures available, it should be possible to process many confectionery products at final solids and some success has been achieved in this respect. The problem appears to be dissolving the sugar at the high solids level and unless temperatures high enough to melt the sugar are used, when caramelization becomes a problem, additional water must be added. The advantages of this type of processing over conventional equipment are thus not yet fully proven.

Paste mixing

Many confectionery products are cold mixed pastes such as lozenges, cream pastes, and truffles. These are usually produced in either Z blade dough mixers or planetary beaters using a dough hook. With products containing both a syrup phase and fat overmixing should be avoided to prevent fat separation.

Plastic forming

The plastic forming operation is widely used in the confectionery industry for converting plastic masses such as boiled sugar or toffee into individual sweets. The process is shown in Figure 11.3.

The product will have been tempered to the correct consistency and flavour ingredients incorporated during the kneading process in the case of boiled sugar, or on a slab, cooling drum or cooling conveyor in the case of toffees. The mass is then fed into a batch roller.

The batch roller consists of a tapered cradle about 2.5 m long, formed by a set of conical rollers which revolve on their axes and spin the sugar mass into a cone. The rollers automatically reverse their direction of rotation at intervals and the whole cradle is enclosed so it can be heated to keep the product at the correct consistency. At the narrow end of the cradle a set of forming wheels is fitted into which the product is fed and emerges

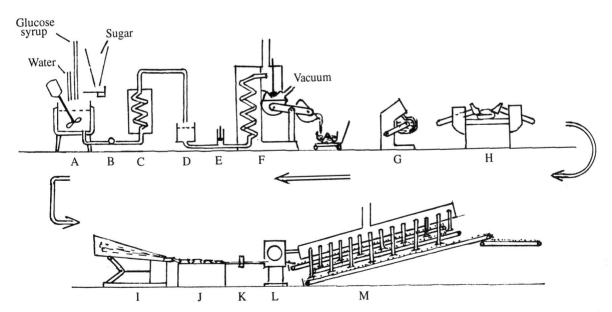

Figure 11.3 Plastic forming process. A, Ingredient mixing; B, feed pump; C, preheating; D, holding vessel; E, metering pump; F, vacuum coil cooker; G, flavour mixer; H, kneader; I, batch roller; J, presizer; K, metal detector; L, forming machine; M, cooler.

as a rope of about 5 mm diameter. As the product is drawn off, the batch roller may be tilted to maintain the feed to the forming wheels. The rope is then fed to presizer.

The presizer is a set of (usually) four pairs of wheels, with semicircular grooves machined around the circumference. As the rope goes through these, it is progressively reduced in diameter and increases in speed. Due to the speed of this operation, some elastic stress remains in the rope and this must be allowed a brief period with no tension so the stress can relax to prevent distortion of the sweet after forming. The rope now passes to the sweet former, of which a number of types are in use.

(i) *Simple machines* have synchronized pairs of revolving cutters which chop the rope into pieces. If successive cutters are set at right angles, the humbug shape is produced.

(ii) *Seamed tablet machines* have a final pair of rope sizing wheels to control the sweet weight, and the rope is fed into a rotary die. The top half of the die opens to receive the rope and closes to cut the sweet which is then formed by compression plungers which form the sides of the die. The die then opens and the sweets discharge on to a conveyor. These machines are relatively gentle in action and are used particularly for boiled sweets with soft centres.

(iii) *Seamless tablet machines* also have a final pair of rope sizing wheels, from which the rope is fed to a cutting roller which chops pieces of rope into troughs that extend from a die ring. Compression plungers then push the piece of sweet into the sweet cavity in the die ring, squeeze it into the shape and eject it onto a conveyor. Earlier machines of this type had small diameter rotary dies and the compression rate was very high. They could only be used for solid boilings. These machines have largely been superseded by machines with larger diameter dies and longer compression periods and this type can handle both solid and centre-filled sweets. The sweets from all these machines are still plastic and are cooled on wire-link belt conveyors fitted with air cooling jets. These conveyors are best located in air conditioned areas, and because boiled sweets when cold are very hygroscopic, and sweets should be wrapped and packed while still warm.

(iv) *Cut-and-wrap machines* are largely used for toffees. The rope from the presizer is fed through forming rolls which create the rectangular or round shape required. This is then chopped into pieces by a high speed knife. Simultaneously, the wrapping paper is fed into the machine, cut to the length required and positioned so that when the toffee piece is pushed into the jaws of the wrapping head it folds the paper round it. The overlapping paper is then tucked in and as the wrapping head revolves forms a tube around the toffee. Twisting heads then grip the ends of the tube and spin two or three turns to complete the wrap. Other versions of these machines produce wraps with folded ends. The machines will handle cellulose films, waxed paper, metallized paper or paper with a foil strip attached. Some types will feed a separate foil strip backed with waxed paper under the outer wrapper. Speeds are now up to 1000 sweets per minute. On emerging from the wrapping machine the toffees are still plastic and are usually passed through cooling tunnels to harden before packing.

Wet crystallizing

Wet crystallizing is the process of building up a thin coherent coating of sugar crystal on the surface of a sweet. This seals the surface, and when used on products such as fondant creams or marzipan, retards drying out, extending the shelf life from a few days to six months or more.

The process utilizes the fact that cold sugar syrups between 67–74% solids will not crystallize unless seeded. This seed is provided by sugar crystal present in the surface of the sweet. In practice a sugar syrup is made up from high quality granulated sugar by stirring in heated tanks, without boiling. It is then transferred to cooled holding tanks and cooled down to about 27–30°C. Meanwhile sweets to be coated are placed in wire mesh baskets and put in trays or tanks. The cooled syrup is then carefully run into the trays or tanks to cover the sweets. The surface is covered to prevent evaporation and the sweets left for up to 12 h for crystallization to develop.

When an adequate coating is formed, the syrup is drained off and the sweets left to drain in a humid atmosphere. If the atmosphere is too cool or dry, the syrup does not drain and forms a fine crystal glaze on the sweets which detracts from their appearance.

It is not possible to be specific about syrup concentrations, temperatures or crystallizing times since they are interdependent and depend on the desired result. In general, high syrup solids and higher crystallizing temperatures develop the coating faster, but the crystal is coarser and the surface rougher.

After use the drained syrup has more sugar added and is reprocessed for further use, but it does gradually invert and develop colour so that three cycles is about the maximum possible before the syrup should be discarded either for recovery or use in other products. The number of cycles can be increased by buffering the syrup with sodium citrate, but even so only about five cycles are possible.

During crystallization, saturated sugar syrup is absorbed by the sweet and care must be taken that centres for crystallizing are firm enough not to collapse during processing. The sweets in the baskets are random packed and touching and the crystallizing bonds them together. When they are dry and separated, this does not impair the protection afforded by the crystal coating. It is imperative that crystallizing rooms are free from vibration and are maintained at constant temperature. If this is not so, spontaneous crystallization can occur in the syrup which settles out as an incoherent coat on the sweets.

Continuous sweet production

Over a number of years the main development in confectionery processing has been the creation of fully continuous production lines requiring less labour and the reduction of process times. Automatic production of feed mixes and continuous cooking have been available for many years, but the problem has been the incorporation of colour, flavour and acid, and forming the base product into finished sweets. These operations were usually carried out batchwise and although such continuous forming machines were effective, they usually had to be fed by hand.

Most of the problems are now solved and continuous production involving no direct manual input is normal for most high tonnage production. Different approaches arise from the consistency of the mass to be handled, which can range from fluid pourable products suitable for depositing, to highly viscous masses which must be handled in the plastic state.

Mixing

Dispersion of colour, flavour and acid in fluid masses requires a liquid acid, which is usually buffered lactic acid and because the low viscosity makes these products more susceptible to crystallization, the degree of agitation must be low.

For boiled sweets a rotating funnel device is used. The boiled syrup flows on to the conical side of the funnel and builds up layers as it rotates. The additives are sprayed on to this and are sandwiched between successive layers as the cone rotates. This is then further mixed by offsetting the discharge from the rotating funnel on to a chute where further layering occurs and the product flows to the depositor.

For toffees, an open trough mixer is used with an axial stirrer designed to minimize air incorporation.

For plastic boiled sweets a totally enclosed tubular mixer is used to prevent the incorporation of air bubbles into the product, and this is followed by a continuous kneading operation which completes the mixing, and also reduces the temperature of the boiled sugar uniformly to give a texture suitable for sweet forming.

Forming

The automatic depositing of sweets is a technique which has been in use for some time, but this was mainly into starch filled trays where the starch has been formed into the mould. The development of a depositor which could handle boiled sweets and toffees, and the introduction of metal and silicone rubber moulds made continuous casting of these products possible. This has since been developed to include centres within the sweets but aeration is still not possible. The sweets are cooled within 15 min and fed directly to wrapping machines.

Another version of this machine produces centres for chocolate covering, which were previously moulded in starch. This machine can deposit a variety of centres simultaneously into the same mould and discharge them in ordered rows suitable for direct feeding to a chocolate enrober within about 12 min. Thus a chocolate covered assortment can be made in one operation. An alternative approach is to deposit a centre within a chocolate coating directly into a mould in one operation.

The plastic forming process for boiled sweets, although more complex, is more versatile than depositing. The flavour mixing unit can also be used to inject and disperse air to give an aerated sugar. This can then be extruded with a non-aerated coloured sugar to produce stripped sweets. Both clear and aerated sweets can be continuously centre filled, or the sugars can be laminated to give layered sweets.

Conditioning

These processes run with little operator control, but they are susceptible to changes in ambient conditions and air conditions and air conditioning is virtually essential. With reduced process times, sweets have to be cooled rapidly requiring low temperatures in cool boxes. This can give rise to condensation if the air in the packing area is not dry. Moisture attack is a major problem since it can cause stickiness, reducing the efficiency of wrapping machines, and can seriously reduce shelf life.

Panning

Panning is an operation which has been used by the pharmaceutical and confectionery industries for centuries. In principle it is the application of many layers of coating to centres tumbling in a revolving pan mounted at about 30°C to the horizontal. Coatings may be sugar syrup (hard panning), glucose syrup dried off by the application of fine sugar (soft panning) or chocolate; see Figure 11.4.

The operation consists of adding enough coating medium to cover the centres completely with no surplus and drying this off either with hot air (hard panning), extra sugar (soft panning) or cold air (for chocolate). Each coat must be thoroughly dried before the application of the next. Panning is a time-consuming operation, but a number of pans can be handled simultaneously by one operator.

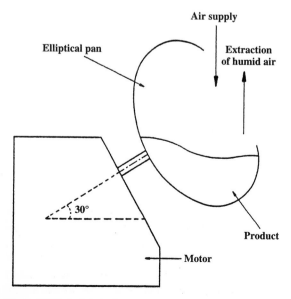

Figure 11.4 A simple dragee pan.

Colouring of hard panned products is carried out in the final few coats of syrup. It is easier to obtain a uniform finish if lake colours are used.

Finished coatings are usually beeswax or carnauba wax for hard panned goods or glazes of gum, shellac or cellulose derivatives on chocolate work. These coatings are applied in finishing pans, which in the case of beeswax are lined with a wax coat. Dry conditions are essential for a good finish.

Silver dragees are built up from graded sugar crystals by hard panning with a low solids (45–50%) sugar syrup. When the desired size is achieved, the balls are coated with a film of gelatin solution containing acetic acid. Coating with silver is carried out by placing the silver leaf in a glass pan and adding the balls in a stream with the pan running. This is a highly specialized operation which requires much experience for success.

Centres containing fat are usually sealed using a coat of gum arabic and starch or sugar to prevent migration into the sugar or chocolate coating.

Modern systems are highly automated and use coating sprays actuated by sensors for the engrossing operation.

Centre filling

Boiled sugar is frequently used as a casing around soft centres. The hand process uses a U-shaped trough with closed ends as a former. The sugar casing, which is frequently but not necessarily pulled, is spread into a sheet to line the trough and the centre then added. The casing is sealed by moistening the edges, usually with a damp cloth, and pressing them together. (The substitution of more hygienic methods than the damp cloth is still far from complete.) The still-plastic product is then placed in a batch roller and passed through a plastic forming machine.

Many centre-filled products are now made continuously. The batch roller has a PTFE-coated tube fitted down the centre and the casing is wrapped around this. As the casing is drawn off from the batch roller, the centre is pumped through the tube continuously. It is important that the feed rates are balanced, that air does not get into the product and that the casing is kept plastic. If these requirements are not met, breakage of the rope can occur in the forming machine which will entail removal of the dies for cleaning. Powder centres such as sherbet can also be fed in this manner, but for this application the tube is wider and has an auger feeder revolving in the centre. It is difficult to include much more than 18% centre in a sugar coating.

Sugar sanding

Sanding is the confectionery term for coating with sugar crystal. It is used mainly on jellies. The jelly sanding machine consists of a wire feed band fitted with a steaming unit. The jellies are fed along the band under the steamer which moistens the surface and are then thrown by a spinning bar into a revolving drum containing high grade crystal sugar. The tumbling action coats a layer of sugar crystals all over the jellies, which are then dried either on trays or by passing through a drying tunnel.

Aeration

Whilst the term aeration specifically implies the incorporation of air, it is also loosely used for the expansion of a confectionery mass by any means, including chemical or mechanical. Chemical aeration is usually the breakdown of sodium bicarbonate by heat or acid to give carbon dioxide in products such as honeycomb, cinder toffee or puff candy. Mechanical aeration includes the following:

(i) *Pulling* or the repetitive stretching and folding of a plastic confectionery mass. This eventually forms a ribbed surface which on folding traps channels of air in the products. Further stretching and folding reduces the diameter of these channels to give fine continuous air passages through the product. Products utilizing this process include seaside rocks, Edinburgh rock, satin and striped boilings and chews.

Skinning. To achieve a satin finish on pulled boiled sugar, a small portion of the batch is left unpulled, spread into a thin sheet and wrapped around the pulled portion. This functions effectively as a varnish coat, but has the added advantages that it helps to delay moisture attack on the pulled sugar surface. The procedure has also been used to reduce the effect of inversion in boiled sweets containing acid.

(ii) *Cracknel aeration* which forms much larger air channels. In the hand operation, the start point is a bag of boiled sugar inflated with compressed air, which is sealed, stretched, cut and joined in parallel. Repetitive stretching, cutting and folding produce a multi-channelled coarsely aerated texture. A similar result can be produced mechanically by extrusion and coalescence of semicircular filaments.

(iii) *Whipping* is normally carried out in a planetary beater using a wire whisk or flat beater and is used for nougat or marshmallow. This process can also be carried out under pressure, but the foam must be allowed to expand gradually.

(iv) *Continuous foam mixing* involves the injection of metered quantities of compressed air into a syrup flowing through a pressurized dispersion head and allowing the resultant foam to expand gradually to atmospheric pressure. This process can also be used on chocolate.

(v) *Vacuum expansion* can be applied to heated, pulled boiled sugar or to chocolate which normally contains enough air and dissolved gas to expand. Products expanded in this way must be allowed to set under vacuum.

(vi) The *incorporation* of other highly aerated products such as puffed rice.

Fondant

While it is possible to make fondant by hand, most processes are continuous and consist of boiling a syrup to about 120°C (248°F) and cooling this rapidly to about 50°C, when it is subjected to violent agitation to cause rapid nucleation and crystallization. Once the process has started, the plant is full of seed crystal and the fondant is virtually fully crystallized as it emerges. Crystal size is reduced by beating at lower temperatures and at higher shear rates, but this increases the plant loading and energy costs.

Due to the latent heat of crystallization of sugar, heat is liberated during beating and many plants have water cooling jackets. After preparation it used to be normal for fondant to be left to cool and mature for 24 h. The benefits of this are dubious when most fondant cream processes redissolve about 50% of the crystal in base fondant and continuous processes have eliminated the maturing stage.

Wrapping

Individual sweet wrapping performs a number of functions, it protects the sweet from contamination, it prevents them sticking together and it is usually decorative. Materials may be waxed paper or plastic film both of which may be metallized or have a foil strip attached. In some cases an understrip of waxed paper with a foil laminate may be included under the main wrap. For products which are prone to cold flow, such as toffees and chews, waxed paper is more effective at preventing distortion than plastic film unless an understrip is used.

Unless it is heat sealed, the immediate wrap does not prevent moisture absorption by the sweet and a further moistureproof protection is required. This may be a tin, a glass or plastic jar, or most frequently a moistureproof film bag produced on a continuous form, fill and seal machine.

PRODUCTS

Boiled sweets

Boiled sweets or high boilings were originally made by boiling a sugar solution with cream of tartar over coke fires to about 149–155°C (300–310°F). The cream of tartar inverted some of the sugar and the resulting mass, comprising a mixture of sucrose, dextrose and fructose at about 97% solids, could be coloured, flavoured and set to a transparent glassy product.

The action of cream of tartar is dependent on time of boiling and temperature and so results were variable. Furthermore, boiled sweets based on invert sugar are very hygroscopic and readily develop stickiness on exposure to the atmosphere. An improvement was to use manufactured invert sugar in controlled amounts, but the introduction of glucose syrup or 'glucose' gave sweets which were less prone to stickiness, and sugar/glucose syrup or sugar/glucose syrup/invert sugar combinations are now universal. The ratios of sugar to glucose can range from 70% sugar and 30% glucose to 50:50 mixtures.

Coke fires later gave place to forced-draught gas fires, but the most suitable heating medium is steam and this requires either very efficient heat exchangers or evaporation under vacuum.

Batch vacuum pans came first, but these have largely been displaced by coil cookers where the syrup is heated by passage through a coil and discharges at about 25–30 inches (635–760 mm) mercury vacuum into a vacuum receiver where the moisture is evaporated. With this type of cooker, very low residual moisture levels, e.g. 1%, are possible. Due to evaporative cooling the product temperature is about 132–138°C (270–280°F).

The other method is to form a very thin film of syrup on the inside of a heated cylinder by a rotating shaft carrying wiper blades. These cookers can readily achieve temperatures of 300°F

(150°C) and can also be operated under modest vacuum for improved evaporation. Discharges from both these types of cooker can be by manual means or pumps.

The sugar from high vacuum cookers is a viscous plastic mass. Added flavours and colours may be partially dispersed through the batch by stirring in before the mass is poured onto the slab or kneading machine table. Acid, normally as hydrated citric acid, may be sprinkled over the molten mass and final dispersion achieved by kneading. The more fluid sugar from atmospheric or low vacuum cookers can be handled in the same manner, but on continuous depositor plants the colour, flavour and acid, normally as buffered lactic acid, are metered into special mixing devices which achieve adequate dispersion in the sugar as it flows from the cooker to the depositor. The plastic masses are usually shaped by a sequence of rope forming, size reduction and shaping in seamed or seamless tablet machines. After cooling to set, the sweets are wrapped in moisture-proof packs as rapidly as possible, preferably while still slightly warm.

Candied fruits

The process of candying is the exchange of water in the fruit for a highly concentrated sugar syrup. The rate of diffusion of the sugar syrup into the fruit is dependent on the temperature of the syrup, the maturity of the fruit, the size of the fruit and the sugar concentration of the syrup.

The chosen fruit should be firm, ripe and free from blemishes. Blanching should be carried out by immersing the fruit in hot water. This process breaks down the fruit tissue and reduces the shrinkage of the fruit during candying. When necessary, the surface of the fruit should be pricked with silver or steel needles. Traditionally, the fruit is loaded into wire baskets and lowered into a low concentration sugar syrup. The sugar concentration of the syrup should be 40% at the start and increased to 75% over a period of 2 weeks. The addition of 10% invert sugar or glucose syrup will produce a more tender product. In the continuous process the fruit, held in baskets, is blanched with weak boiling sugar syrup. This syrup is replaced by a syrup of slightly higher concentration. The concentration of sugar is then slowly increased under vacuum to 75%, while the temperature is raised from 49–82°C (120–180°F). Production by the continuous method reduces process time to one day.

Preserved ginger

The ginger used to produce this delicacy comes from the rhizomes or rootlike edible stems of *Zingiber officinale*. The finest quality is Chinese ginger which lacks the harsh quality and fibrous nature of ginger from other sources, but in recent years Australian ginger has become important. Stem ginger, derived from the side roots, is of a better quality than cargo ginger from the main stems. The root is peeled and then pickled in brine (14–18% total solids) prior to shipment.

On receipt at the processing factory, the root should be chopped to size. A suitable size for retail sale is 60–80 pieces per pound (25–35 per kg). The roots should then be immersed in 40% syrup prepared with a 75:25 sucrose:glucose syrup ratio, for 24 h. The solids content should then be increased by adding sucrose to 58, 70 and 75% total solids for successive 24 h periods. Following draining, the ginger should be dried in a stove and then wet crystallized or sugar sanded.

Caramels and toffees

The name 'toffee' was originally used for products which did not contain milk and were similar to butterscotch. The name is still used in this context for toffee apples but most toffees these days do contain milk. Although some sources regard caramels as having higher milk, fat or moisture levels than toffees, there are no definitions and the two names may be regarded as being interchangeable. Toffees and caramels are made by boiling sugar, glucose syrup, milk (usually condensed milk), vegetable fats and salt. Other ingredients may include cream, butter and various flavouring additions such as treacle or malt extract. Many toffees are still made with brown sugars because of the flavour they impart. Emulsifiers are frequently added, the usual ones being glyceryl monostearate or lecithin.

The inherent flavour is due to the Maillard reaction between the reducing sugars in the glucose and the milk protein. This is also responsible for colour development. The reaction is time and temperature dependent and accelerates at temperatures above 112°C.

Good toffee texture requires thorough emulsification of the fat which must be carried out before cooking commences. The original toffee pans were fitted with contra-rotating stirrers to achieve this, but for large scale production it is usual for the mix to be emulsified in a separate plant. Cooking times should be as short as possible consistent

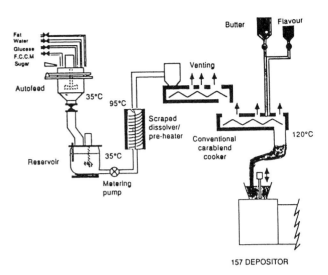

Figure 11.5 Toffee plant.

with adequate caramelization and many manufacturers do not add any additional water to the mix, the amount already in the ingredients being just enough to dissolve the sugar. The final moisture content depends on the texture required but is normally between 6–8%. Boiling temperatures are about 118–126°C.

Continuous plants are now quite common and fall into two main groups. The first type, developed by Tourell in conjunction with Callard and Bowser, effectively reproduces the processing characteristics of the batch process; similar cookers are now made by other manufacturers. Figure 11.5 shows a typical toffee plant.

The premix is first heated through a scraped-surface heat changer and passes to a trough cooker containing a closely fitting rotor designed to reproduce the action of the contra-rotating stirrers of the pan. The cooking time is controlled by feed rate and an adjustable overflow weir which varies the residence time in the cooker. The cooked toffee passes to a similar but smaller unit called a blender where additions of flavours or butter may be made.

A second group of cookers all evaporate the moisture at high speed. These include the Microfilm cooker originally developed for boiled sugar, scraped-surface heat exchangers and plate heat exchangers. In all these cookers the residence time is too short for caramelization to develop and a holding stage of controlled time and temperature is added to allow for this.

Most toffees are shaped either by depositing or by the cut and wrap process. Deposited toffees must flow readily and flatten out in the moulds to facilitate wrapping. The flow characteristics depend upon the protein and sugar contents and flow can be increased by reducing the levels of usage. Unfortunately, the same factors influence shape retention when the toffee is cold and recipes have to be a compromise to suit individual circumstances.

Cut-and-wrap processing shapes toffee in the plastic state after cooling to about 45–50°C when flow is not a problem. Due to the working of the plastic mass, however, there is a tendency for fat to be expressed, and for this reason fat levels are frequently lower than for deposited toffees.

Coconut ice

This is basically a fondant composition which is grained using a very small quantity of seed crystal formed either by rubbing the grain into the syrup or adding a small quantity of fondant. To this is added desiccated coconut and the batch spread on insulated tables to set. It must not be shock chilled. Desiccated coconut as supplied gives a very dry texture and is usually tenderized by the addition of warm glucose or invert sugar syrup which is absorbed into the coconut before use.

Cream paste

Cream paste is a variant of lozenge paste but contains glucose syrup and fat, and/or coconut flour to prevent drying out and hardening. It is not stoved and usually extruded into shape. The most common use of cream paste is the combination with liquorice paste in liquorice allsorts.

Chewing gum

Gum base

A gum base is made from one or more of the following:

(i) *Chicle* is the coagulated latex from the sapodilla tree and was originally the main component of gum base;

(ii) *Jelutong-pontianank* is the latex of the jelutong tree which grows mainly in Malaysia and Sumatra and is used as a partial replacement of chicle;

(iii) *Other resins* – quite high ratios of synthetic products may be used in present day formulations of gum bases, the manufacturers tend to be reticent about disclosing their ingredients.

Manufacture of gum base is a specialized process and most makers of chewing gum or bubble gum buy the base ready prepared.

Chewing gum

The final product is prepared by mixing a sugar syrup, derived from sucrose, high DE glucose syrup and dextrose, with icing sugar and flavouring, plus the gum base at a rate of about 25% of the total batch weight. The gum base may first be washed, dried, sieved and melted into blocks. The blocks are transferred to heated Z-arm mixers and the icing sugar, sugar syrup and flavouring added. Following mixing, the paste is extruded and cut into 5 kg slabs which are allowed to mature for at least 24 h. The slabs are then kneaded at an elevated temperature and extruded into sticks for subsequent wrapping.

Compressed tablets

Two methods, slugging and wet granulation, are used for the manufacture of compressed tablets. In slugging, the powder mix is treated with a lubricant and glidant before precompression into large granules; see Figure 11.6. For wet granulation, sufficient binder solution is added to the dry ingredients to make a malleable paste. The paste is then forced through a wide meshed sieve and dried in an oven at 32–65°C (90–130°F). The dried mixture is treated with a lubricant and flavour and sent for pressing. Pressing is carried out in continuous machines which operate at high

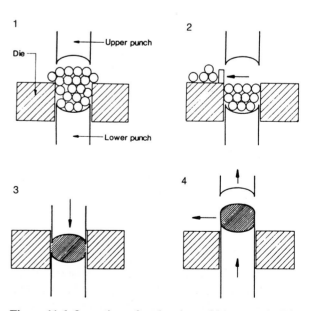

Figure 11.6 Operation of a slugging tablet press. 1, Die is filled with powder; 2, excess powder is removed; 3, powder is compressed between the punches; 4, upper punch is raised and tablet is ejected by lower punch (from LFRA, 1983).

speed. Tablets can be formed either in single or multiple die actions in which the lower and upper dies are brought together under pressure.

The mixture must flow smoothly into the dies and the presence of a certain amount of 'fines' is helpful in developing a good tablet structure. Hardness depends on an even flow of granules into the dies and the amount of pressure used during compression. The amount of binding agent and lubricant used in compounding the recipe also influences the physical properties of the tablet.

For tablet sweets, icing sugar and dextrose are normally used as a base, with gelatin, gum acacia or gum guar as binding materials. Stearic acid and its salts are excellent lubricants.

Base fondant

Base fondant is a suspension of very fine sugar crystals in a syrup phase. Approximately 60% of the product is crystal with a diameter of about 8–10 μm. Whilst it is possible to make fondant by hand, most processes are continuous and consist of boiling a syrup to about 120°C (248°F) and cooling this rapidly to about 50°C, when it is subjected to violent agitation to cause rapid nucleation and crystallization. Once the process has started, the plant is full of seed crystal and the fondant is virtually fully crystallized as it emerges. Crystal size is reduced by beating at lower temperatures and at higher shear rates, but this increases the plant loading and energy costs.

Due to the latent heat of crystallization of sugar, heat is liberated during beating and many plants have water cooling jackets. After preparation it used to be normal for fondant to be left to cool and mature for 24 h. The benefits of this are dubious when most fondant cream processes redissolve about 50% of the crystal in base fondant and continuous processes have eliminated the maturing stage.

Base fondant is used primarily for the manufacture of fondant creams, but is also used for seeding other products such as fudge and chews to create a uniform, fine crystal structure on cooling or during storage.

Fondant creams

Two methods may be used to convert base fondant into creams. In the first the base fondant is reheated to about 70°C when it becomes fluid enough to deposit into moulds after the addition of colour and flavour. The second or so called 'bob' method is to boil a syrup identical in composition to the base fondant to the same solids,

turn off the steam and add the same quantity of cold base fondant. This will give virtually the same product at the same temperature as remelting base fondant but has the advantage that only half the output requires processing into base cream.

Fudge

Fudge is basically a toffee or caramel with a high sugar content which has been deliberately crystallized during processing. It is produced by boiling a caramel batch to 120–123°C, cooling it to about 105°C and adding fondant to initiate crystallization. The quantity of fondant added will affect the final texture. Usually about 5% is required but increasing the quantity will produce finer crystal and a more plastic product. It is then cast or spread on tables to allow crystallization to develop. Like many other confectionery products, fudge should not be subjected to shock temperature changes and should not be cooled too rapidly, since crystallization will be delayed. Shock temperature changes usually lead to mottling or 'starring' of the fudge. This is a visual defect caused by contraction of the syrup phase within the crystal structure. Fudge frequently contains inclusions of nuts, dried fruit or chocolate and these may be added after boiling to assist the cooling process.

Jellies

The main gelling agents used for confectionery are agar, gelatin, gum arabic, pectin and starch. Agar jellies are no longer very popular and gum arabic and starch are largely used for hard gums and pastilles. Gelatin gives a sweet with a very elastic chewy texture, whereas pectin produces a softer eating short texture. Pectin has a superior flavour-release characteristic and has become increasingly popular as a confectionery jelly. Processing varies with the nature of the gelling agent.

Pectin jellies

Close control of pH is required for jellies made with high methoxyl pectin. The procedure is to premix pectin and sugar to aid dispersion and add this with vigorous mixing to hot water containing a buffer salt to maintain pH 5. This prevents hydrolysis of the pectin and inversion of the sugar. This solution is boiled to dissolve the pectin and then the sugars are added. The jelly is boiled to the required solids level, usually about 80%, and after addition of colour and flavour, is acidified to a pH of 3.5. Below pH 3.5 the rate of setting increases, above that pH the gel strength is impaired.

Pectin jellies must be deposited rapidly after acidification. Pressure cookers can be used to eliminate the need for moisture evaporation and the pectin solution containing sufficient sugar to give the desired final solids is pumped continuously through the cooker which may be of either a scraped surface or live steam injection type.

Low methoxyl pectin jellies do not require acidification and setting is caused by the presence of calcium ions. There is normally enough calcium in tap water to achieve this. These non-acid jellies are used mainly for producing imitation Turkish delight.

Gelatin jellies

Gelatin solutions cannot be boiled, so the sugars are dissolved and boiled to the required solids level, usually about 84%. Meanwhile the gelatin is soaked in at least two times its weight of water for at least 2 h. It can then either be added to the boiled syrup or heated to melt the gelatin while the sugar syrup is cooled to about 95°C, when the two solutions are mixed. Incorporated air is removed either under vacuum or by allowing it to rise to form a scum which is removed. Acidification should not be carried out above 60°C.

Agar jellies

Agar is dissolved in water and boiled to achieve solution. When solution is complete the sugars are added and the jelly boiled, normally to about 76–78% solids. This solution must be cooled to about 60°C before acidification and depositing.

Agar requires a large volume of water for complete hydration and solution. Normally at least 35 times the weight of agar is used. Since sugars retard the hydration, it is usual to add the agar to the water and boil rapidly to achieve complete solution. After this the sugars are added and the jelly boiled to the desired solids level, generally between 76–78%. Agar jellies lose gel strength when boiled above 80% solids and gelling power is rapidly destroyed in hot acid solutions. For confectionery jellies, the boiled batch is cooled to 60°C (140°F) before the addition of acid.

Jellies produced from agar alone have a short eating texture and do not carry flavour well. To improve the texture, fruit pulps, jams and other gelling agents are frequently added.

Alginate jellies

Alginate confectionery jellies are rare, the best known being a simulation of a glacé cherry. Gelation is achieved by injection of a syrup containing sodium alginate into a calcium chloride solution which gives a calcium alginate gel. Alginate jellies are irreversible in that they do not melt on heating and do not disperse in the mouth.

Starch jellies

These can be made by boiling a slurry of thin boiling starch, sugar and glucose syrup, but this is a time consuming operation and normally pressure cookers of either the scraped surface or steam injection type are used. These eliminate the need for excess water, and jellies can be processed at the final solids levels.

Liquorice

Liquorice paste

Liquorice paste is compounded from wheat flour, crude sugar, glucose syrup, molasses, liquid caramel, gelatin, block liquorice, salt and aniseed oil. Wheat flour is used at about 25% of the total weight of the cooked batch. Soft wheat flour with a low protein content is more suitable for liquorice manufacture than hard wheat flour. Crude sugars and molasses are needed to develop the full flavour of the confection. The use of salt improves the flavour of the liquorice.

In the conventional process for liquorice paste manufacture, the flour is slurried in one to two times its own weight of water and then pumped or gravity fed into the boiling sugar mass. It is essential that the mixture held in the open steam heated pot is effectively stirred to prevent localized overheating. The batch is then cooked for 1–4 h, depending on whether the paste is to be used in conjunction with other confections or in novelty lines. Because of the long cooking period, it is economically convenient to manufacture large batch sizes of 1 tonne or more. The gloss and texture of liquorice paste are controlled by the extent of gelatinization that has occurred in the wheat flour during boiling. Gelatinization can be controlled by varying the slurry water, the rate of addition of the slurry, the sugar composition of the boiling mass, the type of wheat flour and the boiling time.

Liquorice paste can be manufactured using continuous methods. In a typical layout, a premix is prepared and pumped at 66–71°C (151–160°F) into a scraped film heat exchanger. The heat exchanger consists of a central shaft rotating between the heated walls of a tube. The rotator blades sweep the paste forward to the discharge outlet where the liquorice is extruded at between 110 and 120°C (230–248°F). In conventional liquorice paste manufacture, the finished product is dried by stoving at about 45°C, 55% relative humidity for 12–24 h. Liquorice paste made by a continuous procedure does not need to be stoved.

Liquorice allsorts

Liquorice allsorts are produced by the layering or simultaneous extrusion of liquorice paste and cream paste. The liquorice paste is processed to have a short texture to match that of the cream paste.

Lozenges

Lozenge paste should be produced in small batches from the finest icing sugar and 42 DE glucose syrup, with a solution of gum arabic or gelatin as binder. Mixing should be carried out under standard conditions, particularly with regard to time, in double-armed units. Well mixed paste will retain the impression of the fingers when pressed into a ball and will not break or crumble. Once mixed, the paste should be used quickly to reduce dragging in the dies. Scrap must be used up as soon as possible after production to prevent case hardening. The trayed stamped lozenges should be stoved at between 32 and 43°C (90–109°F) until they are hard enough to be packed. A significant amount of the flavouring oil is lost during the manufacture of the lozenges. The quantity of flavouring should be increased to compensate for the loss taking place during stoving. A small trace of blue colour added to white lozenges will improve the brightness of the product.

Marshmallow

Marshmallow is produced from white sugar, high DE glucose syrup or invert sugar, dextrose and a foaming agent. Foaming agents suitable for marshmallow are egg albumen, gelatin and milk protein. Batch or continuous plant can be used for production and in earlier processes planetary or horizontal beaters were used. These could be either atmospheric or operate under pressure such as by the Morton pressure whisk. Continuous systems have revolutionized the manufacture of

marshmallow and are highly effective in producing stable foams. Typically, air is forced into the mix under pressure and subdivided by passage through a high speed dispersion head, jacketed to prevent heat loss. The quantity of incorporated air, the rate of dispersion and the pressure of all are independently variable.

The marshmallow must not be suddenly exposed to normal atmospheric pressures after being manufactured in a pressurized unit and normally a long discharge pipe is used to allow gradual expansion. Factors affecting the quality of the foam produced in a continuous unit are the viscosity of the syrup, the temperature of the syrup, the pressure of the air, rotor speed, solids concentration of the syrup and the type of foaming agent.

Marshmallow mixes are not easy to deposit in starch at final solids levels of about 76–78% solids and some manufacturers deposit at lower solids and lightly stove the product to dry off excess moisture. Alternatively, marshmallow can be run into trays and cut after setting.

More recent developments include extruded marshmallow, where the mix is cooled after aeration in a scraped-surface heat exchanger and extruded on to a bed of starch as a rope, which after setting is cut into pieces. This product can be packed virtually straight from the extruder. The texture is not as elastic as a deposited mallow.

Marzipan

Marzipan is prepared from cane or beet sugar (sucrose), glucose syrup, almonds and glycerine or sorbitol. The blanched almonds are refined and to these are added the sugar syrup which has previously been concentrated to 90% total solids. Following thorough mixing, the batch should be cooled and again passed through a refiner. At this stage the marzipan can be rolled into the sheets and thence cut into bars or worked into mosaic patterns.

Many smaller manufacturers do not make their own base product but buy almond paste and add extra sugars and other ingredients as extenders. A number of marzipan substitutes are produced using either apricot or peach kernels or soya flour.

In common with all products containing high moisture and sucrose levels, marzipan can ferment. It is important that the non-crystallizing sugar (glucose syrup solids, or dextrose or fructose at the usual levels of addition) is at least 1.5 times the water content to prevent this.

Nougat

Nougat originated in Montelimar in France where it is still a speciality. True nougat is made from honey, egg white, nuts and preserved fruit. These days many nougats are made from sugar and glucose syrup and may not contain honey. The use of sugar allows nougat to be made in grained or ungrained form and some products use gelatin as the aerating agent instead of egg albumen.

Nougat may be made by boiling a sugar syrup to the required solids content, and in the case of egg albumen whipping an aqueous foam as light as possible. The syrup is cooled to 100°C and the foam mixed in using a stirred pan or beating machine. At the end the nuts and fruit are added and the product cast into trays lined with rice paper and allowed to set. It is then cut using rotary knives.

When gelatin is used the soaked gelatin is added to the sugar syrup in a beater and whipped to the required density, then cooled, extruded and cut to size. This type of nougat often contains desiccated coconut. Sometimes to improve cutting, a small quantity of fat and glyceryl monostearate is added to assist the cutting process; this also reduces the tendency for the product to stick to the teeth.

Nut products

The terminology of products based on sugar and nuts is very confused and differs from country to country. Products such as pralines, nougatines and croquantes are all made from roasted nuts and caramelized sugar and may or may not be refined (reduced to a pasty consistency using a refiner). In Germany the name 'nougat' is also used for a product of this type.

Noisettes are similar but contain a proportion of cocoa liquor.

Pralines are pastes made by dry melting sugar to introduce some caramelization and adding roasted nuts, usually almonds or hazelnuts. The mass is poured on to a slab to cool it rapidly, broken up and put through a refiner. Pralines are used as centres, either in chocolate assortments or in boiled sweets.

Croquant is made by the incorporation of chopped nut pieces into dry melted sugar, cooling the mass and forming into pieces whilst still plastic.

Peanut brittle is a similar product except that the boiled sugar is a normal boiled sweet composition containing glucose syrup and the product is usually formed into bars, rings or broken into pieces.

Truffles

Truffles are paste products which may be either syrup or fat based. The syrup based product contains milled sugar or fondant, glucose syrup, condensed milk or milk powder, cocoa powder, chocolate or cocoa liquor, or a compound. The fat-based product contains chocolate, or a compound, milled sugar, glucose syrup, condensed milk and may contain a small quantity of gelatin.

Mixing is carried out in a paste mixer or a planetary beater using a flat blade beater or dough hook. The usual flavour for truffles is rum but many other flavours are used. Traditionally the paste was rolled into balls and covered with cocoa powder or chocolate vermicelli. Currently many chocolate assortments contain a high proportion of truffle centres which are either enrobed or used as filling for moulded sweets.

CHOCOLATE CONFECTIONERY

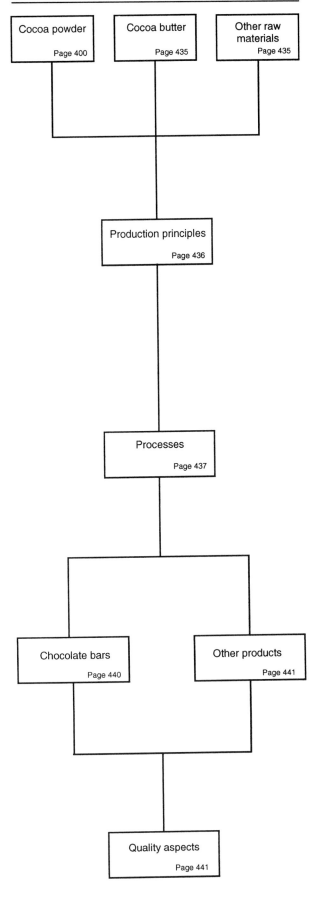

RAW MATERIALS

The production and properties of cocoa beans, cocoa, cocoa butter and chocolate are fully described in Chapter 10.

Cocoa powder

Many speciality cocoas are produced by variations in the roasting and alkalization treatments. Cocoa powder is sometimes used as a flavouring and colouring agent in sugar confectionery and for this purpose may be dispersed in vegetable fat to assist incorporation.

Cocoa butter

Cocoa butter is a natural fat present in cocoa beans. The properties of the fat dictate the behaviour, texture and manufacturing characteristics of the chocolate. Cocoa butter can crystallize in at least four crystal forms, only one of which produces a stable chocolate with a long shelf life (see under PRODUCTION PRINCIPLES). The stable crystal form of cocoa butter melts at 37°C (99°F).

Cocoa butter equivalents

There are a few fats, some natural such as illipe or shea nut, and some manufactured products which are both physically and chemically similar to cocoa butter and whose mixtures with cocoa butter also have similar properties. These fats are called cocoa butter equivalents and can be mixed with chocolate with no adverse effects.

Cocoa butter substitutes

These fats are physically similar to cocoa butter but are chemically different and are used with defatted cocoa, sugar and milk powder to make substitute chocolate or compound coatings. These fats, and coatings made from them, must never be mixed with cocoa butter since they are incompatible and the products become impossible to handle. Some of these fats do not require tempering and can be cooled much more rapidly than cocoa butter. With practice it is not difficult to differentiate between chocolate substitutes and genuine chocolate by taste.

Compounds

Compounds are chocolate compositions where the cocoa butter is wholly or partially replaced by other vegetable fats. A wide range of compounds is available from those compatible with chocolate to some where the vegetable fats are not compatible and fat-reduced cocoa must be used to prevent interaction of the fats.

A few compounds are based on hydrogenated coconut oil. The melting of this fat gives a noticeable cooling effect in the mouth. The shelf life of most compounds is inferior to chocolate and they have found wider use in the bakery than in the confectionery industry.

Emulsifiers

The usual emulsifiers in manufactured chocolates are lecithin, YN (a synthetic phospholipid developed by Cadbury) and polyglycerol polyricinoleate (PGPR). The last is very effective in reducing the yield stress in chocolate and makes it easier to remove ripples in coatings. The function of these emulsifiers in chocolate products is to improve the wetting of the solids by the fat and thus reduce viscosity.

Viscosity reducing agents

Viscosity reducing agents are added during the manufacture of chocolate to improve the flow of the product and to lower the requirement for additional fat. Traditionally lecithin extracted from oilseeds has been used for this purpose, but a range of synthetic material with similar structures to the natural products have been developed.

Lecithin

The efficiency of commercial lecithin varies not only between sources but between different batches from the same oilseed. Purchases of lecithin for use in chocolate manufacture should be based on comparative testing of the viscosity reducing properties when added to a standard chocolate batch. Soya lecithin tends to have superior viscosity reducing properties to lecithin from other sources. Maximum viscosity reduction in chocolate is achieved at about 0.5% of the mixture, beyond this level the viscosity tends to increase again.

Milk products

The main milk products available were considered under SUGAR CONFECTIONERY, but the following may be noted:

Milk powder

In the UK in chocolate manufacture milk powder has been largely or partially replaced by milk crumb which gives a characteristic flavour preferred by some consumers.

Whey powder

Demineralized whey powder is permitted in some countries as a partial replacement for milk powder in milk chocolate.

Milk crumb

Milk crumb is a mix prepared from cocoa beans, milk and sugar. The blend of cocoa beans in the crumb can be made to the specification of the chocolate manufacturer. Crumb is more expensive to use than the traditional raw material milk powder, but the chocolate produced has an enhanced flavour and texture The stages in manufacture of milk crumb are:

(i)   Quality checks on the incoming milk
(ii)   Repasteurization of the milk
(iii)   Mixing with sugar syrup
(iv)   Evaporation to low moisture content.
(v)   Mixing with cocoa mass.
(vi)   Setting period
(vii)   Slicing then holding under vacuum
(viii) Crushing to powder.

Milk crumb can be used alone after refining or with the addition of cocoa butter or other appropriate fats or by blending with extra cocoa mass and other recipe ingredients. The presence of a high level of milk solids limits the process temperature to 50°C (122°F) or less. Milk crumb is used extensively in the manufacture of milk chocolate and introduces a caramelized flavour from Maillard reaction products formed in the crumb process.

PRODUCTION PRINCIPLES

Chocolate is the classic example of a product where stability of shape is achieved by solidification of fat. It is also virtually anhydrous and therefore is not prone to microbial deterioration. Chocolate composition is more subject to legal standards of composition than most other confectionery and the inclusion of vegetable fats which is common in the UK, the USA, Australia and elsewhere has resulted in many arguments in Europe over the definition of chocolate.

Particle size

The particle size of chocolate is important as it affects viscosity and thus cocoa butter usage. It has been said that to eliminate grittiness the particle size should be less than 30 μm, but this is affected by the nature of the solid material, and in fact a proportion of considerably larger particles can be tolerated. This is important in relation to the refining process.

Tempering

Cocoa butter crystallizes in a number of crystal forms or polymorphs. Only one of these is stable at room temperature and is known as Form V (Form five). When handling chocolate for subsequent setting this form must be the only one present as seed crystal. In the process of tempering the chocolate is first heated to about 50°C to melt any crystal present and then cooled to about 27°C to start the crystallization process. At this stage it will contain crystals not only of Form V but other unstable polymorphs as well. On reheating carefully to about 32°C the unstable crystals melt leaving only Form V crystals as seed in the tempered chocolate. The temperatures for cooling and reheating cannot be specified accurately as they depend on the composition of the chocolate being used and have to be established by trial. In smaller operations tempering is carried out manually in tempering kettles, but many manufacturers use fully automatic continuous tempering machines.

Temper meter

The temper meter is an instrument for plotting a cooling curve of tempered fat or chocolate. A standard container is filled with a tempered sample and placed in iced water. A sensitive thermocouple linked to a recorder is inserted in the sample and plots the cooling curve. As the sample cools and crystallization occurs, the latent heat released exceeds the cooling rate and the sample heats up. The degree of supercooling and subsequent temperature rise provide information on the state of temper of the sample.

Viscosity

Molten chocolate is non-Newtonian in viscosity, that is, the apparent viscosity is dependent on the rate of shear. Viscosity is affected mainly by the contents of cocoa butter and any emulsifiers or viscosity reducing agents. These can have a marked effect on yield stress: the choice of emulsifier depends on the type of product required. In recent years there has been an increase in the viscosities at which chocolate is handled, this may result in higher yield stress values which also influences the choice of emulsifier.

Yield stress

There is a minimum force required before molten chocolate will flow, termed the yield stress. Unless the yield stress is high any decorative marks raised on the surface of sweets would flow out and disappear. However, moulding chocolate does not need to have a high yield stress.

Water

Free water dramatically increases the viscosity of chocolate and must not be allowed to get into equipment. For this reason chocolate machinery is rarely washed with water but flushed with cocoa butter.

PRODUCTION PROCESSES

Chocolate manufacture

The manufacture of cocoa mass or cocoa liquor, cocoa powders and cocoa butter are fully described in Chapter 10. The stages involved in the manufacture of chocolate from cocoa mass are:

(i) *Recipe compilation*, carried out in heavy duty mixers or melangeurs,
(ii) *Refining*, size reduction carried out on multiple steel roll refiners with very fine adjustment; the particle size of chocolate is normally about 30 μm,
(iii) *Conching*, agitation under heat to remove unwanted volatile substances and reduce viscosity; final viscosity is adjusted at this stage.

Mixing

In the past the melangeur was virtually the universal chocolate processing unit. It was used for refining, mixing ingredients and conching. The output, however, was relatively small and it has to a large extent been replaced by larger specialist machines. The melangeur consists of a flat bottomed circular dish with either a granite or cast iron revolving base, heated by gas or steam. Two granite rollers are mounted each side of the central pillar and run on the bed as it revolves. The gap between the rollers and the bed is adjustable so that as refining progresses they can be lowered. A system of scrapers deflects the charge under the rollers during each revolution of the bed. Similar machines are used in other industries where they are also known as edge runners.

Refining

In chocolate confectionery, refining is the term used for particle size reduction of the cocoa solids in a mixture with fat, so as to develop a smooth pasty consistency. The smoothness of a chocolate is related to the efficiency of refining.

The equipment is normally of the five-roller variety, with granite rolls for coarse reduction, then steel for fine particle reduction. The steel rolls are capable of delicate adjustment and precision in their manufacture. They are constructed in the form of hollow cylinders with end facing and are cooled by water circulation.

To balance the distribution of frictional heat, the thickness of the steel walls varies along the length of a roller and there may be an extra thickness at the ends. Frictional heat developed at the ends is difficult to remove by cooling water alone, and any preferential expansion there would distort the roll. Such a change of shape would interfere with the fine adjustment of the rolls and cause the discharged chocolate to vary in particle size according to its position along the rolls during its passage through the refiner. Modern rolls are made with slightly conical ends to deal with the same problem.

Statements on the output of refiners may be misleading, as the rate depends on the desired quality of the finished product. Small adjustments of the gap between the rolls can produce large variations in throughput.

The efficiency of a refiner depends on the operational method as well as the quality of design of the machine. The first factor affecting the acceptability of the product is the precision of control over the ratio of speeds between the various rolls. This ratio depends on the type of drive, the gearing of the rolls and the level of control available to the operator. Each roll is run at a different speed, the lowest at the lowest speed and the top roll at the highest speed. In a typical refiner, the variations are: 20 rpm (the bottom roll), 30, 45, 66 and 100 rpm (the top roll). Because of the speed

difference, each roll has a slipping action as it moves in relation to the next roll. Efficient refining is a combination of pressure and the ratio of speed and movement between the rolls.

It is essential that the chocolate is of the right consistency before refining, neither too greasy nor too dry. Output levels are affected if the product is too liquid, particle size reduction is impaired if it is dry. The latter is due to a buildup of paste on the lower roll. Such accumulation places a considerable strain on the refiner, giving excessive wear on the rolls, gears and bearings.

Conching

The main reasons for conching chocolate are:

(i) To reduce viscosity
(ii) To improve flavour
(iii) To improve texture.

During the refining process, an enormous increase in surface area of the solid particles occurs and the resulting chocolate mass is quite dry. The sugar particles are not easily wetted by the cocoa butter and although small are compacted into aggregates.

The original conche was of the longitudinal pattern which consisted of a temperature controlled trough with a granite bed on which a granite roller or rollers moved back and forth. The ends of the trough were shaped so that the chocolate ahead of the roller, on reaching the end of the conche, was thrown back over the roller into the bulk of the chocolate and the process repeated on the return stroke.

To make the dry charge workable, some cocoa butter and lecithin had to be added to achieve sufficient fluidity for operation but from then on the fluidity increased due to the working of the mass. The action of the chocolate being thrown back by the ends of the conche created a large surface area and at the temperatures used, 75–80°C for plain chocolate and 50–60°C for milk chocolate, moisture and the volatile acids such as acetic, propionic and butyric acid were driven off. In addition, some oxidation of flavour components occurred resulting in a reduction in astringency and the development of a mellower flavour. Conching does not reduce particle size, but it is said to round off the corners of the sugar particles.

The longitudinal conche had a processing time of up to 4 days for plain chocolate but only about 1 day for milk chocolate. It was found, however, that the evaporation of unwanted volatile components was faster if the mass could be worked in the dry state before the addition of cocoa butter and lecithin. This was not practical in the longitudinal conche and most conches these days are designed on rotary principles. They frequently have two distinct stages of operation and often are provided with dry air injection to assist the evaporation and oxidation stages. Continuous conches are also now in use which have reduced conching times for milk chocolate to about 4 h.

Complete processing

The McIntyre mixer refiner is virtually a complete chocolate processing unit in one machine. Basically it consists of a horizontal water-jacketed temperature-controlled drum fitted with an internal rotor. The internal design is such that grinding, mixing and the conching operation can all be carried out in the one machine. It is used mainly for relatively small scale chocolate production and for the manufacture of compound.

Enrobing

Originally introduced for chocolate coating, enrobing machines are now also used for cream coating and caramel coating. The procedure is basically similar for all processes but the details vary significantly. The sequence of events is:

(i) Centres are fed on to a feed band and transferred to a wire mesh band which passes through the enrober.
(ii) The coating medium is maintained at a constant temperature and in a controlled condition in an agitated tank and is pumped to a flow pan. The flow pan creates a continuous curtain of coating and feeds a bottoming device which forms a bed of coating which floods the mesh band.
(iii) The centres are passed through this curtain and bed and are covered on all surfaces.
(iv) The centres then pass under a fan which blows off surplus coating which returns to the coating tank.
(v) The centres are then vibrated to remove ripples left by the fan, smooth the coating and drain surplus which also returns to the tank.
(vi) The centres discharge from the enrober on to a cooling conveyor passing over a detailer which is a rapidly spinning rod across the end of the wire band. This breaks the tail that forms as the centre leaves the wire band and either throws it onto the sweet or back into the enrober.

Modern chocolate enrobers are large machines up to 2 m wide, and have elaborate control systems for temperature, air flow (velocity and direction) and amplitude and frequency of vibration. The objective is to control coating ratios closely and maintain the chocolate in the optimum state of temper. Many enrobers are fitted with decorators, either mechanical marking discs or wheels which just touch the liquid chocolate to make patterns, or piping nozzles on a manifold which can be oscillated or moved in a circular orbit leaving a thin piped decoration of the same or a contrasting coating on the sweet. The sweets then pass into a coolbox to set the chocolate.

Prebottomer

It is always difficult to achieve a uniform chocolate coating on the base of the sweets in the enrober and in many cases a prebottomer is used. The enrober feed band is usually about 6 m long and runs on a water cooled table. Before this band is a temperature controlled tank fitted with a wire mesh belt over which the sweets travel. Underneath the top of this belt is a revolving drum which picks up the tempered chocolate from the tank and forces it through the mesh. The centres pass over the chocolate, picking up a layer on the base which is set as the sweets pass over the water cooled feed band. At the enrober, the feed band passes round a knife edge leaving the chocolate bottom adhering to the centre. This is then given a further coat of chocolate in the enrober itself.

Drip feeding

When an enrober is operating on tempered chocolate it is possible to feed cooled but untempered chocolate continuously at the rate at which it is being used. Provided the usage rate is not too high, the tempered chocolate in the enrober tank seeds the untempered chocolate as it is added. The operating conditions may require modification from those where all tempered chocolate is being used.

Decorating

Originally chocolates used to be decorated by fork marking where a pattern was raised on the surface of the molten coating using a wire fork, by piping a design, often in a contrasting chocolate, or by adding a topping such as nuts or crystallized violet petals when the coating was still liquid. Marking and piping have been mechanized for many years, some of the equipment being quite complex.

Spinning

Hollow chocolate items such as Easter eggs or Christmas novelties are frequently produced by spinning. They may be made in two halves or as a single shell. Both are made in split moulds but when two halves are required the mould has a partial divider between the halves. In the process the required amount of tempered chocolate is deposited into the mould which is quickly closed and attached to the spinner which rotates the mould around two axes at right angles simultaneously. The chocolate spreads over all the interior of the mould which is then cooled to set the shell when the mould may be opened to remove the product.

PRODUCTS

Chocolate

Plain chocolate

This is a suspension of finely ground roasted cocoa beans or cocoa mass and sugar particles in cocoa butter. The flavour of chocolate is primarily dependent upon the selection of the cocoa beans and the degree of roasting, but the degree of conching has an effect, and in most cases flavours such as vanillin, ethyl vanillin or spices are added.

In general terms, a plain chocolate will contain approximately four parts of cocoa liquor (ground cocoa beans or cocoa mass) to five parts of sugar. This mixture contains 23% of cocoa butter. A moulding chocolate requires about 28% of fat and an enrobing chocolate about 33%, which means that additional cocoa butter is required. Because cocoa butter is expensive, cocoa butter equivalents may be used as the additional fat, though many countries will not permit the use of these fats in chocolate. Butter oil is sometimes now added to plain chocolate as an antibloom agent. The quantity used is about 4% of the chocolate. Typical formulations of chocolate and the resultant composition are shown in Table 11.7.

Milk chocolate

This is similar to plain, with the addition of dried whole milk powder or milk crumb. Milk chocolate contains approximately one part of cocoa liquor, two parts of whole milk powder or milk crumb equivalent and 4.5 parts of sugar. This mixture contains 15% of fat including butter fat from the milk and thus milk chocolate requires more addi-

Table 11.7 Milk and plain chocolate – constituent percentages by weight

	Sugar	Non-fat milk solids	Non-fat cocoa solids	Fat	Flavouring and emulsifiers	Total
Milk chocolate						
Sugar	46.2	–	–	–	–	46.2
Milk powder	–	15.6	–	5.0	–	20.6
Cocoa mass	–	–	4.9	5.4	–	10.3
Cocoa butter	–	–	–	22.4	–	22.4
Flavouring and emulsifiers	–	–	–	–	0.5	0.5
Totals	46.2	15.6	4.9	32.8	0.5	100.0
Plain chocolate						
Sugar	48.2	–	–	–	–	48.2
Milk powder	–	–	–	–	–	0.0
Cocoa mass	–	–	18.4	20.1	–	38.5
Cocoa butter	–	–	–	12.8	–	12.8
Flavouring and emulsifiers	–	–	–	–	0.5	0.5
Totals	48.2	0.0	18.4	32.9	0.5	100.0

tional fat than plain chocolate. Due to the butter fat content, milk chocolate is less brittle than plain chocolate and more resistant to bloom.

Many recipe variations are possible. Apart from using milk crumb in varying quantities, skimmed milk powder and butter oil can replace whole milk powder and also whey solids can be used. Many of these options are cost related, and in this respect much work has been carried out on the effect of emulsifiers in reducing viscosity and saving cocoa butter.

Chocolate flake

This is produced by passing a specially processed chocolate paste through refining rolls and planing the film from the last roll. By controlling the take off bars of uniform size can be produced.

Chocolate vermicelli, chocolate strands or streusel

This product is made by extruding a tempered chocolate paste through a perforated die plate and setting the strands as they emerge.

Chocolate liqueurs

Two distinct type of chocolate liqueurs are manufactured; these can be classified as 'crusted' and 'crustless' varieties. Crusted liqueurs are produced by depositing a supersaturated liqueur – sugar syrup at a concentration of 71–74% – into bottle shapes made in warm moulding starch. An even crystal shell is produced by turning the deposited shape as soon as a sufficiently thick sugar crust has formed. The thickness of the shell is governed by:

(i)   The concentration of sucrose in solution
(ii) The holding temperatures
(iii) The presence of other ingredients
(iv) Agitation received during manufacture.

True liqueur additions are usually between 25–30%. The crystal shell is double coated with plain chocolate. Crustless liqueurs are prepared by depositing prepared liqueur syrups into preformed chocolate shells. The preformed chocolate mould can be produced in two ways. The first method has the following sequence:

(i)   Filling bottle moulds with chocolate
(ii)   Inversion of the moulds
(iii) Partial draining
(iv)   Return of moulds to the upright position
(v)   Setting the chocolate
(vi)   Filling the chocolate shapes with liqueur syrup.
(vii) Top coating with chocolate.

The second method entails:

(i)   Partial filling of a bottle mould with chocolate
(ii)   Closure of the mould
(iii) Spinning at high speed
(iv)   Puncture of the formed bottle top
(v)   Filling the shapes with liqueur syrup
(vi)   Sealing the top with chocolate.

Crustless liqueurs normally contain a higher percentage of true liqueur but have a shorter shelf life.

Aerated chocolate

This is made by expanding chocolate containing fine air bubbles or carbon dioxide under vacuum and setting it before the vacuum is released. Aerated chocolate does not readily demould and is usually produced in a preformed chocolate shell.

Bars or count lines

Count lines in the food context may be defined as individually wrapped and individually sold bars (or occasionally other shapes) which usually fall into one of three main categories.

(i)   *Confectionery bars.* These are composed principally of the usual confectionery items–chocolate, caramel, toffee, marzipan, sugar-based creams and fondant, and so on, frequently

plus additives specially to characterize the final product. Some components may be whipped or otherwise aerated and the bars may be enrobed with chocolate or couverture or not. They are frequently given a wafer backbone.

(ii) *Cereal or granola bars.* These bars are basically composed of cooked (and frequently expanded) cereals bonded together by various means including the use of honey, sugar syrups, dextrins, hard fats and so on. They may contain many items in addition to cereals such as nuts, dehydrated vegetable and fruit material and other textural and flavour giving adjuvants. Nutritionally they are very much more satisfactory than confectionery bars though to many people may not be as acceptable.

(iii) *Fruit bars.* These are substantially composed of three major items: fig paste, date paste and pastes made from dehydrated vine fruits (raisins, currants, and so on). They are frequently texture improved by the inclusion of 'crunchy' items which may include nuts and cereals. These play a minor part in their formulation, however.

It need hardly be said that these three already wide categories are supplemented by intermediate ones which combine to some extent the characteristics of the parents and that the dividing line between the categories is thus frequently blurred.

QUALITY ASPECTS

Density

Baumé is a scale of density, based on the concentration of salt solutions. Its common usage in confectionery is to define concentration of glucose syrups.

Refractive index

The presence of dissolved substances in water alters the refractive index, which is a measure of its ability to bend a light beam. Since this varies with the concentration of dissolved substance the refractive index of a pure sugar solution can be used to measure the concentration.

Sugar syrups above about 85% solids when measured at room temperature can cement the refractometer prisms together on cooling and cause damage when they are opened. For these syrups the control of solids content is usually done from the boiling temperature.

Refractometers

Refractometers are widely used in the confectionery industry as control instruments for measuring the sugar content of syrups. The scale is usually directly calibrated in percentage sucrose. Corrections may be applied for the presence of other sugars.

The Brix scale

Degrees Brix is the same figure as percentage sugar solids by weight at 20°C. Table 11.8 gives correction factors for other temperatures.

Relative humidity

Relative humidity is defined as the quantity of moisture vapour in air, divided by the quantity of moisture vapour in saturated air at the same temperature and pressure, multiplied by 100. Instruments for measuring relative humidity include hair and paper hygrometers, wet and dry bulb thermometers and now electronic instruments are available using various types of moisture sensors. These are very rapid in response, but need to be calibrated at intervals. For many years the standard instrument has been the sling hygrometer or some form of aspirated wet and dry bulb thermometer used in conjunction with psychrometric tables. The importance of relative humidity lies in the ability of products to absorb moisture from the atmosphere and it is to prevent this that air conditioning is so important in factories.

A number of instruments are available for determining ERH but they are expensive if only occasional determinations are required. Alternatively the Landrock and Proctor method may be used, in which a number of airtight containers contain saturated salt solutions of known relative humidity. By exposing the product in a range of humidities the weight gain or loss is found and by plotting these results the point where no weight change occurs gives the ERH. Although the Landrock and Proctor method does not require equilibrium but only weight loss or gain, opening the container disturbs the humidity and time must be allowed for this to restabilize.

Another method uses the known relative humidity of glycerol solutions, as given in Table 11.9. By placing a quantity of product in a small airtight container, enclosing about two drops of

Table 11.8 Corrections to be applied to Brix readings made at temperatures above and below 68°F (20°C)

	Temp. (°F)	° Brix and correction to be made									
		10	20	25	30	35	40	45	50	55	60
Subtract	40	0.5	0.6	0.7	0.8	0.9	0.9	0.9	0.9	0.9	1.0
correction	50	0.5	0.5	0.5	0.5	0.6	0.6	0.6	0.6	0.6	0.6
	60	0.2	0.2	0.2	0.2	0.2	0.2	0.2	0.2	0.2	0.2
Add	70	0.1	0.2	0.2	0.2	0.2	0.2	0.2	0.2	0.2	0.2
correction	80	0.5	0.6	0.6	0.6	0.6	0.6	0.6	0.6	0.6	0.6
	90	0.9	1.0	1.0	1.0	1.1	1.1	1.1	1.1	1.1	1.0
	100	1.3	1.4	1.5	1.5	1.5	1.5	1.5	1.5	1.5	1.5
	120	2.5	2.6	2.6	2.6	2.6	2.6	2.6	2.6	2.5	2.5
	140	3.8	3.8	3.8	3.8	3.8	3.8	3.7	3.7	3.6	3.6
	160	5.1	5.1	5.1	5.1	5.1	5.0	5.0	4.9	4.8	4.8
	180	6.7	6.5	6.4	6.4	6.3	6.3	6.2	6.1	6.0	5.9
	212	10.0	9.6	9.4	9.3	9.1	8.9	8.7	8.4	8.2	8.1

Table 11.9 Refractive index and ERH of glycerol solutions (adapted from International Critical Tables)

Glycerol (%)	Refractive index	ERH
0	1.333	100.0
10	1.345	96.0
20	1.357	92.0
30	1.371	89.0
40	1.384	86.0
50	1.398	80.0
60	1.413	73.0
70	1.428	62.0
80	1.443	49.0
90	1.458	29.0
100	1.474	0.0

pure glycerol in a small cup made from cooking foil and allowing it to come to equilibrium overnight, the glycerol concentration may be determined by a laboratory refractometer and from this the ERH established. The disadvantage of both these methods is the time taken for equilibrium to be established.

Particle size

Particle size is very important in relation to texture and handling properties. Many methods are available for the determination of particle size and particle size distribution, these include micrometers, the microscope, sedimentation, particle sizing and counting (Coulter counter) and light scattering. In all methods the chocolate must be thinned either with a light oil or a special solvent.

The most convenient micrometer is probably the engineer's anvil micrometer using a flat plunger. The disadvantages are that the method tends to measure largest particle size and it is possible for two particles to get superimposed. Since the particles tend to lay flat it does not usually measure the largest dimension.

The microscope requires a calibrated eyepiece micrometer and again the particles tend to lay flat so the measurement is of the largest dimension. Frequently projection microscopes are used and more recently scanning microscopes have become available which will provide a read out the particle size distribution.

Many sedimentation methods exist. In principle a suspension of chocolate in a light oil or solvent is placed in a tapered sedimentation tube, or centrifuge tube and allowed to settle. After a fixed time the volume of sediment is measured. The method is empirical, but can be improved by taking a number of determinations, pouring off the liquid after various times and examining the sediment by the microscope.

The Coulter counter measures particle volume and counts the number of particles between a range of sizes enabling a distribution curve to be constructed. The chocolate has to be suspended in a suitable electrically conducting solvent and details are provided with the instrument.

Light scattering measures the area of the particles in the path of the light beam which is a low power laser. These instruments provide a direct readout of particle size distribution.

Viscosity

Whilst sophisticated viscometers are essential for complete characterization of flow properties, for routine use simple flow rate, falling ball or torsion instruments are often used in production. The measurement of the viscosity of tempered chocolate is very difficult.

Research

The main centre of research in the UK is at the laboratories of the Leatherhead Food Research Association, LFRA. Work is supported by industrial contributions and sponsorship with some finance from central government. Most major UK companies maintain applied research programmes in their laboratories and fundamental research of interest to the industry is carried out in universities such as Leeds, Reading, Nottingham and Surrey. Research on methods of analysis is carried out at the Laboratory of the Government Chemist

and in the laboratories of the Ministry of Agriculture, Fisheries and Food.

The International Sugar Confectionery Manufacturers Association (ISCMA) was founded in Europe in 1953. An earlier body, the International Office of Cocoa and Chocolate (IOCC), was founded in Belgium in 1930 and through its Cocoa Research Policy Committee, initiates and supports studies for the improvement of cocoa cultivation in producer countries. A joint technical committee of IOCC and ISCMA promotes research of interest to industry, particularly in the area of methods of analysis.

Research in the USA is carried out in the laboratories of major food manufacturers, through sponsorship at universities and in the laboratories of the US Department of Agriculture on the use of new ingredients and methods of analysis. Sponsored work is undertaken by universities on behalf of trade associations such as the Pennsylvania Manufacturing Confectioner's Association.

A similar position applies in Europe, with strong support for cooperatively sponsored research and education in West Germany. Other countries active in research include the Netherlands, France and Russia.

REFERENCE

LFRA (1983) *Research Report no. 432*, Leatherhead Food Research Association.

FURTHER READING

Beckett, S.T. (1993) *Industrial Chocolate Manufacture and Use*, 2nd edn, Chapman & Hall, London.

Chatt, E.M. (1953) *Cocoa*, Interscience, New York.

Cook, L.R. (1963) *Chocolate Production and Use*, Magazines for Industry, New York.

Jackson, E.B. (1995) *Sugar Confectionery Manufacture*, 2nd edn, Chapman & Hall, London.

Kempfe, N.W. (1964) *Technology of Chocolate*, Manufacturing Confectionery Publishing Co., Glen Rock, New Jersey.

Kreiten, K., Joike, H. and Meiners, A. (1983) *Silesia Confiserie Manual No. 3*, Silesia/Gerhard Hanke KG, Neuss.

Lees, R. and Jackson, E.B. (1973) *Sugar Confectionery and Chocolate Manufacture*, Leonard Hill-Blackie, London.

Lees, R. (1980) *A Basic Course in Confectionery*, Specialised Publications Ltd., Surbiton.

Lees, R. (1981) *Faults, Causes and Remedies in the Confectionery Industry*, Specialized Publications Ltd., Surbiton.

Meiners, A. and Joike, H. (1969) *Silesia Confiserie Manual*, Silesia/Gerhard Hanke KG, Neuss.

Minifie, B.W. (1989) *Chocolate, Cocoa and Confectionery Science and Technology*, 3rd edn, AVI, Westport, Connecticut.

Rohan, T.A. (1989) *Processing of Raw Cocoa for the Market*, FAO, Rome.

Stock, K.W. and Meiners, A. (1973) *Silesia Confiserie Manual No. 2 (Panned Goods)*, Silesia/Gerhard Hanke KG, Neuss.

Urquhart, D.H. (1955) *Cocoa*, Longman Green and Co., London.

12 Snack Foods and Breakfast Cereals

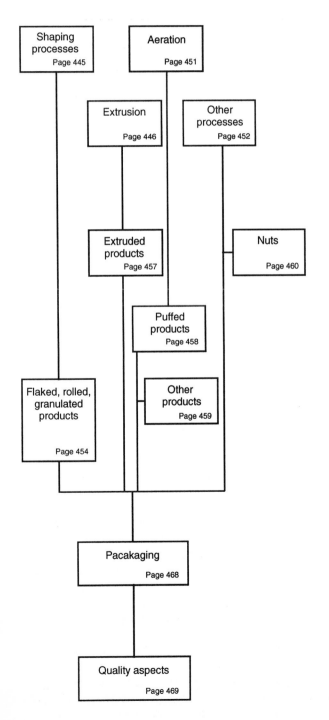

INTRODUCTION

SNACK FOODS

The definition of 'snack foods', which we shall follow in this chapter, is much the same as that used by the grocery trade or included in marketing surveys. The American market research company Frost and Sullivan Inc. suggest the following list:

Conventional snacks:
    Potato chips (UK 'crisps')
    Corn flakes and other forms of corn
    Savoury/extruded products
    Nuts
    Salted/savoury crackers
Newer snacks:
    Instant hot pot snacks and cup soups
    Cereal bars/mixes
    Meat snacks
    Fish snacks
    Novel – ethnic and filled
    Countline confectionery items
    Dairy beverages (milk, yoghurt, fermented milks)
    Sweet baked items including cookies.

Many other foods, however, are typically eaten as snacks in the other sense in which the word is commonly used. This meaning is typically defined in dictionaries as 'a hasty, casual or very light meal'. This would include the traditional mid-morning, mid-afternoon or work-break snack, or the growing multi-snack meals that form the basis of a 'grazing' lifestyle in which three main meals a day can be replaced by up to seven or eight snacks taken largely 'on the hoof'.

No hard and fast rules can be laid down for what is and what is not a snack food. Where a particular product is frequently used as a snack food but is directly derived from a normal, non-snack foodstuff, then it will be given only brief mention in this chapter. Further details can be found in the appropriate chapter relevant to that particular type of food. On the other hand, the many current snack foods which have little or no 'main meal' relevance are dealt with in this chapter.

BREAKFAST FOODS

In the past, when hard manual labour was the lot of most people in the West, the large breakfast gained in popularity. In many countries, it became more or less standardized, for those who could afford it, as a combination of bacon, eggs (usually scrambled or fried), mushrooms, tomatoes, kidneys, fried bread and a few other odds and ends. This was frequently preceded by porridge, and followed by buttered toast and marmalade or other confiture – 'Le petit dejeuner Anglais'. This can still be encountered in many hotels, but has been substantially replaced, both in home and hotel, by a much lighter breakfast consisting basically of some form of cereal, plus frequently fruit (dried, cooked or au-naturel) or fruit juice, and tea or coffee.

Breakfast cereals are produced either from individual varieties of grain or from mixtures thereof; wheat, oats, maize (corn), rye, rice and occasionally, barley, are all employed for this purpose. A range of manufacturing techniques are employed to produce a wide variety of products specifically designed to appeal to all ages and, in many cases, ready to be served at the table directly from the package. A large proportion of these breakfast cereal foods will be dealt with in this chapter; others are referred to elsewhere in this book.

MANUFACTURING TECHNOLOGIES

INTRODUCTION

It can fairly be said that many of the recent major advances in food processing technology have largely emanated from the requirements of the rapidly expanding snack and breakfast foods sectors. The simple basic technologies, referred to in the text of this chapter, are not addressed separately here in any detail. However, the more complex manufacturing techniques, which are essential elements in the conversion of the various raw materials and their mixtures into the desired end products are discussed in some depth. For present purposes, these technologies, and their resulting products, have been classified as follows: shaping processes, extrusion processes, aeration processes and other miscellaneous processes.

SHAPING PROCESSES

Included in this section are descriptions of rolling, flaking and granulating processes. In certain cases, flaking and rolling can be used interchangeably to describe the same operation, as for example, in the flaking or flattening by means of co-rotating rolls of partially precooked cereal grains. In contrast, the terms cannot be interchanged in the case of nuts. Here, rolling is not normally employed; flaking, on the other hand, involves a slicing operation as, for example, in the case of flaked almonds.

Granulation is a very different process, normally based on a form of deliberate partial comminution of otherwise large pieces of the intermediate or finished product. A typical example occurs in the manufacture of 'Grape Nuts', a breakfast cereal which, incidentally, is based neither on grapes nor on nuts.

Flaking (rolling) of cereals

Precooked cereal grains are flaked, either in whole or in part, to achieve the required final dimensions of the finished product. Thickness is normally of particular importance. The principles of this process are the same for all cereals, though the detail is necessarily different. These differences are mentioned later under PRODUCTS.

The grain, or grain pieces, usually corn, wheat, oats (groats), barley (pearled) and rice, are first steamed and, with the exception of oats, usually fully cooked, with a concurrent rise in their moisture content of up to 20%. At the same time other ingredients are added if desired – sugar, salt, malt extract, honey, and so on and the temperature of the cook (99–104°C) leads to the inactivation of any enzymes which are still present in the grains. It may be necessary to hold the cooked grains for a length of time for conditioning and they may be also be partially dried as well. The hot, moist and now well plasticized grains are passed individually between a pair of large heavy steel rolls running at 250–400 rpm, heated to a specific temperature, and with a closely controlled gap between them to ensure uniformity of the eventual flake thickness. The flakes are then dried to an appropriate moisture content.

A similar process may be used to flake extrudates produced from single or mixed flours or semolinas in a cooker extruder (see below under EXTRUSION PROCESSES) or pelletizer. The pieces must, of course, be appropriately conditioned before rolling and then are treated in exactly the same manner as rolled whole cereal grains.

Further processing of the flakes may entail:

(i)   Sugaring – accomplished by a spray of sugar plus honey or another 'non-sticky when dry' component with the sugar.
(ii) Toasting in a very hot oven to cause some blistering (a form of aeration), colouring and some flavouring.
(iii) Addition of flavours by incorporation in sprays or by dusting the flakes

Rolling of cereals has a subdivision, which has been used for many years in the production of shreds and which still has many advantages over the use of the extruder for this purpose. Wheat is the most usual cereal to be shredded and Shredded Wheat has been a popular breakfast item for a long while. It is manufactured by a process with a stage that is unique in breakfast cereal manufacture.

Washed, whole wheat is boiled in salted water for as long as is necessary to cook the contained starch thoroughly and make it fully dextrinized. At this time, the drained grains will have nearly doubled their weight. They are then allowed to cool and achieve (i) a uniformity of moisture content and (ii) a degree of starch retrogradation (i.e. of crystallization) over a lengthy period of up to a day. The wheat is then fed continuously to pairs of counterrotating rolls, one of which is grooved and the other is not. There is no clearance between the rolls and the only way that the wheat can pass between the rollers is through the grooves. 'Doctor' combs are positioned to remove the shreds so formed.

A series of pairs of such rolls feed their shreds on top of each other, any necessary flavouring is added and the mass of shreds is conveyed on a belt where it is cut by deliberately blunted knives which compress and meld the sides and ends of each piece, which can be of any desired size. These now go to a hot oven (about 255°C), where they first expand somewhat before setting and drying, and are finally coloured (browned) at the end of the baking process, which is normally carried out in a typical travelling biscuit oven. The final moisture content of the product is approximately 4%. The flours or semolinas of other cereals may also be used to produce shredded products provided that appropriate modifications are made to the standard wheat process to provide fully acceptable final products.

The term 'flaking', as in 'corn flakes', has been used synonymously so far with the word 'rolling'.

In fact it is not, and another form of flaking is possible and is used in a few products.

Firmly structured food materials, e.g. nuts and some fruits and vegetables 'au naturel', and some cooked and solid, formed products, may be thinly sliced into flakes on what might be termed the 'spokeshave' principle. This employs a very sharp blade of some kind, set at a fixed spacing to give the desired flake thickness. Such methods of flaking usually require specialized manufacturing equipment because of the great variability in the size, shape and nature of the material to be flaked. Almond flakes are an example of product produced by this technique.

Granulating processes

Within the breakfast foods area, there is nowadays a frequent need for the production of granules (e.g. as in 'Grape Nuts'). There are two basic methods of granulating which can be used, as appropriate.

The first method uses the 'pharmaceutical' style of granulator, or the outlet die of a cooker-extruder. A fairly stiff dough of the final cooked product is forced through a sieve having the necessary aperture sizes and breaks off (or is broken or cut off) on release, falls onto a travelling oven belt and is then dried and/or toasted and brought to the necessary final crisp state. The original dough may or may not be aerated.

In the second method final product dough may be sheeted with appropriate smooth rolls and the sheets (usually around 10 mm thick) are baked to a moisture content which makes them readily breakable, but not overly. This is to avoid a wide range of size fragments and the generation of excess fines. The baked sheets are then passed between suitably spaced ribbed steel rollers – or alternatively between spiked rolls – to produce the necessary granules.

There are many variants of these basic methods to serve special needs. Aggregates, which may be mixed in colour and composition, can be produced by loosely gumming particulates together with dextrin, or another vegetable gum, drying them and breaking them up into appropriately sized pieces.

EXTRUSION PROCESSES

Extrusion technology has grown considerably since the 1950s. The principal developments that have taken place include:

(i) The generation, application and control of temperature during the total extrusion process;

(ii) Arising from (i), the possibility of cooking the product in various ways (high temperature, short time or lower temperature, longer time) during the process;

(iii) The possibility of 'puffing' or aerating the product by rapid evolution of steam at the point of extrusion at a temperature above 100°C;

(iv) The possibility of rapid, though usually only partial, drying of product by reason of (iii),

(v) Arising from (i), effective microbiological control of the finished product and destruction of unwanted enzyme activity;

(vi) The possibility of producing fibres as well as the enormous variety of puffed or unpuffed pasta shapes and sizes. This enables particular textures such as meat simulants to be produced, mainly through the use of plasticized proteins as extrudates. 'Baco-bits', which simulate cold, crisp fried bacon, is a typical example of such an extruded product.

These changes could not have taken place without major developments in extruder technology, which have for the most part been as follows:

(i) A change of layout from the original S-shaped format to a straight-through format. This permits better access to and control of all sections of the extruder.

(ii) The development of a wide range of screws or worms having different pitches, with deeper or shallower flights and operating cylinders which may be shallow- or deeply grooved or ungrooved, and of large or small diameter. This results in wide variability in the shear forces produced and the associated energy input.

(iii) The use of jacketed barrels for steam- or hot-water heating, or cold-water or refrigerant cooling. Some designs of extruder also feature hollow screw shafts through which steam can be passed, or electric heating. The use of water or steam injection into the food at various stages of the mixing and extrusion processes is also used to control the temperature.

The basic extruder design, together with the speed of rotation of the screw, the composition and temperature of the mix, and the die design will all dictate the mechanical energy input to the mix. This is partially converted into heat, which may then have to be removed or augmented by the means described above.

(iv) The introduction of twin-screw, twin-cylinder designs.

Nowadays, the food extruder has achieved such a high level of sophistication that it is probably unrivalled in its range of capabilities by any other food-industry process. The things which can normally be accomplished by most modern extrusion plants include:

(i) Metering, mixing and holding the ingredients in a manner appropriate to the desired product.

(ii) Heating, and cooling if necessary, the contents of the extruder barrel. This can be accomplished by absorbing the energy generated in the course of mixing, melding and otherwise manipulating the dough, and by the further addition or removal of heat as required. This process will determine the degree to which the starch present is gelatinized.

(iii) Venting the extruder barrel, with or without vacuum, to control and standardize behaviour and quality of the product on final extrusion.

Increasing sophistication in extruders demands increasing sophistication in their feedstocks. All the complex control systems of such extruders can be rendered ineffectual unless the feedstock is standardized, this standardization not only encompassing the proportional mix of items each with its own quite precise specification, but also limitations on the overall analysis and physical properties of the final mix including of course, particle size(s).

The reasons for this necessity are fourfold:

(i) The particle size of the ingredients must be such as to allow the appropriate flow of mix through the extruder – too small an average particle size (e.g. flour) can cause jamming, whilst too large a size may cause too great a speed of flow and resultant inadequate processing in some areas.

(ii) Some precooking of one or more ingredients may lead to premature plasticizing of the mix and problems arising therefrom.

(iii) In the case of expansion extrusion particularly, an excess of fat can abort the exercise: 3% of fat or oil is the usual top limit here and this limitation also obtains in many non-expansion extrusion situations.

(iv) The overall efficiency (in terms of throughput and energy usage) of the extrusion process can frequently be improved by the addition of GMS (glyceryl monostearate) to the mix. The optimal quantity (frequently of the order of 1.5%) has to be determined experimentally for each specific mix).

Types of extruder

Piston or pump extruders

The simplest form of extrusion is exemplified by the syringe and the spinneret. Here liquids, viscous or otherwise, are extruded through a narrow orifice by pressure, either from a piston in a cylinder or from a pump. In both cases the pressure employed is quite low and no significant amount of work of any kind is applied. The composition or formulation of the liquid has been achieved before it enters the cylinder of the extruder and is unchanged during the operation.

The extrusion of foods started, and still continues, in the context of the butcher's shop. Preminced, spiced and generally formulated sausage-meat paste is caused by a piston to extrude through a nozzle located in the end of a cylinder. This carries a circlet of prepared tubular sausage skin, so that the minced meat, as it extrudes, is provided with appropriate edible containment. Modern large scale sausage making is carried out on the same principle but uses a straightforward continuous screw extruder, which is usually allied with a manufactured rather than a natural skin. Miniature, cocktail or snack sausages, usually about 4 cm long, are made by the same method.

Collet extruder

The Collet extruder was designed in the 1940s for (and now is really only useful for) processing maize grits. It is capable of producing a modest range of variously shaped expanded cooked snacks from low moisture feedstocks. It is a high shear, high pressure, self-heating and self-cooking extruder; the high shear generates a lot of heat in a short length of time. Inlet moisture content is typically 11% (range 9–17%); outlet moisture content 5%; temperature of mix immediately before extrusion 175–180°C; about 0.10% of the input energy to the screw drive motor is converted into heat absorbed by the product.

The products require little drying after extrusion and can readily be sprayed with sugar syrup or other flavouring additives and toasted if desired before consumption.

Pasta press

Most forms of extrusion employed in the preparation of snack foods stem from the pasta press (or extruder) which is itself also used in the manufacture of the first stages of some modern snack foods.

The feed materials for ordinary pasta (spaghetti, macaroni, tagliatelle, etc.) are very simple – wheat semolina and water. The wheat used in the manufacture of good quality pasta is almost exclusively durum or amber durum, the semolina from which eventually gives a product that has a translucent appearance. If significant quantities of other wheat semolinas are used the appearance and eating quality of the product are inferior; the presence of non-durum semolina must be declared on the product label.

The pasta press is a single-screw extruder with three major zones: (i) mixing and conditioning, (ii) plasticizing and (iii) extrusion. In early pasta manufacturing machines these three zones were positioned for compactness one above the other, giving an inverted S flow. Modern pasta presses are designed on a straight-through format.

(i) The first zone receives a metered supply of the durum semolina and water with an overall moisture content of about 22%. This mixture is propelled along a trough by broken screw flights which also act as beaters or mixers.

(ii) The semolina, now appropriately hydrated, drops into the entry of the closed cylinder plasticizer, which incorporates a propelling screw furnished with appropriate flights and walls to ensure adequate shear and moulding. As a result, the semolina reaches the final pre-extrusion state as a uniform, relatively high viscosity plastic dough. The moisture content remains about 22%; about 0.02% of the input energy to the screw drive motor is converted into heat absorbed by the product. Most pasta machines have cold-water cooling jackets to remove the heat generated by this intensive work. Before the final stage, the extruder is vented, frequently with a vacuum outlet, to ensure the absence of air bubbles in the dough, which would interfere with the translucency of the final product and its surface integrity.

(ii) The final stage is a very low shear building up of pressure in the cylinder at the end of which are the nozzles constituting the die from which the extrudate appears. The shape of the nozzles governs the shape of the finished pasta, from simple rods of spaghetti, strips of tagliatelli or tubes of macaroni to the more or less complicated shapes of various short-cut pastas. The temperature of the paste at this stage is about 55°C, well below the boiling point of water; there is therefore no puffing effect as the pressure is brought down to atmospheric.

The paste is then cut off to an appropriate length.

Drying completes the manufacturing operation. (Other aspects of pasta manufacture are dealt with in Chapter 5.)

High pressure shaper

This is a single screw machine used for snack foods which are cooked with or without expansion after leaving the extruder. Inlet moisture content 25%; outlet moisture content 25%; temperature of product immediately before extrusion 90°C; 0.03% of the input energy to the extruder motor is converted to heat absorbed by the product. The products may be part dried and may be expanded in a fryer or an expansion chamber.

Low shear cooker-extruder

This is used mainly for making semi-moist pet foods but could also be used for similar intermediate moisture human foods. The high moisture content of such products limits the amount of energy that can be derived from the screw operation and this necessitates jacket heating. Inlet moisture content 28%; outlet moisture content 25%; temperature of the mix immediately before extrusion 120°C; 0.02% of the input energy to the extruder drive motor is converted into heat in the product.

High shear cooker-extruder

This is the most modern development and is extremely versatile. High shear cooker-extruders can produce products of widely varying size, shape, colour, degree of cooking, and so on, in either solid

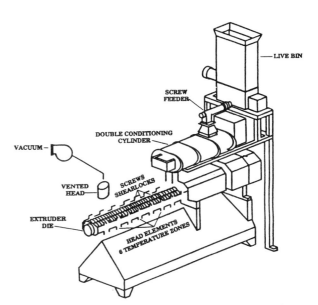

Figure 12.1 Cutaway view of twin-screw extruder (from Booth, 1990).

or expanded (aerated) form. Products from a twin-cylinder machine may be of two colours and flavours, in sandwich form, or of filled round bar shape. The expanded product may be further dried, and can be fat- and flavour-sprayed and dusted.

Among the snack foods made possible by this type of extruder are the so-called flat breads, the expanded-wall tubes or rods of cooked breakfast cereals and snacks, and expansion-extruded products of the type resembling wafer biscuits. Half products for further processing into snack foods may also be produced.

Figures 12.1–12.4 depict typical exterior and

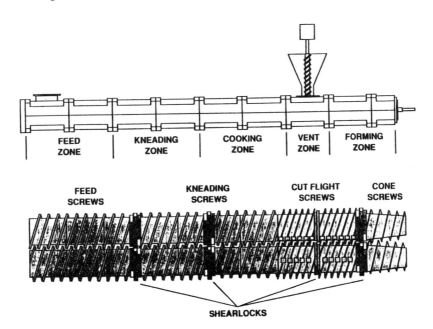

Figure 12.2 Extruder components (from Booth, 1990).

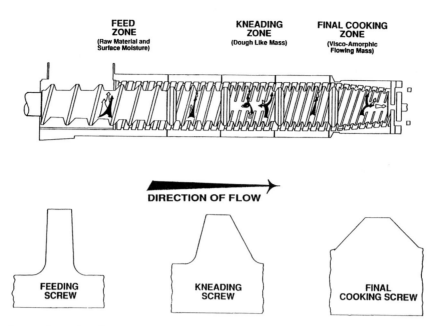

Figure 12.3 Extruder configuration (from Booth, 1990).

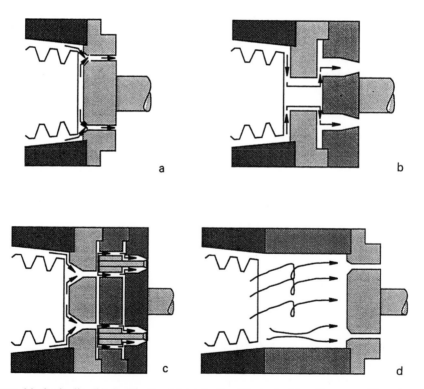

Figure 12.4 Die types: (a) single die, (b) double die, (c) triple die, (d) single die with spacer (from Booth, 1990).

interior layouts of modern extruders. Preliminary mixing and any necessary preconditioning of the feedstocks are not shown: more exhaustive details may be found in the references listed under Further Reading at the end of this chapter.

The machines are normally of twin-screw design. There is frequently the possibility of interchanging parts, particularly screws, cylinders and dies of different geometry. This enables a wide range of raw materials to be processed, with a wide range of moisture contents. Typical operating conditions are: inlet moisture content 15%; outlet moisture content 8%; temperature of product immediately before extrusion 120°C; 0.11% of the input energy to the extruder drive motor is converted to heat absorbed by the product.

Pre-die temperatures are considerably in excess

of 100°C, so the extrudate is normally therefore fully cooked and any starch present is gelatinized. The pre-die temperatures may be varied almost at will and the amounts of expansion and moisture flash-off can be controlled through the effective use of change parts and modern control systems.

AERATION PROCESSES

Aeration consists of changing a non-porous liquid, plastic or solid food into a porous spongy-textured or honeycomb one. The vacuoles may range in size from the microscopic to the relatively large, such as those found in bread, and may be filled with air, carbon dioxide, nitrogen, or occasionally other gases such as nitrous oxide. Aeration can be reversed in some cases by liquid absorption (e.g. trifle).

The means by which aeration is accomplished are quite varied, as is necessary to cope with individual materials and circumstances. They may, however, be classified into various categories such as expansion extrusion; gas injection (e.g. air into ice-cream mix); flash frying of moist material; rapid high temperature baking; the use of yeast fermentation; the use of vacuum together usually with some heat and dehydration; the use of the very hot plate; whipping of slurries, batters and creams; the use of baking powder and other raising agents. The efficacy of the method used depends on the variable characteristics of the material it is desired to aerate. These characteristics include surface tension, fat content and dispersion, whether starch is present and if so, in what form, and the presence of gums, pentosans, albumens and similar natural or synthetic materials.

Many of the foods we eat are aerated – most of these would be very much less attractive to the palate if they were not – and the marketing advantages of an expanded product are, or should be, obvious for the most part. Quite a large proportion of snack and breakfast foods are aerated to a greater or lesser degree though the expanded state is frequently unrecognized by the consumer. For example, the fact that ice-cream is normally half product and half air is not usually appreciated but it has several advantages to the consumer as well as to the producer. Aeration can provide better eating qualities, such as lightness and crispness, greater bulk, a greater surface area in relation to weight and better facilities for flavouring.

The summarize, the methods currently used are as follows:

(i) Fermentation, in which aeration is provided by yeast growth and the evolution of CO_2

(ii) The use of raising agents, comprising baking powders and ammonium bicarbonate or 'stuff', which also depend on CO_2 production

(iii) 'Puffing', in which product containing superheated moisture is subjected to a sudden release of pressure. This is analogous in many ways to (iv) but is normally only applicable to whole cereal grains

(iv) Expansion extrusion, by which superheated product is caused to emerge under pressure from a die and the moisture in it immediately vaporizes

(v) Frying in very hot fat

(vi) Whipping to entrap air

(vii) Vacuum-oven drying

(viii) Air injection (e.g. into ice-creams) or CO_2 injection (Oakes bread process)

(ix) Nitrous oxide (concurrently used as a propellant)

(x) Special cases – hot plate or microwave oven to cook and expand cereal grains.

The relatively new processes of ohmic heating and ultra-high pressure treatment may in due course also offer opportunity for aeration.

The essence of satisfactory aeration is that the product should retain the cellular structure it gains, and this depends on the structural qualities of the ingredients in the food, the way in which they have been affected by the process overall, and the presence or absence of modifying ingredients such as oil or fat. The 'fixing' of the aeration is all important.

All these methods have to be regarded as 'horses for courses' – the formulation or required product frequently narrows choice to one or, at most, two alternatives. This arises because of two major factors

(i) the presence of too much fat in the product may exclude many potential alternatives, and

(ii) the degree of cooking of the product, and where and how it happens, is frequently a deciding factor. In this context, it should be mentioned in passing that the use of cold extrusion in the case of a product which cannot be expansion extruded can frequently be employed as a prelude to aeration by another method.

Aids to aeration, such as appropriate additives to, or the pretreatment of, formulations are widely used. Additives such as guar, carob or xanthan gums are frequently effective, as are the addition of ingredients rich in pentosans, and changes to the nature of the starch included in the formulation.

OTHER PROCESSES

Baking and drying

The oven is one of the most ancient of all food manufacturing facilities. However, the technology has developed considerably over the years. It was originally used for the production of bread, the cooking of meats and for the drying and/or smoking of foods. It now it has many other specialized functions, which include expansion, colouring, flavour development, sterilizing, enzyme destruction and conditioning, to mention but a few. These varied functions have made necessary the development of multifunctional equipment. Many of these types of oven are concerned with the manufacture of snack foods and breakfast foods; their general features are described briefly below.

(i) The heating of ovens is almost always carried out using gas or oil as fuel. Special radiant heat generators and so-called microwave units are also occasionally employed. By appropriate use of these techniques including combinations thereof and of proper control methods, various temperature zones can be created and the requisite heating of the product from without and within may be accomplished. So temperatures can be arranged for immediate rapid heating or an initial slow bake, and for such requirements as toasting, browning or drying. Steam may also be injected into the oven to reduce moisture loss.

(ii) Ovens can be static; the brick oven is still used for some purposes. However, most are usually mechanized and are loaded and unloaded by a mechanically driven steel belt which can have a longitudinal or spiral path. If the product is loaded on trays, then a tower oven with mechanical uplift and descent may be used.

(iii) The vacuum oven, usually used for low temperature dehydration, is a special case but is now engineered so as to be available for continuous throughput operation.

(iv) The drying oven, usually heated by forced air circulation, is available in many forms and is widely used in breakfast and snack food production.

These types of oven are referred to later in relation to specific kinds of product.

Frying

Two forms of frying are possible: (i) that carried out in a pan with a very shallow layer of fat, where the material being fried is frequently turned over so that the whole of its surface is eventually heated by the hot fat as well as by conduction within itself; and (ii) total immersion or deep frying, where the material either floats in or is completely covered by the fat. The 'shallow layer' method of frying is only very infrequently used in food production on the large scale and so only frying by immersion will be considered here. A subdivision of deep frying is termed 'flash frying'. This is widely used in snack food production when a very short term, high temperature cook is needed.

Frying has a number of functions, which are by no means all related to cooking in a medium in which heat transfer is much more rapid than in an oven. The things which are customarily expected to be achieved include:

(i) Total cooking or final cooking of snack or breakfast food items to the eventual condition required.

(ii) Reduction of moisture content and also of any excess fat content of the material in question (e.g. as in pork scratchings).

(iii) Expansion of moisture-containing materials and the 'fixing' of the expansion by protein denaturation, drying of contained dextrinized starch, etc.

(iv) 'Crisping' and colouring the product as necessary (e.g. potato crisps).

(v) Overall improvement in rigidity, strength and mouthfeel of the product.

The quality of the oil used for frying is dealt with in Chapter 8 but it is pertinent to mention here that the flash, fire and smoke points need to be taken into account in view of the necessary temperatures required. Also, the stability of the oil and its nutritional qualities (preferably high in monounsaturates) have to be considered because, inevitably, small amounts of residual fat are usually left on the product. This can frequently be minimized by centrifugation or otherwise, but is usually sufficient to allow flavouring mixes or powders in sufficient quantity to adhere firmly to the product after frying. The relatively low bulk density and large surface area of many products help in this matter. Oils used for frying frequently contain antioxidant stabilizers, both for the sake of maximizing the shelf life of the final product and also extending the working life of the frying oil itself.

Coating

Many snack foods and breakfast foods are coated with a variety of materials, such as sugar, honey, natural or synthetic flavour concentrates, cheese powder, batters and savoury mix powders, in order to make them attractive on the palate and give them an individuality. For example, cereal flakes and expanded cereals may be sprayed with sugar solution and dried off after each application. Up to 50% of the final product may be sugar.

Methods used for coating are very varied. Some examples are as follows:

(i) The coating pan, as used for pharmaceutical tablets and products such as almonds, where a pear-shaped open pan angled at about 45% is slowly revolved. This results in its contents being tumbled, while sprays of, for example, sugar solution are applied alternately with jets of hot or warm drying air until the desired level of application is reached. Such coatings may be 'polished' by a final dusting of icing sugar, talc or similar powder.

(ii) The products travel on a belt which is vibrated at intervals and dusted with appropriate powders; some of the powder adheres and the rest is recirculated. The product needs to be somewhat adhesive for this method to be effective.

(iii) The products may be dipped into or sprayed with a batter, which is then converted into a 'shell' coating by heat and drying. The batter may be smooth cream or particulate in nature and may, in some cases, be aerated itself.

(iv) Products such as a fruit and/or cereal bar or biscuit, may be coated in an 'enrober', through which it is usually carried on a wire mesh belt and deluged from above and below with tempered liquid chocolate or a substitute. The excess drains off and the product then enters a cooling tunnel where the couverture sets.

(v) A 'coated' product can be produced in a twin-screw extrusion operation. A central main core of product is surrounded by an annular extruded 'jacket' which may be of very different composition and style from the core. The Japanese-made Rheon machine, based on a similar approach, will produce multicoated products such as analogues of Scotch eggs.

(vi) Glazing, or the coating of ice on frozen snack products (e.g. prawns) resulting from the application of a cold water spray has the function of preventing dehydration and avoiding toughening and a general deterioration of appearance and mouthfeel.

Mixing

This is a process which is often incorporated within another process as a necessary ancillary (e.g. as in extrusion) but is also recognized as a process in its own right. When employed in this context, it may also incorporate functions other than mixing, such as aeration, emulsification, dough formation and comminution. Consequently, a wide range of mixing equipment has been developed and is available commercially. Some of the principal 'mixing' functions are as described below.

Mixing of solids

Particle size defines the necessary style of mixer – for powders, the ribbon mixer in a trough is one possibility, as is a sealed drum with baffles, rotating on the axis of the drum or end-over-end. Such machines are also suitable for mixing materials of particle size averaging that of granulated sugar. For mixing solids of disparate but larger size (e.g. muesli ingredients) a rotating twin conical 'V' type of mixer is suitable.

Mixing of solids and liquids

Here the required state of the mixture defines the type of plant to be used. For the production of suspensions or solutions, the high speed propeller agitator within an open tank may be adequate. If some degree of comminution is necessary, or lump formation is a problem, then the Silverson type of mixer is needed. Aeration is a feature of some mixers (e.g. in the Chorleywood bread process). If the final product is a dough, the sort of mixer required becomes a more massive Z-blade or similar machine, or the type of integral dough-forming mixer incorporated into extruders.

Mixing of non-miscible liquids

Here the aim, aided by stabilizers, is to produce an emulsion or cream from, for example, an oil–water mixture. A high speed propellant mixing screw may be adequate here, otherwise a homogenizer may be necessary.

Mixing of miscible liquids

This hardly needs mention, merely stirring with paddles or a recirculating pump is usually adequate so long as sufficient time is taken, particularly when the liquids have different specific gravities.

Masa production

A wide range of snacks manufactured and used in USA, Mexico, Central America, and also now in the UK, require a special form of maize as a starting point – masa flour or masa meal. Masa is produced by treating whole grains of maize with lye (lime, caustic soda or KOH) which renders the pericarp easily removable. The temperature of the treatment also partially cooks the maize. Subsequently, the alkali is washed out of the grain, the pericarp removed and the masa can then be used immediately for tortilla or taco manufacture. However, it is mainly dried and ground to provide masa flour. Among other uses, this can be used as a basis for making corn flakes.

Sieving

The mouthfeel of products markedly affects their acceptability. It is therefore necessary to control the perceived granularity, roughness or whatever criterion may be applied to measure this factor. To standardize on the optimum range of particle size, whether as a powder, or as suspended particles in a liquid, it is necessary to sieve many products. Fines go to rework, oversize go back to recomminution.

Chopping and slicing

Here, to a large extent, the processes used are specifically related to individual raw materials or products. There is, however, an overriding necessity, which holds in almost all cases, to keep the knife blades sharp. Blunt knives give rise to fines and improperly cut material and a corresponding financial loss.

Chopping is the easier process – the ubiquitous bowl chopper, properly programmed, can deal with most requirements. Slicing, however, frequently has to be carried out by specially tailored equipment Such a case is that of cut groats, which constitute the highest volume oat product for human consumption, forming as they do the basis of porridge, oat cakes, cereal bars, and so on. The cutter used for this purpose consists of a hollow perforated drum, rotating with a doctor knife close to its outside surface. Groats within the drum project though the holes and are cut, usually into three pieces, which then constitute 'pinhead' oatmeal. These are the bare bones of what is quite a sophisticated operation. Slicing of nuts, e.g. almonds, also requires specialized equipment.

Chilling and freezing

Quite apart from their preservation function, these processes, particularly freezing, are integral parts of the structuring and presentation of some snack foods. The obvious examples are ice-cream and many analogues. The aeration and the physical nature (hard or soft) of ice-cream cannot be accomplished without ancillary freezing.

Brining and curing

Salt is the primary agent in both these processes, which are quite similar. Other agents may be added to the brine for curing purposes: nitrates and/or nitrites, for example, and special flavouring materials.

Brining, which is mostly carried out on meat and fish, results in two essential effects, namely dehydration by osmosis and preservation. Additional effects are on flavour and colour.

Brining and curing may be carried out by straightforward immersion, by 'tumbled' immersion to speed up penetration of the cure into hams and by vacuum and pressure methods.

PRODUCTS

FLAKED, ROLLED AND GRANULATED BREAKFAST PRODUCTS

The line dividing breakfast food products and snack foods is not at all precise – corn flakes, for example, are now being widely advertised as being suitable for any daytime use. Consequently, this section cannot be unambiguously divided into the two areas. A start will, however, be made on what is largely accepted as being the breakfast food area, viz. cereal products. These may, however, be at times combined with items sourced from potatoes and other vegetable (and largely starchy) items and small quantities of other additives.

Corn flakes

These are probably the best known and most popular cereal breakfast product in the West. They arose in the USA from the introduction in the mid-nineteenth century by the 7th Day Adventist Church of flaked corn as a major item of their (vegetarian) diet. Whilst they still manufacture it worldwide today, Mr Kellogg, who introduced his brand a little later, and his successors are now the leading manufacturer in the present galaxy of flaked 'me too' products. Cornflakes normally contain 6% of sugar (based on cereal weight), or the equivalent amount of corn syrup and/or non-nutrient sweetener, a source of malt flavour (malt extract as powder or syrup) and about 2% of salt, together with any other ingredients, which may enhance nutritional value, sustain advertising claims or promotions, or fulfil the specific needs of consumers of varying ages.

Most of the spray coatings frequently applied to flakes, and also to expanded extruded products, contain sugar. A special 'sweetener' mix is required in order to avoid consequent stickiness, especially in humid conditions. This is normally mainly sucrose but additional sugars, frequently emanating from honey, are present to a small extent, up to about 8%. By this means, sugar coatings of up to 50% are possible. Some of the sugar may, of course, be replaced by synthetic sweeteners. A hard, translucent crisp brittle coating results from drying the sprayed-on sugar mix.

The final quality of corn flakes is dependent on whether they are made by the traditional individual-grain rolling method or by the alternative of producing cooked cereal pellets (half products) on a cooker extruder, which are subsequently rolled. The former method produces a much better quality product but the second is more versatile in that it can utilize a variety of feedstocks, such as flour, grits, semolinas and brans, or combinations thereof.

Wheat flakes

Wheat flakes and similarly processed items are frequently seen on breakfast tables. Straight wheat flakes (e.g. 'Force') are made by a method very similar to that for maize (corn) but involve a variety of modifications to accommodate the different grain and starch characteristics. The final flakes, whilst still somewhat adhesive, may be compacted into small oval cakes before drying to produce the product 'Weetabix'.

Rice flakes

The flaking of rice is usually carried out by the whole grain approach as for flaked wheat. Nowadays, however, it is sometimes prepared via the 'half product' route based on continuous cooker-extrusion (see below). Rice presents some stickiness problems after cooking, which takes about an hour at a steam pressure of 1 bar. To avoid such problems, the moisture content of the cooked rice should not exceed 30%.

Different kinds of rice are found to vary in their requisite cooking time and their liability to stickiness, but in any case some degree of aggregation of the grains (lumping) is commonly found after cooking. The delumped, cooked rice is dried to a moisture content at which it will effectively produce whole, regular sized flakes. Flaking is achieved by passing the individual grains between the usual heavy steel rolls without differential drives. These and the associated doctor knives sometimes require Teflon coating to prevent sticking, even after the rice has been duly cooled and tempered for a few hours.

All directly consumed cereal flakes may be toasted to develop colour as well as desired blistering. Additionally, they may be sprayed with various additives before toasting to improve the overall flavour over that resulting from the toasting process. Toasting also serves to decrease the moisture content of the flakes to about 2%, making them ready for cooling and packing.

Oat flakes

Whilst maize, wheat and rice are of a convenient individual grain size to produce suitably sized breakfast flakes of the usual crisp toasted style, the same is not true of oats or barley which are usually consumed in a different way.

Oats, which are widely used in breakfast products, are a rather special for two reasons. In the first place, they are normally only consumed dry and in a fully cooked state in mueslis. Otherwise, they normally require further cooking to produce porridge. A minor exception is that of instant porridge, where the oats are fully precooked before packaging.

Oats for human consumption are bought from the oat miller as groats, shelled oats or as comminuted groats, cut or ground. An integral part of oat milling is kilning. Oats contain about three or four times the amount of fat contained in wheat; they also contain a fat-splitting enzyme (lipase). Unless this is inactivated, the groats will rapidly develop off-flavours from the freed fatty acids, particularly if they are cut or ground. After kilning by the miller, the processed groats, comminuted or not, keep very well within the usual moisture content limitations (about 12–13% maximum moisture). This stabilization is accomplished by raising the moisture content of the groats to around 20%, steaming them to 96°C, maintaining them at this temperature for 2–3 min, and then drying to the necessary 12% moisture or thereabouts before storage, further processing or sale. Stabilization does not take place unless heating is carried out at a moisture content well above that normal for the grain.

For breakfast use, oats are usually supplied in a crushed form, as oat flakes or as rolled oats, rather than as the somewhat unfashionable oatmeal (comminuted groats). However, the finest porridge can only be made from cut or milled oatmeal although this requires long term cooking. The recipe for making porridge is roughly three parts (by volume) of water to one part of oats, plus a generous pinch of salt to get rid of the 'raw cereal' flavour which otherwise obtrudes. The milk to accompany it should always be poured over the finished porridge, or for the purist, be contained in a separate dish. It should never be added to, or form part of, the liquid added initially to the oats before cooking. Forms of porridge may be made from other cereals. In such cases, the dish is frequently known as polenta, especially when made from corn (maize) grits or maize semolina.

Bran flakes

These are made principally from wheat bran and oat bran. Bran is not necessarily of a constant composition because it may be freed of grain endosperm to a greater or lesser degree when it is cleaned in the break section of the flour milling process. During the milling of wheat, of course, the bran is kept slightly moist to preserve its integrity and to avoid any dry shattering or powdering, which would make the flour speckled and lower its value. It ends up with a small part of the endosperm still attached; economics decide how far its removal should be taken.

For breakfast use, bran may be incorporated in any desired proportion and in any desired state (powder or flour to flake) in a cereal base which provides the desired dry and firm final structure. This may be in expanded pieces, extruded unexpanded pieces, flakes or other forms. It may be and is frequently flavoured – sweetened, malted and with added honey. It is sometimes coated.

Bran is frequently incorporated in present day mueslis for its widely publicized health benefit claims.

Potato flakes

These should not be confused with potato chips and crisps, which are made by slicing, but need to be mentioned as exemplifying roller treatment of root vegetable-derived solids in special dough formats. So far, only complex mixtures of potato granules, potato starch and occasionally other potato or vegetable products with a final moisture content of around 50%, have been used for multistage rolling into the thin film necessary for flake manufacture. The final processing includes cutting the potato film into appropriate sizes, followed by rapid heating in an oven or frying. The methods applied here are by no means as straightforward and standardized as those in most other rolled or flaked product production. However, potato in combination with cereals provides a now well-established range of products.

Muesli

Muesli is a product of varying composition, but it is always based on a mixture of flaked cereals. To this base may be added, in a wide range of proportions, dried vine fruits (seedless raisins, sultanas), nuts in variety and varying piece size, other dried fruits (banana slices, chopped dates, figs and prunes), coconut, pine kernels, sunflower and pumpkin seeds, sugar, malt extract, and so on.

There are three problems of a technical kind in producing (and packaging) muesli:

(i) Initial bulk mixing, for example in a rotating apex-over-base cone or double cone mixer, is no problem, but avoiding the subsequent separation of the components right up to final packaging can be. The basis of it is the widely differing specific gravities, particle sizes and smoothness or roughness of the ingredients. A stirring rather than a vibratory or other mixing technique is indicated following the initial bulk mixing.

(ii) Within the package are components of widely varying moisture content and it is necessary to ensure that no significant transfer of moisture occurs from one to another. This could result in nuts becoming tooth breakingly hard and vine fruits becoming as tough as leather. Each case has to be considered separately and the ingredients must be moisture-conditioned before use in a mix.

(iii) In packaging, access or ingress of moisture in any form must be precluded, and warmth and light both shorten shelf life.

Only a few products have been mentioned in this section, but they are typical of the great variety which are now available as clones of their predecessors.

SNACK AND BREAKFAST PRODUCTS PRODUCED BY EXTRUSION

Introduction

A recent UK survey of breakfast foods, mainly based on cereals and listing around 250 brands, when read with prior knowledge of manufacturing methods, indicated that quite a large proportion actually were or could have been made by processes involving extrusion. Some are produced using what corresponds to the 'half product' technique, plus a further frying to provide final aeration, cooking and drying, or plus rolling to flakes. The outcome is that extrusion, whether 'expansion' or otherwise, can be tailored to produce more or less any flight of fancy which the marketer requires in terms of shape, colour, aeration, composition, bulk density and many other factors.

Methods

The variety in these breakfast foods is derived first from the kinds and forms of the basic cereals (and similar normally starchy materials such as potato), from the additives to the base (flavour, colour, comminuted fruit and nut pieces etc.), from the surface additives (sugar, cheese or other coating), from 'toasting' and from other physical treatments, which may be mechanical as well as thermal and designed to influence acceptability to both eye and palate. In the case of snack foods particularly, extrusion may be used twice in the manufacture of many items, particularly when half products (see below) are involved.

The production of extruded snack foods and breakfast foods can be somewhat arbitrarily divided into (i) tempering, mixing, precooking and plasticizing the mix, and subsequently bringing it to whatever extrusion temperature, pressure and moisture content is required at the point of extrusion, and (ii), the actual extrusion, with cut-off into whatever shape is required. This is followed

by the individual operations necessary to finalize the particular product, such as flash frying, toasting or baking, drying, flavouring (including salting), and so on.

Half products

Half products are dried, precooked products that are frequently prepared by extrusion. As their name implies, they rely on subsequent processing, often involving frying or some other form of puffing, to transform them into the final finished product. Like pregelatinized starches, they have a storage life under normal ambient temperature and humidity conditions of many months due to their low fat and nil-enzyme contents. They can be stored by their original manufacturer until required or sold on to a third party for subsequent processing.

Half products are particularly valuable to the relatively small scale producer of snack foods, whose turnover does not justify capital investment in a complex extruder and its ancillaries, nor the employment of the necessary highly skilled personnel to operate it. Their use has further advantages in that the desired mix can be standardized and the final producer does not have to face any of the problems arising from the need to maintain, blend and process inventories of various raw materials.

In view of the fact that a larger than usual proportion of the selling price of cereal- and potato-based snack foods and breakfast foods lies in adding value, the division of manufacturing into the two stages outlined above is economically feasible. The production of half products has several additional advantages:

(i) Expanded extruded snack and breakfast food products have a very low bulk density. The space occupied by them during transit is therefore very expensive and distribution is only economical over relatively short distances. In contrast, half products may be shipped in the same manner as grain and flour.

(ii) Many finished products are required to be light and brittle and are therefore quite fragile. Consequently, subsequent handling must be minimal and gentle. In contrast, the handling and transportation of half products is much easier. Amongst other advantages, this permits flexibility in the choice of location of finished product manufacture.

(iii) The shelf life of the final product is much shorter than that of the half product and final product-demand changes can usually be much more easily accommodated by final product producers than by half product manufacturers.

(iv) A single half product may be used as a basis for a variety of shapes, types and flavours of final product, each of which may be made on its own relatively low capacity line. This arrangement can lead to improved plant utilization (e.g. less downtime on the half product extruder) with consequent lower costs and major savings on final production scheduling and operations.

(v) Manufacturers of half products gain the benefit of bulk buying of ingredients as compared with final product producers.

It can safely be said that the current international market penetration of extruded snack foods and many cereal-based breakfast foods could not have been achieved on such a scale without prior development of the half product concept.

SNACK AND BREAKFAST FOOD PRODUCTS DERIVED FROM SPECIFIC AERATION PROCESSES

Fermentation

The products of concern here are principally those based on pastry and other cereal doughs containing yeast. A variety of biscuits frequently used as snacks are dealt with elsewhere. Typical products are PASTIES (see below), the individual fruit pie, and any base and cover (if present) of the normal or open sandwich. PRETZELS and BAGELS (see below) also begin with fermented dough, as do croissants.

Raising agents

Snacks in which raising agents may be used largely parallel those produced by fermentation, but whilst having their own limitation, are not limited in manufacture by the time factor and nutritional and temperature constraints attendant upon yeast fermentation. Most cakes are made using baking powder (see Chapter 5).

Puffing

This process is primarily used to make a variety of puffed whole cereal grains, wheat and rice particu-

larly, which may then be coated with sugar and other flavourings. A less known but now considerable outlet for such products lies in the use of whole or kibbled puffed cereals as a lightening and bulking ingredient in various snack bars, particularly in those containing dried fruits, which tend otherwise to be heavy and stodgy.

Expansion extrusion

As indicated previously, many 'expanded' breakfast foods and snacks have their origin in expansion extrusion, whether derived directly from the final section of the extruder or from half products.

Frying

The multiple functions of frying – cooking, dehydration and aeration, colouring, coagulation and forming a very thin adhesive layer for powder coating – means that it is widely used independently or as an adjunct to various forms of extrusion. One snack product in which all these functions are combined is the production of PORK 'SCRATCHINGS' (see below) from pork rind. Another snack product, CROUTONS (see below), is made by deep frying specially baked bread. This is staled and cut into variously shaped small pieces, with the possible incorporation of flavour additives before or after the frying.

Whipping to entrap air

This method is used in the smaller scale production of ice-cream just as crystallization and solidification of the basic mix is occurring, in order to achieve the desired 100% overrun, that is halving the specific gravity of the product. It is also used in the manufacture of creams, both natural and synthetic and other forms of toppings and dips connected with a variety of snacks.

Air injection

This is used in large scale ice-cream production and in the manufacture of aerated chocolate and of certain cereal-based bread and confectionery products, many of which can be regarded as snacks. If air is to be replaced by carbon dioxide, then the gas injection often occurs in an Oakes mixer.

Nitrous oxide

This is used as an ingredient in the 'instant' whipped cream, used as a topping for many snacks.

Hot plate and/or microwave heating

Both these techniques can be used to expand an early aerated snack, popcorn. The latter technique give rises to aeration by internal heating and evaporation of moisture from within the product.

PRODUCTS DERIVED FROM OTHER PROCESSES OR COMBINATIONS OF PROCESSES

The 'other processes' included here go beyond those described previously. A few are in widespread use; others are used solely in the manufacture of single products. For convenience, they will be dealt with more or less in order of importance.

Potato crisps and potato chips

Potato crisps (called potato chips outside the UK) hold nearly 50% of the UK snack food market and are of major importance worldwide. The characteristics of the feedstock required are:

(i) Uniform, regularly shaped, medium sized tubers over 40 mm.
(ii) Clean skins with shallow eyes.
(iii) No internal bruising, disease, greening or shrinkage and softening.
(iv) Ideally, a specific gravity of 1.085, equivalent to 21% dry matter.
(v) The reducing sugar content should always be below 0.25% and preferably below 0.10%.

Record and Maris Piper are frequently used as maincrop varieties and Bintje, Home Guard and Red Craigs as earlies.

The factory processes employed are as follows:

(i) Inspection of feedstock and manual removal of obvious trash on a conveyor belt. Checking that stored potatoes have had their sugar content adjusted by 'burning off' any excess (temperature adjustment to store).
(ii) Mechanical washing and stoning of potatoes using brushes, with gravity separation of stones in a vertical flume.

(iii) Steam peeling of potatoes, usually by steam-jet loosening of the surface skin, followed by a dry abrasive skin-removal treatment and the use of brushes and water sprays. This approach minimizes water usage and concentrates the skin residues.
(iv) Inspection and trimming by hand on a conveyor belt.
(v) A holding period to maintain continuity of production. A buffer tank containing a weak solution of SO_2 or sodium metabisulphite is usually provided at this point to avoid enzymic blackening.
(vi) Slicing of the potatoes to the desired thickness using multiple knives.
(vii) Washing slices free of starch and mechanical separation of slivers and nubbins concomitantly with dewatering.
(viii) Blanching of the slices in steam or hot water (160–200°F, 71–93°C), which amounts to a partial precooking. Blanching times depend on slice thickness and blanch temperature and can be from 3 or 4 min upwards.
(ix) Drying the slices by shaking and blasting with air to avoid water carryover and to start the dehydration continued in the next (frying) stage.
(x) Frying or cooking on a conveyor in a thermostatically controlled bath of oil, which is subjected to continuous recirculation, filtration and indirect heating. The cooking time is strictly controlled to ensure consistent product quality and colour. In some processes, the frying may be carried out batchwise in kettles; in this case the product is referred to as 'kettle' or 'kettle-style' chips or crisps.
(xi) Defatting by shaker and by hot air blasts.
(xii) Flavouring and salting by dusting on the seasoning, whilst the defatted crisps are being tumble conveyed
(xiii) Cooling of the finished crisps.
(xiv) Packing of the crisps as soon as possible in small laminate packages designed to prevent the ingress of both oxygen and moisture.

During the processing of the potatoes to crisps, the moisture content of the potato is reduced from 79% (approx.) to 5% (approx.). Of the 95% of dry matter in the crisps, 35–40% is fat. Taking into account peeling, trimming and other small losses in the plant (starch, slivers and nubbins, etc.) and the uptake of fat, the conversion ratio of potatoes to crisps is around 4:1. The processing of potatoes into manufactured products demands very large quantities of water, around 10–14 tonnes of water per tonne of potatoes, according to the type of soil they are grown in, the harvesting conditions and the efficiency or otherwise of the plant equipment. A measure of water recycling is highly desirable.

Crisps are made in a great variety of flavours but the plain variety, presalted or with a salt twist in the pack, is still the most popular. Of the 'flavoured' kind (if one disregards salt as a flavour), the 'salt and vinegar' and 'cheese and onion' varieties are very popular, followed by bacon, chicken, beef and onion, tomato, pickled onion and the branded savoury flavours.

The current preoccupation with lowering the fat content of diets on medical grounds has meant that some fried potato products are produced with reduced residual fat contents. This may be achieved with the aid of vacuum frying, by the use of shorter frying times, or by agitation of the product immediately after frying. The eating quality of the product may not be preferred to that of products with normal fat content.

A related interest, to produce products with increased proportions of unsaturated fat, may be met by the use of appropriate frying oils high in unsaturates, Note, however, the deleterious effect of these oils on keeping quality, due to their greater susceptibility to oxidative rancidity.

Potato chips

Ordinary chipped potatoes or chips (French fries outside the UK) will not be considered here as they do not fall within the snack food classification. However, miniature dehydrated chips are manufactured, packed and used in the same way as crisps. The method of manufacture is identical to that for crisps with only two major exceptions:

(i)   The potatoes of appropriate size are cut in the same way as are crisps; in other words they are sliced in one direction and then at 90° to the first direction. The cross-sectional dimensions are much smaller though.
(ii) The frying is usually carried out in a vacuum fryer so that the necessary degree of dehydration can be achieved without excessive darkening or scorching of the product; the fat temperature in a vacuum fryer is much lower than in an atmospheric fryer.

Nuts

Nuts are used widely as snacks to accompany drinks. The variety accounting for the largest consumption for this purpose is peanuts, but others such as almonds, cashews, hazelnuts (filberts), pecans and Brazils and several others are also used. All except a small proportion of sweet chestnuts, walnuts and hazelnuts are now imported in to the UK, some usually in the shell, others more often (or exclusively) shelled.

With one exception, chestnuts, which have a moisture content that may exceed 50% and a fat content of only 3%, all nuts used as snacks have a low moisture content, usually not exceeding 7%, and a high fat content, up to around 70%. Many are consumed roasted. This treatment may be carried out either by deep frying on a continuous moving belt in the same manner as potato chips or more often nowadays, by 'dry roasting', that is by baking. Again, the high content of fat, some of which migrates to the surface on roasting, facilitates the pick-up of salt. This is frequently applied as the nuts are being tumbled in a rotating drum or coating pan.

Nut oils vary markedly in their fatty acid makeup and hence in their molecular weights, melting points, boiling points and degree of unsaturation. Because peanut oil (arachis oil) is much less unsaturated than the others, it is by its very nature much less liable to oxidative changes (including rancidity) than is, for example, walnut oil. Consequently, under similar circumstances, the shelf life of peanuts is much longer than that of walnuts.

The roasting or kilning of nuts extends their storage potential and their shelf life by destroying lipolytic enzymes, thereby preventing the production of fatty acids and the unwanted and disagreeable flavours arising therefrom. The reduction of moisture, also occasioned by roasting or kilning, has the effect of limiting and slowing down any residual latent enzyme activity in all kinds of nuts.

Nuts are particularly prone to infestation (the flavour, texture and high food value all appear to be most attractive to various forms of insect life) and also, in adverse high moisture conditions, to mould growth. Some moulds are harmless, except insofar as they may impart a mouldy smell or taste, but others are quite dangerous and produce mycotoxins that are poisonous and in some cases carcinogenic. For these reasons, the fumigation of nuts is quite a regular operation, but care must always be taken to ensure that the levels of fumigation residues do not exceed those regarded as safe, and that these residues do not affect the generally delicate nutty flavours.

The large scale storage of nuts is best carried out at a temperature of 4–8°C, preferably in a store with a low oxygen atmosphere. Every precaution should be taken to prevent temperature and humidity changes that could give rise to condensation with its risk of local mould growth.

The growth of mould on nuts under imperfect storage conditions is likely to lead to production of aflatoxins, which are now known to be highly toxic to many animals and presumably also to humans. UK Regulations prescribe a maximum limit in all foodstuffs of 4 µg kg^{-1} (ppb) and similar limits apply in many other countries. The quality control of nuts should normally therefore include a test for aflatoxins.

Various alternatives are employed to extend the shelf life of nuts packed for retail sale. The best (but the most expensive) is the tinplate or aluminium container, which may be nitrogen flushed before sealing and which gives very good physical protection to the nuts. Laminates (including foil laminates) are also used, with particular care being taken to keep oxygen out and moisture in. The principal characteristics required of nuts to be used as snacks are as follows:

(i) There should be little apparent physical damage.
(ii) There should be no musty or 'off' smell, nor smell due to insecticide spraying.
(iii) There must be no obvious signs of mould.
(iv) There must be no obvious signs of infestation.
(v) The nuts must be clean and dry.
(vi) There must be no obvious rancidity in the oil of the nuts.
(vii) The nuts must be plump and not shrunken.

Almonds

The role of almonds as a snack is less than that in confectionery but is nevertheless considerable. In contrast to peanuts, which require cultivation and planting, special harvesting equipment and a reasonably fertile soil with adequate rainfall, the almond tree, with its deep rooting system, is very resistant to drought and can survive and yield on very barren soil. Harvesting is simple; shaking the trees at the appropriate time is perfectly effective. The yield per hectare is much less than that of the peanut, however.

Shelled almonds to be used as snacks are normally processed close to where they are produced. The shell is cracked between rolls and the kernel or nut is separated by sieving and by air lift. The kernels are then briefly blanched at 82–93°C (180–200°F) for 3–5 min. Blanching carried out at over 93°C (200°F) may result in discoloration. The almonds are then cooled and washed with cold water and the skins removed by rubber rollers and nylon brushes. After this, they are dried at around 68°C (145°F) to a moisture content not exceeding 6%. Grading for size and colour follows before packing, usually either in 25 kg fibreboard boxes or in lined sacks. The nuts are best stored at around 4–8°C at a relative humidity which does not permit water uptake nor lead to condensation with ambient temperature changes. Almonds are an expensive commodity and chilled storage is economically feasible.

For use as snacks, almonds are usually roasted and salted and are packaged in a similar manner as peanuts used for the same purpose. Transport and storage of almonds in the shelled state are naturally much more economical than when they are unshelled.

Brazils

Brazil 'nuts' are really the seeds within the fruit of a South American tree. The hard shelled fruit, about 150 mm (5–6 inches) in diameter, contains its seeds, each shaped and arranged like sectors of an orange. Unlike the orange sectors, however, each seed is composed of a hard shell with the kernel closely invested within. There are usually 8–14 nuts within the fruit.

To remove the kernels whole from Brazil nuts effectively requires that they should first be boiled. This also blanches the kernels, which may be peeled in the same way as almonds.

Brazils have a fat content in the range 60–70%; 5–6% of moisture is satisfactory though they may dry out on storage to 3–4%, which can give rise to a rather hard and unattractive texture. Brazil nuts are not usually roasted nor salted; their method of packing is similar to that for PEANUTS (see below). Brazil nuts are unusual in that they are very resistant to infestation.

Cashew nuts

The principal current source of cashew nuts is India, but they are also produced in Brazil and East Africa. The nut is attached to the end of the fruit of the cashew tree, and between the hard shell of the nut and the kernel is a highly alkaline liquid containing phenols – an irritating mixture, possibly designed by nature to inhibit consumption of the nut by wild animals.

The nuts are unusual in being quite sweet and having a unique mouthfeel; they are recognized as a luxury snack and are priced accordingly. In order that they should be given maximum protection against rancidity and staling, they are almost always packed in highly protective containers, tin-plate or heavy gauge plastic drums or similar, which have been purged of oxygen before closure.

For good keeping quality these nuts should not contain more than 6% moisture. Their fat content is about 48% and they have a total sugar content of 22%.

Chestnuts

These nuts are in a completely different category to the varieties previously described. Although they fit the general specification of nuts, having a hull and a shell around the nut itself, the shell is thin and soft and peels off easily. The nut within usually contains over 50% moisture and no more than 3% fat. They are therefore very susceptible to mould growth and to shrinkage due to loss of moisture. Because their value is relatively low *vis-à-vis* the majority of other nuts, it is usually uneconomic to hold them at chilled storage temperatures. Their storage life tends therefore to be short, two or three months under good temperate conditions, where protection is given against predators, excesses of temperature and high humidities and a keen lookout is kept for any signs of mould infection. As a snack, sweet chestnuts are roasted and, unlike the other nuts considered here, are eaten hot.

Coconut

The oil content of desiccated coconut is about 57% and the fibre content about 12%. The moisture content of the coconut should be kept to 3.5% or less if mould development is to be avoided. Rapid changes in the storage conditions may lead to the pick-up of moisture and subsequent deterioration.

In the past coconut has been subject to salmonella infection, and purchasers should ensure that the supplier has carried out all necessary pasteurization procedures. This can be achieved by steam heating at 96°C (205°F) for 8 min, followed by drying or roasting.

Hazelnuts, filberts or cob nuts

This nut is grown widely in Turkey, Italy, Spain and the USA. There are two types, one round and the other flatter. The shell is hard and very brittle and is readily removed by cracking between rolls. Hazelnuts can be blanched and skinned in the same way as ALMONDS (see above) but, in a similar manner to that applied to some irregularly surfaced nuts, may be skinned using lye. The fat content is very high, almost 70%. In order to have a low water activity and hence to keep well the moisture content should not exceed 4%. Like almonds and peanuts, hazelnuts are usually roasted before consumption; this improves shelf life as well as flavour. Packaging is similar to that of PEANUTS and ALMONDS.

Macadamia nuts

These nuts are produced by large evergreen trees, up to 50 feet (15 m) high, native to the tropical and well-watered areas of Australia. They are also grown in Hawaii and in East Africa. The trees produce drooping clusters of up to 20 flowers, which give rise to fruits with leathery outer husks. These eventually split to release the hard shelled round seeds which, in this instance, are classed as nuts. The nuts ripen over a period of months; they are not picked but allowed to drop off the trees. After mechanical shelling, the nuts are washed – the 'floaters' are the best quality – and then usually 'dry roasted' (cf. PEANUTS) and salted. They require hermetic foil laminate packaging to remain in good condition over an extended shelf life.

Peanuts, groundnuts or monkey nuts

The peanut constitutes by far the largest proportion of all nuts consumed as snacks, but its use in this fashion is much less than its use to produce arachis oil and protein-rich peanut meal. It is of course a principal feedstock for vegetable oil-producing plants. It is grown especially in southern USA, in India and in parts of Africa, and is leguminous. The pods, which later become the relatively thin reticulated shells, are formed above ground as is usual with Leguminosae, but are later forced below the surface of the ground by the lengthening flower stem; it is underground that the nuts mature. The whole plant is harvested and, after slow drying, the pods, which have now changed to shells, each containing two or three seeds or nuts, are removed. All that remains is then to decorticate them or remove the shells. From this paragraph it is at once clear that the peanut is not really a nut but is a pea or bean. However, commercially it is considered a nut or an oilseed.

Peanuts for use as snacks are carefully selected for size and quality. Electro-mechanical sorting is carried out to grade the nuts by colour. Some peanuts are consumed raw but the majority are roasted and salted.

Roasting has a stabilizing effect on the peanuts, caused by both enzyme destruction and moisture reduction. The original 'roasted peanuts' were not roasted but fried, being produced by passing the nuts continuously through boiling fat or oil before draining and salting. The newer 'dry' roasted peanuts, on the other hand, are baked without fat or oil being added. If not roasted, peanuts may be blanched, as are almonds.

A variant on peanuts has been developed from soya beans by which the final product closely resembles in flavour and texture split dry roasted peanuts. Details of the process are not available but the 'beany' flavour of the soya has been satisfactorily removed and the imitation is very close to the genuine on almost all counts.

To provide adequate shelf life, packaging films such as polyester/foil and polypropylene laminates are used, which form effective light-, moisture- and oxygen-barriers, are fat-resistant and are particularly effective when flushed with nitrogen. Under these circumstances, the considerably longer shelf life necessary for nuts *vis-à-vis* potato crisps, for example, can be achieved.

With about 28-30% of protein and about 50% of oil or fat, peanuts are very nutritious. It is perhaps unfortunate in this context that the oil is not a good source of unsaturates and that a significant allergy to peanut protein in any form has recently become prevalent.

Pecan nuts

These are a variety of hickory nuts, a genus which comprises inedible as well as edible kinds, and were until relatively recent times only grown in North America. They are now, however, produced in appropriate climates worldwide. This nut is produced in large quantity and is very popular, particularly in the USA, where it originated.

Pecans closely resemble walnuts and the trees also are similar. The nuts however are much more stable and amenable to storage than walnuts. Because the trees have not been subjected to the disastrous disease which killed so many walnut trees, the nuts have in many instances replaced the resulting shortfall in the supply of walnuts. The nuts are harvested by tree shaking followed by 'lawn sweeper' collection. They are then separated from trash and stored for about three weeks before further treatment. Then, if long storage is required, the moisture content of the kernels must be reduced to 4% and the temperature held at 20°C (76°F). Under these conditions, the nuts can be safely stored for a few years. After shelling, the nuts are sold, usually in halves; smaller pieces,

however, have a good market for use in ice-cream, cakes, and so on.

Pine nuts

A few varieties of pine trees produce pine nuts which, from size and edibility considerations, are worth harvesting. They are the seeds of the tree carried within its cones. Harvesting is costly for nuts which are each only about one-sixth of the size of an almond. Therefore, as quantities available are low, the cost of the final roast product is high, even when labour costs are low as in Mexico and China, where much of the available supply originates. The keeping quality of the nuts is low, even when refrigerated and they should be consumed promptly.

Pistachio nuts

The medium-sized trees producing pistachio nuts grow in rather arid semi-tropical conditions. Originally, they were encountered wild in the Middle East, particularly in Turkey and Iran, but nowadays they are also grown in orchards in many parts of the world. As in the case of almonds, the latitude in which the trees are grown largely determines whether or not they will bear fruit. The trees, which are quite decorative, will grow and thrive but not fruit in the UK.

The nuts are unusual in being green in colour. After harvesting, they are dried down to about 5% moisture content and are suitably graded. Then, as required, the nuts are roasted and salted in their shell. They are first soaked in brine for a while and then dry roasted, drying at first and then bringing the temperature up to 120°C (250°F), at which roasting occurs. These go for direct consumption; the remainder are shelled unsalted, then dried and may be roasted before being used commercially in ice-cream or confectionery manufacture.

Walnuts

Walnuts are grown principally in France, Italy, Poland and the USA. An unfortunate outbreak of disease has reduced the number of trees growing in the UK to negligible proportions. At harvest time, the walnut is similar to the almond in its anatomical features, with a thick fleshy hull that readily splits off to reveal the hard shell within. Because of its convoluted shape, the walnut does not lend itself easily to skinning or peeling; it is also difficult to get the nut whole out of the shell and this also increases the cost of shelled nuts.

Walnuts have other drawbacks; they are very prone to infestation and partly because their oil is highly unsaturated, they are also prone to oxidative rancidity. They require kilning at 60–65°C to destroy any infestation and to dry the nuts down to a maximum moisture content of 4% to minimize enzyme activity (particularly of lipase), which would otherwise lead to the lipolytic form of rancidity, with high free fatty acids (FFA).

They do not last as long in store or in retail packs as the majority of other nuts discussed here and are best consumed in as fresh a condition as possible. Unlike other snack nuts, walnuts are not often roasted or salted.

Fruit bars

The term 'fruit bar' is not meant to describe products which have been formulated specifically to provide particular nutritional benefits but snack items that are highly acceptable in terms of overall eating quality. Wide variations are available in the makeup of such bars, which usually start with a base of macerated date flesh, macerated seedless or seeded vine fruits and figs. To these may be added candied citrus peels, dried banana, dried apricot, crystallized ginger, and so on, all of which are incorporated in the very stiff paste, which is produced in a heavy-duty Z-blade mixer. Chopped nuts and some form of cereal, such as rolled oats, may be added to improve mouthfeel and texture. A relatively recent type of additive is that provided by puffed cereal grains (Presco process) which, after granulating, provide a lightening, nutritionally attractive and texturally desirable component.

There are relatively few constraints in the manufacture of such bars. All ingredients may be checked microbiologically, but the widespread use of SO_2 and of blanching is of great assistance in minimizing bacterial, mould and yeast counts, and so selection is made relatively easy. The low moisture and very high sugar content (therefore low water activity) means that no microbiological growth need be expected. Enzyme activity is also restricted for the same reason and so shelf life can be considerable.

The bars are customarily fabricated by rolling out the dough, preferably on PTFE-coated rolls, to the required thickness, then slicing the dough to an appropriate width with circular rotating knives, and finally guillotining it to the desired length. PTFE coating of cutters and of all parts of the machine in contact with the sticky dough is essential. Before wrapping, the bars can be made much less sticky on the outside by the application of powdered talc, lactose or a similar surface-coating powder. Fruit bars may also be enrobed with chocolate or other form of couverture. A particular kind of fruit bar is known as a FRUIT LEATHER – see Chapter 4.

Croutons

Croutons have moved out of their traditional role as additives to soups, and are now found in salads and also as a replacement for crisps and peanuts as a savoury snack with drinks. They are now offered in various flavours in the same way as potato crisps.

The basic raw material in crouton manufacture is a baked raised dough – a form of bread. It may be a typical loaf, or a thin stick-like 'baton', or any other appropriate shape of leavened or otherwise raised bread. It is usually baked specially for the purpose and has an appropriate cellular structure for its end use. An excessive crust thickness is to be avoided as the yield is thereby reduced.

The bread has then to be reduced to the approximate size and shape of the eventual crouton. This cannot be done mechanically until the bread has staled sufficiently or, in other words, the crumb has set by reason of retrogradation of the starch present. The bread sticks can then be sliced into discs or the loaves sliced and cubed, or cut to whatever shape and size are required.

Subsequent preparation of the croutons depends on their prospective end use. If they are to be used in hot soup and to retain their crispness as long as possible, then they require a very high final fat content – 50% or more. If, on the other hand, they are to be used as a snack or in salads, then 20–25% may be adequate. In the first instance deep frying, followed by hot air blowing, draining and centrifuging, may be appropriate; in the second, measured amounts of fat may be applied by spraying, followed by baking to the necessary extent to give a low moisture content (perhaps 5%), a golden colour and a long shelf life.

The croutons may be flavoured by mechanically sprinkling them with appropriate amounts of whatever flavours are required (e.g. cheese, garlic, herbs, etc.) plus salt at an appropriate stage in manufacture, such as following draining after frying, when good adhesion can be obtained. An alternative method of flavouring is, of course, to flavour the bread dough at the outset, but this is somewhat limiting in many respects.

In the same way as all crisp snack foods, croutons need to be protectively packaged, with moisture exclusion being the main criterion. The addition of antioxidant to the fat provides an extended shelf life to croutons.

Crêpes or pancakes

The crêpe or pancake has been a very popular snack in France, possibly for centuries, and its virtues are becoming ever more widely recognized in the UK. In France and North America it has been employed as a suitable base for an exceedingly wide range of toppings or inclusions. These extend from seafood, through minced meats, various forms of cheese and other savoury items, to the only role that it has played for the most part up to now in the UK – that as a basis for sweet or dessert items in Crêpes Suzette and apple pancakes. Spring rolls are an Eastern variant of the same theme.

The basic recipe for crêpes or pancakes consists of plain flour, salt, eggs, milk and butter (or oil). This is satisfactory for both sweet filled and savoury filled crêpes, but those designed exclusively for sweet use are improved by the addition of sugar. A typical recipe for domestic use is:

Plain flour	100 g
Salt	5 g
Eggs	2
Milk	350 g
Butter or oil	15–25 g
Sugar (if required for sweetening)	15–25 g

Because tastes vary, so do ingredient quantities. Economic factors also affect the final choice. In addition to the wide variety of materials that may be folded into crêpes as fillings, such items as grated or powdered cheese or lemon juice are also frequently included in the crêpe mix as well as using them (or a comparable ingredient) in the filling.

Before cooking, the viscosity of the crêpe batter should be checked. It should be quite runny (coffee cream thickness) for the best results, very thin final crêpes, but appreciably thicker for producing the normally encountered and more easily handled routine crêpes. Addition of more or less water just before cooking effects any necessary reduction of viscosity.

Cooking is a matter of a minute or two on an oiled or buttered surface, the crêpe being turned in the pan or on the griddle, so that cooking takes place on both sides until, preferably, a golden colour appears. In certain circumstances pancakes can be frozen and kept for several months, and, even when not frozen, they can be kept for a few days and then reheated. As usual, they are not as good as when freshly made, but may be adequate.

Croissants

See Chapter 5

Bagels

Bagels are snacks derived from Jewish cuisine and are made from fermented dough by a method resembling that used in making pretzels. They are a special type of circular bun with a hole in the middle. A typical bagel recipe (for small scale production) is as follows:

For 12 bagels:

Plain flour	300 g
Salt	10 g
Caster sugar	50 g
Baker's yeast	25 g
Tepid water	100 ml
Egg	1
Caraway seeds	a few

Sieve the flour, salt and half the sugar into a mixing bowl. Suspend the yeast in the tepid water. Add the yeast mixture and the egg to the flour, mix to a stiff dough and knead until smooth. Cover and leave to rise in a warm place for about 1.5 h until the volume is doubled. Knead (or knock back) and divide into 12 pieces. Roll each piece into a sausage-like shape, form it into a ring, and leave it to rise for about 15 min. During this period, bring an appropriately sized pan of water to the boil and add to it the remaining sugar.

Drop the rings in, one by one, and when all are in, reduce the heat and simmer for 5 min. Remove the rings and drain. Place on lightly oiled baking sheets, brush with egg, sprinkle with caraway seeds and bake in a preheated oven at 200°C (400°F) for 20–25 min or until crisp and golden.

Pretzels

Pretzels originated in Germany and Alsace, where they were used as an accompaniment to beer; their salty taste encouraged consumption. Pretzels are now well-known international snacks which are still used to accompany drinks, as well as main items in hand-held snack meals, such as fried chicken joints. Pretzels are frequently much less salty nowadays and come in a variety of sizes. The method of manufacture is rather unusual and is akin to that used for producing the BAGEL (see above). The smallest pretzels may be made by an extrusion process but the larger ones are made substantially as follows.

Pretzel paste is a stiff dough, made with flour of medium strength using 250 ml of warm water to 500 g of flour. To this is added 20 g of baker's yeast, 15 g of salt and 10 g of caraway seeds. The paste needs to be kneaded well and left in a warm place for half to one hour to ferment and aerate. The fermented paste is then moulded into shape (a loose knot) and is placed into a boiling alkaline liquid containing 30 g of sodium bicarbonate (sometimes ammonium bicarbonate is also added) per half litre. When they float they are removed from the liquid, placed on an oven tray or belt, sprinkled with coarse salt and baked to the desired colour. They may also be brushed or sprayed with beaten egg (to provide 'shine') and sprinkled with cumin seeds.

The above small scale preparation method has of course been translated into large scale manufacturing terms in which the dough is extruded into a rope and is mechanically formed into bows and cut off. These are then treated with alkali, boiled and then baked as in the above method. The stick pretzel is made in exactly the same way as the ordinary pretzels, but consists of short lengths of the moulded rope of dough.

Doughnuts

The doughnut manufacturing technology resembles that of the bagel and the pretzel in many respects, except that instead of the final baking, the boiled dough pieces are deep fried. The jam normally found in the centre of the doughnut bun is nowadays usually injected rather than incorporated at the time the bun piece is formed. After draining off the excess fat, the doughnuts are dusted with caster sugar.

See also Chapter 5.

Dips

A dip is a material of fairly stiff creamy consistency, most often consumed cold and used to provide a usually savoury (but sometimes sweet) zest to crisps, breadsticks, biscuits or other rigid usually aerated snack accompaniments to drinks. Other rigid bases which are used with dips are raw vegetables such as carrots and celery sticks, cauliflower, and so on. The dip should not drip after being picked up.

The basis for a dip is variable, ranging from mayonnaise and mayonnaise/salad cream mixtures to sour cream, thick yoghurt and a great variety of other (usually homogenized) mixtures which may contain starches, gums and other thickening agents as well as (usually) fats, emulsifiers and frequently a source of protein as well, e.g. whole egg or egg whites.

Within these bases are incorporated a wide range of additives, usually finely chopped or grated but sometimes in powder or liquid form. The intent is to provide an interesting colour and texture as well as flavour to the usual white or creamy coloured base. Materials frequently used are chives, cheese, garlic, peppers, salami, anchovies, nuts, olives and onions, but the possibilities are legion, in fact practically all the materials seen in hors-d'oeuvres are usable in appropriately comminuted form in dips. In addition to the obvious additions to a dip base, there may also be many less obvious ones such as herbs, flavouring essences, condiments, and so on.

For long keeping it is desirable to blanch some fresh ingredients before use to avoid both the development of usually bitter off-flavours and the liquefaction of the base which may arise from enzyme action.

Sweet dips are less common than savoury ones, but may include in a sweet creamy base, such as sweetened yoghurt or whipped cream, such items as chopped preserved ginger (either syruped or crystallized) or kumquats, and any of the great variety of crystallized fruits and candied citrus peels that are available.

The method of manufacture, the composition, pH, presence or absence of preservatives and the processing methods employed all have to be taken into consideration when the prospective shelf life and the packaging of dips are being considered. A life of six months in an unopened glass pack plus several weeks' refrigerated storage capability after opening is regularly attainable in products of reasonably low pH (e.g. 3.5).

See also Chapter 7.

Fish-based snacks

These can be varied in nature, frequently but not always preserved or specially prepared, and may be dried, salted, fermented, smoked or marinated. They are frequently raw.

Smoked fish is used worldwide both for both snack and main meal purposes; smoked salmon is an appropriate Western example. The salmon fillets or 'sides' are first brined to a level which will make the final product short keeping (e.g. 3–4 weeks, chilled) or longer keeping (a few months), the final levels being about 2.5 and 5% respectively. They are then smoked in a kiln to reduce the moisture content of the fish, using a particular kind of wood or sawdust to produce the required flavour. Moisture reduction is normally 15–25%. This has the effect of 'concentrating' the fish flavour, which in combination with the flavour imparted by the smoke can lead to a most attractive product. Keeping quality is also enhanced by this process.

Another fish component in many snacks is the anchovy, canned fillets of which are widely used, as are various forms of pickled and marinated raw herring fillets such as Bismark herrings and the superlative Scandinavian 'gaffelbiter' or fork pieces. New fish snack variants are constantly being introduced, such as fish 'sticks' and crab 'sticks' to rank with the great variety of prawns, shrimps and other seafoods regularly used in this way. It should perhaps be mentioned that the current 'best seller' snack-lunch sandwich in the UK is the prawn mayonnaise variety.

Toast toppers

The concept of the toast topper was probably derived from Welsh rarebit and the currently available ranges always include one or more cheese-based variants. The toast topper consists of a savoury mixture of ingredients which may be spread on to a piece of toast and then placed under the grill. The final outcome is a savoury snack in which the topper ingredients are grilled or toasted. This usually produces a plasticized, hot and usually quite fatty covering to the toast slice.

The range of ingredients that may be used in toast toppers is very wide. A variety of cheeses provide different textures and flavours, e.g. cheddar, mozzarella (as used in pizzas) and a variety of the Emmenthaler cheese produced in the canton of Valais in Switzerland and customarily used in the making of raclettes. Meat and fish products, particularly cured varieties, are other possibilities. Vegetable and herb flavours are also appropriate, as are flavour enhancers, monosodium glutamate for example, and egg may also feature in the mixes as a more bland ingredient. Overall, a relatively high fat content is desirable in this type of product.

The principal difficulty with commercial toast toppers is that they have to be processed in order to achieve a satisfactory shelf life. Canning or other heat processing sometimes degrades both the texture and flavour of the product to an undesirable degree.

Pasties

Pasties consist of various forms of pastry, duly sheeted and folded, and containing a variety of fillings. The range includes Cornish pasties, Mexican tacos, Indian and Middle East samosas and Chinese spring rolls. These are all eaten widely in Western countries – there are hundreds of similar products consumed as snacks worldwide within their own localities. The pasties and their contents (which may be fish- or meat-containing or vegetarian) are finally baked or fried and can be eaten hot or cold. The Chinese spring rolls tend to differ somewhat from the other three mentioned varieties in that the 'pastry' used is more in the nature of a crèpe.

Vol-au-vents

Vol-au-vent cases come in a wide variety of shapes and sizes, the smallest of which are unashamedly snacks. The cases are made of puff pastry and the methods of manufacture are well known. Miniature vol-au-vent cases were sold like biscuits not very long ago, indicating that they are capable of mass production.

The production of puff pastry has been described in Chapter 5. In order to make vol-au-vents, the puff pastry is rolled out to a thickness of about 12 mm and is then cut into oval, circular or square shapes. A circle about 8 mm deep is then cut in the centre of each. After glazing, the pieces are baked and cooled. The centres are then removed leaving a hole into which filling is placed. the baked top of the removed portion may be placed on top of the filling.

There are many fillings suitable for vol-au-vent cases which can provide a wide variety of snacks for the cocktail party or the buffet snack bar. Such fillings may include chopped chicken in white sauce, mushrooms in a sauce, shrimps or prawns in mayonnaise, and so on.

Pork 'scratchings'

Pieces of fresh pork skin with their considerable proportion of adherent fat are deep fried to such an extent that most of the fat is removed and the remaining skin protein is well cooked and crisped. The product is reminiscent of the crisped outside of a roast pork joint in both texture and flavour, and after removal of as much fat as is possible from it, the flavour is frequently enhanced using a flavouring powder which adheres to the fatty surface. This is a popular bar snack.

Bombay mix

Bombay mix is not a single product but is the name applied in Britain and North America to a range of moderately hot spiced mixtures of nuts, pulses and cereals, commonly prepacked and sold as retail snacks. In India, the term is not apparently used, although products of this type are widely available. Typical ingredients may include potato sticks, rice puffs, noodles of various shapes, peanuts, gram flour, chick peas, lentils, cashews and possibly sultanas, with salt, sugar and a little vegetable oil. The spices usually include turmeric (for the yellow colour), chilli and pepper.

Confectionery bars

These bars are among the most popular of all snack foods. They vary so much in composition that attempts to classify them would be futile. Suffice it to say that they may be chocolate coated (enrobed) or not; may have a biscuit or wafer base or else a composite base of caramel; they may contain added processed cereal, nuts or dried fruits, or mixtures thereof; or they may have aerated or flake chocolate centres, and so on. Nutritionally, they tend to be heavy on sugar and fats, but at least they can prevent the 'low blood glucose' syndrome. They do not normally have any other nutritional pretensions.

See also COUNT LINES, Chapter 10.

Intermediate moisture products

Apart from traditional examples, e.g. cheese, these foods have not yet become well known or developed except in the pet food area. They tend to have moisture contents of around 25% but such low levels of water activity (around 0.6–0.7) that microbiological changes do not take place at normal temperate-zone ambient temperatures. The one example which has made some headway as a 'pub' bar snack is the miniature salami sausage.

The stability of salami is due to a variety of factors. First, it has a low water activity as a result of which bacteria and mould growth is inhibited. Second, it is made mainly of cured meats with a high level of salt, sodium nitrate and some sodium nitrite. Third, it is smoked and simultaneously heated and dried to its around 25% moisture content, which provides additional protection. Fourth, spices are incorporated which themselves have a preservative action; finally, in some instances, the use of preservatives, e.g. sorbic acid or sorbates, may be allowed.

PACKAGING OF BREAKFAST AND SNACK FOODS

The usual criteria applicable to food packaging apply here, but the emphasis may vary. The protection requirements may be summarized as follows:

(i) Mechanical strength – protection against crushing or deformation
(ii) Avoidance of taint or contamination from the package or the environment
(iii) Protection against moisture ingress or egress
(iv) Protection against gaseous or gas-carried oxidants, etc.
(v) Minimal damage caused by light
(vi) Avoidance of microbiological ingress or initial presence

Materials for the packaging of any given product must be chosen with all these factors in mind and particularly the fact that in most cases the contents of the package will be eaten without further cooking.

However, there are in addition other considerations apart from the always present economic one:

(i) Weight, shape and size of the pack for convenience in bulking, transport and shelf stacking, not to mention display value of the pack
(ii) Single use or multiple use of the pack (breakfast foods especially) and facility for re-use
(iii) Ease of opening of the pack, with due regard to reasonable protection against misuse.

Sophisticated product protection is provided by flexible film; board supplies mainly mechanical protection. Data on the protective properties of packaging films, such as may be found in the Tables in Chapter 16, Food Packaging, may be used in conjunction with relative cost data to provide a basis on which the choice of an appropriate film or film laminate may be made. To the list should probably be added metallized films, which can in some cases replace aluminium foil.

The computer is now widely used to determine the optimal packaging format in terms of material, design, size and shape, cost, and so on, for a given shelf life.

QUALITY ASPECTS

A large proportion of breakfast foods and snack foods are formulated. This means that they should provide nutritional advantages to the consumer over and above those which might normally be expected. Such formulation can change and follow the current trend (scientific, fashionable or otherwise) in ingredient popularity. A prime example of this is the way in which oat bran is tending to replace wheat bran, at least in part, because of its reputed effect in reducing blood cholesterol levels and also its apparently superior efficacy as an aperient in many cases.

Formulations may also be altered to include specific additives designed to eliminate any nutritional shortages that may show up in dietary surveys. Recent examples of such changes have been the addition of various minerals such as iron, zinc and calcium. Vitamins, particularly the water-soluble ones, have been incorporated for a long time; a recent addition is folic acid.

The use of fats, naturally occurring and added, in snack foods is another area where both quantity and quality are of great interest to the informed consumer. There is concern over both high levels of monosaturates and total fat content. As with all foodstuffs, there is a general requirement for high quality in terms of both nutrition and hygiene.

REFERENCE

Booth, R.G. (ed.) (1990) *Snack Foods*, AVI and Van Nostrand Reinhold, New York.

FURTHER READING

Frame, N.D. (ed.) (1993) *The Technology of Extrusion Cooking*, Blackie Academic and Professional, Glasgow.

Kent, N.L. and Evers, A.D. (1994) *Technology of Cereals*, Pergamon/Elsevier Science, Oxford.

Maga, J.A. (1984) Cereal-based Snack Foods. In *Handbook of Cereal Science and Technology*, J.L. Klaus and K. Kulp (eds), Marcel Dekker, New York, Chapter 20.

Matz, S.A. (1984) *Snack Food Technology*, 2nd edn, AVI and Van Nostrand Reinhold, New York.

Mercier, C., Linko, P. and Harper, J.M. (1989) *Extrusion Cooking*, American Association of Cereal Chemists Inc., St. Paul, MN.

Woodroof, J.G. (1979) *Tree Nuts*, 2nd edn, AVI and Van Nostrand Reinhold, New York.

Woodroof, J.G. (1983) *Peanuts*, 3rd edn, AVI and Van Nostrand Reinhold, New York.

13 Composite Foods and Ready Meals

INTRODUCTION

In the two centuries since Nicholas Appert filed his revolutionary patent for the preservation of food by heating it in closed containers, the food industry has made enormous progress towards two objectives:

(i) To conserve individual items of food so that they may be kept in good or perfect condition from the times and places where they are relatively abundant until they can be consumed at other times and in other places where they would not otherwise be available;

(ii) To maximize the commercial efficiency of production.

Yet still, almost everywhere in the world, the food which is thus produced and conserved, often on a very large scale, is assembled, prepared and served for consumption on the very small scale. The quantities of food for even a major banquet are small compared with the daily output of any moderate-sized factory; the quantities for the average domestic meal are minute in comparison. It is also almost always the case that once a meal is prepared it is at its best if it is consumed immediately. Significant amounts of food may be left over and served again, sometimes in enterprising and attractive ways, but this is not usually considered as good as enjoying the food when it was first set out.

The modern catering industry in only the last few decades has applied increasingly the principles of food preservation and of the division of labour to the preparation and service of complete meals. In this kind of industrialized process, the stages of assembly and preparation of the multiple components of a meal become increasingly separated both in time and place from the final stage of serving the meal to those who will eat it.

In other chapters of this manual we have considered a wide range of manufactured food items more or less independently of one another. In this chapter we now consider some of the questions which arise when foods of dissimilar properties are assembled together as components of composite meals and when the components have to remain together for longer periods than would commonly be allowed in domestic or small-scale catering.

Development of this modern way of food preparation, both for small scale domestic meals and for large scale catering (which may often include high quality catering), turns on the solutions to two kinds of problem, temperature control and migration of components.

Temperature control

Prepared meals, for reheating

The preparation of these by the manufacturer usually involves sufficient heating to cook the food and render it microbiologically safe. It is necessary to design the product so that at the end of all the preparation processes it reaches optimum eating quality when served at the table; this means that the reheating conditions and any thawing conditions if the product is frozen, must be precisely specified by the manufacturer and properly executed by the caterer.

Intermediate storage

Especially in the case of 'cook–chill' meals, the temperature and maximum time of intermediate storage must be strictly specified by the manufacturer and equally strictly followed by the storage contractor (if any) and the caterer. The chill chain must be kept intact, for obvious reasons of both quality and safety. Similarly with 'cook–freeze' products the cold chain must be maintained and it is important also to specify and control whether and how the product is to be thawed or reheated directly from frozen.

Migration of components

Among the most enjoyable experiences when eating is to encounter widely different sensations of texture and flavour in the mouth at the same time. To present well balanced and contrasting mixtures of sensations is a highly prized aspect of the culinary art. However, the greatest differences in texture are often related to large differences in composition, particularly in moisture content, and when components of different composition are in contact with one another then migration of moisture or other substances is likely to take place. This may change the texture in unwanted ways, for instance when the lettuce in a sandwich becomes limp and the bread turns soggy.

There may be further difficulty from the same cause if migration leads to chemical changes. If, for instance, the movement of moisture from a vegetable into a mayonnaise has the effect of reducing the salt-on-water concentration in the mayonnaise, the keeping quality and the microbiological safety of the system may be impaired.

In the sections which follow we shall draw attention to the technological means adopted by the industry to reconcile these various conflicting requirements.

BREAD PRODUCTS

SANDWICHES AND FILLED ROLLS

This ready-to-eat convenience market has grown exponentially since the mid-1980s to reach an estimated value in the UK alone of about £2 billion in 1996. The products are extremely 'high risk' as they consist of cooked and raw ingredients which have undergone several stages of preparation (e.g. slicing, chopping) and a large amount of handling in constructing the final product. Consequently they need to be manufactured in extremely clean surroundings and the usual safeguards pertaining to 'high risk' or 'high care' production areas must be enforced.

Following production, the ready-to-eat sandwiches and filled rolls are packaged and maintained at chill temperatures to retard the growth of micro-organisms. It is also possible to freeze the products to ensure maximum safety but of course this detracts from the convenience nature of the products. Some packs of sandwiches are gas flushed with atmospheres high in carbon dioxide to extend their shelf life and long life rolls and filled bread products have been produced having shelf lives, at chill temperatures, of up 30 days.

The major components of sandwiches and filled rolls (bread, sliced meats, eggs, salad vegetables, pickles, mayonnaise), when made to contact each other in a sandwich, interact with each other in a physical way, which in turn leads to a deterioration in eating quality. Moisture can migrate directly from a 'wet' component (e.g. tomato) into the relatively dry bread or it can, if the sandwich is enclosed in a pack, escape into the surrounding atmosphere to be absorbed gradually by the outer surfaces of the bread, again producing a soggy texture.

The problem can be minimized by introducing physical barriers to the migration (e.g. butter or fat-based spread) onto the bread surfaces. Alternatively the ingredients of the sandwich can be chosen to have similar water activities to the bread to minimize moisture migration during subsequent storage. Low water activity cloths/ webbing have been introduced into sandwich packs to absorb moisture from the atmosphere and the pack.

Chilling a bread product also causes problems in staling, as bread has a faster staling rate at low temperatures compared to room temperature. However, over the usual shelf life of the sandwich (24–48 h) these effects are small. Bread also goes stale rapidly at temperatures in the range -3 to $-10°C$ but hardly at all at $-20°C$ or below. Bread or sandwiches can therefore be frozen for storage at $-20°C$ provided care is taken to ensure that freezing and thawing are both carried out as rapidly as possible and that the proper storage temperature is maintained. (Unbaked or part baked bread or pastry is not subject to this problem and can be frozen for storage without special difficulty). Microbiological problems associated with sandwich storage at incorrect (high) temperatures are well known.

Clean, high quality raw materials are obviously essential for producing a safe product. To minimize the numbers of micro-organisms on the final product, reputable manufacturers have 'high risk' or 'high care' production and preparation areas where there is a strict regime of personal, equipment and factory cleanliness and associated hygiene checks on the fabric of the production area, the equipment and the workers. Hazard analysis of critical control points (HACCP) or similar risk assessment programmes should be in place in sandwich manufacturing operations in the UK.

The choice of low pH pickles or sauces as part of the filling, and of components with low water activity where this is possible, helps to slow down the growth of certain organisms. Once the sandwiches are packed the methods for minimizing the growth of micro-organisms includes in particular the use of chill temperatures and high carbon dioxide atmospheres.

TOASTED BREAD PRODUCTS

Convenience snack meals containing toasted bread have appeared in a limited range of frozen and chilled goods such as 'Welsh Rarebit' and 'Croque Monsieur'. Partially toasted bread is topped with cheese; the combination will be cooked under the grill before serving, so that the cheese will be melted and slightly browned.

Moisture uptake by the toasted bread from the surrounding atmosphere during storage and thawing, may cause problems of lack of crispness in the toast. Adequate packaging usually alleviates this problem.

Texture changes may occur in the cheese during prolonged storage – the cheese may become hard and rubbery. Correct choice of cheese and processing conditions usually helps to minimize the problem. Fried bread products such as croutons, i.e. small cubes of bread which have been fried, are usually stored at ambient temperatures and are used to garnish prepared soups, bouillons and salads. The product should be crunchy on eating but if it is stored incorrectly in a humid atmosphere, it may pick up moisture and acquire an unacceptable spongy texture. If croutons are stored in the light in the presence of oxygen, they develop rancid flavours due to the oxidation of the oil which they absorbed during frying. Nitrogen flushing prior to closure of the package, followed by storage in light proof, gas impermeable packs (usually aluminium foil lined) helps to alleviate the problem.

There is now a range of flavoured bread snacks, similar to croutons, usually garlic or herb flavoured. These may be used in the same way as croutons or eaten with drinks as an aperitif. They are usually moulded directly from wheat dough to form bite sized pieces, then sprayed or soaked in oil. Again, flavour problems associated with fat rancidity are the major cause of concern when the product is stored at ambient temperatures. Consequently, inert gas flushing and foil lined packages are used to prevent ingress of oxygen and light. Compounds present in the garlic and herb flavours may also act as natural antioxidants and in any case the strong flavours certainly help to some extent to mask the off-flavours arising from fat oxidation.

PIZZA PRODUCTS

Frozen and chilled ready prepared pizzas have increased in popularity rapidly in recent times. The products consist of a flat layer of wheaten dough onto which is spread a thin layer of tomato puree with the other ingredients (e.g. fish, ham, meat, mushrooms, cheese, peppers, olives, pineapple) arranged on top. The whole is cooked rapidly and then frozen or chilled.

As with pies and tarts, moisture migration from the topping into the cooked dough can cause loss of textural quality. This can be a particular problem when pizzas are cooked and transported to customers in home delivery operations. The cheese texture can deteriorate through freezing and thawing, as with Welsh Rarebit and Croque Monsieur, and the correct choice of cheese is necessary to minimize this problem. The traditional cheese for pizza is mozzarella, a soft unripened cheese with a plastic, stringy texture; cheese analogues are now commonly used in addition or instead (see Chapter 3).

OTHER BREAD PRODUCTS

Small market shares of the bread market are occupied by pitta bread and prepared filled pitta bread products. Chilling or freezing helps to preserve the microbiological stability of the filled products. Migration of moisture from filling to bread is the major defect which can occur during chilling or freeze/thawing.

'Burger-in-a-bun' products (cooked meat slices or patties, with or without salad materials and sauces, served in soft bread rolls) and filled croissants also have a small market share. The problems encountered are similar to those with filled rolls or sandwiches.

CREAM-BASED PRODUCTS

Whipped, aerated or clotted cream is incorporated in many desserts and cakes, originally to provide a luxurious treat. Today, among health conscious nations, cream is perceived as a cholesterol-rich food which should be eaten in moderation. Whipped, aerated and clotted cream have high fat and moisture contents and the problems associated with the storage or cream products are associated with these two components.

When cream is in contact with dry sponge cake or meringue base, moisture migrates from the cream. The migration is slow, as the water-in-fat emulsion has to be broken before free moisture can migrate. Migration slows down if the product is frozen, so frozen cream cakes, gateaux and desserts are now commonplace.

The growth of micro-organisms is common in cream bases. Some organisms can cause lipolysis and rancidity in the fat by production of free fatty acids. Other organisms are pathogenic and may multiply in temperature-abused cream-based products causing food poisoning if they are present in large numbers.

Frozen, chilled and ambient-stable dairy desserts are all available today. Frozen dairy desserts include ice creams, yoghurts, mousses and custards. Many now are made with whole nuts, fruits, fruit puree or chocolate, either mixed with the bulk of the dessert or layered above or below the base to give double or triple layers.

The microflora of such desserts are usually the same as those of the ingredients. If fruit is mixed with the dessert then the pH of the dessert is lowered and there is a beneficial effect on the microflora. However, with layered desserts, where there is no mixing of the layers, each layer has its own microbial population and spoilage pattern. Strict control of the temperature of chilled desserts is essential.

If fruit is mixed with chilled yoghurt then curdling may occur due to the reduced pH imparted by the fruit acids. This is overcome in several packaged products in which the fruit is placed in a separate compartment from the yoghurt and the consumer is allowed to mix them immediately before consumption.

MICROWAVE REHEATING

The microwave oven is now widely used by caterers, restaurants and in the home for reheating manufactured ready prepared meals before serving them hot at the table. It is therefore common for the manufacturer of the meal to provide instructions on the outer packaging on how to set about microwave reheating. A number of factors must be taken into account in formulating these instructions to ensure that the meal, as finally served, reaches the table properly cooked and at the right uniform temperature for eating.

Several features of the microwave oven may be encountered, alone or in combination:

(i) A turntable to rotate the food and achieve more even heating

(ii) More than one microwave source to even out the microwave energy within the oven

(iii) An independently heated internal fan-assisted convection oven. The judicious combination of the convection oven with microwave energy can be used to advantage in producing desirable qualities in the reheated product, which would not be possible using microwave energy alone

(iv) An independently heated internal grill; the grill provides radiant heating of the surface of the food to temperatures above 100°C so that surface browning may occur (which cannot be done by microwave energy alone, since microwaves heat the interior of the product and neither the internal nor the surface temperature can rise above the boiling point of water before all the internal water is evaporated

(v) A tunnel microwave in which food travels on a belt through powerful microwave sources. In many cases the tunnel microwave is supplemented with a 'pass-through' infrared convection oven to obtain surface browning. Tunnel microwaves are used in large scale catering operations (e.g. hospitals) or in food processing factories.

It is evident that the rates of heating of products will be different according to the type of oven and any extra heating sources present; preparation instructions should take account of this. Plastic, glass or ceramic containers are poor absorbers of microwave energy and therefore permit the best reheating of their contents; metal containers may absorb so much that the food inside is barely heated at all. Lightweight aluminium dishes may however be used in many applications where the bulk of the container is small relative to that of the contents.

The absorption of microwave energy by a food material, and therefore the heating effect and the temperature reached in a given time, are affected by:

(i) The shape of the material; the most efficient and uniform heating occurs with uniform material of regular shape completely filling the microwave cavity. This is almost impossible to ensure completely but the closer to this ideal the better and the more irregular the shape of the food being heated the more irregular will be the final temperature.

(ii) The composition of the material; large differences in moisture content or salt content among the food components of a dish may lead to differential absorption of the microwave energy, shielding of some components by others and therefore irregular final temperatures.

These variations may make it quite difficult to reheat a meal so that all of the dish is hot enough both for agreeable eating and to ensure microbiological safety, while none of it is overcooked. For best reheating the separate components of a multi-component dish should all be of approximately similar composition and should be arranged in the container like bricks in a wall, with a uniform overall appearance. Neither the flavour nor the appearance of such a product will be the most attractive possible and some compromise between appearance, flavours and heating efficiency must therefore be sought.

Susceptor devices have been developed to provide concentration of microwave energy at certain points (e.g. the surface) of the products being reheated. These thin aluminium-coated cardboard shapes are wrapped around the food (e.g. apple pie) and if positioned correctly in the microwave can cause browning of the surface of the food.

PASTA PRODUCTS

See also Chapter 5

Pasta products come in many forms. They may be canned, dehydrated or fresh. They may be stored at frozen, chilled or ambient temperatures. There is probably no other food from a single common source (wheat) which offers such a variety of shapes and sizes – macaroni, spaghetti, vermicelli, tagliatelle, noodles and lasagne are but a few of the names given to different shapes. A form of spaghetti, known as 'perciatelli' or 'small pierced strands', with a narrow hole running the length of the spaghetti strand allows for rapid reheating in a microwave, with steam produced at the centre of the strand to provide internal heating.

The use of pasta products in manufactured mixtures has increased markedly over the last few years. These include ravioli (pasta envelopes with meat or other fillings), lasagna (sheets of pasta interleaved with cooked meat and a cheese sauce) and 'Pot Noodles' in which pasta shapes are mixed with dried peppers, onion, meat pieces and other ingredients to form a dehydrated snack base.

A possible problem with canned pasta stored for long periods is the migration of flavours from richly flavoured sauces into the pasta which properly should be bland in flavour. Textural changes may also occur, first because the heat process necessary to sterilize the can may be excessive for cooking the pasta and second as a result of low pH such as may occur in, for example a tomato sauce.

PASTRY PRODUCTS

Within this section are included pies, puddings and tarts, which may be frozen, chilled or maintained at room temperature. They include egg custards, fruit pies, quiches, meat pies, en-croute products and cheese cakes. For manufacturing considerations, see PIES, PASTIES, SAUSAGE ROLLS in Chapter 1. We are concerned here mainly with storage effects after manufacture.

The major problem that occurs during storage of filled pastry products is migration of moisture from the filling to the pastry base, sides or top. With only a small degree of moisture migration pastries become soft and lose their crumbly texture. In the extreme, a pastry is left which is heavy, soggy and unappetizing. Moisture migration can also occur to or from the surrounding atmosphere of a refrigerator into the outer surfaces of unprotected pastry. However, this spoilage of product quality is not usually as great as in the case of moisture migration within the product itself and can usually be prevented by use of correct packaging.

Several methods to retard moisture migration from ingredients into pastry have met with success:

(i) 'Blind baking' or partial baking of the pastry case followed by application of a glaze or seal on the cooked pastry, helps to protect the moisture absorbing sites within the pastry. The filling can be added with only small amounts of moisture migrating during storage. However, this process in a factory environment is inconvenient and significantly adds to the processing costs.

(ii) Materials (e.g. starches) which bind water and thereby reduce the overall water activity can be incorporated into the filling to minimize water migration from the filling to the pastry.

(iii) If a filling contains egg which is then cooked, the egg protein is denatured during cooking and water is released which becomes available for migration to the pastry. The use of dried egg, which has been denatured previously in the drying process, is reported to have beneficial effects particularly in egg custards and quiches. Even when the dried egg is reconstituted with water before incorporation into the filling, subsequent cooking does not produce large quantities of free moisture for potential migration from the filling.

A further problem sometimes encountered on storage of pastry products is moisture adsorption at the surface with subsequent mould growth on the moist surface. This is not very common on the outer surface unless the pastry is packed in material with a low water vapour transmission rate but it may occur on inner surfaces especially where there is space between the pastry and the filling. If temperatures during storage are allowed to fluctuate, or if products are not cooled quickly after baking, there will be temperature gradients within the product or between product and packaging. Moisture is then likely to evaporate from the warmer regions and be deposited on the cooler surfaces. Mould growth on the insides of pies, pasties and similar products may easily be caused in this way.

Occasionally mould growth on the surface of pies and pasties packed in plastic wrappers has been noted if the packaging contains pin holes. During storage at different temperatures there is considerable exchange of air above the surface of the pastry with the external atmosphere due to a 'pumping action' and mould spores enter the pack. If adverse storage temperatures are maintained mould growth occurs. Strict quality assurance at the packaging point and stock rotation will minimize this problem.

Another effect on pastry products stored for long periods at room temperature is oxidation of the fat in the dried surfaces of the pastry. 'cardboardy' flavours sometimes result. Keeping the products away from direct light and high temperatures within their shelf life minimizes the problem.

Sometimes taints or odours can be absorbed onto the surface of pastry products from contact with or closeness to strongly smelling substances. Correct packaging or storage usually alleviates these problems.

READY MEALS

These products include ready plated, precooked, frozen whole meals and ambient-stable microwaveable meals which only require thawing (if frozen) and a short time of reheating prior to serving.

FROZEN, READY PLATED, PRECOOKED DISHES

These include roast dinners and other meals in which the components (e.g. meat, potatoes, Yorkshire puddings, vegetables) are precooked and then assembled on a dish prior to freezing and packaging. The preparation and assembly of these meals are generally in 'high risk' or 'high care'

areas as there is no further cooking of the meals prior to consumption.

The major challenge in production, aside from ensuring the microbiological safety of the products, is the exact precooking of the ingredients so that they will be of the correct texture when stored, reheated and consumed. Precise control of cooking operations is required.

Moisture migration from one ingredient to another can take place even at frozen storage temperatures, producing soggy textures in some components after reheating (e.g. the potatoes and Yorkshire puddings in a roast beef dinner). Flavours may migrate from strongly flavoured components to others when they are in close contact under frozen storage conditions.

AMBIENT-STABLE MEALS

These are products which have been processed, therefore cooked, in a can or in a plastic microwaveable container so that when opened reheating is all that is required prior to consumption. Often the meals are highly flavoured with hot spices or acidic sauces to mask the slight caramel or 'warmed-over' flavour which can be detected in blander products. Popular products are various types of chilli con carne and curry

OHMIC HEATING (see also Chapter 15) is a process which can give cooked products in laminated plastic/foil pouches shelf lives of up to 12 months. A current is passed through a moving column of the food. The current passing through the food heats it up quickly to high temperature so that the necessary heat process conditions can be attained with minimum heating-up time. The heating is direct and energy efficient. Under these conditions the flavours and textures within the food are much better maintained than under conventional cannery retorting conditions. There is also minimal 'warmed over' flavour. The heated food is rapidly chilled through tubular heat exchangers and is filled aseptically into pouches or other suitable containers. Products such as boeuf bourgignon and steak and kidney pie filling have been manufactured successfully using this technique. One limitation of the ohmic heating process is its inability to heat up products containing ions which conduct electricity. High conductivity produces low currents and minimal heating. Thus, the technology cannot be applied to salty foods.

One problem with ambient stable meals is that at long storage times, flavours from all the components permeate the bulk of the food and the taste of the product becomes uniform. Individual solid components may have the taste of the whole product, thus detracting from product quality.

SALAD PRODUCTS

Salad products can be split into two main types: (i) those which are packaged as loose mixes of leaf and sometimes root vegetables and (ii) those to which mayonnaise, salad cream or a clear oil-based dressing has been added (e.g. coleslaw). Both types are distributed at chill temperatures.

SALAD INGREDIENTS

The quality of the final product depends to a major degree on the quality and handling of the ingredients and the use of high quality fruits and vegetables is essential to produce good quality salad products. Good hygienic practice, through the use of 'high risk' or 'high care' operations, is essential to prevent food poisoning or spoilage organisms from being introduced into these products which will be eaten without any heat processing.

The crops should be grown under clean conditions, using land and irrigation water free from faecal pollution. To confirm this, crops on arrival at the processing factory should be almost free from *Escherichia coli*, which is an indicator of faecal pollution. The raw materials should be harvested carefully to avoid damage and bruising and should be free from significant blemish, disease or mould. They should be stored under controlled conditions, normally in chill store. Modified atmospheres may be used for longer term storage (see below).

Once the raw materials are removed from store for subsequent processing they should be prepared and used immediately and not stored in semi-prepared form in uncontrolled surroundings. Fruits and vegetables are normally inspected, trimmed to remove any blemish, washed peeled and cut, shredded or diced as required. Where several washes are used, the water should be progressively cleaner, the final rinse being in chlorinated water with 5–15 ppm total residual chlorine. Solutions containing 0.3–1% sodium chloride, 0.1–0.5% citric acid, 0.3–1% ascorbic acid or 2–5% acetic acid may be used to reduce deterioration of cut surfaces. Stronger acid solutions can be used where acidic dressing such as vinaigrette or mayonnaise is to be added.

Pasta and potato are cooked and then cooled before use as salad ingredients. If animal protein foods are used, e.g. ham, sausage, prawns, cheese, egg, they also should be suitably treated to remain stable during the shelf life of the products.

SALADS WITHOUT DRESSINGS

This type of product, the loose mixes of leaf and root vegetables, has increased in popularity over the last few years and can now be found in transparent packs, often under modified atmospheres, on the shelves of most supermarkets. Among the ingredients may be chopped lettuce, Chinese leaf, bean sprouts, cabbage, radiccio, watercress, red and yellow peppers and carrot. Any combination of these may be packaged at chill temperatures but each mixture has its own particular problems in storage. Shelf life of these products is usually 3–5 days at chill temperatures.

Ready-to-eat salads are classed as high risk products and the usual precautions must be observed in salad selection, washing and chopping (see above). Otherwise there will be problems of microbial growth, even at the temperatures of chilled storage.

The major problem for these products during storage is that the vegetable components respire at different rates. The presence of shredded carrot or other root vegetable usually causes most problems, as their respiration rates are very different from those of the green vegetables. Respiration rates may be modfied by MODIFIED ATMOSPHERE PACKAGING (see Chapter 16). Usually mixtures of oxygen/carbon dioxide/nitrogen are used with the proportions of each being accurately matched to the nature of the ingredients. A mixture of 5% O_2/5% CO_2/90% N_2 is frequently chosen for the control of respiration and the growth of micro-organisms. Other gas mixtures are commercially useful especially when used with semi-permeable membrane plastics.

Respiration of vegetables in pack also causes problems through moisture release. The moisture detracts from the visual appearance and overall quality and reduces the product shelf life. Anti-fogging agents on the internal surfaces of the plastic packs can help to minimize the problem of visual appearance; ricinoleate mixtures may be used for this purpose.

At the cut surfaces of the salad vegetables, enzymic browning reactions can occur and cause loss of visual appearance and reduction in shelf life. The reaction produces pink, brown or orange discoloration, particularly on lettuce, and is due to the actions of oxygen, natural enzymes and substrates present in the plant cells, exposed on cutting. It can be minimized by prewashing chopped salad vegetables in a dilute acid (as described above) and by ensuring that chilled storage temperatures are maintained.

Developments in plastics technology have produced materials (semi-permeable membranes) which allow some of the unwanted gases to escape from the pack while retaining those which are beneficial. With such possibilities, the technique of modified atmosphere packaging may still be in its infancy (see Chapter 16).

The incorporation of ethene (ethylene) absorbing agents into plastic packs also opens up possibilities for controlling atmospheres and extending shelf life. Vegetables and fruits which lose their green colour on ripening (e.g. tomatoes, apples, bananas) produce ethene in the ripening process; if ethene can be continually removed from the environment then ripening can be delayed.

Longer shelf lives of 8–10 days at chill temperatures are now attainable with modern technology (e.g. in transcontinental distribution chains). Careful temperature maintenance and monitoring of the raw materials are required for this to be successful.

SALADS WITH DRESSINGS

Salads with dressings have seen a rapid growth in sales due to the development of new salads with pasta, nuts, apples and a wide variety of less common fruits and vegetables, for example cherry, grapes, orange, peach, pineapple, kiwi fruit, mandarin segments, peppers, tomatoes, onions, sultanas, dill and bean sprouts.

Mayonnaise or clear acidic vinegar-based dressings are usually added to 20–30% of the product pack weight. They must be well mixed with the salad product to ensure a thorough acid coating of the ingredients and a product pH 3.5–4.2. Ingredient unit size should be small, to ensure a high surface-to-mass ratio. After mixing the product may be piston-filled into plastic pots or trays.

Salads with well formulated and carefully applied dressings can have product life under chill conditions of 5–18 days depending on the ingredients. During this time water migrates from the fruit and vegetable tissue into the mayonnaise, so diluting it. This is accompanied by migration of oil from oil-based dressings into the vegetable tissue. These migrations raise the pH of the dressing and reduce its viscosity, while the fruit or vegetable tissues become translucent. Both effects

modify the texture of the products and limit their acceptable life. The presence of pH gradients must be recognized and allowed for.

The yeasts *Saccharomyces dairensis* and *S. exiguus* are common spoilage organisms in acid salads. The combination of low pH, organic acid (particularly acetic acid) and chill temperature inhibits the growth and development of food poisoning organisms. Handling during preparation must be kept to a minimum and be of the highest hygienic standard to prevent food poisoning Salmonellae, Campylobacter or Staphylococci from entering the product. Variations in pH due to moisture and other migrations can cause unexpected problems in an otherwise apparently safe, acid food.

The life of the product is normally limited by the growth of yeasts and moulds or by texture changes. In some countries sorbate, benzoate or sulphite may be added as preservatives, particularly against yeast and mould spoilage, to prolong shelf life.

SOUPS

CANNED SOUPS

Canned soup production is to some extent the prerogative of the big canners with high speed production, but it is a useful outlet for any canner with surplus good raw material which would otherwise be wasted, e.g. large carrots, celery unsuitable for canning as hearts, oversized mushrooms and many others. There are also a number of relatively small producers of speciality lines such as game soups, turtle soup and exotic products of many varieties.

Cream of tomato soup remains a popular variety. It is prepared from tomato puree or paste which is imported from Mediterranean countries in cans or large aseptic containers. To this may be added vegetable oil, sugar, modified starch, cream, whey powder, citric acid, salt and spices. The modified starch provides the thickening or stabilizing agent necessary for the creamy texture of the finished product. It is a difficult soup to prepare, liable to curdle if the percentage of protein is too high. To avoid this, milk should be added last and it is sometimes recommended that a proportion of the milk solids be replaced by whey solids. Tomato soup should be emulsified and sieving gives it a final 'gloss'.

A number of canned soups are available in the condensed form, intended to be diluted with equal parts of water or milk. The usual pack is the A1 can giving about 570 ml (1 pint) of soup for serving. Condensed soups can also be used undiluted by the versatile cook in the preparation of many dishes and many of the producers now include recipe ideas on the labels of their canned soups.

The basis of most soups is a good stock. Very few canners now produce stock in the old domestic way, by simmering or pressure cooking meat, bones and vegetables with added herbs and spices; instead they use as a basis meat extracts and other materials of standard colour and flavour. This is a much simpler and more reproducible operation.

Apart from the basic stock most soups will contain either a garnish of recognizable small pieces of the major ingredient or, for example in a mixed vegetable soup, whole small vegetables such as peas and beans and diced or cut larger vegetables such as carrots, potatoes, celery, and so on. Pearl barley or rice are often incorporated and the majority of soups include a thickener. Flavour may be enhanced by yeast extract, hydrolysed protein and to a lesser extent these days by monosodium glutamate.

Since many cream soups have delicate flavours and light colours, raw materials such as starch, milk powder and spices must be chosen to be of good bacteriological quality so as to reduce the necessity for prolonged sterilization.

Soups may be filled hot to avoid further exhausting although steam flow closure may be practised. Heat processing may be in stationary or continuous cookers, with or without agitation, according to the type of product. Starch has an extreme delaying effect on heat penetration; with such products some form of agitating cooker may be essential and for most products they give an improved quality. Processing times and temperatures vary from mild for an acid tomato soup to severe for thick soups of high pH. Processes for cream type and other thick soups need to be carefully evaluated; it is important to review them when any formulation changes are made.

The nature of the internal surface of the cans will depend to a large extent on the product. Low pH soups or those which produce sulphur staining if plain tinplate was used, require fully lacquered cans.

Customer demand has escalated the production of canned and bottled 'cook-in-sauces'. Using soup production technology the products usually contain particles and chunks of the appropriate ingredient. They are used to provide colour and flavour to meats which are subsequently heated in them. Examples are cook-in curry sauces, tomato-based sauces and sweet and sour sauces.

OTHER HEAT PROCESSED LIQUID SOUPS

A growing market niche is that of chilled soups of relatively short shelf life. Soups having delicate and unusual flavours are gently cooked, packed into cartons (aseptically in some areas) and stored under chill conditions. Small volumes, unusual names and ingredients, despite the short shelf lives, give these soups a premium price, 'up-market' image. Soups can be packed in plastic microwaveable pots and recently there has been a marked increase in this convenience product.

DEHYDRATED SOUPS

Within this sector are included powdered soups, powdered instant soups, powdered cook-in-sauces, compressed stock cubes and instant gravy granules. The advantage of this type of product over their liquid equivalents is generally their easier (and therefore lower cost) transportation and storage. Storage times are usually 18–24 months at room temperature.

Disadvantages of the dehydrated products are:

(i) Flavour loss from ingredients during the dehydration processes of manufacture. Freeze drying techniques can reduce this problem.

(ii) Flavour transfer from dried ingredients to other components which are in close proximity in the sealed storage pack.

(iii) Irreversible cellular changes in dehydrated ingredients which cause texture problems when rehydration takes place.

(iv) Moisture migration, even at very low moisture contents, from the ingredient to the atmosphere and back to the surface of the ingredients when there is a change in storage temperature. This can lead to problems with clumping and caking at the surface of the food particles and hence give problems during rehydration. It is essential therefore, to minimize this phenomenon, to reduce the moisture content of each of the ingredients to as low a value as possible. Pack integrity must be maintained to ensure a permanent barrier to ingress of moisture; the packaging is often plastic and/or metal lined for this purpose.

Instant soup powders – where hot water is to be poured onto the powder in a cup – require a different type of modified starch from the traditional powdered soups. When water is added the starch must thicken the product within a matter of seconds rather than minutes. In the case of instant gravy granules, the porous and honeycombed nature of the dried granules provides a high surface area-to-weight ratio, so that dissolution of the granules is rapid when water is added.

Encapsulation technology has enabled the production of dry powders which are surrounded by a fat/gelatin layer. This tends to prevent atmospheric moisture entering the dried particles and minimizes loss of volatiles to the surrounding atmosphere. Shelf lives of the products are enhanced.

14 Miscellaneous Food Ingredients

INTRODUCTION

The subjects in this chapter are food materials which could not be placed conveniently into any one of the other chapters, or might have been placed in several of them. In most cases this is because they have useful properties of wide application, relevant to a variety of food types.

ADDITIVES

An additive is defined in the UK Food Labelling Regulations (1996) as: 'any substance not normally consumed as a food in itself and not normally used as a characteristic ingredient in food, whether or not it has nutritive value, the intentional addition of which to a food for a technological purpose in the manufacture, processing, preparation, treatment, packaging, transport or storage of such food results, or may reasonably be expected to result, in it or its by-products becoming directly or indirectly a component of such foods'. Essentially similar definitions apply in most other countries.

The use of additives is governed by legislation in most countries. Usually, as in the UK, they are governed by a 'positive list' and only the additives on that list are permitted. The maximum levels at which they may be used are laid down in most cases. Particular substances may also be prohibited in certain circumstances, e.g. in foods for babies and young children (in the UK).

In the EU the process of harmonization of different national lists is now completed, with a single list for all countries and each permitted additive identified with an 'E' number.

In the United State the concept of additives 'generally regarded as safe' (GRAS) was brought into the legislation in 1958, to cover additives in use at that time and considered to be safe by recognized experts or from experience based on their common use in food. The current US Regu-

lations, including the list of GRAS substances, are to be found in Chapter 21 of the Federal Register (21CFR), Parts 70–82 and 170–186.

The current categories of additives, as permitted in the EU, are as follows. Each category is referred to later in this chapter; where they are discussed more fully elsewhere in this Manual, the reference is given.

(i)   Antioxidants,
(ii)   Colours,
(iii)   Emulsifiers and stabilizers,
(iv)   Flavourings,
(v)   Mineral hydrocarbons,
(vi)   Miscellaneous additives – acids, anticaking agents, bases, bulking aids, firming agents, flavour modifiers, flour bleaching agents and improvers, glazing agents, humectants, sequestrants,
(vii)   Modified starches,
(viii) Preservatives,
(ix)   Solvents,
(x)   Sweeteners.

Processing aids

Substances used as processing aids – antifoaming agents, liquid freezants, packaging gases, propellants, release agents – are included in Additives Regulations as 'miscellaneous additives' because traces of them are likely to be found in the finished foods produced with their aid. They are not however considered in this chapter.

Pesticides

The legal situation varies somewhat among countries, but although there is widespread concern about the possible effects of pesticide residues in food, pesticides themselves are not usually considered to be food additives and they also are not considered in this chapter.

ACIDS

Acids may be produced in a food or added as an ingredient to preserve it, to give an acidic flavour or to control the acidity of the final product for some specific purpose such as the setting of a jam by controlling the pH. In the EU, certain acids only are permitted for use in conjunction with emulsifiers: see later under EMULSIFIERS.

Acetic acid, E260

$$CH_3CH_2COOH$$

Acetic acid is the principal component of vinegar, produced by the secondary fermentation of beer, wine or other alcoholic liquids. Details of the manufacture and uses of beer vinegars are given in Chapter 8.

Industrial acetic acid used to be made from the distillates produced by the destructive distillation of wood chippings. Now it is usually made synthetically, by the reaction of methanol with carbon monoxide using a rhodium or iridium catalyst.

$$CH_3OH + CO \rightarrow CH_3COOH$$

The methanol and carbon monoxide are produced together by treating natural gas with steam.

'Glacial' acetic acid, of 98% purity, is the usual material of commerce; aqueous solutions of various strengths are also available. Pure acetic acid has specific gravity 1.049 at 20°C, boils at 118°C and freezes at 16.6°C (hence 'glacial', since it solidifies at a very moderate temperature). It is miscible with water in all proportions.

Solutions of 4–5% concentration have direct use in foods as a vinegar substitute: in the UK they must be referred to as 'non-brewed condiment'.

Undissociated acetic acid is strongly inhibitory to the growth of a wide range of micro-organisms, see Chapter 9 for a full discussion.

Citric acid, E330

$$CH_2(COOH)CH(COOH)CH_2COOH$$

Commercial citric acid is manufactured by fermentation of sugar, molasses or corn syrup with *Aspergillus niger*. The product normally sold to the food industry is the monohydrate.

Citric acid is a tribasic acid, has a high solubility in water (59.2% w/w) and is only mildly sour. It softens at 70–75°C (158–167°F) and melts between 135 and 152°C (275 and 306°F). Dry storage conditions are necessary to prevent caking.

It is used for its pleasant acidic flavour and as a chelating agent to control the ill effects of traces of heavy metals. Suitable dosage rates are 0.5% for jellies, 1.0% for boiled sweets and 1.5% for acid drops.

Sodium citrate is a colourless, crystalline material with a slight saline taste. It is extremely useful as a buffer in pectin products, crystallizing syrups and prepared sugar syrups.

Glucono δ-lactone, E575

$$\overset{\displaystyle \ulcorner\ \ \ \ \text{O}\ \ \ \ \urcorner}{COCHOHCHOHCHCH_2OH}$$

This is an internal lactone of gluconic acid. In aqueous solution it is initially neutral, but it hydrolyses slowly to give gluconic acid, COOH-CHOHCHOHCHOHCH_2OH. It therefore finds use in preparations in which slow or controlled acidulation is required.

Lactic acid, E270

d-α-hydroxypropionic acid, $CH_3CH(OH)COOH$

The majority of commercial lactic acid is made by fermentation, the commonest starting material nowadays being liquid whey or one of its derivatives, the ferment being *Lactobacillus bulgaricus* or other *Lactobacillus* species. Some lactic acid is made synthetically from acetaldehyde by way of lactonitrile.

Phosphoric acid, E338

orthophosphoric acid, H_3PO_4

Of all the common food acids, phosphoric acid produces the lowest pH value in solution. It is particularly used as an acidulant in soft drinks. It also is useful as a sequestering agent for the inactivation of metal ions. The various useful phosphate salts are listed later, under PHOSPHATES.

Tartaric acid, E334

$$HOOCCH(OH)CH(OH)COOH$$

The acid occurs naturally in many fruits. Tartaric acid itself is not used directly as a food ingredient, but potassium bitartrate (potassium hydrogen tartrate or potassium acid tartrate, E336(i), KOOC-CH(OH)CH(OH)COOH) is used as an ingredient of raising agents and to initiate the precipitation of tartrates in the stabilization of wines.

Acidity regulators

These are usually weakly alkaline substances, added to control the acidity directly, for instance magnesium carbonate or calcium hydroxide, or to buffer the effect of added acid, for instance sodium citrate.

ANTICAKING AGENTS

These are added to dry products such as milk powders, salt, cocoa products, to ensure good flowing properties. Magnesium carbonate is an example, added to table salt.

ANTIOXIDANTS

When fats are exposed to atmospheric oxygen they gradually become oxidized and develop off-flavours. It may also give rise to deleterious nutritional effects, for example, destruction of essential fatty acids and vitamins. Some fats oxidize more readily than others. Substances which retard the oxidative deterioration of fats are called antioxidants.

Antioxidants function in two ways. They may act as oxygen scavengers, removing oxygen from the oil, or by virtue of their ability to inhibit autoxidation by capturing free radicals. Antioxidants are not effective against hydrolytic rancidity. The antioxidants permitted in the EU are listed in Table 14.1.

In the USA the following are permitted, together with PG, BHA and BHT:

α-Tocopherol
Guaiagum or guaiacol gum
Trihydroxybutylphenone (THBP)
Hydroxymethylbutylphenol (HMBP)
tert-Butylhydroxyquinone (THBQ)

Also permitted in some other countries is:

Nordihydroguaiaretic acid

E300 Ascorbic acid, E301 sodium ascorbate and E302 calcium ascorbate, being strong reducing agents, are also effective antioxidants. E315 Erythorbic acid and E316 sodium erythorbate are

Table 14.1 Antioxidants permitted in the UK

E number	Name		Maximum levels
E310	Propyl gallate	(PG)	200 mg kg^{-1} in fats (BHT 100 mg kg^{-1})
E311	Octyl gallate	(OG)	200 mg in the fat content of various foods
E312	Dodecyl gallate	(DG)	
E320	Butylated hydroxyanisole	(BHA)	25 mg kg^{-1} in dehydrated potato
E321	Butylated hydroxytoluene	(BHT)	400 mg kg^{-1} BHT or BHA in chewing gum, dietary supplements
E315	Erythorbic acid		meat products 500 mg kg^{-1} fish products 1500 mg kg^{-1} (as erythorbic acid)
E316	Sodium erythorbate		

classed as antioxidants in the UK Regulations but are permitted for use only in preserved and semi-preserved (cured) meat and fish products. Their actions in these products are the same as those of ascorbic acid and the ascorbates, see Chapter 1.

E330 Citric acid is not normally classed as an antioxidant but because it forms complexes with traces of transition metals, so neutralizing the catalytic effects which these metals can have on the oxidation process, it can have a significant antioxidant effect.

Tocopherols, particularly α-tocopherol, possess good 'carry-through' properties. 'Carry-through' refers to the ability of certain components to escape destruction on baking or frying. BHA and BHT also possess good carry-through properties during baking, but their volatility gives them reduced effectiveness during frying operations, BHA being less volatile than BHT.

COLOURS

Natural and nature-identical colours

These materials have long been available and they now make up a significant proportion of the colours used in the industry, due to consumer pressure in recent years against the use of artificial colours. The range of colours available is limited and the colour intensity and brightness are often much less than those of alternative synthetic colours. Stability to light is relatively poor in most cases, as may be stability to acids, SO$_2$ preservative or ascorbic acid. Nevertheless there is a range of formulations acceptable for many purposes.

The most commonly used natural and nature-identical colours are:

(i) Yellow, orange and red colours – carotenoids
(ii) Green colours – chlorophyll, including its copper and magnesium salts
(iii) Red and purple colours – anthocyanins

Because of low brightness and colour intensity dosages are generally high. Lakes are not available. Some of the colours are derived from the spices and have some flavour which can be difficult to mask. They are generally more expensive than synthetic colours.

Artificial or synthetic colours

The food manufacturing industry as it developed in the late nineteenth and early twentieth century experimented with and used the synthetic dyestuffs being made available at that time by the chemical industry. Colours were added to foods to enhance their initial appearance or to compensate for deterioration of the natural colours during storage. However the toxic nature of many of the earliest dyestuffs used, particularly the azo dyes, became increasingly recognized and by the mid-twentieth century the use of the azo dyes was generally abandoned. All the other colouring matters then in use were investigated, along with all other food additives, for possible short term and long term toxicological effects; only those whose safety-in-use has been demonstrated to the satisfaction of expert committees in many parts of the world are now permitted to be used.

Some marginal cases remain. Due to consumer pressure in Europe and America, tartrazine and amaranth are now little used, though the evidence against them is not considered sufficient to remove them from the EU permitted list.

Colours may be used in their ordinary form or as 'lakes', where the colour is adsorbed on to calcium or aluminium hydroxide. Lakes are useful, for example, for the dispersion of colours in fats or where moisture migration occurs during processing. Whereas in these cases colour migration may lead to uneven distribution of a soluble dye, the immobilization of the colour on an inert carrier can prevent this and the greater opacity of the lake gives a deeper colour.

Permitted colours

There are now positive lists, all slightly different, of the natural and artificial colours permitted for use in the EU, the USA, Japan, Australia and other countries.

The full EU list is given in Table 14.2, the USA list in Table 14.3.

In each country or region there are also detailed lists of special cases – foodstuffs to which the addition of colour is not permitted, foodstuffs to which only specified colours in specified quantities may be added and colours which may be used

Table 14.2 Colours permitted in the EU for use in foodstuffs

E100	Curcumin
E101 (i)	Riboflavin
(ii)	Riboflavin-5'-phosphate
E102	Tartrazine
E104	Quinoline Yellow
E110	Sunset Yellow FCF
	Orange Yellow S
E120	Cochineal, Carminic Acid, Carmines
E122	Azorubine, Carmoisine
E123	Amaranth
E127	Erythrosine
E128	Red 2G
E129	Allura Red AC
E131	Patent Blue V
E133	Brilliant Blue FCF
E140	Chlorophylls and chlorphyllins:
(i)	Chlorophylls
(ii)	Copper complexes of chlorophyllins
E141	Copper complexes of chlorophylls and chlorophyllins:
(i)	Copper complexes of chlorophylls
(ii)	Copper complexes of chlorophyllins
E142	Green S
E150a	Plain caramel
E150b	Caustic sulphine caramel
E150c	Ammonia caramel
E150d	Sulphite ammonia caramel
E151	Brilliant Black BN, Black PN
E153	Vegatable carbon
E154	Brown FK
E155	Brown HT
E160a	Carotenes
(i)	Mixed carotenes
(ii)	β-Carotene
E160b	Annatto, bixin, norbixin
E160c	Paprika extra, capsanthin, caposrubin
E160d	Lycopene
E160e	β-Apo-8'-carotenal (C 30)
E160f	Ethyl ester of β-apo-8'-carotenic acid (C 30)
E161b	Lutein
E161g	Canthaxanthin
E162	Beetroot red, betanin
E163	Anthocyanins
E170	Calcium carbonate
E171	Titanium dioxide
E172	Iron oxides and hydroxides
E173	Aluminium
E174	Silver
E175	Gold
E180	Litholrubine BK

Aluminium lakes prepared from the above colours are also permitted.

Table 14.3 Colours permitted in the USA for use in foodstuffs

FD&C	Blue No. 1	(Brilliant Blue FCF)
FD&C	Blue No. 2	(Indigo carmine)
FD&C	Green No. 3	(Fast Green FCF)
	Orange B	
	Citrus Red No. 2	
FD&C	Red No. 3	(Erythrosine)
FD&C	Red No. 40	(Allura Red)
FD&C	Yellow No. 5	(Tartrazine)
FD&C	Yellow No. 6	(Sunset Yellow FCF)

only in specified foodstuffs. The situation is very complicated even within a single region such as the EU, and a manufacturer must always check and observe the exact regulations of the country or countries in which it is intended to sell the products.

Caramel

The caramel colours are made in a variety of ways all of which involve heating concentrated solutions of sugar (of various types and origins) with acid or alkali and with or without a basic nitrogen source such as ammonia, ammonium salts or amines. The manufacturing process, still considered as much an art as a science, therefore yields a wide range of caramel classes and types. There are four main types:

(i) *Type I: plain caramel (PC) (E150a)*
     pH range 3.1–3.9; colour intensity 5–80.
     Used mostly for spirits, medicines, biscuits, pastries.
(ii) *Type II: caustic sulphite caramel (CSC) (E150b)*
     pH about 3.1; colour intensity 40–80
     Used only as a flavouring for spirits.
(iii) *Type III: ammonia caramel (AC) (E150c)*
     pH range 4.1–5.7; colour intensity 60–200
     Used in beers, bread and many other foods.
(iv) *Type IV: sulphite ammonia caramel (SAC) (E150d)*
     pH range 3.9–5.8; colour intensity 35–270
     Used for cola and other soft drinks, vermouth, vinegar.

The colour intensity is equivalent to the specific absorption at 610 nm.

Important factors in the choice of a caramel for any specific purpose include the pH, colour intensity and isoelectric point (IEP). The last is particularly important if caramel colours are to be mixed or if the food material contains substances such as tannins in solution, where flocculation could occur if materials of different IEP come together.

Manufactured caramels are not completely stable but deteriorate more or less slowly to insoluble resins without flavour or colouring power. They should be stored at low temperature and for limited periods only; the maximum shelf life of a caramel manufactured and stored under ideal conditions is given as five years. Light and atmospheric oxidation should be avoided, also metal containers other than stainless steel (to avoid catalytic effects).

EMULSIFIERS

Emulsifiers play an important role, as the manufacture of many foodstuffs involves the formation and stability of emulsions of two basically immiscible phases, usually oil (O) and water (W). The use of an emulsifier increases the ease of formation and promotes the stability of an emulsion. The concept of hydrophilic–lipophilic balance (HLB) is now widely used in the selection of emulsifiers for particular purposes. Each emulsifier is assigned an HLB value, which is a number expressing the balance of the number and strength of its polar as compared to its non-polar groups, and this serves as an indication of the emulsifying action it will perform. Whereas the HLB system has proved quite adequate in applications in which only water and oil are involved, the selection of a proper emulsifier for food applications is often considerably complicated by the fact that the emulsions concerned are so complex that the determination of their 'required HLB' is extremely difficult. Nevertheless, the HLB system can be very useful in narrowing the selection from the many emulsifiers available.

In addition to the HLB of an emulsifier, attention may have to be given to its physical nature, which may be dictated by the form of the complete product, and to its chemical composition, minor differences in which can cause vast differences in performance. The method of preparation of an emulsion can also be important; for example, the conditions of operation of a homogenizer or a heat exchanger may lead to either an O/W or a W/O emulsion, from the same original mixture.

In some cases, such as in ice-cream, bread and cake, emulsifiers exert important influences on the product by means of their effect on other ingredients like starch and protein, their unique crystallization characteristics or their indirect induction of interactions lowering the interfacial tension between the two phases. Blends of emulsifiers usually give better results than a single compound, and techniques such as response surface methodology have been used to determine the optimum levels and proportions of several emulsifiers for a blend designed for a specific application, for example in cake mix shortenings. Some of the most common applications of emulsifiers in foodstuffs are:

(i) *Bread*: at 0.1–0.5% (of the flour), emulsifiers improve the loaf volume, crumb structure and tenderness of the bread and markedly decrease its rate of staling. Diacetyltartaric acid (DATA) esters are used together with a small proportion of hard fat in bread improvers.

(ii) *Cakes*: at 1–4% level (of the fat), emulsifiers improve the cake volume, crumb softness, texture, moisture retention, and keeping quality of the cake.

(iii) *Ice-cream*: at 0.1–0.5% level (of the mix), emulsifiers improve the aerating properties (overrun), dryness, texture, body and apparent richness of the ice-cream.

(iv) *Margarine*: at 0.5% level (of the fat), emulsifiers improve the stability, texture and spreadibility of the margarine and some reduce its spattering when used for frying.

(v) *Whipped fillings*: at 2% level (of the fat), emulsifiers improve the whipping characteristics, appearance, volume and stability of the fillings, topping and icings.

(vi) *Choclate*: the main advantage of adding emulsifiers to chocolate is to reduce the chocolate viscosity and thus reduce the fat requirement. Lecithin is the most widely used emulsifier, the optimum level of utilization being about 0.4% of the total mix. Other emulsifiers, such as phosphorylated monoglycerides (both ammonium and sodium salts), polyglycerol polyricinoleate and sucrose dioleate, are useful in the reduction of choclate viscosity. Sucrose dioleate is not at present permitted in the UK for use in chocolate. Sorbitol tristearate and other sorbitol esters are sometimes used to reduce bloom on chocolate.

Stearyl tartrate

This is used as a bread improver, sometimes in combination with tartaric or diacetyl tartaric acid esters (see below).

Partial glycerol esters

The partial esters include any compound formed by incomplete esterification of the hydroxyl groups of glycerol with:

(i) Any single fatty acid or mixture of fatty acids. This class comprises the well-known monoglycerides, diglycerides and their mixtures which were the first synthetic emulsifiers to be used in the food industry. Because of their great diversity of application, stability and relative cheapness, they are still by far the most widely used emulsifiers in foodstuffs. They are used in the preparation of bread, cakes, margarine, ice-cream, whipped toppings, confectionery, cereal products, starch products and frozen and dehydrated foods. Saturated monoglycerides inhibit oiling out of peanut butter.

(ii) Any mixture of fatty acids with one of a series of organic acids. These are modified mono- and diglycerides and are generally more hydrophilic than the parent compounds.

The permitted organic acids are:

(i) *Acetic acid*. The acetoglycerides, which are exceptionally stable in the α-polymorphic form, are particularly effective emulsifiers in high-ratio cakes and in cake mixes designed for single stage preparation. They are also used in edible protective coatings, enrobing fats, spreads, salad oils, pan release agents and lubricants for machinery used to process foods.

(ii) *Lactic acid*. The lactoglycerides, such as glycerolactopalmitate, tend to be stable in the α-polymorphic form and are widely used as emulsifiers in shortenings, both for conventional and high-ratio cakes and in cake mixes. They are particularly effective aerating agents when incorporated in liquid shortenings for cakes and are also used as whipping agents in topping creams. Their other applications include ice-cream mixes, baked goods, bread and confectionery coatings.

(iii) *Citric acid*. These esters are not widely used, their main application being as antispattering agents in margarine. They exert a mildly antioxidant effect when incorporated into oils and fats.

(iv) *Tartaric or diacetyltartaric acid*. These esters improve bread, cakes and other baked goods. They also improve confectionery, margarine and whipped toppings. The diacetyl tartaric (DATA) esters are permitted in the UK and are widely used, but the simple tartaric acid esters are not allowed.

(v) *Phosphoric acid*. These esters are not widely used, their main application being as viscosity reducers in molten chocolate.

Partial esters of polyglycerols

Polyglycerol mixtures, which are made by the alkali-catalysed thermal dehydration of glycerol, consist of chains of glycerol molecules joined together by ether linkages. The remaining free hydroxyl groups are incompletely esterified with:

(i) Any single fatty acid or mixture of fatty acids
(ii) Dimerized fatty acids of soya bean oil
(iii) Interesterified fatty acids of castor oil.

Since the length of the polyglycerol chain, the degree of esterification and the nature of the fatty acids can all be varied independently of each other, a wide range of esters, covering the HLB range, can be prepared. This versatile class of emulsifiers has found application in margarine, peanut butter, ice-cream, confectionery, chocolate, cakes, baked goods, low calorie spreads, toppings and bakery release agents. They retard crystal formation in salad oils. The esters tend to be hydrophilic in character and are therefore particularly effective in synthetic creams and high ratio shortenings.

Propylene glycol esters

Propylene glycol may be completely or partially esterified with:

(i) Any single fatty acid or mixture of fatty acids. These esters tend to crystallize in the α-polymorphic form and are very effective emulsifiers in cake mixes and in plastic and liquid shortenings for cakes. They impart excellent whip stability to dessert toppings, ice-cream and icings made from prepared dry mixes. They are also used in confectionery products.

When a molten equimolar mixture of glycerol monostearate and propylene glycol monostearate is cooled, mixed crystals are formed in which the majority of glycerol ester is in the normally unstable α-polymorphic form. These crystals disperse readily in water and this property is retained for over one year. The aqueous dispersions foam when shaken, and due to this unique property, the mixed crystals exhibit enhanced effectiveness in bread, cake mixed, sponges, whipped sauces, purées, toppings and confectionery.

(ii) Any mixture of fatty acids and lactic acid. Mixtures obtained by the reaction of lactic acid with mixed fatty acid esters by propylene glycol and glycerol are particularly effective as emulsifiers in cake mixes.

(iii) Alginic acid. Propylene glycol alginate is used as an emulsifier in baked goods, brewery products, canned goods, soft drinks, sauces and margarines containing highly unsaturated oils. It is also used as a thickening agent in a wide variety of foods.

Fatty acid esters of sorbitan and their polyoxyethylene derivatives

The sorbitan esters, prepared by the reaction of sorbitol with fatty acids, are more hydrophilic than monoglycerides and their hydrophilicity can be still further increased by condensing them with ethylene oxide. The esters cover a wide HLB range and have found applications in baked goods, whipped toppings, ice-cream, confectionery coatings and beverages. A blend of sorbitan ester and a polyoxyethylene derivative usually gives better results than a single ester.

In chocolate manufacture, sorbitol tristearate and other sorbitol esters are sometimes used to reduce bloom tendency and enhance texture, gloss, keeping qualities and melt-down on the palate. They tend to have more effect in artificial coatings than in real chocolate.

Monostearin sodium acetate

This is an effective antispattering agent for margarine, but is not permitted in the UK or the USA.

Miscellaneous emulsifiers

Other compounds such as cellulose ethers, sodium carboxymethyl cellulose and brominated vegetable oil are stabilizers rather than emulsifiers, and are used in baked goods, frozen foods, sauces, soft drinks, soups, jellies, and so on.

Brominated vegetable oils were removed from the permitted list in the UK in 1970.

The fatty acid esters of sucrose comprise a versatile class of emulsifiers which when pure are entirely non-toxic. Until recently, however, great difficulty was experienced in completely removing the toxic solvents, usually dimethylformamide, used in their manufacture and for this reason they are not permitted in some countries. However, methods of manufacture which do not involve toxic solvents have now been found and the esters are gaining wider acceptance. They have proved to be effective emulsifiers in cakes, biscuits, confectionery, chocolate, bread, margarine, shortenings, and salad oil.

Calcium stearoyl lactylic acid and sodium stearyl fumarate, which have been gaining wider acceptance, are effective improvers for bread and other baked goods.

FLAVOURS AND FLAVOURINGS

Herbs and spices

The term 'spicing' has been used to cover the use of spices, herbs and certain aromatic vegetables to impart odour and flavour to foods.

The following is a convenient classification.

Hot spices: capsicum (chillies), cayenne pepper, black and white peppers, ginger, mustard.

Mild spices: paprika (pimiento), coriander.

Aromatic spices: allspice (pimento), cardamom, cassia, cinnamon, cloves, cumin, dill, fennel, fenugreek, mace, nutmeg.

Herbs: basil, bay, dill, marjoram, mint, tarragon, thyme.

Aromatic vegetables: onions, garlic, shallots, celery.

Although spices and herbs are only used in relatively small proportions, their use can make or mar the finished product. The aim should always be to arrive at a balanced overall odour and flavour effect, complementing and accentuating rather than swamping the flavour of the basic ingredients and without any single spice predominating excessively.

To arrive at a balanced spicing for a given formulation is very much a matter of choice, of experience in blending the different odour and flavour notes to produce the desired result, and of trials to establish the level of incorporation of the selected blend. To a would-be manufacturer lacking the necessary experience, the best advice that can be given is to seek the assistance of one or another of the leading spice houses, who are always willing to submit a range of expertly blended alternative combinations for trials for the specified purpose, and who will suggest levels of addition which can be used as the starting points for experiments.

Having established a standard spice mix for a particular use, it is important to ensure that subsequent supplies show no variation. This is normally achieved by comparative tasting of suitable diluted samples. Additionally, deliveries should be examined visually, and samples can be tested for moisture content and salt or dextrose content (as the case may be). If GLC equipment is available, it can be effectively used to compare consignments with standards.

Spice extracts

Spices and herbs may be incorporated into the product in several forms: as the whole spice or herbs, as ground spice, as essential oils, as oleoresins, as prepared and filtered infusion in vinegar or in marinades. All of these forms have disadvantages. Whole or ground spice is subject to microbiological and other contamination, infestation and deterioration on storage. The direct use of ground spices may result in the presence in the product of specks which can spoil the appearance of some products.

Essential oils (obtained by distillation) and oleoresins (obtained by solvent extraction) are incomplete versions of the total odour and flavour constituents of the corresponding spices, and being highly concentrated, precise weighing and addition are essential. The preparation of infusions and marinades is laborious and time consuming. Furthermore, all of these forms of flavour addition suffer from the natural variability odour and flavour strength of different supplies of any given spice.

The alternative to all of these is spice extracts. These consist of the flavour components of a spice, or a specified mixture of spices, dispersed on one of several types of base, usually salt or dextrose. The flavour strength per unit weight is made to correspond approximately to that of a good quality supply of the spice or mixture of spices involved.

These materials provide assured flavour standardization and virtual microbiological sterility with freedom from tannins, enzyme activity, foreign matter, specks, husks, seed coats, and so on, ingestation and deterioration on storage. Because of their rapid dispersibility, they can be incorporated towards the end of a cooking process, thus minimizing loss of volatiles in the process. Where a number of spices are involved in a formulation, there is a saving of labour and less risk of error, as only one ingredient has to be dispensed instead of several.

Spice and flavouring components may also be supplied in the form of liquid essences, which are similarly easy to use and control.

Flavour enhancers

As the name suggests, these enhance the natural flavour of foods. They have some flavour themselves but their main action appears to be to stimulate the tongue and olfactory organs to the other flavours present. Probably the most well known is monosodium glutamate which is widely used to enhance the flavour of meat products.

GELLING AGENTS AND THICKENERS

STARCHES

Starch occurs in water-insoluble granules and constitutes the reserve carbohydrate of plants. Chemically the starches are glucose polymers in the form of straight chains (amylose) or branched chains (amylopectin).

When an aqueous suspension of starch is heated the granules absorb water, swell and burst, releasing the starch and forming a viscous solution which sets to a gel on cooling. With time, the straight chain amylose molecules of medium length in the gel tend to become oriented together, 'crystallize' and to be precipitated. This causes the gel to weaken and eventually to break, with syneresis or loss of water and separation. The process is called 'retrogradation'. Gels made from starches with amylopectin of longer molecular chain lengths (e.g. potato as opposed to maize) are less prone to retrogradation. Starches with high proportions of amylopectin (e.g. waxy maize as opposed to normal maize) produce weaker gels but are less prone to retrogradation. Generally, cereal starches tend to give opaque dispersions in water, while root starches tend to give clear dispersions.

Corn flour (maize starch)

On heating an aqueous suspension of corn flour the viscosity develops slowly, thinning out after peak viscosity is not very marked or rapid and on cooling it becomes opaque and a gel structure forms. If the gel structure is prevented from forming (e.g. by continuous stirring until cold) or is allowed to form and then destroyed by vigorous stirring or sieving at any time up to about four days after the heating process, a gel will subsequently reform, its gel strength varying inversely as the length of time elapsing between heating and either discontinuing stirring or the destruction of the previously formed gel structure. If however, a gel is allowed to form and stand for at least five days and is then vigorously agitated, no gel structure is subsequently reformed. This combination of properties provides the basis for the successful design of a sauce making process based on cornflour.

Waxy maize

Waxy maize starches are obtained from hybrids of waxy corn, and consist wholly or largely of amylopectin. They show little or no tendency to gel on cooling of heated aqueous suspensions. On heating, however, they tend to pass rapidly through the region of maximum viscosity and to thin out rapidly and to a marked extent, which means they do not have reasonable tolerance of minor variations in the conditions of sauce boiling and it is, therefore, difficult to obtain prescribed results consistently.

Wheat flour

Wheat flour contains gluten as well as wheat starch, but the gluten appears to play no significant part in the stabilization of sauces and thick liquors, and the thickening effect appears to correspond to that of the wheat starch content only. Wheat flour can be used in a similar way to cornflour, but perhaps less satisfactorily.

Sago

Sago starch, obtained from the stem of the sago palm, behaves somewhat similarly to cornflour, but shows less tendency to gel.

Tapioca

Tapioca starch does not show a significant viscosity 'peak' (i.e. the viscosity achieved does not vary markedly with minor variations in heating conditions) and shows little tendency to gel.

Potato starch

Some English speakers call this 'Farina' but in Latin countries that word more usually refers to wheat flour. Potato starch shows a marked viscosity 'peak' with rising temperature, beyond which it thins out rapidly, but shows little or no gelling tendency.

Rice

Rice flour develops a viscosity between that of corn and wheat flours, but with a higher gelatinization temperature; the gel develops on cooling. Rice flour has some popularity due to its stability in acetic acid systems and has been found of value in piccalilli formulations.

Modified starches

Modified starches can be made with a wide range of properties, suitable for many different applications. The main methods of modification are given below. Each can be applied separately and some also in combination which further increases the possible range of properties in the final products.

Pregelatinized starch

This is made by cooking the starch (usually corn or potato starch) rapidly in water and immediately drying it, e.g. by passing a slurry over rollers heated by superheated steam. This produces a powdered, fully gelatinized starch which can be rehydrated immediately in cold water – an 'instant' starch, in fact.

Oxidized starch (E1404)

Starch may be 'stabilized' by oxidation with sodium hypochlorite, peracetic acid or other oxidizing agents. Oxidized starch gives more stable gels than the untreated equivalent.

Acid-thinned starch or thin-boiling starch

Starch is treated with hydrochloric or sulphuric acid at temperatures up to about 60°C, avoiding swelling or gelatinization; sodium chloride or sodium sulphate may be added to inhibit swelling further, the starch is then washed and dried. The modified starch when treated with hot or boiling water goes rapidly into solution, the viscosity of the solutions is low but they set rapidly to give gels of good firm consistency.

Cross-linked and side chain modified starches

The commonest products here are the starch phosphates, E1410–1414, and starch esters, E1414–1442. They are made by treatment with phosphoric acid or acetyl or succinic anhydride. Such starches may be used for instance in the preparation of sauces, where they give remarkably constant viscosity. Prolonged heating (e.g. by pasteurizing) has little effect, there is no tendency to gel and no significant syneresis or change in viscosity on prolonged storage.

The texture and mouthfeel of products made with modified starches may be different from equivalents made with unmodified starch, in ways not dependent on viscosity alone. Relatively small degrees of modification of the starch can produce large differences in properties. The degree of modification or substitution (DS) is defined as the average number of hydroxyl groups replaced by substituents, in each glucose residue; since there are three free hydroxyl groups in each glucose residue, the maximum possible DS is three. In the USA the DS of the modified starches are set by Regulation, mostly between 0.07 and 2 but with some as low as 0.02.

Pectin

The pectin substances occur naturally in the middle lamella of plant tissues and may be regarded as part of the cement which holds the cells together. During the ripening of fruits there is a conversion of insoluble protopectin, a polygalacturonide, into soluble pectin or polygalacturonic acid. When fruit becomes overripe molecular breakdown occurs due to pectolytic enzyme activity, producing shorter chain pectins which have inferior gelling properties.

The main sources of commercial pectin are apple and citrus fruits. In both cases the residues from juice extraction are available as pectin-rich raw material. Extraction is carried out under acid conditions and the resulting extract clarified by enzymic removal of starch and protein, decolorized with activated carbon, filtered through diatomaceous earth and finally concentrated under vacuum. About 5% of pectin represents the limit to which concentration can be taken on account of the extremely high viscosity of the solution. The pectin is precipitated out by the addition of alcohol. Immediate alcohol treatment precipitates high methoxyl pectin. If the alcohol treatment is delayed for some days to allow hydrolysis of the pectin, low methoxyl pectin may be precipitated. Ammonia may be added before the alcohol to yield amidated low methoxyl pectin.

Chemically, pectin is composed mainly of long chains of partially methylated polygalacturonic acid (Figure 14.1).

High methoxyl pectin

High methoxyl (HM) pectins are those in which minimum chain degradation has taken place and about half of the galacturonic acid groups are

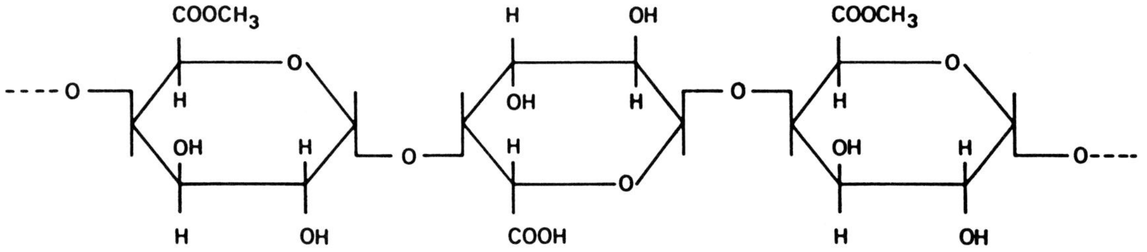

Figure 14.1 Part of pectin molecule.

methylated. They form strong gels with solutions of high sugar content (60–70%) and pH 2.8–3.5.

When HM pectin is dissolved in water it forms a weak acid, due to the tendency of carboxylic acid groups to dissociate. The pH should be controlled by buffering, normally with mixed citric acid and sodium citrate. The sugar present has a further effect, of dehydrating the pectin and reducing its solubility. Cross-linking occurs between the pectin chains to form the gel. Under optimum conditions HM pectin gels set rapidly, above 80°C setting will usually occur in less than 5 min.

Low methoxyl pectins

Low methoxyl (LM) pectins are those in which less than 50%, down to 7%, of the carboxylic groups are methylated. They form gels at sugar concentrations below 60%, in the presence of divalent metallic ions, usually calcium, at pH 3.1–3.5. The gels are formed by cross-linking of the pectin chains through the calcium ions; the gel strength is not greatly dependent on sugar concentration. Setting occurs down to 60°C; the setting rate is solely temperature-dependent and is slower than with HM pectin.

Amidated pectins

In these, about 20% of the carboxylic groups are converted to amide groups. They form gels under similar conditions to low methoxyl pectin.

Amidated low methoxyl pectins

These have advantages in that the gels are thermally reversible and fairly tolerant of variations in calcium level. The calcium ions required for setting may be present in sufficient quantity from the fruits or vegetables used if the water is also hard (i.e. with high contents of calcium or magnesium); if additional calcium is required it may be added as the lactate or citrate.

Pectic acid is the completely demethoxylated acid, possessing no power to form sugar–acid gels.

Pectin is graded according to its setting characteristics, the strength of the gel which is formed and the rate at which the gel sets. The grade is defined as that quantity of sugar which when used with a unit quantity of pectin will produce a gel of a defined standard strength. Thus 1 g of 150 grade pectin when used with 150 g of sugar under specified conditions will produce a 'standard' gel.

In order to produce consistent materials, pectin manufacturers standardize batches of pectin by blending and dilution with sugar. If required for diabetic products, unstandardized pectin should be used in order to ensure the complete absence of added sugar.

Selection of the type of pectin to be used depends on the product. If, for instance, fruit pieces are present in a jelly or a jam, a fast set is required to ensure that setting takes place before fruit flotation or sinking occurs. If a clear jelly is being manufactured, then a slow set is advantageous so that air bubbles can be completely removed to give a clear product. In practice, a compromise must usually be found, with the rate of setting controlled by the combination of pectin type, pH and sugar solids.

Besides its well-known uses in the manufacture of jams, jellies and candies, pectin has numerous other commercial applications:

(i)   glues and mucilage
(ii)   sizing of textiles
(iii)   lacquers and explosives similar to nitrocellulose and nitroacetate
(iv)   stabilization of oil–water emulsions
(v)   stabilization of salves and medicinal jellies
(vi)   improvement of quality of processed cheese
(vii)   improvement of quality of bakery goods
(viii) manufacture of dehydrated fruit juices and beverages.

Powdered pectin should be redissolved in water using a high speed mixer to ensure complete dispersion before use. A solution containing 3–5% of actual pectin is suggested, if a standardized pectin is used the full quantity of material needed may be higher, perhaps up to 10%.

GELATIN

Acid gelatin is produced by the acid treatment of skins, usually pigskin, lime gelatin by alkali or lime treatment of skin or bones. The method of production is not usually important in choosing a gelatin for a food product, the main quality criteria are colour, odour and gel strength.

Gel strength is measured in Bloom units, usually in the range 80–220 Bloom. The lower bloom gelatins give firmer texture but are less elastic than high bloom gelatins.

GUMS

The important criteria in purchasing gums are purity, mesh size and viscosity. Colour is not always a reliable guide to quality.

Gum arabic (Acacia)

Gum arabic is the exudate from *Acacia* trees. It is a good emulsifier and inhibits crystallization in sugar solutions by increasing the viscosity of the mix. It will produce a solution in water up to concentrations of about 50%. The viscosity of a solution of this gum is inversely proportional to the temperature and is adversely affected by acid conditions, e.g. in solutions of pH 4.0 or less. Excessive heat causes the breakdown of the gum and discoloration of the solution. Gum arabic solutions must be filtered through a medium mesh before use. Gum arabic is now available as a spray dried material which is free from contamination and much easier to handle. Because of high and variable prices in the past gum arabic has now been replaced in many of its former applications, for instance by modified starches.

Guar gum

Guar gum is a polymer of mannose and galactose which can be used in conjunction with other gums to improve stability. It is soluble in cold water, has excellent pH stability and is compatible with gelling agents. A solution of this gum will slowly lose its viscous nature when left to stand for some time.

Locust bean or carob gum

The gum swells in cold water and forms a highly viscous solution; mixtures in water should be used hot. Carob gum itself does not gel but in conjunction with agar or carrageen it may greatly improve gel strength and stability.

Gum tragacanth

Gum tragacanth is the exudate from leguminous plants of *Astragalus* spp. It is commonly available in the form of dried whitish ribbons. It swells slowly when added to cold water and should be soaked overnight before use. It has wide pH stability, the maximum viscosity being achieved at pH 5.0. It is heat stable in acid conditions so it is almost ideal for acid sauces. However it is expensive. Gum tragacanth will deteriorate under long storage and stocks should be limited to not more than three months' supply.

Xanthan gum

Xanthan gum is a complex polymer produced by microbial fermentation of glucose. Solutions of xanthan gum are very stable to variations in pH and to a wide range of temperatures.

The gum may be combined with guar gum and locust bean gum to form high viscosity solutions or gels. Xanthan gum can often be used as a substitute for gum tragacanth.

Agar

Agar, or agar-agar, is an extract of various seaweeds. The extraction is carried out by boiling in water, followed by purification and concentration. Once produced in strips, it is now available in powdered form which eliminates the need for soaking before use. Normal usage levels of agar, e.g. in confectionery jellies, are from 1–1.5%. It is soluble in boiling water, the gel sets at 30–40°C.

Alginates

Alginates are salts of alginic acid and are produced from a brown seaweed by digestion with alkali followed by precipitation of alginic acid which is then converted to the sodium salt. The sodium salt is soluble in cold water. Alginic acid and the calcium salt are insoluble. Irreversible gels are formed on the slow addition of a calcium salt to the aqueous solution. The main uses of alginates are as stabilizers and thickeners.

Carrageenan (Irish moss)

Extracted from red seaweeds, κ- and ι-carrageenan give thermally reversible gels in the presence of calcium salts, stronger in the presence of proteins; they have applications in milk and chocolate drinks. λ-Carrageenan does not gel.

GLAZING AGENTS

These are used to improve the surface appearance of products. In the past mineral hydrocarbons were used widely, e.g. for dried fruit. Their use is now discouraged and vegetable oil emulsions have taken their place. Other examples of glazing agents are waxes such as carnauba wax.

MINERAL HYDROCARBONS

These include microcrystalline wax used in some chewing gum, and mineral oil for the dressing of dried fruit. Use of the latter has largely been abandoned voluntarily by the trade. Note also that food machinery is now lubricated almost exclusively with vegetable oils, to avoid any contamination of product with mineral oil.

PHOSPHATES

A large number of phosphate salts have applications in food manufacture. Table 14.4 presents a summary of these and other applications.

For applications of polyphosphates in meat production, see also Chapter 1.

PRESERVATIVES

The preservatives permitted in the EU are listed in Table 14.5. They are permitted in a wide but clearly specified range of foods and drinks, with maximum tolerances laid down in all cases. The preservative permitted in the largest number of foods is sulphur dioxide.

Table 7.2 in Chapter 7 shows quantitative relationships between various practical sources of SO_2 when added to wine; the data are applicable to other food applications.

SOLVENTS

The Regulations relate to solvents which may be used to facilitate the incorporation of ingredients into food. They do not apply to extraction solvents used only during processing, none of which should remain in the finished food.

SALT

Salt is present to some degree in most foods and is added to many. There are three main types of salt used in the European food industry:
(i)   Brine evaporated salt, made by the evaporation of strong brine, usually pumped from deep mines; in Cheshire (UK) the evaporation is done under vacuum, hence the name vacuum salt; this is the commonest kind of salt in use in the UK, being readily available.
(ii) Rock salt, mainly mined from underground deposits, as at Stassfurt in Germany.
(iii) Solar salt, made by evaporation of sea and salt lake water under the action of sun and wind; this is mainly available in Mediterranean and other seaboard countries.

(i) and (ii) are almost 100% chemically pure, (iii) is much less pure, sometimes only 80% NaCl. In addition to mechanical impurities such as dust, sand grains and water, the chief chemical impurities are calcium and magnesium chloride and sulphate, sodium sulphate and carbonate, and traces of heavy metals such as iron and copper. Small quantities of copper (as little as 0.2 ppm) have been shown to cause a characteristic and troublesome brown discoloration in cod.

Solid salt has a specific gravity of 2.17 g cm^{-3}, the bulk density may vary from 0.9 to 1.6 tonne m^{-3} depending on the fineness and the dampness, with an average around 1.2. When pure, salt is hygroscopic in relative humidities above 76%; calcium and magnesium chlorides and sulphates, which are common impurities, increase the hygroscopicity.

The solubility in water at 10°C (50°F) is 26.3 g per 100 g of solution, falling to 23.3 g per 100 g at −21.2°C (−6°F) which is the eutectic point. The freezing points of salt solutions of different concentration are given in Table 14.6. Salt has a nega-

Table 14.4 Summary of food uses of phosphates

Phosphate		Main applications
Orthophosphates		
Monosodium phosphate	NaH_2PO_4	Cheese (emulsifier)
		Meat processing mixtures
		pH regulation
Disodium phosphate	Na_2HPO_4	Cheese etc. (emulsifier, 2–4%)
(anhydrous, dihydrate, dodecahydrate)		Adjustment of Ca/P ratio in milk
Trisodium phosphate	Na_3PO_4	Cheese emulsifier
		Meat processing mixtures
Monoammonium phosphate	$(NH_4)H_2PO_4$	Yeast food
Diammonium phosphate	$(NH_4)_2HPO_4$	–
Acid calcium phosphate	$Ca(H_2PO_4)_2$	Baking powders
(dihydrate)		
Dibasic calcium phosphate	$CaHPO_4$	Nutritional addition of Ca and P
		Gelation of alginates
Tricalcium phosphate	$Ca_3(PO_4)_2$	Anticaking agent for table salt
	$Ca_5OH(PO_4)_3$	
Pyrophosphates		
Disodium dihydrogen phosphate	$Na_2H_2P_2O_7$	Baking powders
Acid sodium pyrophosphate		Cheese (emulsifier)
		Meat processing (antienzymic browning)
		pH regulation
Tetrasodium pyrophosphate	$Na_4P_2O_7$	Gelation in instant milk mixes
Sodium pyrophosphate		Cheese (emulsifier)
		Meat processing mixtures
		pH regulation
		Sequestrant (e.g. in milk processing)
Triphosphates (Tripolyphosphates)		
Sodium tripolyphosphate	$Na_5P_3O_{10}$	Meat processing mixtures
Others		
Sodium polymetaphosphate	$Na_{n+2}P_nO_{3n+1}$	Cheese (emulsifier)
Sodium phosphate glass	$n = 12$ on average	Extraction of pectins from fruit
(CALGON®)		Softens skins of canned peas, beans
Sodium aluminium phosphate,	$Na_3Al_3H_{15}(PO_4)_8$	Cheese (emulsifier)
acidic		Meat processing mixtures
		Blood anticoagulant
		(NB: Use of aluminium salts is
		discouraged in some countries)
Bone phosphate	about 15% P	Nutritional addition of Ca and P
(edible material made from animal bones)		Anticaking agent

tive heat of solution; if 20 parts of salt are mixed with 100 parts of water or wet foodstuff the temperature of the mixture is depressed by 4.8°C (8.6°F).

The strength of a brine is most conveniently measured with a hydrometer. Baumé hydrometers are calibrated to indicate the salt concentration directly. The Twaddell hydrometer is based on a scale such that

$$\text{Specific gravity} = \frac{200 + °\text{Twaddell}}{200}$$

Other hydrometers, called 'salometers' or 'salinometers' have scales to indicate the degree of saturation of the salt solution, from 0 = pure water to 100 = saturated brine. Relationships between these different measures are shown in

Table 14.6. In using any type of hydrometer to determine brine strength, the brine must be at the temperature for which the hydrometer was calibrated.

Hydrometer methods are satisfactory for measuring the strengths of pure salt brines but can be misleading if applied to brines which have been in contact with vegetables, meat or fish, where the specific gravity depends not only on salt concentration but also on soluble solids extracted from the foodstuff. In such cases, chemical determination of the salt content may be necessary.

For making brines for factory use, as in pickling vegetables or curing meat, salt with fine crystals is preferred since it is readily soluble; for dry salting coarse crystals are better. For large scale production of brine, automatic installations are available providing saturated brine which may be

Table 14.5 Preservatives permitted in the EU

E200	Sorbic acid	
E202	Potassium sorbate	
E203	Calcium sorbate	
E210	Benzoic acid	
E211	Sodium benzoate	
E212	Potassium benzoate	
E213	Calcium benzoate	
E214	Ethyl p-hydroxybenzoate	Maximum levels
E215	Sodium ethyl p-hydroxybenzoate	(including some
E216	Propyl p-hydroxybenzoate	permitted in
E217	Sodium propyl p-hydroxybenzoate	combination)
E218	Methyl p-hydroxybenzoate	laid down for a wide
E219	Sodium methyl p-hydroxybenzoate	range of drinks and
		other foodstuffs.
E220	Sulphur dioxide	
E221	Sodium sulphite	
E222	Sodium hydrogen sulphite	
E223	Sodium metabisulphite	
E224	Potassium metabisulphite	
E226	Calcium sulphite	
E227	Calcium hydrogen sulphite	
E228	Potassium hydrogen sulphite	
E230	Biphenyl, diphenyl	For surface
E231	Orthophenyl phenol	treatment of
E232	Sodium orthophenyl phenol	citrus fruits
E233	Thiabenzadole	
E234	Nisin	Cheeses, some
E235	Natamycin	puddings, cured
E239	Hexamethylene tetramine	sausages.
E242	Dimethyl carbonate	Some drinks
E284	Boric acid	Caviar
E285	Sodium tetraborate (borax)	
E249	Potassium nitrite	Cured meats
E250	Sodium nitrite	
E251	Sodium nitrate	Cured meats,
E252	Potassium nitrate	pickled herring,
		some cheese.
E280	Propionic acid	Certain prepacked
E281	Sodium propionate	and energy reduced
E282	Calcium propionate	breads; Christmas
E283	Potassium propionate	pudding.
E1105	Lysozyme	Ripened cheese

diluted to the required concentration at the point of use.

SUGAR

See Chapter 4

SWEETENERS AND SUGAR SUBSTITUTES

The sweeteners permitted in the EU and the USA are shown in Table 4.7. Intense sweeteners are so-called because they have a sweetness many times

that of sugar. They have little or no application in conventional confectionery products, but if bulk sweeteners or polydextrose become established as ingredients, since they are mainly less sweet than sugar, it is possible that additional sweetening may be required.

Acesulfame-K (6-methyl-1,2,3-oxathiazin-4(3H)-one-2,2-dioxide)

Acesulfame-K is synthesized from acetoacetic acid. It is a white powder and is claimed to have unlimited storage life, good solubility and stability in acidic solution. In some applications it can have an aftertaste.

Aspartame

Aspartame is a methylated dipeptide of phenylalanine and aspartic acid. It is less soluble in water than other artificial sweeteners but readily dissolves in dilute acids. It is not very heat or acid stable and under acidic conditions it slowly decomposes to the constituent amino acids. The phenylalanine thus produced can affect sufferers with phenylketonuria, and a warning to this effect should be stated on the product label. Nevertheless, aspartame has achieved wide acceptance in low calorie products because of its excellent sweetness profile. Aspartame in soft drinks is stabilized by the presence of caramel (colour).

Saccharin

Saccharin is still widely used due to its wide availability, low cost and good solubility. Earlier doubts over its safety-in-use have been overcome except in the minds of some consumers. In the USA saccharin if present must be accompanied by a 'health warning' on the label.

Low calorie or 'diet' drinks are sweetened with mixtures of saccharin and acesulfame-K or aspartame. These mixtures have improved flavour profiles compared with those made with saccharin alone, which have a bitter aftertaste.

Cyclamates

Cyclamates have received safety clearance in most countries but are still banned in the USA. They have a cleaner taste than saccharin.

Thaumatin and neohesperidin

Thaumatin and neohesperidin DC are more recent developments and experience of their use is still somewhat limited.

Table 14.6 Properties of salt solutions

(°Salometer)	Salt concentration at 3°C (38°F) (°Baumé)	(°Twaddell)	(%w/w)	Specific gravity (3°C)	Freezing point[a] (°C)	(°F)
0	0.0	0.0	0.000	1.000	0.0	+32.0
10	2.8	4.0	2.631	1.020	−1.5	+29.3
20	5.6	8.0	5.262	1.040	−3.1	+26.4
30	8.2	12.0	7.892	1.060	−5.0	+23.0
40	10.7	16.0	10.523	1.080	−7.0	+19.4
50	13.3	20.2	13.154	1.101	−9.1	+15.6
60	15.7	24.2	15.785	1.121	−11.6	+11.1
70	18.0	28.4	18.416	1.142	−14.5	+5.9
80	20.4	32.8	21.046	1.164	−17.8	+0.0
88.6	22.4	36.6	23.300	1.183	−21.1	−6.0[b]
90	22.7	37.2	23.677	1.186	−18.9	−2.0
99.9	25.0	41.6	26.285	1.208	+0.1	+32.2[c]
100	25.0	41.6	26.308	1.208		

[a] Point at which ice begins to form. [b] Eutectic point. [c] Transition point, NaCl to NaCl.2H$_2$O.
Source: American Meat Institute (1950) and Lange's Handbook of Chemistry (1956).

Table 14.7 Permitted sweeteners

Intense sweeteners

E number	Name	Sweetness (relative to sucrose, approx.)	Usage maximum (mg kg^{-1})
E950	Acesulfame-K	150 times	500
E951	Aspartame	200 times	1000
E952	Cyclamic acid Na cyclamate Ca cyclamate	30 times	500
E954	Saccharin Na saccharin K saccharin Ca saccharin	300 times	500
E957	Thaumatin	3000 times	50
E959	Neohesperidin DC	2000 times	100

Under US legislation (21 CFR 180) the following intense sweeteners are permitted: acesulfame-K, aspartame, saccharin and thaumatin. The latter is permitted in chewing gum only.

Bulk sweeteners

			Quantum satis
E420	Sorbitol	60%	
(i)	Sorbitol		
(ii)	Sorbitol syrup		
E421	Mannitol	60%	
E953	Isomalt	50%	
E965	Maltitol	90%	
(i)	Maltitol		
(ii)	Maltitol syrup		
E966	Lactitol	40%	
E967	Xylitol	90%	

Bulk sweeteners

These are sometimes also referred to as 'sugar replacers'. The bulk sweeteners permitted are shown in Table 14.7. They are polyhydric alcohols (polyols) and have some of the functional properties of sugar or glucose syrups. All have laxative properties to a greater or lesser extent and daily intake needs to be controlled. All are claimed to be less cariogenic than the carbohydrates but not all are non-cariogenic. Xylitol is claimed to help in remineralization of decayed teeth. Some have negative heats of solution and can therefore exert a cooling effect in the mouth. All are considered to have a calorific value of 2.4 kcal g^{-1} (10.02 kJ g^{-1}). All are more expensive than carbohydrates, some considerably more.

Properties considered to be critical when selecting a bulk sweetener include cariogenicity, sweetness equivalent, solubility and viscosity, hygroscopicity, the cooling effect, laxative effect and cost. In most cases the best results are obtained by using mixtures of two or more.

Sorbitol

Sorbitol is a hexahydric alcohol, normally sold as a 70% solution or as a powder with a purity of 99.5%. Certain commercial sorbitols contain mannitol, and it is said that this impurity enhances the non-crystallizing character of the alcohol. The optical rotation $[\alpha]^{25}_D$ of sorbitol is −1.988. It has a sweetness approximately equivalent to that of dextrose.

Sorbitol is used as a sugar replacer in diabetic chocolate and as a humectant in conventional confectionery at levels between 5–10%. Due to its hygroscopic nature it can cause viscosity problems during processing.

Mannitol

Mannitol is a naturally occurring sugar alcohol, produced industrially by the hydrogenation of fructose, which yields three parts sorbitol for every part of mannitol. It is therefore more expensive than sorbitol. In confectionery it is usually used with sorbitol.

Isomalt or palatinit

Isomalt does not occur naturally, it is made from sucrose by enzymatic conversion followed by hydrogenation. It is a sweet, low calorie bulking agent with properties and characteristics similar to those of sucrose, with approximately half the sweetness intensity of sucrose. Isomalt is non-cariogenic and causes insignificant changes in the blood glucose levels in humans. These properties suggest that isomalt is a suitable sweetening agent for diabetic foods, low calorie products and tooth friendly confectionery.

Maltitol

Maltitol occurs naturally and is available in crystalline form, but is used largely in the form of hydrogenated high maltose starch conversion syrup. It is sold by a number of manufacturers under a wide variety of trade names. Product compositions differ slightly between manufacturers.

Lactitol

Lactitol does not occur naturally. It is made by hydrogenation of lactose. It is available as a crystalline monohydrate or dihydrate.

Xylitol

Xylitol occurs naturally but is produced commercially by hydrogenation of xylose produced from cellulose. It is available as a crystalline material which has a pronounced cooling effect when eaten or dissolved in water.

BULKING AGENTS

These are non-calorie or low calorie fillers which can be used in low calorie foods. They are derived either from cellulose, which is insoluble in water and non-calorific, or from polydextrose which is soluble, non-crystallizing and has a calorie value one quarter that of sugar. Functionally polydextrose is similar to glucose syrup.

The polyols (see above), with calorie values of only 2.4 kcal g^{-1} (10.02 kJ g^{-1}), may also be regarded as partial bulking agents.

WATER

Water used as a food ingredient and for the final preparation and cleaning of food ingredients and manufacturing plant should be of potable quanlity. Where a mains water supply is available, water for these purposes should be drawn direct from the main. Other water supplies should be regularly tested to confirm their continuing suitability.

Potability is not, however, the sole requirement of a suitable water. Additional criteria may be required for different products or processes. When used as an ingredient, water should be non-alkaline and free from traces of hydrogen sulphide or iron. Unless there are specific requirements for the presence of calcium or magnesium to provide or maintain textural characteristics in the food, it may be desirable to use soft or softened water.

When the mains supply is not suitable it will be necessary to install treatment plant to adjust the water. Modern water supply practice also gives increasing chances of variability in the water delivered: this should be checked with the water supply authority before purchasing and installing treatment plant.

Hardness of water

The hardness of water is mainly due to the presence of salts of calcium and magnesium held in solution. Certain salts, such as calcium carbonate, are practically insoluble in pure water, but if carbon dioxide is present a certain amount of the carbonate will go into solution as bicarbonate. On boiling this water, the carbon dioxide escapes and the carbonate is thrown down as a cloudy white precipitate. It is usual to state the hardness of water as the equivalent of parts per million of calcium carbonate. The hardness due to salts which are insoluble after boiling is referred to as 'temporary' hardness, while the calcium and magnesium salts which remain in solution constitute the 'permanent' hardness. The hardness of water, particularly the temporary hardness, should be low if clear syrups or brines are to be made, for instance for the canning or bottling of fruits. The degree of hardness is of still greater importance in vegetable canning where the presence of lime in the blanching water or brine may cause toughening of the skins of such products as peas and beans. In blanching and covering green peas, for instance, the effect of lime in the water becomes noticeable if the hardness exceeds about 1000 ppm.

Waters from different districts vary greatly in hardness. In certain places practically pure rain water is obtainable, with hardness not exceeding 3 to 4 ppm, while in chalky country and in some well waters the hardness may rise to 600 or 700 ppm. Canners and bottlers in particular should be certain of the suitability of their water before starting operations, and where necessary water softening plant should be installed.

Effluents

Plants handling fruits and vegetables can use and discharge to waste large quantities of water which may represent a considerable production cost. Also in most countries, trade effluent is subject to strict control with respect to the soluble and insoluble matter it contains. Specialist advice should be sought when necessary about possible treatment of the water to minimize the cost of disposal.

To reduce the amount of water used and therefore the volume of effluent, it should be used more than once wherever possible. One common situation is to have a circulation system for water for can cooling purposes; meticulous control of the bacteriological quality of the cooling water by chlorination is essential. Again, if water is used at various stages of preparation of vegetables, then the same water may be used at two or more stages; the reverse flow method is used, the clean water being used for the first time at a stage nearer to the finished product, and then reused for an earlier stage of preparation. Such methods can save a great deal of water quite safely if properly devised and controlled. If badly applied then vegetables may be inadequately cleaned and, worse, undesirably high bacterial loads may build up. Such process water is often chlorinated as it comes into the factory (in-plant chlorination) and this

has a real advantage in maintaining a hygienic condition of flumes, elevators and other handling equipment.

REFERENCES

American Meat Institute (1950) *Pork Operations*, AMI, Chicago.
Federal Register (21CFR) Title 21 of the Code of Federal Regulations, Parts 70–82 and 170–186.
Lange's Handbook of Chemistry (1956) 9th edn, Handbook Publishers, Sandusky, Ohio.

FURTHER READING

Ash, M. and Ash, I. (1994) *Handbook of Food Additives*, Gower, Aldershot.
Ashurst, P.R. (ed.) (1994) *Food Flavourings*, 2nd edn, Chapman & Hall, London.
Birch, G.G., Green, L.F. and Coulson, C.B. (1971) *Sweetness and Sweeteners*, Chapman & Hall, London.
Branen, A.L., Davidson, P.M. and Salminen, S. (eds) (1990) *Food Additives*, Marcel Dekker, New York.
Burdock, G.A. (1995) *Fenaroli's Handbook of Flavor Ingredients*, 3rd edn, 2 vols, CRC Press, Boca Raton, Florida.
Burdock, G.A. (ed.) (1996) *Encyclopaedia of Food and Color Additives*, CRC Press, Boca Raton, Florida.
Butterworths Food Law (1992), Butterworths, London.
Glicksman, J. (1969) *Gum Technology in the Food Industry*, Academic Press, London.
Hardman, T.M. (ed.) (1989) *Water and Food Quality*, Chapman & Hall, London.
Harris, P. (ed.) (1990) *Food Gels*, Elsevier, Barking.
Heath, H. and Reineccus, G.A., (1986) *Flavour Chemistry and Technology*, Chapman & Hall, London.
Hendry, G.A.F. and Houghton, J.D. (eds) (1996) *Natural Food Colorants*, 2nd edn, Chapman & Hall, London.
Hudson, B.J.F. (ed.) (1990) *Food Antioxidants*, Elsevier, Barking.
Imeson, A. (ed.) (1996) *Thickening and Gelling Agents for Foods*, 2nd edn, Chapman & Hall, London.
Jukes, D.J. (1992) *Food Legislation of the UK: A Concise Guide*, 3rd edn, Butterworths, London.
Lewis, R.J. (1989) *Food Additives Handbook*, Van Nostrand Reinhold, New York.
Marie, S. and Piggott, J.R. (eds) (1991) *Handbook of Sweeteners*, Chapman & Hall, London.
Mayer, D.G. and Kemper, F.H. (eds) (1991) *Acesulfame-K*, Marcel Dekker, New York.
Molins, R.A. (1991) *Phosphates in Food*, CRC Press, Boca Raton, Florida.
Nabors, L.O'B. and Gelardi, R.C. (eds) (1991) *Alternative Sweeteners*, 2nd edn, Marcel Dekker, New York.
Piggott, J.R. and Peterson, A., (1994) *Understanding Natural Flavors*, Chapman & Hall, London.
Prakash, V. (1990) *Leafy Spices*, Wolfe Publishing, London.
Reineccus, G. (1994) *Source Book of Flavours*, Chapman & Hall, London.
Rowell, R.A., Gresty, G.S. and Woodroffe, G.P. (1988) *Practical Food Law Manual*, Sweet and Maxwell, Andover.
Russell, N.J. and Gould, G.W. (eds) (1991) *Food Preservatives*, Chapman & Hall, London.
Smith, J. (ed.) (1991) *Food Additive User's Handbook*, Chapman & Hall, London.
Smith, J. (ed.) (1993) *Technology of Reduced-Additive Foods*, Chapman & Hall, London.
Underriner, E. and Hume, I. (eds) (1993) *Handbook of Industrial Seasonings*, Chapman & Hall, London.

Walter, W.H. (1992) *The Chemistry and Technology of Pectin*, 3rd edn, Academic Press, London.

Wedzicha, B.L. and Wedzicha, G.J. (1996) *Chemistry of Sulfur Dioxide in Foods*, 2nd edn, Chapman & Hall, London.

Whistler, R.L. and BeMiller, J.N. (1993) *Industrial Gums: Polysaccharides and Their Derivatives*, Academic Press, London & San Diego.

For the USA the majority of additives legislation will be found in the appropriate Parts of Title 21 of the Code of Federal Regulations (21CFR), Parts 70–82 and 170–186.

15 Food Preservation Processes

INTRODUCTION

The major technologies used to preserve food products include (i) inactivation of microorganisms, e.g. by heat pasteurization, ionizing radiation and high pressure and (ii) prevention and retardation of the growth of micro-organisms, e.g. by chilling, freezing, acidification, vacuum and modified atmosphere packaging and dehydration.

Some processes, such as freeze drying and pasteurization/chilling involve a combination of two or more processes. Whichever process is used, the correct selection of raw materials and preprocess treatments are essential for achieving a high quality product. The final packaging and storage conditions will determine the shelf life of the product.

CHILLING AND FREEZING

BASICS

Units

Temperature

The two most common scales used are (i) the Celsius (also known as the centigrade) scale, on which the melting point of ice is 0°C and the boiling point of water at atmospheric pressure is 100°C; (ii) the Fahrenheit scale, on which the melting point of ice is 32°F and the boiling point of water at atmospheric pressure is 212°F. The Fahrenheit scale is still extensively used in mechanical engineering and refrigeration work but is gradually being replaced by the Celsius scale.

There are two absolute scales: (iii) the Kelvin scale, corresponding to Celsius, which is used for very low temperature work, where ice melts at 273K and water boils at 373K; the Kelvin scale is the one adopted for the Système International d'Unités, known as SI Units; (iv) the Rankine scale, corresponding to Fahrenheit, where ice melts at 460°R and water boils at 672°R.

The conversion factors are as follows: $1°C = 1K = 1.8°F = 1.6°R$

Units of heat energy

The Btu or British thermal unit is the amount of heat required to raise the temperature of one pound of water through one degree Fahrenheit. This is widely used in the USA, but has now been replaced by SI units in the UK and elsewhere. Other non-SI units employed are the CHU or Centigrade heat unit (the amount of heat required to raise the temperature of one pound of water through one degree Celsius) and the calorie (the mount of heat required to raise the temperature of one gram of water through one degree Celsius).

The SI unit of energy (heat) is the Joule (J) or newton-metre. A more convenient unit is the kilojoule (kJ): the amount of heat required to raise the temperature of 1 kg of water through 1K.

Equivalents to note:

1 kJ	= 239 calories	= 0.948 Btu
1 cal	= 4.184 J	= 0.004 Btu
1 Btu	= 1.055 kJ	= 252 calories
1 therm	= 105.5 MJ	

Units of power (rate of heat flow)

The basic unit is the watt (W), which is defined as 1 joule per second. A more convenient unit is the kilowatt (kW). Some important equivalents in refrigeration are:

1 horsepower (hp)	= 745.7 W
1 Btu h^{-1}	= 0.293 W
1 cal s^{-1}	= 4.187 W

Ton refrigeration

The ton refrigeration is a measure of refrigeration plant performance. It represents the rate of cooling produced when a (US) ton (2000 lb) of ice melts during a 24 h period. In the USA, 1 ton refrigeration is 288 000 Btu per 24 h when the evaporator temperature is 5°F and the condenser 86°F. In SI units, 1 ton refrigeration is 3.54 kW.

Several other definitions of ton refrigeration have been used in the past; thus Wymass Anderson, 322 000 Btu/24 h; Lloyds, 318 080 Btu/24 h; Ice Making, 510 080 Btu/24 h; International, 342 860 Btu/24 h.

Heat transfer

The rate of heat transfer is of particular importance in chilling and freezing processes. Under any given circumstances it is characterized by the heat transfer coefficient U, defined as:

$$Q = UAT,$$

where Q is the rate of heat transfer (in units of e.g. kW), U is the heat transfer coefficient (kW m^{-2} K^{-1}), A is the heat transfer area (m^2), and T is the temperature difference between the cooling (or heating) medium and the product (K).

In still air, heat transfer between the product and the air occurs mainly by natural convection. Typical heat transfer coefficients range between 4 and 8 W m^{-2} K^{-1}. For natural convection to a liquid such as water, values range between 200 and 500. With circulation (forced convection), these can be raised to 900 or more. Where melting ice is used as the heat transfer medium, the value of U is only about 100 W m^{-2} K^{-1} because of irregular contact, air pockets, and the fact that heat is transferred mainly as a result of the melted ice running over the product. Sprays are likewise less efficient than total immersion.

The rate of heat transfer is also favoured by (i) a greater surface area for a given weight, i.e. thinner product; (ii) a greater temperature differences between the product and the medium. For example, brine at −1.5°C (29°F) can typically cool fish from 16 to 2°C (60 to 35°F) in two-thirds of the time taken by brine at 0°C (32°F). The rate of cooling of a warm fish is greater at the beginning of the operation and falls off rapidly as it approaches the temperature of the medium. Again, the smaller the ice particles employed, the faster the cooling. Ice cubes of side 1 cm cool in as little as half the time taken by cubes having a 10 cm side. Also, irrespective of the theoretical thermal requirement of about 1 kg of ice per 4 to 5 kg of fish, as dictated by the value of the latent heat of ice (0.34 J kg^{-1}) and a typical value of the specific heat of fish, a minimum ratio of ice to fish of 1 kg of ice per kg of fish is required to achieve the optimum cooling rate. Lower values lead to appreciably slower cooling.

Of course the rate of transfer of heat from or into a body is determined by its thermal conductivity, and this factor becomes increasingly dominant as the thickness of the product or package increases. The temperature difference between the surface and the medium falls accordingly.

A variety of different types of thermometer are available for measuring temperatures in frozen food applications. Environmental temperatures are often indicated using mercury or liquid-in-glass types. Wet and dry bulb types are used for measuring relative humidity in different environments. The most commonly used electrical device is the thermocouple; the type T device incorporates a copper/copper–nickel alloy (constantan) junction. The temperature may be displayed digitally or the unit used in conjunction with data logging equipment. Some probes are made in the form of corkscrews to permit easy penetration of a block of frozen product to measure the correct core temperature.

Thermocouples are also employed as the sensor in time–temperature integrating devices, which are able to log temperatures and determine, through a computer program, the integrated effect of storage conditions.

Food refrigeration

Refrigeration is one of the main techniques for increasing the shelf life of fresh and manufactured products. It may be used in two principal ways, either chilling of freezing. Chilled food storage temperatures vary from +1 to 5°C, although, in some cases, where freezing or chill damage do not take place, lower temperatures, e.g. −1°C can be employed. The acceptable storage time depends on the product and its composition: milk can be stored for several days, cheese, meat and bacon for considerably longer, several weeks. Frozen foods on the other hand are generally taken as being at temperatures of −10°C or below and are stored at similar temperatures.

Most foods can be satisfactorily frozen; exceptions are milk, shell eggs, salad vegetables and some fruits with high water content. However, the frozen food market has expanded from frozen fruit and vegetables, meat, poultry and fish products, to a wide range of cooked products, including meat, poultry, mixed vegetable and potato products, speciality dishes, pizza and pastry products as well as cream cakes.

Internationally, a class of frozen foods known as 'Quick Frozen' or 'Deep Frozen' are recognized as having the highest quality. These are frozen to

−18°C or below and have been rapidly frozen between the temperatures −1°C and −5°C. In France the term surgeli is used to represent this class of products, whereas products not meeting this specification are referred as congeli. In the latter class are frozen meat (−10°C) and frozen poultry (−12°C). In West Germany the corresponding terms are tiefgefroren (maintained at −18°C) and gefroren (maintained at −12°C).

Chilling

Chilling of foods is carried out to slow down deterioration. Foods will spoil over a period of time due to biochemical, chemical or physical changes, but most importantly due to microbiological growth.

As a general rule, the lower the temperature at which a food is stored above freezing, the slower the deterioration. In practice, chilling is usually taken to mean cooling to within the range +1 to +5°C, which will restrict deterioration.

The techniques used for chilling are very similar to those for producing frozen foods and include batch cabinets and continuous in-line chillers. The former category includes large chilled rooms for conditioning or for chilling meat carcasses and poultry, as well as smaller units for packaged products. These are usually air-blast chillers. In the case of hot cooked products such as meat pies, liquid nitrogen or liquid carbon dioxide sprays may be used. All these include continuous belt or band tunnels and spiral chillers.

Horticultural produce

Chilling is also used as a storage technique for raw horticultural produce to prolong freshness prior to processing or marketing. Hydrocooling with water, ice flakes or solid carbon dioxide is used to remove metabolic field heat and prolong the useful life of the produce.

Hydrocooling is commonly applied to root and other vegetables such as celery, radishes and carrots, whereas apples are air cooled in boxes. Vacuum cooling is also carried out on lettuce, cauliflower, cabbage and strawberries. This relies on the fact that when the surface moisture is removed from a product, the latent heat of evaporation is extracted and the product temperature decreases.

Special gas-tight chilled stores are used for the long term storage of fruits such as apples and pears, which evolve carbon dioxide metabolically after picking. The shelf life is prolonged by careful control of the oxygen to carbon dioxide ratio, usually by scrubbing out the excess carbon dioxide

produced metabolically. Apples can be stored for up to eight months in an atmosphere of up to 8% carbon dioxide and as little as 2% oxygen, provided the temperature remains between −1 and +4°C. The technique is known as controlled atmosphere (CA) storage. When applied to packaged products in which a larger number of gases can be used, it is known as modified atmosphere (MA) storage.

Fish

Fish muscle, although basically similar to meat in composition, possesses a lower level of adenosine triphosphate (ATP) and in consequence the pH declines after death to around 6.4–6.8. When the muscle passes through rigor mortis there is considerable muscular contraction, which results in unsightly appearance of the fish. To prevent this happening it is necessary to chill fish immediately after they are caught at sea; this is usually done by adding granular ice.

Meat

The chilling of carcass meat has enabled meat producing countries in the southern hemisphere to supply the large urban populations of North America and Europe.

Care must be taken not to chill meat too rapidly after slaughter or 'cold shortening' will occur, which in extreme cases will lead to a toughening of the meat beyond acceptable limits. This is related to the loss of ATP and to other post mortem changes in the meat. It can be avoided by ensuring that the temperature does not fall below 10°C (50°F) before the pH falls below 5.5. This process of conditioning takes about 10 h for lamb or beef and a shorter period for poultry and pigs. Newer techniques such as electrical stimulation of recently slaughtered carcasses cause the muscle to use up its reserves of ATP rapidly and reduce the conditioning period.

Beef also requires a period of ageing up to two or three weeks at 4°C (38°F) before it reaches optimum tenderness. Ageing occurs rapidly with pigs and poultry.

Freezing

The freezing process may be divided into three major stages: the first involves the cooling of the food product to the freezing temperature, during which time the sensible heat is removed; the second stage involves the formation of ice and the removal of the latent heat of fusion at a relatively constant temperature around -1 to $2°C$; during the final stage, the frozen product is cooled down to the storage temperature, $-18°C$, or otherwise depending upon the product. Apart from a small degree of supercooling in solid food products or a larger degree in some liquid or syrup products, all foodstuffs with high water contents behave similarly.

The second stage is the most important, since the rate of freezing is the rate determining step. Freezing rates in most modern freezers are sufficiently fast to avoid deterioration of the quality of products. Some quality advantages may be noticeable with ultrarapid freezing; however, storage will reduce the differences. The freezing time is determined by the initial and final temperatures, the temperature of the heat transfer medium, the physical properties of the product, in particular its latent heat of fusion and the overall heat transfer coefficient, which includes the resistance to heat flow of the packaging. Many attempts to predict freezing times from these data have been made. The most widely used and in some ways the simplest is that due to Rudolf Plank. His treatment of the problem includes the assumptions that the initial temperature of the food is the freezing point (i.e. precooling is ignored) and that freezing takes place under steady state conditions. Plank's formula enables the freezing time t to be calculated in terms of the latent heat L, the difference in initial temperatures T, the thickness of the product d, the thermal conductivity k of the frozen product and the overall heat transfer coefficient U between the cooling medium and the product surface:

$$t = (L/T) (P.d/h + R.d^2/k)$$

P and R are constants that depend upon the geometry of the product. Some typical values of P and R are: rectangular slab, 1/2 and 1/8, respectively; sphere, 1/6 and 1/24, respectively.

The mean linear rate at which the ice front moves in a product can be used to classify different processes: slow freezing in blast air cooled rooms, 0.2 cm h^{-1}; quick freezing of retail packages in air blast or plate freezers, 0.5–3 cm h^{-1}; rapid freezing in fluidized bed freezers, 5–10 cm h^{-1}; and ultrarapid freezing in liquid nitrogen sprays, 10–100 cm h^{-1}. For most commercial operations, rates of freezing greater than 5 cm h^{-1} are considered suitable for quick frozen foods. For larger bulk products, e.g. beef carcasses and quarters, mean rates of 0.1 cm h^{-1} over 3–5 days are common.

The mean linear freezing rate, often referred to as the nominal freezing rate, is defined as the shortest distance between the surface and the thermal centre (slowest responding point, equal to half the thickness for regular shaped bodies) divided by the nominal freezing time from 0 to $10°C$ below the initial freezing point (centre temperature).

The time required to reduce the temperature of the food product from the initial ambient temperature to a given value at the thermal centre during freezing is known as the effective freezing time.

REFRIGERATION SYSTEMS

Mechanical refrigeration

This type of refrigeration system (see Figure 15.1) operates on the closed cycle vapour compression principle and consists of a compressor and condenser to liquefy the vapour at high pressure and an expansion valve which allows the liquid to vaporize in the evaporator or cooler. The low pressure saturated vapour is drawn into the suction end of the compressor where it is compressed adiabatically to a vapour whose corresponding saturation temperature exceeds that of the cooling medium. After discharge from the compressor, the superheated high pressure refrigerant vapour is passed through the condenser in which it is converted into a liquid. When this passes through the expansion valve, the pressure of the gas is reduced and the temperature falls. In

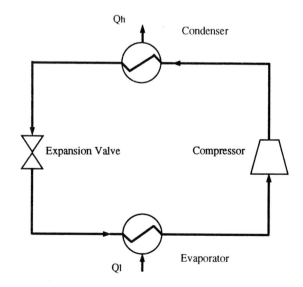

Figure 15.1 Mechanical refrigeration cycle. QI, heat absorbed from product; Qh, heat rejected to environment.

the evaporator the gas absorbs the heat extracted from the air when products are cooled.

Compressors

Four types of vapour compressor are in current use:

(i) Reciprocating compressors, in which the rotary motion of an electric motor is changed into reciprocating motion using an eccentrically mounted crank and rod attached to the piston. Two-, four- and eight-cylinder compressors are commonly available.
(ii) Rotary blade compressors. There are two basic types, one which has blades that rotate with the shaft and the other which has stationary blades. In these compressors, the low pressure vapour from the evaporator is drawn into the opening and fills the space behind the blade as it revolves. The trapped vapour is compressed as the blade revolves and is pushed into the condenser.
(iii) Screw-type compressors. These use a pair of special helical rotors, driven by an electric motor, which trap and compress the vapour as it progresses along the helical thread. This type is used extensively in the food industry because of the limited wear and maintenance required and also because of its efficiency at the required working temperatures.
(iv) Centrifugal or turbo compressors. The vapour is compressed using an impeller consisting of a disc with radial blades which spins rapidly, thereby imparting a centrifugal motion to the vapour. This type of compressor is relatively simple as it requires no valves, pistons or cylinders. Control is easily effected simply by throttling either the suction or discharge of the compressor.

Condensers

This is the part of the closed system where the gaseous refrigerant is cooled to the liquid state at high pressure. It may be water cooled or air cooled. Various types of heat exchanger are available for this purpose, including atmospheric condenser tubes or coils, shell and tube types, concentric pipes and forced air convectors. When air is used for cooling, the pipes are finned so that the surface area for heat transfer is increased and the system becomes more efficient.

Expansion valves

The basic component of the refrigeration compression system is the refrigerant flow control valve,

often known as the expansion valve. It has two functions: (i) to meter the flow of liquid refrigerant into the evaporator at a rate which is commensurate with the rate of vaporization and (ii) to maintain a pressure differential which allows the refrigerant to vaporize under the desired low pressure while, at the same time, condensing at a high pressure in the condenser.

Five basic types are used in practice:

(i) Hand expansion valves. These are needle valves which create differential pressure across on orifice and can be adjusted manually. They are mainly suitable for non-fluctuating load conditions.
(ii) Automatic expansion valves. This type of valve is actuated by a diaphragm which is controlled by the force of an adjustable spring and the evaporator pressure. If the evaporator pressure falls below the required pressure the spring causes the valve to open until the set pressure has been re-established by increased evaporation, and vice versa.
(iii) Thermostatic expansion valves. These are similar to the previous type, except that the diaphragm has an additional balancing pressure of a vapour pressure thermometer, the bulb of which is located at the end of the evaporator tube. Unlike the automatic expansion valve which maintains a constant pressure differential, this type maintains a constant degree of suction superheat at the evaporator outlet. Thermoelectric types are also used. Pressure limiting valves have an additional built in spring which enables a maximum operating pressure to be maintained. This avoids excessive compressor overload.
(iv) Capillary tube. This is one of the simplest methods of controlling flow rate. However, it is essential that the flow capacity of the tube is equal to the pumping capacity of the compressor. Systems using capillary tubes as a restrictive device are inflexible as far as operating conditions are concerned.
(v) Float controls. These are used with flooded evaporator types of refrigerator, and usually consist of a buoyant hollow metal sphere or cylinder which responds to change in liquid level. The increase or decrease in flow rate depends on the position of the float.

Evaporators

The design of an evaporator depends upon the application (air conditioning, freezing, chilling, storage, transport). The following principal types are used:

(i) Bare tubes. These are made of steel (for use with ammonia) or copper (for use with organic refrigerants) and are used in the form of flat zig-zag or oval trombone coils. This type is often used in frozen food storage rooms, suspended from the ceiling.

(ii) Plate surface types. These consist either of tubes enclosed in sheets of metal or are constructed by welding moulded sheets together. They are mainly used in domestic refrigeration systems, less often in transport systems and low temperature storage rooms.

(iii) Finned tubes are tubes with discs welded or fixed on the outside so that the effective heat transfer surface is greatly extended.

Evaporator systems are classified according to the method of liquid feed. The dry expansion type features complete evaporation of liquid, the flooded type is totally immersed in the liquid refrigerant and the liquid overfeed type is filled to an extent that there is more liquid present than can be evaporated. The last type is particularly useful in multiple evaporating systems.

Absorption systems

The absorption refrigeration cycle, often referred to as the Platen–Munters system, employs the evaporation of a volatile refrigerant in order to produce a cold environment, and an adsorbent (either liquid or solid) to produce continuous movement of the refrigerant. In its simplest form, the cycle is operated by heating a solution of ammonia in water, thereby liberating ammonia gas which is condensed to give liquid ammonia. The liquid ammonia passes through the refrigerated compartment where the food is stored and extracts its latent heat by evaporation, thereby producing the cold environment. The process is maintained by absorbing the evaporated ammonia in water and repeating the cycle. The temperature may be varied by including hydrogen gas in the system; the greater the amount of hydrogen, the lower the temperature that may be achieved. To produce a temperature of $-18°C$ ($0°F$), the ammonia must boil at 0.1 MPa, which requires a hydrogen pressure of 2.65 MPa. The total pressure in the system is 2.75 MPa and consequently a robust construction is required (see Figure 15.2).

The main advantage is that the system does not require a compressor, only a supply of heat. It is therefore highly suited to domestic systems. In some versions a pump may be used instead of gravity flow to move the ammonia–water solution from the absorber to the heated generator. A pump is also required when hydrogen is not present to produce the differential pressure.

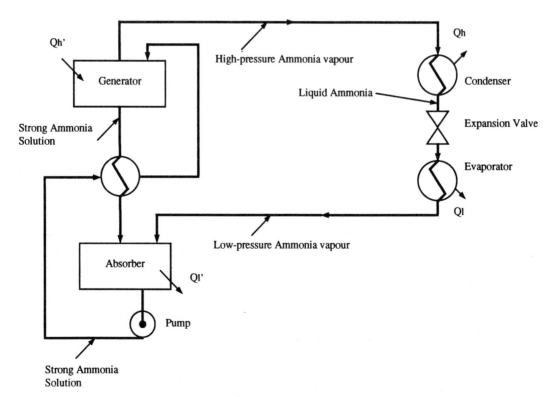

Figure 15.2 Ammonia absorption refrigeration cycle. QI, heat absorbed from product; QI', heat rejected from absorber; Qh, heat rejected to environment; Qh', heat supplied to generator.

An alternative system uses water as the refrigerant and lithium bromide solution as the absorbent. The system will not produce temperatures below 0°C; however, it is useful for chilled food systems. For a temperature of 4°C at the evaporator, the evaporator pressure is 8.13 mbar (7100 kPa gauge).

Refrigerants

A refrigerant is any substance which acts as a cooling agent by absorbing heat from another substance. For the vapour compression cycle, the choice of refrigerant depends upon economic factors and specific physical and chemical properties. The ideal characteristics are that the fluid should be non-flammable, non-explosive and non-toxic, both in the pure state and when mixed with air in any proportion. It should not interact with the materials of construction and lubrication and should be non-toxic if it comes into contact with food.

For economical operation the most important properties of a refrigerant are high latent heat of vaporization, low specific volume of vapour, low compression ratio, low specific heat of the liquid and high specific heat of the vapour. The power required per unit of refrigerating capacity is similar for all common refrigerants, and consequently the major factors which affect the selection of a refrigerant are those which reduce the size of plant, its weight and cost and which permit continuous operation and easy maintenance.

Early refrigerants

Initially, all early refrigeration plants used either ammonia or carbon dioxide. These plants were essentially large and consequently other refrigerants such as sulphur dioxide and methyl chloride were developed for smaller commercial and domestic use. For air conditioning applications, methylene chloride and carbon dioxide were used, because they were non-toxic and considered safe.

CFCs (Freons and Arctons)

Apart from ammonia, which has excellent properties, all the other refrigerants were replaced in modern installation by the fluorocarbons. These were developed in the 1920s especially for refrigeration systems and became extensively used until recent times when their effects on the Earth's ozone layer had become of concern.

The chlorofluoromethanes, i.e. compounds with one carbon atom and containing one of more chlorine and/or fluorine atoms, are R11, R12, R13 and R14, where R stands for refrigerant, and the last number indicates the number of fluorine atoms in the molecular structure, e.g. R11 is trichloromonofluoromethane (CCl_3F), R14 is carbon tetrafluoride (CF_4). They are also known as 'Freons' with numbers identical to R numbers (Du Pont, USA), or 'Arctons' (ICI, UK) with different numbers (A9 = R11; A6 = R12; A3 = R13; A4 = R22; A7 = R21).

HFCs and HCFCs

Research has shown that it is the chlorine in the CFC molecule that attacks the ozone layer. Bromine is even more damaging; therefore all references to chlorinated compounds apply with equal or greater force to brominated compounds. The damaging effect is very much reduced if the molecule is not fully halogenated i.e. contains hydrogen in addition to chlorine, fluorine and carbon. Therefore the fluorocarbon compounds are now designated as:

(i)   CFCs, fully halogenated chlorofluorocarbons, use of which is being phased out as rapidly as possible.
(ii) HFCs, hydrofluorocarbons, which contain no chlorine and as a result have no damaging effect on the ozone layer. Unfortunately there do not appear to be sufficient of these compounds with the appropriate properties to replace all of the CFCs in current use. The HFCs, for example HFC-134a, are all new products.
(iii) Fluorocarbons. These contain only carbon and fluorine; a few of these compounds are available and have no effect on the ozone layer, for example FC-14.
(iv) HCFCs, hydrochlorofluorocarbons, which do contain chlorine but are not fully halogenated. HCFCs do have an adverse effect on the ozone layer, but this is very much less than that of the CFCs. Certain HCFCs, notably HCFC-22, were already known and widely used before the ozone depletion theory was first proposed; others, e.g. HCFC-123, are new. HCFCs and HFCs are sometimes collectively referred to as HFAs, hydrofluoroalkanes.

Recent developments

By the end of 1995 at the latest, production of CFCs in developed countries had effectively

Table 15.1 Summary of phaseout schedules as set out under the Montreal Protocol and European Community Regulations

Refrigerant	Montreal Protocol December 1995	EC Regulation 3093/94
CFCs	75% cut by 1/1/1994 Phase out by 1/1/1996[a]	85% cut by 1/1/1994 Phase out by 1/1/1995[a]
CCl$_4$	75% cut by 1/1/1994 Phase out by 1/1/1996[a]	85% cut by 1/1/1994 Phase out by 1/1/1995[a]
Halons	Phase out by 1/1/1994[a]	Phase out by 1/1/1994[a]
1,1,1-Trichloroethane	50% cut by 1/1/1994 Phase out by 1/1/1996[a]	50% cut by 1/1/1994 Phase out by 1/1/1996[a]
HBFCs	Phase out by 1/1/1006[a]	Phase out by 1/1/1996[a]
Methyl bromide	Freeze at 1991 levels from 1/1/1995 25% cut by 1/1/2001 50% cut by 1/1/2005	Freeze at 1991 levels from 1/1/1995 25% cut by 1/1/1998
HCFCs	Freeze at 2.6% of (CFC consumption + total HCFC consumption in 1989) by 1/1/1995 35% cut by 2004 65% cut by 2010 90% cut by 2015 Phase out by 1/1/2020[b]	Freeze at 2.6% of (CFC consumption + total HCFC consumption in 1989) by 1/1/1995 35% cut by 2004 60% cut by 2007 95% cut by 2013 Phase out by 2015

[a] Provision for possible 'essential use' exemption. [b] 0.5% of consumption allowed until 2030 for the servicing of existing refrigeration and air conditioning equipment.

ceased as the result of international legislation restricting their production and sale for most current applications including refrigeration and air conditioning.

As a member of the European Union, the UK is a joint signatory of the Montreal Protocol, the principal international phaseout agreement. Furthermore, EU member countries have adopted a more stringent phaseout policy than that outlined in the protocol. At the time of going to press the phaseout conditions are as set out in Table 15.1

Insulation

The function of insulation is to retard as effectively as possible the flow of heat into refrigerated spaces. The more efficient the insulation is in performing this duty, the more easily and economically can the desired low temperatures be maintained. The best heat insulator is a vacuum. The next best is probably a gas such as still air, which has a very low thermal conductivity. Under conditions where there is no air circulation, there is no heat transmission resulting from natural convection. Insulating materials such as granulated cork or slag wool subdivide the air space into cells so minute that convection, though not entirely

prevented, is greatly reduced and ceases to be important. This is a consequence of surface resistance to the motion of the air and of the smallness of the temperature difference between opposite cell faces. In addition, the interposition of material between two walls greatly reduces heat transmission by direct radiation.

It is, however, important that the insulating material absorbs a minimum amount of moisture, because its presence substantially increases the conductivity of the material, by 50–100%. This is particularly important for insulated surfaces that are at a temperature below the dewpoint of the surrounding air. In such cases, it is necessary to seal the surface against the penetration of water vapour, which would condense within the material, thereby causing a serious increase in heat flow, possible breakdown of the material, rotting of woodwork and corrosion of metal surfaces.

Insulants in common use include cork, slag wool, cellular expanded rubber, glass wool and aluminium foil. Some properties and applications of each of these are as follows.

Cork

This low conducting, non-hygroscopic material, in particular cork slab (baked and moulded cork), has long been accepted as possessing desirable characteristics for cold storage work and despite the many available alternatives, is still widely employed. The slab cork, which has a thermal conductivity (k value) of $0.07 \text{ W m}^{-2} \text{ K}^{-1}$, is usually erected in two or more courses laid in cement mortar or asphalt. Subsequent courses have their joints broken and are fastened with wooden skewers driven obliquely into the preceding course. Finishing of walls and ceilings may be by a cement plaster or an asphalt emulsion, to which is applied a coat of aluminium enamel or two coats of white enamel. Floor insulation needs more protection. A granolithic finish, carried some 150 mm up to the side walls, laid on top of several cm of good concrete, is frequently employed.

Moulded cork covering for pipes and fittings is made up using the desired thickness and comes in half sections. These are fastened with wire tape, the joints being carefully filled with cork filler or cement and the outside painted with an asphalt paint. The important point is to ensure at all times that there are no cracks through which water can penetrate the insulation. At times it is desirable to add to the covering for pipes and moulded fittings a waterproofing jacket of two-ply roofing paper or a stock 'canvas' strip wound spirally and painted with enough asphalt paint to saturate it.

Slag wool

Slag wool is available as a loose fill and in quilted, mat and slab form, all of which have a k value of $0.06 \text{ W m}^{-2} \text{ K}^{-1}$. When packed at densities less than 160 kg m^{-3}, the loose material is liable to settle down, leaving air spaces through which heat can leak fairly readily.

Cellular expanded rubber

Cellular expanded rubber possesses a very low conductivity ($0.04 \text{ W m}^{-2} \text{ K}^{-1}$), withstands vibration and the action of water. It is light in weight and has a low ignition point.

Glass wool

Low conductivity (a k value of $0.06 \text{ W m}^{-2} \text{ K}^{-1}$) and freedom from capillarity are desirable properties of glass wool. Additionally, it does not disintegrate under the action of vibration, its density does not change with vibration and it is fire-, rot- and vermin-proof. The material is available in loose, quilted, mattress and slab form.

Aluminium foil

Crumpled aluminium foil lowers heat transmission by reducing its radiant and convective components. As is well known, the true conductivity of air is less than any commercial insulating material, but radiation and convection are important additional factors. Air spaces 15 mm thick or less separated by paper reduce convection, but radiation is still a large factor. If the paper is replaced by bright metal the radiation factor is reduced, and if the metal is replaced by crumpled aluminium foil both conduction and convection are also reduced, i.e. areas of contact are lessened and the already small air space is broken up into a number of smaller spaces. The resultant k value is $0.05 \text{ W m}^{-2} \text{ K}^{-1}$. The foil tarnishes with time, it forms an oxide coating, but this does not appear to affect its emissivity, which remains constant at 0.1 or less. Other desirable features are moisture-, odour- and fire-resistance; it is also rot- and vermin-proof.

Polymeric materials

A wide range of polymeric materials, many clad with structural materials, are now used. Some typical examples are:

	k (W m^{-2} K^{-1})
Rigid mineral fibre board	0.034
Polyurethane board	0.023
Polyisocyanate board	0.023
Expanded polystyrene board	0.034

COMMERCIAL FREEZERS

Blast freezers

The most important method of freezing food products involves blowing cold air over the surface of the food until the core temperature is about 10°C below the initial freezing point of the product. The product may be in the form of packaged or unpackaged material, supported on trays or hanging free (e.g. in the case of carcass meat), in a batch system or supported on a continuous moving belt. Continuous systems are either linear or spiral in configuration, often incorporate a simple washing and cleaning system and are surrounded by insulated enclosures. The cold air is recirculated continuously over the bank of refrigerant tubes (usually finned) and the food product.

For free flowing products, e.g. peas, diced vegetables, beans, small fruits or potato chips, the most common method of freezing is to fluidize the product by passing the air upwards through a continuously moving bed of food product on a perforated tray or moving belt.

With very moist food products, the initial stage of freezing involves some moisture removal and hence weight loss. As freezing continues, there is also some loss of ice by sublimation. In recirculated air systems in which the air becomes saturated, good design can result in low weight loss, e.g. 0.5–2.0%.

Fluidized bed freezers

This technique is used for products which can be supported on an upwards blast of cold air. The air velocity required to achieve this state, termed fluidization, is very critical and has to balance the weight of the products. Heat transfer is rapid compared with deep bed freezing of similar products in a conventional air blast freezer. The products from this type of operation are known as individually quick frozen (IQF).

Commercial fluidized bed freezers are continuous tunnels and the product is moved along on either a perforated belt or an inclined perforated plate. The product is often stored in bulk and provided it has been thoroughly frozen, will retain free flowing characteristics and will remain suitable for subsequent packaging in retail size bags. The process is particularly well suited to peas, broad beans, sprouts, diced vegetables, mushrooms, prawns, potato chips (French fries) and soft fruits, e.g. strawberries, raspberries, blackberries bilberries and black- and redcurrants.

Plate freezers

In this method of freezing, the product is tightly compressed between refrigerant cooled, heat transfer surfaces in the form of plates. This is particularly useful for packaged products where good heat transfer may be obtained by close contact and the geometry of the package maintained during freezing. The spacing is usually limited to 75 mm to maintain rapid freezing. It is also used for bulk products such as fish or meat offal.

There are two main types, horizontal and vertical. A typical horizontal freezer has up to 20 plates, and the product may be manually loaded on to trays or, alternatively, automatically loaded, with unfrozen packets displacing frozen ones.

Plate separations of 25–75 mm are the most usual and hydraulic actuators are used to contact the plates and products. The compression forces are about 5–30 kPa, and allowance for expansion of the product has to be made during the freezing operation.

Vertical freezers work on similar principles; however, the methods of handling the products are different. This type of freezer is best suited to unpackaged products such as wet fish, meat and offal. Typical freezing times for fish blocks 75 mm thick are 120 min at $-34°C$ and 90 min at $-40°C$. Plate freezers require about half the time to achieve freezing compared with air blast freezing for the same thickness of product.

Cryogenic freezers

This type of freezing includes direct contact of the refrigerant with food products. In recent years it has become a useful method of freezing without recourse to large capital expenditure on refrigeration plant. The most common refrigerant for this purpose is liquid nitrogen. It is usually sprayed on to product which has been prechilled using the evaporated vapour. The product is carried on a continuously moving band in an insulated tunnel and the evaporated vapour is dispersed by fans, so that it is in direct contact with food entering the tunnel. The product is sprayed with the liquid towards the end of its passage through the tunnel and is finally allowed to equilibrate to obtain a uniform low temperature.

Liquid nitrogen plants are especially used for products which require rapid freezing to maintain texture and flavour, such as soft fruit, tomatoes, shrimps and prawns. They may also be used for rapid cooling of hot meat pies and similar products.

Liquid nitrogen (R728) has a boiling point at atmospheric pressure of $-196°C$ $(-320°F)$. Another direct contact refrigerant is carbon dioxide (R744), which has a boiling point of $-78°C$ $(-109°F)$ and is usually sprayed on to the food in the form of a fine snow from cylinders of liquid carbon dioxide.

Liquid contact freezers

For securely packaged foods such as poultry, direct liquid contact freezing is used. The freezing time for poultry using this method ranges from 30 min to 1 h, depending upon size and the temperature of the freezant. For brines containing calcium chloride this may be as low as $-34°C$ ($-30°F$), for those containing sodium chloride, $-18°C$ ($0°F$).

Ethylene glycol and propylene glycol are also used as direct contact freezants. Aqueous freezants tend to become diluted in use, and care should be taken to maintain the concentration at the required level. The concentrations of aqueous ethylene glycol solutions range from 15–50% vol., corresponding to temperatures of -5.3 to $-35.8°C$. For aqueous propylene glycol solutions, the concentrations range from 5–59% vol. corresponding to temperatures of $-1.7°C$ to $-49.4°C$. Propylene glycol is one of the most extensively used antifreeze agents in the refrigeration industry. It is particularly stable and is non-corrosive.

Scraped-surface heat exchangers

This type of equipment is used for continuous cooling of liquid products such as ice-cream mixture or fruit juice for freeze concentration and slush freezing. It consists basically of a cylindrical annulus constructed from two concentric tubes, through which the product to be cooled passes. The coolant (refrigerant) is passed through the outer shell. The heat transfer surface is prevented from fouling by scraper blades mounted on the rotating inner cylinder and this gives rise to high heat transfer coefficients.

Brine systems

Where it is not possible or convenient to use the expanded refrigerant gas for cooling, a secondary refrigerant is employed. In most cases, a brine, either calcium chloride or sodium chloride, is used for this purpose. Calcium chloride is used for industrial process cooling, product freezing and storage, and applications where temperatures below $-18°C$ are required. The lowest temperature, which is attainable without solid calcium chloride bring precipitated is $-55°C$ with a 29.87% weight solution (eutectic temperature). Sodium chloride solution is used for applications not requiring such low temperatures or where the solution is used in direct contact with packaged food, e.g. chilling and freezing of poultry and meat products. The lowest temperature (eutectic) which may be achieved is $-21.2°C$ with a 23% weight solution.

Freezers for ice manufacture

Natural ice was, until the dawn of mechanical refrigeration, the most important cooling medium. Harvested during the winter in high latitudes, the ice would be transported almost all over the world: ice houses for storing winter ice for summer use were common in large country houses in the eighteenth and nineteenth centuries.

Today, block, plate, tube and flake ice are the commonest type of ice. However, block ice, first made in cans frozen with refrigerated brine has now largely been superseded by other types.

Plate ice is more energy intensive in production than flake ice. The ice builds up on a vertical surface which has liquid refrigerant circulating behind it. Unfrozen water collects in a trough and is recirculated over the refrigerated plate as pre-cooled feed water. Either warm water or hot gas is substituted for the cold refrigerant to harvest the ice.

Tube ice, as its name implies, is produced in the curvature of a tube and has the advantage of sticking less during storage. A film of water is frozen either outside or inside the tube, the refrigerant circulating on the opposite side. At the end of the freezing cycle the ice is removed by introducing hot gas.

Flake ice can be continuously produced by applying water to either side of a refrigerated drum. The ice is opaque. Its production is highly efficient; no surface thawing is required since the ice is removed from the drum by a doctor blade. For many applications flake ice is preferred; it has the highest surface area/volume ratio and thus melts and cools rapidly.

Defrosting

The major problem with all non-insulated freezing systems, especially evaporators, is that the moisture in the surrounding atmosphere is deposited on the surface and ultimately forms a thick layer of ice. This impedes both freezing or chilling and the whole operation becomes very ineffective unless the ice is removed by defrosting. Ice formation is particularly restrictive with shell-and-tube evaporators.

Two principal methods of defrosting are used: (i) natural, involving shut-down or 'off-cycle', which uses the heat in the surrounding air to melt the ice, and (ii) supplementary heat defrosting, which utilizes water, brine, electric heating or the hot gas from the discharge of a compressor. Automatic defrosting is achieved by incorporating a timer in the circuit and the system is closed down for specified periods.

STORAGE OF FROZEN FOODS

Cold stores

Enormous advances have been made in the design, construction and operation of cold stores for frozen food in recent decades. Premoulded panels of insulants such as polyurethane allow swift, inexpensive construction of cold stores with very high insulating properties (k factor). The insulation must be sealed against water vapour. Provision must be made for underfloor heating to prevent the ground freezing, expanding and causing frost heave in the cold store. This is normally effected by using waste heat from the refrigeration system warming oil which is then piped through a grid in the subfloor.

Adequate refrigeration is required to combat heat leakage through the insulation, to extract the heat from machinery, lights, human operators and door openings and to bring the temperature of 'warm' product down to storage temperature.

Loading and unloading vehicles through 'port' doors can greatly reduce the ingress of warm, humid ambient air. These are openings which have the same shape as the vehicle's open doors and against which the vehicle body makes an effective seal. Complete automation of pallet handling obviates the need for human truck operators and allows all lights to be turned off in the cold store, reducing the amount of heat to be extracted from the space.

Storage temperature

A major factor in the success of quick freezing is the small size of the ice crystals formed during the process. If, however, the storage temperature is not right and in particular if fluctuations in temperature are allowed to occur, then the tendency will be for these crystals to fuse together and form larger ones so that one of the main objects of fast freezing will eventually be defeated. Temperature during storage and distribution is therefore as important to the quality of frozen foods as the original freezing process. Immediately after quick freezing, products are put into cold storage where the temperature is maintained constant at about $-29°C$ ($-20°F$). Distribution from the cold stores is accomplished by bulk transport of frozen food

in insulated vehicles, either road or rail, and in some cases by road vehicles possessing their own refrigerating plant. Where insulated vehicles are used without refrigerating plant, the low temperature is usually maintained by use of 'dry ice' (solid carbon dioxide) carried in racks in the vehicle. In retailers' cabinets the temperature should be maintained at or below $-18°C$ ($0°F$). These cabinets have a load line marked on the inside and food should never be stacked above this line. For home storage of frozen foods, domestic refrigerators are now marketed with frozen food compartments with star markings. One star should represent a temperature not above $-6°C$ ($21°F$) and the food will retain its quality for one week; two stars, not above $-12°C$ ($10°F$) for one month; three stars, $-18°C$ ($0°F$) for three months. However, it is always advisable to consult the instructions printed on the frozen food container for verification in case there are abnormal circumstances pertaining to a particular product.

Lloyd's rules

Many owners of public cold stores, especially those situated at ports, have had their installations surveyed by the engineers of the Committee of Lloyd's Register. If the installation is approved, this carries with it the privilege of registration in the Society's Register. The advantages of being in Lloyd's Register are that the store becomes known to many potential clients and intending users are confident that the store has been approved and classified by an impartial, highly experienced and competent authority.

Once classified, the value and location of the store, together with relevant details regarding the temperatures for which the various chambers have been designed and their storage capacity, are printed in the Register book. Approved stores are thereafter subject to an annual reinspection. Twenty-four stores classified in a typical year (1977) were designed to maintain $-30°C$ ($-22°F$); four to hold $-23°C$ ($-10°F$) and one to hold $-20°C$ ($-4°F$).

Precisely what temperature the store is operated at is, of course, subject to the discretion of the operator. The classification temperature certified in the Register is simply a 'notation temperature'. This is the temperature actually achieved during the test with the store closed; it therefore excludes the effects of, for example, product heat load or any heat associated with activity in the store, such as door openings, people, trucks, lighting, etc.

PRODUCT AND QUALITY ASPECTS

Microbiology of frozen foods

The freezing process, although not destroying micro-organisms as heat sterilization does, essentially retards growth, while at the same time effecting a slow loss of viability. Spores and preformed toxins, however, survive the storage process, and damaged bacterial cells may be viable after a long resuscitation period. All this means that the raw material should be as free as possible from organisms, that good hygiene is required in the factory and that cross-contamination must be carefully avoided. It is also important to avoid delays in handling products; for example, *Staphylococcus aureus* grows rapidly on vegetables after blanching, and several outbreaks of food poisoning have resulted from frozen peas, potato chips and meat products such as burgers.

Considering the various groups of micro-organisms, it is important to see what the effect of freezing is in general terms. Most of the food poisoning micro-organisms belong to the group referred to as mesophiles, consisting of sporulating, non-sporulating aerobic and non-aerobic bacteria. Many of these have the property of breaking down proteins and are referred to as having proteolytic properties. In general food poisoning organisms do not grow and produce toxin below 10°C; the main exceptions are *Clostridium botulinum* Type E at 3°C, *Staphylococcus aureus* at 6.7°C, *Salmonella* spp. at 6.7°C and the faecal indicators, which include the coliform group (*Escherichia coli* at 3–5°C, *Aerobacter aerogenes* at 0°C and the Enterococci at 0°C). However, exact minimum growth temperatures also depend upon the product.

A second group of micro-organisms are the psychrotrophs (cold seeking) and psychrophiles (cold tolerant) bacteria which grow at lower temperatures. Most psychrophiles produce larger amounts of enzymes at low temperatures than at high. Proteolysis, lipolysis and alcoholic fermentations can occur as a result of the growth of these micro-organisms. The minimum growth temperatures for psychrotrophs are: *Micrococcus* at −1 to +3°C; *Coliaerogenes* at −1.5 to +1.5°C; *Escherichia coli* at −2°C; and *Lactobacillus* at 0 to −2°C. In foodstuffs of animal origin the most frequently found psychrophilic organisms belong to the genera *Achromobacter*, *Aerobacter*, *Alcaligenes*, *Flavobacterium*, *Micrococcus* and *Pseudomonas* together with the moulds *Aspergillus*, *Cladosporium*, *Mucor* and *Penicillium*. Since most of these organisms are aerobic, advantage can be taken of vacuum and nitrogen packaging for meat, poultry and fish, provided that care is taken to ensure that the temperature does not rise sufficiently to allow *Clostridium botulinum* Type E to grow under the anaerobic conditions.

It is generally considered that below −10 to −12°C, microbial growth ceases, or is at the most extremely slow.

Special care should be taken to ensure that correct storage temperatures are maintained; under no circumstances should products be allowed to thaw out, since thawing temperatures are ideal for the growth of many of the contaminating microorganisms. *Salmonella* spp. survive the freezing process and in the case of thawed poultry and pork products, special care should be taken to ensure that they are thoroughly cooked before being consumed. Fluctuating storage temperatures should be avoided, as these encourage microbial growth.

Nutritional composition

Changes in the nutritive value of food products after freezing and during storage are particularly important, especially since the advent of nutritional labelling. The general status of frozen foods is high compared with other methods of preservation, e.g. dehydration and heat processing. Frozen foods are generally recognized as nutritious and relatively similar to their fresh counterparts as purchased in retail outlets.

In vegetables, the most serious losses are of the thermolabile nutrients (ascorbic acid, thiamin and riboflavin), occurring mainly during blanching and subsequent reheating or cooking.

In meat products, the B vitamins are particularly important in the diet. Ageing of meat causes loss of thiamin and to a lesser extent riboflavin. Freezing and rate of freezing appear to have little direct effect, although some workers recommend very rapid (sharp) freezing at −38°C for 2 h to give satisfactory retention of B vitamins in meat and liver. Storage of meat causes some loss of nutrients, often in the drip, which tends to increase with increased storage life. In the literature the results appear variable, partly due to different analytical techniques. It is not easy to determine whether other parts of the process have had an influence on a particular processing variable.

Product, process and packaging factors

Product, process and packaging (PPP) factors are now regarded as a major part of any study or trials of storage stability. Stability, or keeping quality, graphs plot the logarithm of the number of days stored against the time. The relationships are generally, but not always linear. The concept parallels that of 'time–temperature tolerance' (TTT) (q.v.). The value of tests of this type is that they can be used to determine optimum combinations of PPP factors to give maximum or extended shelf life for frozen products.

Time–temperature tolerance factors

Time–temperature tolerance (TTT) studies for quality changes during frozen food storage were carried out extensively and systematically by the USDA Western Regional Research Laboratory at Albany, California in the 1950s. The products investigated included fruits, vegetables and poultry products. Logarithmic relationships of storage time versus temperature were established for various storage temperatures. The criteria used included taste panel evaluation using triangular tests and chemical indices such as chlorophyll conversion, ascorbic acid content and instrumental measurement of colour.

Various terms have been used in relation to quality parameters. These include just noticeable difference, JND; first noticeable difference, FND; first perceptible difference, FPD; stability time, ST; and high quality life, HQL. Most of these refer to trained tasting panels, and consequently more practical terms such as 'acceptability time' have been proposed. The 'practical storage life' of a product is defined as the maximum time a product would remain acceptable to a discerning consumer.

Freeze/thaw stability

Most frozen foods will suffer some physical deterioration if they are subjected to thawing and refreezing. These are often textural changes brought about by the formation and reformation of ice crystals. Fish and meat both suffer under these circumstances and in the latter case drip is often the result. Ice-cream can suffer from crystal formation which destroys its texture. Other effects of freeze/thaw cycles are protein denaturation, starch retrogradation and emulsion breaking.

It is possible to give foods some protection against damage from freeze/thaw cycles by using certain stabilizers. Polysaccharides such as sucrose, sorbitol, carrageenan and modified starches exhibit such cryoprotective properties.

Microwave heating

Frozen foods can be conveniently heated for domestic consumption in a microwave oven. In the UK, microwaves may be generated at two frequencies, namely at 915 MHz for industrial applications and at 2450 MHz for domestic ovens.

Commerical applications are fairly limited. A large scale butter tempering system, thawing 0.3 m³ blocks from -20 to $+2°C$, has proved to be very efficient. This operates at 915 MHz; its power consumption is 90 kW.

Domestic ovens (range 72 W–1 kW) are available either as single units or combined with radiant/convection type ovens. Many of these are relatively sophisticated with microelectronics control systems for programmable cooking, infrared browning devices and rotating turntables. Many frozen food products, including complete meals, are specially prepared and packaged for microwave heating.

The basic physical principle involved is that microwave energy penetrates the whole foodstuff so that the centre point is heated at the same rate as other points. This type of heating is sometimes called 'volumetric' heating. Microwave heating is extremely rapid, especially with foods containing water, since water, especially near the freezing point, has the capability of vigorously absorbing microwaves. Metallic structures when placed in a microwave field tend to cause arcing and it is consequently difficult to measure temperatures accurately using thermocouples. A technique which uses coloured temperature-sensitive crystals has been developed, the signal being transmitted via an optical fibre out of the microwave field. Plastic or ceramic containers are required for reheating food products.

Thawing

This subject is dealt with under specific product headings in Chapters 1 and 2.

HEAT PROCESSING

BASICS

Introduction

Thermal processing of food involves:

(i)   the preparation of the raw material,
(ii) filling the pack, i.e. metallic can, glass bottle or jar, or plastic container,

(iii) removing the entrained air (exhausting),
(iv) seaming or sealing the pack,
(v) processing, i.e. heating followed by cooling,
(vi) labelling, and
(vii) storage.

This is a generalized scheme, the exact details of which depend on the nature of the product and the pack or container. Canning and bottling are more specific titles, as are the processes known as sterilization and pasteurization.

Aseptic processing

Aseptic processing is essentially the packaging of a sterilized or pasteurized product under sterile conditions, so that no recontamination takes place. The principle is limited to liquid or to semi-liquid foods with some particulate material, which can be sterilized in bulk quantities. This is achieved by heating to the sterilizing temperatures, holding at the sterilizing temperatures for sufficient time, conveying to coolers prior to filling and then filling under aseptic conditions.

Ultra heat treatment

UHT or ultra heat treatment processes are designed to sterilize products by exposure to high temperatures for short times. The principle underlying this method of treatment is to sterilize without adversely affecting the flavour of the product. Details of such processes are given under UHT MILK in Chapter 3.

Acidity and pH

Among the important constituents of fruits and vegetables are many organic acids, which while varying greatly in structure and composition, all have the common property of giving rise to hydrogen ions when dissolved in water. The hydrogen ion concentration is a measure of the intensity of acidity and is expressed as the pH value of the solution. Pure water, which is neutral, has a pH value of 7. Increasing acidity is indicated by a progressively smaller pH value; conversely a pH value above 7 is indicative of an alkaline solution. Fruits tend to have a pH value in the range 2.5 to 4.0, while the vegetables are less acid and lie in a range from about pH 5.0 for certain carrots to about pH 6.5.

Acidity is the most important single factor determining the requirements for an effective and safe sterilization process. It also affects the corrosivity of the product in relation to a metal con-

tainer such as a can and hence influences the choice and specification of container and the shelf life of the product.

The heat processing of a 'high acid' product, defined as one having a pH lower than 4.5, may be effected in boiling water, since only a limited range of micro-organisms can grow in such an acid medium and they may all be controlled by heating at 100°C (212°F). On the other hand, 'low acid' foods (as those with a pH value higher than 4.5 are often described) require exposure to much more severe heat treatments in order to achieve microbiological stability and freedom from health or spoilage risk. Typical examples of such products include vegetables, meat and fish. Water or saturated steam under pressure are normally employed as the heating medium in such cases.

Note that 'low acidity' in canned foods has a quite different significance from 'low acidity' in acid pickles. (See under ACETIC ACID preservation in Chapter 9.)

As far as corrosion is concerned, the pH value and the actual type of acid present may be important. As indicated below, the pH of most canned fruits lies between 2.7 and 4.3, and those of most canned vegetables between 4.7 and 6.3.

pH values of canned fruits

These are as follows: purple plums, 2.7–3.3; loganberries, 2.7–3.1; golden plums, 2.9–3.2; Victoria plums, 2.8–3.3; gooseberries, 2.7–3.3; apples (solid pack), 2.8–3.3; damsons, 2.9–3.4; blackberries, 2.8–3.5; raspberries, 2.8–3.6; blackcurrants, 2.9–3.5; greengages, 3.0–3.5; acid cherries, 3.1–3.4; sub-acid cherries, 3.2–3.4; strawberries, 3.2–3.8; sweet cherries, 3.7–4.4. Tomatoes are intermediate between fruits and vegetables, with pH values 4.1–4.7.

pH values of canned vegetables

These are as follows: carrots, 4.6–5.7; celery, 4.9–5.6; vegetable macedoine, 5.1–5.8; beetroot, 5.0–5.7; dwarf and runner beans, 5.1–5.9; spinach, 5.4–6.3; potatoes, 5.4–6.2; broad beans, 5.9–6.1; peas (fresh green), 5.7–6.4; peas (processed), 5.9–6.3.

The hydrogen ion concentration is an important factor in sterilizing canned foods, bacterial spores being much more easily destroyed in fruits at pH around 3.0 than in vegetables at pH 5.0–6.0. Bacterial spores do not germinate and grow at pH values much on the acid side of pH 4.5. As a result, even if they have not been destroyed in canned fruits, they cannot readily develop in the stored cans. On the other hand, it is absolutely necessary to destroy all harmful organisms in canned vegetables as the conditions of acidity are such that they may develop on storage.

Commercial sterility

This is a term used to describe food which has been so processed that it will not spoil or endanger the health of the consumer when kept under normal storage conditions. It is often used to describe canned foods where the process is known to destroy yeasts, moulds and most bacteria but will not destroy thermophilic spores.

Botulinum process

In order to achieve microbiological stability of canned foods or foods packaged in another manner, it is necessary to submit the sealed cans/containers to a thermal or other process that will destroy, or render inactive, all micro-organisms that could cause toxicity or spoilage. Since absolute sterility, that is the complete destruction of all microorganisms present, is seldom attained, 'sterilizing' is usually not a strictly accurate term and it is common practice in the industry to use the word 'processing' to describe the heat treatment applied. The 'scheduled process' is the time during which the product is maintained at the designated temperature.

For canned products, the most important single factor in determining the magnitude of the heat process necessary to confer stability is the acidity or, more properly, the hydrogen ion concentration or pH. Foods can be divided into various groups on the basis of their pH values, but the most important value from the point of view of heat processing is pH 4.5. It is only above this pH (that is, in less acid conditions) that the most heat resistant pathogenic organism occurring in food, *Clostridium botulinum*, is able to develop. Consequently, the minimum process applied to canned foods with pH values above 4.5 (low acid foods) must be sufficient to eliminate any risk of botulism. This invariably means a high temperature process in steam heated pressure vessels. The only exception to this rule is in the very rare case of a product which contains legally permitted chemical preservatives, such as cured meats, in which case somewhat milder processes are applied. It is, of course, theoretically possible to process any product at atmospheric pressure but the times involved, apart from being absurdly uneconomical, would result in most products being cooked to a pulp.

The pasteurization of acid foods, pH less than 4.5, is normally carried out in boiling water or steam at atmospheric pressure for relatively short times. This may be in batch or continuous tank cookers or round shell types which impart agitation.

Pasteurization

This subject is covered under specific food headings in Chapters 1, 3, 6 and 9.

'Pasteurization' is a term which applies to the milder heat treatments applied to certain products, which kill non-sporing pathogens and prolong the shelf life of the food, but are insufficient to confer indefinite microbiological stability. The pasteurization of milk is a prime example of a heat treatment designed to eliminate pathogens from the food. In canning, heat treatments are usually greater than those that could properly be described as pasteurization, which is applied to products which rely to a large extent on other preservative factors for their stability. Examples include pickles and sauces containing acetic acid or carbonated soft drinks and beer of low pH.

Yeasts

Due to their low heat resistance – a few minutes' exposure to a temperature of 74°C (165°F) is all that is necessary to kill them – yeasts are rarely involved in spoilage of canned foods unless packs are grossly understerilized.

Process evaluation

In order to establish a process for any particular product, it is necessary to measure the rate of heat penetration to the slowest heating point in the can. In conduction heating packs this is normally the geometric centre of the product; where heating is entirely or partly by convection, it is usually a little lower. Heat penetration determinations are made using cans fitted with needle-type thermocouples which record the temperature at their tips, these being located at the slowest heating points. Direct temperature reading, self-recording potentiometric instruments may be used to plot a graph of the temperature versus time. Using the information thus obtained, it is possible to integrate the lethal effect of any process with respect to thermal death times of any particular micro-organism. It is modern practice to calculate this in terms of the destructive effect on *Clostridium botulinum* of an equivalent number of minutes at 121°C (250°F); this is expressed as the F_0 value. So, for example, the heat penetration graph might be plotted for a process of 60 min at 115.5°C (240°F). This is then converted into a graph based on lethal rates (lethalities compared with the effect on *Cl. botulinum* of 1 min at 121°C (250°F)), which is then integrated to give a total value of, say, 8 min. This would mean that the process has an F_0 value of 8, or it is equal in destructive effect on *Cl. botulinum* to 8 min at 121°C (250°F), assuming instantaneous heating and cooling. This expression of process in terms of an F_0 value is, of course, quite arbitrary but it is a useful way of expressing it in simple numerical form and for comparative purposes.

The rate of heat penetration into products is influenced by many factors such as product formulation and homogeneity, fill and residual headspace, ratio of solid to liquid, position and stacking of cans in retorts and amount and type of agitation. Agitation has an effect which is most marked with the 'semi-conduction' heating of products. Those heated by convection usually need no agitation and conduction heated products benefit little unless the headspace 'bubble' can be induced to move through the product, thereby causing a stirring effect.

Having obtained heat penetration data, it is still necessary to determine what lethal effect of F_0 value any particular product requires in order to stipulate a satisfactory process. There is one basic criterion which is that all canned foods above pH 4.5 must be given the minimum safe process or botulinum cook. The only exceptions are those containing inhibitory ingredients such as some cured meats, cakes and condensed milk. For all practical purposes, the botulinum-safe process is one having an F_0 value of 3. This process will not, however, necessarily ensure freedom from spoilage by organisms more heat resistant than *Cl. botulinum* and in order to decide what F_0 value a product requires other information is necessary. This may be obtained by bacteriological examination of raw materials, pack inoculation studies or previous experience. Usually some combination of all three sources of knowledge is used to decide upon a process.

Almost invariably, a process is determined by the requirements of microbiological stability and is laid down as the minimum necessary to achieve this while minimizing the cooking effect on the product. Only extremely rarely is the process needed for adequate cooking greater than the process needed for microbiological stability. Heat penetration data can be mathematically converted

from one can size to another, or from one retort temperature to another.

RETORTING

Equipment

The basic pressure processing vessel is the batch retort, which may be either vertical (see Figure 15.3) or horizontal according to cannery layout and space available. It should be fitted to receive steam and water and also compressed air, when cans requiring pressure cooling are going to be used. Vents should be at least one pipe size larger than inlets. Regular maintenance is necessary to ensure efficiency of retort operation and accuracy of instrumentation.

Crateless retort systems have been developed in which cans are tumble packed into the top of the retort where they fall into a hot water 'cushion'. This water is pumped out prior to processing in steam. The retorts are refilled with water at the end of the process and the cans are discharged through a hydraulically operated door into a cooling canal. Cans of 10 cm and larger diameter are partially cooled prior to discharging to avoid peaking.

Other batch retort systems operate by processing the container either under steam heated water with an air overpressure, where necessary, throughout the process (e.g. Rotomat-type retorts) or in an intimate mixture of steam and compressed air (e.g. Lagarde-type retorts). Such systems have the advantage that they lend them-

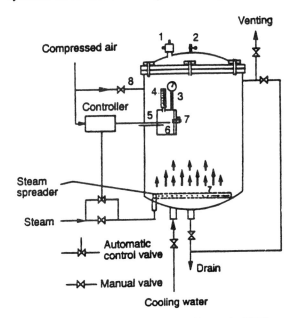

Figure 15.3 Vertical retort (from Holdsworth, 1996).

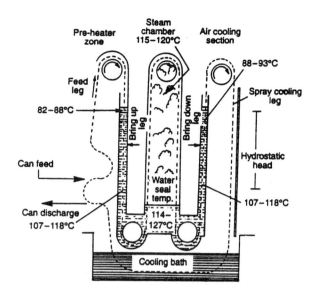

Figure 15.4 Hydrostatic cooker/sterilizer (from Holdsworth, 1996).

selves to the processing of glass containers and certain retortable plastic containers as well as conventional metal cans.

Continuous pressure cookers are those into which cans are introduced through a valve or port and in which a constant steam pressure can be maintained, with consequent advantages in the saving of space, steam and manual handling, and in less variability in treatment of the cans. Continuous cookers are available which impart agitation. These are usually of the reel type, where cans roll on the outer retort casing during about one-third of each revolution. They are most suitable for convection heating packs and those conduction heating products that show increased heat penetration with this type of agitation. The main disadvantages of continuous cookers are, apart from the initial capital expenditure, inflexibility with regard to can size and losses incurred with breakdowns. With modern equipment, however, the latter are rare and provision is made for the rapid freeing of any jams due, for example, to damaged cans.

Hydrostatic sterilizers (see Figure 15.4) have been in use for some years and because of their flexibility and mode of operation are now widely used for large scale canning operations. In these cookers, water columns are used to balance the pressure in the steam chambers, thereby eliminating the needs for transfer valves or ports, and changes in pressure are gradual. The latest hydrostatic sterilizers are capable of handling over a thousand cans per minute. Models which impart can rotation are available.

Dry heat as a means of raising the can tempera-

tures for sterilization has attractions in that heating rates for convection packs may be increased and since this may be done at atmospheric pressure, the equipment may be simplified. This concept is used in the French Steriflamme sterilizer. A disadvantage is that stronger and therefore more expensive cans may be necessary in larger sizes.

Processing procedure

Heat processing consists of three stages: retort come-up (which includes 'venting'), time at retort temperature and cooling. The come-up stage is eliminated in continuous cookers with direct can entry through a transfer valve, and in hydrostatic cookers it is automatically taken care of by cans passing through the first water column. In batch cookers, however, the period of coming-up to retort temperature is very important. It is here that venting, that is eliminating air from the retort (rather like the exhausting of cans), must take place. Air is a very poor thermal conductor and if left in the retort may form pockets, thus insulating some of the cans which will as a result not be properly processed.

Report instrumentation

Retorts are normally controlled by temperature recording instruments, which operate the steam entry valves by compressed air. It is, of course, essential that they function efficiently and accurately. Retorts should also be fitted with pressure gauges and thermometers, which are usually of the mercury-in-glass type but may also be platinum resistance. All these instruments should be regularly checked against each other, and at intervals against an independent standard thermometer. With accurate instruments, disagreement between thermometers and pressure gauges during the actual heating process is a sure sign of air in the retort due to inadequate venting. Appropriate action should be taken to rectify this situation. Unfortunately, the corollary, that agreement indicates effective venting, is not necessarily true since air can gather in pockets which may not coincide with the gauges.

TEMPERATURE CONTROL

It is invariable practice to control all heat sterilization vessels by means of temperature. Various types of temperature controlling equipment are available, operating on the principle of an indicating thermometer or thermocouple located within the retort shell or in an external vented well attached to the retort. This thermometer is connected to a continuous recorder, which in turn controls the operation by compressed air of a valve controlling the flow of steam to the retort. Sufficient steam is allowed intermittently through the valve to maintain the desired temperature. Retorts should also be fitted with pressure gauges so that a constant cross check between pressure and temperature is possible. It is essential that all equipment is maintained in good order and regularly checked for accuracy against a standard thermometer.

In hydrostatic sterilizers, a pressure differential control system automatically adds water or drains off water from the legs to balance the pressure, and hence temperature, inside the steam chamber.

Closing temperature

The temperature of the contents at the time the can is closed is important for several reasons, and therefore should be chosen and controlled carefully. It will affect the final vacuum in the can, may influence shelf life, particularly of fruit packs, and must be standardized if equal heat treatments are to be given to different batches of the same product.

Choice of closing temperature is dependent upon the type of product and its preparation techniques. Hot filling is not always possible and under such conditions some alternative method of obtaining a vacuum in the can should be used.

HEADSPACE IN CANS

Control of headspace is important for economic and technical reasons. Underfilling will contravene Weights and Measures legislation and may cause rusting in the headspace, while overfilling may seriously affect the efficiency of the canning operations. The adequacy of exhausting, particularly if steam injection is employed, may be reduced and this may lead to straining or permanent distortion of the can ends during heat processing. In agitating cookers the movement of the headspace gas is used to provide disturbance of the contents and to assist heat transfer within the can. With acid products, which produce hydrogen by corrosion of

the can during storage, the headspace is the space reservoir for the hydrogen to fill before the can becomes blown. In these and other ways an adequate and controlled headspace is important. The usual target is 5–6 mm in the closed cooled can.

One canned product to which the foregoing does not apply is sweetened condensed milk which is not heat sterilized but relies on its high osmotic pressure to prevent microbial outgrowth. A completely filled container is aimed for with this product, since the presence of an enclosed air atmosphere is conducive to mould growth and 'buttons' of osmophilic moulds may grow on the surface at air bubbles.

The composition of headspace gases in sound cans varies considerably and alters during storage. In the case of canned fruits, the oxygen disappears rapidly (generally within one week of canning) in plain cans and less rapidly in lacquered cans (usually after four to six weeks' storage). At this stage, the gases consist almost entirely of nitrogen and carbon dioxide, the latter usually being present to the extent of 10–35%, except in the case of repacked fruits, such as fruit salads, when it is present in much smaller quantities. Experiments have shown that the tissue gases of fruits contain very little oxygen and consequently the small percentage of this gas found in newly packed cans must be attributed mainly to air entrapped at the time of closure. After the oxygen has almost or entirely disappeared, hydrogen may start to form and in the course of time, the can may become a hydrogen swell.

In the case of canned vegetables, the oxygen content falls gradually when the cans are held in store. The carbon dioxide content will depend on the pretreatment of the vegetables (i.e. blanching and exhausting). Hydrogen swells very seldom form in canned vegetables, but the presence of small amounts of hydrogen in the headspace gases can often be detected by analysis.

As far as milk, cream and most meats are concerned, the carbon dioxide content will depend on the pretreatment. In most cases the percentage of this gas is small. Hydrogen only develops very slowly, and hydrogen swells are practically unknown in these types of canned foods.

VACUUM IN CANS

In general, the presence of an adequate internal vacuum in cans is a sign of good canning practice. Notable exceptions are carbonated beverages, which naturally have an internal pressure and milk products, which have little or no vacuum. Apart from anything else, the presence of even a low internal vacuum ensures that the can ends remain concave during possible fluctuations of temperature during storage, so that there is no possibility of confusion with cans that have spoiled and have their ends distorted by the pressure of carbon dioxide produced by fermentation or bacterial spoilage.

The level of vacuum required depends primarily on the corrosive nature of the product and its susceptibility to oxidative deterioration. Fruits, generally being highly corrosive, require a high vacuum of about 250 N m^{-2} or above, so that there is a reservoir for the hydrogen eventually formed during corrosion. It should be noted, however, that cans of 103 mm diameter or greater, and particularly A10 cans of 157 mm diameter, have an increasing tendency to buckle inwards, or panel, with increasing vacuum. With these can sizes, vacuums above about $250–270 \text{ N m}^{-2}$ should not be exceeded unless the cans are made of special plate or have bodies specially strengthened by beading. Products other than fruit usually do not need such a high vacuum, although it may be beneficial in terms of product quality, particularly colour, and to minimize certain corrosive effects such as internal headspace rusting and black discoloration.

Vacuum in cans is normally measured using a Bourdon-type gauge, which is fitted with a sharp, tapering, hollow needle projecting through a rubber bung. Using hand pressure, the needle can be made to pierce the tinplate whilst the bung acts as a seal. In cans with very small headspaces, the readings so obtained may be several centimetres below the true reading because of the diluting effect of air within the gauge tubing. If accurate readings are required on low headspaces, a gauge known as the FIRA gauge is available which compensates for this effect and allows a true reading to be obtained using a specially designed, simple-to-operate calculator.

It should be remembered that the vacuum gauge is actually a pressure differential gauge and simply measures the difference between the pressure within the container and that of the external atmosphere. Different readings would thus be obtained for the same container when the pressure of the external atmosphere changes (due to, for example, the altitude above sea level or to differing weather conditions).

Vacuum in canned foods may be created by the mechanical removal of air or by heating the product before closing.

Vacuum filling and closing

Vacuum filling

Vacuum filling of solids is essentially reserved for paste-type meat products such as luncheon meat and meat spreads. A vacuum filler is normally coupled to a vacuum mixer and holds the meat product under vacuum prior to extruding it as a solid tube into the empty can, which has previously been evacuated of air in the filling head chamber. Such products are normally filled cold and closing the cans is normally performed by a vacuum closing machine.

Vacuum syruping

Vacuum filling of liquids is normally known as vacuum syruping. The purpose of the vacuum syruper is to remove occluded and entrapped air from products before the addition of the covering liquid. The operating sequence is basically as follows: cans filled with their correct weight of fruit, for example are conveyed into a vacuum chamber where a high vacuum is drawn on the cans. Syrup is then added to the can to a predetermined level, giving a consistent headspace in the cans. The can is finally vented to the atmosphere and leaves the unit to be conveyed to the seamer. Some machines use a double vacuum sequence, the second vacuum being drawn just before the syrup fill is complete, when it serves to draw off excess syrup and ensure a uniform fill.

Apart from removing air from the product, the vacuum syruper offers economies in syrup consumption and consistent headspace; the latter assists in the attainment of consistent in-can vacuum levels using a vacuum seamer or a seamer fitted with steam flow closing equipment.

Vacuum closing

Vacuum closing of cans is achieved using a vacuum seamer, which is, in some cases, preceded by a clinching unit. Basically, the vacuum seaming operation consists of (i) placing or loosely clinching the end on to the can, normally filled with cold product, so that the air may be drawn from the can be the application of an external vacuum, (ii) transferring the can by means of a suitable transfer valve into an evacuated chamber (or chambers) where the air in the can is withdrawn, and (iii) double seaming the can whilst it is still maintained in an evacuated chamber.

Compared with hot filling/heat exhaust techniques, vacuum seaming offers the advantages of being able to handle products which are difficult to heat due to their solid nature or slow conduction heating characteristics, or products which cannot be heated before closing and processing due to their texture etc.

High vacuum packing technique

The high vacuum packing technique has been used for over 50 years in the preparation of canned products such as whole kernel sweet corn, and it is now being considered for a wider range of vegetables and some fruit products. Basically, the technique entails the addition of a very small quantity of water or brine to the product-filled can, and the can is then seamed in a vacuum seamer to give a vacuum of not less than 78 kN m^{-2}. Heat transfer during sterilization, which is now usually carried out in continuous agitating cookers, is achieved by means of the water vapour given off by the boiling brine in the can condensing on the product surface. Because of the high vacuum in the can, the brine boils at a low temperature and, coupled with the rotation of the can in the cooker, rapid heat penetration and low process times are achieved.

The principal advantages claimed for the high vacuum packing technique over conventionally canned foods are: (i) a saving of weight of the filled can by as much as 25–45% due to the reduction in the quantity of brine added, (ii) more rapid reheating of the can contents, when required, prior to consumption, and (iii) higher retention of nutrients and higher organoleptic quality. As regards disadvantages, stronger can bodies using higher strength tinplate and/or beading may be required to increase collapse resistance and so withstand the high vacuum levels achieved.

Exhausting

It is desirable for processed food cans to be closed with a partial vacuum in the can to reduce the strain on the can ends during heat sterilization and cooling and to present the concave end appearance during shelf life even at elevated temperatures. The cans are 'exhausted' in the sense that the air and other entrapped gases are partly removed before they are closed.

In addition to the purely mechanical objectives of exhausting referred to above, there are other important advantages, mostly resulting from the reduction of the oxygen content which is so achieved. Oxygen may cause deterioration of certain canned foods – loss of vitamin C and discoloration of certain lightly coloured products, for example. Perhaps more important, however, is its effect as a promoter or accelerator of can corrosion. Rusting in the headspace, acid attacks on plain (i.e. unlacquered) tin surfaces and on iron exposed by damage occurring to the tinplate during can manufacture and closure, are all affected by the amount of oxygen enclosed in the can. With those fruit packs which produce hydrogen in the can by acid attack on the steel base of the tinplate, the initial vacuum level frequently controls the shelf life. When sufficient hydrogen is produced to give rise to a greater than atmospheric pressure inside the can, the can becomes blown and is unsaleable. Apart from the influence of oxygen in increasing the rate of corrosion, the amount of space available inside the can to be taken up by the hydrogen will define the shelf life before it blows. Hence control of headspace and high vacuum levels are desirable for acid fruit products.

Exhausting can be achieved by several methods and the choice is affected by the nature of the can contents. A thin liquid will rarely occlude gas below its surface and only the headspace air needs consideration. Viscous and semi-solid products may contain considerable entrapped air when filled into the can, while fruit and vegetable tissue may contain carbon dioxide derived from respiratory processes.

For fruits and vegetables the syrup or brine is normally filled as hot as possible (ideally 82°C (180°F) or above) so that air from the headspace is expanded and partially displaced by steam from the hot liquid. For some vegetables, the vacuum obtained in this way may be enough. Can closing is today frequently preceded by steam injection, a much cheaper and more convenient method in which the can-closing machine is equipped with jets of steam which sweep the air from the headspace as the can is closed. The vacuum so obtained is dependent upon the level and upon the smoothness of the surface of the can contents, as well as upon the setting of the injection jets and the steam pressure. For good results, therefore, the filling process, both temperature and volume of contents, must be under close control. Since steam injection (or steam flow closing, as it is often called) does not eliminate gases trapped below the surface, it is frequently used in conjunction with vacuum syruping in the case of products such as halved peaches which may trap air. Steam injection needs more filling-line control and will still give a wider spread of vacuum levels, but cost and convenience recommend it increasingly.

Many meat and fish products are vacuum closed, that is, the double-seaming operation takes place inside a vacuum chamber. The necessity to enclose the can at the seaming head and to draw the necessary vacuum makes the process more complex and slower, but for any semi-solid products it is the method of choice because it can deal with cold filled materials which have air trapped within the product.

CLINCHING

One method of air removal from the can prior to closing that is not frequently used is to heat the filled can and its contents in steam or hot water before double seaming. This is best done with the lid loosely attached so that, while the air may come out of the can, loss of contents is minimized and undesirable dripping of condensate into the can is prevented. This loose attachment of the end is known as clinching and represents the beginning of the double seaming operation with the formation of the cover hook extending only as far as is necessary to hold the lid in position, leaving the end freely rotatable. Special machines are available to effect clinching.

COOLING

Once their contents have been sterilized, it is desirable to cool the cans as rapidly as possible. This is normally effected with cold water, the bacteriological condition of which must be carefully controlled.

With batch retorts, the cooling water is introduced at the end of the steaming cycle. As the steam condenses the pressure surrounding the cans falls rapidly, more rapidly than the pressure inside the cans, causing considerable strain on the can ends. Although the ends are designed with concentric expansion rings to enable them to move in and out, too great a strain may result in permanent deformation or 'peaking'. The strain will be increased if the can is overfilled or if it was inadequately exhausted before closing, and the larger diameter ends have less intrinsic resistance to peaking. With good cannery control, deformation due to poor can filling and closing technique can be avoided but with the larger cans it is necessary to resort to pressure cooling, in which the counterbalancing pressure of steam in the retort is replaced by compressed air before the cooling water is introduced. As cooling proceeds, the air pressure is allowed to fall gradually. Pressure cooling is generally necessary for cans of 10 cm

diameter upwards, and may also be needed for smaller diameters at retorting temperatures above 115.5°C (240°F).

Another feature of cooling in batch retorts is that uneven cooling may occur, with a marked temperature gradient among the cans from bottom to top depending upon the direction of flow of the water. It is usual to flood the retort from the bottom, with an overflow above the level of the cans: if the flow of water is not rapid, then not only will the top cans remain above the water level for some time, but also for much of the cooling cycle the water near the inlet will be much colder than that at the top near the overflow. This effect can be minimized by having an adequate supply of water to fill the retort rapidly and also by spraying water from the top of the retort as well as from the bottom.

The objective in cooling cans is to reduce the temperature of the contents to a level at which cooking and quality deterioration stop, and below the range at which THERMOPHILIC BACTERIA, which may have survived the process, can grow out and spoil the product. Ideally this would mean cooling to about 21–27°C (70–80°F). However, since the cans are wet with the cooling water, there is an advantage in leaving some heat in the cans to accelerate evaporation, since surface water can cause rusting and also adversely affect can-line hygiene and labelling efficiency. A compromise cooling target temperature is therefore chosen, usually about 38°C (100°F). If the baskets of cans are tipped to an angle of about 45° from the horizontal on removal from the retorts, most of the water will drain from the countersunk end and they will dry fairly rapidly at this temperature. In continuous cookers, cooling conditions are somewhat better controlled than in batch cookers and rapid pressure changes and temperature variability are generally more easy to avoid.

COOLING WATER

When cans are cooling from a high temperature they contract considerably and undergo changes in internal pressure. In such circumstance even well-made seams may permit slight inward leaks. It is therefore essential that the water used for cooling should be as near sterile as possible, and in practice, water carrying more than 50–100 micro-organisms per ml is regarded as unsuitable. This means that unless the water comes direct from a reputable drinking water supply and is used once only, it will probably need disinfecting before use. Even good mains water may require pretreatment if it is stored in a tank in the cannery before use, because bacterial buildup may take place under such conditions. Water from deep wells is often of good bacteriological quality, but other natural waters may be heavily contaminated. With the high cost of water, its recirculation and re-use becomes more attractive and careful disinfection is essential.

The disinfectant of choice is chlorine, added either as a gas or as liquid sodium hypochlorite solution. Bacterial destruction is not instantaneous and so at least 20 min contact with an adequate level of chlorine is necessary for the desired effect. Since other organic substances present in the water will react with the chlorine, it is important to ensure that sufficient addition is made to maintain a small residuum of free chlorine right through to the time that the water is used. Too high a residual, apart from being wasteful, can increase the corrosiveness of the water.

Satisfactorily treated water will have a low bacterial count, and such determinations must be carried out when establishing the chlorine dosage routine and at regular intervals as a routine check. However, the day-to-day control will be more conveniently and immediately established by chemical determination of chlorine. Confirmation of correct dosage is obtained by total chlorine determination and the adequacy of this dosage rate is shown by the presence of free chlorine in the water as it enters the can cooler. Both checks may be made with sufficient accuracy by rapid colorimetric tests, e.g. using 'Chlorotex' reagent (British Drug Houses Ltd.), for total chlorine and DPD (diethyl-paraphenylene diamine) for free chlorine.

The minimum contact time of 20 min will be ensured by having sufficient storage capacity and by making certain that the water flow through the tank does not short circuit by placing inlet and outlet points at opposite ends of the tank and inserting baffles to direct the flow if necessary.

In recirculated water systems, a cooling tower is usually incorporated and its construction should be such that it can be readily cleaned. With re-use, the chloride content of the water tends to build up, as do also other soluble residues washed off the cans. Some supply of make-up water into the system is always necessary, but dependent upon the extent of this, the whole supply should be drained and refilled at regular intervals.

CODING

It is proper practice for the canner to imprint a code on the end of the can that is attached so that the product can be identified subsequently. The code normally indicates type of product and date of manufacture but may also include details such as the line on which the product was packed and the shift or time of day. Since one of the uses of the code is to enable isolation of stocks which for some reason are below standard, then the more detail the code contains the more efficiently this can be done.

Coding has traditionally been carried out by embossing the lid. Close control of the depth and regularity of the embossing dies is necessary if, on the one hand, the desired legibility of the code is to be obtained without, at the other extreme, damaging the plate or lacquer coating. Nylon dies are much less likely than metal dies to damage the ends. However, it is increasingly the practice to use ink jet printers to code cans. These have clear advantages: there is no danger to the internal lacquer of the can, the codes produced are very clear and the information possible is increased. Often the can is coded with an expiry date and an accurate time of manufacture.

DRYING

There are several important reasons why cans should be dry before they are removed from retort baskets or before they pass from continuous cookers to labellers and casers. In the first instance, labelling efficiency may be higher with dry cans and the risk of can rusting during storage will be reduced. There are additional hygienic reasons. Wet cans will deposit water on the can tracks leading to and though labellers and case-packers and bacterial buildup, impossible on dry tracks, will take place. If cans run on their double seams along wet infected tracks, leaker spoilage levels will rise, particularly if the cans are roughly handled in the tracks. With wet cans unloaded by hand from retort baskets, the leaker spoilage risk is complicated by the possible involvement of pathogens, staphylococci and salmonellae for example, from the hands of the operatives.

The drying rate of cans in retort baskets will be dependent upon the temperature of the cans (about 35 to 38°C (95 to 100°F) on removal from the retort is desirable) and the efficiency of draining of water from can countersinks by tipping the baskets.

It has been shown that if the cans in the baskets are dipped into a hot detergent bath after removal from the retort, then their subsequent drying rate is greatly increased. On continuous cookers, a can dryer should be sited as near as possible to the cooker outlet, and if this is impracticable, the cans should be conveyed on end-on nylon covered rope conveyors to avoid double-seam infection until they are dry.

ASEPTIC PROCESS

Heating methods

Aseptic canning

The process of aseptic canning was invented by Dr W.M. Martin and developed by the Dole Engineering Company of Redwood City, California. In this process, in effect, the cooking and sterilizing aspects are separated. The product, ready for canning, is heated in a continuous flow to a temperature around 149°C (300°F), at which temperature sterility is achieved in a few seconds. This ultrahigh temperature short time (HTST), also known as the ultra high temperature (UHT) process, especially in the case of milk, is possible because the rate of flow of product through the sterilizer is so adjusted that a only a small quantity is present in the heating and cooling sections. Thus, heat may be introduced and removed from the product very rapidly. High temperature steam is the usual heating medium. This may be injected directly into the product; if necessary, an equivalent amount of water vapour may be flashed off under vacuum at a later stage in the process. Alternatively, the heat from the steam may be applied indirectly through a heat exchanger.

A typical process may be of the order of 149°C (300°F) for 2 or 3 s with the product cooled immediately to around 16°C (60°F) for filling. In the Dole equipment, the cans are presterilized as they pass along to the filler through a tunnel heated with superheated steam or hot gas at a temperature of around 232°C (450°F) (the melting point of tin is the limiting factor in this part of the plant). Filling and double seaming of the cans are carried out under aseptic conditions in a steam-filled chamber, the can ends having received a similar dry heat sterilization to the empty cans in a special magazine attached to the seamer.

In this way heat degradation during sterilization is reduced to a minimum and product of similar quality can be obtained in small or large cans. Because the sterilization is a continuous process, the product must be sufficiently fluid to be pumped through the system. With UHTST processes, care must be taken to ensure that enzyme destruction has been achieved as well as sterility,

otherwise undesirable chemical changes in the product will occur on storage.

The APV 'Jupiter' system has the capability of sterilizing particulate material of up to 32 mm diameter. The solids are processed in an aseptic double-cone pressure vessel and the sterile liquid component is added and mixed with the solids prior to filling. However, problems of filling particles of such size aseptically have still to be overcome.

Other heating methods

Foods for aseptic processing may also be sterilized using HEAT EXCHANGERS, notably plate heat exchangers, tubular heat exchangers and scraped surface heat exchangers. Another option is OHMIC HEATING.

Heat exchangers

Heat exchangers have been in use in the food industry for many years as an efficient means of heating, cooling or both. The following are the most important currently in use.

Plate heat exchangers

These have been in use for many years primarily for liquids such as milk and juice. They are very efficient in operation, take up less space than other alternatives and are less expensive. However, since the product passes through a narrow gap between plates, this type of exchanger is not well suited to handling very viscous products or products containing particles.

Tubular heat exchangers

These are available in a number of configurations which include single tube, multiple concentric tube and parallel tube systems. For many years their use was restricted because the industry favoured plate heat exchangers. However, the advent of aseptic processing has revived interest in this type of heat exchanger. One of their advantages is that they are operated at high fluid velocities, which reduces the possibility of product burning on the walls and permits long product runs before cleaning is necessary. The short residence times for the product also mean that quality is likely to be better.

Scraped-surface heat exchangers

These have already been referred to under COMMERCIAL FREEZERS in relation to their use in ice-cream manufacture. Sometimes called swept-surface heat exchangers, they are expensive and require motive power. They are used for viscous products, such as thick sauces or foods containing particles or fibre. Revolving blades continually scrape the product off the heated tube walls, making them suitable for processing such products.

Ohmic heating

Ohmic heating is based upon the principle of passing an electric current through an electrically conducting product; as with microwave energy, electrical energy is transformed into heat. The advantage over microwave heating is that heat penetration is virtually unlimited. Temperatures of between 120 and 140°C are achievable and at much faster speeds than traditional cooking methods.

Products containing a large proportion of particulates, such as complete ready meals, can be effectively heated in this manner. Thus, ohmic heating is very useful for aseptic processes and for bag-in-box packaging and some very high quality meals with a long shelf life (two years) have been produced by this method.

Ohmic heating systems are not widely used but are likely to become more common in the next few years. In the UK the Government gave clearance in 1992 for their use in the preparation of ready meals; their use is therefore restricted to products of this type at the time of going to press.

FOOD QUALITY ASPECTS

Thermophilic bacteria

As the name implies, this is a group of bacteria which flourish at temperatures above the normal growth range; they have optimum growth temperatures of about 55°C (113°F). Many do not grow at temperatures below about 40°C (104°F) and these are known as 'obligate' thermophiles, while others, so called 'facultative' thermophiles, may have a growth temperature range extending down to 25°C (77°F). Thermophilic spore-forming bacteria are important in canning because their spores have high heat resistance and it may be necessary to extend the heat process excessively to eliminate them from a pack. Fortunately, the obligate thermophiles tend to have the most heat resistant spores and provided that cans are properly

cooled after retorting, the presence of residual spores of this type is of no significance for storage in temperate conditions since the growth temperature range will not be reached.

Spoilage due to thermophiles may lead to blown cans or to so-called 'flat sours' in which, while the contents have gone acid or sour there has been no significant gas production to distend the can.

In view of their heat resistance, destruction of large initial loads of thermophiles by extending the process is normally not acceptable as it may lead to excessive heat damage to the can contents. Attention is therefore paid to keeping the raw material infection low, by choice of materials with low counts of thermophiles (a canners' grade of sugar, for example, with an accredited count, is available) and by attention to plant hygiene in situations where thermophile buildup can occur, such as in blanchers, filler hoppers, and so on. As mentioned in COOLING of cans, adequate cooling of processed cans to about 38°C (100°F) will minimize risks of outgrowth of thermophilic survivors.

Tin in canned foods

In the UK, a legal maximum tin content of 200 ppm of can contents has been recently introduced. Experience shows that with lacquered cans, tin contents rarely exceed 30 to 40 ppm even after prolonged storage. It should also be noted that various other national regulatory bodies throughout the world are reducing, or considering a reduction in the permitted acceptable levels of tin in foods.

While fruit products in plain cans do not often contain more than 200 ppm, this dissolved tin is not evenly distributed throughout the contents, being more concentrated in the solid fruit or vegetable. In plain cans where a large surface of tin is exposed to attack, the presence of oxygen or other oxidizing agents such as nitrates will lead to extensive tin dissolution. Thus, close control of headspace and vacuum levels with canned fruits is essential. Until recently the only serious manifestation of detinning accelerated by nitrates has been in products containing nitrate-cured meat, particularly in an acid medium such as bacon with beans in tomato sauce. Local detinning with discoloration of the product has also been seen with cured meats at normal pH levels. More recently, however, in many parts of the world excessive use of nitrate fertilizers has led to unusually high levels of nitrate in certain vegetables and resulted in rapid and extensive detinning of plain cans which had been used traditionally without trouble. Some acid products have been particularly implicated in this problem.

There have also been one or two isolated outbreaks of heavy and rapid detinning of plain cans by formulated citrus drinks, where the concentrated juice has been diluted for canning with a water abnormally high in nitrate content. In at least two of these outbreaks, the tin content caused illness due to acute gastrointestinal irritation.

Nutritional values of canned foods

The practical canner will wish to know how to avoid unnecessary losses of nutritive material at the various stages of the canning process. Losses may occur at any or all of three stages: (i) between picking or gathering the crop and its arrival in the canning factory; (ii) during the canning process; and (iii) on storage of the cans and subsequent preparation of the contents for the table. The common practice of canning fruits and vegetables within a few hours of picking ensures that the loss of nutritive material is kept to a minimum during the first stage. In this respect, the canner is generally in a much better position than the domestic buyer of 'fresh' fruits or vegetables, but there may be a small loss of vitamin C, particularly if the fruit is ripe or the vegetables are allowed to wilt. The losses during the next two stages are discussed below under the headings of the chief nutritive constituents. No account is taken of the nutritive material discarded during peeling, slicing, podding, and so on. Losses in practice are not likely to exceed the figures given. It will be seen that blanching can be the most serious cause of loss of water-soluble nutrients: in this respect steam blanching is preferable to water blanching.

Sugars

There is no loss of sugars at any stage during the preparation or storage of canned fruits, except for those which are blanched to soften them for filling into 'solid packs'. Up to 30% may be lost during the blanching of fresh peas, sliced green beans and other small vegetables of large surface area; the bulkier vegetables, such as whole carrots, turnips, and so on, may lose about 10% of their sugars during blanching. There is no further loss during the canning processes or on storage, but 30–40% may be wasted if the covering liquid is discarded when the vegetables are prepared for the table.

Starch

There is a slight loss of soluble starch from some vegetables, notably peas and beans, during blanching, but this is too small to be of any account.

Protein

The smaller vegetables, such as fresh peas, sliced green beans, and so on, may lose up to 20% of their protein during blanching, but the bulkier vegetables seldom lose more than 5%. The amount of protein wasted by discarding the covering liquid when preparing vegetables for the table is not known accurately, but is probably about 20–30%.

Mineral substances

Losses here are comparable to those for sugars (see above), arising from water blanching and by dissolution into the covering brine or syrup in the can.

Vitamin A

There is little vitamin A in fruits, but its precursor, carotene, is present in considerable quantities in some vegetables, notably carrots and the leafy green vegetables. This vitamin is practically insoluble in water and is stable at high temperatures. The loss of carotene during canning is negligible.

Vitamin B1

Legumes are a fairly rich source of vitamin B1. The vitamin is soluble in water and is rather unstable at high temperatures. The losses by leaching during blanching are of the order of 25%; the heat treatment probably causes a further loss of 15–25%, except in beans in tomato sauce, where the long process at a high temperature causes further destruction. The usual wastage of about 30% will occur if the covering liquid of canned vegetables is discarded.

Vitamin B2 complex

Riboflavin is soluble in water, but is relatively stable to heat; losses of the order of 25% occur during blanching, but there is little destruction during processing. Wastage will occur if the covering liquid is discarded.

Vitamin C

This is the principal vitamin of fruits and vegetables. It is soluble in water and is susceptible to oxidation, particularly in the presence of oxidases or metallic catalysts. The losses during fruit canning occur mainly by oxidation during cooking; the amount of oxygen in the can is the main controlling factor and a high closing temperature and a small headspace reduce the loss to a minimum. Canned fruits generally retain 65–95% of the vitamin C of the raw fruits. The losses of vitamin C on storage of canned fruits appear to be very small, except in the case of some red or purple berries. In canned vegetables the chief loss occurs during blanching, the average loss at this stage being about 30%. Losses during cooking are generally small, and most vegetables retain at least 50% of the vitamin content of the raw material. Losses during storage appear to be negligible, but wastage through discarding the covering liquid may amount to 30–40%.

DEHYDRATION

BASICS

Introduction

Dehydration as a method of food preservation arrests the elements of natural decay by depriving micro-organisms of the moisture necessary for them to remain active. In addition, chemical reactions within the foodstuff generally are retarded. It is not a sterilizing process and means must be provided to preserve the equilibrium and prevent the foodstuff from regaining moisture until such time as deliberate reconstitution is sought. Dehydration implies more than 'drying' in that the capability must remain of restoring the material rapidly to a condition resembling its original fresh state, in which its required qualities can be recognized, even many years later. Dehydration brings with it a reduction in weight and frequently in volume which gives dried products an advantage during storage and transport. Their convenience in further processing and in the home assures them of a sizeable market.

Moisture content

The (percentage) moisture content of unprocessed foodstuffs is usually given per unit of dry matter and moisture combined, i.e. on a wet basis,. After drying, however, many foodstuffs have their moisture content quoted on a 'dry-weight' basis, i.e. per unit mass of their bone-dry solids. This is convenient as the denominator remains constant whatever the condition. Confusion can sometimes arise unless the basis is made clear.

Psychrometry

In saturated air, the partial pressure of the water vapour is equal to its vapour pressure P_0. In unsaturated air, on the other hand, the partial pressure P is less than P_0 at the same temperature. The ratio P/P_0 (usually expressed as a percentage) is termed the relative humidity. In order that a food material may transfer moisture to the surrounding atmosphere, the vapour pressure of water exerted at its surface must exceed that within the air. In many parts of the world, climatic variations do not allow this to occur for long periods and as a result the moisture content of materials stored in the atmosphere is liable to oscillate. Consequently, artificial methods of drying are necessary. Fresh air must be introduced, preferably at a relatively high temperature, to reduce the ratio P/P_0 to a sufficiently low value to ensure that drying does take place.

At all times, a foodstuff will seek to equilibrate with its environment, and the nearer it is to achieving this, the slower the process becomes. The aim is always to keep the driving force high enough for the process of protection to be faster than the process of deterioration, until safe limits are achieved. It should be borne in mind, however, that any increase in temperature in the meantime may result in deterioration of quality. Information on the moisture carrying properties of air is conveniently portrayed on a psychrometric chart.

A simple thermometer in moving air will indicate the actual or 'dry bulb' temperature. If a second, adjacent thermometer has wet cotton wool or the like wrapped around the sensing bulb, then the water will tend to evaporate in the moving air. This process requires energy (the latent heat of vaporization), which is extracted from the thermometer bulb. The temperature of the bulb falls as a result and this difference in temperature causes sensible heat to flow from the air stream to the wet bulb. When dynamic equilibrium is established, the heat transfer effect will balance the mass transfer effect and the 'wet-bulb temperature' will be attained. The difference between the dry-bulb and wet-bulb temperatures is known as the 'wet-bulb depression' which is a function of the humidity (moisture content) of the air. Its measurement is widely used as a technique for measuring humidity.

Food structure and drying

Drying mechanisms

Provided the material to be dried behaves like the 'sponge' on the wet bulb thermometer described above, this model system can be taken as an imitation of the actual process occurring during the drying of foodstuffs. For this to be true, the moisture within the food must have continuous and ready access to the surface from which it freely evaporates, being fed as by a wick from the interior of the drying piece. Early in the drying process this may be so and under these conditions, 'constant-rate drying' occurs. With most food products, however, the transfer of moisture from the bulk to the surface is hindered by the internal cell structure (see below) and 'falling rate' drying is observed in which the drying rate falls progressively, often in a complex and unpredictable manner, with declining moisture content.

Foodstuffs are mostly cellular and contain dilute solutions of sugars, salts, proteins, and so on, within a labyrinth of cell walls through which the moisture must diffuse to reach the outer surface and escape. This puts a restriction on the rate of migration and hinders the transport of moisture within the bulk of the product. As vapour is picked up from the outer surface, the local cell contents become more concentrated. Diffusion of moisture then occurs from nearby inner cells, the contents of which become more concentrated in turn, and so on. As the diffusion path becomes progressively more difficult, some cells, unable to retain their fullness, collapse and set up stresses which tear at the structure creating fissures which become new and easier paths of moisture egress. Eventually the material exhibits signs of shrinking or shrivelling and, in general, drying gradually slows down.

Other things being equal, drying will proceed faster if the wet bulb depression is large; for a given wet bulb depression, drying will be faster if the dry bulb temperature is high. As drying proceeds, care has to be taken to ensure that the surface temperature of the foodstuff does not rise unduly and it may be necessary to reduce the dry bulb temperature of the surrounding air.

Bound water

The cell contents of a foodstuff are a complex of solutions, which as water is driven off, become more concentrated and can be conceived as reaching a condition where chemical bonds have to be broken to obtain further change. This would not ordinarily be reversible and hence has no proper part in food dehydration. Physically and chemically 'bound' water, which is a rather imprecise term, may be present in amounts in excess of that which is in equilibrium with air at the corresponding temperature and humidity. In contrast, 'free' water may be initially present at the start of the drying process, either at the surface of the product or, more likely, in dilute solutions in cut cells, which present no diffusion barrier. Free moisture can be distinguished from bound moisture in that its removal is reversible.

With many food products, the presence of bound moisture precludes the possibility of drying down to a bone-dry condition. In such circumstances, the product attains a limiting moisture content, known as the equilibrium moisture content (EMC) after extended drying. EMC is a function of both the humidity and temperature of the drying air. The humidity of the air that is in equilibrium with product containing a specified residual moisture content is termed equilibrium relative humidity (ERH).

Case hardening

Internal stresses during drying coupled with migration of solutes may result in the partial sealing of some fruit and some meat surfaces, which slows up considerably the action of further drying and makes the avoidance of 'wet centres' difficult. Diffusion of moisture in both liquid and vapour form has to proceed at rates that will minimize the onset of these effects. Proper preparation and attention to drying procedure for each particular commodity have made this phenomenon uncommon.

Agglomeration

Some commodities may not be acceptable in fine powder form or may be too dusty to handle. Agglomeration is a technique for gathering a number of fine particles together into larger ones. This is usually achieved by rewetting the particles slightly with steam while they are airborne and then redrying them. The resulting open-work granules have a more pleasing and acceptable appearance and tend not to 'cake' on reconstitution. This contrived structuring is particularly useful for modifying the bulk density to achieve 'one spoonful per cup' requirements.

The drying process

Observation of the foodstuff at the start of a batch drying process will show that its temperature initially rises rapidly to the wet bulb temperature of the surrounding air. It will maintain this temperature as long as its moisture content is above the 'critical moisture content', which characterizes the demarcation between the constant rate and falling-rate drying periods. Constant-rate drying occurs under conditions where the moisture is able to evaporate freely from the surface. Many foods, it should be noted, do not exhibit a constant rate drying period; in such cases, falling-rate drying is observed from the outset. Once the moisture content falls to the critical value, the evaporation rate will start to decline as the progress of the moisture to the drying surface is hindered by the internal cellular structure of the food. As a result, the cooling effect of the evaporation process diminishes and the temperature of the solids progressively increases towards the dry bulb temperature of the air.

In general, higher air speeds will give faster drying, as will thinner layers and smaller pieces. For the same 'depression', higher temperatures give faster drying. If the wet bulb temperature is held constant, higher dry bulb temperatures (i.e. increased depression) will also speed up drying. Such rules are simplistic and relate to drying regardless of dehydration. In practice, conditions may be varied during the process, usually in several stages, to suit the particular system. Thus to dehydrate effectively we need to know quite a lot about:

(i) The characteristics of the material being dried:
  - its water activity (see below) at various moisture contents for a given temperature
  - its thermal conductivity
  - its diffusion resistance
  - its effective pore size
  - any markedly significant change points in the drying rate,
(ii) The characteristics of air/water vapour mixture at various temperatures,
(iii) The rehydration quality or reconstitution characteristics expected after defined storage.

Then we can decide: (i) how we must dry, (ii) what form of dryer we shall need and (iii) what speed of drying will be economical.

Water activity

The ratio of the vapour pressure above a solution to that above pure water at the same temperature is commonly called the water activity, $a_w = P/P_0$. Being less than unity, it gives a simple measure of what may be a complex situation in that it indicates the difficulty which further water molecules may have in moving out of solution. However, the term 'activity of water content' has also been applied to other than free surface actions and indeed to very low moisture conditions when erratic molecular migration may occur. It is generally thought of as being related to the availability of water for microbial conversion, as well as enzymatic and chemical reactions, and can thus be used as a guide to the effectiveness of a particular moisture content in maintaining quality.

Normal bacteria are unlikely to grow below $a_w = 0.91$, normal yeasts below 0.88, normal moulds below 0.80, halophilic (salt loving) bacteria below 0.75, xerophilic (dry loving) fungi below 0.65, and osmophilic (sugar tolerant) yeasts below 0.60. A value of a_w of 0.70 at room temperature is usually considered safe for short periods of storage. To achieve a specific a_w, the food may be dried and then subjected to controlled rehumidification, or it may be submerged in a solution having a higher osmotic pressure (e.g. candied fruit).

Intermediate moisture foods

Many foodstuffs in a range of moisture content 15–30% have an a_w value at 20°C in the region 0.65–0.85, and thus are very slow to suffer in an environment typified by (say) 20°C and 70% RH. These are termed intermediate moisture foods and are ordinarily safe from bacteria and yeasts. Moreover, they contain enough moisture to be palatable in this relatively stable semi-moist condition. Consequently, semi-moist foods as a class are of growing interest. However, they are still vulnerable to moulds and so have a limited unprotected shelf life. There is also the possibility of lipid oxidation and non-enzymic browning; in some meat products of a_w 0.85 held at 38°C for several weeks, there is some evidence of reactions which render certain proteins unavailable nutritionally.

DRYING SYSTEMS

Classification of dryers

A wide variety of dryers suitable for drying foodstuffs is available, each with its own particular characteristics. Dryers may be classified as follows: (i) batch versus continuous, (ii) direct, indirect or other heating method or (iii) atmospheric versus vacuum operation.

Batch dryers as their name implies process foodstuffs in individual batches. These dryers are best suited to drying relatively small throughputs and are quite versatile in that they can handle a wide range of different products for which the drying conditions may be varied at will. They are, however, relatively labour intensive. Continuous dryers, on the other hand, are best suited to handling large throughputs of a single product. They are less flexible in general than batch dryers.

Direct dryers in which there is direct contact between the heating medium, normally hot air, and the product are widely used in the food industry. Drying rates are generally high and equipment is available that covers the complete range of production rates and useful drying temperatures. Indirect dryers are used where contact between the product and the drying medium is undesirable for one reason or another. In such dryers, the product is in contact with a heated surface; it may either be still or agitated. The usual heating media are condensing steam or hot water. Indirect dryers are usually less thermally efficient and more costly (like-for-like) than their direct counterparts. Other heating methods employed in special cases include microwave, radiofrequency and induction heating.

The majority of dryers operate at atmospheric pressure. However, vacuum dryers are used where the product is particularly thermally sensitive; an extreme example is the freeze dryer. Naturally, all types of vacuum dryer are indirectly heated.

The choice of dryer type for a particular application is a complex procedure which is best left to a specialist. It should be stressed that dryers not only remove moisture, they often have a major influence on product quality as well. Details of some of the more common types of dryer are given in Table 15.2.

Drying in heated air

Some general guidelines for drying in heated air are:

(i) The rate of air flow should be high in the early stages when water is evaporating freely, but except in the air-lift dryer (see below), it should be insufficient to lift the material being dried; a linear rate of 180–300 m min^{-1} is often suitable for vegetable pieces and 130–260 m min^{-1} may be used for fish or meat.

(ii) In the early stages of vegetable drying, dry bulb air temperatures in the range 90–100°C may be used if the wet–bulb temperature can be kept low – preferably not above 50°C. This induces rapid evaporation of water from the foodstuff, which remains at about the wet bulb temperature as a result of evaporative cooling, so that no heat damage occurs. If carried to excess, very high evaporation rates can cause a form of case hardening, particularly if thick sections (e.g. thicker than 1 cm) or proteinaceous foodstuffs such as meat or fish are dried; with these latter materials, lower dry bulb temperatures and higher humidities tend to be employed.

(iii) In the later stages of drying (below about 15% moisture content), the rate of migration of water to the surface of the pieces is very slow and prolonged drying at relatively low temperature (55°C or below) is needed to bring the moisture content to below 6%. This final stage may be carried out in some form of finishing dryer (see below), where a low rate of air flow may be used, since the amount of water to be evaporated per unit mass remaining is small.

Hot air dryers

Some of the more common types of dryer used to dehydrate foodstuffs are described below.

Air-lift dryers

The air-lift dryer, also known as the pneumatic conveying dryer or flash dryer (see Figure 15.5) is essentially a vertical column up which a stream of heated air flows. The moist material in granular form is fed into the lower part of the dryer and is carried upwards by the air stream as it dries. The dried product is collected in a cyclone and the air normally passes through a secondary cleaning device, such as a wet scrubber, prior to discharge.

The ring dryer is a common variant of the air-lift dryer. In this device the drying solids circulate around a ductwork ring. During the course of each pass, they encounter a centrifugal classifier. The fine dry particles are directed to a cyclone for product collection; the larger, still moist, particles are recirculated with fresh gas for further drying and comminution in a disintegrator.

Mashed potato powder, starch and gluten are typical examples of food products dried in such dryers.

Table 15.2 Types of dryer

Feed form	Handling style	Main feature	Sub-group	Batch or continuous	Mode[a]
Liquid pastes, large or small pieces	Material static, equipment static, process varied	Chamber or cabinet ovens with solid or perforated or mesh shelves or hooks		B	A, V, FD
			Finishing bins	B	
All forms	Material static, equipment moves it through each process stage	Shelf moves	Solid band or belt	C	A,V,FD
			Perforated band or belt	C	A
			Vibrators	C	A,V,FD
		Trucks of trays	Tunnels	C	A,V,FD
Liquids	Material scraped	Thin-film evaporators	Steam, wiped	C	V
			Scraped cylinder	C	A
Pastes	Drum(s)		Film on outside	C	A,V
Liquids	Material sprayed	Spray dryers		C	A,I
		Towers		C	A
Small pieces, powders	Material stirred	Mechanical stirrers	Vanes	C	A
			Cascade shelf		
			Screw conveyor discs		
	Gravity feed	Vertical flow		C	
		Radial flow			
	Tumbled	Louvre dryers		C	
		Belt dryers			
		Trough dryers			
		Sprial dryers			
		Rotating conical chamber		B	
Small pieces, powders	Material blown	Fluidized bed		B,C	A,I
		Pneumatic ring or air lift		C	

[a] A = air; FD = freeze drying; V = vacuum; I = inert gas atmosphere.

Belt trough dryers

The belt trough dryer was developed in the USA to provide partially dried vegetables as part of the process of dehydrofreezing. The dryer comprises an endless wire mesh conveyor belt, supported and transported by roller chain sprockets carried by three parallel drive shafts. Between these the belt forms a trough, the bottom of which rests on a flat-surfaced air grate which slopes upwards in the direction of travel. Heated air is forced upwards via the grate and through the belt. The foodstuff (e.g. diced root vegetables, peas or fruit), is fed on to the trough in a direction parallel with the drive shafts and is raised by the motion of the belt towards the lip of the trough, tumbles back, and in a series of spiral movements, travels along the dryer towards an adjustable discharge weir at the exit end.

Because of the continuous turning over of the bed of foodstuff, unusually high temperatures can be employed; for example, carrot dice of 9 mm edge can be dried to 65% weight reduction in 20 min at dry- and wet-bulb temperatures of 150°C and 45°C, respectively.

Cabinet ovens or chamber dryers

Cabinet ovens are small scale general purpose dryers able to handle anything from cereals to sweetmeats, and can also be used for absolute moisture content determinations. They consist basically of an enclosed chamber fitted with a variable heater, a fan to induce air movement, together with deflectors for air flow adjustment, outlet air louvres and adjustable inlet air louvres.

The larger models are fully instrumented. The process is constantly maintained under close control and materials needing varying drying conditions may be catered for. In the orthodox type of chamber, provision is made for the material either to be spread on trays or hung on hooks or rods according to its nature. The air flow is directed across the material. It is usual to recircu-

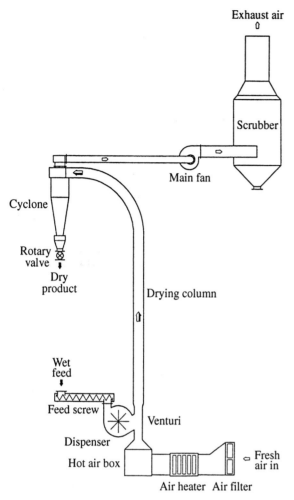

Figure 15.5 Pneumatic conveying (flash) dryer (courtesy of Barr Rosin Ltd).

late a certain percentage of the air, according to the humidity required at the time in the chamber.

Drum dryers or roller dryers

The substance to be dehydrated is spread in a thin film, in liquid or mash form, on single or double drums heated by steam. Drying takes place in a matter of seconds, before one complete revolution of the drum has been made and the dry film is removed from the surface of the dryer by 'doctor' knives or scrapers. Most drum dryers operate under atmospheric conditions, but where the product may be damaged by drying in contact with air, the drums are enclosed in a vacuum chamber.

Drum dryers are commonly used for the dehydration of liquids such as milk, whey and fruit juice, for pastes and slurries (e.g. yeast) and for carefully prepared diced potato, making flakes which reconstitute into a fluffy mash. Precooked meat and vegetable soup powders are also produced on such dryers.

Evaporators

These devices concentrate liquids very effectively; however, they yield a concentrated liquid, as opposed to a dry powder, as product. An evaporator is normally used to preconcentrate a dilute solution (e.g. milk) prior to spray drying.

Finishing dryers

In the final stages of dehydration, the rate of drying is limited by the rate at which liquid water or water vapour can be made to diffuse from the centre of the piece of foodstuff to the outer surface where evaporation take place. Because of the need to limit the temperature to relatively low values in order to avoid heat damage during this phase of the process, lengthy drying times may be required in order to achieve low final moisture contents. Many food manufacturers therefore resort to the use of separate finishing dryers for this purpose, rather than overburdening the primary dryer.

One type of finishing dryer, the bin dryer, is in the form of a cylindrical vessel, 1 to 2.5 m in depth, with a perforated base through which heated air is blown. Most bin dryers are of the batch type and sometimes feature a 'clip-on' trolley-form bin. Others are arranged for continuous passage of the foodstuff. Another form of finishing dryer, the tray dryer, consists of a cabinet containing a series of deep trays, through which the heated air is blown. In all cases relatively low air temperatures are used, say 55°C or less. The drying period may be 3–12 h. Continuous dielectric dryers, in which the heat is provided by microwave or radiofrequency energy are also used as finishing dryers.

Fluidized bed dryers

This type of dryer was originally developed for finish drying of potato granules (a form of mashed potato powder). A bed of partially dried material, of say 12% moisture content, which when static would be 50–60 mm deep, is supported on a horizontal perforated or porous plate. A current of heated air is forced upwards through the plate at a velocity sufficient to separate and support (i.e. fluidize) the individual particles. The material flows through the dryer from the feed inlet at one end to the exit, over a weir, at the other. Because of the mixing action in the 'fluidized' bed, the particles dry evenly and quickly (see Figure 15.6).

This type of dryer can be used to dry peas and

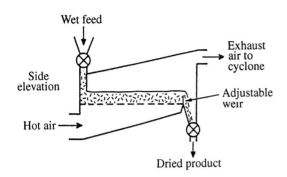

Figure 15.6 Fluidized bed dryer (courtesy of AEA Technology plc).

dice as well as powders. The air velocity and temperature must be adjusted to suit the commodity being dried, and may need to be reduced as the food dries. This is achieved by dividing the plenum chamber into a number of separate sections. Dust and fine airborne particles are separated from the exhausts by cyclones and remixed with the discharged product.

Loft or platform dryers

Loft or platform dryers are one of the earliest and simplest types, and consist of a room or raised chamber with a perforated drying floor on which the material to be dried is laid directly or in mesh-bottomed trays. Heating may be by relatively crude methods employing direct convectional flow of combustion gases, or by more controllable systems in which the flow is controlled by a fan and damper, or indirectly by some form of heat exchanger. These dryers are not notable for producing dehydrated foods for human consumption as control is broad banded and approximate, though certain fruits, e.g. apples, are sometimes dried in this fashion. Vegetables for animal feeding, and grass, clover, grain hops and the like are commonly dried in this type of kiln. If the dryer is controllable at a relatively low temperature, say 45–50°C, it may be used for the drying of culinary herbs such as sage or parsley.

Mobile dryers

These are designed so that the dryer can be taken to the task, rather than taking the materials to the drying plant and are used mainly for agricultural work – grain or grass drying. Often the products of combustion of propane or another fuel, are mixed with copious quantities of excess air for direct contact with the material being dried, while the fan(s) and auger-type conveyors are driven from the 'power-take-off' of the tractor used to pull the dryer.

Batch units can consist of two concentric, horizontal or vertical, perforated cylinders, between which the grain is introduced and agitated, while hot gases pass radially from the inner plenum chamber to the outer wall. Drying over the moisture range applicable takes about 2½ h, including cooling.

Continuous flow units require careful matching of variable speed augers, which transport the grain into and out of a deep chamber where the warm air enters the bed through midway ducts and moves up and down, giving rise to both counter and concurrent flow. Care is required to avoid the problems of dust and seed cracking. Moisture content reduction is in the range 19–14.5%.

Overdraught (cross-flow) dryers

A significant proportion of dehydration today is carried out in some form of overdraught dryer, which may be in the form of a cabinet dryer or a tunnel dryer in which the material is supported in shallow multilayers with the air directed at high speed between and across the surfaces.

In all hot air dryers, large volumes of air are required to supply the heat to the product and to carry away the moisture evaporated. For the sake of heat economy, it is advantageous to recirculate a proportion of the air, although not in sufficient quantities to cause its humidity to be raised to such an extent as would seriously hamper drying. Overdraught dryers must, therefore, include power driven fans of larger capacity, adequate air heating units, air ducts between the heaters and the drying chamber(s) capable of providing an even distribution, recirculation air ducts controlled by dampers for recirculating a proportion of the air from the drying chamber back through the heaters, and intake and exhaust air ducts, which are also controlled by dampers.

In overdraught cabinet dryers, the foodstuff is spread in thin layers on trays, usually at 4.75–7.5 kg m^{-2}, though higher tray loads can be used if the material is respread or turned during the drying period. The trays are arranged in cabinets in which heated air at a controlled temperature and humidity is directed evenly over the trays. Steam-heated batteries most frequently provide the heat.

In large cabinet dryers, the trays are usually supported on trolleys which can be pushed into and out of the cabinet, and also turned around so that the air does not always impact onto the same edge. In some cases, the entire drying process is completed in a single cabinet; in others, the cabinets are arranged so that trolley loads of trays pass from one to the next, with different temperature and humidity conditions in the successive cabinets. In yet other layouts, only part of the dehydration process is carried out in cabinets; it is completed in finishing dryers. Recirculation within the cabinets is practised as far as humidity conditions allow; in multistage cabinet systems air is taken from the later stage cabinets through heaters to the earlier stage cabinets.

Moving the trucks individually through successive cabinet doors is wasteful. As a result, arranging the 'cabinets' in line continuously to form an overdraught tunnel was an obvious development. The best designs are able to combine high capacity and high quality of product. The trolleys are conveyed through the tunnel on guide rails, usually by some mechanical device. There are various systems of airflow over the foodstuff.

(i)   Concurrent flow tunnels – i.e. in the direction of movement of the trucks (termed parallel flow in the USA). The hottest and driest air impinges first on the wettest material and a high rate of evaporation ensues. The air is partly cooled as a result, and becomes cooler as it passes over the trays in successive trucks in the tunnel. At the far end, the dry bulb temperature of the air may have dropped from 90 or 95°C to 60 or 65°C. It will be carrying a heavy load of moisture and will have lost a great deal of its drying effectiveness. As a result, it is not possible to recirculate much of the air and therefore cool air has to be taken into the heaters continuously. Consequently these tunnels are somewhat limited by their relatively high heat (and fuel) consumption.

(ii) Counterflow tunnels – i.e. opposite to the direction of movement of the trucks. In these designs, the heated air is first blown on to the driest material at a temperature low enough not to cause heat damage to the product (usually about 70°C). Owing to evaporative cooling along the tunnel, the air is considerably cooler than this when it reaches the wet end (i.e. the end at which the trucks of wet material are inserted) and evaporation in this stage is relatively slow. Because of their simple design, counterflow tunnels achieved considerable popularity and much of the dried fruit commercially produced and a substantial proportion of dehydrated vegetables processed in the USA, has been dried in this type of equipment.

(iii) Combined tunnels – these are of two types, parallel and centre exhaust tunnels.
      Parallel tunnels. In parallel tunnels, introduced in Britain during World War II, a concurrent flow tunnel and a counter flow tunnel were built side by side. The concurrent flow tunnel became the so-called 'wet' tunnel into which the trolleys containing the scalded vegetables were loaded; when the trolleys had passed through the wet tunnel, they were transferred to the counterflow or 'dry' tunnel. In this original design, the trolleys ran alongside each other in pairs and there was room in each tunnel for seven trolleys. Commonly, pairs of trolleys were inserted at intervals of 30 min, thereby giving a total drying time of 7 h.

Air entered the system by way of an inlet into the recirculation duct of the dry tunnel: it passed through the heater battery and the fan and was forced through the drying section, whence it passed again into the recirculation duct. Part of it was diverted into the wet tunnel section, part recirculated through the dry tunnel system. That part which had been diverted to the wet tunnel joined the recirculation air there, passed through the wet tunnel heater and fan and through the drying section; part was then exhausted and part was recirculated. By appropriate manipulation of dampers quite good control of humidity could be obtained in the tunnel, with substantial economy in the use of heat.

Commonly used temperatures in these tunnels were: wet tunnel inlet end, 99°C dry bulb; exhaust end, 60°C dry bulb, 43°C wet bulb; dry tunnel, inlet end, 65°C.

Other systems of parallel tunnels are in use, but the principle is the same – a 'wet' tunnel or tunnels with concurrent flow and a 'dry' tunnel or tunnels with a counter flow air stream.
      Centre exhaust tunnels. The same general considerations operate in the centre exhaust tunnel, which is a long tunnel with a single large extraction fan near the centre, forcing air to be drawn in through heaters at each end of the tunnel. The trolleys are inserted at one end and moved along in stages, finally to be withdrawn at the other. Consequently, for the first part of the drying period the foodstuff is in a concurrent flow tunnel but after it has passed the central exhaust region, it is in a counter-flow tunnel. Unless separate finishing dryers are to be used, the counter-flow stage will be longer than the concurrent-flow

stage, because of the slowness of removal of vapour from partially dehydrated material.

(iv) *Transverse-flow tunnels*. These are usually combined with interstage reheating of the air and appropriate ducting to provide what is essentially a counterflow system.

This type of dryer can be designed to have a very high evaporation rate, with good thermal efficiency, but the trucks have to fit very snugly in the tunnel to avoid short circuiting the air. It is often followed by a finishing dryer, to which the foodstuff is transferred when its moisture content has fallen to below 20%. The driest air enters where the material is about to emerge, being partially exchanged and mixed with fresh air and reheated between stages so that its vapour-holding capacity is increased as it travels along the tunnel. Such a dryer may contain a load of twelve trolleys, each carrying 180 to 230 kg of wet foodstuff at entry; these trolleys may be moved along one-twelfth of the tunnel at 15 or 20 min intervals, giving a total drying period of 3 or 4 h. The air entering the dryer (at the material discharge end) may be heated to 60 or 65°C, and its temperature may be increased to 90 or 100°C as it travels towards the food-inlet end. This style of dryer is economical in terms of floor space, and the process is more under control, but it incorporates fans and air heaters at a number of points rather than a single fan and heater at one end. Hence, its power rating for the numerous air direction changes can be higher than that for a simpler tunnel.

Puffing dryers

Various ploys which set out deliberately to alter the structure of the foodstuff have been developed. These have a secondary benefit, namely to speed up the final stages of drying. One such system, the foam mat dryer, is suitable for viscous slurries. The slurry is whisked into a stable foam, which is then spread in a thin (6 mm) layer on a perforated belt. The drying air is forced through nozzle-like apertures, creating small craters, which present a greater surface from which drying can be completed in a few minutes.

Another system, designed for solids, places batches of partly dried dice in a pressurized 'gun', sudden release from which causes the dice to expand to a more open-pored texture. A fluidized bed or vibrofluidized bed dryer may also be employed to puff cooked cereal grains.

Radial dryers or turbine dryers

These are used particularly for the drying of grain, which is contained between two concentric vertical, perforated sheet metal, bin-type cylinders. Located within the inner shaft, a succession of circulating fans force air through the material in a horizontal direction, the air returning to the centre of the dryer after passing through heating coils fitted to the outside cylinder. This continues in stages until the air is discharged at the top. It follows, therefore, that cold air coming into the bottom of the dryer cools the dried grain being discharged, while 'green' grain entering at the top is preheated by the exhaust air before the main drying process takes places. The method prevents the surface of the grain becoming hardened before drying is completed.

Rotary dryers and louvre dryers

A rotary dryer consists of a large slowly rotating cylinder fitted with internal flights which keep the material moving so that it neither sticks to the wall of the dryer nor mats together. The cylinder is mounted at a small angle to the horizontal to facilitate movement of the product from the upper (feed) end to the exit. In the direct-fired version, often termed the cascading rotary dryer, hot air flows either concurrently or countercurrently to the solids. The flights lift the drying material and cascade it though the hot gas. In an alternative, indirectly heated design, known as a rotary steam-tube dryer, heating is accomplished by the material coming into contact with steam heated tubes mounted inside the dryer along its length.

The rotary louvre dryer does not contain any flights. Hot air flows through a series of longitudinal louvres and directly contacts the drying solids, which pass through the dryer as a moving bed located on the bottom of the cylinder. Rotary (including louvre) dryers are used for sugar, salt, grain, fishmeal and other substances of a granular nature.

Shuffling or cascade dryers

In these, inclined shelves spaced rather like the rests of a shallow rise stepladder, vibrate with a 'lift and throw' motion. the material (e.g. sugar) shuffles down the slope while heated air passes through the stream between the 'steps'.

Spiral dryers

The spiral dryer is contrived to give a long path in small space for drying grain. It consists of a

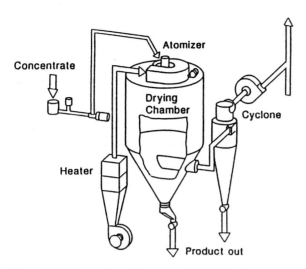

Figure 15.7 Spray dryer (courtesy of A/S Niro Atomizer).

rotating drum with two or three spirals built inside. The material is fed to the inmost spiral by a worm conveyor, travels to the outside wall of the drum via the spirals and is then discharged from the drum casing by scrapers. The hot air used for drying follows the path taken by the materials being dried. The arrangement of the spirals, which are fitted with baffles, ensures that the material being dried is kept in intimate contact with the hot gases. The evaporative capacity of this type of dryer can be 60 kg (water) h^{-1} m^{-3} of cylinder volume.

Spray dryers

The outstanding characteristic of this type of dryer is the exceedingly short time of contact between the material being dried and the drying medium (usually hot air). Liquid is transformed into a dry powder in a few seconds. Such dryers have obvious advantages for materials which are liable to suffer discoloration, oxidation, decomposition, loss of flavour or protein denaturation when heated for any but the shortest time. They have been adopted for a wide variety of products, such as milk, eggs, yeast, sugar solutions, coffee extracts and also vegetable extracts, fruit juices and blood plasma; high throughputs can easily be accommodated.

The general construction of a spray dryer (see Figure 15.7) incorporates four main features:

(i) Heater and at least one fan to produce air at the required temperature and velocity
(ii) Atomizer or jet to produce liquid droplets of the required size
(iii) Chamber in which the liquid droplets are brought into intimate contact with the hot air

(iv) Means for removing the product from the air stream.

The air heater, situated on the pressure side of the intake fan, is designed according to the temperature desired, which may be anything between 90 and 650°C. Steam, hot oil, hot furnace gases or electricity can be employed as the source of heat. The fan should be ample to give the required air velocity within the chamber, usually between 15–30 m s^{-1}. The air is first passed through a filter to remove any airborne impurities.

Perhaps the most important component of a spray dryer is its atomizer device. This has the function of continuously dispersing the liquid in the form of small droplets of uniform size in a controlled direction. The atomizer must be capable of operation for long periods under steady working conditions. Three main types are employed:

(i) The nozzle atomizer (the oldest type) in which the liquid feed is pumped at pressures up to 140 bar (14 MPa), or up to 275 bar (27.5 MPa) for egg dryers and very high velocities through a minute aperture in a disc.
(ii) The two-fluid nozzle, which employs air or sometimes steam as the atomizing medium at pressures in the order of 14 bar (1.4 MPa). This type of atomizer can accommodate only relatively low liquid throughputs.
(iii) The rotary, centrifugal or wheel atomizer, which for many products appears to have certain advantages: clogging and blocking are avoided; it may be operated with suspensions as effectively as with solutions; it gives a product of fairly uniform particle size. Trouble from eddy currents set up by the motion of the rapidly rotating disc is avoided by fine control of the air stream. A variety of rotary atomizers is available. These meet the objective of obtaining the maximum degree of dispersion of the particular droplets. Unless atomization is very uniform, some larger drops are formed which will not dry completely.

In the case of (i) or (ii), multiple nozzles can be mounted within a single spray chamber in order to achieve the desired throughput.

The drying chamber, i.e. that part of the dryer in which the hot air stream and the droplets are brought into intimate contact, has been the focus of a number of constructional variations. In most spray drying applications, however, a lower conical base is surmounted by a cylindrical body. The atomizer is usually located at the top of the cylinder and the product gathers towards the base of the conical hopper from where it is continuously removed.

Chamber connections vary according to the means adopted for air stream distribution. One model, for example, introduces the hot air through an annular air distributor to give uniform rotational and downward velocities in the same direction as that imparted to the droplets by the atomizer. The moisture-laden air exits the chamber through a centrally disposed downwards facing duct, which leads out of the side of the conical chamber base. In this style of spray dryer, cold air is blown over the spray to cool the air in the vicinity of the atomizer. This prevents premature evaporation before full dispersion of the droplets is obtained. There are many variations of this technique.

In contrast, another type of spray dryer has only one air blast, directed centrally upwards towards the jet head, with vanes to create swirling. The laden gas carries the powder out of the cone to one or more separating cyclones. An exhaust fan acting as a back up helps to draw the air out of the chamber.

Chambers for foodstuffs are usually welded in aluminium or stainless steel and of a very fine finish. The dimensions, in particular the diameter, are of great importance. For a given degree of atomization the diameter must be sufficiently large to ensure that the droplets do not hit the wall before they are dry. At the other extreme, it should not be so large that the atomization and matching flow of hot gas are not under control; this would result in a non-uniform interaction between the air and the particles. In general, the minimum diameter to be used will depend on the uniformity and size of the droplets leaving the atomizer. Diameters up to 9 m are not uncommon in the food industry, yielding a particle size in the range 280 to 800 μm. Diameters of less than 1.5 m, although seldom encountered except in pilot plant, give rise to a finer product, 50–100 μm.

The bulk of the powder formed in the chamber falls to the bottom of the conical hopper from where it is removed, usually by a helical screw conveyor. Some of the finer powder, however, is entrained and carried away by the air stream, and various devices have been adopted to recover it. The discharge air may be washed by passing it up a small scrubbing tower down which the feed liquor is caused to flow before being pumped to the atomizer. More often, cyclone separators are employed. In many designs, the gas is exhausted through a set of bag filters emptied by mechanical shakers and a reverse blast of air. In some cases,

where the product is sufficiently valuable, an electrostatic precipitator may be used in conjunction with a cyclone separator.

In normal operation using a stream of hot air, the feed liquor enters at room temperature and the product obtained is a fine uniform powder. In certain cases, however, when the material to be dried is not particularly sensitive to heat, the feed may be introduced in a superheated state and a cold predried air stream employed; 'flash' evaporation and drying take place, with the liquid transferring both sensible and latent heat to the air stream. The product resulting from this method is of a granular nature.

In practice the process makes certain feeds highly viscous. The dried product is found to contain spherical particles with small amounts of air or water entrapped inside. The pressure of the water during drying is usually sufficient to burst the globules, but with very viscous feeds the 'skin' may become too strong for this to happen. This problem can usually be overcome by mixing a small proportion of a liquid of high vapour pressure in with the feed. The resulting extra pressure ruptures the globules, while the additive separates. Thus if maltose (specific gravity 1.242 at 18°C) is spray dried in a N_2 stream at 116°C, and the feed is mixed with 6 volumes percent of ether, the product is a dry powder of specific gravity 0.037.

Because of the wide range of droplet sizes, empirical relationships still guide the performance of spray dryers, which has to be based on the largest droplet being just dried and the smallest being overdried. The wet bulb temperature tends to remain constant.

Through-circulation band or conveyor dryers

In these devices, the foodstuff is transported through the dryer on a conveyor belt, which is normally constructed from stainless steel. The belt is perforated and heated air is blown through the foodstuff. Baffles divide the dryer into individual zones in which the temperature and humidity conditions are separately controlled. The simplest form of conveyor dryer employs only a single band. However, other types feature several bands in a vertical array, travelling at different speeds. This enables the foodstuff to pass quickly and in relatively thin layers through the initial zones of the machine which are the hottest, but after it is partially dried, to travel more slowly and in thicker layers through the subsequent zones which are maintained at lower temperatures.

Through-draught tray dryers

When the air flows upwards or downwards through a bed of foodstuff, the dryer is termed 'through-draught'. This method is commonly applied to finishing bins, fluidized beds, conveyor dryers and a particular type of cabinet, the tray dryer, which has tiers of deep trays. This latter type is rectangular, holding (say) ten trays with mechanical handling, up to 6 m^2 each in area, stacked one above the other in two banks of five. Under each bank there is a battery of heaters. Air is blown upwards through the heaters and trays at about $350 \text{ m}^3 \text{ min}^{-1}$. The temperature of the air immediately above the lower heaters (the 'dry' zone of the cabinet) is about 50°C for 'sensitive' materials and about 70°C just above the upper heater. Temperatures are raised for those foods which will tolerate them.

The operating procedure is sequence staging. A tray is loaded with about 60 kg of prepared vegetables. If the drying time is, say, 4 h, then every 24 min a tray of dry material is removed from the bottom position, the trays in the lower half move down one place and the bottom tray from the upper half of the cabinet is respread. This tray is then inserted into the top position in the lower half. In the top half the four remaining trays move down one place and the fresh tray of prepared material is inserted.

The main disadvantages of through-draught drying are a high usage of labour and a somewhat lower heat economy than for a well-designed cabinet or conveyor dryer; advantages are fuller use of capacity and the opportunity to attend to the product at intermediate stages in the drying cycle.

Tower dryers

This type of dryer was developed in Switzerland. Derived from the spray dryer, it was designed to be more gentle in treatment in order to maintain the fresh characteristics of the liquid food being dried. In the tower dryer, the material is fed into a distributor which delivers it as an even 'rain' of droplets of sizes up to 2 mm diameter at or near the top of a cylindrical chamber, which may be 50 m or more high and 15 m in diameter. The particles fall slowly down through an even, rising current of air (or other gas) of controlled temperature and humidity. The conditions are chosen to bring the foodstuff to the desired moisture content by the time it just reaches the base of the tower, where it is collected. The temperature of the drying gas is much lower than is usual in hot air

dryers (e.g. 35°C is suggested for tomato juices), and the gas may have to be predried. The relatively long time of contact in the rising air current in the tall tower can give rise to the possibility of oxidation.

Also termed 'tower dryers', but really gravity, shaft or cascade dryers, are the tall tubular types for minor moisture removal which allow the (granular) material slowly to descend under its own weight against a current of centrally introduced warm air. The natural interstices between the particles form a porous path often of high resistance, occasionally enhanced by zig-zag divertors.

Vacuum dryers

Dehydration under vacuum has certain special merits. Drying can be carried out at lower temperatures than with air drying and heat damage is minimized. Further, oxidation during drying is virtually impossible.

Principles and features

Operation of dryers processing commercial quantities of foodstuffs as discrete pieces, in systems having a working pressure well below atmospheric, poses special design problems. Obviously, the drying chamber must be sufficiently robust to withstand the considerable external pressure. Also, the pieces depend on conduction and/or radiant forms of heating and tend, therefore, to be spread out in thin exposed layers. If the material is prepared on trays, the doors to the dryer must provide easy and rapid access and are correspondingly large and heavy; they must seal well, as must all other connections, so as not to impose continuous leakage loads on the vacuum pumps. The vapour coming from the material being dried has a high specific volume and this must be reduced before it is vented to the atmosphere. Pull-down time to operating conditions should not be a delaying factor, so the shell should fit compactly around the internal fixtures, without impeding vapour flow.

Earlier evacuated vessels, such as evaporators, were cylindrical and tubular, and therefore inherently strong, but when tray-loaded systems were introduced on a significant scale, heavily ribbed rectangular shapes which were more costly to construct, were resorted to. Again, since ample supplies of condensing water were generally available at below 13°C, corresponding to a vapour pressure of better than 1.6 kPa, and it was considered that vapour from the foodstuffs would be produced at no more than 60°C (vapour pressure 19.7 kPa), then the main engineering concern was to provide a total pressure in the dryer such that the partial pressure of vapour was predominant.

A well-tried means of doing this, handling both condensible and non-condensible gases, is the multistage venturi-type steam driven ejector/compressor, and this was the device adopted in the first large scale vacuum dryers for foods. Holding the dryer at a total pressure of 1.6–2.0 kPa, the first two stages bring the vapours to a condition in which they can be dealt with in a spray condenser, serviced by a wet rotary vacuum pump. The non-condensables are subject to a further two augmention stages.

As drying proceeds, vapour production falls away. Most of the vapour pressure 'downgradient' is spent within the piece, so the chamber pressure tends to fall and the leakage pressure may become predominant. The ejector system, which has a maximum designed capacity for both functions, adapts to this but becomes more costly, relatively, to run unless adjusted. It is of course dependent on the temperature of steam and of available cooling water.

If the two functions, removing water vapour and removing air, are taken separately, then a refrigeration unit backed by a large rotary vane constant volume 'vacuum' pump may be employed. The water vapour condenses as ice on the refrigerated coils, usually positioned in an isolatable section of the evacuated chamber, so that the accumulated moisture may be disposed of by deicing at intervals. The air exhausting pump must be beyond the ice-condenser 'pump' to be protected from contamination by water vapour and must be capable of rapid pull-down and of low (partial) air pressures.

Vacuum band dryers

Vacuum band dryers can be used to dehydrate heat sensitive products, or products which suffer readily from oxidation. They are best suited to handling thick pastes, but small shaped pieces and powders can also be dried in this class of dryer. Examples include chocolate crumb, meat and vegetable extracts, and fruit juices and purees. The foodstuff travels over or between externally heated batteries of surfaces, heated by conduction or infrared radiation. These are all located within a vacuum chamber. The support may consist of a number of endless woven-wire belts, the foodstuff being passed from belt to belt.

Freeze dryers

Freeze drying is dehydration by freezing and sublimation; ice is formed from the natural liquid content of biological material and then this frozen segregate is extracted directly as vapour under carefully controlled conditions of temperature and pressure to leave the fundamental structure rapidly restorable to its original state by the addition of water (see Figure 15.8). Applied to perishable materials such as foods, this process allows them, when suitably protected from moisture, light and oxygen, to be preserved for extremely long periods at ambient temperatures. On reconstitution, their appearance, palatability and quality are normally superior to foods dried by other means and stored under similar condition.

Freeze drying for the medical, veterinary and pharmaceutical fields has been practised on a relatively broad scale for many years, supplying markets which are accustomed to a high production-added-value on top of a high initial cost, the materials involved being principally assessed by the users in terms of viability. Safeguards involving small load cycles and particularly low temperatures are stipulated, which cannot be sustained economically with less expensive commodities, for which precise morphological restoration is not the measure. The process of preservation for graft tissues, plasma, vaccines and antibiotics tends to be cautiously prolonged.

In the case of foodstuffs, much more intense treatment, including deliberate inactivation during pretreatment, can be given without detriment to nutritive value, visual appearance, rapid reconstitution or other factors which make the product attractive to the consumer. The entire process must be economically acceptable within lower ceilings and on a much large scale.

Acceptability

Freeze drying of foods is confined at the present time to a limited number of products, e.g. beverages. From a spate of enthusiasm for freeze drying everything on universally adaptable plant, processors have concentrated their efforts along particular lines, sometimes to produce intermediate products whose appearance, flavour and scrupulously retained texture are sought by further processors, so that the general public encounter them indirectly. Reconstitution, next to cost, has been shown as an overall criterion. To the fastidious process assessor, a long reconstitution time may point to partial drying from the liquid phase, or overdrying in an irreversible manner or incor-

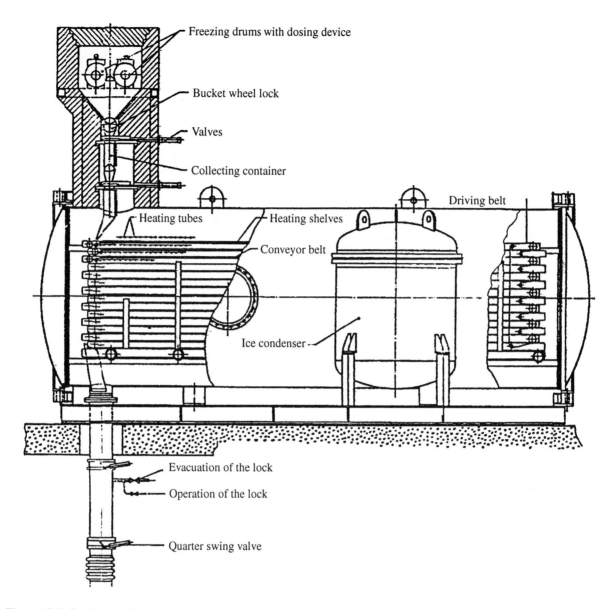

Figure 15.8 Continuous freeze dryer.

rect freezing. In this respect it is important to differentiate between rehydration and reconstitution, the latter embracing the former. Evolution of less expensive plastics as tried and thoroughly tested packaging materials has also done much to bring down the overall costs.

Advantages

While freezing in itself is an action which brings foodstuffs to a state in which decay is halted, preservation depends on maintaining a sequence of environments (the 'cold chain') which denies deterioration the opportunity to resume. The further step of drying enables the foodstuff to be sealed in lightweight packs which are sufficient protection in themselves, needing no particular temperature control nor level. The quality merits are readily discernible by the consumer. Again, while it is difficult to achieve efficient use of energy in other preservation processes, including other forms of drying, freeze drying lends itself to heat pump design concepts which are undoubtedly economical. It is more involved and more demanding (within the actual factory) of initial capital outlay than other means of achieving extended preservation, so that commitment to traditional processes has been an impediment; but the actual process cost is no greater a fraction of total cost and freeze drying has justified itself alone in many cases by the savings it furnishes outside the factory. Major overall attractions lie in low distribution and storage costs, convenience in use and retention of quality; these offset the higher costs attributable to processing.

Assessment is largely organoleptic. The process variations affecting these attributes can be related to a scale of economics. Plant has been devised specially to cater for various forms of foodstuffs and preservation employing freeze drying is now competitive with other methods, albeit in a narrower range than initially contemplated. A process which offers to preserve good appearance carries an inducement to dry solid foodstuffs in as large pieces as possible. Against this, freezing times and drying throughputs demand short paths in at least one dimension.

Freezing

Formation of ice in the food is the basic dehydrating action; the secondary action is to extract the frozen moisture as vapour without disturbing the remaining material. The particular freezing procedure positions and dictates the form of ice and predetermines its accessibility for extraction, whether or not this has been made easier by softening of cell walls, where scalding or partial cooking is permissible.

If the basic structure of the food does not change during the sublimation process, then the internal space available for readmission of water is also predetermined. Thus the primary action of freezing has special significance and has to be undertaken in carefully differing manners appropriate to particular foods.

In deciding the rate of cooling, a balance has to be struck between factors such as undesirable solute concentrations, position of ice crystals, subsequent growth of crystals and formation of internal paths. Very rapid freezing promotes small ice crystals, mostly within cells, from which true sublimation can be difficult or at least slow. Very slow chilling, on the other hand, will tend to create instability due to undercooling, and when crystals do form, they are small at first, but on further cooling, solidification leads to large crystals, mostly outside disrupted or crushed cells. The result may give easy vapour paths but also loss of structure, and perhaps problems in storage, as well as a reduction in final water-holding power. An intermediate condition should be sought (less than 1½ h from ambient through the critical range, i.e. 0 to $-5°C$).

The safe frozen level (as a rule well below $-7°C$) is equally important, but for the complex solutions in the cells is not always clearly defined; the thawing level is usually more definite and

always higher. The procedure must involve cooling to a temperature below the acceptable freezing zone, after which it is generally provident to retrace to a temperature level just below the melting point prior to sublimation. However, prolonged retention in this elevated state may alter the crystalline formation.

Some liquids do not appear to yield ice crystals on initial cooling, but concentrate into a highly viscous state and finally a glass-like solid. It is possible in some instances to change this irreversibly to a crystalline structure by gently warming the material and holding it at a higher temperature. After this transition, the cooling and solidification processes may be continued to completion in a more favourable manner.

Air-blast freezers may be successfully employed on cycles which bring the foods finally to a condition slightly below the predetermined ice temperature at which the drying operation is to be carried out. The material may rise nearer to this level while being placed in the dryer in which it should rapidly be brought under control.

Liquid and semi-liquid foods (e.g. beverages, sauces) are often much more heat sensitive than solid foods. Thick beds are slow to dry; the neatest arrangement is to prefreeze them as thin sheets (e.g. on a drum freezer), which can then be fragmented and heaped in order to strike a balance between easy vapour escape paths and adequate heat transfer.

Evaporative freezing may be employed in certain cases, but as this in effect involves drying from the liquid phase to extract latent heat, it should not be prolonged nor resorted to except in such cases as cooked meats, where denaturation has already taken place and the effects are not easily discerned.

Measurement

In air employed for drying, the partial pressure of the water vapour is not actually measured; it can be deduced indirectly from a knowledge of the dry bulb and wet bulb temperatures. In the high vacuum conditions pertaining to freeze drying, a distilled water icicle will equilibrate to the vapour pressure around it, and its temperature can likewise be used for inferential measurement. Normal vacuum gauges indicate total pressure and with appropriate traps can be arranged to give the partial pressure of air, so that the vapour pressure may be estimated by difference. It is feasible therefore to control freeze drying processes from both temperature data and minute pressure gradients.

As an illustration of what can be achieved, prefrozen beef steaks 15 mm thick, spread in an AFD (accelerated freeze drying) cabinet at 12 kg m^{-2}, have been dried to less than 2% moisture content in less than 6½ h. Processing details for many other products are given in an HMSO (1961) publication.

Dehydrofreezing

This is a hybrid process, as the name implies. Not only is the moisture in the heart of a discrete piece of foodstuff progressively more time consuming to remove in any method of drying, but the air in the fine pores may delay water uptake during reconstitution. Since preservation is paramount in the processor's objective, it may be considered expeditious to perform the easy, early drying stage and then change tactics by quick freezing the remaining moisture.

Dehydrofreezing is a term coined to describe such a composite (deterioration-arresting) process, which at the expense of reliance on the 'cold chain', provides a product, which unlike some completely air-dried material, does not require prolonged soaking before cooking and compares in quality and storage life with wholly quick frozen products.

Dehydration is therefore carried only to the stage of removing about half the water originally present – and if the size of the piece of foodstuff is appropriate (e.g. 9 mm dice, peas), this can be done in under half an hour. This intermediate material is then quick frozen by standard air-blast freezing techniques.

Belt trough and pneumatic dryers have been used. Uniform drying of each and every piece is essential. The shrinkage accompanying the intensive spell of drying means that the pieces require only about half the refrigerated storage space that normal frozen foods would occupy.

Obviously the consumer, unless captive, has to have the reason for the unusual appearance explained, since he is probably expecting a finished product comparable in end-sale-point attractiveness with that from complete quick freezing.

Ancillary equipment – fans and blowers

To avoid unequal or uneven drying, it is a requirement that the heated air operates over a limited range of temperature, so that conditions on its path across or through the bed of material do not alter significantly in terms of its heat supplying and its vapour carrying capacities. It follows that large quantities of air must be driven round the system at a relatively low static pressure. The main resistance to flow that has to be overcome is the heater battery. The air 'blower' is usually sited in the coolest part of the airway, after wet air has been discharged and fresh air drawn in and before the heater.

Propeller fans directly confronting trays on trucks require elaborate air straighteners to obviate air velocity differences between trays and are rarely used now in that fashion. Axial flow fans have their motor in the adverse conditions of the airway, although otherwise they have a high efficiency. Centrifugal fans are now available in high width/diameter ratios and so have low tip speeds and are quiet. The forward-curved-blade type overloads above design pressure. The backward-curved-blade type does not overload, has a steeper pressure curve, high efficiency and is also self-cleaning. Turbo-blowers with a blast gate can be arranged for either constant discharge pressure, constant volume or constant weight. Selection and control of the air mover requires specialist advice.

QUALITY ASPECTS

Selection and pretreatment considerations

The physical nature of the substance to be dried has important bearings on the economics of drying. Thus foodstuffs may conveniently be thought of as mainly liquid (e.g. beverages), mainly pastes (macaroni-and-cheese, spaghetti, tomato sauce) or mainly discrete pieces (steaks, chips, dice, berries). It is a further advantage to look on structure as fibrous (muscle), sac-like (oranges, blueberry) or dense cellular (potato). Composite dishes, if the ingredients are not dried separately, have generally to be treated in a manner that favours the slowest drying component.

In terms of nutritional value, there may be a chosen ripeness. Pulses, grain and berries can be graded to size. Desire for homogeneity and waste-free produce will prompt extra attention to general selection and preparation. Natural external and internal moisture barriers should be removed or overcome, and various procedures have been devised to this end. Examples are: the deep removal of the plug from fruits such as whole strawberries and raspberries, so that drying is mainly independent of vapour passage through the impermeable outer surface; scratching by abrasion whilst frozen of the thick skins of berries and fruits such as blackcurrants; pricking and steam scalding of peas and beans without noticeable damage to appearance; peeling and segmenting of peaches, apples, apricots and other firm flesh fruits; peeling, subdividing and steam scalding of root vegetables; separation of muscle from fat and interconnective tissue; cutting across the fibres to give short easy vapour flow along them; partial cooking to condition all walls, especially in sea-foods like shrimps. In the case of chicken meat, partial cooking not only makes stripping easier but also has in itself a dehydrating effect. Partial cooking of thin pieces of vegetable may take place during blanching. It is pertinent to stress that the product can only have qualities in keeping with the raw material, and varieties of fruit (say) of high colour and flavour, perhaps slightly under-ripe, should therefore be selected.

Compression of dehydrated foods

Dehydration alone brings about a great reduction in weight and through shrinkage a considerable reduction in volume of many commodities. For diverse reasons, the actual bulk of certain dehy-drated foods may be further reduced by compres-sion. An extreme example is cabbage, where the ratios of raw material to packed dehydrated product can be (by weight) 18:1, (by volume (38:1) when compressed, compared to 12.5:1 and 6.9:1, respectively when compression techniques are omitted.

Comparable figures for meat, which when fresh and transported in refrigerated vehicles over long distances, should logically take into consideration the non-payload weight and space of refrigeration equipment and insulation, are as follows: 1 tonne of chilled carcass beef occupies 3.07 m^3 of actual hold space; 1 tonne of frozen carcass beef (which can be stacked) occupies 2.51 m^3; 1 tonne of bone-less frozen beef, 2.23 m^3. The dehydrated equiva-lent in the form of mince weighs about 242 kg; compressed and packed into cans in wooden cases the total weight is 280 kg and the volume 0.357 m^3.

The techniques for compression differ widely according to the nature of the food. Dehydrated scalded vegetables with a high content of soluble solids, for example, are first plasticized by heating for 1 to 2 min in a current of warm air at 60–70°C; the air should preferably be of mod-erate relative humidity to prevent excessive drying. If the pieces are then transferred without loss of heat to a heated mould maintained at 50–60°C, they are compressible into coherent blocks by a pressure of 14 to 27.5 MPa applied

for 10–30 s, the conditions being chosen to give the best results with the vegetable concerned. Blocks with a density 800–1000 kg m^{-3} may be obtained by this method and they have accep-table reconstitution characteristics. The blocks may conveniently be 12 to 18 mm thick, and square, rectangular or circular in shape. Circular blocks of a size to fit neatly into A10 cans (3 l or 3 kg) have been evolved to form a high density catering pack.

Reconstitution

Prior to consumption, preserved dried pieces of food generally require as much as possible of the original moisture to be restored, so that the soluble constituents of the cells are returned to their original state. This is rather more than rehy-dration and a simple yardstick is water-holding power.

Unless reconstitution takes place elaborately under vacuum (e.g. by piercing an evacuated con-tainer submerged in water), gases other than water vapour in the porous dry material will have to be displaced or dissolved. This may delay the uptake of water in the case of sizeable pieces. Floating or sprinkling such pieces on the water of reconstitu-tion is often a faster method than total immersion in that outlets for the air are not closed by surface swelling.

By and large, however, all that may be required is to warm the water. In presenting the product as a convenient food, a balance generally has to be struck between cost conscious processing, which may have caused departure from the original char-acteristics of the foodstuff and speed of reconstitu-tion. If the product has been dried so that it may be mixed conveniently with other ingredients in powder or granular form, then the problem will not arise. Alternatively, the distributor may under-take reconstitution in bulk as an added service to the consumer and sell the product to the public ready to use. Whatever the method, if reconstitu-tion is rapid and the original characteristics are restored, the aims of the process have been satis-fied.

NON-THERMAL METHODS

IRRADIATION

Introduction

Food irradiation originally dates back to 1930 when the first patent was taken out for the process. As a means of preserving food, it is unquestionably effective in killing bacteria, e.g. *Salmonella*, *Campylobacter*, *Shigella* and *Listeria* spp. It can also be used to inhibit sprouting or ripening in vegetables and fruit. Its lack of development over the intervening years has been in part due to the need to satisfy stringent safety criteria in order to convince governments and the public of its acceptability as a process.

Only recently has the irradiation of food for sale been permitted in the UK. The Food (Control of Irradiation) Regulations 1990 now permits irradiation of certain categories of food by persons with irradiation licences provided that the stipulated doses of irradiation are not exceeded. Other countries, including some in the EU (Belgium, France and Denmark), accepted food irradiation earlier for use on a limited range of products. In many countries, spices are the only foods permitted to be irradiated.

The US legislation is to be found under Title 21 Code of Federal Regulations Section 179, where a similar list of permitted foods and maximum losses are given.

This part of this chapter will deal with the current legislation and process involved in this treatment of food.

Current legislation

At the time of going to press, the UK legislation referred to above permits the irradiation of the following categories of food at the maximum radiation doses given:

Item	Maximum dose (kGy)
Fruit	2
Vegetables	1
Cereals	1
Bulbs and tubers	0.2
Spices and condiments	10
Fish and shellfish	3
Poultry	7

No less than 98% by weight of the item must be in the above category for irradiation to be permitted. The food must have been irradiated in the UK or, if imported, it must follow the same requirements and must be accompanied by appropriate documentation. Any food which contains an irradiated ingredient must be labelled accordingly and declare the ingredient as such. Food may only be irradiated by a person holding a licence permitting them to carry out such treatment. At the time of writing only one such licence

has been granted in the UK, to Isotron, Swindon, and this for herbs and spices only.

Detection of irradiated foods

There has been a great deal of consumer resistance to the introduction of irradiated foods and a consequent demand for methods of detection. This has so far been difficult to do because much of the change in a food brought about by irradiation is the same as that brought about by conventional processing. Currently research into three techniques, electron spin resonance, thermoluminescence and the formation of volatile compounds from lipid-containing foods are showing the most promise.

Methods of irradiation

There are two methods of irradiation in use in Europe.

(i) *Gamma rays*. These are produced during the decay of certain radioisotopes, e.g. cobalt-60 or caesium-137.
(ii) *Electron accelerators*. These produce high energy electron beams, accelerated to high speeds.

In both cases, the costs of plant and specialized housing units are very high, currently estimated at between £2 million and £2.5 million. The high resultant costs of treatment to offset such outlays may be prohibitive for many foods. Both methods produce no heat and frozen foods can be safely treated in either way.

The cobalt-60 method is that used in the UK. The radioactive source is contained within stainless steel tubes in a framework to produce a uniform radiation field. This is housed within a concrete cell and the product is conveyed into it and follows a path around the source to allow for even exposure. Exposure time, and hence the dose received by the product, is determined by the conveyor speed. When not in use the source is lowered into a concrete pit or a deep pool of water.

HIGH PRESSURE PROCESSING

The application of high hydrostatic pressures in the range 100–600 MPa at or around ambient temperatures will inactivate not only vegetative micro-organisms and spores but also enzymes and proteins. The main advantages of this technique are that the food will not be heat degraded and the quality should closely resemble that of the raw product.

The equipment required for high pressure processing of a food product consists of a forged monolithic cylindrical vessel made of low alloy steel of high tensile strength. The wall thickness depends on the maximum working pressure, the diameter of the vessel and the number of cycles for which the vessel is designed.

For a vessel to withstand a pressure of around 100 MPa, a diameter of 100 mm and a length of 1000 mm would be required, giving a working volume of 8.5 l. Prestressed vessels are used for higher pressures but these are more expensive than monolithic blocks. Pressures may be generated by water or oil pumps and pressure intensifiers. Before pressurizing the vessel, it is necessary to evacuate the air using a vacuum system. Electric heating bands around the vessel may be used to elevate the temperature.

The first products from this technology were launched in Japan in 1990 and included jams, fruit sauce for yoghurts, fruit jelly, fruit juices and tenderized beef. Developments in Europe and the USA are expected in the future. The technique, with its high capital cost and limited throughput, will only produce high quality, high priced products.

REFERENCES

HMSO (1961) *The Accelerated Freeze Drying (AFD) Method of Food Preservation*, HMSO, London.

Holdsworth, S.D. (1996) *Thermal Processing of Packaged Foods*, Chapman & Hall, London.

FURTHER READING

Baker, C.G.J. (ed.) (1997) *Industrial Drying of Foods*, Blackie Academic & Professional, London & Glasgow.

Blackholly, H. and Thomas, P. (1989) *Food Irradiation*, University of Bradford, Horton Publishing, Bradford.

Brennan, J.G., Butters, J.R., Cowell, N.D. and Lilly, A.E.V. (1990) *Food Engineering Operations*, Applied Science Publishers, London.

Brennan, J.G. (1994) *Food Dehydration – A Dictionary and Guide*, Butterworth Heinemann, Oxford.

Cleland, A.C. (1990) *Food Refrigeration Processes, Analysis, Design and Simulation*, Elsevier Science Publishers, London.

Dalgleish, J.McN. (1990) *Freeze-Drying in the Food Industry*, Elsevier Science Publishers, London.

Diehl, J.F. (1995) *Safety of Irradiated Foods*, 2nd edn, Marcel Dekker, New York.

Fellows, P.J. (1996) *Food Processing Technology – Principles and Practice*, Woodhead, Cambridge.

Footitt, R.J. and Lewis, A.S. (eds) (1995) *The Canning of Fish and Meat*, Blackie Academic & Professional, London & Glasgow.

Gosney, W.B. (1982) *Principles of Refrigeration*, Cambridge University Press, Cambridge.

Gould, G.W. (ed.) (1995) *New Methods of Food Preservation*, Blackie Academic & Professional, London & Glasgow.

Hallowell, E.R. (1980) *Cold Storage and Freezer Storage Manual*, 2nd edn, AVI, Westport, Connecticut.

Heldman, D.R. and Singh, R.P. (1981) *Food Process Engineering*, AVI, Westport, Connecticut.

Herson, A.C. and Hulland, E.D. (1980) *Canned Foods, An Introduction to their Microbiology*, 7th edn, J. and A. Churchill, London.

Holdsworth, S.D. (1983) *The Preservation of Fruit and Vegetable Products*, Macmillan Press, London.

Holdsworth, S.D. (1992) *Processing and Aseptic Packaging of Food*, Elsevier Science Publishers, London.

Josephson, E.S. and Peterson, M.S. (1983) *Preservation of Food by Ionizing Radiation*, 3 volumes, CRC Press, Florida.

Keey, R.B. (1978) *Introduction to Industrial Drying Operations*, Pergamon Press, Oxford.

Kyzlink, V. (1990) *Principles of Food Preservation*, Elsevier Science Publishers, London.

MacCarthy, D. (ed.) (1986) *Concentration and Drying of Foods*, Elsevier Science Publishers, London.

Rees, J.A.G. and Bettison, J. (1990) *Processing and Packaging of Heat Preserved Foods*, Blackie Academic & Professional, Glasgow.

Schwartzberg, H.G. and Rao, M.A. (eds) (1990) *Biotechnology and Food Process Engineering*, Marcel Dekker Inc., New York.

Shapton, D.A. and Shapton, N.F. (1991) *Principles and Practices for the Safe Processing of Foods*, Butterworth-Heinemann, London.

Thorne, S. (1981–1989) *Developments in Food Preservation*, Volumes 1–5, Elsevier Science Publishers, London.

Urbain, W.M. (1986) *Food Irradiation*, Academic Press, New York.

Webb, T. and Lang, T. (1987) *Food Irradiation: The Facts*, Thorsons, Wellingborough.

16 Food Packaging

INTRODUCTION

Food products can be preserved using a variety of methods. For example, they may be acidified, chilled, dried, frozen or heat processed. Whichever method is used, packaging will be required to protect the food and to aid its distribution to the consumer. Packaging can be defined simply as a means of ensuring the safe delivery of a product to the consumer in a sound condition at minimum overall cost. The criteria for the selection of packaging materials are relatively straightforward:

(i) The packaging material should be inert. The packaging regulations in operation in most countries are written such that the packaging material should not transfer its constituents to the foodstuff it contains. However, there are specific instances where some limited movement is desirable, e.g. tin from tinplate containers in some medium and high acid products.

(ii) It should protect the product from adverse environmental conditions, e.g. light, oxygen and moisture, under normal foreseeable conditions of distribution and use. Food products can deteriorate if exposed to such conditions.

Light can change the colour of products by bleaching them over a period time while in storage. Oxygen entering or leaving a food package can affect the shelf life by changing the rate of oxidation of a product or, with modified atmosphere packed products, loss of oxygen can allow the growth of micro-organisms to flourish.

With dried and semi-moist products it is important to maintain the level of moisture within the product to obtain the desired shelf life.

(iii) The package should present the product in an appealing manner and should encourage the consumer to select that particular product rather than another. The use of packaging material, shape, printing technique and colour can be used to achieve the desired effect.

(iv) The package should be easy for the consumer to use. Some packs, for example, are relatively difficult to open. This difficulty should be balanced against the perceived increase in tampering (or grazing) and the need to protect the consumer by the use of tamper-evident devices.

PACKAGING MATERIALS

The packaging industry is diverse, but uses four basic materials, namely glass, paper/board, metals and plastics. Packaging is designed to protect the product as well as supply information to the user about that product. In order to make the most appropriate choice of packaging material, it is important that the product definition includes as much information as possible, covering the degree of protection required. Packaging materials are, in a sense, passive participants in the process of food preservation.

PACKAGING MATERIALS AND THEIR PROPERTIES

Glass

Glass is one of the oldest packaging materials known. It has many advantages. It is clear, rigid and impermeable to gases. It can be made in a variety of shapes and sizes to suit individual customer requirements. The majority of glass used by the food processing industry is known as 'soda-lime' glass, which reflects its ingredients. The basic component of glass is silica derived from sand, flint or quartz. A typical formula for soda-lime glass is:

Silica	SiO_2	68–73%
Calcia	CaO	10–13%
Magnesia	MgO	0.3–3%

Soda	Na_2O	12–15%
Alumina	Al_2O_3	1.5–2%
Ferric oxide	Fe_2O_3	0.05–0.25%

Other minor ingredients may be added for colour.

Glass manufacture is a two-stage process. The mixed raw materials, including cullet (broken glass) are melted in a furnace in the region of 1500°C before being extruded at about 1100°C and cut into gobs. These pass to the moulding and forming machines, where the bottles are formed at about 800°C, after which their temperature drops to about 500°C. At this point, they are given an external 'hot-end' coating made either from tin or titanium. This layer provides a surface to which the 'cold-end' treatment adheres better than it would to a pristine glass surface. Following this coating, the bottle is annealed in the annealing Lehr. The Lehr heats the bottle back up to about 600°C before allowing it cool under controlled conditions down to 100°C over a period of at least one hour. It is at this point that the other coating, referred to as the 'cold-end' coating, is applied. This is an organic material, such as a wax, stearate or polyethylene, which increases the lubricity to aid handling the bottle on a modern production line.

Following the coating process, the glass containers are inspected to ensure that they are free from faults. They are then bulk palletized and dispatched to the warehouse.

The shape or design of glass containers can be made to suit individual customer requirements. The properties of glass can be categorized as mechanical, thermal and optical.

Mechanical

(i) Internal pressure resistance. The internal pressure resistance of glass containers is important, particularly when they are being used for carbonated drinks. The internal pressure can produce stresses at various points on the outer surface, which could lead to the sealed container 'exploding'.

(ii) Vertical load. The design of the container at the shoulder needs to be considered so as to minimize breakage under load.

(iii) Resistance to impact. There are two points to consider with impact resistance. The first is when a moving container meeting a stationary object, i.e. when being dropped. In this instance the container design incorporates strengthened contact points. The second consideration is container-to-container impact, such as when moving containers impact on stationary containers on filling lines. Here again, the impact points are strengthened and the design modified to keep contact to a minimum.

(iv) Resistance to abrasion. The strength of the glass can be reduced by abrasion. Use of the surface treatments described above can provide the necessary resistance, thereby overcoming container failure.

Thermal

The thermal strength of a container is a measure of its ability to withstand rapid changes in temperature. Resistance to thermal failure can be designed in by considering the type of glass, shape of container and the thickness of the glass at specified points within the container. The food processing industry requires glass which will withstand thermal processing at a range of temperatures and pressures.

Optical

The optical properties of glass can be modified by the addition of additives such as metallic oxides, sulphides or selenides. Some of the transition metal oxides will absorb light not only in the visible spectrum but also in the UV and IR regions.

Paper and paper-based products

Paper and board are some of the oldest packaging materials known. Paperboard materials used in the food industry are usually considered in three groups, (i) paper, 250–300 µm; (ii) board, 250 µm +; and (iii) fibreboard. There are no definitive definitions of the thickness of paper and board. Values vary from country to country.

Paper

Paper is produced in a range of grades to provide the customer with the specific requirements needed for the application. It may be laminated with PE (polyethylene) to produce moisture-resistant layers. High quality paper can be used to good effect to produce labels for packaging.

Board (cartonboard)

Board is used to produce both folding and rigid cartons. It consists of a number of plies made from different grades of fibrous material. For example, white-lined chipboard (duplex) has a white surface on one side, made from bleached virgin pulp, with the bulk being composed of 'chip', which is usually grey in colour and is made from a high proportion of waste fibre.

Fibreboard (corrugated board)

Most fibreboard is produced using a multilayer corrugated construction. Using three layers of thin paper is an adaptation of the engineering beam principle of two flat load bearing panels separated by a rigid web. Corrugated board is categorized in three ways: (i) by the thickness and spacing of the corrugated layer, (ii) by the weight in g m^{-2} of the facing layers (the liners) and (iii) by the quality of the papers used.

Metal

The four main metals used in food packaging are aluminium, chromium, steel and tin.

Aluminium

Aluminium is the earth's most abundant metallic constituent. Alumina or aluminium oxide (Al_2O_3) is the only oxide formed by aluminium. Its most common form is bauxite, an impure form of gibbsite ($Al_2O_3.3H_2O$).

A typical aluminium manufacturing process is as follows. Alumina is dissolved in cryolite in carbon lined boxes called pots. A carbon electrode or anode is lowered into the solution and an electric current of 50–150 MA is passed through the mixture to the carbon lining of the pot, which acts as the cathode. The current reduces the alumina to aluminium and oxygen. The aluminium settles at the bottom of the pot while the oxygen combines with the carbon of the anode to form carbon dioxide.

The majority of commercial uses for aluminium require special properties which the pure metal cannot provide. For this reason, alloying agents are added to impart strength, improve forming characteristics and influence corrosion properties.

Chrome

Chromium is used to coat low carbon mild steel equally on both sides with a complex layer of chromium and chromium hydroxides, which is applied by electrodeposition. Chromium coated steel is known as either ECCS (electrolytic chromium/chromium oxide coated steel) or TFS (tin-free steel). It is used primarily as a coating for food packaging which is to contain low and medium acid foods.

Steel

The starting point for tinmill products is the hot-rolled coil supplied by the rolling mill. The steel base may either be ingot cast or, more commonly in the UK, continuous cast. The base steel is cleaned, annealed and temper rolled before being electrolytically coated. The coatings used are either tin or chrome. During annealing, the steel is heated to 600–700°C. This causes recrystallization of the structure within the metal and makes it easier to work. Temper rolling imparts a springiness to the steel and only slightly affects the surface hardness. Tinmill products are sold by area. The unit of area is 100 m^2 and is known as a SITA (Système Internationale Tinplate Area).

Tin

Tin is used for a range of food containers. It is most noticeable in the packaging of high acid foods, such as tomatoes and pineapple segments. Small amounts of tin react with the product during storage, absorbing any oxygen present and maintaining the bright colour of the fruit. Steel coated with tin is known as tinplate; the coating thickness can be different on each side of the plate.

The weight of tin on each surface is expressed in g m^{-2}. For example, E2.8/2.8 has 2.8 g m^{-2} on each surface, giving a total tin coating weight of 5.6 g m^{-2}. Differential coatings, where one surface (the obverse) carries a heavier tin coating than the other surface (the reverse), are identified by marking the more heavily coated surface with continuous parallel lines in the rolling direction. The spacing between the lines identifies the differential coating weights used.

Plastics

The term plastic is used to cover a range of materials based on synthetic or modified natural polymers, which at some stage of manufacture can be formed into the desired shape by flow, aided by heat and pressure. Plastics are of two types, thermoplastics and thermosets.

Thermoplastics are those materials which repeatedly become soft when exposed to heat and harden again when cold and thermosets are those materials which set into a permanent shape when processed under heat and pressure and do not soften when heat and pressure are reapplied.

Plastics are converted into many forms and types of product for packaging uses. Some of these are films; sheeting; tubings – flexible or rigid; coatings – calendered, flow, hot-melt, dip, spray and so on; laminations; additives; mouldings; tapes; cushionings; adhesives; resin coatings; impregnations.

The selection of the correct plastic to use for a defined application can be made after considering the properties required.

Structural plastics

These give rigidity or shape to the container. They define its shape and dimensions, impart mechanical strength, impact resistance and resistance to pressure. Some examples of structural plastics are polypropylene (PP), high density polyethylene (HDPE), poly(vinyl chloride) (PVC), polyester (e.g. polyethylene terephthalate, PET), polystyrene (PS) and polycarbonate (PC).

Barrier plastics

Barrier plastics are used to prevent the ingress or loss of gas or moisture to a sealed container.

(i) Oxygen barriers. Oxygen permeation through the plastic material can affect the flavour and colour of the packed product. Small amounts of oxygen ingress may cause off-flavours to develop in food by oxidative rancidity. Some plastics, which exhibit good gas and odour properties include ethylvinyl alcohol (EVOH), polyvinylidene chloride (PVDC) and some polyesters (PET). PET, while exhibiting some oxygen barrier properties, is a poor oxygen barrier when compared with EVOH and PVDC.

(ii) Moisture barriers. Moisture barriers, apart from protecting the product from moisture gain or loss, are also essential for the protection of the oxygen barrier materials used in some packs, e.g. EVOH. These, while providing an excellent barrier to oxygen, are hygroscopic and will absorb moisture from the product and/or atmosphere, thereby causing a decrease in their effectiveness.

(iii) Light barriers. Plastics can provide for extremes in range of transmission between transparency or opacity. Selected pigmented polymers can be used to control the amount of light transmission through the plastic. This technique cannot completely eliminate light

transmission, but by combining selected polymers can reduce it significantly.

(iv) Gas barriers. Gas transmission through polymers depends on the relative order of magnitude of intermolecular forces. An increasing degree of crystallinity directly decreases the permeability of all molecules. Orientating a crystalline polymer substantially increases the diffusional path with the result that highly orientated polymer films such as PET and nylon, become efficient gas barriers.

The thermoplastics can be further subdivided into polyethylene (PE), polyethylene terephthalate (PET), poly(vinyl chloride) (PVC), polyamide or nylon (PA) and polypropylene (PP).

Metallized films

Metallized films are manufactured by applying very thin layers of aluminium to a plastic film by a process known as vacuum deposition. This results in a material which may have the appearance of aluminium foil. The barrier properties of metallized films are related to the thickness of the deposited metal; the brightness of their metallic finish is affected by the base film.

Metal deposits on films are generally described in terms of light transmission through the film; 4% light transmission provides a good metallic appearance but does not improve the barrier properties of the film significantly. Advances in metallizing technology have resulted in an increase in the functionality of the metallized films. For example, films with less than 1% light transmission having excellent barrier properties can now be produced.

Laminates

Flexible materials are combined on laminating machinery to optimize physical and barrier properties. The concept of bonding is central to any laminating operation. There are three main laminating techniques.

(i) Wet bond laminating This refers to any process in which liquid adhesive is applied to a suitable substrate that is then immediately combined with a second ply to create a laminate.

(ii) Dry bond laminating. A solvent borne adhesive is applied to a substrate that passes through a dryer to evaporate the carrier solvent. The web is then combined with a second substrate in heated pressure nip rollers.

(iii) Thermal or pressure laminating. Thermal laminating can be defined as the bonding of two or more flexible films with one or more of the adhesive types listed below under heat and pressure, without the requirement of a simultaneous application of solvent or solvent-free coating. Some of the applications listed below could be described as either wet bond or dry bond techniques; this makes a description of lamination very difficult.

The 'adhesive' takes the form of one or more of the following:

(i) A previously applied thermoplastic adhesive, which is activated by the heat and pressure of the thermal laminator.
(ii) A special sealing layer on one or both sides of a structured film.
(iii) Adhesive in a dry film form.
(iv) A relatively low melting point film, treated and introduced partially as a sealant.
(v) A 'coated' adhesive on a non-reactive (pressure sensitive) carrier. Some of these adhesives can be utilized at ambient temperatures.

Coating materials

Coatings are used in packaging to protect the food-contact surface from reacting with the food-stuff or, on the outside of the package, to protect the graphics from abrasion. They may also be used to change the properties of the packaging material. Examples include:

(i) Can lacquers. The internal surface of metal containers can be coated with a number of materials, which range from epoxyphenolics through to acrylic and polyester coatings. These coatings are used to protect the internal food-contact surface from reacting with the foodstuff and preventing the can from corroding from the inside.
(ii) Varnish. Varnishes are used on the external (printed) surface of cartonboard and some metal packaging. The primary aim is to protect the graphics/artwork from being damaged during manufacture. On cartonboard, the varnish may be modified to protect the board from moisture ingress in specific applications, such as freezing.
(iii) Coatings on glass. Glass may be coated at both the 'hot-end' and the 'cold-end' to provide protection of the surface and ensure the desired handling properties are achieved.
(iv) Wax. Some cartonboard may be wax coated or dipped to protect the board from ingress of moisture in wet applications. These types of applications are being replaced by either PE linings or twin-walled plastic (PE) sheet.
(v) Polyethylene (PE). Small food cartons for retail applications may have an internal layer of PE. Although this layer may be a lamina-

Table 16.1 Permeability of packaging films to gases

Material	Permeability (ml/m^2/MPa/day)[a]				Ratios to N$_2$ permeability		
	N$_2$ at 30°C	O$_2$ at 30°C	CO$_2$ at 30°C	H$_2$O at 25°C 90% RH	O$_2$/N$_2$	CO$_2$/N$_2$	H$_2$O/N$_2$
PVdC (Saran)	0.07	0.35	1.9	94	5.0	27	1400
PCTFE[b]	0.20	0.66	4.8	19	3.4	24	95
Polyester (Mylar A)	0.33	1.47	10	8700	4.5	31	27 000
Polyamide (nylon 6)	0.67	2.5	10	47 000	3.7	15	70 000
PVC unplasticized	2.7	8.0	6.7	10 000	3.0	25	3800
Cellulose acetate (P912)	19	52	450	500 000	2.8	24	2700
Polythene							
(d=0.954–0.960)	18	71	230	860	3.9	13	47
(d=0.922)	120	360	2300	5300	2.9	19	44
Polystyrene	19	73	50	80 000	3.8	31	4200
(d=0.910)	–	150	610	4500	–	–	–

[a]All permeabilities calculated for 25 μm thick film.
[b]Polychlorofluorethylene.
d=density.
Source: Paine and Paine (1983).

Table 16.2 Resistance of packaging films to various substances

	Resistance[a] to:					
	Water vapour	Gases	Odour	Water	Oil or grease	Chemicals
Polythene (PE)	7	3	3	10	5	10
PVC	2	5	5	9	8	6
PVdC copolymer	9	8	8	10	8	8
Polyester	4	6	8	10	8	10
PT cellulose	0	6	6	5	10	3
MSAT cellulose	5	6	6	6	10	2
MXXT cellulose	6	9	9	6	10	3
PE/MSAT cellulose	8	8	8	6	8	8
Cellulose acetate	1	2	2	6	6	3
Kraft paper	0	0	0	2	0	0
Sulphite paper	0	0	0	2	0	0
Glassine	0	3	3	3	4	2
Waxed glassine	4	5	4	5	6	4
PE coated paper	7	3	3	6	5	7
PVdC coated paper	8	8	8	6	8	7
Al foil, 0.009 mm	10	10	10	10	10	5

[a]Resistance graded on a scale from 0 = no resistance to 10 = impervious.
Source: Paine and Paine (1983).

tion, it is sometimes referred to as a coating due to its thickness.

Coextrusion

A coextruded film is a multilayer film in which each distinct layer is formed by a simultaneous extrusion process through a single die. Coextrusion allows a multifunctional plastic packaging material to be manufactured in one step, rather than the traditional multistep process of coating and lamination.

Summary of functional properties

Significant functional properties of packaging films for food use are summarized in Tables 16.1 and 16.2.

PACKAGING SPECIFICATIONS

A packaging specification is the description of the material, shape, decoration and function to which it is going to be put. It needs to be considered in terms which both the supplier and customer can understand.

Material

The specification should include exact details of the materials to be used. The more detail that can be built into the specification at this stage makes agreement between supplier and end-user easier and helps prevent any misunderstanding.

Shape/dimensions

The dimensions of the packaging need to be specified in units which both parties can understand. The dimensions and desired shape should be achievable both in design terms and also in user terms. Actual size final samples should be seen.

Fitness for purpose

The agreed specifications for the materials and dimensions of the container should ensure that it is suitable for the desired purpose, and that it will not harm the product or be easily damaged during subsequent handling. For instance, it may be possible to design a glass jar in an intricate shape but the result might be a container that cannot be manufactured safely or would be very difficult to handle on a modern high speed filling line.

Decoration

Packaging is generally decorated to convey an image of the product to the consumer. The production of artwork can be complex and needs to be clearly agreed between supplier and customer at all stages. If colours are to be used, then samples should be seen in the correct colour.

Test methods

Testing of packaging materials and finished packages may need to be undertaken to:

(i) Comply with legal requirements
(ii) Satisfy the requirements of the customer to protect the product throughout the distribution chain to the ultimate consumer
(iii) Comply with the agreed customer specification with regard to material selection and artwork/decoration.

Packaging tests can be divided into two main categories:

(i) *Destructive testing.* This is normally undertaken off-line and results in the package being taken apart and examined using a predetermined method. Some typical tests would include container integrity, torque testing of bottle closures, label adhesion and the examination of internal contents of multipackaged containers.
(ii) *Non-destructive testing.* These tests can be divided into on-line and off-line tests. On-line tests could include vacuum testing of glass jars and cans, camera-based tests for label placement and other camera-based measurements. Off-line tests may include the measurement of specific parameters such as package dimensions and colour, including the matching of approved artwork.

While test procedures exist for most types of packaging, it is not within the scope of this chapter to describe every method. The methods used for testing packaging are adequately described in, for example, BSI (British Standards Institution) and ASTM (American Society for Testing and Materials) publications.

PACKAGING SYSTEMS AND EQUIPMENT

PRIMARY PACKAGING

Glass bottles and jars

Glass containers can be divided into two types: narrow-mouth and wide-mouth containers. These are generally referred to as bottles and jars, respectively. Within these broad descriptions are a range of shapes (round, oval, triangular, square, etc.) and a range of sizes from small, e.g. 1 oz (25 g) jam jars, to large, e.g. 25 l bottles.

Metal cans and other containers

Metal containers have a variety of uses, the most common being food and beverage cans. Other uses include containers for biscuits and decorative tins for confectionery. They may be manufactured from either aluminium or steel.

Food cans are rigid and are designed to withstand the large differences in pressure that can develop during processing in a retort to ensure that the contents are 'commercially' sterile. Cans can be stored for up to five years under ambient conditions.

Food cans may be either three-piece or two-piece. The original three-piece cans were manufactured with a lead-based soldered side seam and the ends attached with double seams. Modern manufacturing methods have now replaced the lead soldering method with a welded side seam. Two-piece cans are more modern. They are manufactured from a single disk of metal that is drawn into the desired shape and trimmed to create the flange to which is added the double seamed end after filling. The addition of beading (ridges or flutes) in the can body has enabled the can maker to reduce the metal thickness, thereby maintaining container strength and reducing material costs.

Retort pouches

Retort pouches are used within the food industry to package ambient shelf stable foods. They have traditionally been manufactured from aluminium laminated to polypropylene with an external protective layer of, for example, nylon/aluminium/polypropylene. However, changes in film manufacture and barrier properties have resulted in nonfoil pouches being manufactured for some specialist applications, e.g. microwave oven use. The selection of materials is of importance as the package must maintain its integrity and barrier properties from manufacture through to final use.

Bags

The term bag can be used to describe a range of containers in various materials. They include:

(i) *Paper bags.* There are three basic types, which can be described by the number of layers used in their construction, namely single-ply, two-ply (duplex) and multiwall, which include three or more layers. The strength and vapour barrier properties of multiwall bags can be improved by the inclusion of plastic plies.

(ii) *Plastic bags.* These are available in different shapes and sizes and have replaced paper bags in some applications. Plastic bags may be of a light or heavy gauge depending on the final use of the bag. Light duty bags are generally described in terms of their intended use, e.g. sandwich bags, carrier bags, bin bags, and so on. The manufacture of bags is based on the use of plastic film and heat sealing one or more edges to produce the desired container. They are produced in quantities for use either by industry or by consumers.

(iii) *Net bags.* Net bags are produced from extruded plastic netting using specialized counter rotating dies. The finished net is then cut to the required length to manufacture bags. The ends of the bags may either be heat sealed or tied using metal or plastic clips.

Cartons

Cartons are generally made from sheets of paperboard, which have been cut and scored for bending into desired shapes. These shapes are available in two basic styles, with some special styles being used in selected applications.

Tube constructions are the most common. These cartons have a characteristic fifth panel, which is glue seamed to enable the carton to be folded flat for transportation. The end flaps may either be glued or tucked into place. Generally they are glued in high speed applications.

Tray-style cartons have a solid base with the side panels either being interlocked or glued into place. The lid is then glued into place on the three remaining sides. A typical specialist application would be the collation of bottles or cans in a wrap-round carton, which is glued or locked into place after the wrapping operation. While the majority of cartons are constructed from paperboard, some are made from plastic sheet and may provide product visibility.

Plastic moulded packs

Plastic moulded packs are used for a range of products and can be manufactured by a variety of methods. The large PET bottles are typical of a moulded pack. Moulding is undertaken by two basic methods:

Blow moulding

This is a process to produce hollow objects and can may be undertaken by three different processes.

(i) Injection blow moulding. In this process, the melted plastic is injected into a parison cavity around a core rod. The resulting injection moulded parison, while still hot, is then transferred on the core rod to the bottle blow mould cavity. Air is injected through the core, thereby expanding the parison against the walls of the mould and resulting in a cool container of the desired shape. This process is generally used for small containers.

(ii) Extrusion blow moulding. This uses an extruded tube parison and is generally used for larger containers, some of which may have handles and off-set necks.

(iii) Stretch blow moulding. This technique uses either an injection moulded extruded tube or an extrusion blow moulded preform. Stretch blow moulding is generally used for bottles between 500 ml and 2 l.

Injection moulding

Injection moulding requires hot plastic to be injected at high pressure into a closed mould that may have one or more cavities in the desired shape of the finished article. The mould is then rapidly cooled, the part ejected and the process repeated.

Form-fill-seal packs

Form-fill-seal (FFS) is a technique for making sachets using films based on plastic laminates. Several techniques are available.

Horizontal-form-fill-seal

There are two different techniques for producing horizontal-form-fill-seal (HFFS) packs. The first involves the film being formed into a V-shape before being sealed vertically to form individual pockets. The pockets are filled with product (normally a powder) before the top is sealed and the sachets separated. In the second, the film is formed into a horizontal tube around an indexed conveyor. Product is fed along the conveyor before the cross seals and longitudinal seals are formed. The sachets are then separated. Gas mixtures can be introduced into the sachets to modify the atmosphere within them for specific applications.

Vertical-form-fill seal

With the vertical-form-fill-seal (VFFS) technique, the film is drawn around a tube and a longitudinal seal is made, thereby creating a long tube. Product is fed down the tube in measured quantities before the cross (or horizontal) seal is made. The sachets are separated by cutting and the individual sachets then drop free.

Thermal-form-fill-seal

Thermal-form-fill-seal (TFFS) differs from the above methods in that a plastic film or sheet is fed into the machine and forms the bottom web of the package. The material is heated to soften it before being formed (by vacuum or pressure) into a tray of predetermined shape. The product is placed into the cavity before another film (top web) is heat sealed onto the bottom web. The individual trays are then cut. The trays can be flushed with a modified atmosphere or evacuated, depending on individual customer requirements.

Controlled atmosphere packaging

Controlled atmosphere packaging (CAP) (or storage) involves maintaining the gas mixture within the pack (or storage chamber) at a specific preset composition. Typical applications of controlled atmosphere storage include the transport of bananas from tropical climates to the more temperate climate of Europe and their subsequent storage. Apples are also stored for long periods in controlled atmosphere warehouses.

Modified atmosphere packaging

Modified atmosphere packaging (MAP) involves placing the foodstuff in a sealed pack and either gas flushing with a preselected gas mixture before sealing or allowing the product to change the gas mixture itself. There is no control of the mixture once the pack has been sealed. The selection of the correct plastic film for the application is important if the desired shelf life of the product is to be achieved. The selection of the correct gas mixture is based on prior knowledge.

Fish, meats and poultry do not respire. The gas mixtures selected are based on obtaining the desired shelf life for the product. Generally meats are packed in high oxygen mixtures so as to maintian a bright red colour. For these types of products, a barrier packaging material needs to be selected. This will greatly reduce the rate of permeability of oxygen and other gases, thereby maintaining the gas mix within the pack and extending the shelf life. The high oxygen content may also inhibit the growth of micro-organisms and further increase the shelf life of the product. The oxygen is utilized by the meat and falls over a period of time.

Fruit and vegetables continue to respire after they have been harvested. Preparation such as cutting and slicing can increase this rate of respiration. This results in faster gas changes within the pack. Oxygen is used up and carbon dioxide produced. This raises the possibility of the pack becoming anaerobic and gives rise to the potential for growth of *Clostridium botulinum*, a dangerous food poisoning organism. To counteract this possibility, packaging materials that permit gaseous exchange need to be selected. These may range from microperforatated films to specific films having selected permeation rates. Each application is individual and needs to be discussed with a packaging supplier.

Some products such as cheese absorb the gas (carbon dioxide) within the pack. This results in the pack developing a tight shrink appearance.

Aseptic processing and packaging

Put simply, aseptic processing and packaging involve sterilizing the product in an enclosed system and filling it into sterilized containers within an enclosed sterile chamber. The selection of the packaging material is dependent on the system used to sterilize the material. As most systems utilize heat or hydrogen peroxide, the packaging material must be heat or chemical resistant. Different types of aseptic packaging are available.

Vacuum packaging

Vacuum packing requires all the gas from around the product to be removed. The packaging materials used need to have good barrier properties to prevent the reintroduction of the gases into the pack. Vacuum packing can be considered as a form of modified atmosphere packaging.

Active packaging

Active packaging involves the use of packaging materials that perform a function other than the provision of an inert barrier to outside influences. It was stated in the INTRODUCTION that the packaging should be inert and should not react with the contents. Active packages are an exception to this rule and are specifically designed to interact with their contents.

Active packaging includes:

(i) Modified atmosphere packaging (see above)
(ii) Active films – these may be antimicrobial and based on immobilized enzymes
(iii) Edible films and coatings
(iv) Interactive sachets – such as oxygen absorbers. Other sachets can be used for applications such as removing ethylene or carbon dioxide.
(v) Microwave susceptors, manufactured from metallized films, for reheating products such as pizzas and popcorn.
(vi) TTI (time temperature indicators).

Intelligent packaging

Intelligent packaging is best defined as 'an integral component or inherent property of a pack, product or pack/product configuration which confers intelligence appropriate to the function and use of the product itself'. Intelligent packaging is related to active packaging and initially they both appear to be the same thing. However, intelligent packaging demands more from the pack and responds more than active packaging does.

SECONDARY AND TERTIARY PACKAGING

Secondary and tertiary packaging are additional to the packaging in direct contact with the food. It is used to collate individual packs, provide a surface for information and protect products during transportation from the food manufacturer to the retail sales outlet, e.g. the supermarket.

Collating trays

Collating trays may be made from either plastic (PS) or corrugated board. The function of the tray is to combine generally six or more products into a sales unit.

Outercases/corrugated cartons

Outercases/corrugated cartons are used to protect products during transit from the manufacturer to the retailer and, occasionally, to the consumer. The material used is based on paper but its method of manufacture is designed to increases its strength. Corrugated board is generally made from three layers of material, two external facings and one internal corrugated section. The facings can be made from kraft or semi-chemical materials depending on the application. The internal material is selected for the particular application.

Shrink/stretch wrap

Shrink wrapping employs a thermoplastic film which when heated 'shrinks' back to its original shape. The film is placed around the package to be wrapped and is then heated by means of a heated tunnel or flame gun.

Stretch wrapping, on the other hand, involves the use of a plastic film. Tension is applied to the film, thereby 'stretching' it around the package and allowing its natural resilience to hold the package together tightly.

Pallets

Pallets can be described as a fabricated platform used as a base for assembling, storing, handling and transporting materials and products in a unit load. The majority of pallets have been made from wood but plastic pallets are now being used more within the food industry. Pallets can be made to different sizes depending on specific customer requirements. There are two basic designs of pallet:

(i) Four-way entry pallets. These are block-type, with the blocks spaced so that fork lift trucks and hand trucks can enter the pallet from any of four directions, and
(ii) Two-way entry pallets. These are stringer pallets with solid stringers so that fork lift trucks or other equipment can only enter the pallet from the two opposite ends.

The size of the pallet can be specified according to its particular end use. There are national and international standards determining their construction.

END OF LINE EQUIPMENT

There is a range of equipment that can be placed at the end of production lines, which can check the package contents for contaminants or the pack itself for integrity. Equipment may also be used to apply additions to the package such as tamper evidence or labelling information such as 'best before' dates.

Labellers

Labelling is important for transmitting information about the product to the user. Labelling machinery can be categorized by the label/adhesive/container combination for which it was designed. The machine must be capable of applying the label accurately and consistently. At one extreme, labels may be applied with a simple hand gun/dispenser. At the other, the task may be performed by high speed automatic equipment.

Labels may be of the wet-glue, pressure-sensitive or heat-sensitive types.

(i) *Wet-glue labelling* is the least expensive in terms of label costs. Wet-labelling adhesives can be dextrin-based, casein-based, starch-based, synthetic dispersions and hot melts. Wet-labelling systems all have to perform the same basic function, that is to feed labels one-at-a-time from a magazine, coat the label with adhesive, feed the label to the article to be labelled in the correct position, apply correct pressure to smooth the label onto the article and remove the article from the labelling machine.

(ii) *Pressure-sensitive labellers* have one thing in common. They all require that the label is peeled away from the backing paper. This is carried out by unwinding a reel of die-cut labels and pulling them round a 'stripper' plate. As the label is bent around the plate at a sharp angle, the front edge of the label is pulled away. The container to be labelled passes the stripped label edge thereby transferring the label to the container under light pressure applied by an applicator drum and pressure pad.

(iii) *Heat-seal labelling* has most of the advantages of pressure-sensitive labelling but generally at a lower cost. The heat seal is formed by a plasticized paper which when heated becomes tacky on the underside. The ability of the adhesive to remain tacky after removal from the heat source aids machine design.

Bar coding

Every packaged product on sale in a modern superstore carries a bar code. Every size and variety of package must carry a different number so that the store's computer can recognize and price it correctly. This bar code consists of a group of stripes representing a unique combination of numbers. The pattern of light and dark stripes can be read by a laser scanner, thereby providing the supply chain with both accurate and improved stock management.

Bar codes are produced in a range of sizes which depend on the size of the traded unit. This may vary in size from a small retail pack to a large pallet load. For small packs an EAN-8 in-store number may be allocated whereas an EAN-128 code would be used for pallet loads of product.

The EAN-13 article number is made up as follows:

Country	Company number	Product number	Check digit
XX	NNNNN	nnnnn	C

where XX is the country (UK is 50), NNNNN is the company prefix number used on all company bar codes, nnnnn is the unique product identification issued by the company and C is the check digit calculated to a formula using the previous 12 digits.

Metal detection

There are two basic principles used for the detection of metal in food products, the balanced coil and the magnetic field systems.

The balanced coil system

The balanced coil system is applied except when aluminium containers are to be inspected. In principle, the detector is made up of a three-coil assembly surrounding a tunnel in which a balanced electrical field exists. When metal is introduced into the assembly the balanced state is disturbed. This generates a voltage in the coils, which is amplified and processed to generate a reject signal. Limitations of the systems are:

(i) The sensitivity is proportional to the size of the tunnel. Therefore, all detectors have a limit to how small a particle can be detected

(ii) Any product which is in itself conductive will also generate an unbalanced signal, further limiting the sensitivity, particularly to non-ferrous metals.

The magnetic field system

The magnetic field system (ferrous in foil) consists of a tunnel in which a strong magnetic field is present with a sensing coil surrounding the centre of the tunnel. A particle of magnetic material, even when enclosed in an aluminium container, is magnetized as it is exposed to a magnetic field. When this magnetized particle passes the coil, a voltage is generated across the coil which is amplified to produce a reject signal. Limitations of the system are: (i) it can only detect magnetic materials and (ii) sensitivity is proportional to the size of the tunnel.

Tamper-evident sealing

With ever increasing concern about food safety and malicious contamination, some form of tamper evident device is required. This may vary from container to container.

Membranes

A membrane may be glued or heat-sealed onto the rim or mouth of the container in such a way that it cannot be removed in a single piece. Its removal would leave signs on the sealing surface and the consumer would be aware that the product had previously been opened.

Shrink sleeves

A shrink sleeve may be a label that covers the whole container or just the neck of the bottle or jar. The sleeve is placed over the closure/jar interface and heat is applied to shrink the band into place. Some of these bands have perforations to make them easier to remove.

Tamper-evident bands

These are used on jar and bottle closures, which are made from either metal or plastic. When the closure is removed from the container, the band is detached and drops down the neck of the container as an indication that it has been opened. A typical example is the plastic bands on carbonated soft drinks bottles. These break off when the bottle is opened and remain on the bottle as an indicator.

Pop-up buttons

Some wide-mouth metal closures have a 'pop-up' button in the centre of the closure. The button is manufactured in the up position. However, after filling a vacuum is generated within the jar which is high enough to pull the button down. When the jar is opened, the vacuum is released and there is an audible 'click' as the button 'pops-up' to return to its original position.

Paper labels

There is a range of paper labels in different forms that can be used to indicated tamper evidence. They are attached to both the jar and the closure during the labelling operation and the action of opening the jar tears the label.

Optical systems

These rely on cameras and data processing to capture an image. The image can be simple or complex. They can be used to verify label placement, print details, or product and package colour.

COMPARISON OF PACKAGING SYSTEMS

The comparison of packaging systems is dependent on the end use of the packaged product and the cost of the packaging material within the country in which it is produced. It is therefore outside to scope of the chapter to give comparisons of packaging systems or their costs.

LEGISLATION RELATING TO FOOD PACKAGING

All food-contact materials are covered by specific legal or regulative requirements in the country in which the material is to be used. Packaging is a food-contact material and is regulated both within the UK and the EU.

UK regulations are incorporated into UK law within the Food Safety Act 1990. The regulations include:

(i) The Materials and Articles in Contact with Food Regulations 1987 (SI 1523). These are general regulations and implemented Council Directive 76/893/EEC.

(ii) The Plastic Materials and Articles in Contact with Food Regulations 1992 (SI 3145). These regulations implement Commission Directive EEC/90/128 into UK law. In addition they consolidate the other EU directives relating to plastic materials in that they include:

    (a) 82/711/EEC – laying down the basic rules for testing migration of the constituents of plastic materials and articles intended to come into contact with food;

    (b) 85/572/EEC – laying down the list of simulants to be used for testing migration of constituents of plastic materials and articles intended to come into contact with food.

Within the USA, packaging is controlled by the Food and Drug Administration (FDA) Code of Federal Regulations (CFR) Chapter 21. Other countries throughout the world have their own specific regulations of which users need to make themselves aware.

REFERENCE

Paine, F.A. and Paine, H.Y. (1983) *A Handbook of Food Packaging*, Blackie Academic and Professional, Glasgow.

FURTHER READING

Bakker, M. (ed.) (1986) *The Wiley Encyclopedia of Packaging Technology*, John Wiley & Sons, New York.

Campbell, A.J. (1995) *Guidelines for the Prevention and Control of Foreign Bodies in Food, Guideline No. 5*, Campden & Chorleywood Food Research Association, Chipping Campden, Glos.

HMSO (1992) *Plastic Materials and Articles in Contact With Food Regulations 1992*.

Kadoya, T. (ed.) (1990) *Food Packaging*, Academic Press, San Diego, CA.

Moody, B.E. (1977) *Packaging in Glass*, Hutchinson and Benham, London.

Paine, F.A. (ed.) (1991) *The Packaging User's Handbook*, Blackie Academic & Professional, Glasgow.

Rees, J.A.G. and Bettison, J. (eds) (1991) *Processing and Packaging of Heat Preserved Foods*, Blackie Academic & Professional, Glasgow.

Robertson, G.L. (ed.) (1993) *Food Packaging: Principles and Practice*, Marcel Dekker, New York.

Summers, L. (1992) *Intelligent Packaging*, Centre for Exploitation of Science and Technology, London.

17 Food Factory Design and Operation

INTRODUCTION

The design of food factories is an aspect of the overall food production chain which is often overlooked and regarded as less significant than the more glamorous areas of new product development and sales and marketing. However, the food or drinks factory has a key role in determining the profitability and success of any food or drinks company. It is important to design a factory which is clean and safe in addition to producing a high quality product at a low cost.

The food industry changes very rapidly with product lifecycles ever reducing all the time as consumers, retailers and marketing departments look forward to new products. This places considerable pressures on the food factory and means it has to be a flexible and adaptable facility. The design of the factory is therefore fundamental to the business and its success.

FACTORY DESIGN AND CONSTRUCTION

LOCATION

The location of a factory can be very important and have a significant effect on the performance of the factory and in many cases the quality of food produced in it. It is therefore important to find the most suitable site for the factory. Different companies will have a variety of constraints according to the nature of their business. Some of the more common considerations are described in the following sections.

Site selection criteria

Historically factories have been located in particular regions for specific reasons. A brewery has usually developed close to a good water source and a dairy will have been traditionally located in a rural area in close proximity to good pastures. Labour intensive food products will have been positioned in close proximity to major towns. Some of the key factors governing the location of a site are:

(i) Logistics and proximity of market
(ii) Proximity of ingredients and raw material sources
(iii) Labour requirements and availability
(iv) Services – water, gas, etc.
(v) Financial incentives – grants, land costs, etc.
(vi) Transportation – road and rail links

Having selected the general location for the factory site one needs then to select the actual site on which the building can be constructed. In order to make this decision one needs to consider the following:

(i) Whether a sloping or level site is required
(ii) Access to the location – are roads suitable and is a rail link required?
(iii) Has the land or site been contaminated and what is the underlying geology?
(iv) Will planning permission be granted?
(v) At what level is the water table and what is drainage like?
(vi) What utilities are available – water, gas, electricity, etc.
(vii) Security, in particular access and boundaries.

Although many of the above issues can be resolved they can affect the cost of the overall construction and possibly the operating costs. In order to determine if the site is suitable for the construction of a food factory a survey will have to be carried out to assess the underlying geological strata by obtaining soil samples. The soil samples will be analysed to determine the suitability of the land and in particular what type of foundation would be required. In most cases this survey has already been carried out and the land has already been earmarked for commercial development as an industrial zone. The geological conditions and the previous use of the site can affect the costs of construction. The water table level has to be considered in relation to the obvious threat of flooding. This can be taken into account in the design of the building, but the risk of water contamination through flooding can be of concern. If the product to be manufactured is susceptible to water quality changes then the nature of water supply has to be assessed fully.

The factories or facilities located adjacent to the proposed factory location should also be investigated, as cross-contamination could result from dust, smoke, gaseous emissions, etc. The prevailing wind conditions are important to note, and one needs to evaluate the possible emissions from downwind factories with gaseous emissions that could potentially cause a contamination risk.

Certain food factories have specific requirements of water supply. These include quality, potability, trace elements, quantity and the possibility of contamination. The energy requirements for a factory carrying out a significant amount of freezing or drying could be considerable and one has to assess whether the local electricity supply will be capable of supplying the peak load required, without the factory incurring substantial costs for a new substation or a totally new supply.

Environmental impact

In addition to the food company deciding if a site is suitable to meet its requirements, the impact of the factory on the surrounding area needs to be assessed. The normal method of assessing the impact of a factory on the surroundings is to carry out an environmental impact assessment. In addition one may also wish to carry out lifecycle analysis. The assessment should include the following key issues:

(i)   Waste disposal
(ii) Potential hazards to local community
(iii) Pollution – gas, liquid and solid
(iv) Noise levels – day and night
(v)   Effect of the facility on the use of local raw materials
(vi) Transportation infrastructure

One of the major considerations for the impact of a food factory on the environment is that of waste treatment and waste handling. Most food factories use large amounts of water particularly if they are involved in processing such as fruit and vegetables, dairy, brewery or meat processing. Large quantities of water will be used to wash and for wash down. This waste water could then

Table 17.1 Typical maximum acceptable consent levels

Analysis	Limit (mg l^{-1})
BOD	300
Suspended solids	300
Total dissolved solids	500
Total phosphorus	10
Total nitrogen	40
Oil and grease	100

contain high levels of total solids, biological oxygen demand (BOD) and pesticide residues. Table 17.1 shows some typical maximum consent levels for common effluents.

One of the main measures of the degree of pollution is BOD. To measure BOD, samples are taken from the water stream and diluted to obtain the correct dilution to react. There is then a five day incubation usually at 20°C. Then the amount of oxygen present is determined by means of a sodium thiosulphate titration to a starch–iodine end point. Another measure is that of COD (chemical oxygen demand); this quantifies the amount of organic material that is easily oxidized. This technique also measures the amount of nitrates and sulphites in addition to the organic material. The ratio of BOD/COD is a useful measure. A ratio of above 0.5 refers to a 'degradable' waste and ratios below 0.5 to a 'less biodegradable' waste. The local authorities will monitor or require the food company to monitor discharges and there are consent levels which if exceeded may result in large penalties and legal action may be taken.

Other effluents that require monitoring are phosphates, total solids, grease and oils. One can reduce the effluent load through a number of stages of waste treatment. The first (primary) treatment is to separate the waste. This can be carried out using a number of techniques and in the case of solids one could utilize: screens, sedimentation, flotation, centrifuges, hydrocyclones or ultrafiltration/reverse osmosis.

Primary effluent can either be disposed of or subjected to further (secondary) treatment. Possible options are: land disposal, lagoons or holding ponds, activated sludge systems or activated biological filter.

The tertiary stage of waste treatment which involves chemical treatment to flocculate or coagulate the suspended solids, is often the most expensive. The waste can then be separated by filtering or centrifuging.

Utilities

The major primary utilities used in a food factory are gas, electricity, water and waste handling. In most developed countries and regions utilities do not present a major problem but in developing countries and outlying remote areas they can present a problem and this may result in power being generated on site and the possibility of tankering in water and tankering out waste materials in extreme cases. There are some secondary utilities which include treated waters, steam, hot water, compressed air, central vacuum, refrigeration, nitrogen, carbon dioxide, and so on.

To generate power, hot water and steam there has been a move towards combined heat and power systems (CHP). This type of system uses steam generated from a boiler to drive a turbine to produce electricity. The system can therefore produce steam for heating purposes as well as electricity for driving the equipment. Any excess electricity generated can often be sold to the country's own electricity generating company.

The electricity use pattern in a food manufacturing facility will vary according to the process. Although the major consumption of electricity is in heating and cooling there can also be an appreciable amount used in activities such as mixing, pumping, packaging and conveying materials. There are several key factors involved in the sizing and operation of power supplies. One is to ensure that the peak demand can be met and another is to ensure that the transformer/sub-station is large enough for future expansion of the factory. Energy management is very important in food manufacturing operations and one can use control systems to assist with this. It is useful to size equipment with the cost of electricity in mind and particularly with regard to off-peak tariffs and peak loads.

Gas is an easier utility to manage. One of the key aspects is the sizing of the main supply pipeline to ensure it can supply the required flow rate. If the pipe diameter is sized with future expansion in mind this is not usually a problem. Most gas supplies are normally run direct to the boilerhouse as the main use of gas is usually in steam generation. It is a good idea to size the main header pipe to allow for an additional boiler.

Water is essential to all food factories in one way or another. If water is to be used directly as an ingredient in a product one has to take many more precautions than if it were only to be used for cleaning purposes. As an ingredient the water has to be analysed for its suitability by examining its composition and testing whether it adversely affects the quality of the product. Poor quality waters can create taints in food products.

Because factory water consumption is erratic due to the nature of many food processes a water tank is used to provide a buffer for excess demands. Water often has to be treated before it can be used and treatment varies according to use. Softening is the most common. Soft drinks, beers and similar products require water to be deaerated prior to incorporation as an ingredient otherwise dissolved air can adversely affect the quality, shelf life and flavour of the finished product.

Manpower and resource issues

Most modern factories require a skilled workforce which has been trained in the crafts required. The main skills required in a factory are broad, below is a list of typical functions:

- Production managers
- Line supervisors
- Skilled operatives
- Electricians
- Mechanical engineers
- Fitters
- Packaging engineers
- Maintenance engineers
- Project engineer
- Fork lift truck operators
- Warehouse staff

If the factory is located near a major town or city it is likely that these skills will be available. People are an integral part of the manufacturing process and a major element of total quality management (TQM).

The organization of labour varies, particularly in production departments. Some more modern thinking is to develop team working and use a multidisciplined or multiskilled team approach. A team of workers may be assigned to a production line and they will have full responsibility for the overall performance. Their bonuses and pay may reflect the team's performance. The team will usually have all the required skills to operate and maintain the line or machine and through their cooperation, management and teamwork the task of improving the performance can be undertaken. They will have authority to stop the line or alter production speeds. This approach has also been called worker empowerment where the operator is given more authority and is able to make decisions.

The more traditional approach is to arrange departments according to disciplines where the head of the department will decide the priorities and will be responsible for ensuring jobs are carried out efficiently.

DESIGN PRINCIPLES

Factory design

Food companies differ in their views on the external and internal appearance of the factory and this often reflects the type of food sector they are operating in. Some like to have a highly prestigious fabrication, as a show case, and to reflect the nature of their business, while others consider a low cost 'shed' type construction to be sufficient to meet statutory requirements and their own requirements. Most food factories fall somewhere in between these two extremes.

The fundamental design has to be developed from a brief, which should contain some key criteria such as the size of the factory, future expansion anticipated, multistorey or single storey, access requirements and size of equipment to be housed, drainage requirements and waste treatment, food safety and hygiene aspects, temperature requirements, and so on. It should not be forgotten that the purpose of the building is to protect the materials used in the production of food from the elements and prevent any contamination of the product during the manufacturing process. Therefore, any building needs to provide adequate protection which should include no ingress of dirt, rodent and insect protection and good weatherproofing, including the ability to withstand the most severe weather conditions for the region.

There is always a major debate within companies as to whether the factory should be a single storey building or a multistorey building. A high building can have the advantage of using a gravitational feed system for materials. This can lead to a more fluent operation and the footprint for the building is less than a single level building. However, a multistorey building costs more to construct and can have less flexibility, in addition to being more difficult to manage because it is not possible to observe or hear what is happening on other floors. Historically, flour mills, confectionery factories and biscuit factors have used multilevel buildings.

The traditional design for a food factory is to erect a steel frame and then lay bricks to fill in between the steel framework; cladding is often then used as an external finish. The steel frame can be a portal frame or lattice beams supported on columns. If high sided steel frames are to be used it is usual to reinforce the walls with wind bracing as high sided steel frames can be unstable in high winds. Roofs of steel framed building are usually supported on steel purlings. The roof is best constructed with a single apex. Multiple apex roofs are notorious for leaking.

Other methods used in construction are based on the use of concrete panels. These are sometimes manufactured on site. This method is often referred to as a tilt-up concrete panel construction. This form of construction, more popular in the USA, can lead to very fast and stable construction.

The construction method and design are often dependent on the choice of construction company or architect. Some construction companies have their own style and views on the methods and type of construction to be used, which often relate to costs and previous design experiences.

Hygiene requirements

Factories need to be kept clean in order to prevent the growth of bacteria or other spoilage organisms. The design of the factory can contribute to making that factory both cleanable and easy to clean. The design of walls, floors and ceilings can assist in the task of keeping the factory clean and hygienic.

In order to clean a factory, a number of techniques are employed such as foaming, fogging, general scrub down and cleaning in place (CIP). Cleaning fluids used are normally acid or caustic based. Sterilization is used, in most cases, post cleaning; sanitizers are hydrogen peroxide based.

Foaming, as the name suggests, is a method by which the cleaning chemicals are aerated into a foam and then applied to the surface to be cleaned. This often provides an improved surface contact resulting in increased removal of dirt. The foam is normally left to react for a period of time before being rinsed off.

Fogging is achieved by atomizing the cleaning media into a fine spray which attaches to any airborne solids in addition to settling on equipment and the surfaces to be cleaned. Fogging is used in conjunction with other cleaning methods.

High pressure water hoses are often used in factories and although this method is often very effective in removing major debris, it can create aerosols from contaminated water which settle after cleaning, reinfecting or soiling a previously cleaned area.

Dry cleaning is important and, in particular, the collection of dry wastes using vacuuming systems.

With certain dry powders care has to be taken in their collection since there can often be a potential explosion hazard.

For equipment to be cleaned in place there is a new European standard being developed by the EHEDG (European Hygienic Engineering Design Group, 1992). This group has developed a standard test procedure against which the effectiveness of equipment cleaning procedures can be judged.

Contamination

The food factory has to be designed to prevent contamination of foods. Contaminants can be categorized as shown in Table 17.2.

The risk of the ingress of contaminants can be reduced through the correct design of the building. The design should include air flow controls and filtering where appropriate. Air intakes and outlets should be correctly located to prevent the ingress of fumes, sized correctly and protected to prevent vermin and rodents entering. Doors and other entrances particularly those openings made for services, should be designed to prevent rodents and insects entering. The bottom of door can be made rodent and insect proof by using bristles and metal plates. Masonry around pipework entering the factory should be sealed with cement and a mastic. Other potential entry points such as louvered ventilation systems, vents, roof eaves, building air bricks or vents should be protected.

Contamination may occur from other sources inside the factory such as nuts and bolts, wire, congealed ingredients or lubricants, paint flakes,

Table 17.2 Categories of typical contaminants

Category	Typical contaminant
Foreign bodies	Metal fragments
	Airborne particulates – dust and dirt
	Glass, plastic waste.
Vermin	Birds and bats
	Rodents – rats, mice, voles, squirrels, etc.
Insects	Flies, wasps, bees, moths, etc.
	Beetles, ants, lice, mites, silverfish,
	cockroaches, etc.
Environmental	Odorous gaseous inputs
	Toxic fumes

hair, pens, paper and plastics. There are some technologies available to assist in the detection of materials if they do enter the food processing area. Detection devices are a good method of demonstrating that one is taking all due care and attention in the production of safe foods. Magnets and metal detectors are two of the more common systems used to detect metallic contaminants. Optical sorting systems are used for inspecting regular shaped and coloured products such as peas or grain. X-ray systems are used to detect dense contaminants in some food products such as bones in meat and glass pieces in glass jars.

Figure 17.1 illustrates some safety and contamination prevention design features in a powder mixing plant. There is ventilation above the powder tipping stations. Light fittings are fixed into the ceiling and flush, kick plates are located on the platform along with safety hand railings. Tanks are mounted on plinths to aid cleaning.

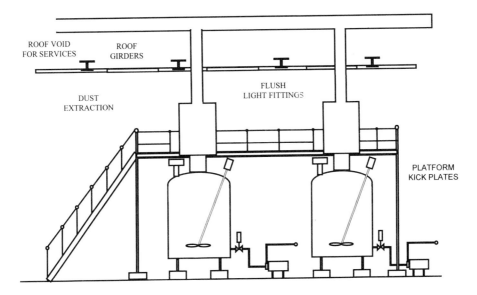

Figure 17.1 Design of a powder mixing plant.

Factory layout

The layout of the factory requires the best arrangement which uses a minimum of costly floorspace, but sufficient space to allow for efficient working. Inefficiencies usually constitute a higher cost than building capital costs. One good method is to design the factory using the straight line principle – ingredients entering at one end and finished goods exiting at the other. This approach cannot always be adhered to, as some processes require large amounts of work-in-progress, long dwell times or cooling systems, therefore making the lines too long and impracticable. In this case designs such as spiralling systems can often make best use of floor space.

Computer-based planning systems can be used allowing simulation of the operation of the factory and a review of the layout prior to committing any costs. These systems can either be based on discrete event analysis (DES) or dynamic computer aided design (CAD) systems. The approach is low cost and an effective way of reducing the risk of a poor layout design. One of the best methods of determining an optimal layout is to investigate the flow of materials and people within the factory. These flows can be used to indicate where there are many cross-overs, potential cross-contamination risks and high levels of movement. In detecting the bottlenecks it is then possible to review new layout scenarios and designs and evaluate them to discover the best options.

There are certain areas within a factory that require partitioning off and separating from the main production areas. The separation necessary is best deduced from an analysis of the risks from a food safety viewpoint. Within the production area there are different levels of risk. Some typical levels of risk are listed in Table 17.3.

A good principle is to separate high risk from low risk areas. The obvious place for partitioning is immediately after filling or sealing the product in its primary pack. After primary packaging there is considerably reduced risk of contamination. Within a chilled or frozen food production facility

Table 17.3 Typical levels of risk

Area	Risk
Storage of certain fresh ingredients	high risk
Fresh food preparation	high risk
Filling of foods into primary packaging	high risk
Secondary and tertiary packaging	low risk
Storage of frozen and ambient foods	low risk

this is usually immediately after cooling. In such cases, the freezer or chiller can act as the partitioning section. For some foods it could be pre- and post-drying that act as a natural barrier. Once the water activity of a product is reduced it may be rendered safe from microbial contamination. In some food operations there is a 'higher risk' area identified which is separated. This area is normally where a food product has been heat treated and is in the process of being packaged and therefore it constitutes a 'high risk' as the food is open to contamination and possible microbial infection. Figure 17.2 shows a typical layout for a pizza factory illustrating some key design principles such as a logical straight line flow of materials.

One of the major problems faced by companies is the changing requirements within the factory due to the continually changing demands on the industry. These changes in products, product mix, size of packs, inspection systems required, and so on, result in larger areas of floor space being required. Usually the greater amount of flexibility there is in the building design, the increased scope there is for a good layout. Items such as columns, drains, and service channels and ducts can, with inadequate design, cause many problems and can introduce constraints.

In addition to the layout of the building and processing equipment there is also the layout of the services, walls and ceiling areas to consider. Drawings and three-dimensional (3D) views often need to be drawn in order to indicate where pipes will be run and where conduits, drains, ducts and cable trays for services laid. In the design phase

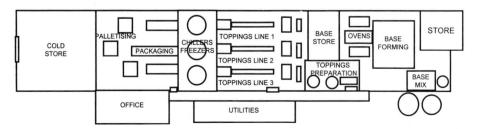

Figure 17.2 Layout of a pizza factory.

one needs to consider all aspects of the spatial requirements. Design tools can facilitate 3D collision avoidance and the overlaying of drawings in three dimensions.

One of the best methods used to separate the services from the processing and manufacturing areas is the provision of a service floor above the processing floor. This can be used to house most of the services which can then be kept away from the factory floor allowing more floor space to be devoted to production. Having services descending from the ceiling allows the floor to be more readily cleaned.

Raw material storage

Raw materials can be delivered in various forms, tankers, sacks, drums, cases, big bags, and so on. Managing and storing raw materials are important as they are often a major component of the cost of the final product. Many companies have adopted just-in-time principles (JIT) but few have managed to achieve this fully. The principles of JIT are to have materials delivered just prior to being required and therefore reduce or eliminate the need for storage on site. The concept in principle is sound but it does not fit well with flexible manufacturing and fluctuating orders which require short response times. Generally JIT has resulted in off-site stores being set up locally by the ingredient or packaging supplier so allowing rapid response. The packaging company can even be located next door, supplying the materials 'through a hole in the wall'.

Many companies still opt to control raw materials by storing adequate quantities on site. Some businesses store seasonal produce and package as and when it is required during the year. Raw materials need to be stored under the correct conditions for that particular item. Some of the key parameters affecting the storage of foods are temperature, humidity, air quality and flow, degree of light and inert or special gas environment. When storing fresh produce such as bananas the conditions and environment are very critical and can have a significant effect on the ripening time.

Certain ingredients (and products) need to be stored in a frozen or chilled state. See Chapter 15 for a description of such storage facilities.

For ingredients that are not susceptible to changes in storage conditions a conventional store can be used. This will consist of racking, allowing storage on pallets and access for a fork lift truck to store and retrieve the ingredients or product. The ingredients can either be stored in set loca-

tions or bar coded so that a more automated and flexible system can be used.

There is a need for special storage facilities in many factories. This can involve the use of storage tanks or silos where materials can be kept under controlled environments and their natural ability to flow can be utilized to pump or blow them to the required usage point. Bulk storage silos normally contain 20 tonnes or more so as to facilitate the off-loading of a complete tanker delivery. Silos and tanks can have different requirements such as inert gas headspaces to prevent oxidation or contamination. Jackets, either hot or cold, can be build into tanks to control the temperature, and agitators can be used to keep mixtures in suspension and homogeneous. With hot jackets the heating medium is normally hot water, heated by steam. For cold jackets a brine solution is normally used.

It is important that ingredients arrive in top quality and are then maintained at a high quality level to be utilized in the process. In order to monitor the required quality, samples should be regularly taken and checked so as to identify any potential deterioration. Raw materials are susceptible to contamination particularly by foreign bodies. There are a number of technologies which can be used to detect and remove foreign matter; these include cyclones, sieves and screens, filters, metal detectors, magnetic separators, vision and colour inspection and X-ray inspection. The technologies can be used to detect and remove potential contaminants such as nuts and bolts, string, hairs, bones, glass and plastics. Most systems are capable of removing very small objects of less than 1 mm in diameter. Retailers often require all dry ingredients to be sieved prior to use as a safety precaution.

Product storage

Many short shelf life products are made to order and shipped directly to the retailer. The retailer or wholesaler will normally have a central or regional warehouse which takes in complete pallets of product. The retailer will then break down the pallets into cases and distribute to the outlets in that region. Many products need to be stored under specific conditions. The main conditions are frozen, chilled, ambient and ambient at controlled humidity.

The storage conditions are essential to ensure that the product is kept in its prime condition for as long as possible. Food products are stored in warehouses on pallets and the pallets are located in a racking system. The level of automation in the warehouse can range from a totally manual system utilizing fork lift trucks to move pallets, to a completely automated computerized system that can automatically store and retrieve pallets. If the warehouse stores large quantities of full pallets of products the system will usually work on a FIFO ('first-in first-out') basis.

Typical warehouses are of a low cost construction, as in most cases they do not add value to the product. The warehouse can be constructed using the racking system, with the racking acting as the frame which can then be either brick lined or just clad. Racking systems are of a modular form and are usually straightforward to expand either upwards or in any horizontal direction to provide an increased number of bays. The usual height of stores is 3–4 racks high but there are some high bay warehouses that are 6–8 racks high. The higher the building becomes the less stable the racking system will be and this affects the level of costs associated with items such as foundations and wind bracing.

Automated stores use stacker cranes which can travel at high speeds to a specific location as directed by a computer and then retrieve the pallet found at that location. The systems that are fully automated are referred to as ASRS systems (Automated Storage and Retrieval Systems). One of the first systems to be fully automated was that of the John Lewis Partnership in the early 1960s.

Racking can be designed so that the bays are a number of pallets deep and the lanes containing the pallets are tilted forward. The full pallets are pushed back into the bay thus allowing them to roll forward under gravity each time one is removed. This type of 'push back' system allows for easy use, but only caters for last-in first-out (LIFO). A LIFO system it is best operated by placing pallets into the back of the racked bay and taking them from the front.

The computer systems used to control the racking systems operate using bar codes located on the pallets identifying what it is and where it is to be stored. It is important in a large store to keep certain goods segregated particularly where there is the possibility of cross-contamination, such as with highly flavoured or scented products. The storage system becomes more complicated when part pallet loads are stored and where 'picking', that is the breakdown of pallets into smaller delivery orders to local stores within a region, is necessary. Picking systems are used in some warehouses where an operator will work from the crane cage and use a display system to direct him to the appropriate bay for the product

and inform him of how many cases of product to pick. Other systems may require pallets to be brought out to a picking area, broken down as required and then the remainder returned to the location in the store. It is important in any store to keep track of the quantity of stored items and where they are located in the store. In this way an inventory can be kept and any losses immediately identified.

Most food products deteriorate with time, therefore stock turnover is usually high. Not all products deteriorate with age, some like cheese will mature with time. Cheese stores usually work on the basis of storing for as long as possible in order to increase the product value. A cheese storage system is somewhat different as cheeses require a specialist warehouse where they can be regularly turned as required in the maturation processes.

INFRASTRUCTURE

The general infrastructure and fabric of a food factory are important and often an area that is neglected or overlooked. The potential problems that can occur from a poorly finished building can cause major long term problems and be expensive. It is vital that the building is finished so as to be fit for the purpose. The building needs to be hygienic, cleanable, maintainable and safe.

External walls

The external cladding of the factory can take many different forms and also depends on the nature of the business. Some of the more common options include metal faced panels, masonry, concrete (precast) and other sheets of laminated materials. There are many factors that influence the selection of finishes such as cost, aesthetics, colour selection, durability and longevity. Some panels can be filled with non-inflammable rock wool, foams or other insulating materials. With filled cavities, the selection of filling is critical as some foams can emit volatile toxic fumes if they catch fire.

Internal walls

The internal wall finish is an important aspect of the factory fabric and one that can potentially cause many problems. Internal walls are normally coated with various materials so as to render them smooth in finish, non-flaking and impervious.

Walls generally require frequent cleaning and should be designed to withstand this.

Traditionally tiles have been preferred. However, although the tile is effective and can be kept clean, the grouting is more difficult to maintain and often cleaning chemicals attack it and cause degradation and potential crevices in which micro-organisms can harbour. The use of cladding has been found to be a more effective means of sealing a wall and the cladding sheets can be joined using a suitable mastic or rubberized sealant. The sheets can be made of a range of materials including stainless steel, polyester, aluminium, glass reinforced plastic and PVC laminates. The specific selection of material will depend on factors such as the level of hygiene required, cleaning chemicals to be used, cost, sealability and the acceptable potential contamination risk.

The corners of walls are often knocked and by strengthening these with metal protectors chipping can be prevented. Partition walls are frequently constructed from foam filled panels. Although this is a low cost method of construction many foams present a major fire risk so advice should be sought on alternatives such as mineral wool and rock wool. The core material will need to be treated to prevent vermin attacking it and fungus and moulds growing in it. Walls generally need to be robust, sealable and sturdy, easy to clean, non-shedding and immune to mould, fungus and parasites, fire safe and adaptable for building alterations. The joins at the floor and ceiling should preferably be rounded giving ease of cleaning and preventing dirt accumulating in crevices.

Ceilings

There are a number of different types of ceilings commonly used. The two main forms are suspended (false) or solid. Suspended ceilings, although commonly used, do have inherent risks as they tend to collect dust, dirt, insects and vermin. If they are to be used they should be dust tight and regularly disinfested.

Ceiling finishes should be such that they can be easily cleaned; lights and other fittings should be recessed so as to present a clean flush surface. The joints between panels are a potential problem area and must provide a good seal. Surfaces where dust can accumulate should be avoided. Lay-in panels can be used which are encapsulated in a special film to provide a flush smooth finish. Alternatively metal faced panels can be used. These can provide sufficient strength to allow one to walk on them. Ceiling tiles that flake must be avoided.

Floors

The surface coatings on floors need to be robust. Significant wear can result from the movement of heavy items, cleaning, fork lift trucks, and so on. The flooring needs to be suitable for food product, ingredients and equipment, durable, free draining, cleanable and hygienic, low cost, adaptable and repairable if damaged.

Thick ceramic tiles, although initially expensive, are often effective as they can be replaced. Epoxy coatings are used extensively and can be low cost; however, they are susceptible to wear and once they have cracked and water enters under the coating it soon lifts. Once it has lifted the whole floor has to be recoated and allowed to set which can have an adverse affect on production. Alternatively an epoxy or polyurethane screed if prepared well and laid properly can provide excellent qualities and finish but can be expensive. Other possible floor materials include concrete, which is easy to lay and economical but has a tendency to crack and wears quickly, and PVC vinyl which creates a good membrane but does not have a high impact strength or abrasion resistance.

Around pillars and supports, tiles or flooring should be curved so as to allow ease of cleaning. Tank legs are best mounted on concrete plinths where possible for the same reason. Doorways require raising to prevent material washing or blowing into the clean area. Depending on the type of factory, foot baths may need to be provided. If these are sunk into the floor they should be easily drained to allow a frequent change of water.

Drainage

The drainage of food factories is always a problem and still provides many debates. For many processes the idea of moving to a completely dry processing floor is attractive as it would result in the food being totally enclosed and protected. Waste would be either minimized or well controlled. This approach is suitable for some food facilities but not all. If one has to wash down and therefore provide drains, there is the consideration as to whether they should be troughs/gullies or pot holes.

Having an open gully type drainage system allows easy inspection and cleaning. The under floor drainage systems are more difficult to maintain because they cannot be seen. However, if a cleaning programme is well managed, hygiene can be controlled. The under floor drains allow more open floor space but require multidirectional falls in the floor, which if large, can result in equipment levelling problems.

Drainage systems should be connected to a trap which can be used to collect the waste prior to discharging into the municipal waste system or to further treatment. This trap or tank allows one to monitor and control the waste and if necessary adjust the pH, skim off fats, and so on. By separating the collection of wastes from each factory area they can be monitored and different wastes segregated. Most of the effort involved in waste treatment is in separating the wastes and concentrating them. Therefore, if different wastes are kept separate, this task is simplified.

Sprinkler systems

Sprinkler systems should be installed in most factories. There are certain areas where systems should not be installed such as in electrical switch rooms where Halon is appropriate. The local fire authority and the building regulations for each specific application should always be consulted. The water used to supply the sprinklers must be separate from other supplies and have its own header tank safely located. The size of the tank should be sufficient to allow a large fire to be extinguished.

Explosion venting

Some food processes can result in explosions if concentrations, particularly of combustible dusts such as sugar, reach critical levels and are ignited. In order to prevent major explosions, 'blow out' panels should be installed. These are special panels or doors that in the case of an explosion will 'blow out' to release the pressure buildup and hence reduce the possibilities of injury to personnel and damage to equipment.

Equipment suppliers will build safety features into their specific machines but additional safety features should be designed into the building structure, for example to vent any explosion directly to the exterior.

HVAC

Heating, ventilating and air conditioning systems (HVAC) are required to some degree in all food factories. Heating will vary according to the size of the facility, the volume of air to be heated and the nature of the materials being handled or processed in the area. Small areas can be heated directly using gas powered heaters. More sensitive areas can be heated using hot air supplied from a central system. The air may need to be filtered and sterilized depending on the microbiological sensitivity of the product being processed. The ducting leading into the room should be regularly cleaned and inspected. Heating systems such as radiators can cause a potential hygiene problem due to the fitments and crevices they can provide.

Ventilation is essential in the majority of factories and should be controlled by sizing the system to give a positive pressure within the processing area allowing used air to be vented outside. In most cases air can be vented to atmosphere, however if any toxic material or obnoxious odours have been collected they should be filtered or treated as appropriate. The air intake and vents should be kept well apart to prevent recirculation. Care should be taken to ensure that the high risk areas contain the highest air pressure so that air flow radiates out from that area and not into it from less clean areas. Air socks are often used to distribute air evenly. Although they are effective they can cause problems if not maintained and cleaned at regular intervals. The need for air conditioning will be dependent on the temperature required within the factory and will dictate the size of refrigeration unit required.

Distribution of utilities

Piped utilities such as steam, water and gas can be distributed from a header pipe on a ring main run around the factory, either at high level or in the roof void. The main header can be sized for the overall capacity. In the case of steam, the pressure in the main header is usually higher than the local pressure, to reduce the size of pipe required and the potential heat losses. The pressure is reduced locally and the pipe sized appropriately for the equipment it is supplying. For water and gas the pipe size reduction causes a pressure drop and the pipe size is calculated to allow the correct flow rate to be delivered at the drop off point.

Electrical power is supplied from a sub-station, usually the factory will have its own, to switchrooms or motor control centres (MCC) where it is fused and distributed to individual items of equipment or to remote MCCs or termination boards and buses. There can be benefits in using a distributed form of motor control and switchrooms as it enables local field cabling to be shorter and thus reduces cabling costs.

Warehousing

The warehousing operation is normally simpler than that of a factory. The basic controls required for a warehouse are those of temperature and humidity. Most warehousing consists of racking for pallets. The pallets are transported in and out using fork lift trucks. The doors in and out of the warehouse may be automatically opened using transponders located on the fork lift truck.

Loading bays are normally located at one side of the warehouse for despatching finished goods from a finished product store or for receiving ingredients into an ingredient store. There are many different kinds of loading bays. The lorry will back up to the bay and the bay will then close around it to provide a seal. This is particularly important when transferring frozen or chilled produce. The fork lift truck can enter directly into the lorry for loading or unloading. For finished produce wheeled cages may be used which are then manually pushed onto the lorry.

MOVEMENT OF MATERIALS

Within food factories materials have to be moved from one location to another. The method of transportation is dependent on the nature and quantity of material to be transferred. Below are described some of the typical methods of transportation.

Conveyors

Conveyors are commonly used in the food industry and come in a number of different shapes, sizes and forms such as:

(i) Belt – plastic, stainless steel, wire mesh, and so on
(ii) Vibrating
(iii) Screw
(iv) Shackle systems and overhead hanging
(v) Bucket conveyors
(vi) Indexing conveyors

Belt conveyors are one of the most common types and are used to convey bottles, cans, ready meals, pizzas, and so on. Different types of belts are used for different products according to the load, durability, hygiene, size and speed.

Some conveyors, such as storeyors, have been adapted to act as storage systems which are used for products such as cereals and snack foods. Vibrating conveyors are usually metallic and are electrically oscillated, the oscillations being controlled such as to cause the material on the conveyor to move along it. The velocity of product along the conveyor can be controlled by altering the frequency of the oscillations.

Screw conveyors are used to elevate or transfer powders. The screw mechanism is turned by a motor, which then transports the product along the spiral. This type of mechanism is used to fill depositing or forming type machines.

Shackle lines are used predominantly in the meat and poultry industry to transport inverted carcasses so that they do not touch the floor. Similar lines are used for transporting trays of products or ingredients. The systems are convenient as they are suspended and do not obstruct at floor level. However, they are generally difficult to clean.

Bucket conveyors are used to supply packaging machines. Products can be counted and positioned in the bucket. They then pass into the filling machine where they are either deposited or wrapped.

The motion of indexing conveyors is intermittent (start–stop). Their function is to deliver objects to certain predetermined locations, hold them there for a given period of time and then to move them forward again. A typical example of their use is to position bottles or other containers beneath a filling head. At this point, the conveyor stops and the bottle remains in this position until it has been filled. The conveyor then starts to move again and continues until the next bottle is accurately positioned beneath the filler. The cycle then repeats itself.

Elevators

Elevators are not widely used; there are several types such as bucket and spiral. Their use is predominently to supply a filling machine with product or a process with ingredients. As the name suggests they are used to raise the material to a particular level.

Pneumatic and vacuum conveying

For the transportation of dry materials which cannot be pumped, such as flour, grains, preservatives, starch and sugar, it is often possible to use pneumatic conveying.

Pneumatic systems usually use air as the conveying medium although inert gases or sterile air can also be used for sensitive materials. The air is blown along a pipe and the product is either drawn into the tube or screw fed into the flow of air. The air then blows the product to the required location. Valving can be used to provide a network of routes to various locations.

As well as air blown systems there are vacuum drawn systems. This type of conveyor uses a pump to pull a vacuum and suck the product from one location to another. Like the pneumatic system it is necessary to ensure there are no leaks in the pipework otherwise the system becomes inefficient.

Pumping

Most liquid products can be transported by pumping them. There is a wide range of pumps available. The selection of the correct pump will depend on factors such as the rheology, density, temperature, flowrate, and so on of the material to be pumped. Some typical pump types include:

(i)   Positive displacement
(ii) Centrifugal
(iii) Liquid ring
(iv) Piston
(v)   Diaphragm
(vi) Screw
(vii) Mono

A pump has to be correctly sized for the required duty otherwise it will have a deleterious affect on the product or not deliver the required duty. A typical pump consists of a motor, drive shaft, seals and bearings, rotor and housing. The rotor is usually constructed of stainless steel as is the housing. The motor may be protected for cleaning purposes by a stainless steel cowling. The seals are very important in food pumps as they protect the food from contamination by the lubricant. Pumps can be run at variable speeds through connecting an inverter or gearbox.

Centrifugal pumps are used for pumping thin, low viscosity liquids and cleaning fluids. Positive displacement pumps are used for pumping thicker viscous materials and those containing small particulates. Piston pumps can pump thick viscous fluids and dry lumpy products such as meats. Diaphragm pumps are used for pumping liquids and for dosing.

Transportation of bulk materials

Other transportation methods for bulk materials are varied and include the use of big bags or tote bins. These containers are used for bulk ingredients or work-in-progress materials. They can be lifted or moved using a fork lift truck.

Fork lift trucks are used to transport materials around most factories. Products and ingredients are usually loaded on pallets which can then be lifted and moved conveniently. The pallets come in different sizes although there has been a move to standardize a 'Euro' pallet made of plastic.

Automated guided vehicles (AGVs), driverless vehicles capable of carrying loads or pulling trailers, are to be found in large modern food factories. Their movement is normally controlled by electric guide wires embedded in the floor, or by optical devices, and is therefore limited to predetermined pathways. However, more recently developed computerized systems permit unconstrained movement of AGVs within the factory.

FACTORY OPERATION

LEGAL ASPECTS

The legal requirements have changed over recent years in Europe and continue to do so in order to bring about harmonization between EC Member States. Outside Europe countries such as the USA have their own system of regulation through the FDA (Food and Drugs Administration) and the USDA (United States Dairy Association). Some of the main European laws relating to food manufacturing are:

In the EU

(i)   EC Food Hygiene Directive 93/43/EEC
(ii)   EC Vertical Directives for milk, meat, fish and egg products

In the UK

(iii) Food Safety Act 1990
(iv)   Food Hygiene Regulations 1970, as amended 1990, 1991
(v)   Environmental Protection Act (EPA) 1990
(vi)   Control of Substances Hazardous to Health (COSHH) Regulations 1988
(vii) Health and Safety at Work Act 1974
(viii) Workplace Health, Safety and Welfare Regulations 1992.

There is also the following UK legislation (UK) that relates to buildings:

(i)   Town and Country Planning Act 1990
(ii) The Building Regulations 1991
(iii) Construction Regulations 1994.

Although there is a good deal of legislation relating to the food factory the information contained in the legislation tends to be rather vague and unclear as to exactly what is acceptable and what is not. The interpretation of many regulations and laws has been through case history which then sets a precedent. The key issue for most food companies is the need to produce quality, safe products within the law. If not, knowledge of the fact soon becomes public and bad publicity is generated which has an immediate adverse affect on that company.

Health and safety at work

Within many countries there is a requirement to meet general health and safety requirements. These regulations require the employer to provide the necessary working environment and equipment to allow employees to work in safety with no adverse affects to their health. Some of the more common aspects of factory safety are:

(i)   Working with dangerous machinery
(ii)   Lifting heavy objects or loads
(iii) Effects of cleaning and other hazardous chemicals
(iv)   Fire risks and procedures
(v)   Ergonomics
(vi)   Minimizing repetitive stress injuries (RSI)
(vii) Ventilation and air conditioning
(viii) Working environment – temperature, humidity, and so on

Within Europe, there is the added requirement of COSHH (Control of Substances Hazardous to Health). This legislation requires companies to maintain a register of all chemicals used within the workplace and to document appropriate procedures for their safe handling and use.

There are specific regulations relating to certain types of machinery. Some of the key points are that safety should be foremost and protective guarding and safety interlocks should be fitted to rotating machinery to prevent danger to the workforce. Under European legislation, new equipment has to be CE marked. This mark is in the form of a kite and certifies that the equipment manufacturer has assessed the risks of operating the particular machine and has incorporated the necessary safeguards. The manufacturer will also hold a technical file demonstrating the development process and safety considerations.

Control of Substances Hazardous to Health (COSHH)

COSHH assessments are required by EU law to be made available to employees to enable them to work safely with chemical substances. To comply with the regulations, employers must undertake an assessment of the work that is carried out by their employees to ensure that they are not improperly exposed to hazardous materials, including dust, fumes, micro-organisms, liquids, gases and solids that could damage their health. The assessment involves an evaluation of the risks to health and of how they could be reduced. Written documentation should include details of composition and potential hazards in use, as well as guidance on how to store and handle the particular chemical, and instructions on what action to take if an accident occurs and the substance is ingested, spilt on the skin, or splashed in the eyes, and so on. Where necessary, appropriate training should also be provided.

In the food industry, most substances, by their very nature, are not hazardous to health and therefore do not require a rigorous COSHH assessment. However, many food factories incorporate laboratories in which chemicals are used for routine analyses, and invariably cleaning chemicals are used for CIP and other cleaning duties. Such materials will probably have to be fully assessed; their manufacturers are legally required to provide specific information to customers for this purpose.

Some food ingredients also need to be covered by COSHH assessments. These are normally concentrates, which are diluted down in the process or are part of a pre-process treatment.

Food Safety Act 1990

The food safety act requires that food is produced in a safe manner. Section 7 states that food shall not be injurious to health. Section 8 states that food has to meet the required safety standard in that it should not be (i) injurious to health, (ii) unfit for human consumption or (iii) contaminated so that it is unreasonable for it to be consumed.

The major portion of responsibility is on the food manufacturer to ensure that safe food is produced. The manufacturer must be able to show that all due care and attention was taken to produce food safely (due diligence). In order to produce safe food the systems have to be in place to do this. The systems, records and actual processes vary according to the type of food being processed. Where there could be contamination measures need to be taken to ensure that those risks are minimized to an acceptable level.

The Act gave increased powers to allow for the emergency closure of food factories if they failed to meet the required hygiene standards. There is also the power to prohibit certain processes or force the closure of factories after a successful prosecution.

Food Hygiene Regulations

The Food Hygiene Regulations in the UK and the EU require that HACCP (hazard analysis and critical control points) is implemented as a method of reducing the risk of food contamination. Certain employees are also required to be trained as food handlers. This means that they must be made fully aware of hygiene issues and practices within the factory and of the danger of handling food after suffering from certain specified illnesses.

Machinery directives

There are many new machinery directives being issued by the EU. Most of the initiatives relate to specific machinery particularly that in packaging and meat processing. The directives describe in most cases the specific features of the type of machine, what safety considerations should be incorporated and the cleanability and hygiene required.

PROCESS AND COST CONTROL

Process control systems

Food processing operations are manual, semi-automatic or fully automated. Those that are automated use computers to control the sequencing of operations either as part of an integrated machine or as a number of items of equipment integrated by the control system. The most common type of sequencing controller is a programmable logic controller (PLC). This type of controller allows the functions required for a sequence of events or actions to be programmed. The sequence can be preprogrammed on a time basis or can use inputs from sensors, operators, and so on to trigger the required actions. The PLC enables cleaning programmes to be written that control the opening and closing of automatic valves and pumps. The programme can include interlocks which prevent the process operations from starting until the section of plant has been cleaned. The software is written in a relay ladder logic structure.

Process controllers can be used to control specific control loops such as temperature, level or pressure. These stand-alone controllers have three-term functions and can also include more sophisticated fuzzy logic control. The three terms, proportional, integral and derivative (PID) can be set manually or the controller can be set to an auto or self-tuning mode. The PID terms are used to set the response of the controller to give the best control and reduce fluctuations. Controllers can be linked into PLC based or DCS (distributed computer system) systems and setpoints and other information transferred electronically.

PLC systems are designed in a modular form with separate cards that slot into a rack assembly. These cards are supplied for various type of control such as analogue input/output, digital input/output, counter and three-term control. If the system is to control a large number of inputs and outputs the factory is divided into sections and separate PLCs are used for each area. The PLCs are then linked together via a network. The factory operator interface or MMI (man machine interface) and data storage can be carried out using a SCADA (supervisory control and data acquisition system). The SCADA system can run on a personal computer or network of personal computers with a PC as a file server.

Production scheduling techniques

Traditionally the scheduling of production in a food factory has been carried out manually with clerks drawing up schedules and often working to a regular pattern. With the changes in the retail market the time from order to delivery has reduced significantly and many companies attempt to operate a JIT system so as to reduce the amount of capital tied up in stock holdings. This has all resulted in scheduling and rescheduling becoming more complex and rapid.

However, with the development of more advanced computers that are able to process vast quantities of data rapidly, quite complex tasks can be carried out. Scheduling of production is often complex particularly where a factory has many product lines and product variants with short production run times.

Standard software formats have been produced to assist and in some cases automate this process. One well known system is material resource planning (MRP). This type of planning system allows the resources needed to make a product such as ingredients, packaging materials, labour, equipment, and so on to be controlled. The system integrates the material requirements for the total production usually on a batch, shift, daily, weekly or monthly basis. Most software systems enable electronic links to suppliers for the automatic replenishment of materials as they are used at the factory.

Recipe management software is available to allow the control of ingredient additions for particular products. As companies produce an ever increasing range of products there is a need to ensure that product ingredients are checked as they are added to the batch mix or continuous process. The use of product and ingredient specification databases allows accessing of ingredient details and requirements by the recipe management system.

The Instrument Society of America (ISA) developed standards for process-based control systems and defined a structure (SP88-MES, Instrument Society of America, 1996). They developed the term manufacturing execution system (MES). This type of system is able to bridge the gap between the SCADA and MRPII software packages and allows an improved level of integration.

Software systems can be integrated. The initial forms of integration were centred around a generic pyramid-type structure as shown in Figure 17.3. This shows a fully automated business system (FABS) in which material resource planning (MRP), the production management information system (PMIS) and the laboratory information management system (LIMS) are linked with the SCADA system and the PLCs on the factory floor.

Many other software systems have now been developed for food businesses, and perhaps the

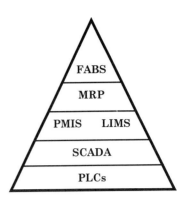

Figure 17.3 Typical pyramid software structure.

structure for the future may be closer to that illustrated in Figure 17.4.

Production cost monitoring and control

The prime objective of the production department within a food company is to produce a product safely, to specification, of constant quality and at the lowest cost. In order to assess manufacturing performance there are a number of techniques and measurements that can be employed. One key measurement is that of efficiency. This can be defined as:

$$E = \frac{P_a}{R_t \times P_t} \times 100$$

where E = efficiency as a percentage, P_a = actual production from a line or piece of equipment over a given time period as a total number of units, R_t

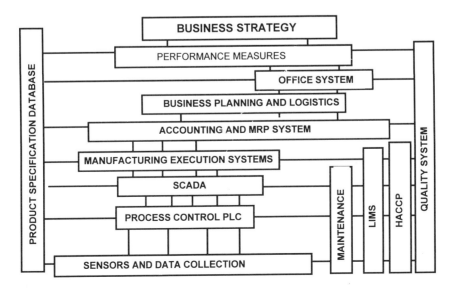

Figure 17.4 A new design for an integration framework.

= given time for which plant was operational and P_t = theoretical maximum production as units/time.

Another measurement is utilization which can be defined as:

$$U = \frac{T_a}{T_t} \times 100$$

where U = utilization as a percentage, T_a = actual time the machine/line was in operation over a given time period, and T_t = maximum time the machine or line could have been in operation over the same given time period.

In order to assess performance, standard costings and performance measures can be derived. The standards are normally set at the maximum achievable levels. The performance of the manufacturing facility can then be assessed against these standards on a real-time basis, hourly, shift, daily, and so on. The difference between the two measures can then be used to judge the level of performance. This technique is called variance analysis.

Variance analysis can be taken further and the costs for each operation in the manufacturing process can be determined and a standard cost for each activity can be derived. This technique is called activity based costing. The process allows an understanding of where costs are expended but more importantly gives a detailed assessment of financial performance for individual operations in the factory.

Another measure used to determine financial performance is return on capital employed (ROCE). This measure is used to calculate how much money is tied up in the assets of an operation compared to the amount of money it is generating as profit. A further measure is the return on investment (ROI) or payback. This is used where additional money is to be invested in the business and the company wishes to know how much additional profit the investment will generate. This then determines how long it will take to recoup that investment.

Benchmarking

Benchmarking is used by many companies in order to improve their performance. The technique involves comparing certain measurable tasks or activities in a company against those same tasks and activities in another; that is, production line, department, subsidiary company, food company or other companies. The companies are then able

to share their findings and investigate why one is better than the other or others. Typical benchmarks could be:

(i) Customer complaints
(ii) Sales order processing
(iii) Invoice processing
(iv) Personnel issues
(v) Number of product rejects
(vi) Machinery performance.

Some of the more generic activities can be benchmarked against a range of companies. Benefits often come through benchmarking clubs where companies share their findings and are able to assist each other to improve their performance. Within a company it is possible to benchmark individual products, production lines, packaging machines or departments.

Minimizing product loss

A traditional method of minimizing product loss is to create a rework product into which all the scraps of material can be incorporated into a product. However, nowadays this is often frowned upon and efforts are focused on designing the systems and factory so as to eliminate product losses.

Within liquid-based food factories such as dairies and breweries product losses during cleaning can be minimized by using sensors to detect the interface between the product and the cleaning fluid. Other methods include using a fixed time period to purge the lines of product or using pressurized sterile air as purge.

In batch production systems losses can be minimized by running one batch after another and scheduling production such that some carryover of product from one product to another is acceptable. This process usually results in lighter mixes being produced first followed by darker mixes.

Waste minimization

With certain products, in particular fruit and vegetables, it is possible to reduce the amount of material brought into the factory in the first place by having more efficient picking systems and on-farm separation processes. The water within the factory can be used several times, with resulting cost savings being achieved. However, as the water is resued, the solids level will increase and the waste will have to be discharged or separated at some point. By reducing the amount of chemicals used on the farm the residues that accumulate in the factory will be reduced.

Energy conservation

Most foods have to be heat treated in order to kill harmful bacteria and pathogens. The cost of heating large quantities of ingredients to relatively high temperatures is high. There are many methods by which energy consumption can be reduced. The most basic consideration is to ensure that an effective and efficient means of energy delivery is available.

Effluent from some factories can provide a means to set up a fermentation process which can generate methane gas. The methane can then be recovered and used as a gas to fuel the boiler system in the factory.

One technique that has been used is 'pinch technology'. This approach involves analysing all the heat sinks and sources within the factory and seeing where energy transfer could take place. One drawback with this approach is that most food factories alter their operations too frequently and therefore the heat sinks and sources are continually changing.

One simple method which has been used is to take the heat recovered from the cooling of products and use it to supply energy to a product about to be heated. This method has been used in pasteurizers for many years and is termed regeneration.

Insurance

Insurance is important to food manufacturing companies, in that there are many risks associated with constructing and operating a food company. These include:

(i) Fire and explosions
(ii) Flooding, subsidence, lightning, etc.
(iii) Injury to workers
(iv) Accidental damage and breakdown of equipment
(v) Theft and malicious tampering
(vi) Contamination of food products and product recalls

Insurance companies are more stringent on the terms and conditions of policies than they have been in the past. Therefore it is within the interests of the manufacturer to ensure the infrastructure and equipment operate with the minimum of risk. Evidence of risk reduction or increased safety measures can often reduce insurance premiums.

To reduce risks frequent audits are a necessity; these can either be internal or carried out by external experts. The inspectors can ensure that all required actions are taken to reduce the risks identified in the audits. HACCP should assist in some of the areas of concern and the principles of HACCP can be applied to minimize risks.

REFERENCES

EHEDG (1992) *A Method for the Assessment of In-Place Cleanability of Food Processing Equipment*, Document 2, Campden and Chorleywood Food and Drinks Research Association, Chipping Campden, Gloucestershire, GL55 6LD, UK
ISA (1996) SP88-MES, Instrument Society of America.
Mortimore, S. and Wallace, C. (1994) *HACCP A Practical Approach*, Chapman & Hall, London.

FURTHER READING

Busvine, J.R. (1985) *Insects and Hygiene*, 3rd edn, Chapman & Hall, London.
Codex Alimentarius Volume A (1988) *Recommended International Code of Practice on the General Principles of Food Hygiene*, FAO/WHO Food Standards Programme, Rome.
Machinery Safety Directive, 89/392/EEC (1989), EDO, Commission of European Communities.

EC Directives and Regulations and UK Acts and Regulations mentioned in the text may be obtained from Her Majesty's Stationery Office, London.

See also
Jukes, D.J. (1993) *Food Legislation of the UK, a Concise Guide*, 3rd edn, Butterworth-Heinemann, Oxford.

18 Quality Assurance and Control Operations

INTRODUCTION

The food manufacturer is in business to make and sell good food at a profit. Consumers will only buy and make repeat purchases if the food is perceived as being good value for money and generally only if they can rely on the product being of a consistent standard. A firm needs therefore to establish a reputation for making a product to a certain standard and to maintain it. The quality standard that the manufacturer wishes to maintain should be clearly defined, preferably in the form of a quality policy statement made at a high level of management. To maintain the highest quality standards requires defined high quality of raw materials, careful supervision of manufacture and a sufficiently high selling price. This will not always be the appropriate objective, as in the majority of situations a lower but nonetheless acceptable standard of quality will be required for the realistically attainable market. This does not mean that variability from the prescribed product standard should be any greater and indeed the converse may apply.

All the staff have a contribution to make in establishing and maintaining quality standards – a moment's inattention can lead to foreign body contamination; faulty personal hygiene could have even more serious consequences. It is therefore highly desirable to build up a knowledgeable workforce trained and motivated to work to the required standards, and to set out in clear form the procedures for maintaining those quality standards. Thus everyone knows what is expected, checks are not omitted and standards do not slip.

A quality policy

The company quality policy should be considered, documented and preferably endorsed at board level, as it will form one of the foundation stones of the company. It will influence sales and profitability as well as manufacturing practice and staff attitudes.

Basis of the control of quality

A minimum standard for quality is set in the UK by the Food Safety Act (1990), its Regulations and other statutory matter, together with numerous Codes of Practice which do not have the official force of law but which must not be ignored. There is a basis for a formal quality policy in ISO 9000. This Standard and its equivalents such as the Australian AS 3900 are generally applicable to all industry but need some extension to cover the particular needs of the food industry.

A more detailed and specific guide, *Good Manufacturing Practice – A Guide to its Responsible Management*, is published by the IFST (1991). In this chapter many of the subjects treated in the Guide are discussed in outline only and readers are strongly advised to consult the Guide itself for further information.

Total quality management (TQM)

The basis of TQM is that every person in the business should be responsible for the quality of his or her work. This responsibility can only be exercised if all workers and managers are aware of the factors that make up the quality of the work and of their effects on the quality of the work of colleagues. Thus the workers at all levels in the business must be trained to recognize quality criteria, to understand the interactions of their work with others and to exercise their responsibilities fully under the supervision of their line managers.

The management of quality becomes a line management responsibility and there is no longer a need for a separate 'quality control' function.

There may well, however, continue to be a need for quality specialists to provide an advisory service on quality matters.

The basis of good manufacturing practice

In the IFST Guide (1991), good manufacturing practice (GMP) is considered as that part of a food and drink control operation aimed at ensuring that products are consistently manufactured to a quality appropriate to their intended use, that is to say, to the qualities intended and expected. It is therefore concerned with both manufacturing and quality control procedures.

Control of quality starts with the selection and purchase of ingredients and packaging, and continues through the manufacturing chain to the quality of the product as it is consumed. It takes in the people, the machinery and the factory together with stores and vehicles. All of these can affect the ultimate quality of the food as it is purchased and consumed. It is up to the manufacturer to ensure that nothing goes wrong.

GMP has two complementary and interacting components: the effective implementation of well-designed manufacturing operations and the effective implementation of a well-designed quality control/quality assurance system. It is not only these two components that must be complementary and interacting, the same must apply to the respective managements of these two functions, with the authority and responsibilities of each clearly defined, agreed and mutually recognized.

Role of quality control

As stated above, if one is operating a perfect TQM system the role of quality control (QC) is greatly diminished, indeed some would say should be unnecessary. Equally, the role of quality assurance is greatly enhanced. Nevertheless in practice most operations still employ an active quality control setup even though the emphasis may have moved to a more on-line position.

The quality system should be audited on a regular basis to ensure it is covering all the necessary quality objectives efficiently and in changing circumstances, and that it is not becoming restrictive or stifling of initiative. The QC department will have a role in quality audits which may be conducted by internal staff or by outside consultants with the appropriate expertise. In most food factories changes are made relatively frequently either as planned progress or to solve short term problems. Whatever the cause it is important that the QC department is informed of all changes and hence remains in a position to advise on their implications for product integrity.

Traceability

A recurring objective throughout all food manufacturing operations should be to be able to relate each batch of product both back to the individual ingredients, their suppliers and the delivery dates, and forward to the packages supplied and if need be their distribution to shops and final consumers.

GENERAL PRINCIPLES

PHILOSOPHY OF CONTROL AND ASSURANCE

Quality has been defined as the composite of those characteristics which differentiate between individual units of product, and which also have significance in determining the degree of acceptability of that unit to a buyer. It is possible to paraphrase this, as a multipurpose measure of the extent to which individual units of a product meet the needs and expectations of the buyer, including that of price.

John Ruskin observed: 'Quality is never an accident, it is always the result of intelligent effort', and this surely is what the reputable manufacturer hopes to apply. While many definitions or descriptions of quality are to be found in the literature, we may identify quality simply as fitness for purpose; fitness, that is, in respect of legal acceptability, value for money, and, of course, organoleptic acceptability. Above all, a product of acceptable quality will perform to the consumer's expectation and satisfaction.

Key principles

The functions of quality control and assurance, for whatever purpose, were enumerated by Kramer and Twigg (1966) under the following headings:

(i)   Establishing specifications or standards
(ii) Development of test procedures and methods
(iii) Development of technically and statistically valid sampling procedures
(iv) Recording and reporting test results

(v) Trouble shooting, or action when standards are not achieved
(vi) Special problems such as customer complaints
(vii) Training personnel in control procedures and hygiene.

Similar philosophies and approaches have been embodied in a number of valuable publications, such as those of Juran *et al.* (1974), and (more specifically in relation to the food industry) Herschdoerfer (1984–86). The philosophy behind the application of QC/QA (quality assurance) is important. Quality control responsibility should be with all concerned in the manufacture of the product and the full commitment of management is an essential component.

In the IFST Guidelines key principles are given. Of quality management, it is said that there should be a comprehensive system, appropriately designed, implemented and resourced to ensure that products will be of the nature and characteristics appropriate to meet the consumers' needs, and that the attainment of that quality objective should have the involvement and commitment of all concerned at all stages of manufacture. This may be said to be the cornerstone of sound QA philosophy, although in practice it is achievable in a variety of ways. The Guidelines go on to cover personnel and training, documentation, premises and equipment, manufacture, rework and product recall, foreign body controls, and various other aspects of GMP over and above quality assurance and control operations *per se*. Application of such principles can be aided by inclusion of such techniques as 'right-first-time' concepts and workforce contribution from 'quality circles' and so on.

Right-first-time

'Right-first-time' is an approach to an overall quality awareness and philosophy which is founded on the concept that although advancing technology may make it possible to manufacture silk purses from sows' ears, it is normally easier and more profitable to start off with silk of good and uniform quality. Strongly expounded by Juran, the approach is one of ensuring that raw materials, plant and equipment, work-in-progress, and so on, are delivered or maintained at a quality standard which minimizes or even eliminates rejection. Not only should the manufacturer produce right-first-time, but to enable this to happen, the manufacturer should receive raw materials which are within the tolerances that allow their unhindered utilization within the specified process.

Further, the involvement of all personnel from purchasing to distribution and sales is expected, under a strongly supportive senior management committed to a defined quality policy which involves a considerable amount of self-assessment by the workforce of their allotted functions in the achievement of prescribed standards. As much responsibility as possible is placed with those who can most directly affect whether the standards are achieved. The right-first-time approach also looks at the identification of problem areas in a disciplined manner and the use of project teams, drawn from a cross-section of persons involved, to detail specific problems and propose solutions or improvements. The Pareto effect is often taken advantage of in this approach, e.g. 80% of the problems with raw materials will result from only 20% of the raw materials with which the factory is supplied.

In inflationary times, where there is significant cost attached to the finance of stocks of raw materials for manufacturing, there may be considerable advantage in minimizing stock levels, often by adopting a 'just-in-time' raw material and work-in-progress supply system. For this to be effective, a right-first-time quality assurance policy is a virtual prerequisite, coupled with appropriate safeguards to prevent the use of unsatisfactory supplies because of pressures to maintain production when nothing else is available.

Quality plan

Fundamental to an effective control system is a quality plan which will make use of hazard analysis of critical control points (HACCP) principles to identify and evaluate the points at which control must be exercised and to build these into a plan for control of all the key factors and stages affecting the quality of the product. In this way reliable control can be achieved at an affordable cost.

The plan should cover every aspect of product development and manufacture from product visualization, through development, raw material procurement and production, to storage, distribution and, to some degree, purchase and consumption of the product. The plan should be recorded in a flexible format so that it may be modified with experience. It should be properly authorized and changes should be controlled and registered. The plan will ascribe responsibilities for carrying out checks and corrective actions and review procedures should be in place.

Management of change

Experience shows that mistakes commonly occur at times of change. It is therefore particularly important to plan changes in advance and to make sure that all the implications of the change have been thought through and communicated to those who are directly concerned and also to those who are thought only to have a peripheral interest.

Emergencies, by their nature, cannot be forecast but it is possible to indicate and plan in advance lines of communication and action which, if followed, are likely to provide a safe and satisfactory solution.

Quality circles

'Quality circles' describes a method of involving employees at all levels in contributing to the maintenance of quality standards and to problem solving. A quality circle comprises a group of people from a variety of disciplines who meet together to discuss quality standards achieved and problems relating to a particular process or production line. These circles increase awareness of quality aspects, generate ideas for further improvement, and in some cases can assist in identifying the source of specific quality defects. They can thus provide opportunities for more employees to contribute to aspects of the operations which affect ultimate product quality.

The operation and administration of quality circles should be through the department responsible for quality assurance and control, to ensure a disciplined and structured approach to the task and the appropriate assessment and actioning of suggestions put forward.

ISO 9000 Quality Standard

This international standard for the management of quality is being widely used in the food industry. The standard was developed primarily with the needs of the engineering industry in mind, but the general principle of systematic documentation is valid in the food industry. It is also being used in catering and hotel business and in service sectors such as food distribution.

ISO 9000 has three parts:

(i)   ISO 9001 – specification for design, manufacture and installation
(ii) ISO 9002 – specification for manufacture and installation
(iii) ISO 9003 – specification for final inspection and test.

The wide acceptance of the Standard by the UK food industry owes much to the availability of Guidelines to its application and to the enthusiasm of most of the first major firms in adopting the Standard. The Standard provides a structured documented system for managing quality that is readily audited and maintained. Well prepared systems provide firm but flexible control without excessive documentation that can be altered to meet changing company requirements and continuously refined with experience. When the Standard is correctly implemented it results in a steady improvement in the standard of quality and a reduction in problems due to unforseen occurrences.

Most early quality systems in food manufacture were written to meet the requirements of ISO 9002 'Specification for manufacture and installation'; more recently it has been recognized that ISO 9001 'Specification for design, manufacture and installation' is often more appropriate.

QUALITY ASPECTS OF THE MANUFACTURING SEQUENCE

Raw materials

In conjunction with the ingredients buyers, QC will have a role in the selection and assessment of suitable suppliers. It will be necessary to evaluate purchasing samples for quality and suitability, and to check that delivered material both conforms to the purchasing specification and matches expectation. Much trouble may be avoided, both for supplier and purchaser, by also obtaining and testing a 'buying' sample of each particular lot in advance. Buying samples are also of value in instances where hard and fast specifications may be difficult to apply, e.g. fresh fruit and vegetables; in such cases the material is purchased on the understanding that it matches the approved sample. Wherever possible, a check should be made that the supplier has an adequate quality control system in operation.

In the case of critical raw materials where quality variation is judged to be either likely or potentially serious, the supplier should be visited to check on standards of manufacture and control and agree on procedures to be adopted in the event of receiving a substandard delivery. Inspection and sampling of bulk lots on suppliers' premises, warehouses or docks can be advantageous where this can be arranged, while occasional inspection of the suppliers' processing operation can often justify itself.

In any event, each consignment received should be inspected and, as time permits, sampled and tested before the bulk is unloaded. More extensive sampling and more thorough testing can be carried out when the consignment is taken into stock. If retained in storage, it should be examined regularly for any signs of deterioration.

When deliveries are made, the warehouse keeper should check the conditions of both the delivery vehicle and its load before commencing off-loading. In some cases the QC department will be notified so that samples can be taken and visual assessment made. Each delivery should be clearly marked so that it can be related to the samples taken for analysis and to the supplier's invoice documents. Wherever possible, the delivery will be placed in a 'hold' area and not taken into production until it has been examined and the samples passed by the QA staff. Packaging materials and supplies of ancillary items such as cleaning materials should be treated in a similar way. The warehouse keeper should maintain a full inventory of stock. Proper stock rotation should be ensured and a regular return of all over-age materials should be made.

Certificate of analysis

The quality checks required on a critical raw material may be too expensive to be carried out on every delivery. It is often possible to obtain from the supplier a certificate of analysis appertaining to the batch from which the delivery was drawn. With occasional cross-checks to ensure reliability, certificates of analysis can eliminate a great deal of duplicate analysis. A 'certificate of compliance' to an agreed specification provides less assurance because it does not give specific information and is less easily cross-checked.

Preparation of raw materials

Many ingredients will require a preparation stage; in some cases it may simply be the removal of paper, cartons and outer wraps to avoid dirt entering the process area; in other cases it will be necessary to peel, wash, trim or standardize to constant composition. Careful inspection of the bulk of the delivery may well be required at this point as well as the rejection of undesirable objects such as stones, insects and metal. There is no complete alternative to manual and visual

inspection but automatic devices such as graders, gravity separators, magnets and metal detectors have a role to play. Where these are in use their proper functioning should be checked frequently and any unusually high rejection rate should be correlated with the particular delivery. The facility to trace back to delivery information should be maintained during the preparation processes.

Blending or mixing

The mixing or blending operation can be critical when it makes up part of the manufacturing operation, as it influences the proper performance of the product through the processing and packaging stages. It governs final eating quality and it affects raw material use and the cost of the product. Because raw materials are mostly natural products their own composition may vary and it will be necessary to bear this in mind during formulation, particularly if the composition of the final product must fall within prescribed limits.

The blending operation may be manual or automatic or a combination of the two. It is worth going to some trouble to make the operation as easy as possible and if possible to introduce a self-checking element so as to minimize errors. Checks on the accuracy of blending should be considered. It may be possible to carry out a simple physical or chemical check for a component such as sugar, salt or acid or it may be desirable to taste the blend or a derivative.

In batch production, the mix may be held while the tests are carried out, which should, of course, be done as rapidly as possible. In a continuous mixing operation there will be no facility to hold material while it is tested and some form of in-line sensor will be necessary to identify variation from standard. A record of the mixing operation must be maintained, showing not only the ingredients used but also relating the delivery batch numbers to product batch numbers. These records should be retained for a period exceeding the likely consumption life of the product.

Processing

During the processing operation the food may be stabilized by, for example, heating, cooling or dehydration, so as to halt or slow natural decay and to transform the food mixture into a form suitable for further processing or immediate packaging for sale. Correct processing may be vital to the production of a food that is safe to eat. It is important to identify the factors that contribute to the safety of the food and to check that these are maintained within prescribed limits. The values actually reached should be recorded, automatically or manually and the records kept on file, cross-referenced to the product batch numbers.

Thermometers, pressure gauges and other measuring instruments should be selected as the most suitable for the purpose. They should be installed so that they measure the required variable sensitively and their accuracies should be checked regularly against calibration standards.

Cross-contamination

In designing a factory and in preparing the quality plan it is important to give careful consideration to steps to avoid processed foods and work in progress becoming recontaminated with spoilage organisms. Such contamination may arise from unprocessed food, humans, other life forms, tools and equipment during manufacture and in storage. Similarly, care must be taken to avoid contamination of any batch of product or of work in progress with material of different composition, such as residues from a previous batch.

Positive release

The concept of positive release is a very useful tool for ensuring that raw materials, work in progress and finished products are only allowed to proceed to the next stage once they have been passed by the QA or QC department. In its simplest form the materials are isolated by placing stickers or tickets on them indicating that they are on hold. Once they have passed all necessary tests these tickets are replaced by ones indicating that they have been passed, and in the case of raw materials, for example, that they can now be used for production. It is essential that such a regime is strictly enforced and that along with the simple ticket system full records of tests carried out are kept.

The canning industry typically uses the positive release principle for the safety of its finished products by the incubation of cans. The cannery will not usually release the product for dispatch until it has passed this test satisfactorily.

Packaging

While packaging materials are, of course, also raw materials, the subject of packaging in the context of quality control and assurance is a wide and complex one. Packaging may need to fulfil a number of functions including:

(i) Containment of the product from producer to consumer
(ii) Protecting the product from the environment
(iii) Providing information to those handling and using the product
(iv) Providing an attractive appearance to aid the marketing of the product and conveying promotional information
(v) Protecting the product from physical damage
(vi) Presenting a convenient form for ease of handling by distributor and consumer
(vii) Being relatively easily disposable and preferably recyclable
(viii) Taking a functional part in the use of the product by the consumer, e.g. teabag filter paper, snackpots, vending cups.

The material also has to interact satisfactorily with the production equipment, mechanical and human, on a cost effective basis and without causing undue downtime, wastage or lack of final integrity. The need for care in defining packaging material specifications and the importance of checking compliance on receipt must be strongly emphasized. The testing of packaging materials is frequently empirical and not all parameters governing satisfactory performance may be identifiable or measurable other than by assessment of performance on the production machinery or in the process to which the material will be subjected. For these reasons the on-line testing of delivery samples as part of the acceptance procedure is to be recommended.

In many cases the packaging material performance and the way it is integrated into the food manufacturing operation are critical to product safety, e.g. the seaming of cans, the sealing of laminates, the application of jar and bottle closures.

It should not be forgotten that packaging extends beyond the individual consumer unit of sale and includes outer cartons, cases, overwraps, pallets and more.

Unit packaging and labelling

During the finishing operations, the product batch or process stream will be broken down into individual smaller lots that will be packaged for sale. This is therefore the last point at which the product can be inspected in bulk, sampled and if necessary recycled or adjusted without loss of packaging. It is a good point at which to locate automatic quality inspection or control equipment. This might be some kind of filter, grading equipment or in-line analytical device, including metal and foreign body detectors.

Once the product is packaged, it is more difficult to sample in a representative fashion; it will however be necessary to check that the product fill is within permitted limits and that the product as a whole is to the standard that the company requires. Samples of the packed product represent the form in which the customer will purchase and use it. It is upon the quality at this point that the producer's reputation depends. The product will be in its package for a considerably longer time than that required for manufacture or for consumption. It is important that the packaging maintains the product at the required quality over the period of the declared shelf life and for a margin in excess of this. This will be assessed by storage trials, taste panels and analytical checks and confirmed by retrospective feedback that anticipated acceptable shelf lives were achieved.

Each product package should be marked so that the production day and possibly the factory, production line and shift can be identified. This code should be related back to the production batch number and the records retained. In this way any package returned for subsequent examination can be related to particular raw material deliveries if necessary.

Storage

The finished product should be stored under good conditions at an appropriate temperature. Care should be exercised to avoid damage to stock at this stage when added value is highest and temptation to pilferers is greatest. As with incoming materials, the warehouse keeper should keep a check on slow moving stock and make a regular return to the appropriate management.

Distribution

Immediately prior to despatch, the packages should be checked to see that they are in good order and to ensure that the package contents match the labelling. Vehicles should be checked before loading to see that they are clean and in good order and that other goods are not being carried that could cause a contamination problem.

QUALITY ASPECTS OF SOME COMMERCIAL OPERATIONS

Imports and exports

Special considerations apply to imports and exports, relating to legislative demands, the market to which the products are directed, the climatic conditions likely to be encountered, the method of shipment and possibly to problems of language and even of literacy regarding the intended method of use by the consumer.

The importer has to satisfy himself that the goods have been produced in accordance with good manufacturing practice and be able to satisfy the importation control authorities that their requirements have been met. In some cases, e.g. meats, there may be a statutory obligation for appropriate certification to have been obtained. Close liaison between the producer and the importer should be established, and an effective method of dealing with and reacting to complaints.

For exporters, there is a particular responsibility for ensuring the adequacy of the packaging and shipping arrangements.

Contract packing

Contract packing is any operation in which all or part of the manufacturing or packaging process is carried out by a producer on behalf of another manufacturer, distributor, wholesaler or retailer. The contract packing itself can be looked at both from the viewpoint of the contract giver and from that of a producer of the goods. We may regard the producer as the contractor and the giver of the contract as the customer.

The contractor

The contractor must ensure that he has a comprehensive understanding of what is required in fulfilment of the contract, and clarification should be sought as necessary in areas of uncertainty, for example, where the greater expertise relating to a particular product or process is that of the customer. It is in the contractor's own interests to seek written agreement for all aspects of the contract, and especially to know what is required in relation to quality control and assurance procedures and records. Any special requirements that the customer has should be identified and agreed.

In contract packaging the contractor is still governed by the legislative requirements of food manufacture, and in most instances of quality control. Indeed there is an obligation for the contractor to carry out the production in accordance with good manufacturing practice.

In most circumstances the contractor is in a position of privilege concerning the buyer's formulae, processes, suppliers, possible promotional activity, and so on, and should treat all such information as confidential. Under no circumstances should all or part of the work involved be passed on to a third party without the prior agreement of the customer.

The customer

The customer has a clear responsibility to ensure that the contractor understands what is required of him, is capable of doing the work required, and has been provided with the appropriate specifications, procedures, documentation, and so on. As far as quality control is concerned, the customer must be sure that the contractor has the appropriate staff and skills necessary, even if this may mean reinforcing those of the contractor from the customer's own resources as part of the contractual arrangement.

All arrangements made between the parties should be confirmed in writing and the customer should ensure right of access to the contractor's premises during production of the product, and to finished product and raw materials, so that the customer's own QC staff can audit as appropriate. It should be borne in mind that in some instances the customer may have greater expertise than the contractor and it is clearly in the interest of both parties for this to be taken into account as information is exchanged. The contractor should be given adequate information on points of contact in an emergency.

HAZARD ANALYSIS OF CRITICAL CONTROL POINTS (HACCP)

Hazard analysis is a technique which has made a substantial contribution to the appraisal of risks from microbial sources, and forms the basis of a critical and disciplined approach to the assessment of dangers that might be anticipated within current knowledge. It is a technique which can be applied just as effectively to the identification of causes of problems in other areas, and for this reason is to be recommended as a generally useful tool in QC. The Food Safety (General Food Hygiene) Regulations 1995 and EC Directive 93/43 require food businesses to identify and control

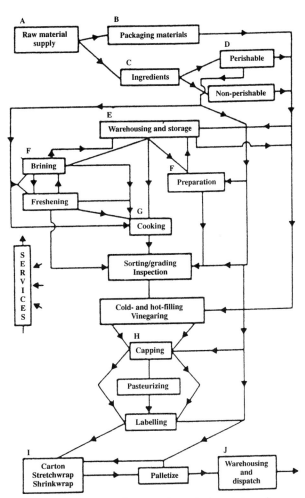

Figure 18.1 Process flow and control points: vinegar pickle production.

food hazards where these are critical to food safety.

In essence, what is required is the logical identification of the areas in which problems are likely to occur (often stimulated by the occurrence of a real life difficulty) followed by a detailed critical appraisal of the events which take place in the area in question. The best starting point is a schematic flow diagram of the process like those shown in Figures 18.1 and 18.2, which show flow diagrams for the production of vinegar pickles. The critical points can then be highlighted and a degree of concern attributed to each. From the results of this, subjected to analysis of variance or other statistical treatments, specifically hazardous points may be identified and appropriate mechanisms of control proposed.

The HACCP process may be defined as: observations and/or tests made to identify actual or potential hazards in operations and to identify critical control points in a process. Control measures are designed and implemented and the control

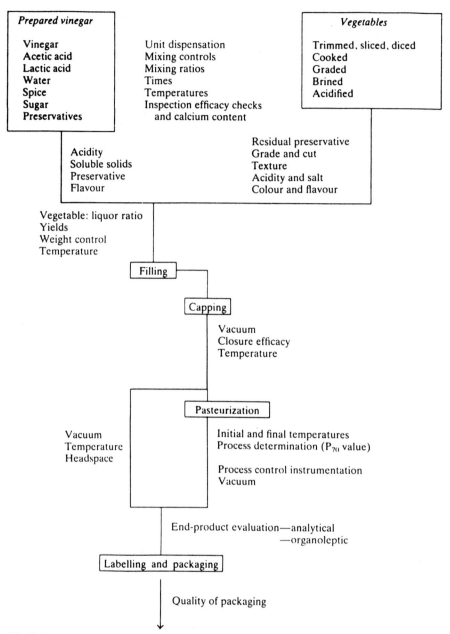

Figure 18.2 Area H of Figure 18.1 – expanded flow diagram and possible assessments required.

points are monitored to ensure that control is maintained.

HACCP is a logical structured process that will consider what may go wrong, how, where and when. The Codex Alimentarius Commission, 1991 has published a logic sequence for applying HACCP (Codex Alimentarius (1995), **IB**, 1.1). The reader in the UK is also recommended to consult the technical manual published by the Campden Food and Drink Research Association (1991) on the subject. This provides worked examples in a variety of situations.

HACCP is of value for evaluating new products and processes with a risk element. The technique is entirely consistent with ISO 9000 Quality Systems with its requirements to use multidisciplinary teams to develop quality plans and process control requirements, to select and use suitable instruments to monitor and control and to record and analyse results and audit the control system.

LEGISLATIVE CONSIDERATIONS

General legislation

Factories and their operations are regulated by the general legislation applicable to all factories. In the UK, the Factories Act 1961 is the backbone of the regulations. In the UK, in addition to the Factories Acts, the Health and Safety at Work Regulations 1974, the Sex Discrimination Act 1975 and others, the food industry is controlled under five primary laws:

(i) Sale of Goods Act 1893 and 1979
(ii) Trade Descriptions Act 1968 and 1972
(iii) Weights and Measures Act 1985
(iv) Food Safety Act 1990
(v) Supply of Machinery (Safety) Regulations 1992, as amended.

Under these Acts most of the ancillary regulations affecting the manufacture and sale of foods and drinks are made.

In simple terms, the Sale of Goods Act requires that articles sold should be fit for the intended purpose so long as that purpose is known to the vendor by virtue of common practice or by specific advice. The Trade Descriptions Act requires that goods and their attributes are fairly and reasonably described, whatever the medium by which the information is conveyed. The Weights and Measures Act and its requirements are covered later under QUANTITY CONTROL. The Supply of Machinery Regulations specify essential Health and Safety requirements for machinery to be used for the manufacture of foodstuffs, including hygienic design. New machinery must carry a CE approval mark.

In the UK there are specialist legal information and updating services, such as those offered by Butterworth's on food law.

Food legislation

The basic food law in the UK is the Food Safety Act 1990. The main cornerstone requirements prohibit the sale of food which is either 'injurious to health', or 'not of the nature, substance or quality demanded by the purchaser'. Other provisions require the registration of all food premises and permit the introduction of Codes of Practice and statutory Orders to embody Regulations originating in the European Union.

A significant change in the 1990 Act was the introduction of the 'due diligence' defence. A vendor or a manufacturer charged with selling defective food may now plead that the cause of the defect was outside his control but that he had 'taken all reasonable care' and 'shown due diligence' to avoid it. For such a defence to succeed evidence will, of course, have to be provided to

show that the quality system was sufficiently careful and was applied with sufficient diligence.

In the USA, the Federal Food, Drug and Cosmetic Act covers similar ground. The emphasis for all food manufacturers must be to ensure good documentary and record keeping procedures. For this reason many are turning to quality systems such as ISO 9000.

Codes of Practice

In the UK, Codes of Practice are considered a more flexible means of controlling standards of quality than strict Regulations and no less effective. They have proved of considerable value to producer, distributor, consumer and enforcement officer alike. The great advantage of Codes of Practice over definitive legislation is that they retain a measure of flexibility which allows more rapid response to changing circumstances, permit operation outside the terms of the Code where it can be justified by a producer to the enforcement authorities or the courts, and constrain development to a lesser degree.

There are many Codes of Practice in existence, dealing with a diversity of subjects from ante- and post-mortem inspection of slaughter animals to the manufacture of vanilla slices. They are published by various national and international bodies, for example, Food and Agriculture Organization, World Health Organization, Codex Alimentarius Commission; local, national, and international trade associations, often in liaison with enforcement authorities or their associations or coordinating bodies. A full list of those current in the UK is given by the IFST (1993).

Those considering entering a market area new to them are advised to check with their local trade association or expert advisor as to whether or not they might be affected by any such Codes of Practice.

Enforcement

The role of enforcement authorities and the manner in which they work varies throughout the world and in this chapter it is intended to deal only with the situation as it exists in the UK. For other countries the reader should refer to expert advice such as is available from the Leatherhead Food RA, or from the appropriate authorities in the country in question.

With regard to the UK, the Food Safety Act

1990 has increased the powers of the local authorities to act. Whereas in the past the authorities could only act if the food or the circumstances in which it was being produced or sold constituted a danger to the consumer, they now have increased powers to prohibit production at all stages and thus take preventive measures to ensure public safety. They have increased powers of entry and rights to examine records and it is the responsibility of the producer to provide these.

HYGIENE

In this section some general guidance is offered. Some product-specific hygiene problems are also discussed. In the UK products must be manufactured in registered premises which comply in design and practice with the UK Food Hygiene Regulations or equivalent legislation in other parts of the world.

The general rule for safe food manufacture is: be quick, be clean, stay cool. Speed is important because foods are derived from natural products and deterioration is inevitable unless action is taken. During processing, foods are exposed and deterioration will be more rapid, so the quicker the food is processed the less will be the deterioration. Avoid leaving food part processed and exposed during production breaks.

Cleanliness is important because dirt carries bacteria, poor personal hygiene can infect the food with toxic bacteria that may multiply in the food, dirty work areas attract vermin, and dirty equipment will transfer infection from poor material to good material.

Cold is important because lower temperatures slow down deterioration. Foods should always be kept as cold as possible without affecting its physical character, and after any heating operation such as sterilizing, baking or frying, the food should be cooled as rapidly as possible to obtain the best quality.

Personal hygiene

People have a great impact on food hygiene. Good personal habits, careful hand washing after using the toilet, washing before starting work and after every production break, keeping hands away from faces, all serve to reduce the risk of contaminating food. Smoking, if permitted at all on the premises, should be restricted to specially designated areas and the eating of sweets, and so on, and the chewing of gum should be forbidden in production areas.

Overalls and other items of protective clothing should be replaced as they become dirty and should only be worn in the changing areas and on the production floor. Hair and beards should be completely covered in nets or snoods.

Items that could drop into food such as jewellery, ear-rings, watches, cuff-links and nail lacquer should not be permitted, and staff who work regularly in manual contact with food should have a routine hand inspection. Poor dental standards could also be a cause for concern.

Employees should all be given elementary food hygiene instruction before they start work and this should be supplemented by further training as they gain experience. Chargehands and supervisors who will be required to give on the job instruction in hygiene should themselves have received adequate training in food hygiene and in instruction techniques.

Personal health can affect food manufacturing standards. Employees should be encouraged to report any problems of diarrhoea or vomiting so that they can be removed to a non-food-contact job. After returning from overseas, a discreet check may be desirable. Any wound or cut must be covered with a waterproof dressing, preferably bright coloured and ideally incorporating a metal strip to aid discovery should it be lost: a check should be made to ensure that it has not been lost during the work period. Any worker with a septic boil, sore or other wound should be moved to a non-food-contact job. Due regard should be taken of the statutory requirements of the Food Hygiene Regulations. Operatives should be trained to avoid unhygienic practices such as picking food or packaging from the floor, moving from a dirty to a clean job without washing and if necessary changing overalls, and using contaminated tools.

Factory structure

The factory building should be bright, clean, well lit and in good repair, so as to encourage workers to recognize that high standards are expected and attainable. The outside of the factory should be kept neat and free from refuse and any lumber that could provide harbourage for pests.

The entrances to the factory, doors, ventilators, drains, service ducts and windows (if any), should be designed to prevent the ingress of birds and of rodents and other mammals. Knockdown insect killers should be located close to entrances, but not in direct line of sight from outside.

Walls should be smooth finished and the surface

strong enough to resist cracking or crazing from chance impact. Cavities behind walls and fittings secured against walls should be avoided, as these provide harbourage for insects. Corners should be rounded and junctions with floors should be well coved. Floors should be smooth enough to clean thoroughly, but rough enough to give a reasonable grip when fat or water may be present. The floor should be laid to an even fall to drains. Floor life will be longer if equipment is not bolted through it and if it is not heavily stressed by overheating or by hard-tyred traffic.

Drains should be laid to a steady fall and designed so that they may be cleaned easily. Consideration should be given to installing roughing filters and primary fat traps at the drain exit from each production section. Blocked drains are a serious hygiene hazard. The cause of the blockage should be identified and rectified, and if necessary production in the affected area should be suspended until the problem is cleared.

The service pipelines and wiring may run around the production area or may need to cross it. Avoid running pipework over any point where food is exposed or above a process vessel. If the pipework is above a false ceiling, then suitable catwalks for access should be provided.

Food machinery – hygienic design

EC Directive 89/332 (in UK, The Supply of Machinery [Safety] Regulations 1992, as amended), sets out basic standards of safety and hygiene, to which all agro–food machinery sold new after 1992 must conform. New equipment may be purchased and installed for food manufacture only when it bears the EU mark certifying that it complies with the requirements of the Directive. The Directive does not apply at present to existing or second hand machines. In practice it is difficult to reconcile the needs for safety and for hygienic design. Manufacturers should consider the effects of raw materials, time, temperature and plant design when setting the hygiene standards required from the plant and deciding on the frequency and method of cleaning required.

Cleaning schedules

A schedule of all the cleaning tasks in the factory should be prepared. For a machine this will identify the extent to which the machine is disassembled by the operator and items that may need to be taken apart by a fitter, the method, equip-

ment and chemicals to be used for cleaning with special attention to any sensitive areas. Then the rinsing and reassembly routines should be identified and how the machine is to be left. Immediately before re-use there may be a further routine to sterilize the machine, or to rinse through or to complete reassembly.

There will be sections in the schedule covering the cleaning of floors, walls, toilets and rest areas specifying equipment and chemicals to be used. Some areas will not be cleaned daily; the schedule will lay down a routine to ensure that all areas are cleaned regularly and that a log is maintained so that this can be confirmed.

Clean-in-place (CIP)

Many food production systems utilize liquid ingredients and therefore have enclosed storage tanks and pipework systems for conveying them. This situation lends itself to CIP systems where, after production has finished and the vessels and pipelines are empty, cleaning materials can be circulated throughout the network. The cleaning materials used depend on the industry but will often involve a non-foaming detergent coupled with perhaps a caustic solution, if fatty residues are present, and/or a sterilizing agent. Final rinsing is also essential. It is essential that consideration is given to the flow rate in the design of such systems and that 'dead' spaces are avoided. It is important that jets are checked and cleaned regularly so that 'blind' areas are not left uncleaned and that tank drains are adequately sized so that there is no buildup of cleaning fluid in them.

CIP systems lend themselves to computerization and fail-safe measures. Typically they will only allow production to recommence if the CIP cycle has been successfully completed. A printed record of the cycle having taken place can be produced.

AUDIT

An audit is a detailed check on the operation of a system, carried out by someone not normally involved in the system. It is important to include in any quality scheme a system of regular audits. Reports of these audits should be followed up to ensure that any recommendations agreed as a result of these are implemented.

One of the strengths of the ISO 9000 Quality Management System is the requirement for regular, planned audits of all aspects of the operation by trained, independent auditors. Audit reports with recommendations for corrective actions serve to keep the documented system and factory practice aligned.

A full formal audit will review all the relevant documentation and procedures, will check on standards being achieved and the suitability of personnel and will make recommendations at a senior level for any changes. Such an audit is time consuming, and for this reason and to obtain an informed but dispassionate judgement, internal or external consultants will usually be involved. In a developing system an annual audit is desirable but in later years the interval between audits may be extended, should expense so require and confidence in the supplier allow.

Factory quality audit

In a larger factory, it is often useful to audit areas of the factory and a team can be set up to carry out a cycle of audits on agreed dates.

These audits will check:

(i)　Documentation and the availability of properly maintained records
(ii)　Standards being achieved and problems being encountered
(iii)　The need to modify aspects of the systems to take account of changing circumstances
(iv)　Opportunities for improvement
(v)　Safety standards
(vi)　Staff training levels in factors affecting quality, especially hygiene
(vii)　Completion of follow-up action from previous audit.

Checklists

The development and use of checklists is a useful tool, particularly when visiting suppliers and where staffing levels are such that different persons may carry out successive quality assurance audits. They provide a convenient way of scoring or rating a supplier and can be modified and amended in the light of experience and changing circumstances. A checklist prepared for a visit to a raw material or finished product supplier might be laid out as in Table 18.1.

REIVEW OF PERFORMANCE AND RESULTS

There should be a commitment to a regular review of performance and results. The information for this review may be assembled by the quality control department but the responsibility of carrying out the review should lie with a senior manager with direct access to the chief executive. The review will take in all aspects of the operation of the quality system including:

(i)　Laboratory results and quality trends
(ii)　Customer complaints
(iii)　Wastage and rejection levels
(iv)　Environmental performance
(v)　Non-conformances and corrective actions
(vi)　Internal and supplier audit reports
(vii)　Concession notes
(viii)　Staff training and performance levels

At the end of the review a report should be prepared for the chief executive recommending actions to be taken to correct faults and to improve the quality system.

STAFF TRAINING

It is essential that staff working in all parts of a food production establishment are properly trained. Aside from the obvious need to train for specific tasks, they should also be trained in food hygiene to an acceptable standard. In the UK both local authorities and special bodies such as the Royal Institute of Public Health and Hygiene run courses suitable for food handlers.

Every food establishment should have a basic set of rules which in the UK should be based on the Food Safety (General Food Hygiene) Regulations, 1995, with which all staff are made familiar. The IFST (1992) publication gives useful guidance. Increasingly all production staff play a role in quality assurance and here too it is vital that staff receive the proper training so that there is understanding of the importance of their role.

LABORATORY PRACTICE

The control laboratory should be properly equipped and staffed with suitably qualified personnel to enable it to carry out the necessary checks and give the required services with the speed and accuracy needed. Special provision for the use of outside resources or other expert service should be made if necessary.

Laboratory systems

Just as there are quality systems for all aspects of food production, a well-run food laboratory

Table 18.1 Checklist for a raw material supplier audit

QUALITY POLICY:	PREVENTION OF CONTAMINATION:	
QUALITY SYSTEMS HACCP: Accreditation: High/low risk policy: Product recall: Complaints: Environmental:	Personnel: Wood: Cross contamination: Glass register: Glass breakage: Covers: Filters/sieves: Equipment logs:	**PEST CONTROL** Contract: Plan: Proofing: Internal baits: External baits: Insect screens: Insect killers: Shatterproof tubes: Pheromone traps: Infestation noted: Fumigation: Records:
MANAGEMENT Structure: Qualifications:	**STORAGE** Raw materials: Segregation: Finished products: Packaging: Temperatures: Pallets: Stock rotation: Rejected stock: Contract storage:	**QUALITY ASSURANCE** Laboratory: QA staff: Traceability: Product recall:
PREMISES EC License: Layout: Roof, ceiling: Walls: Floors: Windows: Doors: Lights: Ventilation: Drainage: Engineers' room: Perimeter: Waste management:	**HYGIENE** Premises: Equipment: Vehicles: CIP: Cleaning equipment: Hygiene team: Cleaning schedules: Cleaning checksheets: Swabs: Records: Audits:	Raw materials Tests done: Certificates: Animal welfare: Pesticides: Authenticity: Water: Packaging: Supplier list: Audits: Positive release:

EQUIPMENT
Appropriate:
Condition:
Paint:
Conveyors:
Lagging:
Materials:
Capacity:
	STAFF
	Number:
	Shift pattern:
	Hygiene rules:
	Hygiene training:
	Hand wash on entry:
	Dress:
	   Laundry contract:
	Jewellery:
	Smoking:
		On-line controls
		Net quantity checks:
		Metal detectors:
		   Detect stainless?:
		   Checked:
		Visual inspection:
		Pack seals:
		Codes:

PROCESS CONTROL Temperatures: Equipment status: Calibration: Recipe control: Re-work: Checksheets: Records:	Lockers: Canteen: Medical: Toilets: Visitors:	Finished products Analytical: Microbiological: Tasting: Positive release: Samples retained: Non-conforming product: Records: Specifications:

should also operate to a quality system to ensure the accuracy and reproducibility of its results. The systems operated and methods used should be documented and should include procedures for the calibration of equipment, staff training, the handling of samples, the test methods used and recording and reporting of results.

The number of food laboratories seeking accreditation for their quality systems is increasing. In the UK the United Kingdom Accreditation Service (UKAS; previously NAMAS) is the body from whom accreditation is commonly sought.

Laboratory safety

Safety at work is the responsibility of the worker, but it is also a management responsibility to see that operators follow safe working procedures, work in a safe environment and are properly trained in safe working procedures. It is important to recognize that potential hazards from sources such as bacterial cultures, powerful chemicals or electrical equipment are present in laboratories and to regulate operations to minimize the risk.

Laboratory audit

Laboratory audits should be carried out at least every six months; in laboratories aiming for the highest standards there will be a permanent audit team operating.

The audit will check:

(i) The selection of analytical methods to see that appropriate methods are specified for use in the laboratory, that these have been adequately checked and that suitable equipment is available.
(ii) The testing of new methods to keep the laboratory up to date and cost effective without over frequent modification of standard methods
(iii) That the specified methods are being used and fully complied with, without short cuts and unauthorized modifications
(iv) That safety standards in the laboratory are being maintained and potential hazards identified and eliminated
(v) That sample receipt, handling and reporting procedures are being followed
(vi) That there is an adequate procedure for checking by the use of blind duplicates and 'spiked' samples
(vii) That operators' skills and training needs are regularly evaluated
(viii) That full value is being obtained by careful selection of samples and sampling points and by analysis of the information available.

Laboratories which are accredited under the UKAS scheme will of necessity carry out a programme of internal audits covering every aspect of their operations over the course of a year. They will also be subject to external audits by UKAS appointed auditors.

ENVIRONMENTAL ASPECTS

Responsible food manufacturers need to consider the possible impact of their operations on the environment, both within and outside their own premises. In the UK the requirements of the

Environmental Management Standard BS 7750 should be considered.

Externally, it is not only smells, noise and effluent that affect the local environment, but also the more subtle influences of transport, services and the movement of personnel. The involvement of the QA staff here is minimal, but an awareness is necessary as it may help anticipate difficulties and may be invaluable in a crisis.

It is also true to say that for some operations particular environments are more suitable than others, climatically and in terms of geographical location. For example, some processing plants need to be in close proximity to the growing area, and some types of production may benefit from a generally low relative humidity environment although the cost of specific air conditioning may not be justifiable. Some foods, particularly fatty foods and those with an extended surface area are particularly vulnerable to taint; off-odours from neighbouring operations and unsuitable materials can lead to absorption of sufficient taint to render the product unsaleable.

Working conditions

As far as the working environment is concerned, the QC department may be called upon to provide both guidance and a monitoring service. Levels of certain gases in factory atmospheres are limited by law or by recommended limits. Fermentation processes or gas flushing might result in the risk of carbon dioxide reaching unacceptable levels, for example, sauce boiling can produce high acetic acid levels, sulphited fruit pulps give rise to sulphur dioxide, and so on. Other in-house operations such as welding, fumigation and running internal combustion engines, can cause problems.

Dust in factories can be a serious hazard. It can lead to pulmonary problems in operators and it is a potential explosion hazard. It is better by far to prevent dust escaping to the environment by good plant design and regular maintenance than to provide a facility to clean up any dust that does escape. All finely divided foods are flammable and could give rise to explosive atmospheres. Noise levels, temperatures, relative humidity and lighting conditions may need to be monitored.

Much of the plant and buildings used in food manufacture is controlled by legislation in the UK, both general legislation for manufacture, such as the Factories Acts, and legislation specific to the food and drink industry such as the Food Hygiene Regulations. The construction and finish of workrooms are prescribed, the provision of cloakrooms, toilets, washing facilities and protective clothing is demanded and numerous other aspects are enumerated. Mandatory requirements apart, enlightened management will provide working conditions which are as good as possible to engender good morale in its staff.

Effluents

Effluents arise in the food industry in all three physical states, solids, liquids and gases. In most situations, the responsibility for the proper control and disposal of these will lie with those responsible for the overall quality control function.

Gaseous effluents include boiler and furnace emissions, exhausts, extraction system discharges and also odours which may cause offence to the public at large. It is as well to be familiar with local as well as national legislation in this respect. Expert advice may be necessary to deal with especially difficult gaseous discharges, treatment ranging from simple scrubbing systems to the installation of catalytic burners in the gas flow.

Except for categories requiring specialist disposal, e.g. oils, solvents and sludges, liquid effluent will normally be aqueous in character and mainly disposed of in one of three ways: direct to the municipal sewage and effluent treatment works; to the municipal works following partial treatment; or to river or sea after full treatment. In most cases the authorities will lay down provisions forbidding the inclusion of toxic wastes and limiting other constituents with any potential to inhibit treatment or to contaminate the water cycle, such as oils and fats, salts, and so on. There will usually be limits on the total volume and rates of discharge, the pH range, settleable and suspended solids content, oxygen demand (expressed as chemical or biological oxygen demand, COD or BOD) and temperature. Local authorities will charge for effluent treatment, normally using a formula based on volume and amount of treatment needed to achieve a standard suitable for discharge into a river, and often related to their basic cost for domestic sewage treatment. Precise conditions may vary in relation to the general type and quality of effluent providing the inflow to any specific treatment plant.

Where the manufacturers undertake treatment on their own account, this will be because it is necessary to meet the minimum quality standard accepted by the municipal works, or because it is more cost-effective for them to undertake full or partial treatment. Typically the most basic treatments undertaken by the food manufacturer involve removal of large particles of solid matter by screening, flotation or settling; provision of oil/fat traps; buffering volume and quality by installing holding capacity to give a more uniform discharge; and removal of finer particles of solid matter by more prolonged settlement. pH correction may also be undertaken and producers of high COD effluents may undertake some biological oxidation by a variety of methods from filter towers to activated sludge systems.

Disposal of solid waste is generally much more straightforward and less likely to cause complications other than as a potential harbourage for pests or as a generator of unpleasant odours. In either case, frequent and hygienic disposal is the solution to most difficulties.

The manufacturer or caterer has a legal duty of care to describe their wastes for disposal correctly and to ensure that they are deposited at a disposal site licensed to receive such wastes.

SPECIFIC MEASURES AND HAZARDS

INSTRUMENTATION

Thermometers

Glass thermometers are obviously undesirable in food factories and in the past vapour pressure and mercury-in-steel instruments were used, frequently linked to recorders. They were slow in response and due to the handling they received, required fairly frequent calibration. The introduction of portable thermocouple instruments was a great improvement, but many instrument manufacturers seriously underestimated the protection needed for the electronics in a factory environment.

The best instruments for accuracy are undoubtedly platinum resistance thermometers, which are now very similar in price to thermocouple instruments, although the probes are more expensive. Fixed thermometers with mains power should be installed wherever possible because these will give greater reliability.

Calibration

Reliable instrumentation is an important aspect of the overall control system. This will apply both to factory and laboratory instruments. It may be useful to differentiate between those instruments which are used for control purposes (e.g. retort thermometers) and those that are used for indication only (e.g. many oven thermometers). Control instruments should be regularly checked and calibrated against 'masters' of proven accuracy. Any product made during a period when the instrument was not known to be within calibration must be regarded as suspect in terms of quality and safety until reviewed by an authorized manager and an informed decision made.

Control instruments should be clearly marked with a reference number and the date when next due for calibration so that they are not used while the calibration is invalid because it is out of date. A register of calibrated instruments should be maintained with details of the required accuracy, calibration method and frequency and reference to the 'master' instruments used to calibrate them.

ORGANOLEPTIC ASSESSMENT

Food is made for eating and the controlled eating of food to validate its acceptability is an important part of quality assurance. Assessment may include appeal to several senses as colour, texture, smell and taste can all be important attributes. Management have a responsibility to ensure that in no circumstances is food or other material tasted which may be found to be unsafe (bacteria, contaminants, natural poisons, etc.).

In some manufacture, informal tasting on the line can be a valuable quality check, but the circumstances and method of such tasting should be carefully controlled to maintain hygiene standards. Off-line tasting should always be confined to a designated tasting area, where samples can be judged safely, impartially and free from external pressures, by informed tasters.

Buying samples need to be assessed by taste for suitability as replacement for existing supplies or for continuity of desired character from season to season. Organoleptic assessment may then be needed for delivery samples to be compared with buying samples. Intermediates and finished products will be judged for continuity of character.

In some cases, the tasting will be of the finished product prepared in the same way as the consumer would for serving, but in other cases a simpler taste medium will be used; for example, spices may be tasted in a boiled starch solution. Analytical tastings will also be used to assess shelf life and to assist performance assessment of different packing systems and changed processes. For this purpose, a trained panel sufficiently large to allow calculation of the statistical significance of the results is desirable.

Routine daily tasting of a range of factory pro-

ducts by a selected panel is a useful way of maintaining awareness of quality throughout the factory. These tastings should be carefully supervised and structured if they are to continue to achieve their objective.

For certain types of product there exist special methods of tasting and assessment which have become industry standards. Examples of these are the Campden Food and Drink Research Association's methods for the assessment of standards of quality of canned fruits and vegetables, frozen fruits and vegetables and so on, and the Torry Taste Panel system for assessing freshness in fish. In the case of the Campden standards the methods include tasting as well as a complete list of other standards, e.g. size grade, defects, foreign material, which enable the assessor to arrive at a numerical value of quality which is translated into alphabetical grades (e.g. Grade A).

Tasting is an important part of product innovation with a need to taste competitive products and numbers of alternative recipes. It is very important that these tastings are controlled and recorded without limiting innovative flair and generation of ideas. Clearly, due regard must be given to safe hygienic preparation of food for tasting and samples must be fit for consumption.

There are many techniques and applications of organoleptic assessment, both for research and development and for market research purposes, some of which, e.g. flavour profile analysis, may be complex both in use and analysis and require particular expertise. They are not generally applicable in day-to-day quality assurance and so are outside the scope of this chapter.

QUANTITY CONTROL

Two basic approaches to the control of weight or of volume are found in national and international legislation, one where the declared quantity is the minimum to be expected by the purchaser and the other where the quantity is, within certain limits, the average amount packed by the producer. There are also circumstances in which products are sold simply by count, and derogations exist for very small packs and for various sorts of portion control packs. The situation in these sectors is relatively straightforward and they are not covered in any detail here.

Minimum quantity

The minimum quantity requirement, such as formed the basis of UK legislation until the implementation of the EC Directive 76/211/EEC, really means that the product as sold must never contain less than the contents declared on the label, and in theory this could be literally enforced, for example, one grain under the declared ounce is an offence. In practice most authorities enforcing legislation of this type are aware of the statistical facts of life, and assuming a normal distribution are likely to regard 'minimum' as something akin to two standard deviations below the mean and regard such distributions as acceptable, gross defects excepted. Hence the manufacturers should normally set their target fill quantity so that it is at least two standard deviations above their contents declaration in the worst operating circumstances. Clearly, appropriate adjustments to this must be made if the fill distribution is not statistically normal and to take account of any checkweighing and sorting operations which are undertaken after filling. With some products and packaging materials, it will also be necessary to consider any quantity changes which take place during storage and distribution.

Average quantity

For manufacturers operating in the European Union and many other markets, it is the 'average quantity' concept which must be complied with. In the UK, most packaged goods are subject to the average system prescribed in the Weights and Measures Act 1985 and its ancillary regulations, which embody the requirements of the EC Directive 76/211/EEC.

This legislation requires that when a group of packages is examined in a defined manner by an authorized inspector, then the average quantity in the group must be the same as or greater than the declared quantity on the individual packages. Whereas the minimum quantity legislation was primarily enforced by checks made at the point of sale, the Average Quantity Law is primarily enforced by inspection at the point of production or importation. Certain limits and tolerances are defined which must be compiled with, and the legislation is enforced by the requirement that when formally tested by an inspector (a reference test), the batch or consignment in question does, in fact, pass that test. If it fails the test, then the sale of that batch is prohibited unless or until correction takes place.

The reference test is carried out on a sample of packages taken either from one batch of product or from one hour's production from a continuous line. The average content of the packages is checked and the number of packages which fall short of the declared quantity by more than a certain level, the tolerable negative error (TNE), is noted. The latter packages are referred to as 'nonstandard' and their number must not exceed 2% of the sample. Packages with a shortfall exceeding twice the TNE are known as 'inadequate', and if any are found in the sample, then the reference test is failed. To ensure that a reference test will not be failed, a packer may opt to measure the quantity in every package to see that no package contains less than the declared quantity. In most cases this is likely to be unacceptably expensive, since the average quantity in the packages will inevitably be significantly greater than the declared quantity. The more usual procedure for packers is to install a control system which satisfies what are known as the 'Three Packer's Rules':

(i) The actual contents of the packages shall be not less, on average, than the nominal quantity

(ii) Not more than 2% of the packages may be non-standard, i.e., have negative errors larger than the TNE specified for the nominal quantity

(iii) No package may be inadequate, i.e., have a negative error larger than twice the specified TNE.

The TNEs are specified in the Regulations, and range from 9% of the declared quantity of packages between 5 and 50 g or ml, to 1% of the declared quantity of packages larger than 15 kg or litres. The higher percentage tolerances for packages containing small amounts recognize the limits of the precision obtainable with normal production equipment.

The average weight legislation is designed to fit the observation that most populations of prepackaged goods have a normal frequency of distribution in statistical terms (see Figure 18.3). Also the

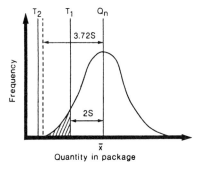

Figure 18.3 Normal frequency distribution as applied to package weight control. Q_n, nominal quantity; $\bar{x}$, mean quantity; S, standard deviation; T_1, Q_n − TNE; T_2, Q_n − $(2 \times$ TNE$)$.

tolerable negative errors given in the Regulations are twice the standard deviations to be expected in products which are 'difficult to fill'. As long as the population is more or less normally distributed, with the standard deviation approximately half the appropriate TNE, then 2.5% of the population will fall between T_1 and T_2 (the area hatched in Figure 18.3, between 1 and 2 TNEs below the average). Hence when the population is normal and the standard deviation is not more than half of the TNE, when Packer's Rule (i) is satisfied then Packer's Rules (ii) and (iii) will also be satisfied. If either of these conditions is not met, a packer will need to take steps to ensure that each of the three rules is satisfied independently.

Unless the packer chooses to measure the quantity of each individual package, the Act and Regulations require that records are kept of the checks which are carried out to maintain quantity control. The form of the records is not specified and is subject to agreement between the packer and the local inspector. It is specified, however, that they must be held for a year and be available for inspection as evidence that checks have been carried out during all production runs.

Packers of any goods in quantities between 5 g or 5 ml and 10 kg or 10 l may voluntarily come under the average system by 'e'-marking the packages and thereby choosing to be controlled by EC Directive 76/211/EEC. Another benefit of 'e'-marking packages is that they can be exported for sale in other Member States without any further metrological control than that exercised in the packing factory.

Although in the UK nearly all prepacked foods are subject to the Weights and Measures Act, there are exceptions, such as for very small quantity packages. These exempted products must satisfy the requirement that each package must contain at least the declared quantity.

Control methods

In considering a quantity control system, there are three acceptable methods if it is impractical to fill every package to the declared quantity under manual control: (i) control by sampling; (ii) control by automatic checkweighing machines; and (iii) for liquids, control by using measuring container bottles.

(i) *Control by sampling*. If a packer opts to control the quantities in his packages by measuring samples of product, he must consider the variability of the performance of the

filling equipment. The first question should be whether the frequency distribution is approximately normal, with a standard deviation less than half of the appropriate TNE. If not, simply ensuring that the average quantity of the samples is above the declared quantity will not guarantee against excessive numbers of non-standard and inadequate packs being produced. In these cases the packer will need to increase the target quantity until the second and third Packer's Rules are satisfied. Alternatively, a checkweigher may be used to remove the underfilled packages, thus shifting the distribution upwards and increasing the average quantity in the remaining packages.

A recommended method of assessing the variability of the performance of filling equipment is given in the UK Department of Trade's (DoT) Code of Practical Guidance for Packers and Importers. This procedure will identify the sources of some of the variability. It also gives the method to be used to calculate the minimum number to which the target quantity needs to be increased to guarantee compliance with Packer's Rules 2 and 3.

When the underlying pattern of filling variability has been established, the minimum target quantity which will satisfy the legal requirements may be calculated. A sampling allowance is almost always necessary when a process is being controlled by sampling so that the quantity control system does not give signals which are too early, so that unnecessary adjustments are made, nor too late, causing large numbers of unsatisfactory packages to be produced before a problem is detected. Appropriate allowances are given in the DoT Code and these help determine the level of the target quantity above the nominal quantity and the amount of checking and control which will be necessary to ensure compliance with the law.

Popular quantity control systems involve plotting results on charts bearing predetermined control limits and taking action when these control limits are exceeded. Three main types of control charts have evolved: original value plots; Shewart charts and cumulative sum charts (Cusum), and further information on their use is to be found in many books on statistical quality control.

(ii) *Control by automatic checkweighing*. Automatic checkweighers measure the gross weight of individual packages on a production line. They check whether the packages are below a certain weight, the set point, and if so automatically remove them. Such devices usually provide for display of information about the characteristics of the production. Simple equipment may provide a digital display of the total throughput and the numbers of packages below or above each set point and record the data. It is usually possible to count the packages outside any set point without necessarily removing them.

Because it is inevitable on a checkweigher that packages are weighed gross, an allowance for the variability of the weight of the packaging material must be included in setting the target weight. A further allowance is required for the 'zone of indecision' of the checkweigher, i.e. the weights at which the checkweigher does not always give the same response. The determinations of both of these allowances are described in the DoT Code.

Some patterns of checkweigher send a feedback electrical signal to an adjustment mechanism, e.g. a servo motor on the filling machine, in response to the data recorded. This can be very effective in controlling average weight where gradual drift occur. Other more sophisticated equipment allows full analysis on a continuous basis and provides full compliance records of the line operation.

(iii) *Control by measuring container*. The UK Measuring Container Bottles (EEC Requirements) Regulations 1977 and EC Directive 75/107/EEC make it legal for bottle manufacturers to sell officially volume controlled containers to packers. These measuring container bottles are specified as having a capacity equal to the nominal capacity at a specified level from the brim. The use of measuring container bottles allows production to be checked in an adequate manner by reference to the liquid level using a template, without having to check the actual volume of the contents.

DATE MARKING AND PRODUCT CODING

In the UK it is now a legal requirement that all foods be date or lot marked in some way (The Food (Lot Marking) Regulations 1992). Effectively, since the Food Labelling Regulations 1984 came into force, most foods have been marked with an indication of minimum durability in the forms of 'Best before', 'Best before end', or 'Use by', dates. There were originally some exemptions to this, such as frozen foods, which have now come within this requirement. There still are some exemptions, such as vinegar, salt and chewing gum, but all these now have to bear some indication of production date or batch number even if in encoded form.

There are several ways of marking products. Marks may be embossed into the surface of the pack as in canned goods, printed on to labels or packaging film and labels may be edge coded by cutting slots into them. However in recent years the most significant development has been the ink jet coder which marks packs with charged particles of ink and is very versatile in what can be printed.

Product coding should be looked upon as a useful tool in any quality system and it is to the advantage of the producer that the product is coded, so that should any complaint arise the producer will be able to identify the production day and possibly the line and operative responsible. Ink jet coding even makes it possible to mark with a precise production time from which it may be possible to investigate fully the cause of the problem if genuine or be absolved from responsibility if the records indicate that such a problem could not have occurred during production. Where a problem is genuine the date marking system should also facilitate traceability of the product from the finished item back through to the ingredients supplier. Where a problem results in a product withdrawal a good coding system can mean the number withdrawn can be narrowed to a much lower number than would otherwise be the case.

CONTROL CHARTS AND SAMPLING SCHEMES

When sufficient data are available it is possible to carry out statistical checks to balance the risk of passing a consignment of material containing too many defectives against the number of samples taken. A double sampling scheme which makes use of a low level of sampling initially, but a higher level of sampling when the proportion of defects rises, can be an economic way of ensuring a satisfactory quality level.

When sufficient data are not available for sound statistical sampling plans to be used, it is useful to plot information as it becomes available in order to identify adverse trends. Sampling rates should then be increased as necessary until normal performance is restored.

In the control of a process or manufacturing operation, the use of control charts to signal the need for change or adjustment is common practice. Such charts require a knowledge of the opera-

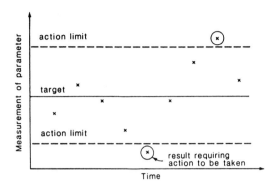

Figure 18.4 A typical control chart.

tion and its variability to allow a target, minimum requirement and range of acceptable variation to be prescribed. These may then be used to control the process within straightforward limits as indicated in Figure 18.4, adjustments being effected when results fall within the action zones.

Cumulative sum charts (Cusum) are valuable for controlling situations where events are likely to alter relatively gradually. They allow adjustments to be made in response to a change in trend, rather than to one or two individual results, and they constantly target the operation back towards the required overall mean.

The use of statistical techniques in control charts and sampling schemes is fully discussed by Herschdoerfer (1984–86) and in many other publications.

PESTS AND PEST CONTROL

Pests may be divided into three groups: insects and other invertebrates, birds, and mammals. In each case it is worth thinking in a number of phases: prevention of access to the factory, elimination from the factory and protection of the food.

Potential breeding of wasps, flies and so on near to the factory should be restricted as far as possible. Refuse should be covered, frequently removed, disinfected if necessary and the bulk for disposal located in a position away from exposed food. Access to the factory can be controlled by not allowing open windows, by screens and air currents at door openings and by keeping all parts of the building in good repair.

Flying insects are attracted by ultraviolet light and this is the basis of commonly used electric knockdown killers. These should be sited close to factory entrances, but so that the light is not seen from outside the building. They should also be sited away from competitive sources of strong light. If infestation does build up in a factory, this can be treated by using high levels of heat under careful supervision but, in general, low temperatures are effective in limiting breeding.

There are circumstances when chemical treatments such as fumigation are necessary to treat infestation, particularly of raw materials imported from warmer countries. Such treatment requires expert attention to ensure that it is sufficient to be effective but not so great as to leave residues that could be unsafe, tainting or illegal.

Raw materials may contain insects present as eggs or in the pupal form. Further development will be limited by low temperatures and low moisture levels. During storage, temperature variation in a stored bulk can cause local buildup of moisture which can lead to some insects hatching and building up metabolic heat resulting in a severe local infestation. All raw materials held in store for long periods should be inspected regularly. Where practicable, material held in hoppers and silos should be recirculated to mix it and prevent buildup of damp patches. It may be desirable to insulate outdoor silos in order to minimize the development of moisture gradients and possible condensation within the silo due to varying ambient temperatures, with the attendant risks.

Birds are very difficult to remove from a factory in an acceptable humane manner and it is well to direct considerable care to their exclusion. Birds try to enter factories for shelter, food and nesting sites. They prefer quiet secluded entries and most species are unlikely to enter at a busy doorway. However, some can enter through quite small gaps at windows, ventilators and damaged roof or walling. Once in, birds can only effectively be removed by licensed operators who will use narcotized baits, mist nets and ultimately air rifles.

The main mammalian pests are rodents, but cats and dogs may also pose problems. Cats were popular for keeping down rodents and discouraging birds and may still be found in boilerhouses, plant rooms and other secluded places where they may be fed on scraps by sympathetic employees. They can be disease carriers and they may foul food process areas. Dogs, apart from guard dogs, are not normally found on factory premises, but they are sometimes carried on transport. If not properly controlled, this could lead to fouling of food or ingredients carried on the vehicle.

In addition to cats and dogs, foxes and rodents could be attracted by factory wastes stored outside the buildings. The waste containers must be strong enough to resist attack and have securely fitted lids.

Rat and mouse populations should be restricted

in the immediate vicinity of the factory buildings by denying them food and harbourage by removing wood, sacking and other materials that could provide warmth and shelter. A well-lit open paved area running right round the buildings will deter rodents, and the first half-metre of the factory wall should be smooth finished to prevent them climbing to find access at higher levels. A mouse can slip through a 10 mm gap, and will gnaw through an aluminium sealing in some circumstances. It is therefore almost inevitable that some rodents will secure entry.

Once in, life should be made as difficult as possible for them by storing all food materials off the floor and half a metre from the walls, and by promptly removing any unwanted waste whether it be food, packaging or engineering materials. Slow moving stock can provide harbourage and should be inspected and moved frequently. All torn ingredient packs should be repaired and used as soon as possible and any spilled material thoroughly cleaned up. Rats need water, and their range can be restricted by denying them access to it.

Baiting for rodents is an expert operation that can be used to confirm their presence as well to poison them. Contractors' staff carrying out bait inspections should be accompanied by factory personnel to note observations and arrange follow-up action. Rodents will eat a small amount of food and will damage a great deal more while eating, but the main damage is done by their excretions.

In hot countries, reptiles can pose special difficulties, particularly as some, such as many lizards, are able to climb almost any surface and even find harbourage in roofs and ceilings.

MICROBIOLOGICAL HAZARDS

Microbiological infection of foods may result in food poisoning, but more often will not have such a severe effect but give rise only to off-flavours and odours, gassing, visual deterioration and so forth. Nonetheless, such problems can still result in loss of business and product withdrawal and present the producer with major problems. It should be said that the microbiological quality of food is entirely controlled by two factors; the way in which the food is handled from raw materials to consumption, and its composition as a medium to support microbial growth.

Bacteria

There is little doubt that poisoning by foodborne bacteria is more frequent than the available statistics indicate, because many relatively mild attacks are not reported and in some outbreaks it is difficult to acquire the necessary evidence. Bacterial food poisoning is much more likely to be associated with large scale food service operations than with food manufacture, but every food manufacturer must be aware of the potential risk and take the appropriate precautions. The increasing market for chilled prepared foods is a particularly vulnerable area if high standards are not maintained. See Chapter 19, Table 19.1 for a summary of the occurrence and properties of the main food poisoning bacteria.

Listeria monocytogenes is a human pathogen which is widely distributed in natural environments; however the disease listeriosis is comparatively rare and rarer still where food is a vehicle of infection. In these rare circumstances it is therefore a foodborne diseases and may be classed with cholera (caused by *Vibrio cholorae*), *Shigella* infections, typhoid fever (caused by *Salmonella typhi*) and paratyphoid fever (caused by *Salmonella paratyphi*).

Nevertheless there has been concern about *Listeria* in recent years because it is able to grow on some foods at refrigeration temperatures. Largely as a result of this the UK Food Hygiene Regulations were amended in 1990 to include temperatures at which certain foods must be stored.

The most common causes of food poisoning are *Salmonella* and *Campylobacter*, but the most feared is *Clostridium botulinum* because of the heat resistance of its spores and because consumption of the toxin that it produces is frequently fatal. Other organisms which may be among the causes of food poisoning and are currently under investigation include *Aeromonas hydrophilia*, *Bacillus subtilis*, *Bacillus licheniformis* and non-proteolytic strains of *Clostridium botulinum*.

Escherichia coli has long been considered a useful indicator of faecal contamination in foods, capable of causing spoilage but not otherwise dangerous in itself. However, verocytotoxic strains (VTEC), causing severe illness with enteric haemorrhage, were first recognized in the early 1980s. These strains are now regarded as important pathogens because of the seriousness of the illness, particularly in children, which can result from them. However, the number of cases discovered remained low until a serious outbreak in Scotland in the autumn of 1996, in which at least 16 elderly patients died. The organism appears to be relatively widespread among foods of animal origin, but it is heat sensitive and destroyed by similar heat processes to those effective against

Salmonella or *Listeria*; it can be adequately controlled by the normal procedures of good hygiene.

The effective prevention of bacterial food poisoning is highly dependent on the commitment and cooperation of the food handling employees, and it is appropriate here to emphasize the value of correct training and retraining, a sound company hygiene and medical policy, and of a happy and dedicated work force, all reflecting a competent and communicating management.

Moulds and yeasts

A number of moulds or fungi have a role in developing the characteristics of certain foods. Some fungi are foods or food ingredients in their own right. In other situations the presence of mould can result in off-flavour becoming apparent in the food itself, with the inevitable consequences for consumer acceptability.

In particular circumstances, moulds growing in foodstuffs or food ingredients may produce mycotoxins, a group of toxic substances with high potency including carcinogenicity. The manufacturer should be aware of these potential hazards and where appropriate provide suitable means of survey and quality control.

Moulds are readily destroyed by the heat treatment given to canned foods. The exceptions to this general statement are *Byssochlamys fulva* and *B. nivea*, which have been responsible for spoilage outbreaks in canned fruit. The ascospores of these moulds are widely distributed and fairly heat resistant and if they survive and develop in the canned fruit they produce enzymes which lead to disintegration of the fruit. They are destroyed rapidly at 91°C (195°F), a temperature which can be achieved more easily in cans without too much heat degradation of the fruit than in bottles with the slower heat 'come-up' cycle. Therefore, spoilage with these moulds has been more common in bottled fruit.

Another mould spoilage potential exists in canned sweetened condensed milk, which is not heat processed, the keeping quality of the product being dependent upon the high solids content. Development of 'buttons' of mould growth has occurred when the cans have not been completely filled and small pockets of entrapped headspace air have coincided with mould spores on the can surface or in the milk. This situation can be controlled by sterilization of the empty cans and ends, good hygiene in the plant and attention to filling efficiency.

Others

Microbiological hazards are not simply confined to those relating to bacteria, yeasts and moulds. Food can act as a viral vector and a vector for parasites, while other microscopic fauna and flora can cause problems. Here again, familiarity with the product group in question and its likely difficulties is important.

HAZARDS TO PLANT AND PROCESSES

Foreign matter

Contamination of ingredients or finished products can arise in many ways, from tainting due to absorption of volatile components of materials such as paints or other finishes to the physical presence of material 'not of the nature or substance demanded' because of pests, carelessness or simply difficulty of detection. Many foreign bodies enter the factory in the raw materials. Careful selection of suppliers will help to minimize this risk but it will be good practice to inspect some raw materials, identified by experience, and to use best available practice to remove foreign matter before the material is taken into use. The foreign bodies removed in this way are likely to be the larger and more easily identified ones but the process reduces the chances of stones and tramp metal causing damage to knives and other delicate machinery, or of large foreign bodies being broken into many smaller ones which will be still more difficult to detect and to remove.

Metals

Some metals in low or trace quantities have important nutritional roles, but higher concentrations can be poisonous. Lead from solder splashes, or from lead pipe in combination with aggressive waters, or from other sources, is undesirable and is subject to legal maximum levels, like other toxic metals. The dissolution of traces of lead from the solder on the side seams of tinplate cans has been a major reason for the widespread introduction of other styles of can over recent years.

Arsenic can occur from combustion products in foods that are dried and may also intrude from its use as a wood preservative or as a fungicide. Cadmium is used to plate springs and some nuts and bolts, but it is soluble in acids and should never be used in food contact situations. Galvanized metals similarly should not be used because of the solubility of zinc in acid conditions.

Aluminium is extensively used in plant, in food packaging and as cladding material, but is soluble in caustic solutions and the pure metal is soft enough to abrade in rubbing contact, leaving black marks on the product. The presence of dissolved aluminium is undesirable, but probably it is not a significant toxicant to most consumers.

Copper, nickel and iron are all oxidation catalysts and quite low levels (0.1 ppm) can lead to quality problems with fats. This is such a problem with copper that all copper and copper alloys should be excluded from any plant processing fatty products. Other metals can cause specific problems in individual circumstances and reference to detailed product information should be made.

Metallic foreign bodies must be detected and removed from all foods. Where there is a possibility of metal being present in the raw material then, as far as is technically feasible, it should be removed before processing commences. Irrespective of whether this is done, the food should be tested after processing to identify and remove any detectable contaminant. Techniques for detecting and removing metals include magnets, metal detectors and X-ray scanners.

Obviously magnets are particularly useful to remove magnetic material from raw materials that are to be finely divided or ground. They should be checked and if not self-cleaning, cleaned regularly and a note kept of the amount of metal extracted, which might correlate with an identifiable supplier.

Metal detectors may be used to detect either the magnetic or the electrically conductive properties of metals. They are considerably more sensitive in the magnetic mode, but this is of no value when the metals to be detected include stainless steel, aluminium and copper alloys. When the detector is set up to detect metals by their conductive property, then all metals may be detected, but the detector sensitivity is lower and the presence of conducting solutions, such as weak salt solutions and the juices in raw meats, may interfere. It may be possible to minimize this interference by examining the food while it is hard frozen. New developments in computer analysis of the signals from metal detectors have considerably enhanced their sensitivity in such difficult applications.

Metal detectors are positionally sensitive because they rely on metal present disrupting the detecting field. Hence they will be most sensitive to spheres of metal or to rod-like objects disposed parallel to the detecting field. Because rod-like objects are more likely to be orientated with the direction of flow of the food, they tend to be sensitive to detection. Any metallic foreign body recovered should be investigated and retained with a record of time and place of detection, material in which found and place of origin of the metal. In the case of small fragments it may be useful to examine and photograph them under a microscope and to determine their composition, for instance by non-destructive X-ray excitation techniques as a help in determining their origin.

X-rays may be used to detect metal particles or other foreign bodies. Until recently such devices were expensive and bulky, limiting their application, but significant improvements have recently occurred and the next decade should see more common use of this technique. Continuous, on-line foreign body detection using X-rays is now commercially available.

It should be remembered that in most situations frequency of real detection is likely to be a trade-off at some level against the frequency of the false alarms which will inevitably occur. It is important to have procedures to check and record the correct functioning of detectors at frequent intervals as part of a quality system.

Glass

The presence of glass in a food factory is fraught with dangers, whether from fragments or splinters of glass in the jars or bottles on receipt or from breakages in use. When glass smashes occur, glass fragments can fly surprisingly long distances from the site of the smash. Furthermore, complaints about glass in foods are particularly emotive, and every effort must be made to minimize the problem.

All glass containers should be inspected for defects or foreign matter and washed and/or air-jet/vacuum cleaned. Glass conveyors should be covered wherever possible. All heating and cooling operations should be carefully conducted to avoid undue thermal shock which could lead to immediate or later breakage. Breakages may occur in the filling and capping operations. In this connection, great care should be taken to avoid the inclusion of any jars or bottles outside the normal size or shape (e.g. with round jars, any showing ovality; or with bottles, any showing off-centre necks), which would be liable to jam and smash in the filling or capping operations. The latter operations should be effectively screened so that, if breakage does occur, flying fragments are prevented from reaching other parts of the production area. In the event of such a breakage, however, all open containers in the vicinity should be rejected; the machine and surroundings should be thoroughly cleaned, and if there is any exposed food, uncovered vessels and so on in the vicinity, these should be regarded as suspect. For this and other reasons, food or vessels should not normally be exposed uncovered.

If glass containers are reused, they should be presorted before washing to remove cracked and grossly contaminated containers and then carefully inspected after washing and immediately before filling. Consideration should be given to filling in windows and relying on artificial light, and also to the replacement of glass thermometers by resistance thermometers (platinum or thermistor). Light fittings should include plastic diffusers which will catch the fragments if a light bulb or tube shatters. Non-glass containers must be found for samples wherever possible. Likewise fragile ceramic products should not be used in production areas.

A log of breakages should be maintained and kept as a record. Glass containers are formed on moulds, and each mould will have an identification number which will appear on the jar or bottle, together with the manufacturer's own identification symbol. The glassware manufacturer will have numbers of moulds in use at any time. If a mould is defective all the jars or bottles made on that mould will have a common defect. All breakage problems should be checked for supplier and mould number in case a common defect can be identified. Notices warning employees of glass hazards should be prominently displayed in all risk areas.

Stones and other solids

Stones and other items of density significantly different from the food may be separated by sieving or by gravity using air or water to suspend the food material. In some instances gravity is not effective, such as when apricots have been dried on sand beds or stones are embedded in potatoes. In such circumstances it is best to monitor supplier performance and to identify the least hazardous source of supply.

Other possible contaminants include carbonaceous matter, building materials and fragments of cleaning equipment and so on. As with all foreign matter, careful analysis of items isolated in production and those arising from consumer complaints should focus attention on where major problems lie and on where preventive effort should be directed.

Organic matter

The range of organic foreign materials that can occur in foods is considerable. There are congeneric materials such as peel on potatoes, shell with walnuts, and bone and gristle with meat which could have been removed by more careful preparation processes. There are other foreign materials brought in with the raw material such as insects, snails, caterpillars, plastic, wood and paper that should have been removed at the preparation stage. There are wrong ingredients, used in error during the formulation of the product and there are foreign materials that may get into the food during production through slipshod control of operations, such as scraps of cleaning equipment, wound dressings, dead flies and wasps and animal droppings. Other items may arise from plant and machinery, such as belting, plastic flakes, rubber, wood and so on. All of these are almost impossible to detect or remove from the processed food. Attention must therefore be concentrated on identifying and eliminating them before the food is processed. In most cases this will require manual inspection and removal.

Inspectors on a raw material line will always remove a proportion of the material they are inspecting. The ratio of inspectors to food must be set so that an acceptable level of removal of unwanted material is achieved without undue wastage of good ingredient. Due account must also be taken of human factors such as fatigue and boredom.

In the grain milling industry each grain is mechanically separated by size and shape to exclude foreign grains; husk and dust are removed by air separators. In the preparation of dried peas and beans, mechanical devices are used to remove seeds with broken shells and insect damage, and colour sorting is used to remove discoloured material.

Detection of an unwanted species of meat can to an extent be achieved by appearance, smell and taste, but to provide the element of certainty needed in pursuing a claim, serum agglutination tests are used, possibly backed up by other sophisticated analytical techniques.

It is quite possible to detect larger pieces of bone in meat by X-ray techniques but it is much more difficult to differentiate smaller ones, for instance rib bones in chicken pieces or bone fragments in mechanically recovered meat. Sieves, size graders and gravity separators all have an important role in removing foreign bodies but some manual inspection will also be desirable.

During processing, organic material can enter

from rodent and insect attack. Insects entering the factory with a raw material as eggs or pupae may hatch and become much more apparent. Wood, bristle and cloth from cleaning material are also a problem as are personal items such as wound dressings and clothing items. All these items are difficult to identify, but it may be possible to remove them by a screen or sieve. Wound dressings are available in a distinctive blue colour and with a metal strip so that they can be detected by metal detectors.

Chemicals and tainting

Chemicals with adverse effects on food quality can contaminate products by carryover from ingredients, e.g. as pesticide residues, by accidental addition or by contact. Apart from the obvious spillages, leakages and human errors, problems can arise through failure to check for erosion of plant materials or absorption of volatiles from paints and resins or other strong smelling items such as cleaning materials and disinfectants.

Foods can quite easily absorb off-odours from materials with which they come in contact or even materials which are stored in the same room. Very low concentrations may be apparent only to the sensitive consumer, but the levels needed for the food to be repugnant to many consumers may also be very low. Fatty foods are particularly but not uniquely vulnerable to taint problems. Specific causes of taint are exemplified below.

(i) Styrene: low levels of styrene have been associated with insulated containers built for the transport of frozen foods and with polystyrene tubs used as food containers.

(ii) Catty taint: has been caused by the presence of low levels of mesityl oxide present as an impurity in the ketonic solvents used in some paint and resin systems.

(iii) Flooring resins: taint problems in some meat factories have been caused by curing agents where resin-based flooring systems have been recently laid. This problem can be exacerbated by washing a new floor with a chlorine containing solution.

(iv) Chlorophenols: this is a wide group of substances, many with disinfectant properties, that cause taint problems even at very low concentrations. As examples, problems have arisen from wood shavings treated with chlorophenols that were used as litter in chicken houses; from a dustbin powder formulation stored in the same warehouse as foodstuffs

and from the use of phenolic drain cleaner followed by a discharge to drain of a chlorine-based sterilant in a food factory.

(v) Printing ink: many inks for the high quality printing associated with food packaging contain solvents that are potentially tainting. These are normally dried off during the printing process, but taint problems have occurred.

(vi) Transport: taint problems have been traced to previous loads carried in transport. This problem is more likely to occur with road and rail tankers and containers.

(vii) Rancidity: very low levels of oxidized fatty acids have an unpleasant taste and odour. Oxidation problems may arise due to defective processing and air leaks, but they are often associated with the catalytic effect of copper alloys coming into contact with the fat.

It is advisable to carry out exposure and taint assessment trials wherever a material of which there is no previous experience is brought into use in the production plant or wherever food is exposed as work-in-progress.

The substantial improvements in crop yield and quality achieved since the 1950s have been due to some extent to the development of a range of highly effective pesticide products. When correctly applied, these are perfectly safe and have demonstrated their efficacy through appropriate trial procedures, screening tests, freedom from taint assessment in the intended product, and so on. However, there may be occasions when high levels are detected on harvested crops and may signal their rejection, and specific screening by analysis may be advisable.

Microbial problems

Consideration must be given by the food manufacturer to the possibility of microbiological problems occurring within the production system. Such problems may range from poor hygiene and cleaning, allowing the buildup of microflora which although non-pathogenic may reduce product quality or shelf life, to disruption of fermentations due to ingress of exogenous strains.

Pesticides

The UK Pesticides (Maximum Residue Level in Food) Regulations 1994, implementing EC Directives 90/642/EEC, 93/57/EEC and 93/58/EEC, detail permissible limits of a range of pesticide residues in certain foods. Encouragingly, the reports of the UK Working Party on Pesticide Residues (1985–1988) show that where residues are found in foods they are generally at low levels.

COMPLAINTS PROCEDURE

Complaint information is of great importance. All complaints, whether they come from customers, shops, salesmen or enforcement authorities, should be notified to a designated senior member of the QC staff. This person should be responsible for making an initial judgement of corrective or other action, or whether a product line might need to be recalled. Routine analysis of the complaints will be carried out by categories (quality, packaging defects, spoilage, foreign bodies, etc.) and byproduct lines. From this analysis comparisons can be made with past experience and possibly with that of others (on the basis of the ratio of complaints to packs sold), and weak points in the company control system can be identified for priority action. It can be valuable to monitor trends of complaints on individual lines within a product group, as this helps smooth out peaks and troughs due to varying levels of sales and consumer uptake due to promotions, seasonality, and so on.

It is well known that consumers in the UK are more likely to make complaints than those elsewhere in Europe, and that they complain more readily about foreign matter than about unsatisfactory quality. However, for every consumer who does complain there may be many more, probably by an order of magnitude, equally dissatisfied who do not register a complaint but may take their custom elsewhere, i.e., switch brands.

Whenever possible the package of the product subject to the complaint should be recovered for examination. Production line and batch information is helpful and in the case of a foreign body, photography, microscopy and non-destructive tests should be used to try to identify the material in as much detail as possible.

Some complaints may arise because of genuine error on the part of the consumer or occasionally because of deliberate action. In some cases it is possible to establish that the cause of the complaint is more likely to have arisen subsequent to the product leaving the control of the manufacturer. As examples: a child's button is more likely to get into the product in a household than in a factory; it is possible to show whether or not a

dead fly has been through the jam-making process; it may be possible to show whether a nail has been in a cooking process; glass can be identified and may be shown to be of a type common in households, e.g. ovenware, but rare in factories. Spurious or unfounded complaints can arise and the manufacturer should be sensitive to such possibilities.

EMERGENCY PROCEDURES

There are a number of situations in which manufacturers may be confronted with an emergency. For example, their product could have been shown to present a possible hazard to health due to a defect in processing or packaging, or their products might have been subject to sabotage or even to the threat of malicious tampering in the marketplace. Any situation of this nature could demand recall of product from sale or distribution and the responsible manufacturer will have made adequate contingency arrangements for such eventualities.

Product recall

No production control operation is perfect and there can be occasions when it becomes apparent that food products are in the distribution chain that are below the procedure's quality standard. In these circumstances product may need to be withdrawn. The company must develop a recall system and confirm that it works, so that when a real emergency arises the recall is rapid and completely effective.

If proper records are kept and it is possible to trace production by clear code marks and to relate production lots back to individual deliveries of ingredients and packaging, then the problems may be narrowed to the recall of a relatively small proportion of the total product distribution. A well-referenced computer record of dispatches can be very useful in these circumstances. A general recall plan is illustrated in Figure. 18.5

Complaints are received by various routes but they should be channelled through one designated manager with the responsibility for making the preliminary decision about whether a potentially serious situation has arisen. When the manager suspects that a general hazard has occurred the chief executive or other designated person should be immediately informed to start the action plan, This will require quarantine of any potential hazard stock within the factory, checking produc-

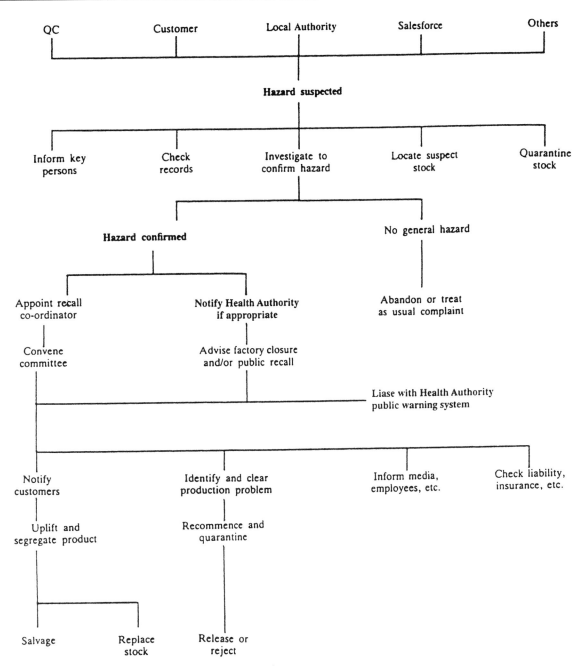

Figure 18.5 Schematic diagram of product recall procedure.

tion records to identify clearly which batches of product may be at risk and tracing stock in the distribution chain. At this point well maintained records and full traceability are of great value.

Investigations will be undertaken to confirm whether a general hazard, requiring recall of all or part of the factory's production, exists. If it is shown that the problem is a limited one then it may be appropriate to stop the general hazard action and take a more limited action as for a routine complaint. If, on the other hand, the existence of a problem justifying further consideration of recall action is established, then a recall coordi-

nator should be appointed; in practice one or more recall coordinators will have been nominated when the plan was formulated and one will be selected and given the necessary authority at this stage. At the same time, the area Health Authority (in the UK, the Department of Health) should be notified of the problem. If they take the view that there is a general public hazard they may advise or require factory closure and they will cooperate with the firm in carrying out a general recall of all products at risk through their Food Hazard Warning System.

The recall coordinator will activate and convene

a recall committee to follow through a number of lines of action. Among these are: notification of staff, major customers, insurers and the media; uplift of product as necessary and its return to a separate quarantine store and continuing investigation to identify as precisely as possible the nature of the problem, its cause and the degree to which it may have affected product. It is evidently helpful if the problem can be narrowed to a particular production line within certain dates or to certain ingredients delivered by known suppliers.

When the problem has been clearly identified, correction work can proceed. It may be possible to salvage saleable material from the quarantined recovered stock. When production restarts, all new product should be quarantined and additional quality checks carried out before it is released for distribution and sale. Throughout the time of the recall operation it is important that a log is maintained of all actions and decisions and that there is a full exchange of information between members of the recall committee and those others who should be kept informed.

Malicious tampering

It may be that a company will be faced with a form of blackmail, allegations of tampering with its product during distribution or on display. The allegations may not be genuine, but the only responsible course is to notify the police and the Health Authority and to take such other action as seems reasonable in accord with the advice given by the public authorities.

Crisis management

The actions that need to be taken in an emergency situation have been outlined above. It behoves the manufacturer to be prepared to face such problems in a disciplined and properly thought out way. Crisis management plans will not only make provision for the appointment of a recall coordinator if necessary, but will include a process for deciding who is to take overall responsibility, who will deal with media enquiries, whether or not a specialist agency should be called in to assist, and so forth. In most cases it is likely that the company will have two immediate objectives – to handle the short term problems as efficaciously as possible and to limit long-term damage to the

minimum. Those with crisis management responsibility may well need authority to override normal line management procedures and crisis management provisions should recognize that problems can arise at any time, 24 h a day and seven days a week. Persons with this special responsibility will need to have ready access to other areas of expertise in the company and may need background training, perhaps with special reference to media implications.

REFERENCES

Campden Food and Drink Research Association (1991) *Guidelines for the Establishment of Hazard Analysis Critical Control Points (HACCP)*, Technical Manual No. 19.

IFST (1991) *Good Manufacturing Practice – A Guide to its Responsible Management*, Institute of Food Science and Technology, London.

IFST (1992) *Food Hygiene Training: A Guide to its Responsible Management*, Institute of Food Science and Technology, London.

IFST (1993) *Listing of Codes of Practice Applicable to Food*, Institute of Food Science and Technology, London.

Herschdoefer, S.M. (ed.) (1984–1986) *Quality Control in the Food Industry*, vols. 1–3, Academic Press, New York.

Juran, J.M., Gryna, F.M. and Bingham, R.S. (1974) *Quality Control Handbook*, McGraw Hill, New York.

Kramer, A. and Twigg, B.A. (1966) *Quality Control for the Food Industry*, AVI, Westport.

Working Party on Pesticide Residues (1985–1988), HMSO, London, ISBN: 0 011 242 8703.

FURTHER READING

British Standards Institution (1983) *Quality Assurance*, BSI Handbook No. 22.

BS 4891 (1972) *A Guide to Quality Assurance*, British Standards Institution, Milton Keynes.

BS 4778 (1979) *Glossary of Terms Used in Quality Assurance (Including Reliability and Maintainability Terms)*, British Standards Institution, Milton Keynes.

BS EN ISO 9000 Series *Quality Systems*, British Standards Institution, Milton Keynes.

Birch, G.G. (1984) *Control of Food Quality and Food Analysis*, Elsevier Applied Science Publishers, Amsterdam.

Brostoff, J. and Challacombe, S.T. (1987) *Food Allergy and Intolerance*, Ballière Tindall, Eastbourne.

Gould, W.A. (1977) *Food Quality Assurance*, AVI, Westport.

Grant, E.L. and Leavenworth, R.S. (1972) *Statistical Quality Control*, 4th edn, McGraw-Hill, New York.

Jowitt, R. (ed.) (1980) *Hygienic Design and Operation of Food Plant*, Ellis-Horwood, Chichester.

Kramer, A. and Twigg, B.A. (1973) *Quality Control for the Food Industry, Vol. 2: Applications*, AVI, Westport.

Mortimore, S.E. and Wallace, C. (1996) *A Practical Approach to HACCP (Training Package)*, Chapman & Hall, London.

Parker, M.E. and Litchfield, J.H. (1982) *Food Plant Sanitation*, Van Nostrand Reinhold, New York.

Price, F. (1984) *Right First Time*, Gower, Aldershot.

Puri, S.C. (1979) *Statistical Quality Control for Food and Agricultural Scientists*, G.K. Hall (Macmillan), London.

Sutherland, J.P., Varnham, A.H. and Evans, M.G. (1986) *A Colour Atlas of Quality Control*, Wolfe Medical, London.

19   Food Issues

INTRODUCTION

The issues which have confronted the food or drink manufacturer in the 1990s, have been indeed multifarious. They have included some which are of true concern for food quality and food safety, but also many which have been matters more of economic or political convenience, or peripheral questions which appear to have been stimulated mostly by minority and sometimes narrowly biased pressure groups.

In this chapter we draw attention to the main issues of public concern today. For the most part we do not enter deeply into the arguments adduced, many of which are of little relevance to practical food manufacture, but we aim to provide background information to assist the manufacturer to decide in each case what kinds of action, if any, should be taken.

NUTRITION

The science of human nutrition – what we need to eat, why and how much – is now fairly well established but legislation in this field has remained in a dynamic situation. While legislation on declaration of nutritional composition has been largely clarified, queries on claims and particularly health claims remain and the reader is well advised to seek up-to-date detailed information from expert sources in those areas in which he or she has particular interest.

THE MAIN NUTRIENTS

The first requirement of the body is energy. The major source of energy in almost all diets is carbohydrate from cereal or root crop starch and sugar. Starch is the most common carbohydrate providing up to half the energy in the diet in industrialized countries and up to 80% in undeveloped parts of the world, derived from cereals and root sources such as cassava, potatoes and yams, while

sucrose, from cane and root beet crops, is the next most available source, contributing from 5 to 20%. Some carbohydrates such as celluloses and pectins (mainly plant cell wall materials) which are largely indigestible by humans, are referred to as 'unavailable carbohydrates' in the literature and are included in the general grouping 'dietary fibre' although they are not necessarily at all fibrous in character.

The available energy from the diet includes that from fat and protein as well as that from carbohydrates. When total food intake is not adequate for the energy needs of the body then the fat and protein reserves are burned off to meet any shortfall.

Fats perform a variety of functions in the diet. They provide a concentrated source of energy, contribute to flavour and texture in foods and carry important fat-soluble vitamins. Fats are composed of fatty acids in various combinations, some of the fatty acids are essential in the diet. Most diets appear to contain these in sufficient quantity not to give rise to problems and dietary deficiency in respect of fats is extremely rare other than when allied with general starvation. The type of fat in the diet is now understood to play a significant part in maintaining health, as discussed below. There is concern that excessive fat content in the adult western world diet could give rise to problems of a cardiovascular nature and to obesity, while excessive preoccupation with low fat diets inflicted upon infants can cause inadequate energy and fat-soluble vitamin intake (see also Chapter 8).

All bodily functions depend on protein and protein-based substances. Some protein constituents can be synthesized in the body while others cannot. Proteins are composed of up to 22 amino acids in various combinations. Eight of these, ten in the case of children, are essential because they cannot be synthesized in the human body. Manufacturers producing foods for special nutritional needs will require access to appropriate expertise in this area as well as on that of minerals, vitamins, and so on (which are covered later).

The minimum physiological need for protein is considered to be 6–7% of the energy intake. An adequate level of protein in the diet is needed to replace nitrogen losses which occur in a variety of ways, for example in urine, from hair and skin loss, and so on. Human milk and hens' eggs have an amino acid composition closest to human requirements, while meat, fish and cow's milk have about 75% of the optimum biological value, soya about 70%, cereals around 50% and pulses 35–50%. Because mixtures of foods complement one another in the total diet, western diets tend to have biological values of around 80%, and less than 70% is rare even in poor countries.

At least 21 minerals have been found to have physiological roles and are therefore dietary essentials. Some are needed in relatively large amounts (e.g. sodium, potassium, calcium), while others are only needed in trace amounts (e.g. zinc, copper, chromium). Trace mineral deficiency is rare in developed countries, except where individuals or groups restrict their diets for whatever reason. Deficiency may be a greater problem in developing countries both because of inadequate overall diets and because soil deficiencies are reflected in the diet. The physiological effects of mineral deficiencies are not dealt with in this volume, but typical examples are the relationships of iodine with goitre, iron with anaemia and magnesium with muscular and nervous system disturbances.

The general role of vitamins is well understood and documented and the consequences of inadequacy are well known. However, unless the food manufacturer is a specialist (e.g. in infant and dietetic foods) or large enough to have in-house expertise, it is wise that any concerns shown by customers or others, about vitamins in particular products, should be checked with competent advisers.

The main points that should be borne in mind are:

(i)   Will any of the processes used in manufacture result in loss of vitamins to a significant degree for the consumer?
(ii)   Could the consumption of this product by any individual result in substitution of another dietary component and thereby alter the vitamin intake of the individual?
(iii) Could the circumstances of distribution and storage result in loss of vitamins to a significant degree?

In the event of a positive answer to any of these questions the manufacturer may be well advised to consider supplementing the vitamin content and/or

more closely controlling the handling and use of the product in question.

Recommended daily amounts

Recommended daily amounts (RDAs), also termed recommended dietary intakes (RDI) or daily allowances, are considered to be sufficient to cover the needs of 'the greater part of the population'. This last phrase is really a piece of statistical pedantry which has found its way into official publications.

Individuals differ in their requirements as measured under experimental conditions; from the limited evidence available the range appears to follow the normal distribution curve. On that basis, the requirements of 95% of the population group would fall within two standard deviations of the mean value. That is to say, the average requirements, as measured, plus 20%, would cover the needs of 97.5% of the population. This is defined as the RDA. There is no evidence to indicate the true needs of the remaining 2.5% of the population.

More than 40 countries have produced tables of RDAs. There are some differences among them, partly because the figures are not so firmly based as to leave no room for opinion and partly because the tables were compiled for different purposes.

Dietary reference values

The COMA Report on dietary reference values (DRVs) (1991), published by the UK Department of Health, was claimed to be the most comprehensive and up-to-date document on the subject in the world. This report established a generic concept of dietary reference values, a blanket term covering all of the requirements described below. Each requirement is intended to refer to healthy people; no allowance is made for special needs imposed by diseases or special medical conditions. The DRVs are interdependent; each depends upon the others being met and no single DRV should be considered in isolation.

Estimated average requirement (EAR) is an assessment of the average requirement or need for food energy, or for a particular nutrient. Obviously, many people will need more than the average and many will need less.

Reference nutrition intake (RNI) is the amount of the nutrient which is sufficient for almost every individual. It follows that this level of intake is therefore much higher than many people need, but by erring on the side of safety it also follows that people consuming the RNI of a nutrient are most unlikely to be consuming insufficient.

Lower reference nutrient intake (LRNI) is the amount of a nutrient considered to be sufficient for the small number of people with low needs. Most people will need to consume more than the LRNI if they are to eat enough. If individuals are consistently eating less than the LRNI, they are at high risk of deficiency of that nutrient.

Safe intake is a term which indicates the intake of a nutrient for which there is at the moment insufficient information to estimate the true requirement. A safe intake is one which is judged to be adequate for almost everyone's needs and not so large as to cause illness or undesirable effects.

The COMA Report (1991) covered energy and 33 nutrients. The DRVs which it gave superseded the previously used RDAs. The number of dietary components for which recommendations were made was higher than before. It is not possible to summarize here all the recommendations for the various groups of people (male and female, young and old, active and sedentary, and so on) but the Department of Health also published *Dietary Reference Values – a Guide* (1991), which summarized the broad content of the report.

For energy, DRVs are given for fats, sugars, starches and non-starch polysaccharides, and as percentages of the total dietary energy. There are no individual DRVs given for fats, sugars, starches and non-starch polysaccharides, but DRVs are provided which indicate that:

(i) saturated fat should average 11% of the food energy intake;
(ii) total fat 35%;
(iii) non-milk extrinsic sugars (primarily sucrose) 11%;
(iv) starch and intrinsic sugars and lactose in milk and milk products 39%.

The figure for non-milk extrinsic factors reflects concern about their role in tooth decay. The Guide also recommended that dietary fibre be redefined as non-starch polysaccharide and set a DRV of 18 g per day.

Nutrition guidelines

A number of diseases have become relatively commonly recognized in recent years in industrialized countries and are often collectively referred to as 'diseases of affluence'. They include coronary heart disease (CHD), various disorders of the bowel, certain forms of cancer, high blood pressure and obesity.

It is clear that numerous factors are involved, including heredity, sex and age, as well as environmental factors including diet. As far as diet is concerned, guidelines for 'nutrition goals' have been published by more than twenty national and international organizations. There is no absolute evidence that changing diets in the directions recommended is of benefit, but there is a general consensus nevertheless that there is enough and growing evidence to make the recommendations.

Two reports were popularized in the UK in the mid-1980s. One from NACNE (National Advisory Committee or Health Education) in 1983, summarized all the advice that had been published earlier but quantified the changes thought to be desirable. The second was the COMA Report (1984). This dealt with diet and cardiovascular disease and came to the conclusion that 'it is more likely than not that the incidence of CHD will be reduced, or its age of onset delayed, by decreasing the dietary intake of saturated fatty acids and total fat ... the evidence falls short of proof'. 'Falls short of proof' epitomizes the problem – it will probably never be possible to state that any one factor is the sole or main cause of CHD; there are numerous risk factors and if any of these can be reduced there is expected to be some overall benefit. Reducing the consumption of fat and sugar will certainly result in a diet of greater nutrient content, and the increase of dietary fibre will be of direct benefit in some bowel disorders and indirect benefit in other disorders of affluence.

Despite the limited evidence that dietary changes will reduce the severity of disease, there is good agreement between the various authorities on the following guidelines.

(i) *Fats* The consumption of fat and especially of saturated fatty acids should be decreased. In the UK and other countries with a high incidence of CHD, the average fat intake is 40% of the total energy intake and it is recommended that this should be reduced to 30–35%.

There is less agreement on polyunsaturates; some authorities point to the effect of polyunsaturates in lowering serum cholesterol levels, others emphasize that the main objective should be to reduce total fat intake, and so polyunsaturates should be used only to replace saturates where fat is necessary in the food.

The COMA Report (1984) recommended reducing total fat consumption in the UK by 17% and saturates by 25%. Their 1991 Report on DRVs recommends a further reduction still, with saturated fats reduced from 15% to 11% of the energy intake.

(ii) *Cholesterol* There is also less agreement about dietary cholesterol. The COMA Report (1984) made no specific recommendation but states that the average British intake of 350–450 mg per day is not excessive and that evidence for an influence of this level of intake on serum cholesterol (and therefore on the incidence of heart disease) is inconclusive. On the other hand, a US report (US Senate 1977) states that the average US intake of 600 mg per day is 'well above the 400 mg limit below which there is a linear relationship with serum cholesterol' and recommends a reduction to 300 mg per day or less (as do several other authorities).

(See also CARDIOVASCULAR DISEASE below)

(iii) *Salts* The average intake of salt at 10–12 g per day is about ten times the physiological need. High intakes of salt have been linked with high blood pressure, and while the evidence is not strong and potassium appears to be involved as well as sodium and chloride, it is considered prudent to reduce the intake of common salt.

A US report in 1977 recommended a reduction to about 3 g per day. They later 'considered that 5 g a day is a more appropriate level of salt intake to recommend at this time to the general population'. In the UK, the COMA Report (1984) recommended only that the intake of salt should not be increased further and that consideration should be given to ways and means of decreasing it.

(iv) *Sugar* Recommendations for sucrose range between those of COMA (see above) to the NACNE recommendation of halving the intake.

(v) *Dietary fibre* It is generally recommended that fat and sugar in the diet should be partially replaced by fibre-rich carbohydrate foods (whole grain cereals and bread, fruits and vegetables) but it is not possible to provide any firm figures (Royal College of Physicians, 1980). The average intake of dietary fibre is about 20 g per day and it has been suggested, without clear evidence, that this should be increased to about 30 g. The most recent COMA recommendation (1984) for non-starch polysaccharides is 18 g as already stated.

Nutrition labelling

At the time of writing the legislative requirements all over the world are still being clarified. However, manufacturers should be aware of the general and increasing need for clear, accurate and informative labelling with regard to nutrition, developing generally along the lines described below.

In the UK and the rest of Europe, declaration of the 'Big Four', i.e. energy, protein, carbohydrate and fat, is used as a minimum where any declaration is made. The inclusion of other nutrition data or claim may signal a requirement for the 'Big Eight', i.e. the Big Four plus sodium, fibre, sugars (as part of the carbohydrate figure) and saturates (as part of the fat figure).

The making of particular claims relating to nutrition or health in the labelling or advertising of any product may signal special legal requirements or constraints, and this is another area in which legislation is under continuing development. Such claims may relate to statements about the presence or absence of particular nutrients or ingredients, about special emphasis given to them, or about the quantity present, e.g. 'low fat', 'light', 'high in fibre'. Special consideration is required in relation to baby, infant and dietetic foods for which specific legislation may and does exist in most countries.

A further aspect which can be related directly or indirectly to nutrition labelling is that concerning dietary intolerances (see below), including the use of symbols such as those of the Coeliac Society, the Vegetarian Society or similar organizations. There may also be religious or moral constraints calling for the avoidance of particular foods or ingredients either completely or unless they have been prepared in a particular manner, e.g. 'organic', 'kosher', 'halal'. In most of such cases there are rules to be followed which the producer must be aware of. Trademark regulations may apply. In some cases an authorizing body may need to carry out an inspection or have a representative present during the actual production.

DIET AND DISEASE

ALLERGIES AND INTOLERANCE

Some food constituents provoke allergic responses in limited sectors of the population. A few examples can be cited, such as phenylalanine found in cheese, wine and chocolate (which is a powerful potentiator for migraines). Crustacea produce a fairly common adverse effect and 'strawberry rash' (due to strawberries) is a well known phenomenon. Tartrazine and possibly other coal tar colours have been associated with hyperactivity, albeit rarely, in children. Monosodium glutamate has been connected with a problem colourfully destribed as 'Chinese restaurant syndrome', and adverse reaction to nuts, especially peanuts, has been widely reported in the media in recent years. Foods which themselves are likely to cause allergic responses, particularly in children, include eggs, milk and wheat products.

A food manufacturer has a clear duty to provide such information as is practicable, so that those who know that they are at risk may avoid particular foods or food components. The first step towards this is undoubtedly to make an open declaration on all food packs, as now required by law in most countries. It may well be supplemented by additional information provided both on and off pack as needs arise and problems become more clearly understood.

ALUMINIUM AND ALZHEIMER'S DISEASE

Alzhiemer's disease is a common form of senile dementia in which abnormally high levels of aluminium become deposited in the brain. Because of the this, attention became focused on the possibility that dietary aluminium intake could be a significant factor, despite evidence that this is not relevant as far as people not susceptible to Alzheimer's disease were concerned. In the disease, plaques containing aluminosilicates develop in the brains of patients, possibly due to the failure of a protective mechanism which normally prevents aluminium salts from entering the brain from the blood.

The current position is that there is no evidence of a link between dietary aluminium and Alzheimer's disease, but that a genetically linked defect results in a less efficient preventive mechanism which allows accumulation of aluminium in the brain as a symptom of the problem.

Aluminium levels in the diet vary from 5 to 25 mg per day, apart from possibly large intakes from pharmaceuticals such as antacids. Most is not absorbed but excreted, mostly in the urine. There appears to be no reason for people in normal health to change their eating habits on this account.

CAFFEINE

Caffeine, present in chicory, cocoa, coffee, cola and tea, is a mild mental and physical stimulant. Caffeine is used for its beneficial therapeutic effects at doses of around 150 mg, about the amount in two to four cups of a typical coffee or tea.

Very high levels of caffeine in the diet can result in sleeplessness and nervousness, although individual reaction is variable. Concerns have been voiced about its long term physiological effects, particularly in relation to cardiovascular disease. These concerns, perhaps coupled with a wish to avoid stimulant beverages near bedtime, have helped to develop a market for decaffeinated or low caffeine drinks.

Those few persons with a hypersensitivity to caffeine should avoid food which contain it. Pregnant women are advised in principle to moderate their consumption of all drugs and this would include caffeine. From considerable research there is no indication that caffeine as normally consumed is harmful to humans in either the short or long term.

CANCER

Cancers of many parts of the human body remain the most common cause of death in western Europe and most developed countries, yet the causes of cancer are still poorly understood. A huge number of substances have been tested for their ability to cause cancers in animal species such as mice and rats, when injected, applied to the skin or ingested; an alarmingly high number are found by one or more or these tests to be capable of causing cancer. The convention is now well established by which the potential risk to humans may be roughly quantified on the assumption that a substance which causes cancer when ingested by experimental animals is likely, therefore, to cause cancer in human consumers.

(i) The 'no-effect' level is established; this is the largest dose of the substance, repeated daily over the lifetime of the animal, fed to animals of the most sensitive of the species tested, which produces no cancers or other untoward physiological effects either in the live animal or when the animal is autopsied at the end of a normal life.

(ii) The 'no-effect' dose is multiplied by 100, to make an (arbitrary) allowance for possible variations in sensitivity among the animals and human species; this is then considered to be the maximum concentration to be tolerated in human food (all doses expressed as mass per unit of body weight of the animal or human concerned).

(iii) If it is likely that the amounts of the substance consumed by humans in an average lifetime could exceed the tolerance level then the substance should not be permitted in human food.

The same approach is taken in assessing the risks of other kinds of illness possibly caused by components of the human diet. The principle of this approach may be generally accepted but its practical application is extremely difficult. There is frequent disagreement among experts in the interpretation of the experimental data. In particular, many challenge the supposition that a substance which causes cancers in animals when fed in the often unnaturally high doses necessary to produce clinical effects, is also likely to cause cancers in humans, at much smaller doses.

A number of substances, for instance cyclamates, have been banned in some but not all countries following assessments of carcinogenicity in this way but the grounds upon which the decisions were made have always been fiercely contended.

CARDIOVASCULAR DISEASE

Cholesterol is a waxy substance produced in the liver whose principal function in the body is to link dietary fats with lipoproteins, thus rendering the fats water-dispersible and enabling their transport through the bloodstream. On metabolism of the fat (or its deposition in the fatty tissues) the cholesterol is liberated and may be deposited on the artery walls, reducing their diameter. This leads to an increase in blood pressure and in the work which must be done by the heart to maintain the circulation of the blood. When cholesterol deposition takes in the blood vessels of the heart, heart disease and heart failure are the likely result.

There is some resorption of deposited cholesterol through the action of low density lipoproteins, but the net amount of deposition increases with the amount of fat in circulation. The full mechanism of the different processes is still not clear, nor is the precise significance in all circumstances of:

(i) the total dietary fat in circulation in the blood;

(ii) the proportion of saturated fats within the total lipids;

(iii) whether there is a specific protective effect from unsaturated dietary fats.

However the consensus of current opinion is that the incidence and severity of heart disease among followers of western-style diets will probably be reduced if consumers reduce (i) the proportion of total fat in the diet and (ii) the proportion of saturated fats in the total fat.

This is the main conclusion of the COMA Report (1984) and of studies in other countries and now forms part of the objectives of the UK government's 'Health of the Nation' campaign.

TRANS FATTY ACIDS AND HEALTH 'RISKS'

Trans fatty acids (TFAs) are natural components of ruminant fats and they are formed during the biohydrogenation that takes place in the rumen. Thus milk, meat and products thereof from cow, mutton and goat contain various amounts of *trans* fatty acids. For example, typical values of *trans* fatty acids of milk fat are in the region of 3 to 6%, and human milk may contain about 4.8% TFAs. The amounts of TFAs are directly related to the quantities of partially hydrogenated food fat consumed by nursing mothers.

Natural unsaturated vegetable oils contain double bonds in the *cis* configuration. This is thermodynamically less stable than the *trans* form, in which the carbon–carbon chains are diametrically opposite to one another. TFAs do not occur in vegetable oils, but they are formed to a small extent during the deodorization/steam deacidification of oils which is carried out at high temperatures. TFAs are the result of geometrical and configuration changes that accompany the hydrogenation process, which is used industrially to improve oxidative stability and functional properties of vegetable and fish oils. Therefore, the dietary intake of TFAs cannot be avoided if our diet includes milk fat, dairy products and/or partially hydrogenated fat used in spreads, margarines, shortenings and foods containing them. The levels and types of *trans* acids depend upon the proportion and type of hydrogenated oil and on the degree of hydrogenation. Especially high levels of TFAs, 30–54%, have been determined in margarines and shortenings containing partly hydrogenated fish oil. Partly hydrogenated frying oil may contain 25–35% TFAs. The most accurate method of determination of the *trans* content in fats is by Fourier transform infrared spectroscopy, while capillary column GC calculation may over-estimate the *trans* content.

The nutritional value of fats containing *trans* bonds has been subject to some argument. The findings of some studies have shown that high intake of TFAs produced adverse effects on the plasma LDL/HDL, low density lipoprotein to high density lipoprotein, cholesterol ratio. In other words high levels of TFAs raise the level of 'bad' or harmful LDL cholesterol and reduce the level of 'good' HDL cholesterol in the bloodstream. Recently, there have been several independent assessments of the nutritional consequences of dietary *trans* fatty acids. One conclusion of the joint meeting of FAO and WHO, in October 1993, on Oils and Fats in Human Nutrition, was uncertain whether the use of *trans* or saturated fatty acids is preferable where such fatty acids are required to formulate food products. The human metabolism can cope with TFAs, but high intake, particularly of those of marine origin, may put a strain on the desaturation/elongation metabolism of linoleic acid.

According to UK nutritionists, the risks of taking *trans* fatty acids have been exaggerated. In an average British diet, the energy from TFAs intake is about 2%, compared with about 16% from saturated fats. This current dietary intake of TFAs is so small that it has been concluded that it poses no risks. Moreover two contradictory research findings have implicated the role of TFAs in heart disease. A European Commission study suggested that TFAs could reduce the risk of heart disease in men. Another study from Southampton medical school has suggested that TFAs might even lower the risk of heart failure.

Recently, the International Life Science expert panel have evaluated critically the available scientific data on *trans* fatty acids. The results of the panel concluded that these was little evidence to link TFAs to coronary heart disease risk. The panel have pointed out that the cholesterolemic effect of partially hydrogenated vegetable oils is dependent on the oil or fat with which it is being compared. These oils lower plasma total and LDL cholesterol levels when substituted for animal fats (e.g. lard and butter) and vegetable fats rich in saturated acids, e.g. lauric oils. However, partly hydrogenated oils raise plasma cholesterol levels when substituted for liquid, unhydrogenated oil such as soya bean oil. Therefore, the overall effect of hydrogenated oil on serum cholesterol level depends on the net change in ingestion of saturated, *trans*, *cis*- and monounsaturated fatty acids. The panel has recommended that further research should be carried out to determine whether TFAs

independently affect plasma lipoprotein cholesterol levels. Nevertheless, in recent years, the fat and oil industry has reacted quickly towards reducing the TFA contents in spreads, margarines and other fatty products. This is being achieved by the application of the existing techniques, for example, interesterification and the new process technology. *Trans* acid free spreads containing 70–75% fat are now commercially available. The plasticity is created through isoelectric dispersion technology which creates a mechanical bond rather chemical bond.

To sum up, the role of TFAs in human health is far from clear. Further research is needed to clarify little known effects of TFAs on blood clotting and blood pressure. From the current evidence, it would be unwise to panic from the 'moderate' consumption of *trans* fatty acids.

COELIAC DISEASE

This disease results from an intolerance of the mucosal cells of the small intestine to gluten, the protein material found in wheat and, to a lesser extent, in oats. Sufferers from the disease must avoid all food containing or made from wheat. As well as obvious foods such as bread and breakfast cereals, this may include beer, snack foods and many others where wheat or wheat products may be ingredients. Many patients can tolerate oat protein, some cannot. In the UK the Coeliac Society maintains a list of branded foods declared by the manufacturers to be free from wheat gluten.

DIABETES

This disease may affect people of any age but is most prevalent among the elderly. The basic cause is a failure in the metabolism of sugars (sucrose and glucose) in the diet. This is commonly but not always due to a failure of the body to produce the insulin which is needed for the glucose hydrolysis. The result is a buildup of free glucose in the blood and consequent disturbance of all the related metabolic pathways. The disease, though not yet curable, can be contained by strict attention to the dietary intake of sugars (in relation to the amount of energy consuming exercise being undertaken), coupled with regular controlled injections of insulin. Any failure to balance the insulin input with the sugar input leads to an excess of one of them and the rapid onset of coma or other very distressing symptoms.

POISONS

A number of food plants contain natural poisons, e.g. bitter almonds (amygdalin), green potatoes (solanine), some varieties of beans (lectins), butter beans (cyanide) and yams (amygdalin). In most cases these poisons are reduced to acceptable levels during the normal preparation of the food, for instance by sufficient heating.

Other foods may be contaminated by the presence of foreign plants that are poisonous, such as potato apples, thorn apples and ergot. These have to be identified at the raw material sampling and inspection stages and appropriate action taken.

MYCOTOXINS

In the early 1960s a large number of turkey poults died of liver damage due to feeding with a meal derived from mould infested groundnuts. The mould, *Aspergillus flavus*, had produced a toxin, subsequently named aflatoxin, in the groundnut kernels during oil extraction and finally appeared in the meal fed to the turkey poults. It was subsequently realized that this is a more general problem and toxic materials produced by the growth of mould on foods are now known as mycotoxins. They are seldom a problem in refined oils and fats, as the neutralization, bleaching and deodorization steps fully remove any mycotoxins present. However, it is possible for unprocessed groundnut oil, favoured in some parts of the world for its nutty flavour, to be contaminated with aflatoxins; peanuts eaten whole might also be contaminated. Legislation varies in different parts of the world on the level of mycotoxins permitted in foodstuffs intended for human consumption and for animal feedingstuffs, the amounts permitted being generally in the range 5 to 30 parts per billion. An excellent review of this topic is given by Diener and Davis (1983).

NITROSAMINES

Nitrosamines are a group of chemical compounds, some of which are now known to be involved in the formation of many kinds of cancer. Nitrosamines have been identified in a number of foods and beverages and may even be formed within the human digestive system. It has been demonstrated that they can be formed during the manufacture or cooking of some nitrite-containing foods and hence, although there is no direct evidence that in the foods normally consumed they are harmful to humans, it is generally accepted as desirable to reduce the level of nitrites and nitrates (potential sources of nitrite) in the diet. This requires attention to a wide range of possible sources such as the nitrites used as curing salts, nitrate in water and vegetables and nitrogen oxides in combustion products in dryers and ovens.

On the other hand, the added nitrite in cured meat products is an essential part of the preservative system. A careful balance is therefore required to ensure the minimum safe residual concentration of nitrite throughout the product, whilst avoiding unnecessary excess (see also Chapter 1).

OBESITY

Obesity aggravates many disorders and it is agreed that people who are greatly overweight ought if possible to lose weight. However, it must be borne in mind that there is no evidence that body weights up to 10% beyond the upper end of the acceptable range, although they may technically count as 'overweight', are medically harmful in themselves.

FOOD SAFETY

Since the late 1980s the biggest single issue in the food industry in the UK, as elsewhere, has undoubtedly been food safety. It demands the continuous attention of every person engaged in any way in the UK food industry today. Food safety, or at least microbiological food safety, does not appear to cause the same level of anxiety in any other well-fed or well-regulated country.

It is not possible, nor shall we attempt here, to answer every demand that is made upon the food industry to prove, in one case after another, its innocence of the charge that through fraud or negligence it makes food which is unsafe for consumers to eat.

We propose instead to list the main areas in which the greatest concerns are expressed and to note, without exhaustive supportive detail, the grounds upon which a food manufacturer may be entitled to reply that good manufacturing practice as applied by the food manufacturer is designed to take full account of the difficulties and to produce the best practicable outcome for all consumers. It may also be of interest to note that the Institute of Food Science and Technology (UK) does produce position statements on issues of the day where resources allow, based on the collective wisdom of its members representing all scientific

MICROBIOLOGICAL FOOD POISONING

The causes of food poisoning and the measures necessary to combat it are quite complex. The reader of this manual with an actual problem to solve will be well advised to consult one of the many good textbooks and working manuals available, some of which are listed at the end of this chapter, and to take professional advice from an appropriately qualified food microbiologist. For an excellent overview of the situation in the UK and the significant implications for food manufacturers, see the Richmond Report (1990).

In 1992 the UK Ministry of Agriculture Fisheries and Food opened 'Food Micromodel' to public access. This is a database of growth curves for the main food poisoning organisms under a wide range of conditions of temperature, pH and water activity (controlled by the salt content), from which reliable predictions of the safe shelf life of food products can be made. (Similar data for all food spoilage organisms are not yet available). Access to Food Micromodel may be gained through one of the UK Government Laboratories or Research Associations designated as 'Expert Centres'. The value of this service to food manufacturers, especially as an aid to the development of new products, will no doubt become clear in the near future.

Table 19.1 gives a brief summary of the most important micro-organisms at the present time, the main symptoms of poisoning, likely sources and broad methods of control or elimination. Other organisms which may cause food poisoning and which have been under investigation include *Aeromonas hydrophilia*, *Bacillus subtilis*, *Bacillus licheniformis* and non-proteolytic forms of *Clostridium botulinum*. In addition, the organism known as *E. coli* 0157 is also causing serious concern where outbreaks occur. Although this is not yet a widespread organism, its effects can very serious, even fatal. See also MICROBIOLOGICAL HAZARDS, Chapter 18.

PESTICIDES AND OTHER AGRICULTURAL CHEMICALS

Pesticides and other agricultural chemicals are essential for the production of ample good quality food for humans and animals. Their use, especially at the late stages of growth and in post-harvest treatments, is strictly controlled. Responsible growers and manufacturers should select pesticides, etc. for effectiveness, low environmental impact, low toxicity and low residue levels. In most western countries the government provides appropriate advice and a monitoring service. In the European Union and its member states, Community legislation is applied and an extensive and on-going surveillance and reporting programme is undertaken. Conditions of use of pesticides are controlled in most countries by national legislation. Internationally agreed maximum levels for pesticide and other residues in specific commodities have also been set by Codex Alimentarius Commission. These maximum permitted residue levels are extremely low and embrace wide margins of safety.

BOVINE SPONGIFORM ENCEPHALOPATHY (BSE)

This, the so-called 'mad cow disease', being a disease of cattle causing degeneration of the brain is, as we go to press with this edition, the prime issue facing UK producers of beef and beef products. The causative organism is considered to be a prion – simplistically a self-replicating protein material. The condition was first diagnosed in England in 1986 and reached epidemic proportions among the cattle population in 1989, after which, following a rigorous policy of slaughter of affected animals, it is anticipated that it will eventually disappear. A major route of infection has been believed to be meat-and-bone meal used in animal feed, infected from diseased carcasses and made without sufficient heat treatment for sterilization (see Chapter 1, MEAT AND BONE MEAL). During the epidemic various UK Government Committees have met and reported and the issues have been debated *ad nauseum* at European level with much confusion in relation to scientific fact *vis-à-vis* political expediency. Severe constraints on the use of beef offals and by-products and their use in animal foods have been applied. Details are not quantified here due to the fast changing scenario, but the IFST is providing a regularly updated position statement (available on the Internet World Wide Web, see above) which may be a useful source of information.

Encephalopathies of various kinds occur in most animal species including humans. BSE in cattle is closely related in symptoms and in pathology to scrapie in sheep, but there is no evidence that either condition may be transferred to the other species in normal circumstances. It is in

interests, these also being accessible on the Internet at http://www.easynet.co.uk/ifst.

Table 19.1 Principal food poisoning organisms

Organism	Symptoms	Distinctive characteristics	Major food sources	Prevention
Staphylococcus aureus	Incubation 1–7 h. Vomiting, diarrhoea, etc. Duration 1–2 days or more depending on the amount of toxin ingested.	Forms toxin, preformed in food. Toxin very heat resistant, organism relatively salt-tolerant.	Foods that are handled, e.g. custards, cream-filled bakery products, sliced meats, dairy products.	Good personal hygiene. Hygienic equipment. Refrigeration. Do not handle.
Salmonella species	Incubation 12–72 h. Pain, diarrhoea, fever, (vomiting), etc. Duration 2–6 days.	In UK, second most commonly reported, after *Campylobacter*. Easily destroyed by heat	Raw meat and poultry, raw eggs and egg products, raw milk.	Avoid cross-contamination of cooked with raw foods. Cook adequately, Refrigeration.
Clostridium perfringens	Incubation 8–22 h. Pain and diarrhoea. Duration 12–24 h.	Anaerobic, forms heat-resistant spores. Large numbers required in food. High growth temperature. Toxin rarely preformed in food.	Cooked meats and meat products, especially foods prepared in bulk, not cooled sufficiently but held warm too long before serving.	Chill cooked foods immediately. Hold hot foods over 60°C. Reheat cooked foods to 75°C.
Clostridium botulinum	Incubation 12–18 h. Mild initially, then severe – can be fatal. Attacks the nervous system.	Anaerobic, forms heat-resistant spores. Forms toxin in food. Toxin destroyed by heat.	Improperly heat-processed low-acid foods. Temperature-abused anaerobic fish packs or cooked uncured meat products.	Destroy spores by severe heating. Prevent growth using combination of preservation factors.
Campylobacter jejuni	Incubation 48 h to over 1 week. Abodminal cramps, fever and bloody diarrhoea. Duration variable	Growth range 30–45°C Unlikely to grow in food but infective dose is low. Sensitive to acid and heat. The leading cause of bacterial gastro-enteritis in the UK.	Raw milk, raw poultry.	Avoid cross-contamination of cooked with raw foods. Cook meat, especially poultry, adequately.
Bacillus cereus	Incubation: emetic toxin 1–5 h, diarrhoea toxin 8–16 h. Duration usually less than 24 h.	Wide growth range, about 10–50°C, optimum 28–35°C. (also some psychrotrophic toxin-forming strains, growth range 4–37°C). Emetic toxin extremely heat stable, diarrhoea toxin destroyed by normal cooking.	Mainly starchy foods, especially fried rice; others include spices, beans, meats, etc.	Chill cooked foods immediately. Hold hot foods over 60°C.
Vibrio para-haemolyticus	Incubation 12–24 h. Diarrhoea, cramps, nausea. Duration 2–5 days.	Salt resistant. Grows at 5–43°C. Large numbers required. Multiplies rapidly. Heat sensitive.	Almost exclusively seafood, especially raw.	Cook and refrigerate seafoods. Eliminate cross-contamination from raw to cooked.
Listeria monocytogenes	Incubation 4–21 days. Flu-like, diarrhoea, meningitis, septicaemia, abortion. High fatality rate. Duration variable.	Growth range −1 to 44°C, grows well at higher traditional refrigeration temperatures. Salt tolerant. Heat sensitive but less so than *Salmonella*. Foodborne infection appears rare.	Wide range of raw foods including raw milk, poultry, meat, seafoods, occasionally chilled products, e.g. mould-ripened chesse, pâté, salads,	Refrigerate below 5°C (ideally <3°C) Restrict shelf life. Heat cooked foods to 70°C for 2 min or equivalent (minimum).
Yersinia enterocolitica	Incubation 1–11 days, typically 2–3 days. Diarrhoea, fever, abdominal pain. Possible appendicitis-like and other symptoms. Duration usually short (2–3 days) unless complications.	Psychrotrophic, growth range 0–44°C. Destroyed by normal cooking.	Raw milk and dairy products, pork, poultry. (NB not all strains isolated are pathogenic.)	Refrigerate below 5°C (ideally <3°C). Cook (reheat) adequately.

After Milner (1995).

fact most unusual for an encephalopathy in one species to infect animals of another species, but under the intense focus on this issue a small number of cases of alleged transferability, one in particular involving a cat, have come to light.

Government and EU policies of slaughter and constraint have appeared to be excessive to many scientists, but continue. The use of bovine offals is very much limited in both the human and animal food chain and readers are recommended to seek up to the minute advice on the situation.

SALMONELLA IN EGGS

Salmonella organisms are among the commonest causes of food poisoning in most countries where reliable records exist and are known to be widespread among poultry flocks. Even so, the degree of concern about them exhibited by the press and public in the UK in the late 1980s and early 1990s was quite disproportionate. That concern was greatly fuelled by public (and Government) misunderstanding of the significance of the report in 1989 of one egg infected internally with *S. enteritidis*. The immediate results were highly damaging for the UK egg industry, but the industry has since almost completely recovered although public concern about the safety of eggs and poultry is resurrected from time to time, often associated with outbreaks of food poisoning attributable to mishandling of the foods in question (see also under EGGS, Chapter 1).

ECONOMIC AND MANUFACTURING ISSUES

EC INTERVENTION POLICY

One well-known effect of the Common Agricultural Policy of the European Community was the overproduction of particular products at times when the subsidy arrangements made it advantageous for farmers to grow those crops. So, for instance, there has been a tendency to get 'mountains' of surpluses such as butter and beef and a 'lake' of surplus wine. These surplus products are usually put into storage and held until new political and economic decisions are made to release them.

Such 'intervention' materials may be released, usually at advantageous prices, to food manufacturers to use as ingredients in their processes. The quality of the materials when released from inter-

vention will depend on the length of time they were in store, their hygienic and other quality attributes at the time they were put into store, the conditions of storage and the quality of the packaging. These important quality factory may or may not have been as well controlled by the EC authorities and their storage agents as they would have been by a responsible food manufacturer. There may be little opportunity for the purchaser to check the quality or to reject if unsatisfactory. The application and control of intervention have undoubtedly improved in recent years, although with political influences it is likely to go on giving rise to problems so long as the current policy is maintained.

EFFLUENTS AND WASTE DISPOSAL

In almost all countries the food manufacturer is closely controlled by existing local and national legislation on the quantities and quality of the wastes (including most notably the waste waters) which may be discharged and the manner of doing so. In many instances the disposal system is paid for or even built by the manufacturer for the local authority. Except where a manufacturer is clearly in default of the regulations, the widely expressed public concerns about the 'green' and environmental aspects of these matters should therefore most properly be addressed to the authorities responsible for the regulations.

IRRADIATION

General information on the use and benefits of irradiation for food preservation, and on the problems and public reactions to proposals to use it, is given in Chapter 15.

RECYCLING OF PACKAGING

Public pressure for the recycling of food packaging varies from country to country and the complex problems raised are dealt with in different ways. The problems, not all of which may be solved together, include:

(i) The greater strength needed in a returnable container compared with a one trip container

(ii) The form of outer container to use – returnable crate, cardboard carton, film overwrapped tray, and so on

(iii) Setting up and organizing the return of containers through the retail system, including the use of returnable crates and the collection of returnable deposits on purchase

(iv) Sorting and separation of returned containers made from different materials, especially different kinds of plastic

(v) Cleaning returned containers before re-use

(vi) Disposal otherwise of containers which are not re-usable.

There are a few long standing examples where glass bottles for particular products are returned, washed and re-used through a well-established system, for instance milk in Britain and table wine in France.

In Denmark it was made obligatory in recent years for all beer and soft drinks on the domestic market to be packed in returnable glass bottles; the use of cans and plastic bottles is prohibited.

In Holland it is now required that all plastic drinks bottles may be returned to the retailer and must be collected from there by the manufacturer; manufacturers may dispose of them as they choose, some now have incineration plants which they use to generate electricity.

In Germany the government encourages manufacturers to instigate 'green spot' schemes – packages marked with the green spot on the label are returnable either to the place of purchase or to central collecting points, where they are taken away by appointed disposal companies. Packaging of all kinds including paper and plastic wrapping can be covered within the schemes.

All of these initiatives have been brought together, along with questions of landfill, incineration, and so on, under the European Union's policy and legislation on packing and packaging waste to be attained by member states during the 1990s. UK legislation was made to come into force from 1997, affecting all businesses producing, filling or selling packaging or packaging materials in excess of 50 tonnes per annum. The legislation deals with recovery, recycling and disposal of packaging and the structure of the system whereby this may be effected as well as registration and enforcement.

TAMPERING AND CONTAMINATION

In the UK since the late 1980s there have been incidents in which would-be extortioners have announced through the press and otherwise that unless large ransoms were paid by named food manufacturers or retailers, foods on the supermarket shelves would be secretly contaminated with poisonous or dangerous materials. One such threat was to put razor blades into jars of a branded baby food and another deliberately to contaminate dairy products with a food poisoning micro-organism. In the event, no ransoms have been paid, few threats have been carried out and the businesses in question have cooperated with the police and the extortioners been caught. The courts have treated these offences extremely seriously and sentences have been generally severe. There have been numerous clumsy small scale 'copy cat' attempts at similar contamination of foods and other goods, all of which have been countered without any ill effects on consumers.

A major consequence of these affairs has been that UK food manufacturers have been pressed by the supermarkets and the authorities alike to pay even closer attention to their systems for the reduction or elimination of all potential sources of contamination of products during manufacture. Particular attention has been paid to contamination by possibly harmful fragments of wood, glass or metal. Measures which are now generally taken in all food factories include the elimination of wooden pallets and all other sources of wood from production departments wherever practicable, the shielding of glass electric lamps with plastic covers, the banning of all glass vessels from production areas and the installation and regular checking of metal detectors at the end of production lines.

All major manufacturers and supermarkets have reviewed, improved and tested their crisis management systems and especially their procedures for the rapid identification of any suspect product in all premises under their control, and recall if necessary of product already sent out to customers. The use of automatic on-line product coding, capable of showing fine details of the production batch and time of filling has been greatly extended. The general introduction of tamper-evident closures on jars, bottles and other re-usable containers has greatly accelerated and traceability of products and their components vastly improved.

OTHER TOPICS

Other topics likely to become 'issues' in the foreseeable future include genetic modification of food-providing organisms such as are used in the cases of vegetarian cheese and tomato purée, biotechnology in general and the continuing evidence of sensitivity to foods and food additives and the advent of so-called 'functional foods'.

LEGAL CONSIDERATIONS

Food manufacture worldwide is unusual as an industrial process in that its products are heavily circumscribed by legislation of one kind or another. Most of the legislation now current in the UK and other EU member states is, of course, derived from EC/EU Regulations and Directives. Many matters are regulated by non-statutory Codes of Practice. The overall purpose of all these laws and codes is to protect the safety of the consumer, to ensure the quality of food and food products and to ensure that the customer is properly informed about what is purchased.

In addition, production plants and their operations are governed by legislation applicable to all kinds of factories. In the UK the Factories Act 1961 contains the basic requirements; the *Short Guide to the Factories Act* (1977) is a useful publication. The Health and Safety at Work Act and Regulations 1974 and the Sex Discrimination Act 1975 are also generally relevant.

BASIC FOOD LEGISLATION

In the UK the food industry is directly subject to four primary laws:

(i)   Sale of Goods Act, 1893 and 1979;
(ii) Trade Descriptions Act, 1968 and 1972;
(iii) Weights and Measures Act, 1963, 1976, 1979 and 1985;
(iv) Food Safety Act, 1990.

Most of ancillary regulations affecting the manufacture and sale of foods and drinks are made under these Acts.

The Sale of Goods Act requires that articles sold should be fit for the intended purpose as long as that purpose is known by the vendor by virtue of common practice or by specific advice. The Trade Descriptions Act requires that goods and their attributes are fairly and reasonably described whatever the medium by which that information is conveyed. The Weights and Measures Act, which embodies EU legislation, and its requirements are covered in Chapter 18.

The legal situation in the USA is similar in principle, with one enabling law, the Federal Food, Drugs and Cosmetics Act (as amended), and detailed Regulations published as the Code of Federal Regulations. These appear under various Titles such as Title 9 (9FCR) – the Meat and Poultry Inspection Regulations; Title 21 (21CFR) – Food Compository Regulations; Title 7 (7CFR)

– the Agricultural Marketing Regulations; and Title 40 (40CFR) – the Pesticide Regulations.

In any matter involving the law there is no substitute for the careful reading of the relevant legal instruments (Acts, Regulations, Orders, and so on) themselves, in particular those of the country or countries where the product may also be made and sold. If legality is in question, or in the face of prosecution, the services of competent and appropriately experienced legal advisors are essential. Specialized advice may also be sought from Trade Associations, Research Associations or consultants. In the UK there are specialist information services on food and related legislation, regularly updated, such as those offered by Butterworth's on Food Law and on Weights and Measures, and Sweet and Maxwell's Practical Food Law Manual as well as various journals such as the Food & Drugs Industry Bulletin and the services offered by the food research associations and a number of independent consultants.

THE FOOD SAFETY ACT 1990 (UK)

The 1990 Act was initially developed to consolidate and modernize the long series of Food Acts and Food and Drugs Acts under which food manufacture and sale in Britain have been governed since 1869. Two fundamental requirements of the earlier legislation were carried forward again – a new Section prohibiting the sale of any food which is injurious to health and Section 14 prohibiting the sale of food which is not of the nature, substance or quality demanded by the purchaser.

One major new offence was introduced under the Act – Section 8 prohibits the sale of any food which 'fails to comply with food safety requirements'. This curious term is not fully defined but is stated to include food 'unfit for human consumption'.

Other important provisions include:

(i)   The introduction of the 'due diligence' defence, elaborated on below
(ii) Provision for Regulations to be made which requiring the obligatory registration of all premises upon which food is handled
(iii) Making food importers legally responsible for the quality of goods brought into the country by them and no longer able to claim 'warranty' from suppliers located outside the jurisdiction of the UK authorities
(iv) Removal of the immunity from prosecution of establishments run by government departments (such as the kitchens in hospitals and

prisons) if they are involved in serving defective food.

The Act, like its predecessors, gave power to Ministers to make subordinate Regulations, such as on composition, labelling, hygiene, and so on, and also to develop statutory Codes of Practice of which there are now more than 20 in existence.

The Food Safety Act also encompasses the requirements of the EU Directive on the Official Control of Foodstuffs 89/397 which first opened the possibility of official in-factory inspection in the UK industry.

The 'due diligence' defence

In the UK under the Food Safety Act the defence of 'having taken all reasonable precautions and exercised due diligence to avoid the commission of the offence by himself or by a person under his control' is now permitted to a food manufacturer (or other vendor) whose food is alleged to contravene the requirements of the Food Safety Act. That is, the accused is entitled to argue that whilst the food might be unsatisfactory in some way, that was not due to his fault because he had foreseen the possibility, taken all reasonable precaution and directed real effort to prevent it from happening.

The key to such defence appears to be the ability to provide adequate and documented evidence of having met both the elements required. To prove 'all reasonable precautions' will involve the identification of what precautions are technically available and a decision about how much is reasonable; that must be decided in advance by the owner of the business or the board of directors, unless or until their view is contradicted by a court. The phrase: '... and all due diligence' seems at first sight to be more straightforward, simply being a requirement to show that reasonable precautions were, in fact, taken. For instance, a reasonable precaution might be to install a metal detector, having established that metallic foreign matter might be a problem, while 'due diligence' could be proved by presenting records to show that the metal detector was continuously in use, was tested by a reliable means with sufficient frequency and when found defective was repaired immediately.

Enforcement authorities

The role of enforcement authorities and the manner of its fulfilment varies throughout the world and in this chapter it is intended to deal only with the situation in the UK. For other countries the reader should refer to expert advice such as is available from, for example, the Leatherhead Food RA, or from the appropriate authorities in the country in question.

With regard to enforcement in the UK, there are broadly three circumstances in which a prosecution under the Food Safety Act may be brought. The first two are where an allegedly defective product is brought to the attention of a local authority, or where an authorized officer to the local authority has purchased an official sample to check that legal requirements are being met. In the first circumstance the complaint might be that the food is mouldy or contains foreign matter or is of poor quality. A prosecution can rely on the single sample.

In the second circumstance three identical samples must be prepared by the authorized officer and sealed. One sample is kept for reference, one is tested on behalf of the local authority and the third is given with an official notice to the retailer who will normally pass it on to the manufacturer claiming 'warranty'. At this stage no allegation of illegality is being made. If the local authority then considers that an offence may have been committed this will usually be discussed with the manufacturer before any notice of prosecution is served. A reasonable attitude to rectification of a technical oversight is less likely to lead to prosecution than will any deliberate attempt to obtain an advantage by misleading or defrauding.

The third circumstance is where on-site inspection in the food distribution chain has revealed some circumstance in apparent conflict with the requirements of the Food Safety Act.

CODES OF PRACTICE

Codes of Practice have proved to be of considerable benefit to producer, consumer and enforcement authority alike. The great advantages of voluntary Codes of Practice over definitive regulations are that they retain a measure of flexibility which allows more rapid response to changing circumstances, they permit operation outside the terms of the Code where that can be justified, for instance, as an honest and common sense reaction to changing circumstances and they constrain development to a lesser degree.

There are many Codes of Practice in existence, dealing with a diversity of subjects from ante- and post-mortem inspection of slaughter animals to the manufacture of vanilla slices. They are published by various national and international bodies, e.g. Food and Agriculture Organization, World Health Organization, Codex Alimentarius Commission, local, national and international trade associations, and so on. They are commonly produced in liaison with enforcement authorities or their professional bodies or coordinating bodies, for example in the UK the Local Authorities' Coordinating Body on Trading Standards (LACOTS). Manufacturers considering entering a new market are advised to check with their own local authority, trade association or expert advisor whether or not they may be affected by any Codes of Practice.

Various Codes of Practice have been produced to give guidance in the UK to the enforcement authorities on the procedures for enforcing the Food Safety Act. The Institute of Food Science and Technology (UK) was commissioned by the UK Government (through the Ministry of Agriculture Fisheries and Food) to identify all the Codes of Practice currently in existence relating to the safety of food and produced a useful listing.

PROFESSIONAL ETHICS

The very existence of such large and complex bodies of national and international food legislation may be seen to rest upon the fact that there is a large and complex food industry in existence. Of course, legislation by itself is quite incapable of producing food – that is the function and the responsibility of the food manufacturing industry and the practitioners of the many professional disciplines employed there – chemists, biologists, microbiologists, nutritionists, food scientists, technologists, engineers and many others. It is here that the role of professional bodies with their interest in validated qualifications, training and personal skills and accompanying codes of professional conduct have an all important role to play.

REFERENCES

COMA (1974) *Diet and Cardiovascular Disease*, Report no. 7, Committee on Medical Aspects of Food Policy, HMSO, London.

COMA (1984) *Diet and Cardiovascular Disease*, Report no. 29, Committee on Medical Aspects of Food Policy, HMSO, London.

COMA (1991) *Dietary Reference Values for the United Kingdom*, Report no. 41, Committee on Medical Aspects of Food Policy, HMSO, London

Department of Health (1991) *Dietary Reference Values – A Guide*, HMSO, London.

Diener, V.L. and Davis, N.D. (1983) Aflatoxins in corn, in Finely, J.W. and Schwass, D.E. (eds), *Xenobiotics in Food and Feeds*, American Chemical Society, Washington, DC.

Factories Act (1977) *A Short Guide*, 2nd edn, HMSO, London.

Milner, J.A. (1995) *Micro-Facts*, 3rd edn, Leatherhead Food R.A.

NACNE (National Council for Nutritional Education) (1983) *Proposals for Nutritional Guidelines for Health Education of Britain*, Health Education Council, London.

Richmond Report (1990) *Microbiological Safety of Food, Part 1*, HMSO, London.

Royal College of Physicians of London (1980) *Medical Aspects of Dietary Fibre*; summary of report, Pitman Medical, Tunbridge Wells, UK.

US Senate (1997) *Dietary Goals for the United States*, Select Committee on Nutrition and Human Need, US Government Printing Office, Washington (1st edn February; 2nd edn December).

FURTHER READING

Bender, A.E. (1990) *Dictionary of Nutrition and Food Technology*, 6th edn, Butterworths, London.

British Nutrition Foundation (1988) *Food Fit to Eat*, Sphere Books, London.

British Nutrition Foundation Task Force (1992) *Unsaturated Fatty Acids: Nutritional and Physiological Significance*, Chapman & Hall, London.

Butterworths Law of Food and Drugs, Butterworths, London.

Department of Health and Social Security (1979) *Recommended Daily Amounts of Energy and Nutrients*, Rpt. No. 15, HMSO, London.

Doeg, C. (1995) *Crisis Management in the Food and Drinks Industry – A Practical Approach*, Chapman & Hall, London.

Eastwood, M., Edwards, C. and Parry, D. (eds) (1992) *Human Nutrition: A Continuing Debate*, Chapman & Hall, London.

Joint FAO/WHO Report (1985) *Energy and Protein Requirements*, WHO Series No. 724, World Health Organization, Geneva.

IFST (1991) *Food and Drink – Good Manufacturing Practice: A Guide to its Responsible Management*, 3rd edn, Institute of Food Science & Technology, London.

Health and Safety Executive (1991) *The Prevention and Control of Legionelliosis (including Legionaires' Disease): Approved Code of Practice*, HMSO, London.

Jukes, D.J. (1992) *Food Legislation of the UK: A Concise Guide*, 3rd edn, Butterworths, London.

Index

Primary references are indicated in **bold**.